U0150789

The Living Record of Science

《自然》学科经典系列

总顾问：李政道（Tsung-Dao Lee）

英方总主编：Sir John Maddox
Sir Philip Campbell
中方总主编：路甬祥

天文学的进程 I

PROGRESS IN ASTRONOMY I

（英汉对照）

主编：武向平

外语教学与研究出版社 · 麦克米伦教育 · 《自然》旗下期刊与服务集合

FOREIGN LANGUAGE TEACHING AND RESEARCH PRESS · MACMILLAN EDUCATION · NATURE PORTFOLIO

北京 BEIJING

图书在版编目（CIP）数据

天文学的进程. I ：英汉对照 / 武向平主编. -- 北京 ：外语教学与研究出版社，2022.8

（《自然》学科经典系列 / 路甬祥等总主编）

ISBN 978-7-5213-3903-1

I. ①天… II. ①武… III. ①天文学－文集－英、汉 IV. ①P1-53

中国版本图书馆 CIP 数据核字（2022）第 146175 号

出 版 人　王　芳
项目统筹　章思英
项目负责　刘晓楠　顾海成
责任编辑　刘晓楠
责任校对　王　菲　白小羽　夏洁媛
封面设计　孙莉明　高　蕾
版式设计　孙莉明
出版发行　外语教学与研究出版社
社　　址　北京市西三环北路 19 号（100089）
网　　址　http://www.fltrp.com
印　　刷　北京华联印刷有限公司
开　　本　787×1092　1/16
印　　张　43.5
版　　次　2022 年 9 月第 1 版　2022 年 9 月第 1 次印刷
书　　号　ISBN 978-7-5213-3903-1
定　　价　568.00 元

购书咨询：（010）88819926　电子邮箱：club@fltrp.com
外研书店：https://waiyants.tmall.com
凡印刷、装订质量问题，请联系我社印制部
联系电话：（010）61207896　电子邮箱：zhijian@fltrp.com
凡侵权、盗版书籍线索，请联系我社法律事务部
举报电话：（010）88817519　电子邮箱：banquan@fltrp.com
物料号：339030001

《自然》学科经典系列

（英汉对照）

天文学的进程

（英汉对照）

主编：武向平

审稿专家（以姓氏笔画为序）

于　涌　　马宇蒨　　王有刚　　邓祖淦　　肖伟科　　何香涛　　沈志侠
尚仁成　　周　江　　蒋世仰

翻译工作组稿人（以姓氏笔画为序）

王耀杨　　刘　明　　沈乃澂　　张　健　　郭红锋　　蔡则怡

翻译人员（以姓氏笔画为序）

王宏彬	王耀杨	史春晖	冯　翀	刘东亮	李任伟	李忠伟
余　恒	汪　浩	沈乃澂	林元章	岳友岭	金世超	周　杰
孟　洁	钱　磊	曹惠来	樊　彬	魏　韧		

校对人员（以姓氏笔画为序）

王　菲	王帅帅	王阳兰	王晓萌	王晓蕾	白小羽	冯　翀
乔萌萌	刘　明	刘东亮	闫　妍	孙　娟	孙瑞静	李　琦
肖　莉	何　铭	何思源	张　帆	张竞凤	张媛媛	周玉凤
宗伟凯	赵凤轩	夏洁媛	顾海成	钱　磊	黄小斌	梁　瑜
葛云霄	韩少卿	曾红芳	蔡　迪			

Foreword by Tsung Dao Lee

We can appreciate the significance of natural science to human life in two aspects. Materially, natural science has achieved many breakthroughs, particularly in the past hundred years or so, which have brought about revolutionary changes to human life. At the same time, the spirit of science has taken an ever-deepening root in the hearts of the people. Instead of alleging that science is omnipotent, the spirit of science emphasizes down-to-earth and scrupulous research, and critical and creative courage. More importantly, it stands for the dedication to working for the wellbeing of humankind. This is perhaps more meaningful than scientific and technological achievements themselves, which may be closely related to specific backgrounds of the times. The spirit of science, on the other hand, constitutes a most valuable and constant component of humankind's spiritual civilization.

In this sense, *Nature: The Living Record of Science* presents not only the historical paths of the various fields of natural science for almost a century and a half, but also the unremitting spirit of numerous scientists in their pursuit of truth. One of the most influential science journals in the whole world, *Nature*, reflects a general picture of different branches of science in different stages of development. It has also reported many of the most important discoveries in modern science. The collection of papers in this series includes breakthroughs such as the special theory of relativity, the maturing of quantum mechanics and the mapping of the human genome sequence. In addition, the editors have not shunned papers which were proved to be wrong after publication. Included also are the academic debates over the relevant topics. This speaks volumes of their vision and broadmindedness. Arduous is the road of science; behind any success are countless failures unknown to outsiders. But such failures have laid the foundation for success in later times and thus should not be forgotten. The comprehensive and thoughtful coverage of these volumes will enable readers to gain a better understanding of the achievements that have tremendously promoted the progress of science and technology, the evolution of key and cutting-edge issues of the relevant fields, the inspiration brought about by academic controversies, the efforts and hardships behind these achievements, and the true meaning of the spirit of science.

China now enjoys unprecedented opportunities for the development of science and technology. At the policy level, the state has created a fine environment for scientific research by formulating medium- and long-term development programs. As for science and technology, development in the past decades has built up a solid foundation of research and a rich pool of talent. Some major topics at present include how to introduce the cream of academic research from abroad, to promote Sino-foreign exchange in science and technology, to further promote the spirit of science, and to raise China's development in this respect to the advanced international level. The co-publication of *Nature: The Living Record of Science* by the Foreign Language Teaching and Research

李政道序

如何认识自然科学对人类生活的意义，可以从两个方面来分析：一是物质层面，尤其是近百年来，自然科学取得了很多跨越性的发展，给人类生活带来了许多革命性的变化；二是精神层面，科学精神日益深入人心，这种科学精神并不是认为科学万能、科学可以解决一切问题，它应该是一种老老实实、严谨缜密、又勇于批判和创造的精神，更重要的是，它具有一种坚持为人类福祉而奋斗的信念。这种科学精神可能比物质意义上的科技成就更重要，因为技术进步的影响可能与时代具体的背景有密切关系，但科学精神却永远是人类精神文明中最可宝贵的一部分。

从这个意义上，这套《〈自然〉百年科学经典》丛书的出版，不仅为读者呈现了一个多世纪以来自然科学各个领域发展的历史轨迹，更重要的是，它展现了无数科学家在追求真理的过程中艰难求索、百折不回的精神世界。《自然》作为全世界最有影响力的科学期刊之一，反映了各个学科在不同发展阶段的概貌，报道了现代科学中最重要的发现。这套丛书的可贵之处在于，它不仅汇聚了狭义相对论的提出、量子理论的成熟、人类基因组测序完成这些具有开创性和突破性的大事件、大成就，还将一些后来被证明是错误的文章囊括进来，并展现了围绕同一论题进行的学术争鸣，这是一种难得的眼光和胸怀。科学之路是艰辛的，成功背后有更多不为人知的失败，前人的失败是我们今日成功的基石，这些努力不应该被忘记。因此，《〈自然〉百年科学经典》这套丛书不但能让读者了解对人类科技进步有着巨大贡献的科学成果，以及科学中的焦点和前沿问题的演变轨迹，更能使有志于科学研究的人感受到思想激辩带来的火花和收获背后的艰苦努力，帮助他们理解科学精神的真意。

当前，中国科学技术的发展面临着历史上前所未有的机遇，国家已经制定了中长期科学和技术发展纲要，为科学研究创造了良好的制度环境，同时中国的科学技术经过多年的积累也已经具备了很好的理论和人才基础。如何进一步引进国外的学术精华，促进中外科技交流，使科学精神深入人心，使中国的科技水平迅速提升至世界前列就成为这一阶段的重要课题。因此，外语教学与研究出版社和麦克米伦出

Press, Macmillan Publishers Limited and the Nature Publishing Group will prove to be a huge contribution to the country's relevant endeavors. I sincerely wish for its success.

Science is a cause that does not have a finishing line, which is exactly the eternal charm of science and the source of inspiration for scientists to explore new frontiers. It is a cause worthy of our uttermost exertion.

T. D. Lee

Editor's note: The foreword was originally written for the ten-volume *Nature: The Living Record of Science.*

版集团合作出版这套《〈自然〉百年科学经典》丛书，对中国的科技发展可谓贡献巨大，我衷心希望这套丛书的出版获得极大成功，促进全民族的科技振兴。

科学的事业永无止境。这是科学的永恒魅力所在，也是我们砥砺自身、不断求索的动力所在。这样的事业，值得我们全力以赴。

李政道

编者注：此篇原为《〈自然〉百年科学经典》（十卷本）的序。

Foreword by Lu Yongxiang

Since the birth of modern science, and in particular throughout the 20th century, we have continuously deepened our understanding of Nature, and developed more means and methods to make use of natural resources. Technological innovation and industrial progress have become decisive factors in promoting unprecedented development of productive forces and the progress of society, and have greatly improved the mode of production and the way we live.

The 20th century witnessed many revolutions in science. The establishment and development of quantum theory and the theory of relativity have changed our concept of time and space, and have given us a unified understanding of matter and energy. They served as a theoretical foundation upon which a series of major scientific discoveries and technological inventions were made. The discovery of the structure of DNA transformed our understanding of heredity and helped to unify our vision of the biological world. As a corner-stone in biology, DNA research has exerted a far-reaching influence on modern agriculture and medicine. The development of information science has provided a theoretical basis for computer science, communication technology, intelligent manufacturing, understanding of human cognition, and even economic and social studies. The theory of continental drift and plate tectonics has had important implications for seismology, geology of ore deposits, palaeontology, and palaeoclimatology. New understandings about the cosmos have enabled us to know in general terms, and also in many details, how elementary particles and chemical elements were formed, and how this led to the formation of molecules and the appearance of life, and even the origin and evolution of the entire universe.

The 20th century also witnessed revolutions in technology. Breakthroughs in fundamental research, coupled to the stimulus of market forces, have led to unparalleled technological achievements. Energy, materials, information, aviation and aeronautics, and biological medicine have undergone dramatic changes. Specifically, new energy technologies have helped to promote social development; new materials technologies promote the growth of manufacturing and industrial prosperity; information technology has ushered in the Internet and the pervasive role of computing; aviation and aeronautical technology has broadened our vision and mobility, and has ultimately led to the exploration of the universe beyond our planet; and improvements in medical and biological technology have enabled people to live much better, healthier lives.

Outstanding achievements in science and technology made in China during its long history have contributed to the survival, development and continuation of the Chinese nation. The country remained ahead of Europe for several hundred years before the 15th century. As Joseph Needham's studies demonstrated, a great many discoveries and innovations in understanding or practical capability—from the shape of snowflakes to the art of cartography, the circulation of the blood, the invention of paper and sericulture

路甬祥序

自近代科学诞生以来，特别是 20 世纪以来，随着人类对自然的认识不断加深，随着人类利用自然资源的手段与方法不断丰富，技术创新、产业进步已成为推动生产力空前发展和人类社会进步的决定性因素，极大地改变了人类的生产与生活方式，使人类社会发生了显著的变化。

20 世纪是科学革命的世纪。量子理论和相对论的创立与发展，改变了人类的时空观和对物质与能量统一性的认识，成为了 20 世纪一系列重大科学发现和技术发明的理论基石；DNA 双螺旋结构模型的建立，标志着人类在揭示生命遗传奥秘方面迈出了具有里程碑意义的一步，奠定了生物技术的基础，对现代农业和医学的发展产生了深远影响；信息科学的发展为计算机科学、通信技术、智能制造提供了知识源泉，并为人类认知、经济学和社会学研究等提供了理论基础；大陆漂移学说和板块构造理论，对地震学、矿床学、古生物地质学、古气候学具有重要的指导作用；新的宇宙演化观念的建立为人们勾画出了基本粒子和化学元素的产生、分子的形成和生命的出现，乃至整个宇宙的起源和演化的图景。

20 世纪也是技术革命的世纪。基础研究的重大突破和市场的强劲拉动，使人类在技术领域获得了前所未有的成就，能源、材料、信息、航空航天、生物医学等领域发生了全新变化。新能源技术为人类社会发展提供了多元化的动力；新材料技术为人类生活和科技进步提供了丰富的物质材料基础，推动了制造业的发展和工业的繁荣；信息技术使人类迈入了信息和网络时代；航空航天技术拓展了人类的活动空间和视野；医学与生物技术的进展极大地提高了人类的生活质量和健康水平。

历史上，中国曾经创造出辉煌的科学技术，支撑了中华民族的生存、发展和延续。在 15 世纪之前的数百年里，中国的科技水平曾遥遥领先于欧洲。李约瑟博士曾经指出，从雪花的形状到绘图的艺术、血液循环、造纸、养蚕，包括更有名的指南针和

and, most famously, of compasses and gunpowder—were first made in China. The Four Great Inventions in ancient China have influenced the development process of the world. Ancient Chinese astronomical records are still used today by astronomers seeking to understand astrophysical phenomena. Thus Chinese as well as other long-standing civilizations in the world deserve to be credited as important sources of modern science and technology.

Scientific and technological revolutions in 17th and 18th century Europe, the First and Second Industrial Revolutions in the 18th and 19th centuries, and the spread of modern science education and knowledge sped up the modernization process of the West. During these centuries, China lagged behind.

Defeat in the Opium War (1840–1842) served as strong warning to the ancient Chinese empire. Around and after the time of the launch of *Nature* in 1869, elite intellectuals in China had come to see the importance that science and technology had towards the country's development. Many scholars went to study in Western higher education and research institutions, and some made outstanding contributions to science. Many students who had completed their studies and research in the West returned to China, and their work, together with that of home colleagues, laid the foundation for the development of modern science and technology in the country.

In the six decades since the founding of the People's Republic of China, the country has made a series of achievements in science and technology. Chinese scientists independently developed the atomic bomb, the hydrogen bomb and artificial satellite within a short period of time. The continental oil generation theory led to the discovery of the Daqing oil field in the northeast. Chinese scientists also succeeded in synthesizing bovine insulin, the first protein to be made by synthetic chemical methods. The development and popularization of hybrid rice strains have significantly increased the yields from rice cultivation, benefiting hundreds of millions of people across the world. Breakthroughs in many other fields, such as materials science, aeronautics and life science, all represent China's progress in modern science and technology.

As the Chinese economy continues to enjoy rapid growth, scientific research is also producing increasing results. Many of these important results have been published in first-class international science journals such as *Nature*. This has expanded the influence of Chinese science research, and promoted exchange and cooperation between Chinese scientists with colleagues in other countries. All these indicate that China has become a significant global force in science and technology and that greater progress is expected in the future.

Science journals, which developed alongside modern science, play an essential role in faithfully recording the path of science, as well as spreading and promoting modern science. Such journals report academic development in a timely manner, provide a platform for scientists to exchange ideas and methods, explore the future direction of science, stimulate academic debates, promote academic prosperity, and help the public

火药，都是首先由中国人发现或发明的。中国的“四大发明”影响了世界的发展进程，古代中国的天文记录至今仍为天文学家在研究天体物理现象时所使用。中华文明同其他悠久的人类文明一样，成为了近代科学技术的重要源泉。

但我们也要清醒地看到，发生在17~18世纪欧洲的科学革命、18~19世纪的第一次和第二次工业革命，以及现代科学教育与知识的传播，加快了西方现代化的进程，同时也拉大了中国与西方的差距。

鸦片战争的失败给古老的中华帝国敲响了警钟。就在《自然》创刊前后，中国的一批精英分子看到了科学技术对于国家发展的重要性，一批批中国学子到西方高校及研究机构学习，其中一些人在科学领域作出了杰出的贡献。同时，一大批留学生回国，同国内的知识分子一道，为现代科学技术在中国的发展奠定了基础。

新中国成立60年来，中国在科学技术方面取得了一系列成就。在很短的时间里，独立自主地研制出“两弹一星”；在陆相生油理论指导下，发现了大庆油田；成功合成了牛胰岛素，这是世界上第一个通过化学方法人工合成的蛋白质；杂交水稻研发及其品种的普及，显著提高了水稻产量，造福了全世界几亿人。中国人在材料科学、航天、生命科学等许多领域，也取得了一批重要成果。这些都展现了中国在现代科技领域所取得的巨大进步。

当前，中国经济持续快速增长，科研产出日益增加，中国的许多重要成果已经发表在像《自然》这样的世界一流的科技期刊上，扩大了中国科学研究的影响，推动了中国科学家和国外同行的交流与合作。现在，中国已成为世界重要的科技力量。可以预见，在未来，中国将在科学和技术方面取得更大的进步。

伴随着现代科学产生的科技期刊，忠实地记录了科学发展的轨迹，在传播和促进现代科学的发展方面发挥了重要的作用。科技期刊及时地报道学术进展，交流科学思想和方法，探讨未来发展方向，以带动学术争鸣与繁荣，促进公众对科学的理解。中国在推动科技进步的同时，应更加重视科技期刊的发展，学习包括《自然》在内

to better understand science. While promoting science and technology, China should place greater emphasis on the betterment of science journals. We should draw on the philosophies and methods of leading science journals such as *Nature*, improve the standards of digital access, and enable some of our own science journals to extend their impact beyond China in the not too distant future so that they can serve as an advanced platform for the development of science and technology in our country.

In the 20th century, *Nature* published many remarkable discoveries in disciplines such as biology, geoscience, environmental science, materials science, and physics. The selection and publication of the best of the more than 100,000 articles in *Nature* over the past 150 years or so in English-Chinese bilingual format is a highly meaningful joint undertaking by the Foreign Language Teaching and Research Press, Macmillan Publishers Limited and the Nature Publishing Group. I believe that *Nature: The Living Record of Science* will help bridge cultural differences, promote international cooperation in science and technology, prove to be high-standard readings for its intended large audience, and play a positive role in improving scientific and technological research in our country. I fully endorse and support the project.

The volumes offer a picture of the course of science for nearly 150 years, from which we can explore how science develops, draw inspiration for new ideas and wisdom, and learn from the unremitting spirit of scientists in research. Reading these articles is like vicariously experiencing the great discoveries by scientific giants in the past, which will enable us to see wider, think deeper, work better, and aim higher. I believe this collection will also help interested readers from other walks of life to gain a better understanding of and care more about science, thus increasing their respect for and confidence in science.

I should like to take this opportunity to express my appreciation for the vision and joint efforts of Foreign Language Teaching and Research Press, Macmillan Publishers Limited and the Nature Publishing Group in bringing forth this monumental work, and my thanks to all the translators, reviewers and editors for their exertions in maintaining its high quality.

President of Chinese Academy of Sciences

Editor's note: The foreword was originally written for the ten-volume *Nature: The Living Record of Science.*

的世界先进科技期刊的办刊理念和方法，提高期刊的数字化水平，使中国的一些科技期刊早日具备世界影响力，为中国科学技术的发展创建高水平的平台。

20世纪的生物学、地球科学、环境科学、材料科学和物理学等领域的许多重大发现，都被记录在《自然》上。外语教学与研究出版社、麦克米伦出版集团和自然出版集团携手合作，从《自然》创刊近一百五十年来发表过的十万余篇论文中撷取精华，并译成中文，以双语的形式呈现，纂为《〈自然〉百年科学经典》丛书。我认为这是一项很有意义的工作，并相信本套丛书的出版将跨越不同的文化，促进国际间的科技交流，向广大中国读者提供高水平的科学技术知识文献，为提升我国科学技术研发水平发挥积极的作用。我赞成并积极支持此项工作。

丛书将带领我们回顾近一百五十年来科学的发展历程，从中探索科学发展的规律，寻求思想和智慧的启迪，感受科学家们百折不挠的钻研精神。阅读这套丛书，读者可以重温科学史上一些科学巨匠作出重大科学发现的历程，拓宽视野，拓展思路，提升科研能力，提高科学道德。我相信，这套丛书一定能成为社会各界的良师益友，增强他们对科学的了解与热情，加深他们对科学的尊重与信心。

借此机会向外语教学与研究出版社、麦克米伦出版集团、自然出版集团策划出版本丛书的眼光和魄力表示赞赏，对翻译者、审校者和编辑者为保证丛书质量付出的辛苦劳动表示感谢。

是为序。

路甬祥

中国科学院院长

编者注：此篇原为《〈自然〉百年科学经典》（十卷本）的序。

Foreword by Wu Xiangping

Progress in Astronomy, a selected collection of milestone astronomy papers spanning the last hundred years or so is the epitome of magnificent history of modern astronomy. Five score years ago, astronomers were pondering questions such as the nature of the nebulae and the energy mechanism of the Sun. One hundred years later, we have seen the remnants of the primordial fireball of the Big Bang, and captured black holes, the cage of everything in the universe. Indeed, the immense progress in astronomy over the past century, enabled by a truly remarkable series of scientific insights and achievements, has radically and fundamentally changed our view of the universe, and promoted the progress of human civilization.

Not surprisingly, *Progress in Astronomy* features many milestone works by Nobel Prize winners and seminal contributors to our discipline: Arthur Eddington, who first observationally confirmed the predicted deflection of light during the eclipse by General Relativity; George Gamow, who calculated the present temperature of the Big Bang fireball; Fred Hoyle, who first used the term "Big Bang" in a public speech; Riccardo Giacconi, who first discovered the cosmic X-ray sources; Martin Rees, who sits as a crossbencher in the House of Lords as Baron Rees of Ludlow in recognition of his remarkable contributions; Saul Perlmutter, who made the epoch-making discovery that the expansion of the universe is accelerating; Andrea Ghez, the first female astronomer to win the Nobel Prize, who confirmed the existence of a supermassive compact object at the centre of the Milky Way; and Stephen Hawking, famed for studies of black holes and his popular science book *A Brief History of Time*. These names and achievements have already gone down in the annals of scientific discovery.

While reading through *Progress in Astronomy*, I was delighted to see the paper which influenced my doctoral dissertation and also my research career for the following two decades. In 1979, Dennis Walsh and colleagues reported in *Nature* the discovery of the first double imaging of gravitational lensing, which provides a novel and powerful tool for probe of the universe. Seven years later, in 1986 when I started my research career in astrophysics, their paper shed new light on my doctoral research on *Gravitational Lensing in the Universe*. I might also mention that two papers published in *Nature* in 1993 claimed the first detection of dark matter in the Milky Way through micro-lensing effect of background stars. I might be the first to explain the phenomenon as the result of star-star lensing in the background galaxy, the Large Magellanic Cloud, which was confirmed by subsequent observations. Dark matter in our Galaxy and also in the Universe has still remained as a mystery today.

武向平序

跟随《〈自然〉学科经典系列——天文学的进程》跨越百年的论文，领略近代天文学史波澜壮阔的画卷，思绪万千，心潮起伏。一方面被先驱者苦苦求索天地万象的精神所折服，另一方面感叹人类面对浩瀚宇宙的那些敬畏和向往。百年前，天文学家还在拷问自己，“星云在哪里？”“太阳的能量从哪里来？”百年后，天文学家已经看到宇宙大爆炸的火球、捉捕到宇宙万物的牢笼——黑洞。的确，百年间天文学的巨大进步和辉煌成绩，改变了人类对大千世界的认知，推动了人类文明的进步。

浏览《天文学的进程》收录的论文，数位诺奖得主的大名赫然醒目，诸多耳濡目染的开拓者呼之欲出：第一个验证广义相对论预言的爱丁顿，预测了宇宙大爆炸火球遗迹温度的伽莫夫，第一个给宇宙“大爆炸”冠名的霍伊尔，第一个发现宇宙X射电源的贾科尼，以天文学研究成就获拉德罗男爵的马丁·里斯，发现宇宙加速膨胀的珀尔马特，以确定银河系中心存在超大质量致密天体（黑洞）首位获得诺奖的女天文学家盖兹，以及坐在轮椅上研究黑洞而妇孺皆知的霍金。他们的名字和他们的成就永远地铭刻在了科学发现的史册中，树立起一座座闪光和敬仰的丰碑。

欣喜地看到，收录在《天文学的进程》一篇论文曾启发了我博士论文的选题，进而影响了我随后二十年的科研生涯。1979年，瓦尔氏等人在《自然》宣布，第一次在宇宙中观测到了广义相对论预言的引力透镜双像，从此开启了利用引力透镜探索宇宙的新时代。七年后的1986年，当我初次迈进天文学的大门，盲无目标地寻找着天文学研究的切入点，《自然》杂志上首次发现引力透镜双像的文章为我点燃了求索的灯塔，引导我最终完成“宇宙中的引力透镜效应”之博士论文。还要特别提及的是，1993年《自然》同期发表的两篇论文宣称发现了银河系暗晕的微引力透镜现象，从而找到暗物质的证据，轰动一时。年轻气盛的我大概是第一个站出来的质疑者，并被随后的观测肯定。

In recent years, a large number of new discoveries and astonishing astronomical phenomena have emerged thanks to construction and operation of many ground- and space-based large-scale astronomical facilities, and *Nature* alone could hardly undertake the mission of reporting on each of these findings. Hence *Nature Astronomy* was launched in 2016, symbolizing that astronomy has entered into another golden age of development. I do hope that the publication of *Progress in Astronomy* can serve not only as the mirror of scientific history, but also as the source of inspiration for us to meet the continuing challenges of exploring the universe.

Wu Xiangping
Academician of CAS

近年间，随着世界各国投入巨资兴建诸多大型天文观测设备，《自然》杂志的主刊已经无法承载逐一报道大量天文学新发现的使命。2016年起，《自然·天文学》创刊，标志着天文学的研究已经迈入黄金发展时期。愿此次出版的《天文学的进程》在带给我们回溯历史的同时，启迪我们迎接探索宇宙的更大挑战，创造天文学研究的更大辉煌。

中科院院士

2022年4月5日

Contents
目录

1932

1933

1934

1935

1965

1966

1968

1969

1971

1972

Volume I

(1869-1972)

The Recent Total Eclipse of the Sun

J. N. Lockyer

Editor's Note

The first Editor of *Nature*, J. Norman Lockyer, had a broad interest in astronomy but particularly in the constitution of the Sun. His chosen technique was that of spectroscopy, by which means he aimed to identify the chemical constituents of the Sun. His distinctive achievement was to have identified spectroscopic lines corresponding to a then unknown atom which eventually turned out to be helium. In the year of *Nature*'s founding, he was concerned to gather evidence about the nature of the Sun's corona—the Sun's plasma "atmosphere", extending millions of kilometres beyond its surface, visible only during a total solar eclipse. His article on page 14 of the first issue of *Nature* advertises his belief that the Sun's corona is not a part of the sun at all but a phenomenon caused either by the atmosphere of the Earth or that of the Moon (which at the time was supposed to be capable of retaining an atmosphere similar to that of the Earth).

IF our American cousins in general hesitate to visit our little island, lest, as some of them have put it, they should fall over the edge; those more astronomically inclined may very fairly decline, on the ground that it is a spot where the sun steadily refuses to be eclipsed. This is the more tantalising, because the Americans have just observed their third eclipse this century, and already I have been invited to another, which will be visible in Colorado, four days' journey from Boston (I suppose I am right in reckoning from Boston?) on July 29, 1878.

Thanks to the accounts in *Silliman's Journal* and the *Philosophical Magazine*, and to the kindness of Professors Winlock and Morton, who have sent me some exquisite photographs, I have a sufficient idea of the observations of this third eclipse, which happened on the 7th of August last, to make me anxious to know very much more about them—an idea sufficient also, I think, to justify some remarks here on what we already know.

A few words are necessary to show the work that had to be done.

An eclipse of the sun, so beautiful and yet so terrible to the mass of mankind, is of especial value to the astronomer, because at such times the dark body of the moon, far outside our atmosphere, cuts off the sun's light from it, and round the place occupied by the moon and moon-eclipsed sun there is therefore none of the glare which we usually see—a glare caused by the reflection of the sun's light by our atmosphere. If, then, there were anything surrounding the sun ordinarily hidden from us by this glare, we ought to see it during eclipses.

近期的日全食

洛克耶

编者按

诺曼·洛克耶是《自然》的第一任主编，他对天文学的诸多领域都有浓厚的兴趣，尤其关注太阳的组成。他通过光谱技术研究太阳以确定太阳的化学组成。他最突出的成就是分辨出了一些谱线，这些谱线对应于一种当时人们还不知道的原子，这种原子后来被确认为氦。在《自然》创刊的这一年，他关心的是收集有关日冕（仅在日全食发生时可见的太阳的等离子体“大气”，它从太阳的表面向上延伸达数百万公里）本质的证据。在这篇发表于《自然》第1期第14页的文章中，他提出了这样的观点：日冕并不是太阳自身的一部分，而是由地球大气或月球大气（当时人们认为月球可以像地球一样保持住一个大气层）引起的一种特殊现象。

如果说我们的美国表兄弟因为担心会从边缘掉到海里（就像他们中的某些人说的那样）而不愿意造访我们这座小岛，那么那些对天文学有强烈兴趣的人可能就完全不愿意来了，因为在我们不列颠岛这样的小地方基本上就看不到日食。这实在是很让人干着急的事情，美国人刚刚已经观测了他们本世纪的第3次日食，而我们在英格兰还一次都没有看到。好在我已经被邀请去观察另一次日食了，这次日食将于1878年7月29日在美国科罗拉多州发生，那里距离波士顿有4天的路程（我想我对从波士顿到那里的行程估算应该是正确的吧？）。

多亏《斯灵曼杂志》和《哲学杂志》的报道，以及温洛克教授和莫顿教授慷慨相送的一些精美照片，我对去年8月7日发生的第3次日食的观测结果已经有了比较充分的了解，这使我渴望知道更多观测信息；此外，我觉得我也有足够的准备在这里评述我们已经知道的一些说法。

首先有必要用一点篇幅来介绍一下已经做过的工作。

日食是一种非常美丽但也令广大民众感到十分恐慌的现象，然而它对于天文学家有着特殊的价值，因为在这些时候，不能发光而又远离地球大气层的月球切断了来自太阳的光线，从而在月球周围以及被月亮蚀掉的太阳的周围都不再有我们平常能看到的那种由于地球大气层对太阳光的反射而产生的炫目的光芒。所以，太阳周围如果有什么东西在平时会因为这种光芒而被掩盖的话，那么在日食的时候我们就应该能够看到。

In point of fact, strange things are seen. There is a strange halo of pearly light visible, called the corona, and there are strange red things, which have been called red flames or red prominences, visible nearer the edge of the moon—or of the sun which lies behind it.

Now, although we might, as I have pointed out, have these things revealed to us during eclipses if they belonged to the sun, it does not follow that they belong to the sun because we see them. Halley, a century and a half ago, was, I believe, the first person to insist that they were appearances due to the moon's atmosphere, and it is only within the last decade that modern science has shown to everybody's satisfaction—by photographing them, and showing that they were eclipsed as the sun was eclipsed, and did not travel with the moon—that the red prominences really do belong to the sun.

The evidence, with regard to the corona, was not quite so clear; but I do not think I shall be contradicted when I say, that prior to the Indian eclipse last year the general notion was that the corona was nothing more nor less than the atmosphere of the sun, and that the prominences were things floating in that atmosphere.

While astronomers had thus been slowly feeling their way, the labours of Wollaston, Herschel, Fox Talbot, Wheatstone, Kirchhoff, and Bunsen, were providing them with an instrument of tremendous power, which was to expand their knowledge with a suddenness almost startling, and give them previously undreamt-of powers of research. I allude to the spectroscope, which was first successfully used to examine the red flames during the eclipse of last year. That the red flames were composed of hydrogen, and that the spectroscope enabled us to study them day by day, were facts acquired to science independently by two observers many thousand miles apart.

The red flames were "settled", then, to a certain extent; but what about the corona?

After I had been at work for some time on the new method of observing the red flames, and after Dr. Frankland and myself had very carefully studied the hydrogen spectrum under previously untried conditions, we came to the conclusion that the spectroscopic evidence brought forward, both in the observatory and in the laboratory, was against any such extensive atmosphere as the corona had been imagined to indicate; and we communicated our conclusion to the Royal Society. Since that time, I confess, the conviction that the corona is nothing else than an effect due to the passage of sunlight through our own atmosphere near the moon's place has been growing stronger and stronger; but there was always this consideration to be borne in mind, namely, that as the spectroscopic evidence depends mainly upon the brilliancy of the lines, that evidence was in a certain sense negative only, as the glare might defeat the spectroscope with an un-eclipsed sun in the coronal regions, where the temperature and pressure are lower than in the red-flame region.

The great point to be settled then, in America, was, what is the corona? And there were many less ones. For instance, by sweeping round the sun with the spectroscope, both before

事实上，人们的确发现了一些奇怪的事情。在月球的边缘处，也可能是在月球后面的太阳的边缘处，可以看到有一个珍珠色的发光晕和一些奇特的红色物质，这个发光晕被称作日冕，而那些红色物质则被称作红色火焰或者日珥。

就像我已经说过的，如果这些东西属于太阳，那么在日食时我们就有可能看到它们，但并不能因为我们能看到它们而得出它们属于太阳的结论。我相信，一个半世纪以前的哈雷是坚持认为这些现象都是由月球大气造成的第一人。而直到近十年，现代科学才给出了令人满意的答案：日珥确确实实属于太阳。这是因为，通过摄影技术，人们发现，在日食发生时日珥也被蚀去了，它并不随着月球的移动而移动。

虽然关于日冕的证据还不是十分清楚，但我想应该不会有人质疑下面这个说法：在去年的印度日食之前，人们一般认为日冕就是太阳的大气，而日珥则是在这个大气上漂浮着的物质。

在天文学家缓慢地摸索前行之时，沃拉斯顿、赫歇尔、福克斯·塔尔博特、惠斯通、基尔霍夫和本生所做的工作给他们提供了一个强有力的仪器，它以惊人的速度拓宽天文学家的知识并给他们以前难以想象的研究能力，这就是光谱仪。在去年日食期间，天文学家第一次成功地将光谱仪用在对日珥的研究上。两个相隔数千英里的观测者独立得到了这样的科学事实：日珥是由氢组成的；我们每天都可以利用光谱仪对日珥进行研究。

因此，从某种程度上来说，日珥的问题已经“得到解决”了，但是日冕又如何呢？

我用观测日珥的这种新方法工作了一段时间，另外我还和弗兰克兰在各种以前未曾尝试过的条件下对氢的光谱进行了非常仔细的研究，在这之后，我们得出了这样的结论：尽管此前人们总是设想存在日冕这样一层非常厚的大气，但是我们在实验室和天文台得到的光谱学证据都反对这一点。我们也向皇家学会提交了这一结论。从那时开始，我越来越确信日冕只是太阳光在通过靠近月球一侧的地球大气层时产生的一种效应。不过也有一个问题一直困扰着我：既然光谱学证据主要依赖于谱线的亮度，那么我们的证据就只是在一定程度上否定了以前的设想，因为强烈的太阳光可能会影响光谱仪对没有发生日蚀的日冕区（温度和压强都低于日珥区）的检测。

对于下一次美国的日食，需要解决的最大问题是，什么是日冕？除此之外，还有许多比较小的问题。比如，通过在日食前和日食后用光谱仪绕太阳扫描一周，或

and after the eclipse, and observing the prominences with the telescope merely during the eclipse, we should get a sort of key to the strange cypher band called the spectrum, which might prove of inestimable value, not only in the future, but in a proper understanding of all the telescopic observations of the past. We should, in fact, be thus able to translate the language of the spectroscope. Again, by observing the spectrum of the same prominence both before and during, or during and after the eclipse, the effect of the glare on the visibility of the lines could be determined—but I confess I should not like to be the observer charged with such a task.

What, then, is the evidence furnished by the American observers on the nature of the corona? It is *bizarre* and puzzling to the last degree! The most definite statement on the subject is, that it is nothing more nor less than a *permanent solar aurora*! the announcement being founded on the fact, that three bright lines remained visible after the image of a prominence had been moved away from the slit, and that one (if not all) of these lines is coincident with a line (or lines) noticed in the spectrum of the aurora borealis by Professor Winloch.

Now it so happens that among the lines which I have observed up to the present time—some forty in number—this line is among those which I have most frequently recorded: it is, in fact, the first iron line which makes its appearance in the part of the spectrum I generally study when the iron vapour is thrown into the chromosphere. Hence I think that I should always see it if the corona were a permanent solar aurora, and gave out this as its brightest line; and on this ground alone I should hesitate to regard the question as settled, were the new hypothesis less startling than it is. The position of the line is approximately shown in the woodcut (Fig. 1) near E, together with the other lines more frequently seen.

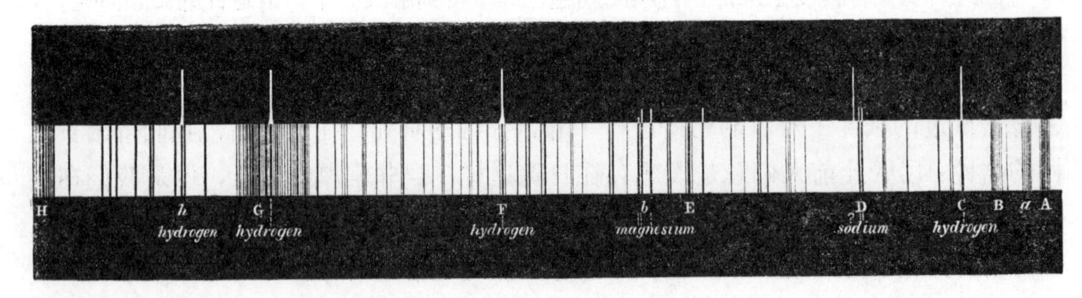

Fig. 1. Showing the solar spectrum, with the principal Fraunhofer lines, and above it the bright-line spectrum of a prominence containing magnesium, sodium, and iron vapour at its base.

It is only fair, however, to Professor Young, to whom is due this important observation, to add that Professor Harkness also declares for one bright line in the spectrum of the corona, but at the same time he, Professor Pickering, and indeed others, state its spectrum to be also continuous, a remark hard to understand unless we suppose the slit to have been wide, and the light faint, in either of which cases final conclusions can hardly be drawn either way.

者只是在日食时用望远镜来观测日珥，我们都可以得到一类解开太阳的奇怪光谱带的办法。应该说，这对于正确地理解过去用望远镜观测得到的结果和将来进行的研究都具有不可估量的价值。事实上，这样我们应该就能够翻译光谱仪的语言。另外，通过在日食前和日食中，或者日食中和日食后观测同一日珥的光谱，我们可能就可以确定强烈的太阳光对谱线可见性的影响。不过，我觉得自己并不是能够胜任这一工作的观测者。

那么美国的观测者们关于日冕的本质提供了什么样的证据呢？实际情况真是极其**荒诞**而又莫名其妙！关于这个问题最明确的说法是：日冕是一种**永久性的太阳极光**！得出这个结论的事实基础是：将来自日珥的光线从狭缝中移走后，仍然能看到3条明亮的谱线，而且其中一条谱线（如果不是全部的话）与温洛克教授在北极光的光谱中观察到的一条（或几条）谱线相吻合。

现在看来，在迄今为止我观察过的所有谱线（大约40条）中，这条谱线是我最常记录到的谱线之一。实际上，这是我通常研究的部分光谱中的第一条铁线，它是在铁蒸气被抛入色球时产生的。因此我认为，如果日冕是永久性的太阳极光，并且这条铁线是它发出的最亮谱线的话，那么我应该总能够看到这条谱线。如果不是后来出现了更令人惊奇的新假说，仅仅依靠这一点我想我是不会认为这个问题已经得到解决的。这条铁线的位置大约在E点附近的刻线（图1）处，图中还显示了其他一些更为常见的谱线。

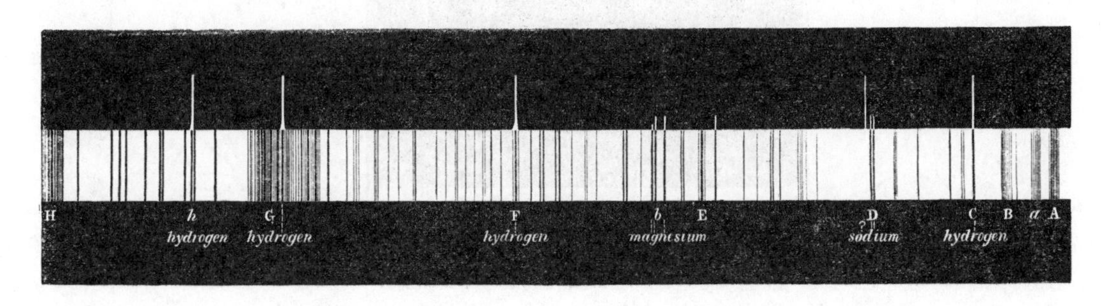

图1. 包括主要的夫琅和费谱线的太阳光谱，其上方是日珥所包含的镁和钠的亮线光谱，其底部则是铁蒸气的光谱。

另外还有一点，不过应该由发现上述重要观测结果的扬教授来补充才是公平的，这就是：哈克尼斯教授也曾宣称在日冕的光谱中发现了一条明亮的谱线，但与此同时，他、皮克林教授以及其他一些人又声称，日冕的光谱也是连续的。这实在是一个令人难以理解的论断，除非我们假设狭缝比较宽或者光线比较微弱，但无论是哪种情况，我们都无法得出上述的最终结论。

So much, then, for the spectroscopic evidence with which we are at present acquainted on the most important point. The results of the other attacks on the same point are equally curious and perplexing. Formerly, a favourite argument has been that because the light of the corona is polarised; therefore it is solar. The American observers state that the light is *not* polarised—a conclusion, as M. Faye has well put it, "very embarrassing for Science". Further,—stranger still if possible,—it is stated that another line of inquiry goes to show that, after all, Halley may be right, and that the corona may really be due to a lunar atmosphere.

I think I have said enough to show that the question of the corona is by no means settled, and that the new method has by no means superseded the necessity of carefully studying eclipses; in fact, their observation has become of much greater importance than before; and I earnestly hope that all future eclipses in the civilised area in the old world will be observed with as great earnestness as the last one was in the new. Certainly, never before was an eclipsed sun so thoroughly tortured with all the instruments of Science. Several hundred photographs were taken, with a perfection of finish which may be gathered from the accompanying reproduction of one of them.

Fig. 2. Copy of a photograph of the Eclipse of August 7, obtained by Professor Mortor's party

The Government, the Railway and other companies, and private persons threw themselves into the work with marvellous earnestness and skill; and the result was that the line of totality was almost one continuous observatory, from the Pacific to the Atlantic. We read in *Silliman's Journal*, "There seems to have been scarcely a town of any considerable magnitude along the entire line, which was not garrisoned by observers, having some special astronomical problem in view." This was as it should have been, and the American Government and men of science must be congratulated on the noble example they have shown to us, and the food for future thought and work they have accumulated.

以上就是我们目前在最重要的问题上获得的光谱学证据。关于这一问题的其他一些结果也同样古怪而又令人困惑。之前有个广受欢迎的说法认为：日冕的光是偏振的，所以它就是太阳光。但美国的观测者们声称这些光**不是**偏振的。就像费伊所说，这实在是一个“让科学界异常尴尬”的结论。另外更奇怪的是（如果是真的话，就更奇怪了），据称有调查表明可能哈雷终究是对的，日冕可能真的是由月球的大气产生的。

我想我说了这么多已经足以表明，日冕的问题还没有得到解决，并且新方法也完全没有取代对日食进行仔细观测的必要性。事实上，天文学家的观测变得比以前更加重要。我非常真诚地希望，对于此后将在旧大陆的文明地区发生的所有日食，我们也能像对最近这次发生在新大陆上的日食一样进行非常认真的观测。无疑，以前人们从来没有像这次这样用所有的科学仪器来彻底地拷问日食。在这次日食中，人们拍摄了几百张照片，我复制了其中一张附在文中，从这张图可以看到拍照工作做得十分完美。

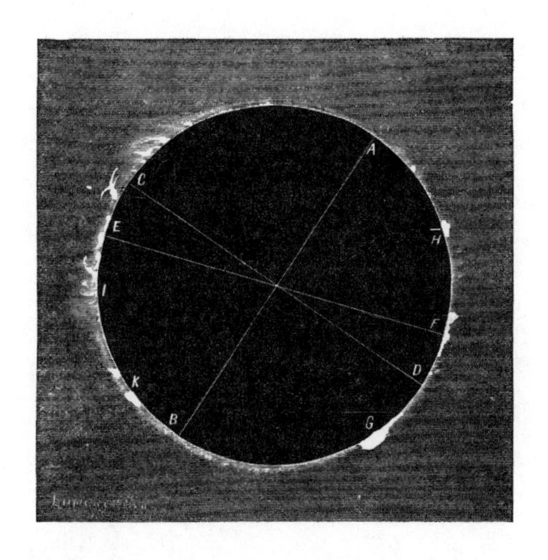

图 2. 莫顿教授的观测队拍摄的一张照片（8月7日的日食）的拷贝

政府、铁路和其他公司，以及许多个人都以极大的热情和娴熟的技巧投入到了这项工作中。结果是从太平洋到大西洋的整个观测线几乎成了一个连续的天文台。我们在《斯灵曼杂志》上读到，“在整条观测线上，几乎所有有点名头的城镇中都驻扎着一些正在思索某些特殊天文学问题的观测者。”这一切好像本来就该如此。美国政府和科学工作者们给我们树立了一个光辉的榜样，也积累了一些有助于进一步研究的证据和思路，应该为此而向他们致敬。

* * *

Since writing the above, I find the following independent testimony in favour of Dr. Frankland's and my own notion of the corona in the *Astronomische Nachrichten*, from the pen of Dr. Gould. He says:—"Its form varied continually, and I obtained drawings for three epochs at intervals of one minute. It was very irregular in form, and in no apparent relation with the protuberances on the sun, or the position of the moon. Indeed, there were many phenomena which would almost lead to the belief that it was an atmospheric rather than a cosmical phenomenon. One of the beams was at least 30′ long."

(**1**, 14-15; 1869)

*　　*　　*

写完以上这些后，我在《天文通报》上看到了谷德博士独立发现的一些能够支持弗兰克兰博士以及我自己关于日冕观点的证据。谷德博士称："它的形状在不断变化，我得到了间隔一分钟的 3 个不同时刻的图像。可以看到，它的形状很不规则，与太阳表面的突起或月球上的某些位置都没有明显的联系。事实上，有很多现象都能够推出这是一种大气现象而非宇宙现象的结论。其中一条光束至少有 30 角分长。"

（汪浩 翻译；蒋世仰 审稿）

Spectroscopic Observations of the Sun

Editor's Note

This paper reports some of the first detailed observations of solar flares, then called "solar protuberances". These had been discovered just ten years before, and were associated with sunspots. American astronomer C. A. Young describes an immense cloudy mass on the Sun's edge, which he estimates to be 22,500 miles high (three times the Earth's diameter) and 1,350,000 miles wide. His spectroscopic observations of the feature show it to be predominantly hydrogen and provide information about velocities of this material. The Doppler shift of the hydrogen emission line implies that the material was moving at roughly 55 miles per second. The feature persisted for five minutes before dissolving into several lumps.

PROFESSOR C. A. Young, of Dartmouth, U.S., has communicated to the October number of the *Journal of the Franklin Institute* the following important observations of solar protuberances, which entirely endorse the work done by Mr. Lockyer in this country. We are enabled to place them thus early before our readers by the kindness of Professor Morton.

September 4th, 1869.—Prominences were noted on the sun's limb at 3 p.m. today in the following positions, angles reckoned from North point to the East:—

1. +70° to +100°, very straggling, not very bright.

2. −10°, large and diffuse.

3. −90°, small, but pretty bright.

September 13th, 1869.—The following protuberances were noted today.

1. Between +80° and +110°, a long straggling range of protuberances, whose form was as in Fig. 1. I dare not profess any very extreme accuracy in the drawings, not being a practised draughtsman, but the sketch gives a very fair idea of the number, form, and arrangement of the immense cloudy mass, whose height was about 50″ and its length 330″ (22,500 miles by 1,350,000). The points *a* and *b* were very bright.

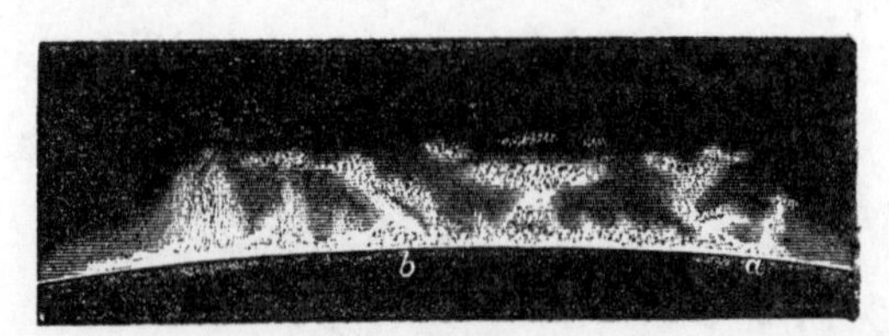

Fig. 1

观测太阳光谱

编者按

这篇文章报道了一些关于太阳耀斑的最早的详细研究。这种当时还被称为“太阳突出物”的现象是10年前才发现的，它与太阳黑子相关联。美国天文学家扬在本文中描述了在太阳边缘出现的巨大云块状物质，他估计这些物质高达22,500英里（地球直径的3倍），宽达1,350,000英里。他对这一现象进行的光谱学研究表明，该物质中的主要成分应该是氢，他还给出了这种物质运动速度的信息。氢发射谱线的多普勒频移意味着，这种物质移动的速度大约是每秒55英里。这种特征持续存在5分钟后，这个巨大云块状物质逐渐分解成了几个团块。

来自美国达特茅斯大学的扬教授已经在10月份的《富兰克林研究院院刊》上发表了关于太阳突出物的一些重要观测结果，这些重要发现是对本国的洛克耶先生所做工作的充分肯定。多亏莫顿教授的热心相助，我们才可以早早地把这些工作呈现给读者。

1869年9月4日，当天下午3点，我们在太阳边缘的下列位置发现了日珥，角度由北向东，分别是：

1. +70° ~ +100°，非常发散并且不太明亮。

2. –10°，大且弥散。

3. –90°，小但是相当明亮。

1869年9月13日，当天发现了下面的日珥：

1. +80°~+110°，有一长条发散的日珥，形状如图1所示。由于不是经验丰富的绘图员，我不敢说图中的描绘绝对精确，但从草图中完全可以分辨出巨大云块的数量、形状和排列。这个不透明的巨大云块高约50角秒，长330角秒（22,500英里 × 1,350,000英里），并且 a 点和 b 点非常明亮。

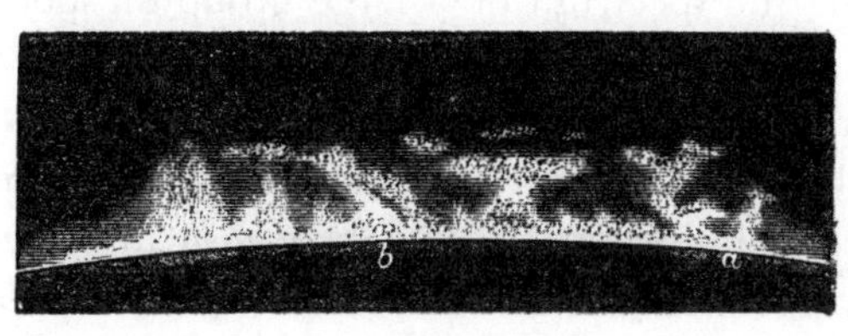

图1

2. +135° small, but very bright at the base, of this form (Fig. 2).

Fig. 2

3. −85° of this form (Fig. 3).

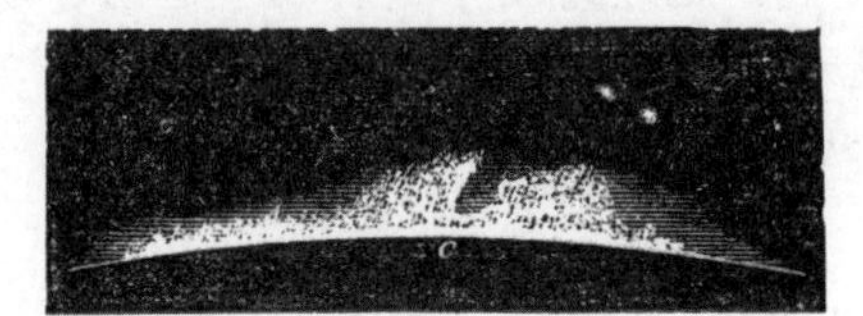

Fig. 3

The dark spot, marked *c*, was very curious, reminding one strongly of the so-called fish-mouth in the nebula of Orion. I saw no change in it for 20 minutes. On the other hand, the first series mentioned were changing rapidly, so that at five o'clock the sketch which was drawn at two was quite inapplicable, only the general features remaining unaltered.

4. −128°, about 20″ high, forked, as in Fig. 4.

Fig. 4

The structure was *cirrus* in every one but No.3, which seemed more like a mass of cumulus.

Today, for the first time, I saw b_1 reversed in the chromosphere when the slit was tangent to disc; 1,474 was easy; the new line at 2,602 cannot be detected as yet.

At 2.25, while examining the spectrum of a large group of spots near the sun's western limb, my attention was drawn to a peculiar double *knobbishness* of the F line (on the sun's disc, not at the edge), represented by Fig. 5, *a*, at the point *e*. In a very few moments a brilliant spot replaced the knobs, not merely interrupting and reversing the dark line, but blazing like a star near the horizon, only with blue instead of red light; it remained for about two minutes, disappearing, unfortunately, while I was examining the sun's image

2. +135°，小，但底部非常明亮，形状如图 2 所示。

图 2

3. –85°，形状如图 3 所示。

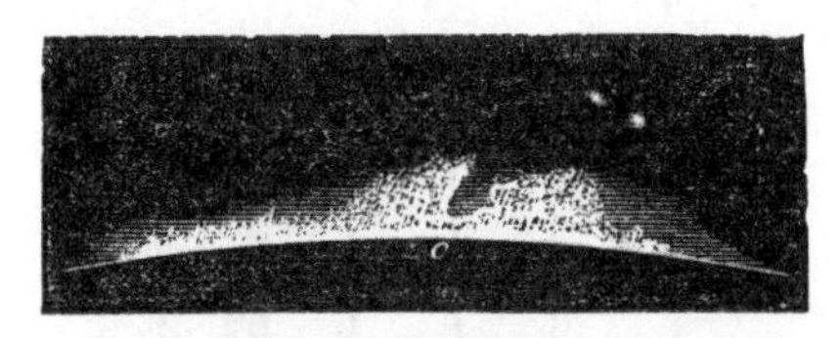

图 3

标注为 c 的暗黑子非常奇妙，很容易让人想起猎户座大星云中被称作鱼嘴的形状。我观察了 20 分钟，没有看到什么变化。另一方面，前述一系列现象却变化得非常迅速，所以在 2 点钟画的草图到 5 点钟时已经不适用了，但总的特征仍然没有改变。

4. –128°，高约 20 角秒，叉状，如图 4 所示。

图 4

除了第 3 个外，其他日珥的结构都是“卷云”，而第 3 个则更像是一块积云。

今天是我第一次看到当狭缝与日面相切时 b_1 在色层球中发生反转，检测 1,474 线相对容易，而在 2,602 的新线到现在还不能被检测出来。

在 2 点 25 分时，当观测靠近太阳西部边缘附近的一大片黑子的光谱时，我注意到了在 F 线处（在日面上，而不是在边缘）有一个奇特的双**球柄结构**，由图 5 中 a 上的 e 点表示。没过多久，一个明亮的亮点取代了球柄结构的突出物，不但打断和反转了那条暗线，而且亮得像一颗在地平线附近闪耀的恒星，但是闪着蓝色而不是

upon the graduated screen at the slit, in order to fix its position, which was at −82.5, about 43″ from the edge of the limb, about 15″ inside of the inner edge of the spot-cluster. I do not know, therefore, whether it disappeared instantaneously or gradually, but presume the latter. Fig. 5, *b*, attempts to give an idea of the appearance. When I returned to the eye-piece, I saw what is represented at Fig. 5, *c*, &c. On the upper (more refrangible) edge of F there seemed to hang a little black mote, making a *barb*, whose point reached nearly to the faint iron line just above F. As given on Ångström's atlas, the wavelength of F is 486.07, while that of the iron line referred to is 485.92 (the units being millionths of a millimetre). This shows an absolute change of 0.15 in the wavelength, or a fraction of its whole amount, represented by the decimal 0.00030, and would indicate an advancing velocity of about 55.5 miles per second in the mass of hydrogen whose absorption produced this barbed displacement.

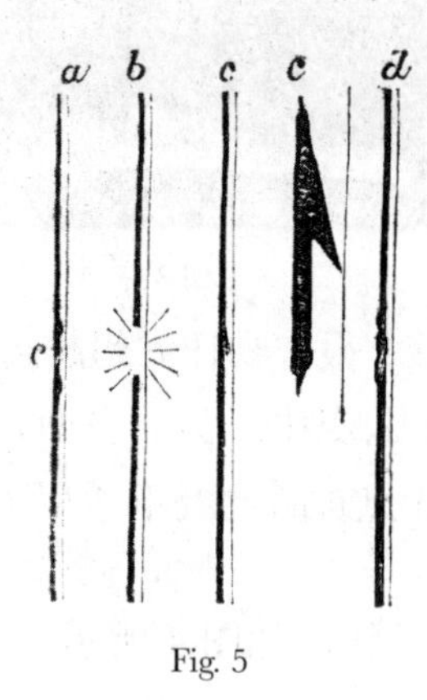

Fig. 5

The barb continued visible for about five minutes, gradually resolving itself into three small lumps, one on the upper, and two on the lower line, Fig. 5, *d*. In about ten minutes more, the F line resumed its usual appearance. I did not examine the C line, as I did not wish to disturb the adjustments and risk losing some of the curious changes going on under my eye.

After the close of this strange phenomenon, I examined, with our large telescope of 6-inch aperture, the neighbourhood in which this took place, and found a very small spot exceedingly close to, if not actually *at*, the place. This was at 2.45. At 5.30 it had grown considerably.

Undoubtedly, the phenomenon seen was the same referred to by Mr. Lockyer when he speaks of often seeing the bright lines of the prominences not only at the sun's limb but on his disc. It is the only time I have had the good fortune to see it as yet.

(**1**, 172-173; 1869)

红色的光芒。这种现象只持续了约 2 分钟就消失了，遗憾的是当时我正在查看狭缝处刻度屏上的太阳的像，想把它的位置固定住，这时位置是 –82.5，大约离突出物的边缘约 43 角秒，在黑子云内缘内部约 15 角秒处。因此我不知道它到底是瞬间消失的还是逐渐消失的，但我认为是后者。图 5 中的 *b* 试图给出这一现象的概貌。当我再次用目镜观察的时候，我看到了图 5 中的 *c* 等所示的图像。在 F 折射较大的上端，似乎挂着一小块黑的微粒，构成一个**鱼钩状**，钩的顶端接近 F 上方那条微弱的铁谱线。在波长表中查到，F 的波长为 486.07，而铁谱线的波长为 485.92（单位是百万分之一毫米）。这说明波长的确改变了 0.15，或者用变化量占原始值的比例表示，即小数 0.00030，这也表明其吸收谱线产生鱼钩状位移的这部分氢的速度提高到了每秒约 55.5 英里。

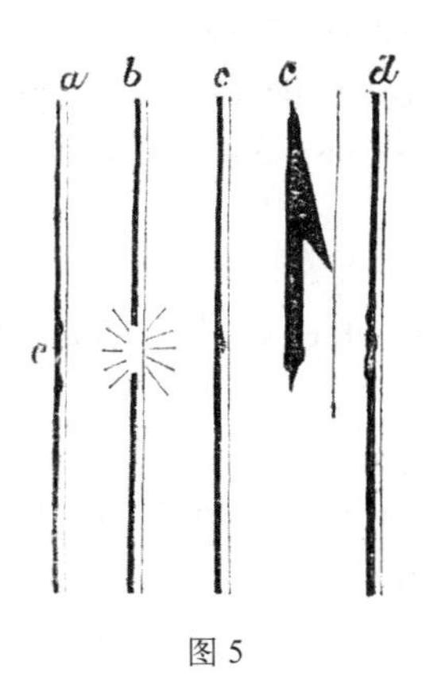

图 5

这个鱼钩持续可见约 5 分钟，之后逐渐分为 3 个小块，一块在上面，另两块相对靠下，如图 5 的 *d* 所示。大约 10 分钟之后，F 线恢复到它之前的样子。我没有检查 C 线，因为我不想冒着失掉正发生在眼皮底下的一些奇特变化的风险去改变调好的位置。

在这个奇特的现象结束之后，我用直径为 6 英寸的大望远镜观测了此现象发生处的周边，发现了一个非常小的点，即使不是恰好**在此现象发生处**，也是在离得特别近的位置。这是发生在 2 点 45 分的事，到 5 点 30 分它已经变得相当大了。

毫无疑问，上述观测到的现象与洛克耶先生提及的不仅在太阳边缘，而且在日面上都能经常观测到的日珥的亮线是一回事。但到目前为止，只有这一次我有幸看到了这种现象。

（刘东亮 翻译；蒋世仰 审稿）

Spectroscopic Observations of the Sun

Editor's Note

This editorial reports on several papers read recently at the Royal Society, pertaining to studies probing the constitution and dynamics of the Sun with the new technique of spectroscopy. *Nature*'s founder and editor J. Norman Lockyer notes that velocities of around 100 miles per second had been detected for matter in the chromosphere. He and others argued that the photosphere and chromosphere together constitute the atmosphere of the Sun, contrary to some alternative proposals. The papers discuss detailed spectroscopic clues about structure and composition obtained from a range of elements, and reflect the great and sudden observational progress on a topic not long before considered inaccessible to empirical science.

ROYAL Society, December 9.—Dr. W. A. Miller, V.P., in the chair. The following papers were read: —

"Spectroscopic Observation of the Sun" —No. V. By J. Norman Lockyer, F.R.S.

The author first referred to several new facts of importance as follows:

"I. The extreme rates of movement in the chromosphere observed up to the present time are—

Vertical movement	40 miles a second
Horizontal or cyclonic movement	120 miles a second

"II. I have carefully observed the chromosphere when spots have been near the limb. The spots have sometimes been accompanied by prominences, at other times they have not been so accompanied. Such observations show that we may have spots visible without prominences in the same region, and a prominences without spots; but I do not say that a spot is not accompanied by a prominence, *at some stage of its life*, or that it does not result from some action which, in the majority of cases, is accompanied by a prominence.

"III. At times, when a prominence is seen bright on the sun itself, the bright F line varies considerably, both in thickness and brilliancy, within the thickness of the dark line. The appearances presented are exactly as if we were looking at the prominences through a grating.

"IV. Bright prominences, when seen above spots on the disk, if built up of other substances

观测太阳光谱

编者按

这篇评论报道了最近在皇家学会宣读的几篇论文，这些论文涉及的都是人们利用新的光谱技术对太阳组成及其动力学进行的探索研究。《自然》的创始人及主编诺曼·洛克耶指出，探测结果表明色球层中物质的运动速度大约是每秒100英里。他和其他一些科学家都认为光球层和色球层一起组成了太阳的大气层，这与另外一些说法相悖。这些文章讨论了从一系列元素得出的一些关于太阳结构和组成的详细光谱线索，同时也反映了在不久之前还被认为是无法通过实证科学解决的问题在最近已经取得的重大而迅速的观测进展。

皇家学会，12月9日——副会长米勒博士主持会议，会议中宣读了下面的文章：

《观测太阳光谱》——第5篇，作者是诺曼·洛克耶（皇家学会会员）。

作者第一次提到了如下一些重要的新发现：

“I. 到目前为止观测到的色球层中物质运动的最大速率是：

纵向运动	每秒40英里
横向运动或回旋运动	每秒120英里

“II. 当黑子出现在日面边缘时，我很仔细地观测了色球层。黑子有时候与日珥相伴出现，另一些时候又不是。这项观测表明，在同一区域内我们可能只看到黑子而没有看到日珥，或者只看到日珥而没有看到黑子。但是我并没有说黑子**在它生命周期中的某个阶段**不会伴随着日珥的出现，也不意味着，黑子的产生不是由某个在大多数情况下与日珥相伴的过程造成的。

“III. 有时候，当在太阳上看到一个明亮的日珥时，明亮的F线将在宽度和色度上发生显著的变化，但其宽度不会超过暗线宽度。其呈现出的恰似我们透过光栅观察日珥时的表现。

“IV. 在日面上看到黑子群时出现的明亮日珥，如果包括除氢以外的其他物质，

besides hydrogen, are indicated by the bright lines of those substances in addition to the lines of hydrogen. The bright lines are then seen very thin, situated centrally (or nearly so) on the broad absorption-bands caused by the underlying less-luminous vapours of the same substances.

"V. I have at last detected an absorption-line corresponding to the orange line in the chromosphere. Father Secchi states* that there is a line corresponding to it much brighter than the rest of the spectrum. My observation would seem to indicate that he has observed a bright line less refrangible than the one in question, which bright line is at times excessively brilliant. It requires absolutely perfect atmospheric conditions to see it in the ordinary solar spectrum. It is best seen in a spot-spectrum when the spot is partially covered by a bright prominence.

"VI. In the neighbourhood of spots the F bright line is sometimes observed considerably widened out in several places, as if the spectroscope were analysing injections of hydrogen at great pressure in very limited regions into the chromosphere.

"VII. The brilliancy of the bright lines visible in the ordinary solar spectrum is extremely variable. One of them, at 1871.5, and another, at 1529.5 of Kirchhoff's scale, I have detected in the chromosphere at the same time that they were brilliant in the ordinary solar spectrum.

"VIII. Alterations of wavelength have been detected in the sodium-, magnesium-, and iron-lines in a spot-spectrum. In the case of the last substance, the lines in which the alteration was detected were *not* those observed when iron (if we accept them to be due to iron alone) is injected into the chromosphere.

"IX. When the chromosphere is observed with a tangential slit, the F bright line close to the sun's limb shows traces of absorption, which gradually diminish as the higher strata of the chromosphere are brought on to the slit, until the absorption-line finally thins out and entirely disappears. The lines of other substances thus observed do not show this absorption.

"X. During the most recent observations, I have been able to detect traces of magnesium and iron in nearly all solar latitudes in the chromosphere. If this be not merely the result of the good definition lately, it would indicate an increased general photospheric disturbance as the maximum sunspot period is approached. Moreover, I suspect that the chromosphere has lost somewhat of its height."

The author appends a list of the bright lines, the position of which in the chromosphere have been determined absolutely, with the dates of discovery, remarking that in the case of C and F his observations were anticipated by M. Janssen:—

* *Comptes Rendus*, 1869, I sem, p. 358.

则谱线中除了氢线之外还应当有其他物质产生的亮线。这时看到的亮线都非常细，位于宽吸收谱带区域的中心（或接近中心），这是由该物质底层一些低亮度的蒸气发射出来的。

“V. 我终于发现了一条与色球层上橘黄色谱线相对应的吸收谱线。塞奇神父说过*，有一条比光谱中其他谱线亮很多的谱线与之相对应。我的观测似乎表明，他观测到的那条亮线，折射率比有争议的那条谱线小，这就是为什么亮线有时候会显得格外明亮的原因。必须在非常好的大气条件下才能在普通的太阳光谱中看到它，尤其是当一个黑子被亮的日珥遮挡住一部分的时候，在黑子光谱中看得更清楚。

“VI. 在黑子周围观察到 F 亮线有时候在一些地方明显加宽，就好像分析氢气在很大压力下注入色球层上很小的区域时的光谱一样。

“VII. 可视亮线的亮度在普通太阳光谱中的变化非常大。在普通太阳光谱中有两条明亮的谱线，我在色球层中也同时观察到了，按照基尔霍夫标度，其中一条在 1871.5，另一条在 1529.5。

“VIII. 在一个黑子光谱中，我发现钠线、镁线和铁线的波长发生了改变。对于铁线，检测到波长发生改变的谱线并**不是**当铁（如果我们认为它们就是单一物质铁发出的）射入色球层时观测到的谱线。

“IX. 当用一个切向夹缝观测色球层时，在日面边缘附近的 F 亮线显示出吸收的迹象，而当夹缝对准色球层的较高位置时，F 亮线逐渐减弱，直到吸收线越来越窄，然后全部消失。其他物质的吸收谱线并没有表现出这种现象。

“X. 在最近的观测活动中，我在太阳色球层的几乎所有纬度都检测到了少量的镁和铁。如果这不仅仅是因为最近透镜清晰度较好的缘故，那就说明当太阳黑子极大期到来的时候，光球层受到了更大的扰动。另外，我猜测色球层的高度也有所降低。”

作者附了一个亮线的明细表，这些亮线在色球层里的位置已完全确定，其中还列出了发现它们的日期，并指出让森曾预测到了他观测到的 C 线和 F 线：

* 《法国科学院院刊》，1869年，第1期，第358页。

Hydrogen		
C.	October 20, 1868	
F.	October 20, 1868	
near D.	October 20, 1868*	
[* *Hydrogen?*—G. G. S.]		
near G.	December 22, 1868	
h.	March 14, 1869	
Sodium		
D.	February 28, 1869	
Barium		
1989.5†	March 14, 1869	
2031.2	July 5, 1869	
Magnesium and included line		
b^1, b^2, b^3, b^4	February 21, 1869	
Other Lines		
Iron	1,474	June 6, 1869
?	1515.5	June 6, 1869
Bright line	1529.5	July 5, 1869
?	1567.5	March 6, 1869
?	1613.8	June 6
Iron	1867.0	June 26
Bright line	1871.5	,,
Iron	2001.5	,,
?	2003.4	,,
? band or line near black line, very delicate ...	2054.0	July 5

Other lines besides these have been seen at different times; but their positions have not been determined absolutely.

The author points out that taking iron as an instance, and assuming that the iron-lines mapped by Ångström and Kirchhoff are due to iron only, he has only been able, up to the present time, to detect three lines out of the total number (460) in the spectrum of the lower regions of the chromosphere,—a fact full of promise as regards the possible results of future laboratory work. The same remark applies to magnesium and barium.

The paper then proceeded as follows:—

* *Comptes Rendus*, 1869, I sem, p. 358.

† This reference is to Kirchhoff's scale.

氢	
C.	10月20日，1868年
F.	10月20日，1868年
接近D.	10月20日，1868年*
[*氢？—G. G. S.]	
接近G.	12月22日，1868年
h.	3月14日，1869年
钠	
D.	2月28日，1869年
钡	
1989.5†	3月14日，1869年
2031.2	7月5日，1869年
镁及相关谱线	
b^1 b^2 b^3 b^4	2月21日，1869年
其他谱线	
铁线	1,474　6月6日，1869年
?	1515.5　6月6日，1869年
亮线	1529.5　7月5日，1869年
?	1567.5　3月6日，1869年
?	1613.8　6月6日
铁线	1867.0　6月26日
亮线	1871.5　6月26日
铁线	2001.5　6月26日
?	2003.4　6月26日
? 靠近暗线的谱带或谱线，非常细	2054.0　7月5日

除上述谱线外，其他谱线也被观测了许多次，但是它们的位置仍然没有被严格确定。

作者指出，以铁为例，如果埃斯特朗和基尔霍夫图谱中标注的铁线是由铁这一种元素的吸收造成的，那么到目前为止，他只能在色球层较低的区域中检测到全部460条谱线中的3条——也许未来的观测工作中可能会观测到全部。镁和钡也如此。

文章接着写道：

*《法国科学院院刊》，1869年，第1期，第358页。

† 参照基尔霍夫标度。

"Dr. Frankland and myself have determined that the widening out of the sodium-line in the spectrum of a spot which I pointed out in 1866, and then stated to be possibly an evidence of greater absorption, indicates a greater absorption due to greater pressure.

"The continuous widening out of the sodium-line in a spot must therefore be regarded as furnishing an additional argument (if one were now needed) in favour of the theory of the physical constitution of the sun first put forward by Dr. Frankland and myself—namely, that the chromosphere and the photosphere form the true atmosphere of the sun, and that under ordinary circumstances the absorption is continuous from the top of the chromosphere to the bottom of the photosphere, at whatever depth from the bottom of the spot that bottom may be assumed to be.

"This theory was based upon all our observations made from 1866 up to the time at which it was communicated to the Royal Society and the Paris Academy of Sciences, and has been strengthened by all our subsequent work; but several announcements made by Father Secchi to the Paris Academy of Sciences and other learned bodies are so opposed to it, and differ so much from my own observations, that it is necessary that I should refer to them, and give my reasons for still thinking that the theory above referred to is not in disaccord with facts.

"Father Secchi states that the chromosphere is often separated from the photosphere, and that between the chromosphere and the photosphere there exists a stratum giving a continuous spectrum, which he considers to be the base of the solar atmosphere, and in which he thinks that the inversion of the spectrum takes place.

"With regard to the first assertion, I may first state that all the observations I have made have led me to a contrary conclusion. Secondly, in an instrument of comparatively small dispersive power, such as that employed by Father Secchi, in which the widening out of the F line at the base of the chromosphere is not clearly indicated, it is almost impossible to determine, by means of the spectroscope, whether the chromosphere rests on the sun or not, as the chromosphere is an envelope and we are not dealing merely with a section. But an instrument of great dispersive power can at once settle the question; for since the F line widens out with pressure, and as the pressure increases as the sun is approached, the continuous curvature of the F line must indicate really the spectrum of a section; and if the chromosphere were suspended merely at a certain height above the photosphere, we should not get a widening due to pressure; but we always do get such a widening.

"With regard to the second assertion, I would remark that if such a continuous-spectrum-giving envelope existed, I entirely fail to see how it could be regarded as a region of selective absorption. Secondly, my observations have indicated no such stratum, although injections of sodium, magnesium, &c. into the chromosphere not exceeding the limit of the sun's limb by 2″ have been regularly observed for several months past. Today I have even detected a low level of barium in the chromosphere not 1″ high. This indicates, I think, that my instrument is not lacking in delicacy; and as I have never seen

“弗兰克兰博士和我已经确定了我在1866年提出的黑子光谱中钠线的展宽效应，并称这可能是存在更强吸收作用的证据，说明较大的压力导致强的吸收。

“在一个黑子光谱中钠线的持续展宽可以被看作是一个附加的证据（如果现在需要一个证据的话），这支持由弗兰克兰博士和我首先提出的有关太阳物质组成的理论，即色球层和光球层组成了太阳的大气，而且在通常情况下，从色球层顶部到光球层底部距离黑子底部（这一底部可以是假定的）任意深度处，吸收作用是连续进行的。

“这个理论基于我们从1866年到现在的所有观测结果，曾报告给皇家学会和巴黎科学院，并已经得到我们所有后续观测结果的进一步验证。但是塞奇神父在巴黎科学院和其他学术机构公布的一些言论极力反对这个理论，而且他的观点与我们的观测结果有非常大的差异，因此我有必要提及这些，并给出我一直认为上述理论与事实相符的理由。

“塞奇神父宣称色球层经常与光球层分离，且在色球层和光球层之间存在一个能发出连续光谱的层，他认为这个层是太阳大气的底部，并且认为在这一层中光谱会发生反转。

“对于他的第一个论断，我首先要说的是，我所有的观测结果都把我引向了一个相反的结论。其次，用色散率较小的仪器，例如塞奇神父使用的那种，不能清楚地看到色球层底部F线的展宽效应，用光谱仪几乎不可能分辨出色球层到底是不是依附在太阳上，因为色球层是一个包层，而且我们要考虑的不仅仅是一个截面。大色散率的仪器可以马上解决这一问题，因为既然F线由于压力作用而展宽，越接近太阳内部的地方压力越大，所以F线的连续弯曲说明这是一个截面上的光谱，而如果色球层只是悬浮在光球层上方的某一特定高度处，我们就不会看到由于压力而导致的展宽效应，但事实上我们总是能观察到这样的展宽。

“对于第二个论断，我想说的是，如果这样一个产生连续光谱的包层确实存在，那么我实在看不出来它怎么能被看作是具有选择吸收的区域。其次，我的观测结果表明没有这样的一层，尽管在前几个月里经常观测到钠、镁等射入色球层的深度没有超过日面边缘2角秒。今天我在色球层中不到1角秒的高度处检测到了少量的钡。我觉得这表明我的仪器精度很高，而且在我的仪器状态良好的条件下，我从未看到过哪怕是一个近似连续的光谱，我倾向于将观测结果归结为仪器误差。这个现象可

anything approaching to a continuous spectrum when my instrument has been in perfect adjustment, I am inclined to attribute the observation to some instrumental error. Such a phenomenon might arise from a local injection of solid or liquid particles into the chromosphere, if such injection were possible. But I have never seen such an injection. If such an occurrence could be observed, it would at once settle that part of Dr. Frankland's and my own theory, which regards the chromosphere as the last layer of the solar atmosphere; and if it were possible to accept Father Secchi's observation, the point would be settled in our favour.

"The sodium experiments to which I have referred, however, and the widening out of the lines in the spot-spectra, clearly indicate, I think, that the base of the atmosphere is below the spot and not above it. I therefore cannot accept Father Secchi's statement as being final against another part of the theory to which I have referred—a conclusion which Father Secchi himself seems to accept in other communications.

"Father Secchi remarks also that the F line is produced by the absorption of other bodies besides hydrogen, because it never disappears. This conclusion is also negatived by my observations; for it has very often been observed to disappear altogether and to be replaced by a bright line. At times, as I pointed out to the Royal Society some months ago, when a violent storm is going on accompanied by rapid elevations and depressions of the prominences, there is a black line on the less-refrangible side of the bright one; but this is a phenomenon due to a change of wavelength caused by a rapid motion of the hydrogen.

"With regard to the observation of spot-spectra, I find that every increase of dispersive power renders the phenomenon much more clear, and at the same time more simple. The selective absorption I discovered in 1866 comes out in its most intense form, but without any of the more complicated accompaniments described by Father Secchi. I find, however, that by using three prisms this complexity vanishes to a great extent. We get portions of the spectrum here and there abnormally bright, which have given rise doubtless to some of the statements of the distinguished Roman observer; but the bright lines, properly so-called, are as variable as they are in any other part of the disk, but not much more so. I quite agree that the 'interpretation' of sun-spot phenomena to which Father Secchi has referred, which ascribes the appearances to anything but selective plus general absorption, is erroneous. But as I was not aware that it had ever been propounded, I can only refer to my own prior papers in support of my assertion which were communicated to the Royal Society some three years ago."

"Researches on Gaseous Spectra in relation to the Physical Constitution of the Sun, Stars, and Nebulae."—Third Note. By E. Frankland, F.R.S., and J. Norman Lockyer, F.R.S.

The authors remark that it has been pointed out by Mr. Lockyer that the vapours of magnesium, iron, &c., are sometimes injected into the sun's chromosphere, and are then rendered sensible by their bright spectral lines. (*Proc. Roy. Soc.*, vol. XVII, p. 351)

能是由局部区域的固体或者液体微粒注入色球层引起的，如果这种注入过程可能的话。但我从未见到过这种注入过程。当然如果这个过程能够被观测到的话，它将立刻证明弗兰克兰博士和我的理论是正确的，即认为色球层是太阳大气的最后一层，而即使认为塞奇神父的观察结果是可以接受的，那也将印证我们的观点。

“但是，对于以前提到的有关钠的实验和黑子光谱中出现的谱线展宽，我认为都明确地说明了大气层的底部在黑子之下而不是在黑子之上，因此我不认为塞奇神父的报告是对我提出的理论中的另一部分观点的彻底否定——塞奇神父自己似乎在其他报告中也是表示同意这一点的。

“塞奇神父还指出，由于 F 线从不消失，所以它是氢以外的其他物质的吸收产生的。这个结论也被我的观测结果否定了；因为我经常观察到它完全消失，并被一条亮线取代。正如我几个月前在皇家学会上所说的，当强烈的太阳风暴来临的时候，伴随着日珥快速的升降，在亮线折射率较小的一侧有时会有一条暗线，但这种现象是由于氢的快速运动导致的波长改变造成的。

“在观察黑子光谱的时候，我发现色散率的提高每次都能使这个现象更加明显，同时也更简单。我在 1866 年发现的选择性吸收谱线达到了它的最强状态，但没有产生塞奇神父所说的其他更复杂的现象。然而，我发现使用三棱镜后，这种复杂的现象基本上就会消失。我们得到了一些有的地方异常明亮的光谱，这使得我们对这位杰出的罗马观测者的一些论断深信不疑。但是那些亮线，严格地说是所谓的亮线，和它们在日面上任何其他位置时一样变化。我认为塞奇神父对黑子现象的‘解释’是不正确的，即认为这些现象绝不是选择性吸收和一般吸收共同作用的结果。但由于我不知道这个观点是否曾经被提出过，我只能参考我自己以前的文章来支持我 3 年前向皇家学会提出的论点。”

《关于太阳、恒星以及星云物质组成的气态光谱研究》——第 3 篇，作者是弗兰克兰（皇家学会会员）和诺曼·洛克耶（皇家学会会员）。

作者谈到，洛克耶先生曾经指出：镁、铁等的蒸气有时候会注入太阳的色球层，从而产生明亮的谱线而被人察觉到。（《皇家学会学报》，第 17 卷，第 351 页）

2. It has also been shown (1) that these vapours, for the most part, attain only a very low elevation in the chromosphere, and (2) that on rare occasions the magnesium vapour is observed like a cloud separated from the photosphere.

3. It was further established on the 14th of March, 1869, and a drawing was sent to the Royal Society indicating, that when the magnesium vapour is thus injected, the spectral lines do not all attain the same height.

Thus, of the *b* lines, b^1 and b^2 are of nearly equal height, but b^4 is much shorter.

4. It has since been discovered that of the 450 iron lines observed by Ångström, only a very few are indicated in the spectrum of the chromosphere when iron vapour is injected into it.

5. The authors' experiments on hydrogen and nitrogen enabled them at once to connect these phenomena, always assuming that the great bulk of the absorption to which the Fraunhofer lines are due takes place in the photosphere itself.

It was only necessary, in fact, to assume that, as in the case of hydrogen and nitrogen, the spectrum became simpler where the density and temperature were less, to account at once for the reduction in the number of lines visible in those regions where, on the authors' theory, the pressure and temperature of the absorbing vapours of the sun are at their minimum.

6. It became important, therefore, to test the truth of this assumption by some laboratory experiments, the preliminary results of which are communicated in this note.

The spark was taken in air between two magnesium poles, so separated that the magnesium spectrum did not extend from pole to pole, but was visible only for a little distance, indicated by the atmosphere of magnesium vapour round each pole.

The disappearance of the *b* lines was then examined, and it was *found that they behaved exactly as they do on the sun*. Of the three lines, the most refrangible was the shortest; and shorter than this were other lines, *which Mr. Lockyer has not detected in the spectrum of the chromosphere*.

This preliminary experiment, therefore, quite justified the assumption, and must be regarded as strengthening the theory on which the assumption was based, namely, that the bulk of the absorption takes place in the photosphere, and that it and the chromosphere form the true atmosphere of the sun. In fact, had the experiment been made in hydrogen instead of in air, the phenomena indicated by the telescope would have been almost perfectly reproduced; for each increase in the temperature of the spark caused the magnesium vapour to extend further from the pole, and where the lines disappeared a band was observed surmounting them, which is possibly connected with one which at times is observed in the spectrum of the chromosphere itself when the magnesium lines are not visible.

(**1**, 195-197; 1869)

2．这也表明：(1) 这些气体中的绝大部分都处于色球层的底部；(2) 在极少数情况下，镁蒸气看上去像是一块与光球层分离的云。

3．1869 年 3 月 14 日向皇家学会提交了一幅草图，进一步说明当镁蒸气注入的时候，并不是所有谱线都达到了同一高度。

比如，对于 b 线而言，b^1 和 b^2 的高度近似相等，但 b^4 短很多。

4．在埃斯特朗先前发现的 450 条铁线中，只有极少数会在铁蒸气注入色球层时出现在色球层的光谱中。

5．作者关于氢和氮的实验可以立刻与这些现象联系起来，这通常要假设大部分产生夫琅和费线的吸收发生于光球层内部。

实际上，就像在氢和氮的实例中一样，只需要假定在密度和温度较低的地方光谱会变得更简单，就可以马上用作者的理论解释为什么可见谱线的数量会在太阳内部吸收气体的压力和温度最小的地方下降。

6．因此，用实验来验证这一假设的真实性就显得尤为重要，初步的实验结果通报如下：

当在两个镁电极之间的空气中产生电火花时，镁谱线并非从一极延伸到另一极，而是只在一个很短的距离内可见，说明镁蒸气只位于两个电极附近。

接下来检测到了 b 线的消失，**发现它们的表现与它们在太阳上的表现完全相同。**在这 3 条线中，折射率最大的也是最短的，而比这条更短的是**洛克耶先生在色球层光谱中没有探测到的一些其他的谱线。**

因此这个初步的实验完全证明了这个假设，而且强化了该假设所基于的理论，即大部分吸收发生于光球层，且光球层和色球层组成了太阳真正的大气。实际上，如果这个实验是在氢气中而不是空气中进行，望远镜中观察到的现象就会几乎完美地重现。因为火花温度的每一次增加，都会使镁蒸气扩展到离电极更远的地方，并且在谱线消失的地方可以看到其上有一个谱带。这可能与镁线不可见的时候在色球层光谱中看到的现象有关联。

（刘东亮 翻译；邓祖淦 审稿）

Are Any of the Nebulae Star-systems?

R. A. Proctor

Editor's Note

In 1870, the nature of many extended astronomical objects generally referred to as "nebulae", apparently devoid of stars, remained mysterious. Some astronomers, notably William and John Herschel, had argued that many of these objects were probably extragalactic star systems: galaxies much like our own, but too far away for individual stars to be resolved. But Richard Proctor here argues that the balance of evidence pointed instead to nebulae being extended objects within our own galaxy. Today we know that these "nebulae" include a wide range of astrophysical objects, including some star clusters in our own galaxy, but also other galaxies, galactic clusters, interstellar clouds of dust and gas, and remnants of supernovae.

THIS may seem a bold question, for it is commonly believed that Sir William and Sir John Herschel—the Ajax and the Achilles of the astronomical host—have long since proved that many of the nebulae are star-systems. If we inquire, however, into what the Herschels have done and said, we shall find that not only have they not proved this point, but that the younger Herschel, at any rate, has expressed an opinion rather unfavourable than otherwise to the theory that the nebulae are galaxies in any sense resembling our own sidereal system.

Sir William Herschel, by his noble plan of star-gauging, proved that the stars aggregate along a certain zone, which in one direction is double. He argued, therefore, that presuming a general equality to exist among the stars and among the distances separating them from each other, the figure of the sidereal system resembles that of a cloven disc. And as the only system from which he could form a probable judgment—I mean the planetary system—presented to him a number of bodies, widely separated from each other and each a globe of considerable importance, he reasoned from analogy that similar relations exist in the sidereal spaces. This being so, his cloven disc theory of the sidereal system seemed satisfactorily established.

Then, of course, those nebulae which exhibit a multitude of minute points of light very close together, and those other nebulae which, while not thus resolvable into minute points, yet in other respects resemble those which are, came naturally to be looked upon as distinct from the sidereal system. The analogy of this system, in fact, pointed to them as external star-systems, resembling it in all important respects.

Then there were certain other objects, which seemed to present no analogy either to the sidereal system or to separate stars. These objects Sir Wm. Herschel considered to belong to our sidereal system; for he could not put them outside its range without looking on

所有的星云都是恒星系统吗？

普洛克特

编者按

1870 年，许多延展型天体的本质仍然是个谜。这些天体通常被称为“星云”，其中似乎看不见恒星。包括著名的威廉·赫歇尔和约翰·赫歇尔在内的许多天文学家们认为，这些天体中大部分可能是类似银河系的河外恒星系统，不过因为过于遥远而无法分辨单个恒星。不过，理查德·普洛克特在这篇文章中指出，各方面的证据综合起来反而更支持星云是银河系内的延展型天体。现在我们知道，这些“星云”所涵盖的天体类型非常广泛，不仅包括一些银河系中的星团，还包括其他一些星系、星系团、由气体和尘埃组成的星际星云，以及超新星遗迹。

这个问题提得也许有些鲁莽，因为威廉·赫歇尔爵士和约翰·赫歇尔爵士——天文学界的埃阿斯和阿喀琉斯——在很久以前就已经证明了很多星云都是恒星系统。然而如果我们深入研究赫歇尔父子说过的话和他们做过的事情，我们就会发现他们俩不但没有证实这个观点，并且小赫歇尔至少还表达了一些不赞同的看法，而没有说明星云就是与我们所在恒星系统类似的星系这一观点。

威廉·赫歇尔爵士凭借他闻名于世的恒星标定法证明了恒星会在一定区域内聚集，并在某个方向上数目加倍。因此他认为，假设恒星和恒星间相隔的距离是均匀分布的，则恒星系统的外形类似于一个裂开的盘。由于他只能根据一个系统——我指的是行星系统——作出可能的判断，这唯一的系统展示在他面前的是许多相距甚远的行星，每一个行星的地位都很重要，于是他由类推法得出相似的关系也存在于恒星之间。这样，他关于恒星系统是一个裂开的盘的理论也就似乎圆满地建立起来了。

另外，有些星云看上去是由许多相距很近的小亮点组成的，而在另一些星云中则无法分辨出微小的光点，但在其他方面与前一种星云相似，后者自然就被看作是有别于恒星系统。实际上，这种星云与恒星系统在所有的重要特征上都很相似，这种相似性向他们表明这种星云是银河系外的恒星系统。

还有某些其他的天体，它们看起来既不像恒星系统也不像一个单独的恒星。威廉·赫歇尔爵士认为，这种天体属于我们的恒星系统，因为，他不能把它们置于

them as objects *sui generis*, which would have been to abandon the argument from analogy. In order to explain their appearance, he suggested that they might be gaseous bodies, by whose condensation stars would one day be formed.

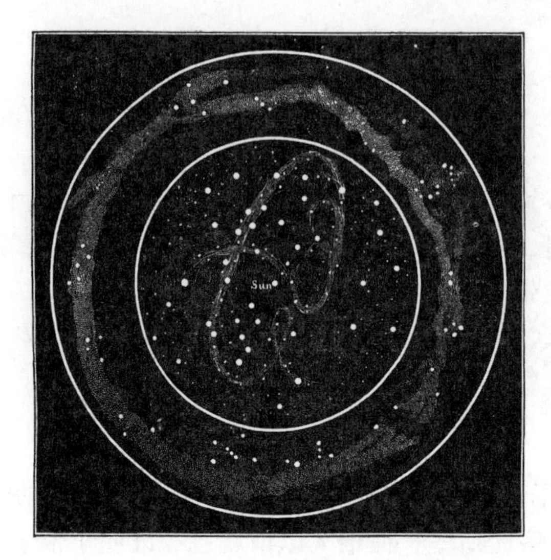

The value of Sir Wm. Herschel's work is not in the least affected even if science have to reject every one of these opinions. He himself held them with a light hand; he had once held other opinions; and he was gradually modifying these. Had he seen one sound reason for rejecting any or all of them he would have done so instantly. For it belonged to the strength of his character that he was never fettered by his own opinions, as weak men commonly are.

Sir John Herschel did for the southern heavens what his father had done for the northern. He completely surveyed and gauged them. It is commonly believed that the results of his labours fully confirmed the opinions which his father had looked upon as probable.

Let us see if this is so.

Sir W. Herschel thought the Milky Way indicated that the sidereal system has the figure of a cloven disc; Sir John Herschel judges rather that the sidereal system has the figure of a flattened ring. Sir Wm. Herschel thought the stellar nebulae are probably external galaxies; Sir John gives reasons for believing that they lie within our system, and Whewell considered that these reasons amount to absolute proof.

It has been further believed and stated that the researches of the elder Struve go far to confirm the opinions put forward by Sir W. Herschel as probable.

Let us inquire how far this is true.

我们的恒星系统范围之外，除非把它们视作自成一类的天体系统，但这样又必须抛弃关于相似性的原则。为了解释这种天体系统的外貌，他认为这些天体系统可能是一些气体，总有一天这些气体会凝聚成恒星。

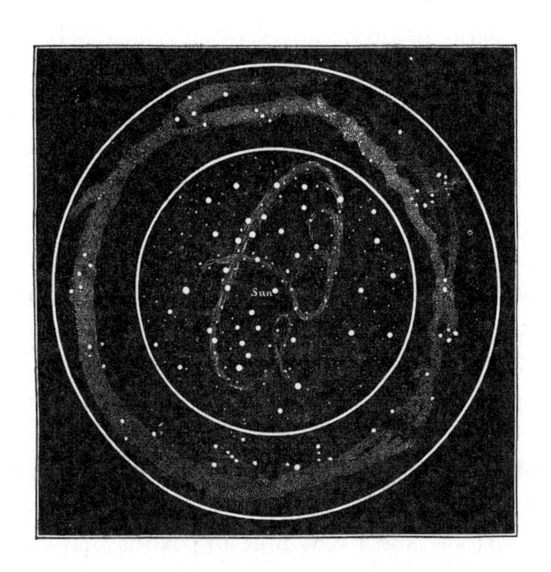

威廉·赫歇尔爵士工作的价值绝对不会因为科学否定了他在这方面的所有观点而受影响。在这些问题上，他的态度灵活多变，他也曾有过其他观点，并逐渐地对这些观点进行修正。假如有可靠的理由让他相信应该放弃他的某些或是全部观点，他就会立即那样做。因为他从来都不会像那些软弱的人们一样被自己的观点束缚，这就是他人格中的闪光之处。

约翰·赫歇尔爵士对南天所做的工作与他的父亲对北天所做的工作一样。他全面地观测并标定了南天的恒星。人们普遍认为，他的研究成果充分肯定了他父亲认为很有可能的观点。

让我们来看一下事实是否如此。

威廉·赫歇尔爵士认为，银河系是一个裂开的盘状恒星系统。约翰·赫歇尔爵士判断，这个恒星系统更像是一个被压平的环。威廉·赫歇尔爵士认为，恒星星云可能是外面的星系。约翰·赫歇尔爵士则给出了一些理由，认为它们处于我们星系内部，休厄尔认为这些理由已经构成了确凿的证据。

人们已经确信，老斯特鲁韦的研究工作在很大程度上证明威廉·赫歇尔爵士提出的观点很可能是正确的。

那就让我们来研究一下他的论断在多大程度上是正确的。

Struve found that the numbers of stars of given magnitudes exhibit nearly the same proportion in different directions. Thus supposing that in a given direction there are three times as many stars of a certain magnitude as there are of the next highest magnitude, then in other directions, also, the same relation is observed. This is a very striking law; but to make it serve as a proof of the opinion which Sir William Herschel had put forward as probable, it would be necessary that another law should be exhibited. For clearly, if that opinion were just, it would be easy to calculate what the relation should be between stars of different magnitudes. Had Struve been able to show that the numbers actually seen corresponded to the relations thus calculated, he would have gone far to render that view certain which Herschel always spoke of as merely an assumption.

But Struve found no such law of stellar distribution. On the contrary, he found a law so different, that in order to force the facts into agreement with Sir William Herschel's views about the sidereal system, he had to invent his famous theory of the extinction of light in traversing space. Now, according to this theory, we cannot see to the limits of our sidereal system, even though we could increase the powers of our telescopes a million-fold; so that if the theory is true, the question which heads this paper is at once disposed of. Obviously, we cannot see galaxies beyond the sidereal system if we cannot see to the limits of that system. And I may note in passing that (independently of Struve's theory) the most powerful telescopes cannot render visible the most distant stars of our sidereal scheme; so that if the nebulae are really external galaxies, the stars we see in them must be enormously greater than those in our galaxies, supposing Herschel was right in thinking these tolerably uniform in magnitude.

Before proceeding to exhibit the evidence which has led me to the conviction that the nebulae belong to our sidereal system, I may mention some reasons for believing that if Sir William Herschel's labours in the sidereal heavens were to be begun now, not only would he not have been led to adopt as probable the view on which he formed his opinions, but he would have rejected it as opposed to known analogies.

He had argued that because the planetary system exhibits a definite number of bodies separated by wide distances, therefore analogy should lead us to regard the sidereal system as similarly constituted, though on a much larger scale. This was perfectly just. Despite the various differences which no one recognised more clearly than he did, this view was the only one he could safely adopt for his guidance, ninety years ago.

But would not he have been the first to reject that view if he had known what we now know of the solar system? If he had known that besides the primary planets, there are hundreds of minute bodies forming a zone between the orbits of Mars and Jupiter; that the rings of Saturn are formed of a multitude of minute satellites; that innumerable meteor-systems circle in orbits of every conceivable degree of eccentricity; that near the sun these systems grow denser and denser; that the comets of the solar system must be counted by millions on millions; that, in fine, every conceivable form of matter, every conceivable degree of aggregation, and every conceivable variety of size, exists within the

斯特鲁韦发现，一定星等的恒星数目在不同方向上所占的比例大致相同。因此，如果假定在一个给定方向上某个星等的恒星数目是比该星等亮一等的恒星数目的3倍，那么在其他方向上，这个关系也会成立。这是一个非常值得注意的定律，但是把它作为证明威廉·赫歇尔爵士提出的论点的证据是不够的，还需要另一条定律。显然，如果那个论点是正确的，将很容易推算出不同星等的恒星之间是什么样的关系。如果斯特鲁韦能够说明实际观测到的数据符合计算得出的关系，那他也就可以确定那个赫歇尔常常只是作为一项假设而提及的论点。

但是斯特鲁韦并没有发现这样一个恒星分布定律。相反地，他发现了一个完全不同的定律，为了强行使实际情况能与威廉·赫歇尔爵士提出的恒星系统的观点保持一致，他不得不发明那个著名的光穿越星际空间时的消光理论。根据这个理论，即便我们把望远镜的放大倍率提高一百万倍，我们也不可能看见自己星系的边界，所以如果这个理论是对的，那么这篇文章标题中提出的那个问题就可以立即迎刃而解了。显然，如果我们无法看见自己星系的边界，也就不能看见外面的星系了。或许我可以顺便提一下（与斯特鲁韦的理论无关），用目前放大倍率最高的望远镜也看不到我们所在星系中离我们最远的恒星，所以如果星云真的是银河系以外的星系，假设赫歇尔认为它们具有差不多相同的星等的观点是对的，那么我们在星云中看见的恒星肯定比我们自己星系中的恒星大许多。

在向你们展示那些使我确信星云属于我们所在星系的证据之前，我先陈述一些理由来说明，如果威廉·赫歇尔爵士现在才开始他对恒星系统的工作的话，那么他不但不可能接受他自己提出的那个观点，还会因为它与现在已知的类推观点相悖而抛弃那个观点。

他曾经争辩说，行星系统是由彼此之间分隔较远的确定数目的星体组成的，因此我们可以类推出恒星系统也具有相似的结构，只不过是在一个更大的尺度上罢了。这是非常合理的。在90年前这是他唯一可以放心采用的观点，尽管对于这两种系统的各种差别没有人比他认识得更清楚。

但是如果他可以了解到我们现在所掌握的关于太阳系的知识，他会不会第一个抛弃他那个观点呢？如果他知道，除了几个主要的大行星之外，在火星轨道和木星轨道之间有一个由数百颗小星体构成的带；土星环是由许多微小的卫星聚集而成；无数的流星群可以在任意可想象到的偏心率轨道上环绕运行；越靠近太阳，星体的密度越大；太阳系中的彗星数以百万计；总之，任何可以想象到的物态、任何可以想象到的凝聚程度、任何可以想象到的尺度的物质在太阳系范围内都存在，难道他能以此类推出恒星系统中只有离散的恒星和正在形成恒星的物质吗？

limits of the solar system,—would he, then, have been led by analogy to recognise in the sidereal system only discrete stars and masses forming into stars?

From a careful study of all that Sir William Herschel has written, I feel certain, that in the case I have imagined, he would have been prepared, even before commencing his labours, to expect precisely that variety of matter, size, and aggregation, which modern observations, rightly understood, prove actually to exist within the range of the sidereal system.

The Herschels, father and son, discovered about 4,500 nebulae. Other observers have brought up the number to about 5,400. When these are divided into classes, it appears that some 4,500 must be looked on as irresolvable into discrete points of light. But of these the greater proportion so far resemble resolvable nebulae as to lead to the belief that increase of optical power alone is wanting to resolve them.

Taking these irresolvable nebulae, however, as we find them, and marking down their places over the celestial sphere, we recognise certain peculiarities in their arrangement. In the northern heavens they gather into a clustering group as far as possible from the Milky Way. In the southern heavens they form into streams, which run out from a region nearly opposite the northern cluster of nebulae; but the *extremities* of the streams are the region where nebulae are most closely crowded. The Milky Way is almost clear of nebulae.

This withdrawal of the nebulae from the Milky Way has been accepted by many as clearly indicating that there is no association between them and the sidereal system. The opinion of the Herschels, if they had been led to pronounce definitively on this point, would have been different, however; for the younger Herschel quotes (as agreeing with it) a remark of his father's to the effect that the peculiar position of the northern nebular group is not accidental. If not accidental, it can only be due to some association between the nebular group and the galaxy. Every other conceivable explanation will be found to make the relation merely apparent—that is, accidental, which neither of the Herschels admit.

But yet stronger evidence of association exists; evidence which I do not hesitate to speak of as incontrovertible. Space will only permit me to treat it very briefly.

There is a certain well-marked stream of nebulae in the southern heavens leading to a well-marked cluster of nebulae. There is an equally well-marked stream of stars leading to an equally well-marked cluster of stars. The nebular stream agrees in position with the star-stream, and the probability is small that this coincidence is accidental. The nebular cluster agrees in position with the star-cluster, and the probability is still smaller that this second coincidence is accidental. Such are the separate chances. It will be seen at once, therefore, how small the chance is that both coincidences are accidental.

The cluster here referred to is the greater of the celebrated Magellanic Clouds. When it is added that the evidence is repeated point for point in the case of the lesser Magellanic

在仔细研究了威廉·赫歇尔爵士的所有著作之后，我确实感到，如果情况正如我已经设想过的那样，那么他甚至可能会在开始他的工作之前就已经做好明确地预期各种不同物质、尺度和凝聚度的存在的准备，这些已被现代观测手段证实确实存在于恒星系统之中。

赫歇尔父子发现了大约4,500个星云。其他观测者又把这个数字提升到了5,400。当对这些星云进行分类时，大约有4,500个星云是不能分辨出离散光点的。但是这些星云中的大多数都非常类似于那些可以从中分辨出离散光点的星云，这使人们确信只要提高望远镜的放大倍率就可以分辨它们。

然而，当挑出这些不能从中分辨出光点的星云，并把发现它们的位置标在天球上时，我们发现这些星云的排布有些奇特。在北天，它们在尽可能远离银河的地方聚成一团。在南天，它们形成许多星云流，发端于与北方星云团正对的位置附近，但是在这些星云流的**尽头**则是星云最密集之处。银河区域内几乎没有星云。

很多人认为，银河区域没有星云已经清楚地说明星云和我们所在的恒星系统没有关联。然而如果要求赫歇尔父子在这个问题上明确表态，那他们的观点就会与此不同；因为小赫歇尔提到过（以赞同的口吻），他父亲曾评论说北天区星云团的特殊位置并非出于偶然。如果不是偶然现象，那只能是由于星云团与我们的星系之间存在相关性。所有能找到的其他解释只不过是让它们之间的关系更明显而已——即赫歇尔父子都不能接受偶然性的存在。

但是有更强有力的证据表明相关性的存在。我可以毫不犹豫地说这个证据是不容置疑的。但是由于版面所限，我只能对此做简要的介绍。

在南天有一个十分明显的星云流，一直通向一个十分明显的星云团。同样存在一个十分明显的恒星流一直通向同样十分明显的恒星团。星云流的位置与恒星流的位置相符，这种一致性源于巧合的可能性是很小的。星云团在位置上也与恒星团相符，这第二种一致性源于巧合的可能性更小。这两个事件是相对独立的。于是我们立刻可以想到，这两个事件都源于巧合的概率会是多么小。

这里提到的星团就是著名的大麦哲伦星云。当我们把小麦哲伦星云的情况也考虑进来的时候，该现象又一次重复出现了，这说明这种相关性的存在是毋庸置疑的。

Cloud, the indications of association appear overwhelmingly convincing. If the nebulae really are associated in this manner with fixed stars, the question which heads this paper is disposed of at once.

But there is yet further evidence.

The nebulae pass by insensible gradations from clusters less and less easily resolvable, to nebulae properly so called, but still resolvable, and so to irresolvable nebulae. Now clusters are found not only to aggregate in a general manner near the Milky Way, but in some cases (on which Sir John Herschel has dwelt with particular force) to exhibit the clearest possible signs of belonging to that zone. If they then belong to the Milky Way, can any good reason be given for believing that the various other classes of nebulae are not associated with the sidereal scheme? Where should the line be drawn?

Again, some of the nebulae are gaseous, and all the gaseous nebulae exhibit the same spectrum. Now, two classes of gaseous nebulae, the planetary and the irregular nebulae, exhibit a marked preference for the Milky Way, and therefore we must admit the probability that they, at any rate, belong to the sidereal scheme. But then a large proportion of the irresolvable nebulae are also gaseous, and as they are formed of the same gases, we see good reason for believing that they also must belong to our galaxy. This, however, brings in all the nebulae, since the recent detection by Lieut. Herschel of the same bright lines in or rather on the continuous spectrum of a star-cluster, shows the great probability which exists that with more powerful spectroscopes all the nebulae may be found to exhibit these bright lines, that is, to contain these particular gases. I pass over the facts, that many nebulae are found to be closely associated with stars, and that if any doubt could remain as to the association being real and not apparent, it would be removed by a picture of the nebula M 17, as seen in Mr. Lassell's great reflector at Malta. The reader will be more interested by the following quotation, which I extract (by permission) from a letter of Sir John Herschel's:—

"A remark which the structure of Magellanic Clouds has often suggested to me has been strongly recalled by what you say of the inclusion of every variety of nebulous or clustering forms within the galaxy, viz., that if such be the case—*i.e.* if these forms belong to, and form part and parcel of the Galactic system—then that system includes *within itself miniatures of itself* on an almost infinitely reduced scale; and what evidence, then, have we that there exists a universe beyond—unless a sort of argument from analogy, that the Galaxy with all its contents may be but one of these miniatures of a more vast universe, and that there may, in that universe of other systems on a scale as vast as our galaxy, be the analogues of those other nebulous and clustering forms which are not miniatures of our galaxy?"

It will be seen that, while Sir John Herschel is quite ready (should the evidence require it) to adopt altogether new views about the nebulae, he is not ready to forego the grandeur

如果星云真的以这种方式与固定的恒星相联系，那么这篇文章标题中的问题就会马上迎刃而解。

但是这里还有更进一步的证据。

通过难以察觉的渐变，星云从非常不容易分辨的星团，逐渐过渡到严格来说可被称为星云但仍然可以分辨的阶段，最终成为不可分辨的星云。现在发现，星团不仅以普通的方式在银河系附近聚集，而且在一些情况下（约翰·赫歇尔爵士在研究这个问题时曾认为存在一种特殊的作用力）还很明白地显示出可能属于那个区域的迹象。如果它们是属于银河系的，那么有什么好的理由可以使我们相信其他类型的星云与我们的恒星系统没有关系呢？这个界线应该划在哪里呢？

此外，有一些星云是气态的，并且所有的气态星云都具有相同的光谱。有两种气态星云，即行星状星云和不规则星云，具有明显的与银河系相似的谱线特征，因此我们必须承认这两种星云有可能属于我们所在的这个恒星系统。但是很大一部分不可分辨的星云也是气态的，由于它们也是由同样的气体组成的，因此我们就有充分的理由相信这些星云也属于我们所在的星系。因为赫歇尔中尉最近在一个恒星团的连续光谱中，或者更确切地说是在连续光谱上，发现了相同的亮线，这表明使用分辨率更高的分光镜很可能会让我们在所有星云的光谱中都能找到这些亮线，也就是说这些星云中都含有同样的气体。我在此略去了一些论据，诸如人们发现很多星云与恒星具有密切的相关性，如果有人对星云与恒星间关系的真实性和明确性表示怀疑，那么拉塞尔先生在马耳他使用大型反射式望远镜对 M17 星云进行观测所得到的一幅照片应该可以消除这方面的疑问。读者可能会对下面我从约翰·赫歇尔爵士的一封信中摘录出来（已获同意）的片段更感兴趣：

“你所提出的银河系中每一种星云和团状结构中包含的物质总使我回忆起麦哲伦星云的结构给我的启发，即，如果事实确实如此——也就是说，假如这些星云和团状结构属于或者组成了银河系的一部分或一块——那么这个系统内部就包含了许多在尺度上无限缩小的**它自身的微结构**。但是，如果不是通过类推法，我们又有什么证据能够证明在星系外部的宇宙的存在，银河系连同它内部所包含的一切物质只不过是更大宇宙中众多微结构中的一个，以及在其他与银河系同样大小的星系中也许会有与我们的星系类似但又绝非与我们星系的微结构一样的星云和团状物呢？”

由此可见，虽然约翰·赫歇尔爵士十分乐意（需要一些证据）全盘接受关于星云的新观点，但他不愿破坏他与他父亲建立的神圣的宇宙观，正是这个宇宙观使他

of those noble views of the universe which he and his father have established, thereby earning the well-deserved gratitude of every lover of astronomy.

And then with regard to the actual form of our galaxy or Milky Way, the figure introduced shows that its apparent one as projected on the heavens may really be due to an arrangement differing both from the cloven disc or flattened ring, a point to which I shall return in a subsequent article.

(**1**, 331-333; 1870)

们理所当然地赢得了每一位天文爱好者的感激。

有关我们所在星系或银河系的形状，文中插入了一张银河系在天球上的投影图，图中显示的排列确实既不像是裂开的盘，也不是被压平的环。我会在以后的文章中讲述这个问题。

（史春晖 翻译；邓祖淦 审稿）

Where Are the Nebulae?

H. Spencer

Editor's Note

Stimulated by Proctor's arguments on the nature of nebulae, the polymath Herbert Spencer notes here that he too had previously questioned the popular view that nebulae were distant galaxies like our own. He points out that some astronomers had noted correlations between the apparent locations of some nebulae and stars in our local galaxy, which should not exist if nebulae were much more distant. Spencer also argues that the relationship between the apparent luminosity of astronomical objects and their distance also works against the idea. In general, larger nebulae should be closer, and therefore easier to resolve into independent stars, whereas in practice, on the contrary, smaller nebulae were more easily resolvable.

MR. Proctor's interesting paper in your last number reminded me of an essay on "The Nebular Hypothesis", originally published in 1858, and re-published, along with others, in a volume in 1863 ("Essays: Scientific, Political, and Speculative". Second Series), in which I had occasion to discuss the question he raises. In that essay I ventured to call in question the inference drawn from the revelations of Lord Rosse's telescope, that nebulae are remote sidereal systems—an inference at that time generally accepted in the scientific world. On referring back to this essay, I find that, besides sundry of the reasons enumerated by Mr. Proctor for rejecting this inference, I have pointed out one which he has omitted.

Here are some of the passages:—

"'The spaces which precede or which follow simple nebulae,' says Arago, 'and, *à fortiori*, groups of nebulae, contain generally few stars. Herschel found this rule to be invariable. Thus, every time that, during a short interval, no star approached, in virtue of the diurnal motion, to place itself in the field of his motionless telescope, he was accustomed to say to the secretary who assisted him, "Prepare to write; nebulae are about to arrive."' How does this fact consist with the hypothesis that nebulae are remote galaxies? If there were but one nebula, it would be a curious coincidence were this one nebula so placed in the distant regions of space as to agree in direction with a starless spot in our own sidereal system? If there were but two nebulae, and both were so placed, the coincidence would be excessively strange. What, then, shall we say on finding that they are habitually so placed? (the last five words replace some that are possibly a little too strong) . . . When to the fact that the general mass of nebulae are antithetical in position to the general mass of stars, we add the fact that local regions of nebulae are regions where stars are scarce, and the further fact that single nebulae are habitually found in comparatively starless spots, does not the proof of a physical connection become overwhelming?"

星云在哪里？

斯宾塞

编者按

在普洛克特关于星云本质的论述的启发下，博学的赫伯特·斯宾塞发表了这篇文章，文中他提到，先前他也质疑过星云是类似于银河系的遥远星系这一流行的观点。斯宾塞指出，有些天文学家已经注意到一些星云的视位置与我们所在的银河系中某些恒星位置之间的相关性，如果星云是非常遥远的类似银河系的星系，那么这种现象就不该存在。斯宾塞还指出，天体视光度与其距离之间的关系也不支持流行的观点。简言之，越大的星云应该越近，从而应该越容易分辨出独立的恒星，但是实际情况与此完全相反，倒是越小的星云中越容易分辨出恒星。

普洛克特先生在上一期《自然》中发表了一篇关于星云的饶有趣味的文章，这使我想起我的一篇题为《星云假说》的论文，这篇关于星云的论文最初发表于1858年，后来又与其他文章一起再次发表于1863年的某部书上（《论文集：科学、政治与推理》，第2部），我在该书中也讨论了相同的问题。当时科学界普遍认为星云是遥远的恒星系统，这一观点是根据罗斯勋爵用望远镜观测到的新结果推出的，我在自己的论文中对这一推论提出了异议。再次回顾普洛克特先生的文章，我发现除了普洛克特先生在他的文章中列举的各种反驳以上推论的理由之外，我曾经指出过的一条原因被他遗漏了。

以下几个段落是从我的文章中节选出来的：

“阿拉戈说：‘在单个星云或更准确地说在星云群前后的星际空间，一般只包含很少的恒星。赫歇尔发现这一规则是永恒不变的。因此，每当由于周日运动，在一个比较短的时间间隔内没有恒星进入他的固定望远镜的视场的时候，他总是习惯对助手说：“准备记录，星云马上就要来了。”’这一现象怎么能够和星云是遥远的星系这一假说一致呢？如果只存在一个星云，为什么它在遥远宇宙中的位置恰好分布于我们所在恒星系统中没有恒星的方向上呢？如果只存在两个星云，而二者的位置都是如此，这岂不是更令人不可思议吗？那么，我们将如何解释星云为什么总处于这样的特殊位置呢？（最后5个词替换掉了以前可能有点偏激的措辞）……考虑到大部分星云处在与绝大多数恒星相反的位置上这一事实，再加上恒星的数量在星云所在区域相当稀少，以及单个星云通常位于几乎没有恒星的地方这些事实，难道这些证明它们之间具有物理关联的证据还不够充分吗？”

The reasonings of Humboldt and others proceeded upon the tacit assumption that differences of apparent magnitude among the stars result mainly from differences of distance. The necessary corollaries from this assumption I compared with the hypothesis that the nebulae are remote sidereal systems in the following passage:—

"In round numbers, the distance of Sirius from the earth is a million times the distance of the earth from the sun; and according to the hypothesis, the distance of a nebula is something like a million times the distance of Sirius. Now, our own 'starry island, or nebula', as Humboldt calls it, 'forms a lens-shaped, flattened, and everywhere-detached stratum, whose major axis is estimated at seven or eight hundred, and its minor axis at a hundred and fifty times the distance of Sirius from the earth.' And since it is concluded that our solar system is near the centre of this aggregation, it follows that our distance from the remotest parts of it is about four hundred distances of Sirius. But the stars forming these remotest parts are not individually visible, even through telescopes of the highest power. How, then, can such telescopes make individually visible the stars of a nebula which is a million times the distance of Sirius? The implication is, that a star rendered invisible by distance becomes visible if taken two thousand five hundred times further off!"

This startling incongruity being deducible if the argument proceeds on the assumption that differences of apparent magnitude among the stars result mainly from differences of distance, I have gone on to consider what must be inferred if this assumption is not true; observing that "awkwardly enough, its truth and its untruth are alike fatal to the conclusions of those who argue after the manner of Humboldt. Note the alternatives":—

"On the one hand, what follows from the untruth of the assumption? If apparent largeness of stars is not due to comparative nearness, and their successively smaller sizes to their greater and greater degrees of remoteness, what becomes of the inferences respecting the dimensions of our sidereal system and the distances of the nebulae? If, as has lately been shown, the almost invisible star, 61 Cygni, has a greater parallax than α Cygni, though, according to an estimate based on Sir W. Herschel's assumption, it should be about twelve times more distant—if, as it turns out, there exist telescopic stars which are nearer to us than Sirius, of what worth is the conclusion that the nebulae are very remote, because their component luminous masses are made visible only by high telescopic powers? . . . On the other hand, what follows if the truth of the assumption be granted? The arguments used to justify this assumption in the case of the stars, equally justify it in the case of the nebulae. It cannot be contended that, on the average, the *apparent* sizes of the stars indicate their distances, without its being admitted that, on the average, the *apparent* sizes of the nebulae indicate their distances—that, generally speaking, the larger are the nearer, and the smaller are the more distant. Mark, now, the necessary inference respecting their resolvability. The largest or nearest nebulae will be most easily resolved into stars; the successively smaller will be successively more difficult of resolution; and the irresolvable ones will be the smaller ones. This, however, is exactly the reverse of the fact. The largest nebulae are either wholly irresolvable, or but partially resolvable under the highest telescopic powers; while a great proportion of quite small nebulae are easily resolvable by far less powerful telescopes."

洪堡和其他研究者的推理基于以下默认的假设：各个恒星视星等的不同主要源于距离的差异。我对这一假设和认为星云是遥远的恒星系统的假说进行过比较。下面的段落给出了上述假设的一些必然的推论：

“如果以整数计算，天狼星与地球之间的距离是地球与太阳距离的100万倍；根据星云假说，一个星云离我们的距离大约是我们与天狼星之间距离的100万倍。这样，洪堡所谓的‘恒星岛或星云’就会‘形成扁平的、具有透镜形状的、处处分离的层，其长轴大约是天狼星与地球之间距离的700或800倍，其短轴则大约是天狼星与地球之间距离的150倍。’既然认为我们的太阳系靠近这一聚集体的中心，那么我们与聚集体最远端的距离大约是我们与天狼星的距离的400倍。然而即使借助目前威力最大的望远镜，也不能逐个分辨出位于聚集体最远端的恒星。那么，用这样的望远镜如何才能分辨出这些100万倍于天狼星距离的星云中的单个恒星呢？这意味着，一颗由于距离过远而无法观测到的恒星，在将其距离拉远2,500倍后，反而可以观测到了！”

如果承认恒星之间视星等的区别主要源于距离的差异，那么从以上结果得到的推论会存在惊人的矛盾。因此，我继续思考，如果该假设不成立，那么究竟可以推出什么。我也注意到“情况非常尴尬，无论是否正确，由洪堡的方法得出的结论都是同样致命的。只能考虑其他可能的选择”：

“一方面，如果这一假设不成立我们会得到什么结论呢？如果一颗看上去体积巨大的恒星不是由于离我们很近，而看上去尺寸较小的恒星也不是由于它们非常非常遥远，那么我们根据太阳系的尺寸和星云的距离又能得出什么推论呢？就像最近已经被指出的，如果天鹅座61这颗几乎不可见的恒星比天鹅座α星的视差更大，虽然按照赫歇尔爵士的假设推算，天鹅座61离我们的距离约为后者的12倍——但是如果事实证明存在可以用望远镜观测到的比天狼星离我们更近的恒星，那么因为只有通过高倍望远镜才能观测到星云中的发光物质而认为星云非常遥远又有什么价值呢？……另一方面，如果假设成立又能得到什么结论呢？用于证明该假设在恒星情况下成立的论据，同样也可以用于星云。既然认为恒星的**视**尺寸一般可以表明它们与我们的距离的远近，那么就必须承认，星云的**视**尺寸也同样对应于它们的距离——即一般而言，尺寸越大表明距离越近，尺寸越小表明距离越远。下面考虑由星云的可分辨性得到的推论。最大或者最近的星云将最容易从中分辨出恒星，而越小的星云则越难从中分辨出各个恒星，那些无法分辨的星云将会是尺寸较小的星云。然而，这与事实刚好相反。即使用最高放大倍率的望远镜观测，那些最大的星云不是完全不可分辨就是只能部分分辨，而大部分尺寸很小的星云，反而可以借助倍率低得多的望远镜轻易地分辨出来。”

At the time when these passages were written, spectrum-analysis had not yielded the conclusive proof which we now possess, that many nebulae consist of matter in a diffused form. But quite apart from the evidence yielded by spectrum-analysis, it seems to me that the incongruities and contradictions which may be evolved from the hypothesis that nebulae are remote sidereal systems, amply suffice to show that hypothesis to be untenable.

(**1**, 359-360; 1870)

Herbert Spencer: 37, Queen's Gardens, Jan. 31.

在写下以上段落的时候，光谱分析尚未得出结论性的证据，即我们现在知道的许多星云由弥散物质组成。但是除了光谱分析的结果可以作为证据以外，我认为由星云是遥远的恒星系统这一假说推导出来的自相矛盾的结果就足以表明该假说是站不住脚的了。

（金世超 翻译；邓祖淦 审稿）

The Physical Constitution of the Sun

Editor's Note

What gives the Sun its heat? Hermann von Helmholtz and William Thomson had recently speculated on the physical constitution of the Sun, and this editorial explores one of their ideas concerning its energy source. Their so-called "contraction theory" offered a conceivable explanation for the output of both solar light and heat. If the sun were gradually contracting, they supposed, while maintaining its temperature, then it would lose energy each year. Helmholtz had shown that a contraction of as little as 0.1 percent in diameter per year would liberate more than 20,000 times the observed solar energy. The hypothesis seemed to give plausible estimates of energy release for a wide range of materials that might make up the Sun.

DR. Gould has addressed an important letter on the above subject to the *Journal of the Frankland Institute*. In the first part he refers to the new light recently thrown on the sun's physical constitution by the observations of Mr. Lockyer, and agrees with him and Dr. Frankland, both as to the absorption taking place in the chromosphere and photosphere itself, and also as to the possible telluric origin of the corona.

He then proceeds with regard to the probable age of the sun:—

"The researches of Helmholtz and Thomson regarding the age of the sun as a source of cosmical heat have shown us limits within which, in the absence of more decisive evidence, we must restrict our theories as to the length of time during which he has warmed the earth. The contraction-theory has been most ably discussed by these eminent physicists, and seems to afford the only satisfactory mode of accounting for the solar light and heat, now that we know both that the meteors generally revolve in cometary orbits, and that the habitability of the earth, as well as the apparent unchanged mutual attraction of the planets, bears testimony to the incorrectness of the meteoric theory. From Pouillet's data (derived from experiments which ought to be repeated in some year when the solar spots are at a minimum) Helmholtz has shown that, even were the sun's density uniform, a contraction of 1/10 percent in his diameter would evolve 20,000 times the present annual supply of solar heat. But when the sun was hotter the same proportional contraction would have evolved yet more heat; so that we must consider the above estimate as a minimum.

"The expansibility of hydrogen gas for 100 °C is 0.3661. No gas appears to have so small a coefficient as 0.360, which would correspond to a linear expansion of 0.108. The expansibility of glass, the smallest known, I believe, even for a solid, is about 1/150 part as great; say 0.00244 in volume, or 0.00081 linear. Therefore for glass even, a contraction of 1 percent in diameter would imply a fall of temperature by 1,230 °C, and a mean specific heat of 218. This seems certainly a minimum value.

关于太阳的物质组成

编者按

太阳的热量是从哪里来的？这篇评论的作者从赫尔曼·冯·亥姆霍兹和威廉·汤姆孙最近对太阳物质组成的研究中发现他们提出的观点中有一个与太阳的能量来源有关。他们提出的“收缩论”为太阳光和热的输出提供了一个可信的解释。他们猜测：如果太阳真的在不断收缩，那么在温度不发生变化时，每年也会损失一定的能量。亥姆霍兹曾经证明：哪怕太阳直径每年只收缩 0.1%，也会释放出比观测到的太阳辐射能高 20,000 倍以上的能量。他们的假说对组成太阳的各种可能的物质释放的能量给出了看似可靠的估计。

谷德博士已经在《弗兰克兰研究院院刊》上发表了一篇有关上述论题的重要报道。在文章的第一部分，他参考了由洛克耶先生的观测结果引发的关于太阳物质组成的最新观点，并且对洛克耶和弗兰克兰博士关于太阳色球层和光球层本身的吸收作用以及日冕源于地球的观点表示赞同。

而后他进一步分析了太阳的大致年龄：

“亥姆霍兹和汤姆孙关于太阳作为宇宙热源的时限的结果给我们提供了一个研究范围，在缺少更明确证据的前提下，我们必须把我们关于太阳照射地球的时间长度的理论限制在这一时间范围内。这些杰出的物理学家已经对收缩论进行了非常深入的讨论，并提出了一个看上去能够解释太阳光和热的唯一令人满意的模式，因为我们知道流星通常沿着彗星轨道运转，也知道地球的可居住性，以及行星之间总是相互吸引，这些都证明流星理论是错误的。根据普耶的观测数据（来自一些实验，我们应该在若干年后当太阳黑子数达到极小值时重复这些实验），亥姆霍兹指出：即便假设太阳密度是均匀的，那么它的直径收缩 1/1,000 也将导致太阳的年辐射量达到目前的 20,000 倍。但如果太阳比现在更热的话，同样比例的收缩将使太阳放出更多的热；因此我们必须把上面的估计看作是一个最小值。

“氢气在 100℃时的膨胀系数是 0.3661，任何气体的膨胀系数都不可能小于 0.360，这相当于线性膨胀系数为 0.108。玻璃的膨胀系数约为氢气的 1/150，据我所知这是最小的，即使对于固体也是一样的，即体积膨胀系数为 0.00244，线性膨胀系数为 0.00081。因此，即使是玻璃，直径收缩 1% 也意味着温度下降 1,230℃且比热达到 218。看起来这肯定是最小值。

"But if we suppose the expansion coefficient to be as large as that of hydrogen, a contraction of 1 percent would correspond to a change of temperature by 8.2 °C or a mean specific heat of 32,700, if equivalent to 20,000 years' supply. This is out of the question.

"Now Thomson has computed that bodies smaller than the sun, falling from a state of relative rest at mutual distances which are large in comparison with their diameters, and forming a globe equal to the sun, would generate 20,000 times the present annual supply. This would be greater did we consider the unquestionable increase of the sun's density towards his centre. And since it seems out of the question that resistance and previous minor impacts could have consumed more than one-half the heat, he inferred ten million times a year's supply to be the lowest, and one hundred million times to be the highest, estimate of the sun's initial heat.

"Now we have every reason for the belief that radiation is proportional to temperature. Assuming this and taking the temperature of the sun's photosphere as 14,000 °C,
10,000,000 times the present annual supply would be radiated
in 3,650,000 years if the specific heat were 218,
in 7,280,000 ,, ,, ,, ,, 1,000.
100,000,000 times the present annual supply would be radiated
in 8,250,000 years if the specific heat were 218,
in 25,500,000 ,, ,, ,, ,, 1,000
500,000,000 times the present annual supply would be radiated
in 11,700,000 years if the specific heat were 218,
in 38,900,000 ,, ,, ,, ,, 1,000

"For vapours, other than hydrogen, the greatest known specific heat, so far as I am aware, is 0.508 (ammonia); and hydrogen, which has less than 3.5, cannot form any considerable portion of the sun's mass.* A specific heat so high as 1,000 seems altogether out of the question; yet it will be seen that, even on this supposition, an amount of initial heat equal to 500,000,000 the present annual supply, would have been radiated in less than forty million years, were the sun's radiative capacity proportional to his temperature. Taking the more probable age, 10,000,000 years, we should find 226 million times the present annual supply to have been radiated within this period if the specific heat were not greater than 218; and even were the specific heat 1,000, the total radiation would have been eighteen million times a year's radiation at present.

"Thus the limit given by Thomson, although so vastly below that afforded by the speculations of some geologists, would appear itself to demand a considerable additional reduction. And I cannot see how we can well suppose the sun in its present form to have

* It seems to form certainly not more than the 18, 000th part of the mass of the earth.

“但是，如果我们假设膨胀系数和氢一样大，那么 1%的收缩相当于温度变化 8.2℃，或者平均比热为 32,700，等价于 20,000 年的能量供给。这是不可能的。

“现在，汤姆孙通过计算得出，一些比太阳小的天体，当它们之间的距离与它们的直径相比很大时，会从相对静止的状态开始聚集并形成一个类似于太阳的球体，那么这些天体产生的能量将是目前年辐射量的 20,000 倍。如果我们认为太阳的密度越靠近中心一定越大的话，这个数还会更大一些。因为阻力和之前的小碰撞不可能消耗一半以上的热量，他估计太阳最初的热量至少是年热辐射量的 1,000 万倍，最多是 1 亿倍。

“现在，我们有充分的理由相信辐射正比于温度。按照这个假设并假定太阳光球层的温度为 14,000℃，则：

<table>
<tr><td>如果比热为 218，</td><td>太阳在 3,650,000 年内</td><td rowspan="2">辐射的热量将是目前年辐射量的10,000,000倍。</td></tr>
<tr><td>如果比热为 1,000，</td><td>太阳在 7,280,000 年内</td></tr>
<tr><td>如果比热为 218，</td><td>太阳在 8,250,000 年内</td><td rowspan="2">辐射的热量将是目前年辐射量的100,000,000倍。</td></tr>
<tr><td>如果比热为 1,000，</td><td>太阳在 25,500,000 年内</td></tr>
<tr><td>如果比热为 218，</td><td>太阳在 11,700,000 年内</td><td rowspan="2">辐射的热量将是目前年辐射量的500,000,000倍。</td></tr>
<tr><td>如果比热为 1,000，</td><td>太阳在 38,900,000 年内</td></tr>
</table>

“对于气体而言，除了氢气之外，就我所知，已知的最小比热是 0.508（氨气）；氢气的比热低于 3.5，不可能是组成太阳大气的主要成分。* 看起来比热高达 1,000 是完全不可能的；即便就按这样的假设，那么，如果太阳的辐射量与它的温度成正比的话，500,000,000 倍于目前年辐射量的初始热量也将在不到 4,000 万年的时间里被辐射掉。在 1,000 万年这个更可能的时间长度内，我们会发现，如果比热不高于 218，那么将有 2.26 亿倍于目前年辐射量的热量在这段时间内被消耗掉；即使比热能够达到 1,000，总辐射量也将是目前年热辐射量的 1,800 万倍。

“因此，汤姆孙给出的极限虽然已经大大低于一些地质学家的推测，但该值似乎仍然需要一定程度的下调。我认为按照目前的情况来看，太阳向外辐射的时间不可能超过 2,000 万年，300 万年或 400 万年可能是一个更合理的估计，除非我们假设在

* 氢在地球中的含量肯定不超过地球质量的1/18,000。

radiated heat for more than twenty millions of years, while three or four millions would seem to be a far more probable estimate, unless the thermic laws be totally different in those exalted temperatures which we must suppose to have existed at some past epoch.

"The very great diversity of the limiting values for the specific heat seems to afford ample scope for every needful allowance on account of the natural action of the particles within the body of the sun, even conceding to this the immense effect (analogous to the increase of specific heat) which has been assigned to it by some investigators. Even did we conceive a primitive heat equal to 200,000,000 times the amount now yearly radiated, and a specific heat 10,000 times as great as is possessed by any known gaseous body excepting hydrogen, we could not deduce so long a period as 80,000,000 of years for the past duration of the sun's heat."

(**2**, 34; 1870)

过去曾经存在过一段温度很高的时期，在那样的高温下热学定律完全不同。

“即使承认这个曾被一些研究者考虑过的巨大效应（类似于比热的增加），比热极限值之间的巨大差异也能为太阳内部粒子在各种必要情形下的自然作用提供极大的可能性。就算我们假设太阳的初始热量等于其目前年辐射量的200,000,000倍，比热是除氢以外现在已知的任意一种气体的10,000倍，那也不可能推出太阳在过去的80,000,000年中一直在辐射自己的热量。”

（魏韧 翻译；邓祖淦 审稿）

Spectroscopic Observations of the Sun

J. N. Lockyer

Editor's Note

Here *Nature*'s editor J. Norman Lockyer describes his recent presentation at the Royal Society on spectroscopic observations of the Sun. Lockyer used spectroscopy to deduce the Sun's composition, and the preceding year he had discovered a new element—helium—by this means. Doppler shifts in atomic emission lines enabled him to deduce motions of material at the Sun's surface, and in this way Lockyer observed the dynamical behaviour of sunspots and solar eruptions. Here he records his latest observations on that topic.

ROYAL Society, May 19.—"Spectroscopic Observations of the Sun". No. VI. By J. Norman Lockyer, F.R.S.

The weather has lately been fine enough and the sun high enough, during my available observation-time, to enable me to resume work. The crop of new facts is not very large, not so large as it would have been had I been working with a strip of the sun, say fifty miles or a hundred miles wide, instead of one considerably over 1,000—indeed, nearer 2,000 in width; but in addition to the new facts obtained, I have very largely strengthened my former observations, so that the many hours I have spent in watching phenomena, now perfectly familiar to me, have not been absolutely lost.

The negative results which Dr. Frankland and myself have obtained in our laboratory-work in the matter of the yellow bright line, near D, in the spectrum of the chromosphere being a hydrogen line, led me to make a special series of observations on that line, with a view of differentiating it, if possible, from the line C.

It had been remarked, some time ago, by Prof. Zollner, that the yellow line was often less high in a prominence than the C line; this, however, is no evidence (bearing in mind our results with regard to magnesium). The proofs I have now to lay before the Royal Society are of a different order, and are, I take it, conclusive:—

1. With a tangential slit I have seen the yellow line bright below the chromosphere, while the C line has been dark; the two lines being in the same field of view.

2. In the case of a bright prominence over a spot on the disc, the C and F lines have been seen bright, while the yellow line has been invisible.

3. In a high-pressure injection of hydrogen, the motion indicated by change of wavelength has been less in the case of the yellow line than in the case of C and F.

4. In a similar quiescent injection the pressure indicated has been less.

观测太阳光谱

洛克耶

编者按

在这篇文章中，《自然》的主编诺曼·洛克耶阐述了他最近在英国皇家学会所作的关于观测太阳光谱的报告。洛克耶根据光谱推测出了太阳的化学组成，此前一年他还用这种方法发现了一种新元素——氦。根据原子发射谱线的多普勒频移他推断出物质在太阳表面的移动，并用这种方式观测到了太阳黑子和太阳爆发的动力学特征。在这篇文章中，他给出了相关的最新观测结果。

皇家学会，5 月 19 日——《观测太阳光谱》，第 6 篇，作者为诺曼·洛克耶，英国皇家学会会员。

最近天气一直很好，而且太阳也达到了足够的高度，这使我在可用于观测的时间内可以重新开始我的工作。观察到的新现象不是很多，如果我进行观测研究的是太阳的某一狭长范围，比方说是 50~100 英里的高度范围，而不是远大于 1,000 英里——实际上接近 2,000 英里的高度，则得到的新的结果将比现在多得多。不过，除了观测到一些新现象外，我还极大地拓展了我以前的观测工作，因此，我虽然花了很多时间去观测那些现在已经很熟知的现象，但也并非是白白浪费了时间。

弗兰克兰博士和我本人在实验室工作中得到靠近 D 线的黄色明亮谱线，这在色球层光谱中属于氢线的部分中没有找到。这促使我对该谱线进行了一系列的观测，尝试性地从另一个角度，把它同 C 线区分开来。

不久前，策尔纳教授已经注意到，在日珥中这条黄线通常不及 C 线高；但这没有被证实（请注意我们的结果只是对镁而言）。我现在提交给皇家天文学会的一些观测证据，思路有所不同但我认为至关重要：

1. 当利用一个切向狭缝观察时，我看到色球层底部有一条明亮的黄线，但是 C 线却是暗的，虽然这两条线都处在同一个视场中。

2. 当日面上有明亮的日珥遮住黑子时，看到的 C 线和 F 线都很亮，却看不到黄线。

3. 在氢的高压喷流中，由波长变化表示的该运动，在黄线的情况下不及在 C 线和 F 线的情况下大。

4. 在近似静止的喷流中，显示的压力变小了。

5. In one case the C line was seen long and unbroken, while the yellow line was equally long, but broken.

The circumstance that this line is so rarely seen dark upon the sun makes me suspect a connection between it and the line at 5,015 Ångström, which is also a bright line, and often is seen bright in the chromosphere, and then higher than the sodium and magnesium lines, when they are visible at the same time; and the question arises, must we not attribute these lines to a substance which exists at a higher temperature than those mixed with it, and to one of very great levity? for its absorption line remains invisible, as a rule, in spot-spectra.

I have been able to make a series of observations on the fine spot which was visible when I commenced them on April 10th, not far from the centre of its path over the disc. At this time, the spot, as I judged by the almost entire absence of indications of general absorption in the penumbral regions, was shallow, and this has happened to many of the spots seen lately. A few hours' observation showed that it was getting deeper apparently, and that the umbrae were enlarging and increasing in number, as if a general down-sinking were taking place; but clouds came over, and the observations were interrupted.

By the next day (April 11) the spot had certainly developed, and now there was a magnificently bright prominence, completely over the darkest mass of umbra, the prominence being fed from the penumbra or very close to it, a fact indicated by greater brilliancy than in the bright C and F lines.

April 12. The prominence was persistent.

April 15. Spot nearing the limb, prominence still persistent over spot. At eleven I saw no prominence of importance on the limb, but about an hour afterwards I was absolutely startled by a prominence not, I think, depending upon the spot I have referred to, but certainly near it, more than 2′ high, showing a tremendous motion towards the eye. There were light clouds, which reflected to me the solar spectrum, and I therefore saw the black C line at the same time. The prominence C line (on which changes of wavelength are not so well visible as in the F line) was only coincident with the absorption-line for a few seconds of arc!

Ten minutes afterwards the thickness of the line towards the right was all the indication of motion I got. In another ten minutes the bright and dark lines were coincident.

And shortly afterwards what motion there was was towards the red!

I pointed out to the Royal Society, now more than a year age*, that the largest prominences, as seen at any one time, are not necessarily those in which either the intensest action or the most rapid change is going on. From the observations made on this and the following day, I think that we may divide prominences into two classes:—

---

* *Proc. Roy. Soc.* 1869, p. 354, Mar. 17.

5. 只有一次观测中，C 线看起来长且没有间断，而黄线虽然与 C 线具有相同的长度，但中间有间断。

太阳光谱中这条线很暗的时候是非常少的，这使我怀疑这条谱线是否与 5,015 Å 的谱线有联系。这条 5,015 Å 的谱线也是一条亮线，经常可以在色球层的光谱中看到，而且当它与钠线和镁线同时出现时，它的亮度高于钠线和镁线。于是问题就产生了，难道我们就不能认为，产生这些谱线的物质与其周围混合在一起的物质相比温度更高而且非常不稳定吗？因为它的吸收谱线在黑子光谱中通常是看不见的。

4 月 10 日，我开始对一个小黑子进行一系列的观测，这个黑子距离其通过日面路径的中点不远。在这次观测中，黑子半影区内几乎完全没有出现吸收现象，我据此判断这是一个很浅的黑子。最近观测到的很多黑子都是这样。几个小时的观测过程中，这个黑子显然在逐渐变深，本影部分变大，且数目增加，就好像有某种下沉运动正在进行；但是这时有云飘过来，我的观测被迫停止。

等到了第二天（4 月 11 日），黑子显然已经变大了，而且这时出现了一个非常明亮的日珥，完全遮住了黑子本影最暗的区域，这个日珥从黑子的半影或者是很靠近半影的地方出现，这一点根据其亮度比明亮 C 线和 F 线更高可知。

4 月 12 日，日珥仍然停留在那里。

4 月 15 日，黑子接近日面的边缘，日珥仍然在黑子上方。到 11 时，在日面边缘处我没有看到明显的日珥，但是大约一个小时以后，我就完全被一个日珥惊呆了，这个日珥与我观测的那个黑子没有关系，但是确实和它靠得很近，其高度超过 2 角秒，并且朝视线方向有巨大的运动。这时天上有薄云，它影响了太阳的光谱，这时我完全看不到 C 线。日珥光谱中的 C 线（在这个波段上的波长变化不像 F 线那么明显）与吸收谱线的重合只有几个角秒。

10 分钟后，谱线向右侧展宽，这就是我提到的运动的全部征兆。再过 10 分钟，亮线与暗线又重合在一起了。

过了很短的时间，那里的运动移向了红端！

一年多以前我曾向皇家学会报告过 *，在任何一次观测中，最大的日珥未必运动得最强烈或变化得最快。根据这次和第二天的观测，我认为可以将日珥分为两类：

*《皇家学会学报》，1869 年 3 月 17 日，第 354 页。

1. Those in which great action is going on, lower vapours being injected; in the majority of cases these are not high, they last only a short time—are throbs, and are often renewed, and are not seen so frequently near the sun's poles as near the equator. They often accompany spots, but are not limited to them. These are the intensely bright prominences of the American photographs.

2. Those which are perfectly tranquil, so far as wavelength evidence goes. They are often high, are persistent, and not very bright. These do not, as a rule, accompany spots. These are the "radiance" and dull prominences shown in the American photographs.

I now return to my observations of the spot. On the 16th, the last of the many umbrae was close to the limb, and the most violent action was indicated occasionally. I was working with the C line, and certainly never saw such rapid changes of wavelength before. The motion was chiefly horizontal, or nearly so, and this was probably the reason why, in spite of the great action, the prominences, three or four of which were shut out, never rose very high.

I append some drawings, made, at my request, by an artist, Mr. Holiday, who happened to be with me, and who had never seen my instrument or the solar spectrum widely dispersed before. I attach great importance to them, as they are the untrained observations of a keen judge of form.

The appearances were at times extraordinary and new to me. The hydrogen shot out rapidly, scintillating as it went, and suddenly here and there the bright line, broad and badly defined, would be pierced, as it were, by a line of intensely brilliant light parallel to the length of the spectrum, and at times the whole prominence spectrum was built up of bright lines so arranged, indicating that the prominence itself was built up of single discharges, shot out from the region near the limb with a velocity sometimes amounting to 100 miles a second. After this had gone on for a time, the prominence mounted, and the cyclonic motion became evident; for away from the sun, as shown in my sketch, the separate masses were travelling away from the eye; then gradually a background of less luminous hydrogen was formed, moving with various velocities, and on this background the separate "bombs" appeared (I was working with a vertical spectrum) like exquisitely jewelled ear-rings.

It soon became evident that the region of the chromosphere just behind that in which the prominence arose, was being driven back with a velocity something like twenty miles a second, the back-rush being so local that, with the small image I am unfortunately compelled to use, both the moving and rigid portions were included in the thickness of the slit. I saw the two absorption-lines overlap.

These observations were of great importance to me; for the rapid action enabled me to put together several phenomena I was perfectly familiar with separately, and see their connected meaning.

1. 一类是活动性强，气体喷射较低的日珥。在大多数情况下，这种日珥不高，仅仅持续比较短的时间，并且会有跳动，形状经常翻新，这种日珥在太阳两极区域出现的频率没有在赤道附近出现的频率高。它们经常与黑子一起出现，但也不全是这样。美国人拍摄的照片中那些极其明亮的日珥就属于这类。

2. 另一类日珥非常稳定，这从其谱线的波长总是不变可以看出。这类日珥通常很高、持续时间很长，但不是特别亮。原则上，它们并不伴随黑子出现。这些就是美国人拍摄的照片里那些“深粉红色”并且暗淡的日珥。

我现在要回到我对黑子的观测上来。到 16 日，许多持续的本影区已经接近于日面边缘了，非常剧烈的活动只是偶尔才出现。我主要观测的是 C 线，以前从未出现过如此快速的波长变化。其运动方向基本上是水平的，或者是接近水平的。这也许就是为什么有三四个日珥，虽然运动剧烈，向外抛射，但总也升不太高的原因。

我附上了一些插图，这些插图是我邀请艺术家霍利迪先生画的，他当时恰好与我在一起，但他以前从来没有见过我的观测仪器和这么庞杂的太阳光谱。我非常重视这些插图，因为它们出自一位未受过专业训练但具有敏锐判断力的观察者之手。

有几次我观察到了一些反常的新现象。氢物质快速向外喷射，同时发出闪光，突然，到处都有谱线很宽的亮线，而且难以辨认，就像是被一束与波长方向平行的特别强的光穿过一样。有时候整个日珥光谱都由这样的亮线组成，这表明日珥本身是由单一的放电形成的，从靠近日面边缘的区域向外抛射，速度有时可以达到每秒 100 英里。当这些现象过去一段时间以后，日珥又出现了，并显示出气旋式的运动；从太阳上向外喷出，看上去是一些分散运动的物质，如我的附图所示。然后逐渐地，形成了一个不太明亮的氢的背景，并以各种不同的速度运动。在这个背景上，呈现出一系列分离的“爆炸”（我的光谱是垂直的），就像镶着宝石的精美耳环一样。

很快我又发现，日珥出现之处背后的色球区域又以每秒约 20 英里的速度被拉回来，这种反冲仅限于局部区域，我手头只有一幅小图，但又不得不用它，在这张图中，运动部分和固定部分都处在狭缝宽度内。我看到了两条吸收谱线的重叠。

这些观测结果对我来说是很重要的；因为这些快速运动可以让我将一些我非常熟悉的分立现象联系在一起，看一看把它们联系起来有什么意义。

They may be summarised as follows, and it will be seen that they teach us much concerning the nature of prominences. When the air is perfectly tranquil in the neighbourhood of a large spot, or, indeed, generally in any part of the disc, we see absorption-lines running along the whole length of the spectrum, crossing the Frauenhofer lines, and they vary in depth of shade and breadth according as we have pore, corrugation, or spot under the corresponding part of the slit—a pore, in fact, is a spot. Here and there, where the spectrum is brightest (where a bright point of facula is under the slit) we suddenly see this appearance—an interesting bright lozenge of light. This I take to be due to bright hydrogen at a greater pressure than ordinary, and this then is the reason of the intensely bright points seen in ranges of faculae observed near the limb.

The appearance of this lozenge in the spectroscope, which indicates a diminution of pressure round its central portion, is the signal for some, and often all, of the following phenomena:—

1. A thinning and strange variation in the visibility and thickness of the hydrogen absorption-line under observation.

2. The appearance of other lozenges in the same locality.

3. The more or less decided formation of a bright prominence on the disk.

4. If near the limb, this prominence may extend beyond it, and its motion-form will then become more easy of observation. In such cases the motion is cyclonic in the majority of cases, and generally very rapid, and—another feature of a solar storm—the photospheric vapours are torn up with the intensely bright hydrogen, the number of bright lines visible determining the depth from which the vapours are torn, and varying almost directly with the amount of motion indicated.

Here, then, we have, I think, the chain that connects the prominences with the brighter points of the faculae.

These lozenge-shaped appearances, which were observed close to the spot on the 16th, were accompanied by the "throbs" of the eruption, to which I have before referred; while Mr. Holiday was with me—a space of two hours—there were two outbursts, separated by a space of almost rest, and each outburst consisting of a series of discharges, as I have shown. I subsequently witnessed a third outburst. The phenomena observed on all three were the same in kind.

On this day I was so anxious to watch the various motion-forms of the hydrogen-lines, that I did not use the tangential slit. This I did the next day (the 17th of April) in the same region, when similar eruptions were visible, though the spot was no longer visible.

下面将这些现象综合起来，从中我们会看到这些现象将告诉我们更多有关日珥的特性。在大的黑子周围，或者一般来说，实际上在日面的任何一部分，只要气体处于完全的静止状态，我们会看到吸收谱线沿着整个光谱的波长方向移动，横穿夫琅和费线，吸收谱线的明暗程度和宽度都会发生变化，这取决于狭缝下面对应的是小黑点、褶皱还是黑子，小黑点其实也是黑子。在光谱最亮的地方（在狭缝下面，该位置有一个明亮光斑），我们突然发现了一个有趣的菱形亮光。我认为这是由于明亮的氢处在高于正常压力的状态下所致，这也就是我们在日面边缘附近的光斑中看见亮点的原因。

分光镜中出现的这种菱形表明它中心位置周围压力的减小，它的出现标志着下述某些现象的发生，而且往往是下述各种现象的同时发生。

1. 在观测期间，氢吸收谱线的能见度和宽度出现了微弱而奇异的变化。

2. 在同一位置出现了其他菱形。

3. 可以比较肯定地确认在日面上形成了一个明亮日珥。

4. 如果靠近日面边缘，这个日珥会延伸到日面以外，这样它的运动形态就更容易被观测。在这种情况下，这类日珥运动大多会呈气旋状，运动速度一般都很快，另外光球大气被撕裂，同时伴有非常明亮的氢线（这正是太阳风暴的另一特征），可见的明亮谱线的数目反映了气体被撕裂位置的深度，并且这个数目的变化直接反映了那里的运动状况。

我认为我们现在已经建立了日珥与光斑中较明亮点之间的联系。

我们在 16 日观测到的在太阳黑子附近出现的菱形亮点是伴随着“脉动”式的喷发出现的，我在前面已经提到过这种脉动；霍利迪先生当时正和我在一起，两个小时内我们看到了两次喷发，在两次喷发之间有一段几乎完全宁静的间歇期。每次喷发都伴有一系列放电现象，就像我前面所说的那样。随后我又见证了第三次喷发。这三次喷发都属于同一类现象。

那天我因为急于观测对应于各种不同运动形式的氢线而没来得及使用切向的狭缝。第二天（4 月 17 日），我使用切向狭缝对同一区域进行了观测，这时看到了同样的喷发现象，但那个黑子已经看不见了。

Judge of my surprise and delight, when upon sweeping along the spectrum, I found hundreds of the Frauenhofer lines beautifully bright at the base of the prominence!!!

The complication of the chromosphere spectrum was greatest in the regions more refrangible than C, from E to long past *b*, and near F, and high pressure iron vapour was one of the chief causes of the phenomenon.

I have before stated to the Royal Society that I have seen the chromosphere full of lines; but the fulness then was as emptiness compared with the observation to which I now refer.

A more convincing proof of the theory of the solar constitution, put forward by Dr. Frankland and myself, could scarcely have been furnished. This observation not only endorses all my former work in this direction, but it tends to show the shallowness of the region on which many of the more important solar phenomena take place, as well as its exact locality.

The appearance of the F line, with a tangential slit at the base of the prominence, included two of the lozenge-shaped brilliant spots to which I have before referred; they were more elongated than usual—an effect of pressure, I hold, greater pressure and therefore greater complication of the chromosphere spectrum; this complication is almost impossible of observation on the disc.

It is noteworthy that in another prominence, on the same side of the sun, although the action was great, the erupted materials were simple, *i.e.*, only sodium and magnesium, and that a moderate alteration of wavelength in these vapours was obvious. Besides these observations on the 17th, I also availed myself of the pureness of the air to examine telescopically the two spots on the disc, which the spectroscope reported tranquil as to up and down rushes. I saw every cloud-dome in their neighbourhood perfectly, and I saw these domes drawn out, by horizontal currents, doubtless, in the penumbrae, while on the floors of the spots, here and there, were similar cloud-masses, the distribution of which varied from time to time, the spectrum of these masses resembling that of their fellows on the general surface of the sun.

I have before stated that the region of a spot comprised by the penumbrae appears to be shallower in the spots I have observed lately (we are now nearing the maximum period of sun spots); I have further to remark that I have evidence that the chromosphere is also shallower than it was in 1868.

I am now making special observations on these two points, as I consider that many important conclusions may be drawn from them.

(**2**, 131-132; 1870)

当我沿着光谱扫视时，发现在日珥底部出现了几百条非常明亮的夫琅和费线！！！可想而知我是多么地惊喜。

色球层的光谱在折射率大于 C 线的区域最为复杂——从 E 线经过 *b* 一直延伸到 F 线附近。高压的铁蒸气是产生这种谱线的主要因素之一。

我以前曾向皇家学会陈述过我看到了色球层的所有谱线，但是那些谱线与我现在观测到的谱线数目相比是完全不值一提的。

由弗兰克兰博士和我提出的关于太阳结构理论的更有说服力的证据可以算是刚刚完成。这项观测不仅支持了以前我在这个方向上的所有工作成果，同时还表明太阳上很多更加重要的现象都是发生在活动区的浅层，并且完全是局部性的。

当我使用切向狭缝观察日珥底部时，发现 F 线处有两个前面提到过的菱形亮点；它们比通常情况下出现的亮点长一些，我认为这是由于压力的影响，压力越高，色球层的光谱越复杂；在日面上几乎不可能观察到这么复杂的光谱。

值得一提的是，在太阳的同一侧有另一个日珥，虽然它的活动性很强，但喷出的物质成分却比较简单，只有钠和镁，在这些气体中谱线波长的不断变化也很明显。除了这些在 17 日得到的观测结果以外，我还在大气比较纯净的时候用望远镜仔细观察了日面上的两个黑子，分光仪显示向上的和向下的活动都是平稳的。我仔细观察了这两个黑子附近的每一个云状隆起结构，发现在黑子半影部分的隆起结构无疑是被水平方向的气流拉长了，而在黑子的底层，到处是类似的云状物，其分布情况不时地发生变化，这些云状物的光谱与位于太阳表面的云状物的光谱是类似的。

我以前曾陈述过，在我最近观察过的黑子中，黑子的半影区域看上去比原来更浅（我们现在临近太阳黑子活动的极大期）；我还要进一步说明，有证据表明现在的色球层厚度比 1868 年时薄一些。

我现在正在对上述的两点情况进行专门的观测，因为我认为根据这些观测可以得出许多重要的结论。

（史春晖 翻译；何香涛 审稿）

The Source of Solar Energy

R. P. Greg

Editor's Note

With no knowledge of nuclear reactions, scientists in the late nineteenth century speculated freely and imaginatively on the source of energy driving the Sun. One idea was that continually infalling meteors might supply the energy through impacts. Yet here Robert Greg points out that this hypothesis conflicts with most of what was then known about meteors. The vast majority of these are objects in regular orbit around the Sun, and barely come any closer to the Sun than does the Earth, so there is no reason to expect them to fall into it. Indeed, do any meteors do so? Some meteors almost certainly do, says Greg, but surely not enough to account for the enormous energy of the Sun.

IT is, I think, rather unfortunate that Mr. Proctor, in his recent work entitled "More Worlds than One", should have re-advocated the earlier and now discarded views of Sir W. Thomson concerning the source of solar heat or energy by *meteoric percussion*. That theory, however ingenious as advanced by the physicist, is surely hardly one to be admitted by the astronomer. Nothing less than an intense desire or necessity for finding some solution to the problem, whence or how the solar heat is maintained, could have encouraged scientific men seriously to advance or support so plausible and unsatisfactory a doctrine, or one, when examined, so little supported by what we really know either of meteors or of nature's laws. Having given much attention to *meteoric* astronomy, may I be permitted briefly to state what I hold are serious and practical objections to the validity of the meteoric or dynamical theory as applied to the conservation of solar heat and energy.

1. Because meteors and aerolites are known to impinge and strike the earth in her orbit, *ergo*, as I understand Mr. Proctor, numbers infinitely greater must no doubt be constantly rushing into the sun, as a body at once far larger, and much nearer to myriads of such bodies than the earth herself; but which, at a much smaller distance, are more likely to be drawn into the sun. Now, all that we really do know about meteors amounts to this, that by far the greater number of shooting stars visible in our atmosphere, in size no larger than a bean, and really separated from each other by thousands of miles, belong to fixed and definite systems or rings, having fixed radiant points for certain epochs or periods, showing clearly that these bodies are revolving round the sun, in courses as true and regular as the planets themselves, and are no more eddying or rushing into the sun, merely because they are so insignificant, than is the earth herself. Having projected upon celestial charts the apparent courses or tracks of nearly 5,000 meteors, observed during every part of the year, I feel I am justified in stating that not more than seven or eight percent of the shooting stars observed on any clear night throughout the year, are *sporadic*, or do not belong to meteor systems at present known to us. More than one hundred meteor systems

太阳能量的来源

格雷格

编者按

19世纪后期的科学家还不知道核反应为何物，他们可以自由地发挥自己的想象力去推测太阳能量的来源。其中的一个想法是不断陨落的流星通过撞击太阳而提供能量。不过罗伯特·格雷格在这篇文章里指出，这一假说不符合大部分当时所知的有关流星的理论。绝大多数流星都是以固定轨道围绕太阳运转的天体，它们几乎不可能比地球更靠近太阳，所以没有理由相信它们会落到太阳上。是否真的有流星落到太阳上呢？格雷格说，也许有些流星确实落到太阳上了，但肯定不足以用来解释太阳的巨大能量。

普洛克特先生在他最近的题为《不止一个世界》的著作中，再次推崇了现在已经被抛弃了的汤姆孙爵士早期的理论，我认为这是相当令人遗憾的。汤姆孙爵士认为，太阳的热或者能量是由**流星撞击**形成的。然而由物理学家提出的这一创新理论，对于天文学家而言显然是难以接受的。太阳的热量究竟源自哪里，又是如何维持的呢？正是寻觅这一问题的答案的迫切需要和强烈愿望激励了科学家，使他们认真地提出或者支持某种理论，这里就有上述那种看似合理实际上很难令人满意的学说，另外还有一种经检验发现很难被我们知晓的关于流星和自然定律的知识支持的学说。我对**流星**天文学下过不少工夫，请允许我再简要地对我坚持的观点加以阐述：无论从重要性上还是实用性上，用流星及其动力学理论去解释太阳上热量和能量的守恒都是不够的。

1. 众所周知，流星和陨石在地球的轨道内可能会撞击地球，因此，在我看来普洛克特先生的说法也就意味着：数量极其庞大的流星和陨石确定无疑会持续不断地撞击在太阳上，因为和地球相比，太阳更为巨大，也更接近这些小天体。不过，当距离很小的时候，这些流星和陨石很可能会坠入太阳内部。然而，迄今为止我们对流星的全部了解都表明：从大气中划过的大量流星，用肉眼看起来仅仅宛若豆子大小，彼此相隔数千英里，各自具有自己固定的运动系统，在某些时段还会呈现出固定的辐射点。这些都清楚地表明：它们和其他行星一样在规则的轨道上绕着太阳旋转，而仅仅因为它们实在是有些微不足道，使得它们和地球一样并不会漩涡式地或者直线式地冲向太阳。我曾经把近5,000颗常年都能观察到的流星的视轨迹投影在天球上，结果表明，对于所有那些在全年任何一个晴朗的夜晚都能观察到的流星，其中最多只有7%或8%的流星是**偶然出现的**或者不属于任何已知的流星群。目前

are now recognised, several of which appear most certainly to be connected with known comets; and from a paper I have just received from Professor Schiaparelli, of Milan, it would appear that the approximate average *perihelion* distance for 44 of these meteor systems is not less than 0.7, the earth's distance from the sun being 1.0; whilst of these 44 systems, only 4, or about 10 percent have their *perihelion* distance under 0.1, that is, approach the sun nearer than nine millions of miles! Now, it is pretty well admitted that meteors are intimately connected with comet systems, yet out of some 200 comets, the elements of whose orbits have been calculated with tolerable precision, only 5 percent have their *perihelion* distance under 0.1. The same argument holds good also for planets, whose numbers also diminish after a certain considerable mean distance from the sun. Are these facts, then, in accordance with the notion that meteoric bodies either increase in number as we approach the sun, or that meteors are so constantly losing their senses, or sense of gravity, as to be ever rushing into or against the sun? I might almost ask, do *any* meteors rush or fall into the sun? Is it probable that the *mass* of all "the countless myriads of meteors" in the solar system exceeds that of a single planet? whether that of Mercury or Jupiter does not much signify. When we take into consideration the gigantic amount of meteoric deposits required to maintain the solar heat for hundreds of millions of years, in the meteoric theory, surely the supply of meteors would long since have been exhausted, were the supply at least confined merely to the meteors under a mean distance of 0.1 belonging to our own solar system! The argument, to begin with, is in a great degree fallacious, *e.g.*, because meteors frequently strike the earth, they must, it is argued, strike the sun in vastly greater numbers, and with far greater velocities. But it is forgotten that the meteors themselves, like the earth, are revolving round the sun as a common centre, in regular orbits, and only by accident, as it were, come into mutual collision, just as the tail of a comet might pass through the system of Jupiter and his satellites; while to the end of time neither the earth nor the meteors need necessarily come into contact with the sun.

2. But it is not merely meteors belonging to the solar system which are taxed to provide fuel for our sun; *space* itself may be filled with meteors ready to impinge upon the sun. The arguments against this are: (1) judging from analogy as well as from facts, comparatively few meteors are *sporadic*, consequently the majority cannot belong to stellar space, but to our own system; (2) granting that space itself is really more or less filled with meteors, these would not necessarily rush straight into the sun, unless, as would very unlikely be the case, they had no proper motion of their own. They might be drawn into or enter our system, it is true, but, according to Schiaparelli, only to circulate like comets in definite orbits.

3. The *zodiacal light* is another victim to the emergencies of the *meteoric* theory of solar energy. Whether composed of myriads of small meteors, or merely a nebulous appendage, or atmospheric emanation belonging to the sun, is it credible that for hundreds of millions of years there could, physically speaking, be sufficient material in the zodiacal light to maintain the sun's heat and supply all the fuel required? Has it ever yet been proved that the entire mass of matter constituting the zodiacal light, is either composed of matter in a solid state, or, if it were, that its mass would be equal to that of our own earth? If

人们已经确认了100多个流星群，基本上可以确信其中一些是和已知彗星有关联的。从我刚刚收到的来自米兰的斯基亚帕雷利教授的文章可知，如果将地球与太阳之间的距离定义为1个单位，那么这些流星群中有44个的平均**近日点**距离不小于0.7个单位。另外，在这44个群中只有4个群（即大约10%）的**近日点**距离小于0.1个单位，也就是说与太阳的距离小于900万英里！目前，在很大程度上人们已经认可了流星与彗星系统密切相关的观点，尽管对于200多颗已经被颇为精确地计算出轨道参数的彗星来说，其中只有5%的彗星的**近日点**距离小于0.1个单位。对于行星来说情况也是如此，在与太阳的间距超出某一个平均距离后，行星的数目也会减少。有观点认为，或者是越靠近太阳流星体的数量越多，或者是流星体不断失去万有引力的影响而冲向太阳，或者与太阳背道而驰，但以上的事实与这些观点中的哪一个一致呢？我还想要问，真的有**任何**流星群冲向或者坠入了太阳吗？太阳系中所有的“数不尽的流星”的**质量总和**能超过一颗行星的质量吗？这颗行星是水星还是木星无关紧要。当我们按照流星理论来考虑维持太阳数亿年的发热所必需的巨大数量的流星沉积时，如果我们将流星的供给最小程度地局限在太阳系中距离太阳的平均距离小于0.1个单位的范围内，那么很显然流星的供给在很早很早以前就已经被消耗殆尽了！这一理论从一开始就是相当荒谬的。比如，这一理论声称，因为流星频繁地撞击地球，所以也一定有数量更为庞大的流星以极快的速度撞击太阳。但是该理论却忽略了流星本身也是像地球一样在规则的轨道上围绕太阳这一公共中心旋转，只是偶尔才会与其他星体发生碰撞，这就像彗尾偶尔可能会扫过木星及其卫星组成的系统一样，但是无论什么时候，地球和流星都绝不会与太阳发生接触。

2. 但是也许，为太阳提供燃料的不只是太阳系内的流星，**太空**中可能到处都有会撞击太阳的流星。不过，以下的证据是反对这一观点的：(1) 从事实和推理可以判断，只有极少数流星是**偶现**流星，绝大多数流星不可能属于其他恒星空间，只能是属于我们的太阳系；(2) 即使太空中确实散布着一些流星，那么它们也不太可能直接冲向太阳，除非在极特殊的情况下，这些流星没有自己固有的运动方向。这样的流星或许有可能被吸引或进入到我们的太阳系内，但是，根据斯基亚帕雷利的观点，它们只能是像彗星一样在一定的轨道上绕太阳运行。

3. 关于太阳能量的**流星**理论出现后，**黄道光**就成了另一个受害者。不管它是由无数小流星组成的，还是由某种星云物质或者来自太阳的大气组成的，对于黄道光中存在的物质能够维持数亿年间太阳的发热，并且为太阳提供全部所需燃料，我们能够相信吗？另外，是否已经有人证明过组成黄道光的所有物质都是固态的？即便如此，它们的总质量是否能抵得上我们地球的质量？如果黄道光是由许多彼此独立

composed of separate meteors, are they not each individually revolving round the sun, rather than occupied in being gradually drawn into it as a vortex? *

Of course I do not say that meteors are *never* drawn into the sun, or that they may not occasionally and by accident enter the solar atmosphere; I have merely endeavoured to show that, from what we really do know about meteors and the laws of nature, it is highly improbable that our sun could derive, in sufficient quantity, a needful supply of fuel from meteoric sources. The comet of 1843, which approached the sun within 550,000 miles, was not sensibly deflected from its course; it is just possible that so small a thing as an aërolite might at that distance have been drawn into the sun; but is it not also possible, from what we know of comet and meteor systems, it may be wisely ordained that the smaller bodies of our solar system, such as meteors, do not as a rule approach the sun too closely; and they probably do not, if their *perihelia* distances are rarely under 10,000,000 of miles?

Aërolites are doubtless of larger size and weight than shooting stars, and as far as is yet known, not so regular in their appearance as shooting stars; but even with that class of phenomena, we notice a certain degree of periodicity in *maxima* and *minima* for certain times of the year, tending to show that they also may be subject to regular laws, and not fall so frequently or promiscuously upon the sun's surface as has been sometimes supposed. If they do not fall in vastly greater numbers, area for area, upon the sun than they do upon our earth, certainly the dynamical effect would be very minute! I may here also observe that even these bodies generally fall to the earth without being consumed, and with a very moderate velocity; their original cosmical velocity having been lost before reaching the surface of the earth. In the case of an aërolite falling upon the sun's surface, its original velocity may similarly have been gradually checked in its passage through the solar atmosphere, and a considerable amount therefore of the calculated mechanical effect lost. Small meteors would probably be consumed thousands of miles from the real body of the sun, seeing that the sun's inflamed atmosphere is now known to extend at times some 50,000 miles. It might almost be a question whether the sun's proper heat may not even be greater than that caused by the simple friction of a meteor through the solar atmosphere!

I merely allude to these minor matters, however, in order to point out some of the numerous uncertainties and difficulties connected with this meteoric or mechanical theory of the origin and conservation of solar heat, in addition to those already alluded to, bearing more especially upon the astronomical bearings of the question. For the present it must still remain a mystery, whence or how the solar heat is maintained, or to what extent really wasted.

(**2**, 255; 1870)

Robert P. Greg: Prestwich, Manchester, July 11.

---

* We beg to refer our readers to Jones' and Liais' observations of the Zodiacal Light. They certainly have not received the attention in this country that they deserve.—ED.

的流星组成的话，那么这些流星为什么不会各自独立地围绕太阳运行，而非要旋涡式地逐步坠落到太阳里面去呢？*

当然，我并不是说流星**从来都不会**坠落到太阳里，也不是说他们不可能偶然地进入太阳大气。我只是想尽力说明，根据我们确确实实掌握的关于流星和自然定律的知识，太阳所需的数量巨大的燃料真的不太可能来自流星。1843 年，彗星靠近太阳时与太阳的间距小于 550,000 英里，但它的轨道并没有发生明显的偏转，在这样的距离下可能只有像陨石那么小的物体才会被拽入太阳吧。但是根据我们对彗星和流星群的了解，太阳系中像流星这样的小天体并不会十分靠近太阳。如果这些小天体的**近日点**距离很少小于 10,000,000 英里的话，那它们怎么可能会是太阳燃料的主要来源呢？

就目前所知，**陨石**的尺寸和重量无疑比流星大，但不像流星那样有规律地出现。我们注意到，陨石这类现象在一年内出现的频率有**大**有**小**，存在一定程度的周期性，这说明陨石现象可能也是遵循一定的规律的，而不是像人们曾经猜测的那样频繁而又杂乱地坠落到太阳表面。如果坠落在太阳上的陨石数量并不比坠落在地球上的陨石数量大很多的话，那么产生的动力学影响必将是非常微小！这里我还想指出，陨石坠落到地球上时一般并没有燃烧耗尽，并且落地的速度是比较慢的，在到达地球表面之前，它们原初的宇宙速度已经被消耗了。对于坠落到太阳表面的陨石来说，情况可能是类似的，在陨石通过太阳大气层时它们的速度就会逐渐降低，因此原先估算的力学效应也随之大幅降低了。我们知道太阳燃烧的大气有时会向外延伸 50,000 英里以上，因此小流星很可能在距离太阳实体数千英里之外就已经燃烧耗尽了。难道太阳本身的热量还不及流星穿越太阳大气时摩擦产生的热量吗？这恐怕是个问题吧！

这里，我仅仅提到了一些细节以指出用流星及其动力学理论解释太阳能量的来源和守恒时存在各种不确定因素和困难，除此之外，先前已经有人对该理论提出过一些疑问，我在这里则更多的是从天文学的角度对该理论提出质疑。至于太阳的热量究竟源自哪里以及是怎样维持和被消耗的，现在仍然是个谜。

（金世超 翻译；何香涛 审稿）

* 请读者参考琼斯和莱斯对于黄道光的观测，可惜他们并没有在这个国家受到应有的重视。——编辑注。

The Source of Solar Energy

R. A. Proctor

Editor's Note

Richard Proctor writes here to correct an earlier essay of a Mr. Greg, which he felt misrepresented his views regarding the energy source of the Sun. Greg implied incorrectly that Proctor believed infalling meteors to be the Sun's sole source of energy, and to act by impact at the solar surface. Proctor here claims that he has no firm opinion on the matter, although many facts do point to infalling meteors as an energy source. They would still deliver their kinetic energy, he says, regardless of whether they reach the Sun in a solid, fluid or vaporous state. Proctor also argues that many facts support the conjecture that the zodiacal light is best explained by a mass of bodies in the neighbourhood of the Sun.

MR. Greg ascribes to me views I do not hold, and then employs my own reasoning to overthrow them. He must have formed his conceptions of my theories from Prof. Pritchard's critique of my "Other Worlds"—a most unreliable source.

To begin with,—I do *not* believe that the solar heat supply is solely derived from the downfall of meteors. I have impressed this very clearly at p. 54 of my "Other Worlds".

I do not believe that *any part whatever* of the solar heat supply is derived from meteoric *percussion*, nor that any meteor ever comes within tens of thousands of miles of the sun's surface in the solid state.

Mr. Greg is very careful to show me that the meteor-systems encountered by the earth cannot fall into the sun. I dwell on this very fact at p. 203 of "Other Worlds"—I say, *totidem verbis*, that no known meteoric system can form a hail of meteors upon the sun. "It is forgotten," says Mr. Greg, "that the meteors themselves revolve round the sun," &c. If *he* has at any time forgotten this, I certainly have not.

"Has it ever been proved," he asks me, "that the entire mass of meteors constituting the zodiacal light, is either composed of matter in a solid state, or, if it were, that its mass would be equal to that of our own earth?" I answer, as Mr. Greg would—"No, it has not been proved, nor is it by any means probable."

There is nothing new to me in Mr. Greg's letter, and little which I have not described myself long ago in the *Intellectual Observer and Student* of 1867, 1868, and 1869. To suppose that I should venture to treat at all of meteoric astronomy, in ignorance of such elementary facts—the very ABC of the science—is not complimentary. Mr. Greg might,

太阳能量的来源

普洛克特

编者按

理查德·普洛克特写这篇文章的目的是为了纠正格雷格先生在以前发表的一篇论文中的错误观点，他认为格雷格先生误解了他对太阳能量来源问题的认识。格雷格先生错误地指出普洛克特认定陨落的流星是太阳能量的唯一来源，而且流星还会与太阳表面发生碰撞。普洛克特在这里声明，他在该问题上还没有十分明确的看法，尽管有许多事实确实证明了陨落的流星可以作为一种能量来源。他说，不管流星是以固态、液态或气态方式降落到太阳上，它们都会释放自己的动能。普洛克特还争辩说，许多事实都支持黄道光形成的最佳解释是它来自太阳附近的一些小天体。

格雷格先生将一些本来不属于我的观点强加在我身上，然后又运用我的推理去推翻这些观点。他一定是从普里查德教授对我的《其他世界》一书的评论中形成了对我的理论的看法，然而这一评论是站不住脚的。

首先，我**不**认为太阳热量的供给完全来自坠落的流星。对于这一点，我在《其他世界》一书的第 54 页已经非常清楚地阐述过了。

其次，我不认为会有**任何一部分**太阳热量源自流星**撞击**，也不认为任何流星会以固态的形式到达距离太阳表面数万英里的范围内。

格雷格先生非常仔细地向我说明了与地球遭遇的流星群不可能坠入太阳。其实，我在《其他世界》一书的第 203 页已经对这一事实进行了详述。在那段叙述中，我正是这样指出的，没有任何已知的流星群可以在太阳表面形成流星雨。格雷格先生在他的文章中提到“流星本身围绕太阳旋转的事实被遗忘了”，即使**他**在某些时候忘了这一点，我也不会忘的。

他还问我“是否已经有人证明过组成黄道光的所有流星都是固态的？即便如此，它们的总质量是否能抵得上我们地球的质量？”我的回答是“没有，这一点还没有被证明过，而且现在还没有办法去证明这一点”，也许格雷格先生的答案也是如此吧。

对我来说，格雷格先生的文章中并没有什么新内容，几乎全部都是很久以前我在 1867、1868 和 1869 年的《聪明的观察家与学生》上阐述过的。假如有人认为我是在完全不了解如此基本的科学常识的情况下无知无畏地研究流星天文学的话，那

at least, have examined what I have written before assigning to me the absurdities he attacks so successfully.

The fact is, this matter of the solar energy only comes in *par parenthèse* in my "Other Worlds". I express no confident opinion whatever about it. I point to some deductions from known facts, and respecting *them* express a certain feeling of confidence. It is not my fault (nor, indeed, can I blame Mr. Greg) if Prof. Pritchard has tacked my words "I am certain" (used with reference to reliable inferences) to a theory respecting which I have distinctly written, that "I should not care positively to assert" its truth. Even that theory is not the absurd one attacked (very properly) by Mr. Greg.

For the rest, most of Mr. Greg's letter is sufficiently accurate, but there are two mistakes in it.

1. We have abundant evidence that the density of the aggregation of cometic perihelia increases rapidly near the sun. For example, whereas between limits of distance 40,000,000 and 60,000,000 miles from the sun this density is represented by the number 1.06, it is represented by the number 1.67 for limits 20,000,000 and 40,000,000 miles, and by the number 8.65 within the distance 20,000,000 miles. The evidence derived from this observed increase of aggregation is not affected by what we know of those cometic or meteoric systems whose orbits nearly intersect the earth's (for they must form but the minutest fraction of the total number) nor by the observed minimum perihelion distance of cometic orbits (for observed comets are but the minutest fraction of the total number).

2. It makes no difference whatever as regards the force-supply of the solar system, whether the substance of a meteor reaches the sun in the solid, fluid, or vaporous state. Given that the substance of a meteor, moving at one time with a certain velocity at a certain distance from the sun, is at another time (after whatever processes) brought to rest upon or within the sun's substance, then either the "force-equivalent" of its motion has been already distributed or the substance of the meteor is in a condition to distribute that "force-equivalent" mediately or directly. In other words, either heat and light have been already distributed, or the central energy has been recruited to the full extent corresponding to the mass, motion, and original distance of the meteor.

I may express here my agreement with the opinion of the Editor of *Nature* that the observations made on the zodiacal light by Lieut. Jones and M. Liais ought to be taken into account in any theory of that mysterious object. Taken in conjunction with the other known phenomena of the zodiacal light, they admit of but one interpretation as to the position, dimensions, and general characteristics of the object. Taken alone, we might infer from them that the zodiacal light is a ring of bodies or vapours travelling around the earth (at a considerable distance); other phenomena suggest that the zodiacal light is a disc of bodies or vapours travelling around the sun; yet others suggest that the zodiacal light is a phenomenon of our own atmosphere. But the only theory which accounts at once

我要说我实在很不欢迎这样的臆测。格雷格先生将那些谬论强加于我并进行如此成功的攻击之前，至少应该先检查一下我到底写过什么吧。

事实上，在我的《其他世界》一书中只是**附带性**地谈了一些关于太阳能量的事情。我只是根据已知事实提出了一些推论，对于**它们**我并不确信，也没有十足的把握。如果普里查德教授将我说过的“我确信”（这个词是用来表述一些可靠的推论的，而且使用时都附注了参考文献）强加到某个我确实写出来过但“我并没有担保”其正确性的理论上的话，那这并不是我的过错（当然，我也不会责备格雷格先生），不过即便这个被普里查德教授强行认为是我确信的理论也并不是格雷格先生（非常合理地）攻击过的那条荒谬的理论。

除此之外，格雷格先生的文章中大部分内容都是非常准确的，不过其中也有两处错误：

1. 我们有充分的证据表明，随着彗星近日点靠近太阳，其聚集密度迅速增大。例如，在距离太阳 40,000,000~60,000,000 英里的范围内，这一密度为 1.06；在 20,000,000~40,000,000 英里的范围内，这一密度为 1.67；而在 20,000,000 英里以内，这一密度增加到 8.65。这些观测得到的聚集度的增加，并不受我们已知的某些彗星或流星群的轨道几乎与地球轨道相交的影响（这样的群肯定会出现，但只占总数中极小的一部分），也与观测到的彗星轨道的最小近地点的距离无关（人们观测到的彗星仅仅是全部彗星中极小的一部分）。

2. 从考虑太阳系的支撑力的角度来看，组成流星的物质到底是以固态、液态还是气态形式到达太阳并没有什么区别。给定某种流星物质状态，在距太阳某一距离以一定的速度运动，而在另一时刻（不管经历了什么过程）被吸引到太阳上的，甚至成为太阳的组成物质，那么，流星运动的“等效力”要么已经转化，要么流星的组成物质正处在间接或直接地转化为这种“等效力”的过程中。换句话说，要么已经转化成光和热，要么中心能量已经有所增加，但总是与由流星的质量、运动以及初始距离确定的能量总量相一致。

《自然》的编辑认为，关于黄道光这种神秘物体的任何理论都应该考虑到琼斯中尉和里阿斯对黄道光的观测结果，对此我表示同意。结合其他关于黄道光的已知现象，他们认为对于黄道光这种物体的位置、尺度以及一般特性只有一种解释。单独来看，根据他们的观测结果我们可能会推测出黄道光是由围绕地球运行的天体或者蒸气组成的圆环（在距离地球相当远的位置上），但另一些现象则暗示黄道光是由围绕太阳运行的天体或蒸气组成的圆盘，同时还有其他现象表明黄道光是由地球大气本身造成的一种现象。不过只有一种理论可以同时解释人们观测到的**所有**现象，那就是：黄道光不过是由于太阳周围连续存在许多小天体碎块或各种蒸气（诸

for *all* observed phenomena, is that which regards the zodiacal light as simply due to the continual presence in the sun's neighbourhood of bodies or vapours (meteoric or cometic, or both) which come there from very far beyond the earth's orbit, and pass away again on their eccentric orbits. A disc thus formed of continually varying constituents would shift in position, and would wax and wane in extent as well as splendour, precisely as the zodiacal light is observed to do.

(**2**, 275-276; 1870)

如流星或彗星，或者两者同时存在）而形成的，这些小天体碎块和蒸气来自非常遥远的地方，远远超出了地球轨道的范围，之后，他们会沿着自己的偏心轨道远离而去。这样一个由持续变化的成分形成的圆盘，其位置自然会发生漂移，而其大小和绽放出的光辉也会时增时减，这和人们观测到的黄道光的表现一模一样。

（金世超 翻译；何香涛 审稿）

The Source of Solar Energy

R. A. Proctor

Editor's Note

Still feeling misinterpreted, Richard Proctor tries once more to clarify his views. He asserts that he believed infalling meteors to supply a portion, but not all, of the Sun's energy. Moreover, he says these meteors are not from the same source as those which strike the Earth, as these cannot reach the Sun. Mr. Greg also made a mistake, Proctor pointed out, in his considerations of how an infalling body might affect the energy of the Sun. Were a large mass of solid iron to enter into the Sun, it would not decrease the Sun's overall energy, as Greg had asserted, though it might decrease the solar temperature. The total solar energy would still increase, just as it would if the earth itself fell into the Sun.

MR. Greg still misses my meaning. I do believe that meteors supply a portion of the solar energy, and I also believe that they fall in enormous quantities into the sun; what I do not believe is that the whole solar energy is derived from meteors, or that any meteors fall in a solid state upon the sun (whose surface is also certainly not solid, even if any part of his mass be).

Mr. Greg's reasoning only proves what I have already pointed out, that none of the meteor systems our earth encounters can supply a meteoric downfall on the sun. This is, however, so obvious as to need no enforcing.

The reasoning by which I show that enormous quantities of meteors must fall upon the sun is wholly untouched by Mr. Greg's arguments, and is, so far as I can see, simply incontrovertible.

Surely Mr. Greg is not in earnest in saying that there would be a loss of solar energy if a large mass of iron fell on the sun before it was quite melted (any conceivable mass would, by the way, be vaporised), *because the sun would have to melt the portion which remained solid*. That solar energy would be consumed in the process is true enough; but if Mr. Greg supposes that the total solar energy would be diminished, he altogether misapprehends the whole subject he is dealing with. If the action of the solar energy in changing the condition of matter forming (as the imagined meteorite would) part of the sun's substance had to be counted as loss of energy, the sun would be extinguished in a very short time indeed. Such processes involve exchange, not loss.

If the earth could be placed on the sun's surface, the action of the sun in melting and vaporising the earth, and producing the dissociation of all compound bodies in the earth's substance, would involve an enormous expenditure of energy, yet the solar energy,

太阳能量的来源

普洛克特

编者按

理查德·普洛克特还是认为他的观点被别人误解了，他又一次发表文章澄清自己的观点。他宣称自己的观点是陨落的流星是太阳能量的一部分来源，而不是全部。他还认为这些流星与撞击地球的流星的来源不同，因为撞击地球的流星无法到达太阳。普洛克特指出，格雷格先生在考虑下落物如何影响太阳的能量时犯了一个错误。如果大量的固态铁落到太阳上，尽管它们可能会降低太阳的温度，但不会像格雷格所说的那样降低太阳的整体能量。太阳的总能量是会升高的，就像如果地球坠入太阳也会提升太阳的能量一样。

格雷格先生仍然误解了我的意思。我确实相信太阳能量的一部分来自流星，并且我也相信有数量庞大的流星坠入太阳。但是，我不相信太阳的全部能量都是来自流星，也不相信流星会以固态形式坠落在太阳上（即使太阳的一部分物质是固态的，但其表面肯定不是固态的）。

格雷格先生的推理只是证明了我先前已经指出的，即地球遭遇的流星群不可能形成太阳上的流星雨。然而这一道理太明显了，没必要再次强调。

我还通过推理指出必然有数量庞大的流星坠落在太阳上，但是格雷格先生在申辩中完全没有谈及这一条。据我看来，这一推理是无可置疑的。

显然，格雷格先生并没有坦诚地指出：如果有大量的铁在熔化之前坠落到太阳上（顺便提一下，任何可以想象到的物质都会被汽化），太阳的能量必然会有损失，**因为太阳必然会熔化掉那些还是固态的物质**。在这一过程中太阳的能量会被消耗，这是完全正确的。但是，如果格雷格先生由此而认为太阳的总能量也会减少，那他就完全误解了他所讨论的整个内容。如果那些即将成为太阳的一部分的物质在发生状态改变（那些假想中的流星就会经历这一过程）时会造成太阳能量的损失，那么太阳肯定会在很短的时间内熄灭。因此，这样的过程中会有能量的交换，但太阳的能量不会损失。

假设把地球放到太阳表面，那么，在太阳使得地球熔化、汽化，并使组成地球的所有物质发生分解的过程中，肯定会有巨大的能量消耗。但是，从整体上来看太阳的能量应该会有所增加的，即便我们不考虑地球可以为太阳提供燃料这一事实。

considered as a whole, would be recruited, even apart from the fact that the earth would serve as fuel. The absolute temperature of the sun would, I grant, be diminished in this imaginary case, though quite inappreciably, but his total heat would be increased by whatever heat exists in the earth's substance.

Apart from this, however, if the minimum velocity with which a meteor or other body can reach the sun, is such as would—if wholly applied to heating the body—completely melt it, then the size of the body makes no difference whatever in the result. The meteor might not be melted if enormously large, but in that case the balance of heat would be communicated to the sun. In reality, of course, the heat corresponding to meteoric motion near the sun is very far greater than is here implied.

But I really must apologise for bringing before your readers considerations depending on the most elementary laws of the conservation of energy.

(**2**, 315; 1870)

我承认，在这个虚构的例子中太阳的绝对温度可能会下降，不过下降的幅度应该是非常微小的，而太阳的总热量应该会由于地球物质中存在的热量而增加。

然而除此之外，如果一颗流星或者其他天体到达太阳表面时的最低速度用来加热该天体就足以使其完全熔化的话，那么天体本身的尺寸差异对于结果就没有什么影响了。如果流星过于巨大的话，也许有可能不会被完全熔化，但这种情况下太阳仍会达到热平衡。当然，实际情况是流星靠近太阳时，其运动对应的热量远比这里假设的大得多。

不过在这里，我也确实要向贵刊的读者们表示歉意，以上只是我基于能量守恒的一些最基本定律所作的一些个人思考。

（金世超 翻译；何香涛 审稿）

The Coming Transits of Venus*

Editor's Note

This comment points out that the next transit of the planet Venus across the Sun's disk is expected in four years. The author calls for scientists to begin developing plans for observations and cooperation. One key challenge is to improve on the error of as much as twenty seconds still present in estimates of the moment when the planet "touches" the edge of the Sun, an error that came from the fuzzy boundaries of the Sun's atmosphere.

TRANSITS of Venus over the disc of the sun have more than any other celestial phenomena occupied the attention and called forth the energies of the astronomical world. In the last century they furnished the only means known of learning the distance of the sun with an approach to accuracy, and were therefore looked for with an interest corresponding to the importance of this element. Although other methods of arriving at this knowledge with equal accuracy are now known, the rarity of the phenomenon in question insures for it an amount of attention which no other system of observation can command. As the rival method, that of observations of Mars at favourable times, requires, equally with this, the general co-operation of astronomers, the power of securing this co-operation does in itself give the Transits of Venus an advantage they would not otherwise possess.

Although the next transit does not occur for four years, the preliminary arrangements for its observation are already being made by the governmental and scientific organisations of Europe. It is not likely that our Government will be backward in furnishing the means to enable its astronomers to take part in this work. The principal dangers are, I apprehend, those of setting out with insufficient preparation, with unmatured plans of observation, and without a good system of co-operation among the several parties. For this reason I beg leave to call the attention of the Academy to a discussion of the measures by which we may hope for an accurate result.

In planning determinations of the solar parallax from the Transits of Venus, it has hitherto been the custom to depend entirely upon the observations of the internal contact of the limbs of the sun and planet proposed by Halley. It is a little remarkable, that while astronomical observations in general have attained a degree of accuracy wholly unthought of in the time of Halley, this particular observation has never been made with a precision at all approaching that which Halley believed that he himself had actually attained. In his paper he states that he was sure of the time of the internal contact of Mercury and the

* Substance of a paper read before the Thirteenth Annual Session of the American Academy of Sciences, held at Washington, by Prof. Simon Newcomb. (The original paper was illustrated by diagrams.)

即将到来的金星凌日*

编者按

这篇文章指出下一次金星凌日（金星划过日面）将发生在4年后。作者呼吁科学家们应该开始着手制定观测和协作的计划。一直以来，由于太阳大气的边界比较模糊，所以在判断行星“接触”日面边界的时刻上一直存在长达20秒的误差。对此作出改进是一件极具挑战的事情。

金星从日面划过而发生的凌日现象比其他任何天文现象都更能引起人们的关注，天文界投入了巨大的精力去研究这一现象。在上个世纪，研究凌日现象是精确获得日地距离的唯一方法，凌日现象的这种重要性使得人们对它的出现相当期待。尽管现在也有其他的方法能以同样的精度确定日地距离，但金星凌日这种天象的罕见性使它受到了比其他观测更多的关注。和观测金星凌日一样，对火星的某些特殊观测同样需要天文学家们的集体协作，而确保这次协作的能力为观测金星凌日提供了在火星的观测中不曾拥有的优势。

尽管下一次凌日是在4年后，但是欧洲的政府和科学机构已经开始着手准备观测的前期工作了。在给天文学家们提供所需的各种支持以便他们能够参与这项观测方面，本国政府不见得会落后。我最担心的问题是出发前的准备工作不够充分，观测计划不够完善，以及各观测组之间不能密切协作。因此我恳请科学院能对如何观测才能期望得到精确的结果进行讨论给予关注。

利用金星凌日确定太阳视差的方法最早是由哈雷提出的，这种方法依赖于对太阳和行星内切的观测。这里尚有一点值得商榷，就是他在论文中说他确定的水星和太阳内切的时间可以精确到一秒以内。尽管现在的天文观测已经达到了哈雷时代无法想象的精度，但实际测量依然没能达到哈雷认为他自己曾经实际达到过的精度。最近一次（1868年11月）对水星凌日的观测，内切时间仍有几秒的误差。而且大家也都知道在对1769年6月的上一次金星凌日观测时，并没有以预期的精度确定太

* 这是在华盛顿举行的美国科学院第13次年会上西蒙·纽科姆教授所宣读文章的要点（原始文章有图示）。

sun within a second. The latest observations of a transit of Mercury, made in November 1868, are, as we shall presently see, uncertain by several seconds. It is also well known that the observations of the last transit of Venus, that of June 1769, failed to fix the solar parallax with the certainty which was looked for, the result of the standard discussion being now known to be erroneous by one-thirtieth of its entire amount. One of the first steps to carry out the object of the present paper will be an inquiry into the causes of this failure, and into the different views which have been held respecting it.

The discrepancies which have always been found in the class of observations referred to, when the results of different observers have been compared, has been generally attributed to the effect of irradiation. The phenomenon of irradiation presents itself in this form: When we view a bright body, projected upon a dark ground, the apparent contour of the bright body projects beyond its actual contour. The highest phenomenal generalisation of irradiation which I am aware of having been reached is this: A lucid point, however viewed, presents itself to the sense, not as a mathematical point, but as a surface of appreciable extent. A bright body being composed of an infinity of lucid points, its apparent enlargement is an evident result of the law just cited.

[The speaker here drew a number of diagrams for the purpose of illustrating his theory.]

The following diagrams show the effect of this law upon the time of interval contact of a planet with the disc of the sun. The planet being supposed to approach the solar disc, Fig. 1 shows the geometrical form of a portion of the apparent surface of the sun, or the phenomenon as it would be if there were no irradiation immediately before the moment of internal contact. Fig. 2 shows the corresponding appearance immediately after the contact. To indicate the effect of irradiation, or to show the phenomenon as it will actually appear on the theory of irradiation, we have only to draw an infinity of minute circles for each point of the sun's disc visible around the planet to indicate the apparent phenomenon. The effect of this is shown in Figs. 1*a* and 2*a*. The exceedingly thin thread shown in Fig. 1 is thus thickened as in 1*a*, and the sharp cusps of Fig. 2 are rounded off as shown in Fig. 2*a*. The apparent radius of the planet is diminished by an amount equal to the radius of the circle of irradiation, and the radius of the sun is increased by the same amount. Comparing Figs. 1*a* and 2*a*, it will be seen that the moment of internal contact is marked by the formation of a ligament, or "black drop," between the limbs of the sun and the planet. This formation is of so marked a character that it has been generally supposed there could be little doubt of the moment of its occurrence. The remarks of the observers have given colour to this supposition, the black drop being generally described as appearing suddenly at a definite moment.

Examining Fig. 2*a*, it will be seen that the planet still appears entirely within the disc of the sun. The geometrical circle which bounds the latter, and that which bounds the planet, instead of touching, are separated by an amount equal to double the irradiation. And, when they finally do touch, neither of them will be visible at the point of contact. The estimate of the moment of contact must therefore be very rough, the means of estimating

阳的视差，而现在知道了当时的结果有 1/30 的误差。因此实现本文目标的第一步就是研究这次失败的原因，以及关于这次失败的各种不同的观点。

在比较不同观测者的结果时，经常会发现存在差异，这通常会被认为是由一种辐射效应引起的。这种辐射效应通常表述为；在观察投射到昏暗背景上的明亮物体时，它的轮廓看起来比实际的大，我能想到的最明显的辐射效应就是一个亮点，无论怎么看，它呈现出的都不是一个数学上的点，而是有扩展的圆面。按照这个规律，一个由无限多个亮点组成的物体看起来也应该有更大的轮廓。

[报告人这时拿出几张图来解释他的理论。]

这几张图给出了按照这个规律当行星和日面内切时应呈现的效果。假如行星接近日面，图 1 是不存在辐射效应时内切前太阳表面一部分的几何图形，图 2 是刚内切后相应的图像。要表示出辐射效应，展示由辐射理论得到的实际景象，只需在行星周围的日面上的每一点画出小圆，效果如图 1*a*、2*a* 所示，图 1 中的细线在图 1*a* 中变粗了，图 2 中相对明显的切点在图 2*a* 中变得圆滑，行星的可视半径也减小到辐射圆半径的大小，而太阳的半径增加了同样多的量。比较图 1*a* 和 2*a*，可以看出内切时刻的标示是太阳和行星边缘处形成的一条切线，也就是常说的“黑滴”现象。通常认为这种“黑滴”现象具有十分明显的特征，以至于一般认为，它的发生时刻是无可置疑的，当然观测者的描述也给这个解释增添了许多色彩，他们通常认为“黑滴”是在某个确切的时刻突然出现的。

检查图 2*a* 可以发现，行星看起来仍然全在日面中，但太阳的实际轮廓与行星的并没有相切，而是在两倍辐射效应半径处分开了。当它们真正相接时，在切点处都看不见。因此对内切时间的估计成了一件相当困难的事情，估计的精度比用通常的准线测微计得到的精度低得多。在实际观测中，眼睛必须持续观测两个圆的切点，

being far less accurate than those afforded by a common filar micrometer. In the actual case the eye has to continue the two circles to the point of contact by estimation, through a distance depending on the amount of irradiation, while measures with a micrometer are made by actual contact of a wire with a disc. Such estimates have, therefore, been generally rejected by investigators, not only from their necessary inaccuracy, but because the time of "apparent contact" depends upon the amount of irradiation, which varies with the observer and the telescope. If there is no irradiation at all, the time of apparent contact and that of true contact will be the same, as shown in Fig. 2, while, when the cusps are enlarged by irradiation, apparent contact will not occur until the planet has moved through a space equal to double the irradiation.

Let us return to the phenomenon at actual contact. According to the theory as it has been presented, the formation or rupture of the black ligament connecting the dark body of Venus with the dark ground of the sky is a well-marked phenomenon, occurring at the moment of true internal contact. This was, I believe, the received theory until Wolf and André made their experiments on artificial transits in the autumn and winter of 1868 and 1869. They announced, as a result of these experiments, that the formation of the ligament was not contemporaneous with the occurrence of internal contact, but followed it at the ingress of the planet, and preceded it at egress. In other words, it appeared while the thread of light was still complete. They furthermore announced that with a good telescope the ligament did not appear at all, but the thread of light between Venus and the sun broke off by becoming indefinitely thin.

The result is not difficult to account for. Irradiation has already been described as a spreading of the light emitted from each point of the surface viewed, so that every such point appears as a small circle. The obvious effect of this spreading is a dilution of the light emitted by a luminous thread, whenever the diameter of the thread is less than that of the circle of irradiation. In consequence of this dilution, the thread may be invisible while it is really of sensible thickness, a given amount of light producing a greater effect on the eye the more it is concentrated. Since the thread of light must seem to break when it becomes invisible at its thinnest point, the formation or rupture of the thread marks, not the moment of actual contact, but the moment at which the thread of light becomes so thick as to be visible, or so thin as to be invisible. The greater the irradiation, and the worse the definition, the thicker will be the thread at this moment.

An interesting observation, illustrative of this point, was made by Liais at Rio Janciro, during the transit of Mercury of November 1, 1868. He had two telescopes, one much smaller than the other. He watched the planet in the small one till it seemed to touch the disc of the sun, then looking into the large one, he saw a thread of light distinctly between the planet and the sun, and they did not really touch until several seconds later.

而切点是通过由辐射量大小决定的一段距离来判断的，同时用测微计对日面相切进行测量。这样的目测显然不够精确，依赖于辐射度的“视觉接触时间”，也会因不同的观测者和望远镜而不同，因此这种方法不被研究者接受。如果完全没有辐射效应，视觉接触时间就会如图 2 所示和实际相切时间一致，而有了被辐射效应放大的交切点，行星运行到相当于两倍辐射效应的距离处才会出现可分辨的视觉接触。

现在来看实际的相切情况。根据上文的理论，在真正的内切发生时，金星黑色轮廓与昏暗的天空背景间黑色接点的形成或破裂是很显著的。在沃尔夫和安德烈于 1868 年和 1869 年的秋天和冬天进行一系列人造凌日实验之前，这一直是标准理论，至少我个人是这么认为的。他们的实验结果表明切线并不是在内切发生时形成的，而是在行星入凌之后或者出凌之前。换句话说，它在光线仍然完整时就出现了。他们进一步指出，这种现象在好的望远镜中根本不会出现，取而代之的只是在金星和太阳之间的光线会变得无限细而消失。

这个结果不难解释。辐射是由被观测表面上每点发出的光扩展而成的，因此每个点看起来都是个小圆。每当光线直径小于辐射圆时，这种明显的扩散效应是一种由明亮的光线发生散射而变弱的效应，正是因为这种扩散，切线在仍然很宽时就看不见了，一定量的光汇聚得越集中，眼睛观察到的这种效应就越明显。光线一定是在它最细的地方断裂，因此它的形成和断裂都不是发生在实际相切的时候，而是发生在光线粗得可以看见了，或者细得看不见了的时候。辐射效应越强，清晰度就越差，光线出现时也就越粗。

在 1868 年 11 月 1 日水星凌日时，里阿斯在里约热内卢进行了一项有趣的观测，其中就展示了这一点。他有两个望远镜，其中一个比另一个口径小很多（口径相差很多）。开始用小望远镜观测，在看到行星接触日面后立刻改用大口径的观测，他发现这时水星和太阳之间仍有光线存在，直到几秒后才相切。

Reference to the figures will make it clear that there is no generic difference between the phenomenon commonly called the rupture of the black drop and that of the formation of the thread of light. If the bright cusps are much rounded, as in Fig. 1*a*, the appearance between them is necessarily that of a drop, while if they are seen in their true sharpness, as in Fig. 1, the form of the drop will not appear. It has been shown that with different instruments the phenomenon of contact may exhibit every gradation between these extremes. The only well-defined phenomenon which all can see is the meeting of the bright cusps and the consequent formation of the thread of light at ingress, and the rupture of the thread at egress.

To recapitulate our conclusions—

1. The movement of observed internal contact at ingress is that at which the thread of light between Venus and the sun becomes thick enough to be visible.
2. The least visible thickness varies with the observer and the instrument, and, perhaps, with the state of the atmosphere.
3. The apparent initial thickness of the thread varies with the irradiation of the telescope.

Two questions are now to be discussed. The observed times varying with the observer and the instrument, we must know how wide the variation may be. If it be wide enough to render uncertain the results of observation, we shall inquire how its injurious effects may be obviated.

The first question can be decided only by comparison of the observations of different observers upon one and the same phenomenon. For such comparison I shall select the observations of the egress of Mercury on the occasion of its last transit over the disc of the sun. This selection is made for the reason that this egress was observed by a great number of experienced observers with the best instruments, while former transits, whether of Venus or Mercury, have been observed less extensively or at a time when practical astronomy was far from its present state of perfection, and that the transit in question would therefore furnish much better data of judging what we might expect in future observations. The comparison was made in the following way:—I selected from the "Astronomische Nachrichten", the "Monthly Notices of the Royal Astronomical Society", and the "Comptes Rendus", all the observations of internal contact at egress which there was reason to believe related to the breaking of the thread of light, and which were made at stations of known longitude. Each observation was then reduced to Greenwich time, and to the centre of the earth.

From these comparisons it appears that the contact was first seen by Le Verrier, at Marseilles, at two seconds before nine o'clock, Greenwich time. In one second more it was seen by Rayet at Paris, Oppolzer at Vienna, Lynn at Greenwich, and Kaiser at Leyden. The times, noted by twenty other observers, are scattered very evenly over the following fifteen seconds. Kam and Kaiser, at Leyden, did not see the contact until nineteen and twenty-four seconds past nine.

借助这些图示，我们可以清楚地知道黑滴断裂和光线形成之间没有什么不同，如果明亮的边缘像图 1*a* 中显示的那样再圆一点，它们之间就会出现黑滴一样的图像，而当它们像图 1 中的那么尖锐时，则不会有黑滴出现。就像上文提到的那样，用不同的设备可以看见这两种极端情况之间的任何阶段。唯一能够很好确定的现象就是亮切线相遇及入凌时光线的形成，以及出凌时光线的断裂。

下面总结我们的结论；

1、观测到的入凌时的内切运动是处于金星和太阳间的光线强到能够被看出的位置。

2、最低可见度随观测者和仪器的不同而有所不同，可能还受大气状况的影响。

3、光线的初始视强度与望远镜的辐射度有关。

下面讨论两个问题。既然观测到的入凌时间与观测者和观测仪器都有关，我们就必须知道；（1）这种误差的幅度到底有多大？（2）如果幅度太大，影响到观测结果的确定，那我们如何才能避免这种误差？

第一个问题只能通过比较不同观测者对同一天象的观测结果来确定。为进行这种比较，我选择了上次水星凌日时出凌的观测数据。之所以这样选择是因为：首先，之前的无论是水星还是金星凌日，要么观测范围太小，要么当时的实测天文学还没有发展到现在的高度，而相对于这些情况，这次对水星凌日的观测都有所改善，并且它是由许多经验丰富的观测人员用最先进的仪器记录下的观测结果；其次，我们所讨论的这次凌日观测也会为以后的观测结果提供参考数据。比较是这样进行的：我从《天文通报》《皇家天文学会月刊》以及《法国科学院院刊》中找出并统计与光线断裂有关的出凌时的内切数据，这些数据的观测经度已知，所有观测时间均归算到格林尼治时间，地点归算到地球中心，之后加以比较。

比较后我们发现，位于马赛的勒威耶在格林尼治时间差 2 秒 9 点时最早看到内切，在随后的一秒内，巴黎的拉耶、维也纳的奥波尔策、格林尼治的林恩、莱顿的凯泽也都看到了，另外 20 个观测者记录的时间平均分布在接下来的 15 秒内，而莱顿的卡姆和凯泽直到 9 点 19 秒和 24 秒时才分别看到。

It thus appears that among the best observers, using the best instruments, there is a difference exceeding twenty seconds between the times of noting contact. This difference corresponds to more than a second of arc, so that really these observations were scarcely made with more accuracy than measures under favourable circumstances with a micrometer, and are not therefore to be relied on. But a great addition to the accuracy of the determination could be made by measures of the distance of the cusps, while the planet was entering upon the disc of the sun. It would tend greatly to the accuracy of the results, if the observers should meet beforehand with the telescopes they were actually to use in observing the transit and make observations in common on artificial transits. It would be a comparatively simple operation to erect an artificial representation of the sun's disc at the distance of a few hundred yards, and to have an artificial planet moved over it by clock-work. The actual time of contact could be determined by electricity, and the relative positions of the planet and the disc by actual measurement. With this apparatus it would be easy to determine the personal errors to which each observer was liable, and these errors would approximately represent those of the observations of actual transit.

Still it would be very unsafe to trust mainly to any determination of internal contact. Understanding the uncertainty of such determinations, the German astronomers have proposed to trust to measures with a heliometer, made while the planet is crossing the disc. The use of a sufficient number of heliometers would be both difficult and expensive, and I think we have an entirely satisfactory substitute in photography. Indeed, Mr. De la Rue has proposed to determine the moment of internal contact by photography. But the result would be subject to the same uncertainty which affects optical observations—the photograph which first shows contact will not be that taken when the thread of light between Venus and the sun's disc was first completed, but the first taken after it became thick enough to affect the plate, and this thickness is more variable and uncertain than the thickness necessary to affect the eye. We know very well that a haziness of the sky which very slightly diminishes the apparent brilliancy of the sun, will very materially cut off the actinic rays, and the photographic plate has not the power of adjustment which the eye has.

But, although we cannot determine contacts by photography, I conceive that we may thereby be able to measure the distance of the centres of Venus and the sun with great accuracy. Having a photograph of the sun with Venus on its disc, we can, with a suitable micrometer, fix the position of the centre of each body with great precision. We can then measure the distance of the centers in inches with corresponding precision. All we then want is the value in arc of an inch on the photograph plate. This determination is not without difficulty. It will not do to trust the measured diameters of the images of the sun, because they are affected by irradiation, just as the optical image is. If the plates were nearly of the same size, and the ratio of the diameters of Venus and the sun the same in both plates, it would be safe to assume that they were equally affected by irradiation. But should any show itself, it would not be safe to assume that the light of the sun encroached equally upon the dark ground of Venus and upon the sky, because it is so much fainter near the border.

可以看到即便是最好的观测者使用最好的观测设备，记录到的内切时间也会存在超过 20 秒的误差，这相当于一弧秒多的视差，因此这样的观测不会比在好的条件下用千分尺测得的结果更精确，也并不可靠。但是在行星进入太阳圆面时测量切线的距离会极大地提高精度，让观测者们预先用凌日时将要使用的望远镜观测人造凌日也会提高精度。比较简单的方法是在几百码远的地方竖一个人造太阳面，然后用发条装置带动人造行星从表面经过，由电子装置确定真实的相切时刻，而行星和圆面的相对位置可以通过实际测量得到。用这样一套装置能够很容易地确定每个观测人员的个人误差，这些误差大致能够代表实际观测时的误差。

但主要依赖于内切时间的测定仍是不保险的。德国天文学家们意识到这种测量的不确定性，并提出在行星划过日面时用太阳仪进行测量。但是用足够多的太阳仪进行测量会很困难，成本也会很高，我们可以用照相法代替。其实德拉鲁先生提出过用照相法确定内切时刻的方案，但由于所有的光学观测都有同样的不确定性，结果仍会受到影响。第一张显示相切图像的照片记录的其实不是金星和日面之间的光线完全消失的瞬间，而是光线暗到刚好能影响底片成像的时刻。这种暗度比眼睛感受到的更加不确定，而且大气会降低太阳的可视亮度，同样也会削弱底片上的光化学反应效应，而照相底片也不像人眼那样能够自动适应环境。

虽然无法用照相法确定内切时间，但我们依然可以以很高的精度来测量两个天体中心的距离。拍摄一张金星影像位于日面内的太阳照片，就能精确地用千分尺确定二者中心的位置，然后以合适的精度测量出两个中心之间以英寸为单位的距离。接下来就需要知道照片上的一英寸对应多少弧度。做这件事情并不容易，不能简单地相信测量出的太阳影像的直径，因为这是光学图像，会受到辐射效应的影响。如果照相底片大小相近，且它们上面的太阳和金星图像的直径比相同，则可以假定两张图片所受的辐射效应影响相同。但是不论图像看起来如何，都不应当假定日光对金星的暗背景和对天空的影响是一样的，因为太阳光在接近边缘处暗得多。

If the photographic telescope were furnished with clock-work, it would be advisable to take several photographs of the Pleiades belt, before and after the transit, to furnish an accurate standard of comparison free from the danger of systematic error. There is little doubt that if the telescopes and operators practise together, either before or after the transit, data may be obtained for a satisfactory solution of the problem in question.

To attain the object of the present paper, it is not necessary to enter into details respecting choice of stations and plans of observation. I have endeavoured to show that no valuable result is to be expected from hastily-organised and hurriedly-equipped expeditions; that every step in planning the observations requires careful consideration, and that in all the preparatory arrangements we should make haste very slowly. I make this presentation with hope that the Academy will take such action in the matter as may seem proper and desirable.

(**2**, 343-345; 1870)

如果照相用的望远镜装了钟表的机械装置，可以考虑在凌日之前和之后对昴星团天区先拍几张照片作为比对标准，从而消除系统误差。如果观测者能够在凌日前后多和望远镜进行磨合，则能够获得更令人满意的数据，从而解决以上这些问题。

我报告的目标并不是要深入到选址和制定观测计划之类的细节中，而是要尽力向大家表明，一个仓促组织、临时装备的队伍是不会取得有价值的结果的，计划中的每一步都需要认真考虑，初步准备工作也都需要慎重考虑。我这次的介绍就是希望科学院能够在此事上准备充分，不负众望。

（余恒 翻译；蒋世仰 审稿）

Fuel of the Sun

J. J. Murphy

Editor's Note

Here Joseph John Murphy argues that perpetually infalling meteors keep the Sun hot on geological timescales. Such meteor-like phenomena have been observed, he says, citing two meteor-like bodies bright enough to stand out against the Sun's disc, which were seen to move across the Sun from west to east before disappearing. If there is a constant supply of meteors, says Murphy, wouldn't they, like all bodies in the solar system, move around the Sun from west to east, and be found in greater numbers near the Sun's equatorial regions, making these hotter than its poles? Murphy is wrong about the source of solar energy, which comes from nuclear fusion, but his speculations testify to the magnitude of the puzzle.

I am not mathematician enough to form any opinion on the merits of the controversy as to the "fuel of the sun"; that is to say, I am not able to decide whether it is consistent with the conditions of the equilibrium of the solar system that the sun's heat should have been kept up through the ages of geological time by the falling in of meteors. But I wish to state some evidence which proves that meteors are constantly falling in, though it does not touch the question whether this source is sufficient to account for the whole or any large part of the total supply of heat radiated away by the sun.

In the first place, the meteors have been seen. On Sept. 1, 1859, Mr. Carrington and another observer simultaneously observed two meteor-like bodies, of such brightness as to be bright against the sun's disc, suddenly appear, move rapidly across the sun from west to east, and disappear.

The fact that their motion was from west to east is important. If the supply of meteors to the sun is constant and tolerably regular, it is scarcely possible to doubt that the meteors, like the entire solar system, move round the sun from west to east, and occupy a space of the form of a very oblate spheroid, having its equator nearly coincident with the sun's equator.

If this is the case, the meteors ought to fall in greater numbers near the sun's equator than near his poles, making the equator hotter than the poles. Such is the fact. Secchi, without having any theory to support, has ascertained that the sun's equator is sensibly hotter than his poles. The instrument used was an electric thermo-multiplier, and the indications show, not the ratio, but the difference of the heat from the two sources compared.

太阳的燃料

墨菲

编者按

约瑟夫·约翰·墨菲在这篇文章中列举理由以证明不断陨落的流星在地质时间尺度中一直在维持着太阳的热度。他说，这种类似于流星的现象曾经被人观察到过，据说有两个像流星一样的天体亮得可以在日面上显现出来，在消失前它们从太阳的东边移动到西边。墨菲说，如果流星以固定的数量不断地落到太阳上，它们就会像太阳系中的所有其他天体一样，围绕太阳自西向东运动，在太阳的赤道附近数量比较集中，以至于使这些区域的温度高于两极。墨菲的观点当然是错误的，现在大家都知道太阳的能量来自核聚变反应，但是从他的推理中我们可以看出人们曾被这个难题深深困扰过。

我并不是个数学家，因此不足以评论关于“太阳的燃料”的争论的价值。换句话说，我不能确定“由于流星的坠落，太阳的热量在漫长的地质年代中应该一直维持在某个水平”这一观点与太阳系的平衡条件是否相一致。不过，我还是想阐述一些能够证明流星确实在持续不断地坠落到太阳上的证据，至于太阳热辐射的全部或大部分能量是否来源于此，那就是和这些证据无关的另一个问题了。

首先，人们已经观测到了流星。1859 年 9 月 1 日，卡林顿先生和另一位观测者同时观测到两个类似流星的天体，在日面的背景上仍然显得很明亮，它们自西向东迅速地横跨过太阳，突然出现然后又突然消失。

流星自西向东运动的事实是非常重要的。如果坠落到太阳上的流星是持续不断而且坠落密度差不多是均匀的话，那么几乎不用怀疑：流星肯定会像整个太阳系一样围绕着太阳自西向东运动，从空间上来说它们应该是非常扁圆的球体，而且其赤道面应该几乎和太阳的赤道面相重合。

如果事实的确如此，那么大量的流星就会坠落到太阳的赤道附近而不是两极，从而使太阳的赤道比两极更热。实际上，事实正是如此。在没有任何理论支撑的情况下，塞奇已经通过实验确认了太阳的赤道明显比两极热。实验中使用了电热倍增器，该仪器显示的结果是来自太阳赤道和两极这两个热源的热量差值，而不是其比率。

It can scarcely he doubted that the meteors must enter the sun's atmosphere with a velocity not much less than that of a planet, revolving at the distance at which they enter. We know that the sun's rotatory motion is incomparably less than this, and consequently the meteors, revolving from west to east, ought to make the sun's atmosphere move round his body in the same direction, and with greater velocity in the equatorial regions, where most meteors fall in. This is what is observed. Mr. Carrington, also without any theory to support, has shown that the motion of the solar spots from west to east is most rapid in the latitudes nearest the equator. We cannot compare the motion of the spots with that of the sun's body, as we do not see his body. But the fact that the motion from west to east is most rapid in the equatorial latitudes proves that these motions are not due to any cause like that which produces trade-winds and "counter-trades" of our planet; for, supposing the sun or any planet to rotate from west to east, in any circulation that could be produced in its atmosphere by unequal heating at different latitudes, the relative motions of the atmospheric currents in high and low latitudes would be similar to that of the trade-winds and "counter-trades", and opposite to that which the motions of the spots indicate in the atmosphere of the sun. This will be true at all depths in the atmosphere.

(**2**, 451; 1870)

毫无疑问，流星必然是以一定的速度进入太阳大气，这一速度与行星的速度相比并不会小很多,流星进入太阳大气后会在相应的距离上围绕太阳旋转。我们也知道，太阳的自转与这种旋转相比慢得多，因此，绕着太阳自西向东旋转的流星应该会带动太阳大气也绕着太阳实体沿相同方向旋转，而且在大多数流星坠入其中的太阳赤道区域，太阳大气绕转的速度会更大。观测到的结果正是如此。同样是在没有任何理论支持的情况下，卡林顿先生已经指出，太阳上纬度最靠近赤道的区域中太阳黑子自西向东运动的速度最快。由于看不到太阳实体，因此我们无法比较太阳实体的运动和黑子的运动，但是，太阳黑子自西向东的运动在赤道附近最为迅速这一事实就能证明，这些运动并不是由造成我们星球上的信风或反信风现象的那类原因引起的。这是因为，假设太阳或行星都是自西向东旋转的话，那么由于不同纬度上热量的不平衡而引起其大气发生的任何循环中，气流在高纬度和低纬度的相对运动都将与信风和“反信风”的运动情况类似，而与黑子运动所暗示的太阳大气的运动情况相反。对于太阳大气来说，在任何高度上都存在这一矛盾。

（金世超 翻译；蒋世仰 审稿）

Periodicity of Sun-spots*

Editor's Note

Rudolph Wolf's observations on the variations of sunspot activity had both estimated their average period—roughly 11.1 years—and shown an interesting contrast between the increasing and decreasing parts of the cycle. Wolf had found that, on average, activity increased over 3.7 years but then decreased more slowly over 7.4 years. Here other scientists present new observations that largely agree with Wolf's, but which also add important details, such as a fixed ratio of about 1 : 2.1 between the duration of any increasing period and the period of the immediately preceding decrease. Modern mathematical methods show that sunspots in fact cycle irregularly, with an average periodicity of about 10.5 years.

IN the short account of some recent investigations by Prof. Wolf and M. Fritz on sun-spot phenomena, which has been published lately in the "Proceedings of the Royal Society" (No. 127, 1871), it was pointed out that some of Wolf's conclusions were not quite borne out by the results which we have given in our last paper on Solar Physics in the *Philosophical Transactions* for 1870, pp. 389–496. A closer inquiry into the cause of this discrepancy has led us to what appears a definite law, connecting numerically the two branches of the periodic sun-spot curve, viz., the time during which there is a regular diminution of spot-production, and the time during which there is a constant increase.

It will be well, for the sake of clearness, to allude here again, as briefly as possible, to Prof. Wolf's results before stating those at which we have arrived.

Prof. Wolf has previously devoted the greater part of his laborious researches to a precise determination of the mean *length* of the whole sun-spot period, but latterly he has justly recognised the importance of obtaining some knowledge of the average character of the periodic increase and decrease. Hence he has, as far as he has been able to do so by existing series of observations, and his peculiar and ingenious method of rendering observations made at different times and by different observers comparable with each other, endeavoured to investigate more closely the nature of the periodic sun-spot curve, by tabulating and graphically representing the monthly means taken during two and a half years before and after the minimum. and applying this method to five distinct minimum epochs, which he has fixed by the following years:—

* Abstract of paper read before the Royal Society December 21, 1871. "On some recent Researches in Solar Physics, and a Law regulating the time of duration of the Sun-spot Period". By Warren De La Rue, F. R. S., Balfour Stewart, F. R. S., and Benjamin Loewy, F. R. A. S.

太阳黑子的周期*

编者按

鲁道夫·沃尔夫在对太阳黑子活动性的观察过程中不仅估算出了它们的平均周期——约 11.1 年，还发现太阳黑子活动性的上升周期和下降周期之间的比值存在一定的规律。沃尔夫发现，平均地说，太阳黑子活动性的上升周期为 3.7 年，而下降过程比上升过程慢，平均周期是 7.4 年。本文中指出：其他科学家的最新观测结果也在很大程度上证实了沃尔夫的基本论点，并得到了一些很重要的数据，比如任一上升周期和随后的下降周期之间的比值固定为 1 : 2.1。现代数学方法证明太阳黑子活动周期实际上并不是恒定的，平均周期大约是 10.5 年。

在沃尔夫教授和弗里茨先生最近发表在《皇家学会学报》(第 127 期，1871 年)上的关于太阳黑子现象的近期研究简报中，我们发现沃尔夫教授的有些结论与我们在《自然科学会报》(1870 年，第 389～496 页）上最新发表的有关太阳物理方面的研究结果并不十分相符。通过对这种差异的原因做进一步研究，我们似乎找到了更加明确的规律，从而在数值上建立起太阳黑子周期曲线中的两条分支之间，也就是黑子数目逐渐减少阶段与黑子数目持续增加阶段之间的联系。

在阐述我们自己的结论之前，为了清楚起见，在这里再次简要地介绍一下沃尔夫教授的研究结果。

沃尔夫教授早先曾主要致力于精确测定太阳黑子周期平均**长度**的工作，但之后他意识到获得一些与太阳黑子数目周期性增减的一般特性有关的知识也是非常重要的。因此他利用现有的观测数据，并用自己独创的特殊方法使在不同时间段由不同观测者观测得到的数据可以相互比较，为的是更准确地研究太阳黑子曲线的周期特性。他用图和表来表示最小值前后两年半时间内每月观测到的黑子数量的平均值，并将这种方法应用于 5 个最小值周期中，这 5 个最小值所在的年份如下：

---

* 本文摘自由沃伦·德拉鲁(皇家学会会员)、鲍尔弗·斯图尔特(皇家学会会员)和本杰明·洛伊(皇家天文学会会员)在 1871 年 12 月 21 日召开的皇家学会会议上宣读的题为《近期太阳物理与太阳黑子周期规律的研究》的文章。

1823.2
1833.8
1844.0
1856.2
1867.2

In a table he gives their mean numbers, expressing the solar activity, arranged in various columns; and arrives at the following results:—

(1) It is shown now with greater precision than was previously possible, that the curve of sun-spots ascends with greater rapidity than it descends. The fact is shown in the subjoined diagram, which it may be of interest to compare with the curves given previously by ourselves in the above-mentioned place. The zero-point in this diagram corresponds to the minimum of each period; the abscissae give the time before and after it, viz., two and a half years, or thirty months; the ordinates express the amount of spot-production in numbers of an arbitrary scale. The two finely dotted curves are intended to show the actual character of a portion of two periods only, viz., those which had their minima in 1823.2 and 1867.2; the strongly dotted curve, however, gives the mean of all periods (five) over which the investigation extends.

(2) Denoting by x the number of years during which the curve ascends, and presuming that the behaviour is approximately the same throughout the whole period of 11.1 years as during the five years investigated, we have the proportion

$$\frac{x}{11.1 - x} = \frac{1}{2},$$

whence

$$x = 3.7,$$

or the average duration of an ascent is 3.7 years, that of a descent 7.4 years.

(3) The character of a single period may essentially differ from the mean, but on the whole it appears that a $\left\{\begin{matrix}\text{retarded}\\ \text{accelerated}\end{matrix}\right\}$ descent corresponds to a $\left\{\begin{matrix}\text{retarded}\\ \text{accelerated}\end{matrix}\right\}$ ascent. Thus the minimum of 1844.0 behaved very normally; but that of 1856.2, and still more that of 1823.2, shown in the following diagram, presents a retarded ascent and descent; on the other hand, the minimum of 1833.8, and still more in that of 1867.2, also shown in the diagram, both ascent and descent are accelerated.

Finally Prof. Wolf arranged in the manner shown in the following table the successive minima and maxima, in order to arrive at some generalisation which might enable him to foretell the general character and length of a future period. Taking the absolute differences in time of every two successive maxima, and the mean differences of every two alternating minima, he shows that the greatest acceleration of both maximum and minimum happens

1823.2
1833.8
1844.0
1856.2
1867.2

在一张表的不同列里他列出了各月太阳黑子数目的平均值，以说明太阳的活动情况，并且得出了以下一些结论：

(1) 现在这张曲线图比以往更精确地显示出，太阳黑子数量上升的速度高于下降的速度。这个事实也可以从附表中看出，而这也可以与我们之前得出的结论（在前面提到的论文中）进行比较。图的零点对应于每个周期的最小值，横坐标给出了最小值之前与之后的时间，即，两年半或者 30 个月。纵坐标则用任意选定的尺度标示黑子产生的数量。两条由细点组成的曲线只显示出了最小值分别在 1823.2 和 1867.2 的两个周期中部分时间的实际特征。而那条由粗点组成的曲线则表示研究工作延伸的所有（5 个）周期内的平均值。

(2) 用 x 来表示曲线上升的年份数，并且假设在 5 年观察期内太阳黑子的活动规律与整个周期 11.1 年内的规律大致相同，则我们可以列出下面的比例式：

$$\frac{x}{11.1-x}=\frac{1}{2},$$

从而得到

$$x=3.7,$$

或者说曲线上升期的平均年份为 3.7 年，下降期的平均年份为 7.4 年。

(3) 对于某一个具体的周期而言，可能和平均值有一定的偏差；但对于整体而言，一个$\left\{\begin{matrix}\text{减速}\\ \text{加速}\end{matrix}\right\}$的下降对应着一个$\left\{\begin{matrix}\text{减速}\\ \text{加速}\end{matrix}\right\}$的上升。所以最小值为 1844.0 的曲线非常正常；最小值为 1856.2 和 1823.2 的曲线上升和下降的速度都比较慢，尤其是 1823.2，见下图；另一方面，最小值为 1833.8 和 1867.2 的曲线则又显示出了上升和下降速度都很快的趋势，尤其是 1867.2，见下图。

最后，沃尔夫教授用下表中的方式列出了几个连续的最小值和最大值，为的是得到一些规律性的结论，使他能够预言太阳黑子的一般特征和未来周期的长度。通过研究与每两个连续最大值对应的时间绝对差和与每两个次近邻最小值对应的时间平均差，他得出了这样的结论：黑子最大值和最小值的最大加速过程是同时发生的。

together. This result strengthens our own conclusions, to be immediately stated, by new evidence, as it is derived from observations antecedent to the time over which our researches extend.

Minima	Differences of alternating Minima	Means	Maxima	Differences of successive Maxima
1810.5				
			1816.8	
1823.2	23.3	11.65		12.7
			1829.5	
1833.8	20.8	10.4		7.7
			1837.2	
1844.0	22.4	11.2		11.4
			1848.6	
1856.2	23.2	11.6		11.6
			1860.2	
1867.2				

From this Prof. Wolf predicts for the present period a very accelerated maximum—a prediction which seems likely to be fulfilled.

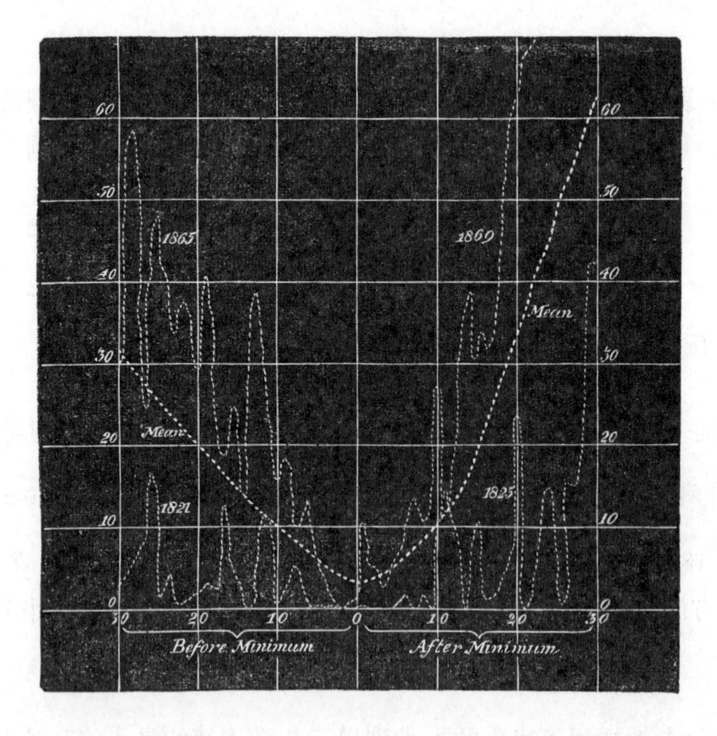

Comparing now M. Wolf's results with our own, it must not be overlooked, in judging of the agreement or discrepancy of these two independently obtained sets, that our facts have been derived from the actual measurement and subsequent calculation of the spotted area from day to day since 1833, recorded by Schwabe, Carrington, and the Kew

通过对以前观察结果的新一轮论证，这一结果也证实了下面马上要谈到的我们自己的论点。

最小值	次近邻最小值的差	平均值	最大值	连续最大值的差
1810.5				
			1816.8	
1823.2	23.3	11.65		12.7
			1829.5	
1833.8	20.8	10.4		7.7
			1837.2	
1844.0	22.4	11.2		11.4
			1848.6	
1856.2	23.2	11.6		11.6
			1860.2	
1867.2				

沃尔夫教授由此预言，我们现在的这个周期是一个加速上升到极大的周期，这个预言似乎很可能会实现。

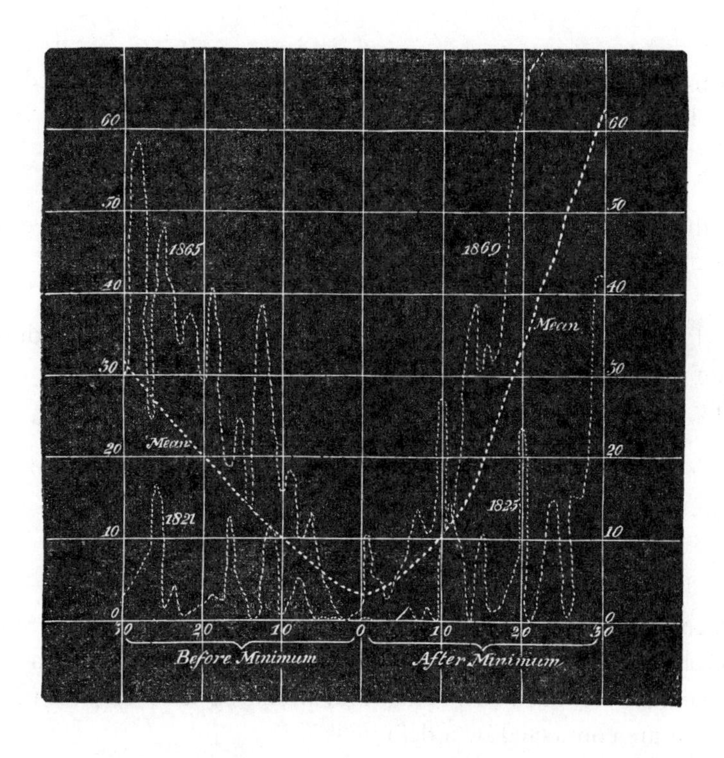

现在比较一下沃尔夫先生的结果和我们自己的结果，不要过分看重对这两个独立研究得出的结论的相同点和不同点的判断，因为我们不能忽视以下事实：我们的结论来自 1833 年以后每天对黑子区域的实际测量和计算，而这些测量数据是由施

solar photograms, which measurements are expressed as millionths of the sun's visible hemisphere, while the conclusions of M. Wolf are founded on certain "relative numbers", which give the amount of observed spots on an arbitrary scale, chiefly designed to make observations made at different times and by various observers comparable with each other. This will obviously, in addition to the sources of error to which our own method is liable, introduce an amount of uncertainty arising from errors of estimation, and the possibility of using for a whole series an erroneous factor of reduction. Nevertheless we shall find a very close agreement in various important results, and this seems a sufficient proof of the great value and reliability of M. Wolf's "relative numbers", especially for times previous to the commencement of regular sun observations.

The following is a comparison of the date of periodic epochs, as fixed by ourselves and M. Wolf: —

Minima epochs	I	II	III	IV
De La Rue, Stewart, and Loewy	1833.92	1843.75	1856.31	1867.12
Rudolf Wolf	1833.8	1844.0	1856.2	1867.2
Maxima epochs	I	II	III	
De La Rue, Stewart, and Loewy	1836.98	1847.87	1859.69	
Rudolf Wolf	1837.2	1846.6	1860.2	

It will be seen from this comparison that only one appreciable difference occurs, viz., in the maximum of 1847, which M. Wolf fixes nearly one and a quarter years before our date.

The mean length of a period is found by us to be 11.07 years, which agrees very well with M. Wolf's value, viz., 11.1 years.

We found the following times for the duration of increase of spots during the three periods, and for the corresponding decrease, or for ascent and descent of the graphic curve, beginning with the minimum of 1833: —

	Time of ascent	Time of descent
I	3.06 years	6.77 years
II	4.12 ,,	8.44 ,,
III	3.37 ,,	7.43 ,,
Mean	3.52 ,,	7.55 ,,

Prof. Wolf gives 3.7 years and 7.4 years for the ascent and descent respectively; and considering that he derived these numbers only from an investigation of a portion of each period, the agreement is indeed surprising, and would by itself suggest that the times of ascent and descent are connected by a definite law.

M. Wolf has expressed in general terms the following law with reference to this relation of increase and decrease of spots: —

瓦贝和卡林顿两人以及邱天文台的太阳照相法所记录的，并以可见太阳半球的百万分之一为单位进行测量；而沃尔夫先生的论断是建立在某些“相对数”的基础上的，他用任选的尺度给出了观测到的黑子总数，主要为了使不同时间，不同观测者的观测结果可以相互比较。很显然，沃尔夫的方法除了存在那些和我们的方法一样的错误源以外，还引入了由估算方法不正确产生的大量不确定性，而且他可能为整个推算过程选了一个错误的因子。然而我们发现两者的重要结论是非常一致的，而这似乎为沃尔夫教授的“相对数”的巨大价值和可靠性提供了强有力的佐证，尤其是在对太阳的常规观测开始之前。

下面是一些对黑子周期历元时期的比较，综合了我们和沃尔夫教授的数据：

最小值时期	I	II	III	IV
德拉鲁，斯图尔特和洛伊	1833.92	1843.75	1856.31	1867.12
鲁道夫·沃尔夫	1833.8	1844.0	1856.2	1867.2
最大值时期	I	II	III	
德拉鲁，斯图尔特和洛伊	1836.98	1847.87	1859.69	
鲁道夫·沃尔夫	1837.2	1846.6	1860.2	

从这些数据的比较中我们可以看出只有一处有明显的差异，即 1847 年的最大值，沃尔夫确定的时间大概比我们的提前了 1.25 年。

我们发现一个周期的平均长度为 11.07 年，与沃尔夫先生的周期 11.1 年非常接近。

我们找出黑子数量增长的 3 个时间段和黑子数量下降的 3 个时间段，对应于图中曲线中的上升阶段和下降阶段，最小值出现在 1833 年。

	上升时间	下降时间
I	3.06 年	6.77 年
II	4.12 年	8.44 年
III	3.37 年	7.43 年
平均	3.52 年	7.55 年

沃尔夫教授分别给出 3.7 年的上升期和 7.4 年的下降期，考虑到他仅仅是在考查每个周期的部分时间后得出的这些数字，这种一致性确实令人惊讶，这本身也说明了黑子增多期和减少期确实会与某些特定的规律相联系。

沃尔夫先生用大众化的语言描述了关于黑子数量增长期和下降期之间的规律：

"The character of a single period may essentially differ from the mean behaviour, but on the whole it appears that a $\left\{\begin{matrix}\text{retarded}\\ \text{accelerated}\end{matrix}\right\}$ descent corresponds to a $\left\{\begin{matrix}\text{retarded}\\ \text{accelerated}\end{matrix}\right\}$ ascent."

We, on the other hand, have, by an inspection of our curves (*vide Phil. Trans.* 1870, p. 393), been induced to make the following remark on the same question: —

"We see that the second curve, which was no longer in period as a whole than either of the other two, manifests this excess in each of its branches, that is to say, its left or ascending branch is larger as a whole than the same branch of the two other curves, and the same takes place for the second or descending branch. On the other hand, the maximum of this curve is not so high as that of either of the other two—in fact, the curve has the appearance as if it were pressed down from above and pressed out laterally so as to lose in elevation what it gains in time."

Although both statements appear to lead up to the same conclusion—viz., that ascent and descent are connected by law—still they differ essentially in this respect, that if A, B, C represent the three following consecutive events, descent, ascent, descent, Prof. Wolf's law refers to the connection between A and B, while our remark refers to B and C. We consider two successive minima as the beginning and end of a single period, while M. Wolf, at least in this particular research, places the minimum within the period, and compares the descent from the preceding maximum with the ascent to the next one.

We have considered the connection thus indicated of sufficient importance to apply to it the following test. If, using the previous notation, a definite relation exists between A and B, the *ratio* of the times which the events occupy in every epoch ought to be approximately constant; similarly with respect to B and C; and this ratio should not be influenced by the *absolute* duration of the two successive events. It is clear that the greater uniformity of these ratios will be a test of their interdependence. The following is the result of the comparison: —

a. Prof. Wolf's law: comparison of A and B

	Periods	Duration of descent (A)	Periods	Duration of ascent (B)
I	1829.5 to 1833.8	4.3 years	1833.8 to 1837.2	3.4 years
II	1837.2 to 1844.0	6.8 „	1844.0 to 1846.6	2.0 „
III	1846.6 to 1856.2	9.6 „	1856.2 to 1860.2	4.0 „

“对于某一个具体的周期而言可能和平均值有一定的偏差，但对于整体而言一个$\left\{\begin{matrix}\text{减速}\\ \text{加速}\end{matrix}\right\}$的下降对应着一个$\left\{\begin{matrix}\text{减速}\\ \text{加速}\end{matrix}\right\}$的上升。”

另一方面，通过研究我们自己的曲线（参阅《自然科学会报》，1870 年，第 393 页），针对上述问题，我们可以得出以下一些结论：

“我们看到第二条曲线不像另外两条曲线那样具有一个整体的周期，它的每一条分支都要长一些，也就是说它的左分支或者上升分支从总体上看要比另外两条曲线的对应分支大，而另一个分支或者说下降分支也遵循同样的规律。另一方面，这条曲线的最大值也没有另两条曲线高——实际上，这条曲线就好像被从顶端压下来然后从两侧被拉伸了，以至于失去了它应有的高度。”

虽然，似乎这两方面的陈述都能导出同样的结论，即上升和下降都有一定的规律性，但是我们之间又有本质上的不同，如果用 A，B，C 来分别代表三个连续的发展阶段，下降，上升，下降。沃尔夫教授的规则适用于 A 与 B 之间的关系，而我们的结论则适用于 B 与 C 之间的联系。我们用两个连续的最小值作为一个周期的起点和终点，而沃尔夫先生，至少在这项特殊的研究中，把这个最小值放在一个周期的内部，并且比较了之前从最大值下降到这个最小值的过程及从这个最小值上升至下一个最大值的过程。

我们认为这种联系用下面的测试来检验是非常必要的。假如，按照以前的想法，A 和 B 之间存在某种确定的联系，则每个周期中两个发展阶段（指 A 和 B）所经历的时间之**比**应该是近似恒定的数。类似的关系也应适用于 B 和 C 之间。这个比值不会受到两个相继的发展阶段持续时间的**绝对值**的影响。很显然比值越恒定越说明这种关系是存在的，下面列出的是对比的结果：

a. 沃尔夫教授的定律：A 与 B 的比值

	周期	下降期（A）	周期	上升期（B）
I	1829.5~1833.8	4.3 年	1833.8~1837.2	3.4 年
II	1837.2~1844.0	6.8 年	1844.0~1846.6	2.0 年
III	1846.6~1856.2	9.6 年	1856.2~1860.2	4.0 年

	Ratio $\frac{A}{B}$		Difference from mean
I	1.265		− 0.728
II	2.615	Mean 2.093	+ 0.522
III	2.400		+ 0.307

These differences from the mean are so considerable that in the present state of the inquiry a connection between any descent and the immediately *succeeding* ascent appears highly improbable. A very new and apparently important relation seems, however, to result from a similar comparison of any ascent and the immediately succeeding descent, or between B and C.

b. Comparison of B and C

	Periods	Duration of ascent (B)	Periods	Duration of descent (C)
I	1833.92 to 1836.98	3.06 years	1836.98 to 1843.75	6.77 years
II	1843.75 to 1847.87	4.12 ,,	1847.87 to 1856.31	8.44 ,,
III	1856.31 to 1859.69	3.38 ,,	1859.69 to 1867.12	7.43 ,,

	Ratio $\frac{C}{B}$		Difference from mean
I	2.212		+ 0.061
II	2.044	Mean 2.151	− 1.107
III	2.198		+ 0.047

(**5**, 192-194; 1872)

	比率 $\frac{A}{B}$		与均值的差
I	1.265	均值 2.093	− 0.728
II	2.615		+ 0.522
III	2.400		+ 0.307

这些比值与平均值的差异如此之大，以至于在这样的情况下研究下降期与**紧随其后的**上升期之间的关系几乎不可能。但是一种新发现的、重要的联系，却可以用类似的方法（即一个上升期与随之而来的下降期之间的比率，或者说 B 与 C 之间的比率）推算出来。

b. B 与 C 的比值

	周期	上升期（B）	周期	下降期（C）
I	1833.92~1836.98	3.06 年	1836.98~1843.75	6.77 年
II	1843.75~1847.87	4.12 年	1847.87~1856.31	8.44 年
III	1856.31~1859.69	3.38 年	1859.69~1867.12	7.43 年

	比率 $\frac{C}{B}$		与均值的差
I	2.212	均值 2.151	+ 0.016
II	2.044		− 0.107
III	2.198		+ 0.047

（刘东亮 翻译；蒋世仰 审稿）

The Connexion of Mass with Luminosity for Stars

J. Larmor

Editor's Note

Here the physicist Joseph Larmor writes to *Nature* to discuss important new work of Edward Milne on the topic of stellar interiors, and the relationship of stellar structure to luminosity. Given that we know very little about matter at the extremely high densities likely in stellar interiors, Larmor notes, Milne's work is a laudable attempt to work from basic principles, such as those of thermodynamics. The results suggest that a star's surface properties must be strongly constrained by the physics deep inside the interior, thereby offering a possible explanation for the empirical fact that the luminosity of a star depends for the most part only on its mass.

VERY remarkable and fruitful correlations have in recent years been detected, mainly at Mount Wilson, between the magnitudes of stars and their spectroscopic characteristics. The interpretation that would naturally present itself is that magnitude can enter into relation with the radiative phenomena of the surface atmosphere only through the intensity of gravity at the surface, which when great flattens down a steady atmosphere far more than proportionately. But if, following Eddington's empirical relation, total radiation of a star is a function of its mass alone, there must be more than this involved; for the radius of the star persists in this relation when expressed in terms of intensities of surface radiation and of gravity, the former determining the temperature roughly by itself. Modern hypothesis, which treats confidently of an "electron gas" with an atomic weight, as Ramsay boldly and prematurely proposed long ago, and subject to the Maxwell-Boltzmann exponential energy formula for statistics of distribution, and to its consequences for the theory of dissociation of mixed gases in relation to pressure and temperature, has on the initiative mainly of Saha led to promising applications to stellar atmospheres, which are held to be of densities low enough at any rate not to forbid this mode of treatment.

It would seem then to be necessary to conclude that these empirical spectroscopic relations on the surface require that the stellar atmosphere must be dominated to some degree by the remote steady interior of the star. Accordingly, tentative theories of the internal constitution of the stars and their flux of radiation have been developed in much detail. With Eddington the stars are perfect gases right down to the centre, though the density may there be hundreds of times that of platinum, as has apparently been verified for the case of the companion of Sirius—the high density involving the view at one time not unfamiliar that two atoms can occupy the same space, if the picturesque conception of atoms "stripped" irrevocably to the bone is to be avoided; and the energy emitted as radiation would come from a dissociation or destruction of matter according to a law involving temperature. On the other hand, it is insisted on by Jeans that the necessary radioactivity for the very long evolutions that are contemplated must be of constant and

恒星的质量与发光度之间的关系

拉莫尔

编者按

物理学家约瑟夫·拉莫尔在《自然》上发表的这篇文章，论述了爱德华·米耳恩在恒星内部结构以及内部结构与发光度之间的关系这一问题上的新成果。拉莫尔指出：因为我们对恒星内部处于超高密度的物质状态所知甚少，所以米耳恩尝试用一些基本原理（如热力学定律）进行的研究是值得称道的。米耳恩的研究结果表明，恒星的表面性质与它的内部物理状态有很强的相关性，这为解释恒星的发光度在很大程度上只取决于恒星质量这一经验事实提供了可能。

近几年的天文观测发现，恒星的绝对星等与它们的光谱特征之间存在很多明显的关联，这些观测主要是在威尔逊山进行的。对于这一点很自然的解释是，恒星的绝对星等只有通过位于表面处的引力强度才能和表面大气的辐射现象联系起来，引力强度在稳定的大气中下降很快，远不是成比例的关系。但是如果根据爱丁顿的经验关系式，一个恒星的总辐射量只是其质量的函数，则这其中一定还有另外的影响因素；因为在表面辐射强度和引力强度的关系式中含有恒星半径这个量，而凭借表面辐射强度本身就能大致确定恒星的温度。主要由萨哈倡导的一个最新的假设已经成功地解决了恒星大气方面的许多问题，一般认为恒星大气的密度很低，绝对不会影响这种处理方式的应用。该假设大胆地采用了拉姆齐在很久以前提出的冒险且欠成熟的假设——给“电子气”赋予一个原子量，其统计分布遵从指数形式的麦克斯韦–玻尔兹曼能量公式，并由此得到了混合气体的离解与压强和温度有关的理论。

接下来我们有必要假设，若要得到恒星表面的这些经验性的光谱关系就要求深藏在下面的稳定的恒星内部结构在某些程度上支配着恒星的大气。根据这一点，人们已经将关于恒星内部结构和辐射通量的试验性理论发展到很精细的程度了。根据爱丁顿的观点，恒星从表面到中心都可以看作是理想气体，尽管中心区的密度也许能达到铂的几百倍。这一点可以在天狼星的伴星上得到很好的证明——为了避免出现原子被不可逆地“离解”到只剩下骨架的独特概念，人们曾经用两个原子共同占据同一个空间这个并不陌生的观点来解释高密度的存在；根据一个与温度有关的定律，以辐射形式放出的能量来自物质的离解或毁灭过程。另一方面，琼斯坚持认为，在预期的漫长演化过程中，那些必要辐射的强度肯定是稳定而确切的，否则恒星将

absolute intensity, else the star would explode: and he has essayed to regard the star as "liquid" in his investigations, apparently, however, implying a very imperfect gas rather than a special phase with its surface of sharp transition. There are other theories of less statical type.

A determined effort to shed off all such special hypotheses has been published very recently by Milne (*Monthly Notices R.A.S.* for November, pp. 17–53), which accordingly invites close attention and scrutiny. The procedure is the natural one, to try to make continuity between the gases of the atmosphere subject to laws more or less already formulated, and a dense interior about which as little is to be assumed as can be helped. He holds that it suffices merely to consider laws of internal density that are in mechanical equilibrium radially under internal pressure P, of which the fraction $(1-\beta)P$ is pressure of the internal field of radiation. He does not find it necessary to consider how this field of radiation of pressure $(1-\beta)P$ is sustained against loss by outward flux: for if he can arrive at results in terms of surface values that are valid for all such equilibrated densities whether otherwise possible or not, they must hold good for the one that follows the actual law of distribution whatever it may be.

The essential feature, so far as a reader can extract the gist from the complication of formulas that seems to be inherent in these discussions, appears to be that the coefficient β, while increasing rapidly downward in the atmosphere in a manner which can be regarded as known, suddenly rises when a photospheric level is reached, altering with steep gradient until a nearly constant value of β is soon attained for the interior of the star: and the same must apply only in less degree to the density ρ. The condition of mere mechanical equilibrium of the interior is found to express the pressure at the interface between atmosphere and photosphere in terms of values at the centre and one quantity C arising from an integral along the radius involving the arbitrarily assumed law of density. The expression for the atmospheric pressure at the interface involves the same constants in such way that on equating the pressures on the two sides of the interface they divide out of the result and only C remains. This C is held, in the light unforeseen of comparison with facts, to be in some degree a characteristic constant for all the stars, and thus may be the new element beyond surface values, and without assuming anything about their interiors, that the law as formulated requires.

This seems to be right enough in a general way, were it not that the formula for C involves the gradient of density within the star close to the interface, and thus its value must be very substantially changed, in absence of some verification to the contrary, by a very slight radial displacement of the surface which is chosen for that interface. For inside the photosphere ρ is as θ^3, while P which is continuous across the interface is as $\theta^4\phi(\theta)$: so C^{-1} is as the value of $P^{-1}(d\rho/dr)^2$ in which the second factor is the internal gradient, at the surface. If this consideration be correct it would appear that it is not legitimate to connect the chromosphere with the interior across a sharp boundary surface, as if they were different phases of matter like a liquid and its vapour. This conclusion would involve that the formula itself for C cannot be well founded: and the reason can be assigned, that

会发生爆炸。他在研究中试图假设恒星是“液体”的，这显然意味着恒星是一种非常不理想的气体，而不是处于一种在表面处变化很突然的特殊状态。另外还有一些其他的弱静态型理论。

米尔恩在最近发表的一篇文章（《皇家天文学会月刊》，11 月，第 17~53 页）中，坚决地剔除了所有这些特殊的假说，因而引发了人们的密切关注和深入思考。他很自然地试图让大气层中的气体与致密的内部之间保持连续性，其中，大气层中气体遵守的定律已经在前面或多或少地提到过了，而致密的内层则几乎没有什么理论可以参照。他提出，如果仅考虑内压为 P 时沿半径方向处于力学平衡的内部密度的变化规律，内压的一部分 $(1-\beta)P$ 源自内部辐射场，则这种连续性是可能存在的。米尔恩认为，没有必要考虑这个压强为 $(1-\beta)P$ 的辐射场如何克服向外的通量损失来维持自身，因为不管是否存在其他可能性，只要他能够得到对所有平衡态密度都有效的表面值，这些结果都会满足一个真实的分布规律，无论其形式如何。

只要读者能够从以上讨论中似乎不可避免的复杂公式中抽取出要点，就会看到基本特征是，系数 β 的值以一种我们已经知道的方式在恒星大气层中由外向内迅速增长，当到达光球层时突然增加，然后以很陡的梯度发生变化，直到很快在恒星内部达到一个基本恒定的值；密度 ρ 的情况肯定也是如此，只是在程度上会弱一些。内部的力学平衡条件可以表示为大气层和光球层间分界面上的压强，这个压强是中心物理量和 C 值的函数，其中 C 是由任意假定的密度分布沿半径方向积分得到的。大气压强在界面处的表达式中也包含同样的常数，在化简结果以使界面两侧的压强相等之后，其他常数都消失了，结果中只有 C 保留了下来。C 在某种意义上可以被认为是所有恒星的一个特征常数，这是在与实际情况的对比中没有料到的。因此它可能是除表面值之外的新要素，上述原理需要这些要素而不必考虑它们的内涵。

从一般意义上说上述做法已经足够完美了，如果不是因为 C 的表达式中包含恒星内部靠近分界面处的密度梯度，则在没有相反证据的前提下，只要所选的分界面沿半径方向有微小的位移，C 的数值必然会发生很大的变化。如果把光球层内的 ρ 看作 θ^3，把在分界面处数值连续变化的 P 看作 $\theta^4\phi(\theta)$，则 C^{-1} 就等于 $P^{-1}(d\rho/dr)^2$，其中第二个因子是在表面处的内部梯度。如果上述结果是正确的，那么将色球层与相隔一个突变边界面的恒星内部看成一个整体就显得不太合理，它们就像同一种物质的液态和气态一样属于不同的相。由此可见关于 C 的表达式不可能是合理的，理

the transition from Milne's formula (21) to (22) is invalid because the interior gradient of ϕ at the interface is very large and cannot be neglected even when multiplied by θ. Apparently one can only assert that the mass of the star involves the value of dP/dr within the photosphere and other quantities relating to the centre of the star, and the luminosity involves the value of P outside it, while the pressure P is continuous across a transition but not dP/dr.

In any case, perhaps not much stress would be laid on the deduction. The formula is regarded probably by its author as essentially an empirical result. When the value of C had been adapted to two prominent stars, the sun and Capella, it turned out in his hands, as he relates, to his astonishment, that it was a universal constant the same for all stars, and if so perhaps not connected with their interior constitutions at all.

(**125**, 273-274; 1930)

Joseph Larmor: Cambridge, Jan. 18.

由是从米尔恩的公式（21）推不出公式（22），因为内部梯度 ϕ 在分界面处非常大，即便是乘以 θ 以后也不能被忽略。显然我们只能说恒星的质量包含在光球层内的 dP/dr 值以及其他与恒星中心相联系的量中，发光度包含光球层外的 P 值，而穿过交界处时保持连续的是压强 P 而不是 dP/dr。

不管怎么说，也许对推导过程没有给予足够的重视。作者可能认为这个公式基本上是依赖经验结果得到的。当 C 值被应用于太阳和五车二这两颗著名的恒星时，作者发现，出乎他的意料，C 是一个对于所有恒星都相同的普适常数。如果真是这样，那么也许 C 值与恒星的内部结构根本没有关联。

（史春晖 翻译；蒋世仰 审稿）

Discovery of a Trans-Neptunian Planet

A. C. D. Crommelin

Editor's Note

Nature **reports that astronomers at the Lowell Observatory in Flagstaff Arizona have for seven weeks been observing an object which appears to be a planet in orbit beyond Neptune. Its behaviour agrees fairly closely with predictions based on anomalies in the orbit of Uranus, and its size, based on one visual observation, seems intermediate between that of the Earth and Uranus. The article notes that the gravitation of this object might also account for a deviation of several days in the arrival of Halley's comet in each of its last two returns. The planet, soon named Pluto, is today no longer considered to be a proper planet, but rather a "dwarf planet"—an asteroid-like object comprised of frozen material.**

ON the evening of Mar. 13 (an appropriate date, being the anniversary of the discovery of Uranus in 1781, and Mar. 14 being the birthday of the late Prof. Percival Lowell) a message was received from Prof. Harlow Shapley, director of Harvard Observatory, announcing that the astronomers at the Lowell Observatory, Flagstaff, Arizona, had been observing for seven weeks an object of the fifteenth magnitude the motion of which conformed with that of a planet outside Neptune, and agreed fairly closely with that of one of the hypothetical planets the elements of which had been inferred by the late Prof. Percival Lowell from a study of the small residuals between theory and observation in the positions of Uranus. That planet was better suited than Neptune for the study, since the latter had not been observed long enough to obtain the unperturbed elements.

Lowell's hypothetical planet had mean distance 43.0, eccentricity 0.202, longitude of perihelion 204°, mass 6.5 times that of the earth, period 282 years, longitude 84° at the date 1914–1915. Its position at the present time would be in the middle of Gemini, agreeing well with the observed place, which on Mar. 12 at 3h U. T. was 7 seconds of time west of δ Geminorum; the position of the star was R.A. 7h 15m 57.33s, north decl. 22° 6′ 52.2″, longitude 107.5°. This star is only 11′ south of the ecliptic, making it likely that the new planet has a small inclination. As regards the size of the body, the message states that it is intermediate between the earth and Uranus, implying perhaps a diameter of some 16,000 miles. A lower albedo than that of Neptune seems probable, to account for the faintness of the body. It appears from a New York telegram that at least one visual observation of the planet has been obtained, from which the estimate of size may have been deduced.

Mention should also be made of the predictions of Prof. W. H. Pickering; one of these, made in 1919 (*Harvard Annals*, vol. 61), gives the following elements; Epoch 1920; longitude 97.8°; distance 55.1, period 409 years; mean annual motion 0.880°; longitude of perihelion 280°; perihelion passage 1720, eccentricity 0.31; perihelion distance 38, mass twice earth's, present

发现海外行星

克罗姆林

编者按

据《自然》报道，天文学家在亚利桑那州弗拉格斯塔夫市的洛威尔天文台已经对一个看似是在海王星轨道外运行的行星进行了 7 周的观测。这颗行星的特征非常接近于人们根据天王星轨道的反常现象所作的预测，它的目测大小似乎介于地球和天王星之间。这篇文章指出，该天体的引力可能是使哈雷彗星最近两次回归地球的时间产生数天偏差的原因。这颗行星很快被命名为冥王星，现在它已不再被人们看作是一颗行星，而是一颗“矮行星”——一个由超低温物质组成的小行星。

3 月 13 日晚（此日恰为 1781 年发现天王星的纪念日，而 3 月 14 日是已故天文学家珀西瓦尔·洛威尔的诞辰）从哈佛大学天文台台长哈洛·沙普利那里收到了一则消息，亚利桑那州弗拉格斯塔夫市洛威尔天文台的天文学家已经对一个星等为 15 等的天体进行了长达 7 周的观测，结果发现该天体的运动与海王星外的一颗行星同步，且与一颗假想行星的情况相当吻合。通过研究天王星的理论位置和观测位置之间的细小差别，已故的珀西瓦尔·洛威尔教授推断出了假想行星的根素。鉴于尚未对海王星观测足够长的时间来获得无扰动根素，相比之下这颗行星会更适合进行研究。

洛威尔假想行星的平均距离为 43.0 个天文单位，偏心率为 0.202，近日点的黄经为 204°，质量是地球质量的 6.5 倍，周期为 282 年，1914~1915 年的黄经为 84°。当前时刻该行星的位置应在双子座的中间，而在 3 月 12 日世界时 3 点观测到该星位于双子座 δ 星西部 7s 赤经的位置，因而观测与预言相吻合。该星的赤经是 7h 15m 57.33s, 赤纬是 + 22° 6′ 52.2″，黄经是 107.5°。该星位于黄道南部 11′ 的位置，很可能是一颗有很小倾角的行星。消息中写道，该行星的大小介于地球和天王星之间，直径约为 16,000 英里。该星的反照率低于海王星，因而星体本身极为暗弱。从来自纽约的电报看，目前已经至少获得过一次关于该星的目视观测，从而可能已经推测出了该星的大小。

这里需要提及的是皮克林教授的预言；他在 1919 年的一个预言（《哈佛年鉴》，第 61 卷）中给出的根素如下：历元 1920 年；黄经 97.8°；距离 55.1 个天文单位，周期 409 年；平均周年运动 0.880°；近日点的黄经 280°；1720 年通过近日点，偏心

annual motion 0.489°. This prediction gives the longitude for 1930 as 103°, which is within five degrees of the truth; actually it was in longitude 108° at discovery. Prof. Pickering's later prediction is further from the truth, making the longitude about 131°.

Gaillot and Lau also made predictions; like the other computers they noted that there were two positions, about 180° apart, that would satisfy the residuals almost equally well. Taking the position nearest to the discovered body, Lau gave longitude 153°, distance 75, epoch 1900. Gaillot gave longitude 108°, distance 66, epoch 1900. The latter is not very far from the truth; with a circular orbit, the longitude in 1930 resulting from Gaillot's orbit would be 128°, some 20° too great. Gaillot performed the useful work of revising Le Verrier's theory of Uranus, thus giving more trustworthy residuals. Lowell pointed out that the residuals of Uranus that led to the discovery of Neptune amounted to 133″, while those available in the present research did not exceed 4.5″; yet even in the case of Neptune the elements of the true orbit differed widely from the predicted ones, though the direction of the disturbing body was given fairly well. He noted that in the present case it would be wholly unwarrantable to expect the precision of a rifle bullet; if that of a shot-gun is obtained, the computor has done his work well.

Another method of obtaining provisional distances of unknown planets is derived from periodic comets; the mean period of the comets of Neptune's family is 71 years; it is pointed out in the article on comets ("Encyc. Brit." 14th edition, vol. 6, p. 102) that there is a group of five comets the mean period of which is 137 years; as stated there, "This family gives some ground for suspecting the existence of an extra-Neptunian planet with period about 335 years, and distance 48.2 units." This seems to be in fair accord with the new discovery, but probably the distance is nearer 45 than 48. Comets also suggest another still more remote planet, with period about 1,000 years, a suggestion which has also been made by Prof. G. Forbes and by Prof. W. H. Pickering.

The question has been asked, "Does the new planet conform to Bode's law?" It is difficult to assign a definite meaning to this question, since Bode's law broke down badly in the case of Neptune; Neptune's predicted distance was 38.8, its actual distance 30.1. For Bode's law, each new distance ought to be almost double the preceding one; the constant term of the law becomes negligible when the distance is great. For the extension of the terms given by the law we might (1) ignore Neptune as an interloper and take the next distance as double that of Uranus, giving 38.5 units; (2) we might take the next distance as four times that of Uranus, which would give 77 units; or (3) we might take the next distance as double that of Neptune or 60 units; none of these values is good, but (1) is the nearest to what we suppose to be the distance. Probably the best course is to assume that after Uranus the law changes; each new distance is then 1.5 times the preceding one; on this assumption, the hypothetical planet with distance 100 and period 1,000 years would be the next but one after the Lowell planet.

The low albedo of the new planet might be explicable if its temperature were much lower than that of Neptune. Owing to its smaller size, it would have lost more of its primitive

率 0.31；近日点距离 38 个天文单位，质量是地球的 2 倍，目前的周年运动为 0.489°。该预言认为 1930 年行星的黄经为 103°，该值与真实值相差不到 5°；事实上，发现时它的黄经为 108°。皮克林教授后来的预言与真实值相差更远，他预言的黄经达到 131°。

加约和劳做了一些预言；和其他天文学家一样，他们认为在远离 180° 的两个位置能够很好地满足残差。选取离被发现天体最近的位置，劳给出黄经 153°，距离 75 个天文单位，历元 1900 年。加约给出黄经 108 °，距离 66 个天文单位，历元 1900 年。后者与真实值相差不远；对于圆形轨道，由加约的轨道得到 1930 年该星的黄经是 128°，比真实值大 20°。加约修正了勒威耶关于天王星的理论工作，给出了更可靠的残差。洛威尔指出导致发现海王星的天王星残差是 133″，然而在目前的研究中能够得到的残差小于 4.5″；然而即使在海王星的情况中，虽然精确地给出了扰动天体的方向，真实轨道根素也远远不同于预言结果。他还指出在目前的情况下要达到来复枪子弹那样的精度是完全不可能的；如果有更好的观测仪器，那么计算机可以将他的工作处理得更好。

另一种获得未知行星临时距离的方法来自周期彗星；海王星家族的彗星平均周期是 71 年；一篇关于彗星的文章(《不列颠百科全书》, 第 14 版, 第 6 卷, 第 102 页)中指出一个由 5 颗彗星组成的团组的平均周期是 137 年；文章中说“这个家族提供了某种基础, 可以假定存在周期大约为 335 年, 距离为 48.2 个天文单位的海外行星。”这似乎与目前的新发现非常吻合，但是距离是接近 45 个天文单位而不是 48 个天文单位。福布斯和皮克林教授提出，彗星暗示了其他更远的、周期约为 1,000 年的行星的存在。

曾有人问到“新的行星满足波得定律吗？”由于海王星已经严重打破了波得定律，因而这个问题已不再有意义。海王星的预言距离是 38.8 个天文单位，实际距离是 30.1 个天文单位。根据波得定律，每一个新的天体距离都必须是前一个天体距离的 2 倍；当距离很远时可以忽略定律的常数项。对于定律的延展项，(1) 我们可以把海王星作为闯入者而忽略，下一个距离取天王星距离的 2 倍，得到 38.5 个天文单位；(2) 我们也可以把下一个距离取天王星距离的 4 倍，得到 77 个天文单位；(3) 或者我们还可以把下一个距离取成海王星的 2 倍即 60 个天文单位。这里的任何一个值都不太好，但是 (1) 的取值最接近我们估计的距离。那么在天王星之后，波得定律不再满足，这可能是最可取的方法；而每一个新的距离是前一个天体距离的 1.5 倍；基于这种假定，距离为 100 个天文单位、周期为 1,000 年的天体将是继洛威尔行星后的又一颗行星。

如果新行星的温度远低于海王星，那么就可以解释这类行星的低反照率。由于该行星很小，它将失去更多自身的原始热量而且只能吸收来自太阳能量的一半；从

heat, and would only receive half as much from the sun; hence its gases might be reduced to a liquid form, with great reduction of their volume. This would result in a relatively smaller disc than the one that might be inferred from its mass.

Some further particulars of the discovery are given by the New York correspondent of the *Times* in the issue for Mar. 15. Quoting an announcement which had been received there from the Lowell Observatory, it is stated that the planet was discovered on Jan. 21 on a plate taken with the Lawrence Lowell telescope; it has since been carefully followed, having been observed photographically by Mr. C. O. Lampland with the large Lowell reflector, and visually with the 24-inch refractor by various members of the staff. The observers estimate the distance of the planet from the sun as 45 units, which would give a period of 302 years, and mean annual motion of 1.2 degrees.

At discovery, the planet was about a week past opposition, and retrograding at the rate of about 1′ per day; this has now declined to 0.5′ per day, and the planet will be stationary in April. It should be possible to follow it until the middle of May, when the sun will interfere with observation until the autumn.

The details of the Lowell Observatory positions have not yet come to hand; when they do, it will be possible to derive sufficiently good elements to deduce ephemerides for preceding years. There are many plates that may contain images of the planet; those taken by the late Mr. Franklin Adams in his chart of the heavens, those taken of the region round Jupiter some twelve years ago for the positions of the outer satellites, and those taken at Königstuhl and elsewhere in the search for minor planets; these all show objects down to magnitude 15. If early images should be found, they will accelerate the determination of good elements of the new planet; in the case of Uranus, observations were found going back nearly a century before discovery, and in that of Neptune they went back fifty-one years. In the present case, forty years is the most that can be hoped for, and probably very few photographs showing objects of magnitude 15 are available before the beginning of this century.

One of the most difficult problems will be to find the mass of the new body; in Neptune's case, Lassell discovered the satellite a few months after the planet was found, and the mass was thus determined. It is to be feared, however, that the new planet would not have any satellite brighter than magnitude 21. Stars of this magnitude have been photographed with the 100-inch reflector at Mount Wilson, but it is doubtful whether it could be done within a few seconds of arc of a much brighter body. Failing the detection of a satellite, the mass can only be deduced from a rediscussion of the residuals of Uranus and Neptune; new tables of these planets will ultimately be called for, but that task must wait until the orbit of the new body is known fairly exactly.

The perturbations of Halley's comet will also require revision; at each of the last two returns, there has been a discordance of two or three days between the predicted and observed dates of perihelion passage; it will be interesting to see whether the introduction

而行星上的气体将会变成液态，体积也大大减小。这样它的盘面将小于由质量推算出的盘面的大小。

《泰晤士报》的纽约通讯员在3月15日的一期上刊载了发现新行星的更进一步的细节。文章引用了洛威尔天文台的一段宣告，说行星是在1月21日劳伦斯·洛威尔望远镜拍摄的一张底片中发现的；此后天文学家即对该星进行跟踪，兰普兰德先生用大洛威尔反射望远镜对该星进行了照相观测，其他的工作人员用口径24英寸的折射望远镜进行目视观测。观测者们推断该星与太阳的距离是45个天文单位，由此周期应当是302年，平均周年运动1.2°。

发现之初，该行星刚通过对冲大约一星期，每天退行1′；目前退行速度减小为每天0.5′，并将在4月保持静止（留）。在5月中旬之前可以对该星进行跟踪观测，此后直到秋天来临之前这段时间内，太阳都会影响观测。

洛威尔天文台还未对观测结果进行详细处理；一旦进行处理，那么将会获得足够精确的轨道根素进而推导出前几年的天文历表。有很多底片可能包含这个行星的图像，如已故的富兰克林·亚当斯先生的天空星图，12年前拍摄木星周围卫星的底片，以及那些在柯尼希斯施图尔山和其他地方拍摄的搜寻小行星的底片；这些底片都能够显示暗到15等的天体。如果可以找到更早期的图像将可以加速新行星轨道根素的确定；对于天王星，观测资料可以上溯到天王星发现前近一个世纪，而对于海王星则可以追溯到51年前。对于目前这颗行星至多追溯到40年前，本世纪之前可能很少有能够记录到15等星的照相底片。

最困难的一个问题是确定新发现天体的质量，对于海王星，在发现该行星几个月后拉塞尔就发现了它的卫星，从而确定了海王星的质量。然而，这颗新的行星的任何一个卫星的星等可能都将暗于21等。在威尔逊山用100英寸的反射式望远镜曾经拍摄过21星等的星，但是并不能确定使用该望远镜能否在几弧秒的范围内拍摄到这样的天体。由于不能探测到卫星，那么该行星的质量只能从对天王星和海王星的残差的再讨论中推导出来。而要进行这种再讨论推导就必须有关于新行星的新的位置数据表，而要得到这个表就必须有这个新行星的精确的轨道根素。

哈雷彗星的扰动也需要修正；在最近两次回归中，通过近日点日期的预言值与观测值之间都有两三天的差别；研究结果是否会由于引入新天体的扰动而有所改进

of the perturbations of the new body effects an improvement. The late Mr. S. A. Saunder made the suggestion at the time of the last apparition of the comet that an unknown planet might be the cause of the discordance, but it was not then possible to carry the suggestion further. The discovery of a new planet therefore opens a large field of work for mathematical astronomers. It will also appeal to students of cosmogony; Sir James Jeans, in an article in the *Observer* for Mar. 16, suggests that it may represent the extreme tip of the cigar-shaped filament thrown off from the sun by the passage of another star close to it. It would have been the first planet to cool down and solidify; he says, "As a consequence of this, it will probably prove to be unattended by satellites."

(**125**, 450-451; 1930)

将是一个有趣的课题。在彗星最近一次出现时，已故的桑德先生提出未知行星的存在将可能导致这种差异，但是并未对这个想法进行更进一步的研究。一颗新行星的发现则为数学天文学家提供了更广阔的工作空间。这也将吸引更多研究宇宙演化的学者投入其中；詹姆斯·金斯先生在 3 月 16 日《观察家报》的一篇文章中写道：这个新行星可能代表离太阳较近的另外一个恒星通过时太阳抛射出来的雪茄状纤维丝的顶端。它将是冷却凝固形成的第一颗行星；他说“作为这个现象的结果，将可能证明这颗新的行星没有卫星。”

（王宏彬 翻译；蒋世仰 审稿）

Lowell's Prediction of a Trans-Neptunian Planet

J. Jackson

Editor's Note

John Jackson of the UK Royal Observatory here discusses the possibility that the orbit of the newly discovered planet had been accurately predicted by the astronomer and polymath Percival Lowell. If true, Jackson suggests, then Lowell deserves high admiration, for the problem, while in principle similar to the prediction of Neptune, is in detail immeasurably more difficult. Jackson reviews Lowell's methods, in which, by hypothesizing a planet of particular mass and orbit, he could reduce the unexplained motion of Uranus by some 70%. However, the brightness of the observed planet is about ten times higher than predicted. Astronomers since have determined that Lowell's calculations were in error, and that Pluto was discovered principally through a painstaking empirical search of the sky.

THE reported discovery of a planet exterior to Neptune naturally arouses the interest of the general public. It will be of importance in theories concerning the genesis of the solar system as to how far it falls into line with the other planets as regards distance, mass, eccentricity and inclination of orbit, and presence or absence of satellites. Its physical appearance will be beyond observation. To those interested in dynamical astronomy, it may be of some interest to consider the data which led to its discovery and to make some comparison with the corresponding facts relating to Neptune.

If the planet which has been reported approximately follows the orbit predicted by Dr. Percival Lowell, the prediction and the discovery will demand the highest admiration which we can bestow. It is true that the problem as regards its general form is a repetition of that solved by Leverrier, Adams, and Galle more than eighty years ago; but its practical difficulty is of quite a different order of magnitude. In short, this discovery, if it turns out to be actually Lowell's predicted planet, was extremely difficult—while Neptune was in fact crying out to be found. Let us look at the actual data.

Uranus was discovered in 1781 by Herschel. Scrutiny of old records showed that it had been observed about a score of times dating back to 1690. The fact that Lemonnier observed it eight times within a month, including four consecutive days, without detecting its character, should be a lesson to anyone who makes observations without examining them. In 1820 Bouvard found that the old and the new observations could not be reconciled, and in constructing his tables boldly rejected the early observations, but the tables rapidly went from bad to worse; the residuals amounted to 20″ in 1830, 90″ in 1840, and to 120″ in 1844. Adams used in his first approximation data up to 1840, Leverrier data up to 1845. Now Uranus had passed Neptune in 1822. As the relative motion is about 2° a year, it means that for most of the time covered by the prediscovery observations the perturbations were very small, while from the fact that the difference

洛威尔对海外行星的预言

杰克逊

编者按

英国皇家天文台的约翰·杰克逊在此讨论了这颗新发现的行星的轨道已经被天文学家、博学者珀西瓦尔·洛威尔精确预测的可能性。在杰克逊看来，如果事实真的如此，那么洛威尔理应得到很高的荣誉，因为虽然计算的原则类似于海王星，但具体过程更加困难。杰克逊介绍了洛威尔的方法，洛威尔通过假设另一个具有一定质量和轨道的行星解释了天王星中 70% 的不明运动。可是，这颗行星的实际亮度比预测值高 10 倍。天文学家们后来认为洛威尔的计算存在错误，而冥王星的发现主要是通过有经验的观测者在太空中艰苦搜索得到的。

已报道的海王星外行星的发现引起了公众广泛的兴趣。该行星对于与太阳系的形成相关的理论来说将是至关重要的，诸如该星在线距离、质量、偏心率和轨道倾斜度，以及是否存在卫星等方面是否与其他行星的规律相符合的问题。该星的物理外貌是不能观测到的。如果对动力天文学感兴趣，可以研究一下发现该行星的数据并将它与海王星的相关资料进行比较。

如果报道的行星大致符合珀西瓦尔·洛威尔博士预言的轨道，那么这个预言和发现将是令人瞩目的。的确，就其一般形式而言，这个问题确实是 80 年前已经由勒威耶、亚当斯和伽勒解决了的问题的重现，但是实际研究中存在的困难在于数量级的差异。简而言之，如果这次发现的新行星的确是洛威尔预言的行星，那么这一发现是极其困难的，而海王星的发现其实是非常容易的。下面让我们看看实际的数据资料。

天王星是赫歇尔在 1781 年发现的。详细审查原始记录会发现，早在 1690 年就已经观测到了这颗星。勒莫尼耶一个月内 8 次观测天王星并且有 4 天是连续观测，但并未探测到它的特征，每个进行观测但未进行详细检测的观测者都应该引以为戒。1820 年布瓦尔发现新旧观测数据并不一致，他在建数据表的时候大胆舍弃了早期的观测资料，但是数据表的结果更糟了；1830 年残差达 20″，1840 年达 90″，到 1844 年则达到 120″。而亚当斯直到 1840 年使用的还是他的第一个近似数据，勒威耶则直到 1845 年还在用。1822 年天王星越过海王星。由于每年的相对运动只有 2°，那么在发现前观测所覆盖的大部分时间内，扰动是非常小的。然而考虑到行星

between the heliocentric distances is much smaller than expected from Bode's law, the perturbations at the time of conjunction were relatively large. Consequently the prediction of the longitude of the disturbing body was very easy, while the determination of the other elements were correspondingly difficult. The fact was that the simple hypothesis of the existence of an exterior planet with any sort of guess as to size and shape of orbit would suffice to predict the longitude. In other words, most of the residuals could be closely satisfied provided that substantially correct values of the longitude of the planet and its attractive force $m\left(\frac{1}{\Delta^2}+\frac{1}{r^2}\right)$ were used. Both Leverrier and Adams easily found values of these quantities, and Galle had no difficulty in detecting the planet.

We now turn to Lowell's "Memoir on a Trans-Neptunian Planet", published in 1915. The observational basis is the outstanding residuals in the motion of Uranus during two centuries, that is, rather more than two revolutions of that planet round the sun, of somewhat less than two revolutions relative to the predicted planet and of about one relative to Neptune. The following are the values of the observed residuals of Leverrier's and of Gaillot's theories taken from Lowell's memoir.

	Leverrier	Gaillot		Leverrier	Gaillot
1709	. .	+2.14″	1855	. .	−0.50″
1753	+5.52″	+4.45	1858	+0.50″	−0.20
1769	+4.77	+2.47	1861	. .	−0.36
1783	−3.30	−0.96	1864	+0.25	+0.18
1787	−5.12	−1.20	1867	. .	+1.20
1792	−3.50	+0.10	1870	−0.50	+1.32
1796	−1.88	−0.69	1873	. .	+0.75
1803	+0.40	−1.19	1876	−1.65	−0.50
1812	+2.00	−0.77	1879	. .	+0.58
1817	+0.50	−0.60	1882	−2.88	+0.52
1820	−0.75	−2.37	1885	. .	−0.17
1827	−2.10	+2.00	1888	−4.22	−0.85
1837	−1.10	−1.22	1891	. .	−1.11
1840	+0.63	+0.78	1894	−5.63	−0.50
1843	. .	+0.74	1897	. .	+0.35
1846	+0.38	−1.40	1900	−4.32	+1.00
1849	. .	−0.25	1903	−3.00	+0.65
1852	−1.17	−0.95	1907	. .	+0.25
			1910	. .	+1.10

The residuals show remarkable differences between the two theories, but Lowell deduced that the residuals exceeded their probable errors four or five times. The problem was to find from these residuals corrections to the elements of the orbit and to find the mass and the elements of the disturbing body. It might almost appear hopeless when we consider that the residuals must be affected by errors in the accepted masses of the known planets. There can be no doubt, however, that the masses adopted by Gaillot for Jupiter, Saturn, and Neptune are very accurate. Lowell's procedure was to adopt a value of the semimajor axis of the unknown body, and a complete series of values for its longitude, and then select the value of the longitude for which the sum of the squares of the residuals was

的日心距比由波得定律估算出的值小得多，那么在合发生时扰动将相对大一些。这样就可以较容易地确定扰动天体的黄经，然而确定该天体的其他根素就相对困难一些。实际上简单的外行星存在的假定，不管对轨道的大小和形状作何种猜测，都足以预言行星的黄经。换句话说，只要充分地利用行星黄经的正确值以及它的引力 $m\left(\frac{1}{\Delta^2}+\frac{1}{r^2}\right)$，那么大部分残差都可以得到很好的满足。勒威耶和亚当斯都很容易地找到了这些值，伽勒也很容易地探测到了行星。

下面我们看一下洛威尔 1915 年发表的《海外行星回忆录》。观测的主要内容是两个世纪中在那颗行星绕太阳旋转两圈多的过程中天王星运动中出现的残差，这种运动相对于预言中的行星不到两圈，相对于海王星大约为一圈。下面是来自洛威尔回忆录中相对于勒威耶理论和加约理论的观测残差值。

	勒威耶	加约		勒威耶	加约
1709	. .	+2.14″	1855	. .	–0.50″
1753	+5.52″	+4.45	1858	+0.50″	–0.20
1769	+4.77	+2.47	1861	. .	–0.36
1783	–3.30	–0.96	1864	+0.25	+0.18
1787	–5.12	–1.20	1867	. .	+1.20
1792	–3.50	+0.10	1870	–0.50	+1.32
1796	–1.88	–0.69	1873	. .	+0.75
1803	+0.40	–1.19	1876	–1.65	–0.50
1812	+2.00	–0.77	1879	. .	+0.58
1817	+0.50	–0.60	1882	–2.88	+0.52
1820	–0.75	–2.37	1885	. .	–0.17
1827	–2.10	+2.00	1888	–4.22	–0.85
1837	–1.10	–1.22	1891	. .	–1.11
1840	+0.63	+0.78	1894	–5.63	–0.50
1843	. .	+0.74	1897	. .	+0.35
1846	+0.38	–1.40	1900	–4.32	+1.00
1849	. .	–0.25	1903	–3.00	+0.65
1852	–1.17	–0.95	1907	. .	+0.25
			1910	. .	+1.10

这两种理论得到的残差值有着明显的差别，但是洛威尔推导发现残差与他们估测的误差相比超出 4~5 倍。现在的问题是，从这些残差的改正中找到轨道根素和扰动天体的质量及根素。当我们考虑到残差必然会受到已知行星质量误差的影响时，那么这些问题的解决就几乎是不太可能的了。毫无疑问，加约引用的木星、土星和海王星的质量都是非常精确的。洛威尔的做法是，引入未知天体的一个半长轴值以及该天体一系列完整的黄经值，然后选择残差平方和最小的黄经。用各种平均距离值（即未知行星的半长轴值）重复进行这一过程，直到变量的取值使得残差最小为

a minimum. The process was repeated with various values of the mean distance until values of the variables were found giving minimum residuals. The process was of course very laborious, but Lowell carried it through with great perseverance. The following extract from his final summary may be quoted: "By the most rigorous method, that of least squares throughout, taking the perturbative action through the first powers of the eccentricities, the outstanding squares of the residuals from 1750 to 1903 have been reduced 71 percent by the admission of an outside disturbing body."

The inclusion of further terms, of additional years and of the squares of the eccentricity, do not alter the results by any substantial amount. Lowell considered that the remaining irregularities could be explained by errors of observation. No trustworthy results could be found from the residuals in latitude so that the inclination of the orbit to the ecliptic could not be deduced, but Lowell considered that it might be of the order of 10°.

As the solution really depends on the difference of the attraction of the unknown planet on Uranus and on the sun, there are two possible solutions in which the longitudes differ by about 180°. The following elements are for the solution satisfying most nearly the position of the newly found body.

Heliocentric longitude on 1914, July	84.0°
Semimajor axis	43.0
Mass in terms of the sun's mass	1/50,000
Eccentricity	0.202
Longitude of perihelion	203.8°

This gives the longitude at the present time as about 104° compared with 107° of the new planet. The predicted magnitude was 12 to 13 or about ten times brighter than the observed; and a disc of more than 1″ was predicted. This is a rather serious discordance.

The smallness of the residuals indicated that the forces were small. The mass given above is only 0.4 of the mass of Neptune. At mean conjunction, the attraction of the predicted planet on Uranus would be only one-fifteenth of the attraction of Neptune in a similar position, and in addition it would last for a shorter time on account of the more rapid relative motion.

The discovery of a minor planet of the fifteenth magnitude is an everyday occurrence. The planet reveals itself by a decided motion relative to the stars in the course of taking a photograph. For a planet in the predicted orbit, the motion shown (mostly due to the earth's motion) would in the most favourable circumstances not be more than 2″ or 3″ an hour, and it would probably need a trail of at least 5″ for the planet to be detected. On the other hand, photographs taken on successive days would show decided motion, but the labour of finding the planet in a region containing many thousands of stars from separate photographs would be very great. Probably the Lowell observers have come across several minor planets before they were rewarded by the discovery of the very distant planet.

止。这无疑是一项非常艰苦的工作，但是洛威尔却坚持不懈地完成了。下面是从他的文章中引出的一句话，“利用最严密的方法，即最小二乘法，引入外界的一个扰动天体计算偏心率一次方的扰动行为，已经把 1750~1903 年天体明显的残差平方减少了 71%”。

进一步引入其他因素，加长年代以及使用偏心率的平方，并没有使结果有任何实质上的改变。洛威尔认为剩下的不规则性可以用观测误差来解释。从黄纬的残差中不能找到可信的结果，因而不能推导出轨道与黄道的倾角，但是洛威尔认为这个倾角可能是 10°。

由于问题的解直接取决于未知行星对天王星和太阳的引力差别，所以可能有两种使黄经相差 180° 的解。下面这些根素就是满足新发现天体条件的最近的解。

1914年7月太阳中心黄经	84.0°
半长轴	43.0
以太阳质量为单位表示的质量	1/50,000
偏心率	0.202
近日点黄经	203.8°

与这一新行星的当前黄经 107° 相比，据此根素得出的其当前黄经大约是 104°。预测星等为 12~13 等，或者说比观测到的星亮 10 倍；而且还预测到了一个大于 1″ 的盘。这里产生了很严重的矛盾。

残差小意味着力小。上面给出的质量仅是海王星质量的 0.4 倍。在平均会合期，预测行星给天王星的引力仅是同一位置海王星所施加引力的 1/15，并且由于较快的相对运动，这一状态持续的时间很短。

发现一个 15 等的小行星是件很常见的事情。在拍照过程中行星以一种相对于恒星确定的运动显露出来。对于预测轨道上的行星，一般在最好的环境下显现出的运动每小时不超过 2″ 或 3″（主要是由于地球的运动），对于可以被探测到的行星，至少需要 5″ 的踪迹。另一方面，连续几天的拍摄可以得到行星确定的运动，但是从分立的照片中在包含数千颗恒星的区域里找到这颗行星将是一件困难的工作。恐怕洛威尔天文台的观测者们在发现这颗遥远行星之前已经偶遇了很多小行星。

Astronomers all the world over will naturally look forward with great interest to see how nearly the newly discovered body moves in the orbit predicted by Lowell, and are anxiously waiting for further details of the observations.

(**125**, 451-452; 1930)

全世界的天文学家对新发现的这个天体的运动离洛威尔预测的轨迹有多近都非常感兴趣，而且充满好奇地期待着关于这一观测结果更详细的进展。

（王宏彬 翻译；蒋世仰 审稿）

Stellar Structure and the Origin of Stellar Energy

E. A. Milne

Editor's Note

Arthur Eddington had recently proposed a model for the structure of stellar interiors, which implied that the mass of a star determines its luminosity in a unique way. Here the English astronomer Edward Milne disputes Eddington's claim. Milne's alternative model predicts that stars should have an extremely dense core—the most likely setting, it was felt, for the processes giving stars their energy—surrounded by a gaseous body of lower density. The core temperatures in his theory could be as high as 10^{11} degrees, some 10,000 times higher than Eddington's theory suggests. Milne's theory made little use of the principles of quantum theory or emerging nuclear physics, but illustrated a growing interest in understanding stellar structure from first principles.

THE generally accepted theory of the internal conditions in stars, due to Sir A. S. Eddington, depends largely on a special solution of the fundamental equations, and according to this a definite calculable luminosity is associated with a given mass. If this were the only solution of the equations it would conflict, as I have repeatedly shown in recent papers, with the obvious physical considerations which show that we can build up a given mass in equilibrium so as to have an *arbitrary* luminosity (not too large) whatever the assumed physical properties of the material. I have recently noticed that the fundamental equations possess a whole family of solutions, corresponding to arbitrarily assigned luminosity for given mass. These solutions show immediately that Eddington's solution is a special solution and corresponds to an unstable distribution of mass. In the stable distributions the density and temperature tend to very high values as the centre is approached, theoretically becoming infinite if the classical gas laws held to unlimited compressibility.

The physical properties of the stable configurations can be described as follows. Suppose a star is built up according to Eddington's solution with his value of the rate of internal generation of energy. Let the rate of internal generation of energy diminish ever so slightly. Then the density distribution suffers a remarkable change. The mass suffers an intense concentration towards its centre, the external radius not necessarily being changed. The star tends to precipitate itself at its centre, to crystallise out so to speak, forming a core or nucleus of very dense material. The star tends to generate a kind of "white-dwarf" at its centre, surrounded of course by a gaseous distribution of more familiar type; the star is like a yolk in an egg. In this configuration the density and temperature are prevented from assuming infinite values by the failure of the classical gas laws, but they reach values incomparably higher than current estimates. For example, it seems probable (though the following estimates are subject to revision) that the central temperature exceeds 10^{11} degrees, in comparison with the current estimates of the order of 10^7 degrees; and the density may run up to the maximum density of which ionised matter is capable.

恒星的结构和恒星能量的起源

米尔恩

编者按

阿瑟·爱丁顿最近提出了一种恒星内部结构的模型，该模型认为恒星的质量唯一地决定了恒星的光度。在这篇文章中，英国天文学家爱德华·米尔恩对爱丁顿的观点提出了异议。米尔恩提出的另一种模型预言，恒星应该具有一个非常致密的核心，在这个核心中极有可能进行着为恒星提供能量的反应，而核心的周围被低密度的气体包围。按照米尔恩的理论，恒星核心的温度可能会高达 10^{11} 度，这比根据爱丁顿的理论估计出来的值高出 10,000 倍。米尔恩的理论几乎没有用到任何量子理论和新兴的核物理学的原理，不过这表明了人们从基本原则出发解释恒星结构的兴趣正在增长。

目前，由爱丁顿爵士提出的恒星内部结构理论被大家普遍接受，该理论主要基于基本方程的特殊解，由此计算出的恒星的光度与恒星质量密切相关。然而，正如我在最近几篇文章中反复强调的那样，如果这是方程的唯一解，那么就会与一些显而易见的物理学原理相矛盾，比如，一颗质量一定的处于平衡态的恒星，可以具有**任意的**光度（只要不是太大），而且与物质的物理特性无关。我最近发现，基本方程可以得出一组解，对于同一质量，会解出任意的光度。这组解显而易见地说明了，爱丁顿的解只是一个特解，并且对应着质量的不稳定分布。在稳定分布的情况下，中心区域附近的密度和温度会变得非常高；若经典的气体定律在无限压缩的条件下仍然适用，则密度和温度在理论上可以达到无穷大。

这种稳定结构的物理特性如下所述：假如一颗恒星是按照爱丁顿的解和他给出的内部产能率而构造的，则只要该产能率略微减小，密度分布就会发生明显的变化，质量向中心紧密聚集，而外部半径并不一定会相应地发生变化。随着这颗恒星中的物质逐渐向中心沉积，就会形成一个致密的核，甚至会达到结晶。恒星在其中心形成了一种“白矮星”，其周围分布着常见的气态物质，这时整颗恒星就像一个鸡蛋中的蛋黄。在这种结构中，由于经典气体定律的失效，密度与温度不再是无穷大，但仍然远高于现行的估计值。比如，中心温度很可能（虽然下面的估计值曾被修正）高于 10^{11} 度，而目前的估计值为 10^7 度；而中心密度可能达到了电离物质所能具有的极限密度。

The unstable density distribution of Eddington's model (curve *A*) and the stable density distribution of actual stars (curves *B*) are indicated roughly in Fig. 1, which is not drawn to scale. It may be mentioned that the instability is of a radically different kind from that discussed by Sir James Jeans. He concluded that perfect-gas stars of Eddington's model were *vibrationally* unstable. In my investigations, the instability of Eddington's model arises from any slight departure of the rate of generation of energy below the critical value found by Eddington. The perfect-gas distribution of my solutions is perfectly stable, but the density necessarily increases until degeneracy or imperfect compressibility takes control.

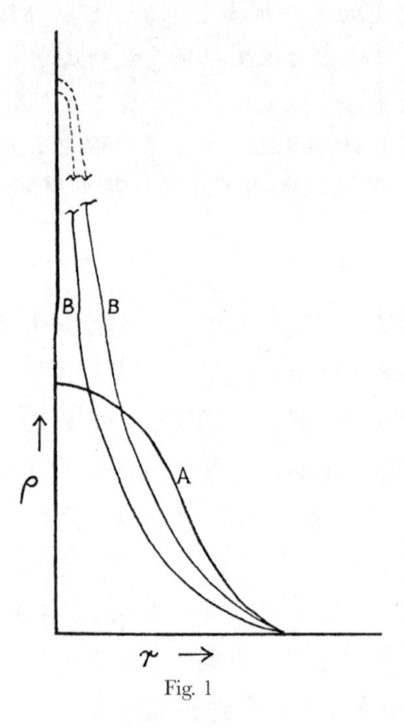

Fig. 1

The consequences amount to a complete revolution in our picture of the internal constitution of the stars. In the intensely hot, intensely dense nucleus, the temperatures and densities are high enough for the transformation of matter into radiation to take place with ease. It is to this nucleus that we must look for the origin of stellar energy, a nucleus the existence of which has previously been unsuspected. The difficulties previously felt as to stellar conditions being sufficiently drastic to permit the evolution of energy largely disappear. Many of the cherished results of current investigations of the interiors of stars must be abandoned; current estimates of central temperature, central density, the current theory of pulsating stars, the current view that high mass necessarily implies high radiation pressure, the supposed method of deducing opacity of stellar material from observed masses and luminosities, the supposed proof of the observed mass-luminosity correlation—all these require serious modification.

The new results are not a speculation. They are derived by taking the observed mass and luminosity of a star, and finding the restrictions these impose on the possible density

图 1 粗略地显示了爱丁顿模型中的非稳定密度分布（曲线 A），以及实际恒星的稳态密度分布（曲线 B），该示意图未标刻度。需要说明的是，这里的非稳定性完全不同于詹姆斯·金斯爵士的论述。詹姆斯·金斯认为，爱丁顿模型中的理想气体恒星处于非稳定的**振动**状态。而我的研究表明，爱丁顿模型的非稳定性源于产能率略低于爱丁顿给出的临界值。在我的解中，理想气体分布是完全稳定的，但是密度会不断增加，直到简并或非理想压缩起主导作用。

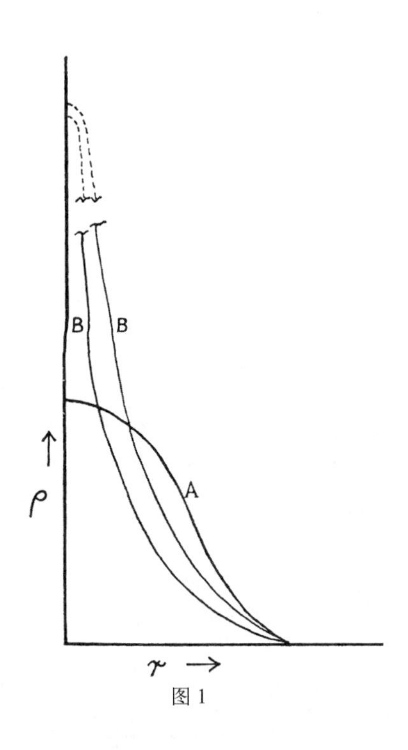

图 1

这些结果对于我们理解恒星内部结构具有革命性的意义。在极端炽热和致密的恒星核中，足够高的温度和压力可以轻易地使物质转化为辐射能。面对这样的核心，我们必须寻找恒星能量的来源，而核的存在以前并不为人所知。以前人们在考虑恒星内部必须存在极端条件才能完成能量演化时遇到的困难现在已经基本解决。然而，我们不得不放弃目前流行的许多关于恒星内部的珍贵研究结果：目前对于恒星中心温度、中心密度的估计，目前关于脉动星的理论，目前关于高质量必然导致高辐射压的观点，通过观测恒星的质量和光度得出恒星物质不透明度的方法，以及对观测到的质–光关系的证明——所有这些成果，都需要进行认真的修正。

这些新得到的结果不是猜测，而是通过测量恒星的质量和光度并给定一些限制条件后推导出来的，这些限制条件要求密度分布必须同恒星的质量和光度相容。通

distributions compatible with this mass and luminosity. By integrating the fundamental equations from the boundary inwards, we are inevitably led to high central temperatures and densities. So long as the classical gas laws persist, the solution is one of the family with a central singularity (infinities in ρ and T), and it is only the ultimate failure of the gas laws which rounds off the distribution with a finite though very large central ρ and T.

(**126**, 238; 1930)

E. A. Milne: Wadham College, Oxford, July 29.

过对基本方程从边界向内积分，我们必然会得到较高的中心温度和密度。只要经典气体定律仍然适用，就可以得到其中一个具有中心奇点的解（相应的密度 ρ 和温度 T 为无穷大）；而只有当气体定律最终不再适用时，才能够形成中心密度 ρ 和温度 T 虽然很高但不是无穷大的分布。

（金世超 翻译；何香涛 审稿）

Stellar Structure and the Origin of Stellar Energy*

E. A. Milne

Editor's Note

Here the astronomer Edward Milne offers an extensive review of new thinking on stellar structure. Astronomers then knew of two distinct classes of stars, one having densities roughly comparable to matter on Earth, and another, known as white dwarfs, with densities roughly 100,000 times greater than water. Physicist Ralph Fowler had recently shown that white dwarfs are prevented from further gravitational collapse by an effective pressure produced by the prohibition in quantum theory of electrons to occupy the same quantum state. Without the benefit of an understanding of nuclear fusion, attempts to link the process of energy generation within a star, involving the conversion of mass to radiation, to the natural evolution of stars, were at that stage severely hampered.

PERHAPS the most striking general characteristic of the stars is that they can be divided into two groups of widely differing densities. In the first group, which comprises the majority of the known stars, the densities are of "terrestrial" order of magnitude; that is to say, their mean densities are of the order of the known densities of gases, liquids, and solids. They range from one-millionth of that of water to ten or, in rare cases, perhaps fifty times that of water. In the second group the densities are of the order of 100,000 times that of water. Of the second group, the "white dwarfs", only a few examples are known, but they are all near-by stars, and it is generally agreed that they must be of very frequent occurrence in Nature, though difficult of discovery owing to their faintness. Whether stars exist of intermediate density remains for future observation. The possibility of the existence of matter in this dense state offers no difficulty. As pointed out by Eddington, we simply have to suppose the atoms ionised down to free electrons and bare nuclei. At these high densities the matter will form a degenerate gas, as first pointed out by R. H. Fowler. But this leaves entirely unsolved the question of why, under stellar conditions, matter sometimes takes up the "normal" density and sometimes the high density. Owing to the probable great frequency of occurrence of dense stars, it might reasonably be asked of any theory of stellar constitution hat it should account for dense stars in an unforced way.

There are two main theories of stellar structure at the present moment. That of Sir James Jeans accounts for the existence of giants, dwarfs, and white dwarfs, but only at the cost of *ad hoc* hypotheses quite outside physics. It assumes stars to contain atoms of atomic weight higher than that observed on earth, and it assumes them to be relentlessly disappearing in the form of radiation; it appeals to discontinuous changes of state consequent on successive ionisations, for which there is little warrant. I think it is true to say that the majority of astronomers do not accept this theory.

---

* Substance of lectures delivered at the Royal Institution on Dec. 2 and Dec. 9, 1930.

恒星的构造和恒星能量的起源*

爱德华·米耳恩

编者按

在本文中，天文学家爱德华·米耳恩全面回顾了关于恒星构造的新理论。当时的天文学家已经知道有两种完全不同的恒星，其中一种的密度与地球上物质的密度相当，而另一种就是被称为白矮星的恒星，其密度约为水的100,000倍。物理学家拉尔夫·福勒最近表明：由于量子理论中禁止电子占据同一量子态而产生的有效压力使白矮星无法进一步发生引力坍缩。当时在不了解核聚变反应的情况下，试图把恒星内部质量向辐射转化的能量产生过程与恒星的自然演变过程联系起来是相当困难的。

也许恒星最为显著的特点就是它们可以分为密度差异很大的两类。第一类恒星，包括我们已知的大多数恒星，其密度所在的量级与地球上的物质相当，也就是说，其平均密度与大家都知道的气体、液体和固体的密度相近。这些恒星的密度从水密度的100万分之一到水密度的10倍，在某些极特殊的情况下可能达到水密度的50倍。对于第二类恒星，其密度的量级相当于水密度的100,000倍。第二类恒星中有“白矮星”，现在只知道几个这样的例子，但它们距离地球都很近，大家普遍认为这类恒星必定会频繁地出现在宇宙中，只是由于其亮度微弱难以发现。至于是否存在密度介于以上两类之间的恒星，仍然有待进一步观测。致密物质的存在在解释上不存在困难。正如爱丁顿曾经指出的那样，我们只需假设原子离解为自由电子和裸露的原子核，就可以解释密度问题。福勒最早指出，在这样高的密度下，物质将形成简并气体。但这完全没有解答为什么在恒星中物质的密度有时是“正常的”，而有时又非常高。因为致密恒星很可能是大量存在的，所以任何关于恒星构造的理论应该能够合理地解释出致密恒星是自然形成的。

目前，关于恒星构造的理论主要有两种。詹姆斯·金斯爵士的理论解释了巨星、矮星和白矮星的存在，但需要引入一些与物理学毫不相干的特殊假设。该理论假设恒星中包含的原子所具有的原子量大于在地球上的观测值，而且这些原子在以辐射形式不断地被消耗；持续电离会引起状态的非连续变化，而这一点是缺乏根据的。我认为事实上大多数天文学家并不接受这一理论。

* 本文取自作者于1930年12月2日和12月9日在英国皇家研究院发表的演讲。

The theory of Sir Arthur Eddington does not claim to account for the observed division of stars into dense stars and stars of ordinary density; nor does it establish the division of ordinary stars into giants and dwarfs. On the other hand, it claims to establish what is known as the mass-luminosity law from considerations of equilibrium only, that is, without introducing anything connected with the physics of the generation of energy. It claims to show that the observed fact that the brighter stars are the more massive can be deduced from the conditions expressing that the star is in a steady state, mechanically and thermally. It does this by making the hypothesis that the stars (giants and ordinary dwarfs) consist of perfect gas. Closer consideration of the actual formulae used by the theory shows that it scarcely bears out the claims made for it by its originator. The "formula for the luminosity" of a star makes the luminosity very nearly proportional to its effective temperature, and so the so-called proof of the mass-luminosity law involves a semi-empirical element, namely, an appeal to the observed effective temperatures of the stars, for the observed values of which the theory fails to account. Another difficulty encountered by the theory is that it makes the interiors of the more luminous (giant) stars cooler than those of the fainter stars, and it makes the interiors of both too cool for the temperature to have any appreciable influence on the rate of generation of energy, by stimulating, for example, the production of radioactive elements or the conversion of matter to radiation.

The claim to establish the mass-luminosity law from mere equilibrium considerations cannot, however, be sustained for a moment. We may regard a star in a steady state as a system provided with an internal heating apparatus (the source of energy). It adjusts itself—state of aggregation, density distribution, temperature distribution—until the surface emission equals the internal generation of energy L. But provided the luminosity L is not too large (in order that the mass shall not burst under radiation pressure), it is clear that a given mass M can adjust itself to suit any arbitrary value of L. If, starting with one steady state, we then alter L (upwards or downwards) by altering the rate of supply of energy, the star will simply heat up or cool down until the surface emission is equal to the new volume of L—precisely like an electric fire. L and M are thus independent variables so far as steady-state considerations are concerned. The fact that L and M show a degree of correlation in Nature must be connected with facts of an altogether different order, namely, with the physics of energy-generation. It is essential to recognise the difference between the formal independence of L and M as regards steady-state considerations and the observed correlation of L with M in Nature. The observed mass-luminosity law must depend on the circumstance that in some way the more massive star contrives to provide itself with a stronger set of sources. The claim to establish the mass-luminosity law from equilibrium considerations only appears to me a philosophical blunder. Further, it is unphilosophical to *assume* the interior of a gas to be a perfect gas; either knowledge of the interior is for ever unattainable or we should be able to infer it from the observable outer layers.

When we dispense with the perfect gas hypothesis and at the same time recognise the independence of L and M as regards steady-state considerations, it is found that a rational

阿瑟·爱丁顿爵士并没有声称他的理论可以解释观察到的恒星可分为致密恒星和普通密度恒星，也没有确认普通恒星能够分为巨星和矮星。另一方面，这一理论认为，不考虑任何与能量生成相关的物理过程，仅从平衡出发，就可以推出“质量—光度”定律。他提出，恒星越亮其质量越大这一观测事实，可以从满足恒星处于力学、热学稳定状态的条件中推导出来。该理论是通过假设恒星（巨星和普通矮星）由理想气体构成而得到以上结论的。进一步考虑该理论中使用的实际公式发现，这一理论根本没有证实创立人为它所作的假设。恒星的“光度公式”说明恒星的光度非常接近于与它的有效温度成正比；而质量—光度定律的所谓证据事实上包含了半经验的因素，即需要求助于实测的恒星有效温度，因为观测值与理论值不符合。该理论遇到的另一个困难是，它使更亮恒星（巨星）的内部温度低于那些更暗淡的恒星；而且两者的内部温度都不足以使能量生成速率受到可观的影响，以激发像生成放射性元素或者使物质转化为辐射这样的过程。

无论如何，仅仅根据平衡过程就确立质量—光度定律是站不住脚的。我们可以将一颗处于稳态的恒星看作是一个具有内在加热装置（能量源）的系统。该系统可以调整自己的聚集状态、密度分布以及温度分布，直至表面辐射等于内部产生的能量 L。如果光度 L 不是非常高（这样，物质就不会在辐射压作用下爆发），一个质量为 M 的系统显然可以通过调整自身以适应任意的光度值 L。对于处于稳态的恒星系统，我们可以通过改变能量供给速率来改变 L 值（增大或减小），这颗恒星的温度将随之升高或降低，直至表面辐射等于新的 L 值，其原理与电暖炉十分相似。因此，对于稳态系统而言，光度 L 与质量 M 是两个独立的变量。自然界中光度 L 与质量 M 具有一定关联性的事实，一定与一个完全不同的原因有关，也就是说与能量生成的物理过程有关。有必要意识到 L 和 M 在稳态系统中的独立性与 L 和 M 在自然界中的关联性之间存在的矛盾。观测到的质量—光度定律必须依赖于以下条件，即质量较大的恒星可以通过某种方式为自己提供更多的能源。在我看来，仅通过平衡过程就确立质量—光度定律的看法是科学上的大谬。此外，无论我们对恒星内部的认识是永远无法达到的，还是应该能从观测到的外层推出内层的情况，**假设**气体内部为理想气体都不具有科学性。

如果我们摒弃理想气体假说，并承认 L 和 M 在稳态情况下相互独立，那么就会发现对恒星结构的合理分析自然而然地解释了致密恒星的存在，而无须引入特殊的

analysis of stellar structure automatically accounts for the existence of dense stars without special hypothesis. Further, it shows, as common sense would lead us to expect, that the more luminous stars must have the hotter interiors. Here the temperatures are found to range up to 10^{10} degrees or higher, depending on luminosity—a temperature sufficient to stimulate the conversion of matter into radiation. In addition, it shows that the central regions of stars must be very dense, ranging up to 10^7 grams cm.$^{-3}$ or higher. Thus the difficulties met by earlier theories fall away as soon as the ground is cleared philosophically.

The foregoing ideas suggest the following as the fundamental problems of stellar structure: (1) What are the configurations of equilibrium of a prescribed mass M as its luminosity L ranges from 0 upwards, M remaining constant? (2) What is the effective temperature T_e associated with a given pair (M, L) in a steady state? (3) What is the value of L which will actually occur for the physical conditions disclosed by the answer to problem (1)?

We observe that the outer parts of a star are gaseous. Consequently we can solve the problem of the state of any actual star by integrating the equations of equilibrium *from the boundary inwards*; we are entitled to assume the gas laws to go on holding until we find that the conditions are incompatible with them. We then change to a new equation of state, and carry on as before. We change our equation of state as often as may be necessary until we arrive at the centre.

The answer to the first of the problems formulated above has been worked out, for certain types of source-distribution and opacity, by the method of inward integration. The results are sufficiently alike to be taken as affording insight into the nature of stellar structure in general, and are as follows. For a given mass M, of prescribed opacity, there exist two critical luminosities L_1 and L_0 ($L_1>L_0$) such that for $L>L_1$ no configurations of equilibrium exist; for $L_1>L>L_0$ the density and temperature increase very rapidly as the center is approached $\left(T \propto r^{-1}\left(\log\frac{\text{const.}}{r}\right)^{-\frac{1}{2}}\right)$, so that in the centre there is a region of very high temperatures and densities where the gas laws are violated; for $L=L_0$ a diffuse perfect gas configuration is possible; for $L_0>L>0$ the only perfect gas configuration is a hollow shell provided with an internal, rigid supporting surface of spherical shape. Since in Nature no internal supporting surface is provided, to find the actual configuration when $L_0>L>0$ we construct the artificially supported hollow configuration and then remove the supporting surface. The mass must collapse, and collapse will proceed until a steady-state is attained in which, except for a gaseous outer fringe, the gas laws are violated. Such configurations may be termed "collapsed". Configurations for which $L_1>L>L_0$ may be termed "centrally-condensed". The physical origin of the different types of configuration is simply the varying effect of light-pressure. For $L=L_0$ the light-pressure due to L is just sufficient to distend the star against its self-gravity and maintain it in the form of a perfect gas. For $L_1>L>L_0$ light-pressure is so high that for equilibrium to be maintained gravity at any given distance from the centre must be assisted by concentrating as much matter as possible inside the sphere in question; when this process is carried out for all spheres, we get a central condensation. For $L_0>L>0$, light-pressure due to L is so low that the mass

假说。另外，常识也告诉我们，明亮的恒星一定具有较高的内部温度。人们发现其温度大约为 10^{10} 度或更高，由光度大小决定——应达到足以使物质转化为辐射能的温度。此外，这表明恒星的中心区域会非常致密，密度达到 10^7 g/cm^3 或更高。因此，只要科学地摒弃一些基础假设，早期理论遇到的困难就可以得到很好的解决。

上述观点提出了以下关于恒星结构的基本问题：(1) 质量 M 值一定，在光度 L 从 0 开始逐渐增加的过程中平衡结构是怎样的？ (2) 处于稳定状态时与特定 M、L 相关的有效温度 T_e 是多少？ (3) 在满足问题 (1) 答案的物理条件下，实际的光度值 L 是多少？

我们注意到恒星的外部是气态的，因此通过对平衡方程**从边界向内积分**，我们可以解出任意恒星所处的状态；我们有理由认为可以一直应用气体定律，直到我们发现条件与它不再相容。然后我们再转换到一个新的状态方程，仍用以前的方式处理。我们按照需要不断改变状态方程直至积分到中心。

这样上述第一个问题的答案就可以解决，对于某些类型的源分布和不透明度，可以采用向内积分的方法。由此得到的结果具有足够的相似性，从而可以理解恒星构造的总体特性，具体结果如下。对于给定的质量 M 和不透明度，存在两个临界光度值 L_1 和 L_0 ($L_1>L_0$)：当 $L>L_1$ 时，不存在平衡结构；当 $L_1>L>L_0$ 时，在不断接近中心的过程中，密度和温度急剧增加 $\left(T\propto r^{-1}\left(\log\frac{\text{常数}}{r}\right)^{-\frac{1}{2}}\right)$，所以中心处是高温高密度区域，气体定律不再适用；对于 $L=L_0$ 的情况，可能存在扩散的理想气体结构；当 $L_0>L>0$ 时，理想气体结构只能是一个内部具有刚性球形支撑面的空壳。在自然界中不存在这样的内部支撑表面，为了复原 $L_0>L>0$ 时的实际结构，我们人为地构造了一个中空的支撑结构，然后把支撑面去除。那么质量一定会发生塌陷，塌陷会一直持续直至达到新的稳定状态；在此状态下，除了气态的外部边界以外，其他区域均不符合气体定律。这样的结构可以被称为“塌陷”。而 $L_1>L>L_0$ 时的结构可以被称为“中心凝聚”。不同结构类型的物理根源仅在于光压的变化。对于 $L=L_0$ 的情况，由 L 产生的光压刚好可以对抗恒星自身的引力，从而保持理想气体的状态。对于 $L_1>L>L_0$ 的情况，由于光压非常高，为了维持平衡态，在距离中心任意远处，引力肯定会因聚集了球面内尽可能多的物质而增加；当该过程在所有球面内同时发生的时候，就会导致中心凝聚。对于 $L_0>L>0$ 的情况，由 L 产生的光压非常低，以致于理想状态下的物质无法支撑自身的重量，从而发生塌陷直到气体定律不再适用。由于 L 的微小变动，$L=L_0$ 时对应的扩散结构是不稳定的。

cannot support itself against its own weight in the form of perfect gas, and collapse sets in until the gas-laws are disobeyed. The diffuse configurations $L=L_0$ are unstable with respect to small changes of L.

For collapsed or centrally-condensed configurations the center will be occupied by a gas in a degenerate state. When the mean densities or effective temperatures of collapsed configurations are calculated, using the Fermi-Dirac statistics for the degenerate gas, they are found to agree with the observed order of magnitude for white dwarfs. Thus, collapsed configurations may be identified with white dwarfs. A white dwarf is thus a dense star simply because its luminosity is too low, and its light-pressure accordingly too low, for it to support its own mass against its own gravity. From another point of view the calculation affords an observational verification of the numerical value of the "degenerate gas constant" the coefficient K in the degenerate gas law $p = K\rho^{\frac{5}{3}}$, and so a check on the Fermi-Dirac statistics.

If collapsed configurations may be identified with white dwarfs, centrally-condensed configurations may be provisionally identified with ordinary giants and dwarfs, though the full determination of the properties of centrally-condensed configurations awaits the construction of certain tables. Centrally-condensed configurations appear to have the properties that as L decreases from L_1 to L_0 the effective temperature rises to a maximum and then decreases again. This would correspond to the observed division into giants and dwarfs. I give this deduction with some caution, as it is not yet demonstrated rigorously in the absence of the tables above mentioned.

A point not yet settled is the question of the continuity of the series of centrally-condensed configurations with the collapsed configurations (Figs. 1 and 2). There are indications that as L passes through L_0 from above to below, the external radius of the configuration may decrease discontinuously, the gaseous envelope collapsing on to the dense core. If this is confirmed, it would follow that a star, when its steady-state luminosity L falls through a certain critical value (depending on its mass), exhibits the phenomena of a nova or temporary star. For it would have to disengage a large amount of gravitational potential energy in a short time, so that the actual emission would undergo a temporary increase, falling again to a value just below its previous value. It would be highly interesting to have observational data as to the densities of a nova before and after the outburst. The early-type spectrum of the later stage of a nova may indeed be taken to indicate a high effective temperature, and so a small radius and high density, in accordance with our prediction.

塌陷或中心凝聚结构的中心将充满处于简并态的电子气。根据简并气体的费米–狄拉克统计，我们可以计算出塌陷结构的平均密度和有效温度，该结果与白矮星的实测数量级相符合。因此，白矮星很可能具有塌陷结构。白矮星之所以属于致密恒星只是因为它的光度过低，相应的光压也很小，以至于其重量无法对抗自身引力的缘故。从另一个角度看，上面的计算也可以验证简并气体常数的数值，即简并气体定律 $p = K\rho^{\frac{5}{3}}$ 中的系数 K，并同时检验了费米–狄拉克统计。

如果认为白矮星具有塌陷结构，则可以暂时假设普通巨星和矮星具有中心凝聚结构，虽然为了完全确定中心凝聚结构的属性，尚需要构建特定的资料表。中心凝聚结构看起来具有以下特点，随着 L 从 L_1 减小到 L_0，有效温度先升到一个最大值然后再下降。这与观测到的巨星和矮星一致。我以审慎的态度提出这个结论，因为在缺乏上面提到的资料表的情况下，尚不能得到严格的证明。

现在，尚未解决的问题是带有塌陷结构的中心凝聚结构的连续性（如图 1 和图 2 所示）。一些迹象表明，当光度 L 从高往低经过 L_0 的时候，该结构的外半径可能会不连续地下降，气体外壳塌陷到致密的内核上。如果这一点得到确认，那么当一颗恒星的稳态光度 L 下降到某一个临界值（取决于星体质量）时，该星体就会发生新星或暂星现象。由于它要在很短的时间内释放大量的引力势能，因此实际的辐射会出现暂时的增加，而后再降低到比先前略小的值。人们在新星爆发前后观察到的密度值变化很有趣。一个新星晚期的早型光谱确实可以说明其具有较高的有效温度、较小的半径和较高的密度，这与我们的预测相符。

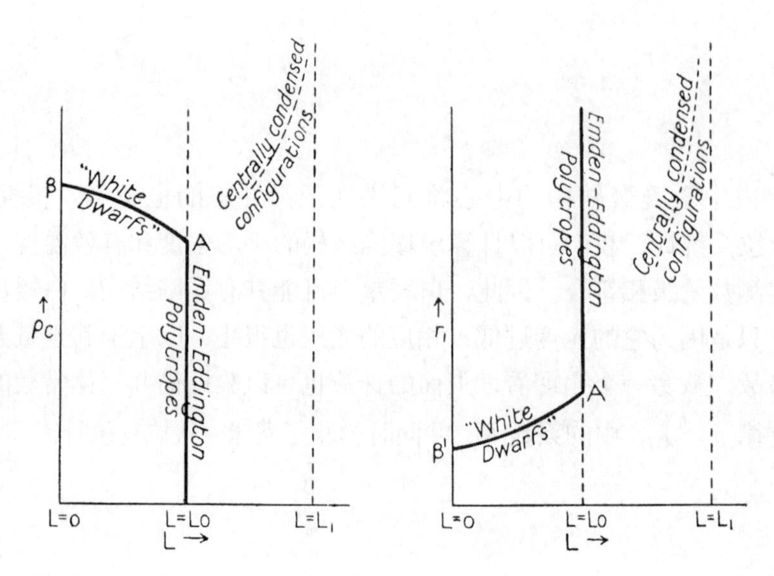

Figs. 1. and 2. The linear series of steady-state configurations of a mass M, of prescribed opacity, as its luminosity L varies. (The "white dwarfs" are the "collapsed" configurations of the general theory. The "Emden–Eddington polytropes" are the gaseous diffuse configurations of the general theory; they are unstable, in general, with regard to deviations of L on either side of the value L_0. The "centrally-condensed" series has not been fully worked out—it awaits the construction of certain tables—but it may be provisionally identified with stars in the state of giants and ordinary dwarfs. The diagram is to be understood as classificatory, not evolutionary.) (ρ_c=central density, r_1=external radius.)

The important point about all the foregoing analysis is that it involves at no stage any special properties of matter or special assumptions. The observed features of the stars are thus found to depend only on the most general properties of matter in association with light-pressure.

A question logically distinct from these is the origin of stellar energy. Here we require to know something of the physics of energy-generation. The following suggestions are frankly of a speculative character. Let us assume, in accordance with a hypothesis first made by Jeans (not his later hypothesis of super-radioactive atoms), that protons and electrons can unite to form radiation. Then thermodynamic considerations show that the process must be reversible—photons can generate matter. We know that matter at ordinary temperatures is stable. Hence we may postulate the existence of a critical temperature above which the process can go on in either direction. Suppose this critical temperature has been passed at 10^{11} degrees. Calculation then shows that at 10^{11} degrees almost the whole of the *mass* in an *enclosure* would be in the form of radiation; and further, that lowering of the temperature of the enclosure would result in more of the surviving matter present disappearing in the form of radiation. The process is in fact the thermodynamic opposite of evaporation: steam condenses to water with emission of energy, and the process is accordingly encouraged by cooling; matter "evaporates" (to radiation) with emission of energy, and the process is encouraged by cooling. Now, the centre of a star is a sort of thermodynamic enclosure with a slight leak. It follows that if (as the steady-state theory indicates) the central region of a fairly luminous star is at a temperature of 10^{11}

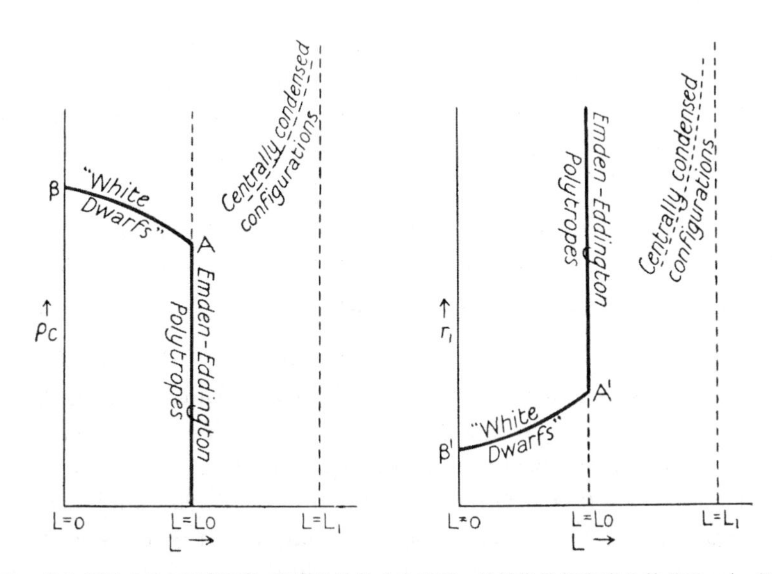

图 1 与图 2. 给定质量 M 与不透明度，稳态结构的中心密度、外部半径与光度 L 的关系。（一般认为，"白矮星"具有"塌陷"结构；"埃姆登－爱丁顿多层球"具有气态扩散结构，通常在 L_0 的两侧，与 L 值的任何偏离都会造成不稳定。由于需要建立特定的资料表，所以"中心凝聚"结构尚未完全解出，但是暂时可以认为该系列与巨星和普通矮星所处的状态一样。以上两图只是为了分类，并不表示演化。）（ρ_c= 中心密度，r_1= 外部半径）

在上述分析中，非常重要的一点是，我们没有引入物质的任何特殊属性或者某些特别的假设。因而观测到的恒星特征仅仅依赖于与光压相关的、最一般的物质属性。

当然，恒星能量的起源问题有别于上述问题。现在我们需要知道一些有关能量生成的物理规则，坦白地说，下面的分析仍具有推测的性质。让我们根据由金斯先生首先提出的假说（不是指他后来提出的超放射性原子假说）假设质子和电子在结合时能产生辐射。而热力学要求这一过程必须是可逆的，即光子也可以生成物质。我们知道，物质在常温下是稳定的。因此我们可以假定一个临界温度的存在，高于这个温度，则转化过程向两个方向都可以进行。假设这个临界温度已经达到了 10^{11} 度。计算表明，当温度达到 10^{11} 度的时候，一个星体**内部**的所有**质量**都将转化为辐射能；当恒星内部温度降低时，将导致有更多的剩余物质以辐射能的形式消耗掉。事实上在热力学中，这一过程与蒸发过程是相反的：蒸汽凝聚为水时要放出能量，所以该过程是由冷却驱动的；而物质"蒸发"（指的是辐射过程）时也要释放能量，这一过程也是受冷却驱动的。这样，恒星的中心就会成为一种稍微有一点泄漏的热力学封闭系统。于是假设（见稳态理论中的说明）在一颗相当明亮的恒星的中心区域，温度达到了 10^{11} 度且有很高的密度，那么事实上中心区域就会成为一个拥有极高辐射

degrees and a high density, then this central region is effectively a reservoir of very dense radiant energy, with a mere sprinkling of ordinary matter present. Natural cooling of this reservoir provides the star's emission to space, and the reservoir is itself maintained by the conversion of matter into radiation inside it and on its confines.

Calculations based on this idea are consistent with the usually accepted evolutionary time scale, and predict a rate of "generation" of energy ε per gram of the right order, namely, $\varepsilon = -\frac{4c^2}{T}\frac{dT}{dt}$ throughout the mass of the core, where $-dT/dt$ is the rate of cooling. The rate of loss of mass is given by the usual formula $\frac{dM}{dt} = -\frac{L}{c^2}$. By combination of these formulae it is found possible in principle to establish a relation linking M with T_c (the central temperature); this is the additional relation which, by expression of T_c in terms of L and M by means of the steady-state theory, must lead in due course to a mass-luminosity correlation. Whether it agrees with the observed mass luminosity law remains for future investigation, but it is a final satisfaction that, after first considering L and M as independent variables, we are able to use the equilibrium configurations thus disclosed to arrive in outline at a solution of the problem of the actual correlation of mass and luminosity in Nature. It is to be noted that the star's generation of energy is naturally non-explosive, for it is simply a consequence of the natural tendency of the star to cool. The star behaves, in fact, simply like a freely cooling body containing a central region of very high specific heat—namely, a pool of intense radiant energy, which is gradually drained away though partially reinforced by the conversion of matter. From this point of view, it is not that a star descends an evolutionary path because its rate of generation of energy slackens; it is rather that the act of evolving and the act of radiating energy are identical.

These suggestions as to the origin of stellar energy and the mode of stellar evolution are not to be pressed. They are to be sharply distinguished from the steady-state theory, which by the rational process of proceeding from the known stellar exterior step by step into the unknown interior indicates an inevitable series of configurations which correspond to the observed bifurcation of celestial objects into "ordinary" stars and "dense" stars.

NOTE.—The fundamental result of the rational method of analysis of stellar structure described in the foregoing article is the division of configurations into two types, the "collapsed" and the "centrally-condensed". The existence of these two types can be demonstrated without complicated mathematics by the following argument. Let r_1 be the radius of a configuration, arbitrarily assigned beforehand. Let us endeavour to construct a gaseous configuration with this radius. If such a configuration be capable of being constructed, let us in imagination take a journey inwards to the centre, starting from the boundary. Let M be the total mass, $M(r)$ the "surviving" mass left inside the sphere of radius r when we have reached the distance r from the centre. Then $M-M(r)$ is the mass already traversed. Consider now the influence of light-pressure. If L is large, light-pressure will be large and will balance an appreciable fraction of gravity, and accordingly the density-gradient will be small. But if L is small, light-pressure will be small, and the density-gradient will be large. Thus, when L is large, we shall have traversed a smaller mass

能的存储器，其中只存在非常少量的普通物质。该存储器的自然冷却过程使恒星能够向外界空间辐射能量,而存储器本身则可以通过内部和边缘物质转化为辐射能来维持。

根据以上分析进行的计算与通常认可的演化时间尺度相符，计算结果预测了核心区域中每克质量的能量“生成”速率 ε 的正确量级，即 $\varepsilon=-\frac{4c^2}{T}\frac{dT}{dt}$，其中 $-dT/dt$ 代表冷却速率。质量损失速率由通用公式 $\frac{dM}{dt}=-\frac{L}{c^2}$ 给出。将以上公式联合考虑，可以发现原则上在 M 与 T_c（中心温度）之间可以建立某种相关性；这是另外的一个关系式——利用稳态理论将 T_c 表达为带有 L 与 M 的式子，通过这一关系必然可以推导出质量—光度之间的关联性。至于是否与观测到的质量—光度定律一致，尚需进一步研究；但这是具有决定意义的结果，首先将 L 与 M 视为独立变量，然后我们就能够利用由此得到的平衡结构，勾画出自然界中实际存在的质量—光度关联性的大致轮廓。需要注意的是，恒星能量的生成是自然地非爆炸的，因为这只不过是恒星逐渐自然冷却的结果。事实上，恒星就像是一个自然冷却的天体，中心区域含有高热源，即聚集着大量的辐射能；尽管由于物质的转化可以补充一部分能量，但整体上是在渐渐枯竭。根据以上观点可以认为，与其说恒星向下延伸一个演化程是由于能量生成速率的放缓，不如说演化行为与能量辐射行为是同一的。

以上关于恒星能量起源和恒星演化模式的观点并非生搬硬套。这些想法与稳态理论极为不同，它通过理性分析从已知的恒星外部一步步推导到未知的内部，从而说明必然存在一系列结构，这与人们观察到的“普通”恒星和“致密”恒星之间的差别相符。

注意：上述文章描述了如何用理性的方法分析恒星结构，该方法得到的基本观点是，恒星按照结构的不同可分为“塌陷”和“中心凝聚”两种类型。通过以下的推导可以证明这两种类型的存在且不需要复杂的数学计算。令 r_1 表示一个结构的半径,这个结构是预先任意给定的。现在让我们用这个半径尽力去构造一个气态的结构。如果这样一个结构可以建成，我们不妨想象自己从边界开始向中心做一次旅行。令 M 表示总质量，$M(r)$ 表示当我们到达与中心的距离为 r 的地方时，“未走过的”半径为 r 的球体的质量。那么 $M-M(r)$ 就表示已走过的区域的质量。下面考虑光压的作用。如果 L 很大，则光压也会很大，并将抵消相当一部分引力，因而密度梯度会很小。但是如果 L 很小，光压也很小，而密度梯度将会很大。因此，当 L 较大时，我们从外层半径 r_1 到 r 之间所走过的区域的质量 $M-M(r)$ 将小于 L 较小的情况。相应地，当

M–$M(r)$ in the shell between r_1 and r than when L is small. Consequently, when L is large, $M(r)$ will be larger than when L is small. In other words, as we journey inwards, when L is small we "consume our mass" faster than when L is large. If L is sufficiently small, we may have consumed our whole mass M before we arrive at the centre; in that case the only configuration of radius r_1 and mass M is a hollow shell internally supported by a rigid spherical surface. If L is sufficiently large, we shall, however, tend to have an appreciable mass $M(r)$ surviving unconsumed however near we approach the centre, and this surviving mass $M(r)$ can only be packed inside r at the cost of high density with violation of the gas laws. Thus these configurations for large L must be centrally condensed. For small L, on the other hand, no configurations of radius r_1 and mass M, unsupported, can exist, and the actual configurations must be "collapsed" ones. "Collapsed" configurations prove to be much more nearly homogeneous than "centrally-condensed" ones.

(**127**, 16-18 & 27; 1931)

L 较大时，$M(r)$ 也将大于 L 较小的情况。换言之，当我们向内走的时候，如果 L 较小，则我们“走过的质量”要比 L 较大时多。如果 L 特别小，我们可能在到达中心之前就走完了全部质量 M；在这种情况下，半径 r_1 和质量 M 组成的结构只能是一个内部由刚性球面支撑的中空壳体。反之，如果 L 足够大，则无论我们多么靠近中心，未走过的剩余质量 $M(r)$ 仍会很大；这些质量只能聚集在半径为 r 的区域之内，具有非常高的密度，气体定律也不再适用。因此具有较大 L 值的结构一定是中心凝聚的；另一方面，对于较小的 L，半径为 r_1 质量为 M 的非支撑结构是不可能存在的，实际结构只能是“塌陷”结构。已证明，“塌陷”结构比“中心凝聚结构”要均匀得多。

（金世超 翻译；蒋世仰 审稿）

Stellar Structure

Editor's Note

In a recent review in *Nature* of new theories of stellar structure, Edward Milne dismissed the views of James Jeans as requiring *ad hoc* hypotheses outside known physics. Here Jeans writes in response. He did not, he claims, simply hypothesise the existence in stars of atomic mass greater than anything on Earth, but drew this possibility from the best possible fit to the latest observed relationship between the luminosity of known stars and their temperature. Jeans also comments on Milne's mathematical treatment of stellar structure, concluding that he has been led to his views by unnecessary presuppositions, and that anything novel in his theory will vanish as soon as these are relaxed.

IN *Nature* of Jan. 3, Prof. Milne writes that my theory of stellar structure accounts for the existence of giant, dwarf, and white dwarf stars, "only at the cost of *ad hoc* hypotheses quite outside physics. It assumes stars to contain atoms of atomic weight higher than that observed on earth, and it assumes them to be relentlessly disappearing in the form of radiation; it appeals to discontinuous changes of state consequent on successive ionisations, for which there is little warrant."

This seems to me a mass of highly concentrated inaccuracy. My actual hypotheses are that the stars we observe in the sky must be stable and need not be gaseous; it seems odd to describe these as "*ad hoc* hypotheses quite outside physics". What Milne describes as "assumptions" are inferences after the main results have been obtained. After the Russell diagram has been obtained, an atomic number of 95 seems to give the *best* fit with observation, but I could have "assumed" a far lower number and obtained quite a *good* fit. Incidentally, no other theory gives any fit at all—or even anything to fit. The appeal "to discontinuous changes of state" appears to be a highly coloured description of the fact that I find that bands of stability and instability alternate.

I find it hard to believe that other astronomers understand my theory as little as does Prof. Milne. If they do, he is no doubt right in saying that the majority do not accept it.

I am not writing to challenge Milne's inaccuracies, so much as to ask whether his new theory must not ultimately prove identical with my own theory, which "the majority of astronomers do not accept". We start out on the same road, by not seeing eye-to-eye with Sir Arthur Eddington. Milne's recent discoveries—that mass and luminosity are independent, that the mass-luminosity relation is a happy (or unhappy) accident, etc.—are merely old familiar landmarks on a road which I travelled and described fully in *Nature* and *Mon. Not. R.A.S.* more than ten years ago.

恒星的构造

编者按

在最近《自然》杂志关于恒星构造新理论的回顾文章中，爱德华·米耳恩否定了詹姆斯·金斯的理论，因为他的理论需要用到一些物理学领域之外的特殊假设。而在这篇文章中，金斯反驳了这一观点。他声称自己并没有无端地假设恒星内存在的原子的质量高于地球上的所有物质，而是从最近观测到的已知恒星光度和温度之间关系的最佳拟合中推出来的。金斯还评价了米耳恩对恒星结构的数学处理，认为他的结论是从几个没有必要的假设中推出来的，一旦没有了这些假设，米耳恩理论中的新奇之处也就荡然无存了。

在 1 月 3 日出版的《自然》杂志中，米耳恩教授提到了我用来解释巨星、矮星和白矮星的存在的理论，他是这么说的“但这一理论需要引入一些与物理学毫不相干的特殊假设。该理论假设恒星中包含的原子所具有的原子量大于在地球上的观测值，而且这些原子在以辐射形式不断地被消耗；持续电离会引起状态的非连续变化，而这一点是缺乏根据的。”

在我看来，以上理解是非常不准确的。实际上我的观点是，我们在天空中观测到的恒星必须是稳定的，但未必一定是气态的；把我的观点说成是“与物理学毫不相干的特殊假设”令人感到很奇怪。米耳恩所称的“假设”其实是一些基于主要结果的推论。在有了罗素图之后，我发现当原子序数为 95 时，可以对观测结果做出**最佳**的拟合；但是我其实可以“假设”一个低得多的原子序数，同样可以得到**很好**的拟合。需要顺便提及的是，没有任何其他理论可以对上述结果进行拟合，甚至连可以用来拟合的根据都没有。米耳恩在文中还提到，我的理论要求“状态的非连续变化”，这显然是对我发现稳定带和非稳定带交替出现这一事实的歪曲。

我很难相信其他天文学家对我理论的理解也会像米耳恩教授那样浅薄。如果真是这样，那么他所说的多数人都不会接受这一理论无疑将是正确的。

我写这篇文章的目的并不是为了挑战米耳恩教授的错误，更不是为了质疑他的新理论是否有必要一定不同于我的所谓“大多数天文学家并不接受”的理论。我们的起点相同，即我们都不完全同意阿瑟·爱丁顿爵士的观点。米耳恩最近的发现——质量和光度是相互独立的，而质量—光度之间的关联是一个令人愉快（或不愉快）的偶然等等，都只不过是我以前有过的阶段性想法，十余年前，我在《自然》杂志和《皇家天文学会月刊》上已对此进行过全面的阐述。

Milne and I part company on the question of stellar boundary conditions. The classical Emden solution starts with a finite density at the star's centre and integrates outwards. Eddington, and then I, followed Emden in thinking that *Nature* was bound to look after the boundary conditions somehow or other. Then I noticed ("Astronomy and Cosmogony", §§ 80, 81) that what appears to be a single solution in a star's interior, spreads out into a whole tassel near the photosphere; it proves to be merely an asymptote to a whole family of solutions which correspond to different boundary conditions. This shows that any boundary condition can be satisfied, so that "the influence of the special conditions which prevail at the surface soon disappears as we pass inwards into the star" (l.c. § 80).

This fundamental point can be illustrated by a simple model suggested by Milne (*M. N.*, **90**, p. 53). Represent a star by a sphere of copper s with a heating coil at its centre, and its photospheric layers by s', a thin coat of asbestos, paint, or other substance. Varying the substance of s' represents varying the photospheric conditions of a star; we want to find how interior conditions depend on s'. Milne solves the problem wrongly, as I think, and concludes that the sphere coated with asbestos will be "much hotter inside" than one wholly of copper, "for it is jacketed by a bad conductor". As every plumber knows, this is not so; only a thick layer of asbestos will make things much hotter inside. It is, I think, the same with the stars. Whatever the photospheric conditions, the photospheric layers are not thick enough to make any real difference.

Thus Milne's involved procedure of integrating inwards, getting infinite or zero density, and then letting masses of unsupported gas crash to finite densities, seems to me all unnecessary; he could have assumed a finite central density to begin with and integrated outwards, and as this is the exact procedure followed in my theory I cannot see how our final results can be different—unless one of us makes a mistake in analysis or arithmetic.

It is, of course, also the procedure followed by Eddington in his classic papers. Where I differ from Eddington is not on general questions of procedure; it is that he thinks a star's centre must be gaseous, whereas I do not. Also, we differ as to whether his very restricted model with $\kappa\eta$ constant is a very good or a very bad model. Milne appears to have followed Eddington so far in tying himself down to this particular restriction. Unless I am greatly mistaken, all that is essentially novel in his theory (as shown in Figs. 1 and 2 in his *Nature* article) will vanish to nothing as soon as he frees himself from this impossible and misleading restriction.

J. H. Jeans

(**127**, 89; 1931)

* * *

米耳恩和我在恒星边界条件的问题上存在分歧。埃姆登的经典解法是从恒星中心的有限密度处开始向外积分。爱丁顿，以及后来的我，都继承了埃姆登的思路，认为物质世界必然会以某种方式寻找边界条件。然后我注意到（《天文学与宇宙演化论》，§§ 80，81）：看似是恒星内部的单一解，可以向外扩展到光球层附近的所有流苏结构；事实上可以证明该解只不过是与不同边界条件对应的一组解的渐进表达。这说明任何边界条件都可以被满足，所以“当我们向恒星内部积分的时候，在表面处占优势的特殊条件很快就会失去影响”（上述引文 § 80）。

这一点可以用米耳恩提出的简单模型来进行解释（《皇家天文学会月刊》，第 90 卷，第 53 页）。不妨让一个铜球 s 代表恒星，在其中心用线圈加热，并在该铜球表面覆上薄的石棉层、涂料或其他物质以代表光球层 s'。改变组成 s' 的物质就意味着改变恒星光球层的条件；我们希望知道 s' 是如何影响内部条件的。我认为，米耳恩解决该问题的方法是错误的，他认为覆盖着石棉的铜球会比只有铜球本身时“内部热很多”，“因为前者覆有热的不良导体”。每个水管工都清楚，事实并非如此；只有很厚的石棉层才会使内部更热。我认为这和恒星的情况是一样的。无论光球层条件如何，光球层的厚度都不足以造成任何真正意义上的区别。

因此，米耳恩采取向内积分的方法，由此得到无限的密度或零密度，并令没有支撑的气体物质塌陷为有限的密度，这些在我看来都是没有必要的。他本可以假定一个有限的中心密度，然后由中心开始向外积分，这就是我的理论所采用的方法。我不认为我们得到的最终结果会有所不同——除非我们俩之中的一个在分析或计算上出现错误。

当然，爱丁顿在他的经典论文中也采用了这一方法。我与爱丁顿的不同之处并非源于方法上的差异，而在于他认为一个恒星的中心必须是气态的，我却不这么认为。同时我们在常数为 $\kappa\eta$ 的爱丁顿受限模型究竟是一个好模型还是一个差模型的问题上也有分歧。米耳恩看起来同意爱丁顿的观点，因此也采用了这一特殊限制。除非是我犯了很大的错误，否则一旦米耳恩放弃这个根本不成立且令人误入歧途的限制，他理论中的那些新颖之处也会自然消失（参见米耳恩在《自然》杂志上发表的文章中的图 1 和图 2）。

金斯

* * *

Referring to Sir James Jeans's letter in *Nature* of Jan. 17, p. 89, I may say that I fully acknowledged in my paper of November 1929 (*Mon. Not. Roy. Ast. Soc.*, **90**, p. 20) that Sir James was the first to recognise the principle that the mass M and luminosity L of a star are independent variables as regards steady state considerations. On p. 53 of that same paper (a page of which Jeans himself quotes in another connexion) I made a general reference of obligation to his work. In my last paper (*Mon. Not. Roy. Ast. Soc.,* **91**) I build on Jeans's permanent contributions to science in three places, mentioning him by name (pp. 4, 9, 51). I could not, however, adduce any of the specific results of his theory of stellar equilibrium in support of my conclusions, for they are totally different; and I could not contrast his results with mine without venturing to discuss his mathematics.

I cannot assent to Jeans's mathematics, because his theory of stellar equilibrium is in formal contradiction with his own (L, M) independence principle. It is an immediate consequence of this principle that for a given mass M in equilibrium the ratio λ of gas pressure to radiation pressure may have any value whatever between zero and infinity, depending on the arbitrarily assigned L. This is fundamental in my analysis. According to Jeans ("Astronomy and Cosmogony", pp. 88, 89) λ is small for large masses and large for small masses, and is calculable in terms of M (p. 97). Jeans may claim the principle, but his theory is not consistent with it.

The point of my analysis is the construction of configurations which satisfy the (L, M) independence principle, even for models for which $\kappa\eta$ is constant. The special properties attributed to these models by both Jeans and Eddington then disappear, and the new general properties which emerge (explaining as they do why some stars are very dense and others not) are shared by other models, since they depend only upon the occurrence of the central singularity r=0 in a certain system of differential equations. Jeans uses throughout Emden's solutions, which possess no singularities.

As regards the branching-out of solutions near the boundary of a star, Jeans is considering a variety of models. *For any one model*, the solution is unique up to the boundary. The work of Mr. Fowler, Mr. Fairclough, and myself published in *Mon. Not. Roy. Ast. Soc.*, **91** (November 1930), discusses the family of such solutions arising from Emden's equation; with any definite configuration, of arbitrarily assigned mass, luminosity, and opacity, there is associated *one* member of the family of solutions, selected by a boundary condition which ensures that the boundary layers, of the prescribed opacity, enclose M and radiate L.

E. A. Milne

(**127**, 269; 1931)

J. H. Jeans: Cleveland Lodge, Dorking, Jan. 5.
E. A. Milne: Wadham College, Oxford, Feb. 5.

看罢詹姆斯·金斯爵士在1月17日《自然》杂志第89页发表的快报,我要说的是,我在1929年11月的一篇论文(《英国皇家天文学会月刊》第90卷,第20页)中就已经完全承认了詹姆斯爵士是第一个认识到稳态下恒星的质量 M 与光度 L 是两个独立变量的人。在同一篇文章的第53页(金斯本人曾在其他地方引用过这一页的内容),我对他的工作进行了全面的介绍。在我最近的一篇论文(《英国皇家天文学会月刊》第91卷)中,我三度(第4、9、51页)提到了金斯的名字,赞扬了他对自然科学的永恒贡献。但我不能通过引用由他的恒星平衡理论推出的特殊结果来支持我的论点,因为它们是完全不同的两回事;在没有仔细分析他的数学推导过程之前,我无法把他的结论与我的进行比较。

我不能认同金斯的数学推导,因为他的恒星平衡理论与他所说的(L、M)相互独立原则存在形式上的矛盾。根据这一原则可以直接推出,对于平衡态中的一个给定质量 M,气体压强与辐射压强的比值 λ 可以取从0到无穷大中的任意值,具体大小由任意给定的 L 值决定。这在我的分析中是最基本的。按照金斯(《天文学与宇宙演化论》第88、89页)的说法,质量较大时 λ 值较小,质量较小时 λ 值较大,且 λ 值可以用 M 的关系式计算出来(第97页)。金斯可以提出这个原则,不过他的理论与这个原则并不一致。

我在分析中的关键点是构造满足(L、M)相互独立原则的关系式,也包括 $\kappa\eta$ 是常量的模型。如此一来,金斯和爱丁顿所创建的模型中的一些特殊性质就会消失,新出现的一般属性(它们能解释为什么有些恒星非常致密,另一些则不是)可以和其他模型共享,因为它们只取决于在一个确定的微分方程系统中是否出现中心奇点 $r=0$。金斯完全照搬了没有奇点的埃姆登解法。

至于方程解在恒星边界附近的拓展性,金斯正在考虑各种模型的适用性。**对于任意一个模型而言**,其解一直到边界处都是唯一的。福勒先生、费尔克拉夫先生和我在共同发表于《英国皇家天文学会月刊》第91卷(1930年11月)的论文中讨论了由埃姆登方程得到的一组解;对于任意一个给定质量、光度和不透明度的确定体系,这组解中都存在**一个**与之相关且满足边界条件的解,使得具有指定不透明度的边界层能够封入质量 M 和光度 L。

米耳恩

(王锋 翻译;蒋世仰 审稿)

Stellar Structure

Editor's Note

Edward Milne's theory of the structure of stars provoked strong disagreement among British astronomers, which was the subject of a discussion organised by the Royal Astronomical Society on 9 January 1931. The discussion itself, reported here, was inconclusive, although events later showed that Milne's approach correctly predicted the very high densities and temperatures at the centre of stars like the Sun.

THE investigation of the structure of the stars, which has for long been a subject of disagreement between Sir James Jeans and Sir Arthur Eddington, has now entered on a new phase through the work of Prof. E. A. Milne. A long paper on this matter by Prof. Milne, which appears in the November number of the *Monthly Notices of the Royal Astronomical Society* together with related papers by Messrs. R. H. Fowler, N. Fairclough, and T. G. Cowling, was introduced to the Society in November last, but no time was available for discussion. The whole of the meeting of the Royal Astronomical Society on Friday, Jan. 9, was accordingly devoted to a debate on the subject, which was briefly opened by Prof. Milne, and in which many astronomers and mathematicians took part.

Prof. Milne's views were outlined in *Nature* of Jan. 3, p. 16. In opening the discussion, he stated that the pioneer work in the subject had been done by Sir Arthur Eddington, and what he himself had done was to rationalise Eddington's theory by clearing it of *ad hoc* hypotheses. He could not accept Sir James Jeans's theory, because it depended on unlikely extrapolations of the laws of physics. He considered that the mass-luminosity law which Eddington claimed to have established was not a possible deduction from the fundamental formulae. It could follow only from the addition of a missing equation expressing the unknown dependence on physical conditions of the rate of generation of energy in stars. He had, therefore, begun his investigations with observed quantities, representing conditions at the surface of the star, and had worked inwards, without making any assumptions unwarranted by observation.

Sir Arthur Eddington, who followed, pointed out that disagreements in physical discussions could be of two kinds: first, those which depended on the adoption of different assumptions or hypotheses in the absence of definite knowledge; and secondly, disagreements on the logical or mathematical deductions from given premises. He thought that the disagreement between Sir James Jeans and himself was entirely of the first kind; but between Prof. Milne and himself there appeared to be a mathematical disagreement, which was very unfortunate. He did not approach Prof. Milne's latest work in a spirit of opposition. He regarded it as quite a permissible attempt to improve the existing theory in points where room for improvement undoubtedly existed. The main feature of Milne's

恒星的构造

编者按

爱德华·米耳恩关于恒星构造的理论在英国天文学界引起了极大的争议，这即是1931年1月9日皇家天文学会举办的讨论会的主要论题。尽管后来人们发现米耳恩的理论正确地预测到了太阳等恒星中心的极高温度和极高密度，但在这次讨论会上并没有形成定论。

恒星构造问题是詹姆斯·金斯爵士和阿瑟·爱丁顿爵士长期争执不下的论题，现在由于米耳恩教授的工作而有了新的转机。米耳恩教授发表在《英国皇家天文学会月刊》1930年11月号上关于这一主题的长篇论文，以及福勒、费尔克拉夫和考林等人的有关论文已经在去年11月被提交给了该学会，但尚没有时间就这一主题进行讨论。因此在今年1月9日（星期五）举行的皇家天文学会大会上专门就这一主题进行了讨论。米耳恩教授用简短的发言拉开了大会的序幕，许多天文学家和数学家都出席了本次会议。

米耳恩教授的观点已发表在1月3日的《自然》杂志上（第16页）。米耳恩在讨论会开始时谈到阿瑟·爱丁顿爵士对这一问题已做了先驱性的工作，而他所做的事情不过是澄清了一些特殊的假设，从而使爱丁顿的理论更加合理化。他不能接受詹姆斯·金斯爵士的理论，因为该理论依赖于物理定律的推论，而这一推论并不可靠。他认为爱丁顿宣称已建立的质量－光度定律并非是从某些基本公式推导出来的，而只能是来自一个附加的缺失方程，这个方程与影响恒星产能率的物理条件之间的依赖关系是未知的。因此，他的研究是从代表恒星表面物理条件的可观测量开始的，然后向恒星内部延伸，并没有作出任何不能被观测所证实的假定。

接着阿瑟·爱丁顿爵士指出，物理学中讨论的分歧可以分为两种：一种是由于在缺乏确定知识的情况下采用了不同的假定，另一种则是由于从给定前提出发的逻辑或数学推导不同。他认为自己与詹姆斯·金斯爵士之间的分歧完全属于第一种；而自己与米耳恩教授的分歧似乎是数学上的分歧，这是很不幸的。他并没有用反对的态度去看待米耳恩教授的最新研究。他认为米耳恩的工作是改进现有理论的一种可行的尝试，这个理论显然存在改进的余地。米耳恩的恒星模型的主要特征是：大部分物质集中在具有很高密度的炽热核心区，并由稀薄的外层包围着。爱丁顿自己

picture of the stars was that the mass was largely concentrated in a dense hot core with a surrounding rarefied envelope. His own theory gave a much more gradual and uniform increase of density from the centre outwards. For the sun it gave a central density of 70, whereas Milne's theory gave a value of about 700,000. The matter there was therefore, according to Milne, not in the condition of a perfect gas, and hence the mass–luminosity relation could not be deduced. There were certainly loopholes in that relation—it did not apply to a star composed mainly of hydrogen, for example—but he did not believe that Milne's theory had sealed up those loopholes. Milne himself had confessedly not been able to satisfy himself that the equations for his stars possessed solutions, and Sir Arthur thought that it was *a priori* unlikely that they did. If such solutions existed, however, it still remained to decide between the two widely different central densities. This might be done by considering the intrinsic opacity in the interior, or its average if it was variable. An appeal on this ground, however, would show that both theories were wrong—that was the long-outstanding discrepancy between the physical and astronomical opacities of matter. Both theories required a larger opacity than would follow from current physical theory, but he had found that the discrepancy on Milne's theory was far greater than that on his own. A great deal depended on how large a mass was concentrated at the centre. If it was merely a point-source of extremely high temperature and density, with the rest of the star following a distribution of density almost identical with that of his own theory, he might be willing to accept the modification as a useful addition to that theory: it would provide a source of stellar energy. The large amount of mass, however, which Milne placed around the centre, inevitably made worse the existing discrepancy with regard to the opacity. Prof. Milne had objected to his theory being called a hypothesis, regarding it rather as an inevitable conclusion. It was inevitable only if the premises were granted, and Sir Arthur had been denying them for fourteen years.

Sir James Jeans found himself in almost complete agreement with what Sir Arthur Eddington had said, because Sir Arthur had not touched on the points on which they differed. If Sir Arthur, Prof. Milne, and himself all suddenly became infallible and omnipotent as mathematicians, he and Sir Arthur would still differ on the same points as before, but Prof. Milne would agree with one or the other accordingly as he considered the stars to be gaseous inside or otherwise. There was nothing new in Milne's theory; all that was accurate in it had been done previously by Eddington or himself. He had shown long ago that if you integrated outwards from the centre of a star you got definite results right up to the boundary, but that there an infinite number of solutions were possible, and Nature itself decided which was the actual one. When, therefore, Milne reversed the process by integrating from the boundary inwards, it did not matter with which of those solutions his initial data corresponded; he would necessarily arrive at the same conditions in the interior. The work, however, would be much more cumbersome. He agreed with Eddington about the opacity discrepancy, but considered that the case had not been put strongly enough. The factor 10 in the luminosity, by which Eddington's own theory deviated from observation, was much greater than was permissible. Expressed in terms of volume, it meant that the sun should be as big as Antares. Since Milne's theory made matters worse still, he considered that that theory was ruled out of court entirely.

的理论则给出恒星密度由中心向外逐渐均匀增大的模型。对于太阳，爱丁顿理论给出的中心密度是 70，而米耳恩理论给出的太阳中心密度约为 700,000。因此，米耳恩认为，太阳中心区的气态物质已不是理想气体，所以不能推导出质量 - 光度关系。这个关系肯定存在漏洞，例如不适用于主要由氢组成的恒星，但他也不相信米耳恩的理论已填补了这些漏洞。米耳恩自己承认他不能证实他的恒星方程组存在解，而阿瑟爵士也认为不太可能有解。不过如果这些解存在，仍然需要在两个差别很大的恒星中心密度之间作出选择。这或许可以根据恒星内部固有的不透明度，或者当不透明度变化时，用平均不透明度来进行判断。不过这种诉求会表明两种理论都是错误的——这是因为长期以来人们对于物质的物理不透明度和天文不透明度一直存在分歧。两种理论均要求有更大的不透明度，并且都超出了现有物理理论的最大值，但是爱丁顿发现米耳恩理论与现有物理理论的差异远远大于他自己的理论。这在很大程度上取决于有多少物质被集中到恒星的中心区。如果只有恒星的中心点具有极高的温度和密度，而恒星其余部分的密度分布几乎与他自己的理论相同，那他或许愿意把这种修正看作是对自己理论的有益补充：这个核将提供一种恒星能源。然而米耳恩把大量的物质安置到恒星中心区，就必定会使已有的关于不透明度的分歧进一步扩大。米耳恩教授反对人们把他的理论称为假说，他认为这样的结果是必然的。这样的结果只有在承认相应前提的条件下才会是必然的，而阿瑟爵士对这些前提持否定态度已有 14 年之久了。

詹姆斯 · 金斯爵士觉得自己几乎完全同意阿瑟 · 爱丁顿爵士的说法，因为阿瑟爵士并未谈到他们之间的分歧。如果阿瑟爵士、米耳恩教授和他自己突然变得像数学家那样确实可靠并且无所不能，那么他同阿瑟爵士之间还会像以前那样保持同样的分歧，而米耳恩教授将会与上述二人中的一位相一致，到底与谁一致取决于米耳恩认为恒星内部是否为气态。米耳恩的理论并不是新东西，其中所有正确的部分都来自爱丁顿或他本人以前的研究成果。他在很早之前就曾指出，如果从恒星中心向外作积分，那么你得到的解一直到边界处都是确定的，但可能存在不定数目的解，而大自然将会决定哪一个解是符合实际的。所以当米耳恩反过来从恒星边界向恒星内部积分时，他的初始数据对应的到底是哪一个解就变得无关紧要了；他必然在恒星内部得到同样的条件。不过这样工作就变得非常繁琐。他同意爱丁顿关于不透明度存在分歧的意见，但认为这一点尚未引起足够的重视。爱丁顿从自己理论中得到的光度与观测值相差 10 倍，远超过误差允许的范围。若用体积表示，则意味着太阳将有心宿二（译者注：天蝎星座 α 星，也称大火，是一颗红巨星）那样大。鉴于米耳恩的理论把问题变得更糟，他认为这个理论根本不值得考虑。

Dr. W. M. Smart directed attention to the fact that the problem was an idealised one, since we knew nothing at all about the interior of a star from observation. Assumptions had to be made, and it should not be forgotten that they were assumptions. He considered the stars themselves were the final umpires in the matter. Sir Arthur Eddington had suggested the opacity as the criterion by which the umpires gave their decision; he would suggest one of several other criteria. Sir James Jeans's theory had been very successful in explaining the formation of spectroscopic binaries by fission, but he did not think that Prof. Milne's centrally condensed stars would provide a satisfactory explanation.

Prof. F. A. Lindemann hoped that Prof. Milne would maintain his theory, because it was the only one which gave really high temperatures in the stars. The temperatures required by Sir Arthur Eddington were, on thermodynamical grounds, incapable of permitting the conversion of matter into radiation. It was all very well to say that Milne's conclusions were inevitable only if you presupposed something. Something had to be presupposed, and Eddington's *ad hoc* assumptions did not meet the facts of the generation of energy. He considered that the discrepancy of a million-fold in the densities of stars according to Eddington's theory was sufficiently large to make the increase to several million-fold in Milne's theory of no significance in deciding between the theories.

Sir Oliver Lodge remarked that what struck him most in the discussion was the remarkable agreement between the three protagonists. They were agreed on knowledge of fundamental importance which was not available twenty years ago, and, in comparison with that, the differences were unimportant. His idea of dealing with the problem would be to start with what we know—the surface temperature of the sun, its rate of generation of energy, the fact that energy comes from the disintegration of atoms. Prof. Lindemann had assured us that the last-named process required a very high temperature. Let that temperature at the centre be assumed. Then we knew the relation between radiation pressure and gravitational pressure, and with regard to the opacity the Compton effect would give us some information. From all these data a mathematician might work out some definite result. He was fascinated by Milne's theory of a nova resulting from the collapse of a star. Such a possibility had not occurred to him, but it seemed to work out.

Several other speakers contributed to the discussion, which concluded with brief comments by Prof. Milne. He did not agree that his work covered the same ground as Sir James Jeans's. He had started from the surface and worked inwards because it was only the surface that we could observe, and he had avoided assumptions which were necessary if you started in an unknown region.

Although it could not be said that the discussion led to a greater measure of agreement between the various speakers, it undoubtedly helped, by bringing together different methods of dealing with the question, to focus the nature of the problem more clearly in the minds of those present.

(**127**, 130-131; 1931)

斯马特博士注意到这是一个理想化的问题，因为我们看不到恒星内部的任何东西。因此，人们必须采用一些假定，但是不要忘记它们仅仅是假定。他认为最终的裁判依然是恒星本身。阿瑟·爱丁顿爵士曾指出，裁判作出判决的判据是不透明度，而他宁愿从另外几个判据中选择一个。詹姆斯·金斯爵士的理论非常成功地用分裂解释了分光双星的形成，但他不认为米耳恩教授的中心高密度恒星能够提供满意的解释。

林德曼教授希望米耳恩教授坚持自己的理论，因为这是唯一一个认为恒星内部温度确实很高的理论。根据热力学原理，阿瑟·爱丁顿爵士所需的温度不能使物质热到辐射出能量。可以合理地认为米耳恩的结论只有在某种前提下才是成立的。某种假定是必需的，但爱丁顿的特殊假定与恒星能量的产生事实不相吻合。他认为既然根据爱丁顿理论得到的恒星密度有一百万倍的差异，那么在米耳恩的理论中把这个差异扩大到几百万倍也就无关紧要了，这不足以作为判定这两种理论优劣的依据。

奥利弗·洛奇爵士表示这次讨论会中令他印象最深刻的是三位主角之间的不寻常的一致性。他们均同意这些 20 年前还不知晓的知识的重要性，而与此相比，分歧是次要的。对于这个问题的处理，他的想法是应该从我们知道的事实出发，即太阳表面的温度、它的产能率以及能量来自原子衰变的事实。林德曼教授使我们确信原子衰变需要非常高的温度。假定这一温度即为恒星的中心温度，于是我们就可以知道辐射压力与重力之间的关系。至于不透明度，康普顿效应会向我们提供一些信息。从这些资料出发，数学家也许会得到某种确定的结果。他对米耳恩关于新星起源于恒星塌陷的理论非常感兴趣，他没有想到会有这样的可能性，但结果似乎就是如此。

还有其他一些参加者在讨论会上发了言，最后由米耳恩教授作简要的评论。他不认为自己的研究与詹姆斯·金斯爵士的研究是建立在同一个基础上的。他的研究是从恒星表面出发向恒星内部延伸，因为我们只能观测到恒星的表面，他避免作出假定，而如果从未知的区域开始研究，这些假定就是必需的了。

尽管不能说各位发言者在这次讨论会上已经达成了更广泛的共识，不过借这个机会将处理问题的不同方法放到一起进行交流，的确有助于把与会者的思想明确地统一到问题的本质上去。

（林元章 翻译；蒋世仰 审稿）

Stellar Structure

F. A. Lindemann

Editor's Note

The British physicist Frederick Lindemann had earlier suggested that particles within stellar interiors would need kinetic energies comparable to their relativistic rest energy if the annihilation of matter were to be the source of stellar energy. Here he writes to change his view. Equilibrium thermodynamics, he notes, does not control the processes generating energy. Rather, the rate of energy release follows from the concentration of particles, and from the details of the rare circumstances required to initiate key reactions. These details would not necessarily demand high velocities for the particles. Lindemann's new analysis meant that nuclear reactions could indeed fuel stellar processes at temperatures well below 1.1 × 10^{13} K, as his earlier arguments had suggested.

IT has frequently been stated, by myself amongst others, that it is necessary to assume in the inside of stars a temperature of the order of mc^2/k in order to explain the generation of energy by the annihilation of matter, m being the mass destroyed in each process, c the velocity of light, and k Boltzmann's constant. This letter is written in order to make clear that this assumption is not necessary. It is perfectly true that the equilibrium constant of a process, subject to the laws of thermodynamics, is of the order $e^{-\varepsilon/kT}$, ε the energy of the process in this case, of course, being equal to mc^2. The equilibrium constant does not, however, determine the generation of energy. What one is concerned with in the case of a star is the rate at which energy is produced; in other words, if one presupposes the simplest process of annihilation, the rate at which protons and electrons disappear in the form of radiation. This is analogous to the rate of chemical reaction, not to the equilibrium constant of a reversible reaction.

In most chemical processes the rate of reaction is governed by the number of molecules activated per second, which again depends upon the number of particles the energy of which exceeds a certain value, say, ε_1. This number is proportional to $e^{-\varepsilon_1/kT}$, and one therefore finds, roughly speaking, that the rates of reaction are negligible unless the temperature is such that the mean energy of the molecule is comparable with the energy of excitation. Since this activation energy is usually of the same order as the energy of the reaction, the conclusion is often extended without great inaccuracy to the total energy of the process.

In the case under consideration, the annihilation of protons and electrons, it seems difficult to imagine any form of excitation, and the rate at which it proceeds can therefore scarcely depend upon a function of this type. Presumably, in such collisions as are effective, certain circumstances, which occur but rarely, have to be fulfilled. When these are fulfilled, and they may not be such as require any high velocities, matter is converted into radiation; in

恒星的构造

弗雷德里克·林德曼

编者按

英国物理学家弗雷德里克·林德曼早先曾经指出：如果恒星的能量来源于物质的湮灭过程，那么恒星内部粒子所需的动能就应该与相对论中的静止能量差不多。不过他在本文中改变了这种看法。他认为平衡态热力学不能决定能量的生成过程。更确切地说，能量释放速率是由粒子的浓度，以及引发关键反应所需的特有环境中的具体要素决定的。这些具体要素并不要求粒子具有很高的速度。林德曼的新观点是，核反应在低于 1.1×10^{13} K 的温度下的确可以在恒星内部发生，他在早期的理论中也曾暗示过这一点。

我和其他人都常常提起：为了解释由物质湮灭所引发的能量的产生，有必要假设恒星内部的温度达到 mc^2/k 数量级，其中 m 为在每一个过程中损失的质量，c 为光速，k 是玻耳兹曼常数。我写这个快报是为了说明这个假设并非必需。根据热力学定律，一个过程的平衡常数约为 $e^{-\varepsilon/kT}$，在这种情况下 ε 是该过程的能量，等于 mc^2。但是平衡常数不能决定能量的产生过程。人们关心的是一个恒星中的能量生成速率；换言之，对最简单的湮灭过程而言，就是质子和电子以辐射形式消失的速率。这类似于化学反应的速率，而不是可逆反应的平衡常数。

在大多数化学作用中，反应速率是由每秒内被激活的分子数目决定的，而这个数目又与能量超过某一特定值（如 ε_1）的粒子数多少相关。因为该值正比于 $e^{-\varepsilon_1/kT}$，所以我们可以粗略地认为，如果温度没有升高到使分子的平均能量接近于激发能，反应速率就可以忽略不计。因为这种激活能通常和反应能具有相同的数量级，所以我们把激活能看作是反应过程总能量的惯用作法就不会产生太大的偏差。

对于质子和电子的湮灭，很难想象有任何形式的激发存在，因此湮灭速率不太会由这类作用所决定。引发湮灭的有效碰撞可能必须满足某种很少发生但具有决定意义的条件。当这些条件得到满足的时候，也许并不需要太高的速度，物质就能被

the vast majority of cases, a collision has no such result. If this view is correct, the rate of annihilation, and therefore the rate of generation of energy, will depend in the first instance on the number of collisions per second, which of course varies with the density and with something like the square root of the temperature; and in the second instance, upon the special circumstances which render a collision effective, and which may, or may not, depend upon the temperature. In either event the simple exponential expression is not applicable, and the conclusion that matter can only be annihilated and energy produced in stars where interiors are at temperatures of the order mc^2/k, that is, 1.1×10^{13} degrees, is valid.

It would be true if the matter–radiation equilibrium has been attained and any further production required a change in the equilibrium constant. It is incorrect if the system has not reached equilibrium, for in this case thermodynamical reasoning is insufficient to determine the rate at which equilibrium will be approached.

(**127**, 269; 1931)

F. A. Lindemann: Clarendon Laboratory, Oxford, Feb. 5.

转化成辐射；在绝大多数情况下，一次碰撞不会引发湮灭。如果这个观点是正确的，那么湮灭速率，也就是能量产生的速率，首先将取决于每秒内的碰撞次数，碰撞次数必然随密度和温度的平方根而变化；其次取决于引发有效碰撞的特殊条件，这个条件可能与温度有关，也可能与温度无关。简单的指数型表达式对两者均不适用，但以下结论是有效的：只有当恒星内部温度达到 mc^2/k 数量级，也就是 1.1×10^{13} 度时，才可能发生通过湮灭而产生能量的过程。

如果物质-辐射已经达到了平衡，要想再发生任何反应都需要平衡常数的改变，这是毋庸置疑的。但是，如果系统没有达到平衡，上述表述就是错误的，因为在平衡未达到之前，热力学条件不足以影响将达到平衡的速率。

（王锋 翻译；蒋世仰 审稿）

The End of the World: from the Standpoint of Mathematical Physics*

A. S. Eddington

Editor's Note

This supplement contains a somewhat light-hearted address on "the end of the world" that Arthur Eddington delivered at the Mathematical Association. One must first ask "which end?", he says. Space itself may have no end: current cosmology suggested a universe shaped like the surface of a sphere, finite but without edges. As for time, the second law of thermodynamics suggests that the entire universe will eventually reach a state of thermodynamic equilibrium marked by complete disorganisation. However, if time is infinite then every conceivable fluctuation in the universe's particles will happen, temporarily disturbing this equilibrium. Eddington ends by predicting how the world will really end: as a ball of radiation growing ever larger, roughly doubling its size every 1,500 million years.

THE world—or space-time—is a four-dimensional continuum, and consequently offers a choice of a great many directions in which we might start off to look for an end; and it is by no means easy to describe "from the standpoint of mathematical physics" the direction in which I intend to go. I have therefore to examine at some length the preliminary question, Which end?

Spherical Space

We no longer look for an end to the world in its space dimensions. We have reason to believe that so far as its space dimensions are concerned the world is of spherical type. If we proceed in any direction in space we do not come to an end of space, nor do we continue on to infinity; but, after travelling a certain distance (not inconceivably great), we find ourselves back at our starting-point, having "gone round the world". A continuum with this property is said to be finite but unbounded. The surface of a sphere is an example of a finite but unbounded two-dimensional continuum; our actual three-dimensional space is believed to have the same kind of connectivity, but naturally the extra dimension makes it more difficult to picture. If we attempt to picture spherical space, we have to keep in mind that it is the *surface* of the sphere that is the analogue of our three-dimensional space; the inside and the outside of the sphere are fictitious elements in the picture which have no analogue in the actual world.

We have recently learnt, mainly through the work of Prof. Lemaître, that this spherical space is expanding rather rapidly. In fact, if we wish to travel round the world and get

* Presidential address to the Mathematical Association, delivered on Jan. 5.

以数学物理的视角看宇宙的终点*

阿瑟·爱丁顿

编者按

本文的内容是阿瑟·爱丁顿在数学协会上发表的一篇关于“宇宙的终点”的非正式讲话。他说，人们肯定首先要问“宇宙的终点是时间方向上的终点还是空间方向上的终点？”空间本身也许不存在终点：现代宇宙学认为宇宙类似于一个球体的表面，有限但无界。至于时间，热力学第二定律预言整个宇宙最终将达到一个以完全无序为标志的热平衡状态。然而，如果时间是无限的，那么宇宙粒子的每一次可能的涨落都会短暂地打破这种平衡。爱丁顿在结束语中预言了宇宙终结的实际方式：宇宙作为一个充满辐射的球不断地变大，每15亿年其大小约膨胀一倍。

宇宙或者说时空是一个四维连续区，因此我们可以从很多不同的角度来讨论它的终点。毫无疑问“以数学物理的视角”来描述以下我试图展开讨论的内容绝非易事。所以我必须相当仔细地考虑这个最初的问题：宇宙的终点是时间方向上的终点还是空间方向上的终点？

球面空间

我们不再寻找宇宙在空间方向上的终点。因为我们有理由相信，宇宙的空间部分具有球面结构。如果朝着空间中的任何一个方向一直走下去，我们不会到达空间的终点，也不会走到无穷远处；但是，当我们走了一段距离（并不是无法想象的远）以后，我们发现自己又回到了原来的出发点，相当于“绕了宇宙一圈”。我们把具有这种特性的连续区说成是有限但无界。球的表面即为有限无界二维连续区的一个例子；我们生活的三维空间被认为具有同样的连通性，但是由于比二维空间多出了一个空间维度，这使得我们很难用图形把这样的三维连续体表示出来。如果我们试图画出一个三维球面空间的话，我们必须要记住，只是这个三维球面对应于我们的三维空间；而这个球面的里面和外面都是我们虚构出来的，因而并不与现实的宇宙相对应。

最近，主要通过勒迈特教授的工作，我们已经获知，我们生活的这个三维球面空间正在快速地膨胀着。事实上，如果我们想环绕宇宙空间一周而回到出发点的话，

* 这篇文章来自爱丁顿于1月5日在数学协会上发表的讲话。

back to our starting-point, we shall have to move faster than light; because, whilst we are loitering on the way, the track ahead of us is lengthening. It is like trying to run a race in which the finishing-tape is moving ahead faster than the runners. We can picture the stars and galaxies as embedded in the surface of a rubber balloon which is being steadily inflated; so that, apart from their individual motions and the effects of their ordinary gravitational attraction on one another, celestial objects are becoming farther and farther apart simply by the inflation. It is probable that the spiral nebulae are so distant that they are very little affected by mutual gravitation and exhibit the inflation effect in its pure form. It has been known for some years that they are scattering apart rather rapidly, and we accept their measured rate of recession as a determination of the rate of expansion of the world.

From the astronomical data it appears that the original radius of space was 1,200 million light years. Remembering that distances of celestial objects up to several million light years have actually been measured, that does not seem overwhelmingly great. At that radius the mutual attraction of the matter in the world was just sufficient to hold it together and check the tendency to expand. But this equilibrium was unstable. An expansion began, slow at first; but the more widely the matter was scattered the less able was the mutual gravitation to check the expansion. We do not know the radius of space today, but I should estimate that it is not less than ten times the original radius.

At present our numerical results depend on astronomical observations of the speed of scattering apart of the spiral nebulae. But I believe that theory is well on the way to obtaining the same results independently of astronomical observation. Out of the recession of the spiral nebulae we can determine not only the original radius of the universe but also the total mass of the universe, and hence the total number of protons in the world. I find this number to be either 7×10^{78} or 14×10^{78}.* I believe that this number is very closely connected with the ratio of the electrostatic and the gravitational units of force, and, apart from a numerical coefficient, is equal to the square of the ratio. If F is the ratio of the electrical attraction between a proton and electron to their gravitational attraction, we find $F^2 = 5.3\times10^{78}$. There are theoretical reasons for believing that the total number of particles in the world is αF^2, where α is a simple geometrical factor (perhaps involving π). It ought to be possible before long to find a theoretical value of α, and so make a complete connexion between the observed rate of expansion of the universe and the ratio of electrical and gravitational forces.

Signposts for Time

I must not dally over space any longer but must turn to time. The world is closed in its space dimensions but is open in both directions in its time dimension. Proceeding from "here" in any direction in space we ultimately come back to "here"; but proceeding from

* This ambiguity is inseparable from the operation of counting the number of particles in finite but unbounded space. It is impossible to tell whether the protons have been counted once or twice over.

我们的行进速度必须要比光速还快；这是因为我们在路上行进的同时，我们前面的路途也在变长。就好像参加了一场终点线向前移动得比运动员跑步速度还快的赛跑一样。想象一下恒星和星系都镶嵌在一个橡胶气球的表面，而这个气球正在持续不断地膨胀着；这样，如果不考虑恒星和星系各自的运动以及它们之间的万有引力，这些天体之间的距离将由于宇宙空间的膨胀而变得越来越大。由于旋涡星云彼此相距很远，以至于它们几乎不会受到相互作用的引力影响，因而它们的形状很可能就体现了宇宙膨胀的效果。几年前我们就已经知道它们正在迅速地相互分散远离，我们可以通过测量它们之间的退行速度来推算宇宙的膨胀速度。

天文学数据表明宇宙空间最初的半径大约为12亿光年。要知道，远至几百万光年的天体实际上已经被我们观测到了，所以12亿光年也并非是不可想象的距离。在那种尺度下，宇宙中物质间相互作用的引力刚好可以把物质聚合在一起与宇宙空间的膨胀趋势相平衡。但是这种平衡并不稳定。于是宇宙开始膨胀，最初膨胀得比较慢；但随着物质之间的距离越来越远，它们之间的引力也越来越小，以至于越来越没有办法抑制宇宙的膨胀。我们不知道宇宙现在的半径，但我估计当今宇宙的半径应该至少是最初半径的10倍。

现在，我们的计算结果依赖于旋涡星云间分离速度的天文观测数据。但是我相信，我们也能从理论上顺利地得出不依赖天文观测数据的相同计算结果。从旋涡星云的退行速度我们不但可以推算出宇宙最初的尺度，还可以估算出宇宙总的质量，进而得到宇宙中总的质子数。我得出的宇宙中的总质子数为 7×10^{78} 或 14×10^{78}。* 我相信这个数值与静电力和引力的比值有密切的关系，并且只与这个比值的平方相差一个常数因子。如果 F 是质子和电子之间的静电力与万有引力的比值，我们得到 $F^2 = 5.3 \times 10^{78}$。理论上我们有理由相信宇宙中的粒子总数是 αF^2，其中 α 是一个简单的几何因子（也许和 π 有关）。可能不久我们就能得到 α 的理论值，从而给出宇宙膨胀速度的观测值同静电力与引力的比值之间的完整关系。

时间的指示牌

我不能继续讨论跟空间有关的问题，而必须开始讨论与时间有关的内容了。宇宙在空间上是闭合的，但是在时间尺度的两个方向上却都是开放的。从“这里”朝着空间中的任意方向出发，最后我们都会回到“这里”；但是从“现在”出发，无论

* 这个结果的不确定性与在有限无界的空间中计算粒子数有不可分割的联系。因为我们无法分辨这些质子到底被计算过一次还是两次。

"now" towards the future or the past we shall never come across "now" again. There is no bending round of time to bring us back to the moment we started from. In mathematics this difference is provided for by the symbol $\sqrt{-1}$, just as the same symbol crops up in distinguishing a closed ellipse and an open hyperbola.

If, then, we are looking for an end of the world—or, instead of an end, an indefinite continuation for ever and ever—we must start off in one of the two time directions. How shall we decide which of these two directions to take? It is an important question. Imagine yourself in some unfamiliar part of space-time so as not to be biased by conventional landmarks or traditional standards of reference. There ought to be a signpost with one arm marked "To the future" and the other arm marked "To the past". My first business is to find this signpost, for if I make a mistake and go the wrong way I shall lead you to what is no doubt an "end of the world", but it will be that end which is more usually described as the *beginning*.

In ordinary life the signpost is provided by consciousness. Or perhaps it would be truer to say that consciousness does not bother about signposts; but wherever it finds itself it goes off on urgent business in a particular direction, and the physicist meekly accepts its lead and labels the course it takes "To the future". It is an important question whether consciousness in selecting its direction is guided by anything in the physical world. If it is guided, we ought to be able to find directly what it is in the physical world which makes it a one-way street for conscious beings. The view is sometimes held that the "going on of time" does not exist in the physical world at all and is a purely subjective impression. According to that view, the difference between past and future in the material universe has no more significance than the difference between right and left. The fact that experience presents space-time as a cinematograph film which is always unrolled in a particular direction is not a property or peculiarity of the film (that is, the physical world) but of the way it is inserted into the cinematograph (that is, consciousness). In fact, the one-way traffic in time arises from the way our material bodies are geared on to our consciousness:

> "Nature has made our gears in such a way
> That we can never get into reverse".

If this view is right, "the going on of time" should be dropped out of our picture of the physical universe. Just as we have dropped the old geocentric outlook and other idiosyncrasies of our circumstances as observers, so we must drop the dynamic presentation of events which is no part of the universe itself but is introduced in our peculiar mode of apprehending it. In particular, we must be careful not to treat a past-to-future presentation of events as truer or more significant than a future-to-past presentation. We must, of course, drop the theory of evolution, or at least set alongside it a theory of anti-evolution as equally significant.

是向着未来还是向着过去前进，我们将永远不可能再回到“现在”。在这里并没有弯曲的时间回路能把我们带回到原来出发的时刻。从数学上说这种差别是由记号 $\sqrt{-1}$ 造成的，正如闭合椭圆和开放双曲线之间的区别也是由这个记号造成的一样。

于是，如果我们要找寻宇宙的终点——其实并不是终点，而是一种永无休止的持续，那么我们就必须从时间的两个方向中选一个作为开始的方向。但是我们如何决定应该选取哪一个时间方向呢？这是个很重要的问题。想象一下你处在时空中的某个陌生的部分，在那里你不会受到常规标志或传统参考标准的影响。但是那里应该有一个时间的指示牌，一边写着“通向未来”，而另一边写着“通向过去”。我要做的第一件事情就是要先找出这个指示牌，否则一旦我弄错了方向，那么无疑，我还是会把你们带到“宇宙的终点”,不过这个“终点”更通常地是被描述为宇宙的**起点**。

在日常生活中，我们的意识充当了时间的指示牌。或者更准确地说，我们的意识根本不关心所谓的时间指示牌；但是无论什么地方，在紧急情况下意识总是朝着某一个特定的方向行进思考，而物理学家也温和地接受了意识的引导并把它的行进方向标记为“通向未来”的方向。这里有一个重要的问题，意识在选择它的行进方向时是否也受到物理世界中的某种事物的影响？如果确实如此，那么我们应该能够直接找出究竟是物理世界中的什么因素使得有意识的人类认为时间是沿单向车道行进的。有观点认为“时间的流逝”只是一种主观的想法而不是物理世界中真实存在的现象。按照这个观点，物质世界中过去和未来的区别只不过相当于左和右之间的区别而已。事实上我们的经验显示时空就像一组电影胶片，它总是按照某种特定的方向放映，但这并不是由胶片（物理时空）本身的性质或特点所决定的，而是与胶片插入放映机的方式（意识）有关。这种观点认为，把事物储存到我们意识中的方式使我们认为时间具有单向性：

“大自然给我们配备了这样的意识思维方式
使得我们永远不能倒退”。

如果以上的观点是对的，那么在我们关于物理世界的图像中就应该抛弃“时间的流逝”这种观点。就像我们放弃旧的时空观以及其他对周围事物观察得到的个人看法一样，我们必须抛弃事物是动态呈现的这种观念，因为这并不是世界本身的性质而只是我们人类在认识物理世界时引入的一种人类特有的理解方式。我们要特别注意的是，如果真是这样，那么我们就不能先入为主地认为“过去到未来”的呈现比“未来到过去”的呈现更正确或更重要。当然，我们也必须放弃进化论，或者至少要同时建立一种具有同样重要性的反进化论。

If anyone holds this view, I have no argument to bring against him. I can only say to him, "You are a teacher whose duty it is to inculcate in youthful minds a true and balanced outlook. But you teach (or without protest allow your colleagues to teach) the utterly one-sided doctrine of evolution. You teach it not as a colourless schedule of facts but as though there were something significant, perhaps even morally inspiring, in the progress from formless chaos to perfected adaptation. This is dishonest; you should also treat it from the equally significant point of view of anti-evolution and discourse on the progress from future to past. Show how from the diverse forms of life existing today Nature anti-evolved forms which were more and more unfitted to survive, until she reached the sublime crudity of the palaeozoic forms. Show how from the solar system Nature anti-evolved a chaotic nebula. Show how, in the course of progress from future to past, Nature took a universe which, with all its faults, is not such a bad effort of architecture and—in short, made a hash of it."

Entropy and Disorganisation

Leaving aside the guidance of consciousness, we have found it possible to discover a kind of signpost for time in the physical world. The signpost is of rather a curious character, and I would scarcely venture to say that the discovery of the signpost amounts to the same thing as the discovery of an objective "going on of time" in the universe. But at any rate it serves to discriminate past and future, whereas there is no corresponding objective distinction of left and right. The distinction is provided by a certain measurable quantity called entropy. Take an isolated system and measure its entropy S at two instants t_1 and t_2. We want to know whether t_1 is earlier or later than t_2 without employing the intuition of consciousness, which is too disreputable a witness to trust in mathematical physics. The rule is that the instant which corresponds to the greater entropy is the later. In mathematical form

$$dS/dt \text{ is always positive.}$$

This is the famous second law of thermodynamics.

Entropy is a very peculiar conception, quite unlike the conceptions ordinarily employed in the classical scheme of physics. We may most conveniently describe it as the measure of disorganisation of a system. Accordingly, our signpost for time resolves itself into the law that disorganisation increases from past to future. It is one of the most curious features of the development of physics that the entropy outlook grew up quietly alongside the ordinary analytical outlook for a great many years. Until recently it always "played second fiddle"; it was convenient for getting practical results, but it did not pretend to convey the most penetrating insight. But now it is making a bid for supremacy, and I think there is little doubt that it will ultimately drive out its rival.

There are some important points to emphasise. First, there is no other independent signpost for time; so that if we discredit or "explain away" this property of entropy, the distinction of past and future in the physical world will disappear altogether. Secondly,

如果有谁坚持时间的行进具有单向性这种观点，我没有任何办法说他是错的。我只能对他说："你是一个致力于把一个真实均衡的世界观灌输给年青学生的老师。但是你所教授的（或没有阻止你的同事去教授的）是完全片面的进化论学说。你并没有告诉学生进化论只是单纯事实的罗列，而是让他们认为似乎从混乱无形到完美有序的过程才是重要的，才真正会是鼓舞人心的。你这样做其实是不诚实的；你应该把反进化论也放在同样重要的位置并讲述从未来到过去的演化过程。你应该向他们展示：现今自然界存在的丰富多彩的生命形态是如何反进化到越来越不适合生存，直到古生代最原始的形态的；自然界是如何从太阳系反进化成混沌的星云；以及，随着从未来到过去的时间进程，自然界是如何选择了一个充满问题的宇宙，而这些问题并不是宇宙本身的构造不好，总而言之，最初的宇宙是一团糨糊。"

熵和无序度

撇开意识的引导性不谈，我们已经发现在物理世界里有可能找到一种时间的指示牌。这个指示牌具有非常奇特的性质，当然，我还不敢说这个指示牌的发现与当年在宇宙中发现客观的"时间流逝"有一样的重要性。但是不管怎样，这个指示牌可以用来区别过去和未来，而空间上的左和右却没有相应的客观区分。我们可以用某种称为熵的可测量量来标志这种时间方向上的差别。假设有一个孤立的系统，我们在两个不同的时刻 t_1 和 t_2 分别测量它的熵 S。因为意识这个概念在数学物理领域里并不是可靠的证据，所以我们想在不靠意识直觉的情况下知道 t_1 究竟比 t_2 早还是晚。这里用到的规则是：熵越大，时间越晚。在数学形式中为：

dS/dt 总是大于零。

这就是著名的热力学第二定律。

熵是个很特殊的概念，与经典物理学常用的概念很不一样。我们可以把熵简单地描述为：熵是衡量系统无序性的物理量。于是，根据熵这个时间的指示牌，我们得到了这样的规律：从过去到未来，无序性是增加的。物理学发展的最奇妙的特点之一就是：多年来，与其他常规的分析方法一样，我们对于熵的认知一直都在悄悄地发展着。直到最近，熵在科学研究中仍只是"居于次要位置"。由熵可以很方便地得到实用的结果，但并不能由此表明人们对熵有了深刻的洞察力。不过现在熵正在谋求更大的发展，并且我认为，毫无疑问，它最终会超过它的对手。

这里还需要强调几个重要的地方。首先，世界上没有其他独立的时间指示牌；所以如果我们不信任或者"通过辩解来消除"熵作为时间指示牌的这个性质，那么物理学关于过去和未来的差别也会随之消失。第二，检验熵的实验结果应该都是一

the test works consistently; isolated systems in different parts of the universe agree in giving the same direction of time. Thirdly, in applying the test we must make certain that our system is strictly isolated. Evolution teaches us that more and more highly organised systems develop as time goes on; but this does not contradict the conclusion that on the whole there is a loss of organisation. It is partly a question of definition of organisation; from the evolutionary point of view it is quality rather than quantity of organisation that is noticed. But, in any case, the high organisation of these systems is obtained by draining organisation from other systems with which they come in contact. A human being as he grows from past to future becomes more and more highly organised—at least, he fondly imagines so. But if we make an isolated system of him, that is to say, if we cut off his supply of food and drink and air, he speedily attains a state which everyone would recognise as "a state of disorganisation".

It is possible for the disorganisation of a system to become complete. The state then reached is called thermodynamic equilibrium. The entropy can increase no further, and, since the second law of thermodynamics forbids a decrease, it remains constant. Our signpost for time disappears; and so far as that system is concerned, time ceases to go on. That does not mean that time ceases to exist; it exists and extends just as space exists and extends, but there is no longer any one-way property. It is like a one-way street on which there is never any traffic.

Let us return to our signpost. Ahead there is ever-increasing disorganisation. Although the sum total of organisation is diminishing, certain parts of the universe are exhibiting a more and more highly specialised organisation; that is the phenomenon of evolution. But ultimately this must be swallowed up in the advancing tide of chance and chaos, and the whole universe will reach a state of complete disorganisation—a uniform featureless mass in thermodynamic equilibrium. This is the end of the world. Time will *extend* on and on, presumably to infinity. But there will be no definable sense in which it can be said to *go* on. Consciousness will obviously have disappeared from the physical world before thermodynamical equilibrium is reached, and dS/dt having vanished, there will remain nothing to point out a direction in time.

The Beginning of Time

It is more interesting to look in the opposite direction—towards the past. Following time backwards, we find more and more organisation in the world. If we are not stopped earlier, we must come to a time when the matter and energy of the world had the maximum possible organisation. To go back further is impossible. We have come to an abrupt end of space-time—only we generally call it the "beginning".

I have no "philosophical axe to grind" in this discussion. Philosophically, the notion of a beginning of the present order of Nature is repugnant to me. I am simply stating the

致的；宇宙中任何地方的孤立系统都应该给出相同的时间方向。第三，在进行这类实验时，我们必须保证我们的系统是严格孤立的系统。进化论告诉我们，随着时间的推移，越来越高的有序系统产生了；但是这与整体上有序性的减少并不矛盾。这在一定程度上是有序性的定义问题；从进化的角度看，需要注意的是有序性的质而非量。但无论如何，有序性很高的系统是通过吸取其他与之接触的系统的有序性来实现的。一个人从过去到未来总是变得越来越有序，至少他自己愿意这么认为。但是如果我们使他成为一个孤立的系统，即切断他的饮食和空气供应，那么他很快就会达到一种大家都认同的“无序状态”。

系统可能达到完全的无序状态，我们称之为热力学平衡态。此时，熵不能再继续增加了，而热力学第二定律又不允许它减少，于是熵就只能保持为一个常数。这时，我们的时间指示牌也消失了，因而对于这样的一个系统，时间变得固定不动了。但这并不表示时间不存在；时间仍然像空间一样存在并延续着，但是它已经不再具有任何单向性了。或者说，单行道仍然存在，但不再有汽车在上面行驶了。

让我们回到时间指示牌的讨论上来。通过前面的讨论我们已经知道，无序性是不断增加的。虽然总的有序性正在减少，但是宇宙中的某些部分却展示出越来越高的特殊有序性；这就是进化现象。但是这些有序的系统最终会被不断增加的机遇与混沌所吞没，然后整个宇宙会达到一种完全无序的状态，即变成热力学平衡态下的一堆毫无特征的均匀物质。这就是宇宙的终点。时间仍然会永远地**延续**下去，可能没有尽头。但是时间的这种**延续**没有了明确的意义。很明显，在热力学平衡态即将到来之前，物理世界中已经不存在意识了，dS/dt 也成为零，再也没有一种指示牌能告诉我们，哪个时间方向通向未来，而哪个又是通向过去的。

时间的起点

如果我们朝着“时间流逝”相反的方向，即朝着过去的方向看，我们将会发现更加有趣的现象。当我们沿着“时间流逝”的方向往回走时，会发现宇宙的有序性越来越大。如果我们一直不停地走下去，就会到达一个物质和能量的有序度都为允许的最高限的时刻。这时，再想进一步走下去已经不可能了。我们已经来到了时空戛然而止的终点——只不过我们通常称之为“起点”。

在这里的讨论中我没有“涉足哲学”的意思。我并不认同哲学中那些与现在自然界秩序的起点有关的概念。我只是想说明物理规律中现有的基本概念使我们陷入

dilemma to which our present fundamental conception of physical law leads us. I see no way round it; but whether future developments of science will find an escape I cannot predict. The dilemma is this:—Surveying our surroundings, we find them to be far from a "fortuitous concourse of atoms". The picture of the world, as drawn in existing physical theories, shows arrangement of the individual elements for which the odds are multillions to 1 against an origin by chance. Some people would like to call this non-random feature of the world purpose or design; but I will call it non-committally anti-chance. We are unwilling to admit in physics that anti-chance plays any part in the reactions between the systems of billions of atoms and quanta that we study; and indeed all our experimental evidence goes to show that these are governed by the laws of chance. Accordingly, we sweep anti-chance out of the laws of physics—out of the differential equations. Naturally, therefore, it reappears in the boundary conditions, for it must be got into the scheme somewhere. By sweeping it far enough away from the sphere of our current physical problems, we fancy we have got rid of it. It is only when some of us are so misguided as to try to get back billions of years into the past that we find the sweepings all piled up like a high wall and forming a boundary—a beginning of time—which we cannot climb over.

A way out of the dilemma has been proposed which seems to have found favour with a number of scientific workers. I oppose it because I think it is untenable, not because of any desire to retain the present dilemma. I should like to find a genuine loophole. But that does not alter my conviction that the loophole that is at present being advocated is a blind alley. I must first deal with a minor criticism.

I have sometimes been taken to task for not sufficiently emphasising in my discussion of these problems that the results about entropy are a matter of probability, not of certainty. I said above that if we observe a system at two instants, the instant corresponding to the greater entropy will be the later. Strictly speaking, I ought to have said that for a smallish system the chances are, say, 10^{20} to 1, that it is the later. Some critics seem to have been shocked at my lax morality in making such a statement, when I was well aware of the 1 in 10^{20} chance of its being wrong. Let me make a confession. I have in the past twenty-five years written a good many papers and books, broadcasting a large number of statements about the physical world. I fear that for not many of these statements is the risk of error so small as 1 in 10^{20}. Except in the domain of pure mathematics, the trustworthiness of my conclusions is usually to be rated at nearer 10 to 1 than 10^{20} to 1; even that may be unduly boastful. I do not think it would be for the benefit of the world that no statement should be allowed to be made if there were a 1 in 10^{20} chance of its being untrue; conversation would languish somewhat. The only persons entitled to open their mouths would presumably be the pure mathematicians.

Fluctuations

The loophole to which I referred depends on the occurrence of chance fluctuations. If we have a number of particles moving about at random, they will in the course of time go through every possible configuration, so that even the most orderly, the most non-

了困境。我没有办法解决这个矛盾；也不能预料未来科学的发展能否避开这个矛盾。这个矛盾是这样的：考虑我们周围的物质，我们发现它们远不是“原子的偶然集合”。现有的物理理论所描述的物理世界图像表明，这些单个元素的分布情况是偶然出现的概率仅仅是10的10次方的10次方分之一。有人喜欢把这种现象称之为宇宙意图或设计的非随机特性；但是，我要把它称为不受约束的反概率性。在物理学领域内，我们不愿意承认反概率性在我们所研究的包含数以十亿的原子和量子系统的相互作用中发挥了作用；而且实际上我们所有的实验结果也证实了所有这些现象都是由随机定律决定的。于是，我们就可以把反概率性从物理定律（微分方程）中剔除了。因为它总要出现在物理框架中的某个地方，所以它会很自然地在方程的边界条件里再次现身。通过把它置于我们现有的物理问题的范围之外，我们幻想着我们已经摆脱了这个困难。只有当我们中的某些人被误导着试图穿越几十亿年的时间回到过去时，我们才会发现被我们剔除的所有反概率性现象都堆在那里像一堵高墙一样形成了一个不可逾越的边界，这即是时间的起点。

目前已经有人提出了一种可以解决以上矛盾的方法，这个方法似乎还得到了一些科学工作者的支持。然而，我不同意那种解释，因为我觉得它是站不住脚的，并不是因为我想让这个矛盾继续存在下去。我更愿意去找一个确切的着眼点，但是我确信现在鼓吹的着眼点其实是个死胡同。我必须首先回应对我的一些小的批评意见。

有时，我会因为没有在我所讨论的内容中充分强调熵是概率性而非确定性的物理量而受到批评。如前所述，如果我们在不同的时刻观察同一个系统，那么熵较大的时刻将是较晚的时刻。严格来讲，我应该如此表述：对于一个较小的系统，上述结论正确的可能性为$10^{20}:1$。虽然我清楚地知道这个结论出错的概率是$1/10^{20}$,但是，由于我在陈述上述结论时不够严谨，有些批评者仍会为此感到震惊。坦白地说：在过去的25年里，我撰写了很多文章和书籍，对物理现象做了大量的解释，恐怕我所做的各种解释中没有多少结论的出错概率会小于$1/10^{20}$。在纯数学领域之外，我的结论的正误比估计更接近$10:1$，而不是$10^{20}:1$；尽管如此，还是自负不已。我认为出错率为$1/10^{20}$的结论对我们认识世界并不会有什么坏处；而我们应该稍微搁置关于出错概率的讨论。如果按照反对者的说法，这个世界上唯一能发表言论的就只可能是纯粹的数学家了。

涨 落

我前面所说的着眼点依赖于概率的涨落。我们考虑一群随机运动的粒子，随着时间的推移它们会经历任何可能的状态，所以只要我们等待足够长的时间，即使是

chance configuration, will occur by chance if only we wait long enough. When the world has reached complete disorganisation (thermodynamic equilibrium) there is still infinite time ahead of it, and its elements will thus have opportunity to take up every possible configuration again and again. If we wait long enough, a number of atoms will, just by chance, arrange themselves in systems as they are at present arranged in this room; and, just by chance, the same sound-waves will come from one of these systems of atoms as are at present emerging from my lips; they will strike the ears of other systems of atoms, arranged just by chance to resemble you, and in the same stages of attention or somnolence. This mock Mathematical Association meeting must be repeated many times over—an infinite number of times, in fact—before t reaches $+\infty$. Do not ask me whether I expect you to believe that this will really happen.*

"Logic is logic. That's all I say."

So, after the world has reached thermodynamical equilibrium the entropy remains steady at its maximum value, except that "once in a blue moon" the absurdly small chance comes off and the entropy drops appreciably below its maximum value. When this fluctuation has died out, there will again be a very long wait for another coincidence giving another fluctuation. It will take multillions of years, but we have all infinity of time before us. There is no limit to the amount of the fluctuation, and if we wait long enough we shall come across a big fluctuation which will take the world as far from thermodynamical equilibrium as it is at the present moment. If we wait for an enormously longer time, during which this huge fluctuation is repeated untold numbers of times, there will occur a still larger fluctuation which will take the world as far from thermodynamical equilibrium as it was one second ago.

The suggestion is that we are now on the downward slope of one of these fluctuations. It has quite a pleasant subtlety. Is it chance that we happen to be running down the slope and not toiling up the slope? Not at all. So far as the physical universe is concerned, we have *defined* the direction of time as the direction from greater to less organisation, so that, on whichever side of the mountain we stand, our signpost will point downhill. In fact, on this theory, the going on of time is not a property of time in general, but is a property of the slope of the fluctuation on which we are standing. Again, although the theory postulates a universe involving an extremely improbable coincidence, it provides an infinite time during which the most improbable coincidence might occur. Nevertheless, I feel sure that the argument is fallacious.

If we put a kettle of water on the fire there is a chance that the water will freeze. If mankind goes on putting kettles on the fire until $t = \infty$, the chance will one day come off and the individual concerned will be somewhat surprised to find a lump of ice in his kettle. But it will not happen to *me*. Even if tomorrow the phenomenon occurs before my eyes, I shall not explain it this way. I would much sooner believe in interference by a demon than

* I am hopeful that the doctrine of the "expanding universe" will intervene to prevent its happening.

最有秩序的、概率上最不可能的状态也会偶然出现。当宇宙达到完全的无序状态（热力学平衡态）后，时间仍然会无限地存在，宇宙中的元素将有机会反复经历各种可能的状态。如果我们观察足够长的时间，我们会发现，系统中一些原子所处的状态结构可能会偶然地与我们现在这个空间中的原子一样；同样偶然地，原子所组成的系统中的某一个系统出现的声波将可能与我口中现在发出的声波一样；而不管你是处于清醒还是昏昏欲睡的状态，这些声波都将去冲击由其他原子所组成的偶然与你的耳朵类似的系统。这个模拟的数学协会报告会将会在时间 t 达到无穷大之前重复很多次，实际上是无穷多次。别问我是否期望你们相信这些真的会发生。*

“逻辑就是逻辑，这就是我所说的全部。”

因此，当宇宙到达热力学平衡时，熵就会稳定地处在它的最大值上，除非出现一个千载难逢的极小的概率，使得熵从最大值回落到明显比最大值小的值。这个涨落消失之后，要等很长时间才会碰巧发生下一次涨落。尽管也许要等 10 的 10 次方的 10 次方年的时间，但是不用担心，我们拥有无限延续的时间。这种涨落的大小并没有什么限制，如果我们等待足够久，也许我们可以碰上一次大的涨落，使宇宙远离热力学平衡态，变成和我们现在这个宇宙一样的状态。如果我们等候更长的时间，其间会有数不清的类似的大涨落发生，也许还会有一次较大的涨落使宇宙远离热力学平衡态，变回到一秒钟以前的状态。

有人提出，我们现在正处在某个涨落的下坡过程中。这种提法存在令人兴奋的微妙之处。我们的宇宙刚好处在涨落的下坡过程而非往上的爬坡过程，这是一种巧合吗？完全不是。就物理世界而言，我们已经**定义**了时间的方向为有序度减少的方向，因此，无论我们站在山坡的哪一边，我们的指示牌都是指向下坡的。事实上，在这个理论里，总的来说时间的流逝并不是时间本身的性质，而是我们所处的那个涨落的山坡的性质。尽管这个理论假设宇宙包含了一个极不可能发生的概率事件，但同时该理论却提供了无限长的时间使得最不可能发生的概率事件最终总能发生。无论怎样，我个人觉得以上的说法是不合理的。

如果我们把一壶水放到火上，这壶水有可能结冰吗？如果一个人把一壶水放到火上无限长的时间，某一天这个人可能会惊讶地发现，他壶里的水居然结冰了。但这类事情不可能发生在**我**身上。即使将来有一天这种事情真的在我眼前发生了，我也不会用上面那样的方式去解释它。我宁愿相信这是一个魔鬼干的，而不是小概率

---

* 我希望“宇宙膨胀”说会阻止它的发生。

in a coincidence of that kind coming off; and in doing so I shall be acting as a rational scientist. The reason why I do not at present believe that devils interfere with my cooking arrangements and other business, is because I have become convinced by experience that Nature obeys certain uniformities which we call laws. I am convinced because these laws have been tested over and over again. But it is possible that every single observation from the beginning of science which has been used as a test, has just happened to fit in with the law by a chance coincidence. It would be an improbable coincidence, but I think not quite so improbable as the coincidence involved in my kettle of water freezing. So if the event happens and I can think of no other explanation, I shall have to choose between two highly improbable coincidences: (*a*) that there are no laws of Nature and that the apparent uniformities so far observed are merely coincidences; (*b*) that the event is entirely in accordance with the accepted laws of Nature, but that an improbable coincidence has happened. I choose the former because mathematical calculation indicates that it is the less improbable. I reckon a sufficiently improbable coincidence as something much more disastrous than a violation of the laws of Nature; because my whole reason for accepting the laws of Nature rests on the assumption that improbable coincidences do not happen—at least, that they do not happen in my experience.*

Similarly, if logic predicts that a mock meeting of the Mathematical Association will occur just by a fortuitous arrangement of atoms before $t = \infty$, I reply that I cannot possibly accept that as being the explanation of a meeting of the Mathematical Association in $t = 1931$. We must be a little careful over this, because there is a trap for the unwary. The year 1931 is not an absolutely random date between $t = -\infty$ and $t = +\infty$. We must not argue that because for only $1/x$th of time between $t = -\infty$ and $t = \infty$ a fluctuation as great as the present one is in operation, therefore the chances are x to 1 against such a fluctuation occurring in the year 1931. For the purposes of the present discussion, the important characteristic of the year 1931 is that it belongs to a period during which there exist in the universe beings capable of speculating about the universe and its fluctuations. Now I think it is clear that such creatures could not exist in a universe in thermodynamical equilibrium. A considerable degree of deviation is required to permit of living beings. Therefore it is perfectly fair for supporters of this suggestion to wipe out of account all those multillions of years during which the fluctuations are less than the minimum required to permit of the development and existence of mathematical physicists. That greatly diminishes x, but the odds are still overpowering. The *crude* assertion would be that (unless we admit something which is not chance in the architecture of the universe) it is practically certain that at any assigned date the universe will be almost in the state of maximum disorganisation. The *amended* assertion is that (unless we admit something which is not chance in the architecture of the universe) it is practically certain that a universe containing mathematical physicists will at any assigned date be in the state of maximum disorganisation which is not inconsistent with the existence of such creatures. I think it is quite clear that neither the original nor the amended version applies. We are thus driven

* No doubt "extremely improbable" coincidences occur to all of us, but the improbability is of an utterly different order of magnitude from that concerned in the present discussion.

事件发生了；而这样做的时候，我总该更像是一个理性的科学家吧。我现在之所以不相信魔鬼会干预我的烹调事务或者其他事情，是因为日常经验使我相信自然界会遵循一定的统一性，即物理定律。我相信这些定律是因为它们已经被反复验证过。当然，也有可能所有这些用来验证物理定律的实验一开始就恰好满足这些定律只是一种巧合。这或许是个难以置信的巧合，但是我相信这种巧合不会像我壶里的水发生结冰的巧合那样难以置信。一旦这种事件发生了，而我又找不到其他合适的解释，那么我只能在以下两种极为难以置信的巧合之间做出选择：(*a*) 这个世界上并不存在物理定律，迄今为止观察到的一致性仅仅是巧合。(*b*) 这个事件是符合物理定律的，只不过难以置信的巧合事件发生了。我更愿意接受前一种解释，因为数学计算的结果表明它的可能性更高一点。相比于违反物理定律而言，我把难以置信的巧合事件的发生看成是更加糟糕的事情；因为我之所以接受存在于自然界中的物理定律只是因为我相信难以置信的巧合事件不会发生——至少在我有生之年不会发生。*

同样，如果逻辑上预言在 $t=\infty$ 之前仅仅是由于原子的随机排列而出现了一个模拟的数学协会的报告会，那我要说的是，我不能接受这就是今天 $t=1931$ 年的这个数学协会报告会发生的理由。我们必须小心这一点，因为这里有个不小心就容易陷入的圈套。在 $t=-\infty$ 到 $t=+\infty$ 之间，1931 年并不是一个完全随机的时间。我们不能因为在 $t=-\infty$ 到 $t=+\infty$ 之间只是在第 $1/x$ 个时间里有一个与今天的数学报告会一样的涨落发生了，就说在 1931 年里发生这种涨落的概率是 $1/x$。从我们现在的讨论情况来看，1931 年的重要性就在于在这段时间里宇宙中出现了能够思考宇宙以及它的涨落的人类。现在我认为有一点是很清楚的，就是人类不可能在处于热力学平衡态下的宇宙中生存。要让生物能够生存下去，宇宙就必须在很大程度上偏离平衡态。因此，对于涨落论的支持者来说，不对在 10 的 10 次方的 10 次方年这段时间中发生涨落的概率比数学物理学家成长和存在的最小概率还小的情况进行说明是完全合理的。这就大大缩小了 x 的值，但是问题仍然存在。**最初**的假设应该是这样的（除非我们承认某些状态在宇宙的结构中不可能发生）：几乎可以肯定地说宇宙在给定的任一时间都将近似处于最无序的状态。**改进**后的说法是（除非我们承认某些状态在宇宙的结构中不可能发生）：几乎可以肯定地说有数学物理学家存在的宇宙在给定的任一时间都将处于最无序的状态，而这个状态也适合人类的生存。我想，我们已经很清楚地看出以上两种说法都是站不住脚的。于是我们被迫接受反概率论；而对待它的最好办法显然是，正如之前所说过的，把所有反概率性现象整理在一起并堆

* 毫无疑问，“极不可能”发生的巧合会在我们面前发生，只是这种巧合事件的发生概率与我们在此讨论的巧合事件有完全不同的数量级。

to admit anti-chance; and apparently the best thing we can do with it is to sweep it up into a heap at the beginning of time, as I have already described.

The connexion between our entropy signpost and that dynamic quality of time which we describe as "going on" or "becoming" leads to very difficult questions which I cannot discuss here. The puzzle is that the signpost seems so utterly different from the thing of which it is supposed to be the sign. The one thing on which I have to insist is that, apart from consciousness, the increase of entropy is the only trace that we can find of a one-way direction of time. I was once asked a ribald question: How does an electron (which has not the resource of consciousness) remember which way time is going? Why should it not inadvertently turn round and, so to speak, face time the other way? Does it have to calculate which way entropy is increasing in order to keep itself straight? I am inclined to think that an electron does do something of that sort. For an electric charge to face the opposite way in time is the same thing as to change the sign of the charge. So if an electron mistook the way time was going it would turn into a positive charge. Now, it has been one of the troubles of Dr. P. A. M. Dirac that in the mathematical calculations based on his wave equation the electrons do sometimes forget themselves in this way. As he puts it, there is a finite chance of the charge changing sign after an encounter. You must understand that they only do this in the mathematical problems, not in real life. It seems to me there is good reason for this. A mathematical problem deals with, say, four electric charges at the most; that is about as many as a calculator would care to take on. Accordingly, the unfortunate electron in the problem has to make out the direction of past to future by watching the organisation of three other charges. Naturally, it is deceived sometimes by chance coincidences which may easily happen when there are only three particles concerned; and so it has a good chance of facing the wrong way and becoming a positive charge. But in any real experiment we work with apparatus containing billions of particles—ample to give the electron its bearings with certainty. Dirac's theory predicts things which never happen, simply because it is applied to problems which never occur in Nature. When it is applied to four particles alone in the universe, the analysis very properly brings out the fact that in such a system there could be no steady one-way direction of time, and vagaries would occur which are guarded against in our actual universe consisting of about 10^{79} particles.

Heisenberg's Principle

A discussion of the properties of time would be incomplete without a reference to the principle of indeterminacy, which was formulated by Heisenberg in 1927 and has been generally accepted. It had already been realised that theoretical physics was drifting away from a deterministic basis; Heisenberg's principle delivered the knock-out blow, for it actually postulated a certain measure of indeterminacy or unpredictability of the future as a fundamental law of the universe. This change of view seems to make the progress of time a much more genuine thing than it used to be in classical physics. Each passing moment brings into the world something new—something which is not merely a mathematical extrapolation of what was already there.

积在时间的起点上。

熵这个指示牌和被称作“流逝”或“发展”的时间动态特征之间的关系导致了一些非常困难的问题，在这里我不可能做详细的讨论。其中的困难之处在于熵这个指示牌似乎和我们预期的时间标记非常不同。我需要强调的是，除了意识以外，熵的增加是我们发现时间具有单向性的唯一线索。曾经有人向我提出一个粗鄙的问题：一个电子（它没有意识）是如何记住时间的流向的？为什么它不会无意中改变方向，即，朝向时间的另一个方向？它需要事先计算好朝哪个方向熵会增加，然后再决定往哪个方向前进吗？我倾向于认为电子确实会做这样的判断。对一个电荷来说，朝相反的时间方向意味着它的电荷符号要发生变化。所以，如果一个电子错误地选择了时间方向，那么它就会变成一个正电荷。这跟狄拉克博士在以电子波动方程为基础进行数学计算时碰到的一个麻烦是一样的，在他的计算里电子真的会走错方向。就像狄拉克所发现的，在一次碰撞之后，电子有一定的概率可以改变电荷的符号。你必须要明白的是，这仅仅是数学的计算结果，并未在现实生活中发现。之所以是这样，我认为似乎有很合理的理由可以解释。假设我们考虑一个最多涉及四个电荷的数学问题，这个问题是任何一个计算器都能够处理的。于是，其中一个倒霉的电子不得不通过观察其他三个电荷的有序性来判断时间从过去到将来的流向。很自然，当我们只考虑三个电子的情况时，很容易发生偶然的巧合使得电子弄错方向；所以电子有很好的机会可以因为弄错了时间的方向而变成一个正电荷。但是在现实实验中，我们所用的仪器会包含几十亿个粒子——多到足够让电子准确地确定出时间的方向。狄拉克理论预言的情况从来就没有发生过，因为与之相适的问题从来没有出现在现实世界中。如果将狄拉克理论应用于只有四个电子的宇宙中，那么经过分析很可能会给出这样的结论：在这个系统中不会存在一个稳定的单向的时间方向，奇特的事情会不断地发生，而这些在我们这个包含了约 10^{79} 个粒子的真实宇宙中是完全不可能发生的。

海森堡原理

如果我们不考虑不确定性原理的话，那么我们对时间的讨论将是不完整的。这个原理是海森堡在 1927 年用公式表达出来并已得到一致认可的。人们已经意识到理论物理学正慢慢地偏离确定性这个基础；海森堡不确定性原理对人们的思维产生了巨大的冲击，因为这个原理作为物理世界的基本定律实际上认为未来理应具有一定程度上的不确定性或不可预测性。相比于经典物理学，这种观念上的转变使得时间的进展具有了更加真实的意义。每一个瞬间过后，世界都可能增添一些新事物——这些事物不可能单纯使用数学方法从过去已经发生的事件中推测出来。

The deterministic view which held sway for at least two centuries was that if we had complete data as to the state of the whole universe during, say, the first minute of the year 1600, it would be merely a mathematical exercise to deduce everything that has happened or will happen at any date in the future or past. The future would be determined by the present as the solution of a differential equation is determined by the boundary conditions. To understand the new view, it is necessary to realise that there is a risk of begging the question when we use the phrase "complete data". All our knowledge of the physical world is inferential. I have no direct acquaintance with my pen as an object in the physical world; I infer its existence and properties from the light waves which fall on my eyes, the pressure waves which travel up my muscles, and so on.

Precisely the same scheme of inference leads us to infer the existence of things in the past. Just as I infer a physical object, namely, my pen, as the cause of certain visual sensations now, so I may infer an infection some days ago as the cause of an attack of measles. If we follow out this principle completely we shall infer causes in the year 1600 for all the events which we know to have happened in 1930. At first sight it would seem that these inferred causes have just as much status in the physical world as my fountain pen, which is likewise an inferred cause. So the determinist thinks he has me in a cleft stick. If the scientific worker poking about in the universe in 1600 comes across these causes, then he has all the data for making a correct prediction for 1930; if he does not, then he clearly has not complete knowledge of the universe in 1600, for these causes have as much right to the status of physical entities as any of our other inferences.

I need scarcely stop to show how this begs the question by arbitrarily prescribing what we should deem to be complete knowledge of the universe in 1600, irrespective of whether there is any conceivable way in which this knowledge could be obtained at the time. What Heisenberg discovered was that (at least in a wide range of phenomena embracing the whole of atomic physics and electron theory) there is a provision of Nature that just half of the data demanded by our determinist friend might with sufficient diligence be collected by the investigators in 1600, and that complete knowledge of this half would automatically exclude all knowledge of the other half. It is an odd arrangement, because you can take your choice which half you will find out; you can know either half but not both halves. Or you can make a compromise and know both halves imperfectly, that is, with some margin of uncertainty. But the rule is definite. The data are linked in pairs and the more accurately you measure one member of the pair the less accurately you can measure the other member.

Both halves are necessary for a complete prediction of the future, although, of course, by judiciously choosing the type of event we predict we can often make safe prophecies. For example, the principle of indeterminacy will obviously not interfere with my prediction that during the coming year zero will turn up approximately $\frac{1}{37}$ of the total number of times the roulette ball is spun at Monte Carlo. All our successful predictions in physics and astronomy are on examination found to depend on this device of eliminating the inherent uncertainty of the future by averaging.

确定性的观点占主导地位至少有两个世纪了，确定论认为如果我们知道了某个时刻宇宙状态的完整数据，比如说 1600 年第一分钟的完整数据，那么单纯的数学推算就可以告诉我们这个世界以前是什么样子以及未来将会是什么样子。未来是由现在所决定的，因为一个微分方程的解由边界条件决定。要理解这个全新的观点，我们必须意识到我们一开始提出的“完整数据”这个概念是存在回避问题实质的风险的。而我们对物理世界的所有认识都是推论性的。我的钢笔并不是作为一个在物理世界中存在的物体使我直接认识到的；我是通过从笔上反射进我眼睛里的光以及握笔时在我肌肉中传播的压力波等感知它的存在和性质的。

我们对过去事物的感知也遵循着完全一样的模式。就像现在我可以通过我的视觉感受去推断一个客观的物体，即我的钢笔一样，我也可以从患上麻疹这个情况推断出几天前应该受到了感染。如果我们完全依照这个规律，就可以就 1930 年已经发生的全部事件去推断在 1600 年使它们发生的原因。乍一看似乎这些推断出来的原因与钢笔（也是一种推断出来的原因）的存在状态一样对我们认识物理世界具有同样重要的作用。于是确定论者们认为这令我陷入了进退两难的境地。如果某个科学工作者在 1600 年的世界中四处探寻并且恰好找到了这些事情的起因，那么他就有完整的数据来准确地预测 1930 年发生的事情；如果他做不到，那么他肯定没有完全认识 1600 年的宇宙，因为这些起因跟我们在其他推论过程中得出的起因一样都是些物理实体。

我还需要继续说明他们是如何为了回避问题的实质而武断地规定哪些内容应该被视为是 1600 年的世界的完整数据，而不考虑那时是否存在可以想象的方式去获取这些数据。海森堡发现（至少在包括整个原子物理学和电子理论的广泛物理现象中），按照自然界的规律，在我们支持确定论的朋友们所需的数据中，只有一半有可能被 1600 年的研究者通过不懈的努力收集到，而对这一半数据的完整认识将使我们自动失去了得到另一半数据的机会。这是个很奇怪的约定，因为你可以选择要了解哪一半数据；你可以知道任何一半的数据，但就是不能同时知道全部的数据。或者你也可以做出妥协而选择不完全地知道这两个一半的全部数据，即有不确定的部分。这个物理规律是确定的：数据是成对出现的，你对一对数据中的一个测量得越精确，则对另一个的测量结果就会越不精确。

当然，想要预测未来，全部数据的两个部分都是必要的，但是如果我们谨慎地选择预测的对象，我们还是可以做出一些可靠的预言的。例如，不确定性原理不会对我预测未来一年蒙特卡洛轮盘赌上的数字零出现的次数造成影响，这个次数大概是小球总的旋转次数的 $\frac{1}{37}$。实践证明，所有物理学和天文学上的成功预测都是通过这种取平均的办法来消除未来的内在不确定性的。

As an illustration, let us consider the simplest type of prediction. Suppose we have a particle, say an electron, moving undisturbed with uniform velocity. If we know its position now and its velocity, it is a simple matter to predict its position at some particular future instant. Heisenberg's principle asserts that the position and velocity are paired data; that is to say, although there is no limit to the accuracy with which we might get to know the position and no limit to the accuracy with which we might get to know the velocity, we cannot get to know both. So our attempt at an accurate prediction of the future position of the particle is frustrated. We can, if we like, observe the position now and the position at the future instant with the utmost accuracy (since these are not paired data) and then calculate what has been the velocity in the meantime. Suppose that we use this velocity together with the original position to compute the second position. Our result will be quite correct, and we shall be true prophets—after the event.

This principle is so fully incorporated into modern physics that in wave mechanics the electron is actually pictured in a way which exhibits this "interference" of position and velocity. To attribute to it exact position and velocity simultaneously would be inconsistent with the picture. Thus, according to our present outlook, the absence of one half of the data of prediction is not to be counted as ignorance; the data are lacking because they do not come into the world until it is too late to make the prediction. They come into existence when the event is accomplished.

I suppose that to justify my title I ought to conclude with a prophecy as to what the end of the world will be like. I confess I am not very keen on the task. I half thought of taking refuge in the excuse that, having just explained that the future is unpredictable, I ought not to be expected to predict it. But I am afraid that someone would point out that the excuse is a thin one, because all that is required is a computation of averages and that type of prediction is not forbidden by the principle of indeterminacy. It used to be thought that in the end all the matter of the universe would collect into one rather dense ball at uniform temperature; but the doctrine of spherical space, and more especially the recent results as to the expansion of the universe, have changed that. There are one or two unsettled points which prevent a definite conclusion, so I will content myself with stating one of several possibilities. It is widely thought that matter slowly changes into radiation. If so, it would seem that the universe will ultimately become a ball of radiation growing ever larger, the radiation becoming thinner and passing into longer and longer wavelengths. About every 1,500 million years it will double its radius, and its size will go on expanding in this way in geometrical progression for ever.

(**127**, 447-453; 1931)

作为一个例证，让我们来考虑一个最简单的预测。假设我们有一个粒子，例如电子，没有受到任何扰动而做匀速直线运动。如果我们知道它现在的位置和速度，那么预测它未来某个时刻的位置是一件很简单的事情。海森堡不确定性原理告诉我们，位置和速度是结成对的数据；也就是说，我们可以无限精确地知道它的位置，也可以无限精确地知道它的速度，但是我们就是不能同时精确地知道这两个量。所以我们在试图准确预测这个粒子未来某个时刻对应的位置时遇到了困难。如果我们愿意的话，我们可以观察这个粒子现在时刻和未来时刻的精确位置（因为它们不是结成对的数据），然后计算出这段时间的平均速度。假如当我们由这个平均速度和最初时刻的位置来计算未来时刻的位置时，我们发现结果完全正确，于是我们成了真正的预言家——其实只是事后诸葛亮。

这个不确定性原理已经完全融入了现代物理学，在波动力学中的确可以认为在电子的位置和速度之间就表现出了那种“相干关系”。由于这种相干关系，同时精确地知道电子的位置和速度是与以上的物理图像相矛盾的。因此，按照我们现在的观点，缺少预测未来所需数据中的一半算不上是信息不灵通；数据不完整是因为在我们做出预测之前它们并不存在。它们是在我们的测量行为发生之时才出现的。

我认为为了充分说明我的主题，我应当用一个预言来作为对宇宙终点可能会是什么样子这个问题的总结。我承认我并不十分情愿回答这个问题。我甚至想，既然我刚才已经解释了未来是不可预测的，我就可以以此为借口拒绝回答这个问题。但恐怕有人会说，你这个借口太勉强，你所需要做的只是计算一下平均值，而那种类型的预言并不违背不确定性原理。过去我们认为宇宙中的所有物质最终会聚集到一起变成一个温度均匀、密度极大的圆球；但是球面空间的学说，尤其是由此得到的宇宙膨胀的最新结论改变了以前的看法。现在还存在一两个未得到解决的问题，这使得我不能得出明确的结论，所以我在这里只是给出其中的一种可能性。大家普遍认为物质会慢慢地转变成辐射。如果是那样的话，整个宇宙就会变成一个不断膨胀的充满辐射的球，辐射会变得越来越弱，相应的波长也变得越来越长。大概每过 15 亿年，宇宙的半径就会增加一倍，宇宙的大小将会以这种几何级数的增长方式永远地膨胀下去。

（沈乃澂 翻译；张元仲 审稿）

Stellar Structure

H. N. Russell and R. d'E. Atkinson

Editor's Note

Henry Norris Russell and Robert d'Escourt Atkinson were both distinguished astronomers, who here give one of the first accounts of the properties of the stars known as white dwarves. In particular, they were concerned to reconcile the very high temperatures of the surface of these stars with the fact that their output of radiation is usually small compared with the Sun. Their conclusion was that white dwarves were indeed much smaller than the Sun. They also introduced the idea that the matter of which these stars consist might, with time, become degenerate, the atoms splitting into electrons and nuclei which separately form very compact atomic structures. Degenerate stars of this kind are now most clearly typified by neutron stars, discovered only in the 1960s.

ZANSTRA'S recent determination of the temperatures of the *O*-stars in planetary nebulae[1] makes it appear extremely likely that these stars are all generically in the "white dwarf" class, with mean densities far above anything known on the earth. He has found temperatures between 30,000° and 100,000° for about twenty of these objects, and yet their luminosities are comparatively small. For a fairly typical nebular nucleus we may assume a photographic magnitude of 12.5 with a parallax of 0.002″, making the absolute magnitude +4, or about eight magnitudes fainter than typical galactic *O*-stars. It is, of course, true that the luminous *efficiency* falls off with rising temperature in this range, but an increase in temperature at constant radius must always involve an increase in brightness, proportional, even in the farthest part of the Rayleigh–Jeans region, to at least the first power of the temperature. Thus these stars must be of very small radius.

Zanstra obtains his figures from the difference between the measured brightness of the star and that of the nebula, which latter is assumed to be excited by the main (Schumann region) radiation of the star, and to convert all of it to long wave-lengths. The nebula thus performs the correction from visual to bolometric magnitude for us, as it were; if it does not do so completely, the star must be still hotter than is calculated. The correction is found to vary between 2 and 7 or 7.5 magnitudes; in an average case it would be about 5 magnitudes, and the temperature would be rather more than 55,000°. Such a star would then be about 5.9 magnitudes brighter than the sun bolometrically, with ten times its surface temperature; this means a radius 1/43 of the sun's, so that even with the sun's mass its density would be more than 100,000 gm./c.c.

The masses are, however, certainly greater. In a typical planetary, the line-of-sight rotational velocity can be taken as 5 km./sec. at 6″ from the nucleus, and with a parallax of 0.002″ this gives a mass 80 times the sun's.[2] If the gas is partly supported by radiation

恒星的构造

亨利·罗素，罗伯特·阿特金森

编者按

亨利·诺里什·罗素和罗伯特·阿特金森是两位著名的天文学家，他们在本文中讨论了白矮星的性质，这篇文章也是最早报告这方面研究的文章之一。特别是他们致力于解决白矮星表面温度高而辐射却比太阳低的矛盾，他们的结论是白矮星实际上比太阳小得多。他们还提出这些恒星中的物质也许会随着时间推移而变成简并状态的思想，即原子中的电子和原子核分离开来，并且各自形成了非常致密的结构。20 世纪 60 年代发现的中子星就是这类简并星中最典型的代表。

赞斯特拉最近对行星状星云中的 *O* 型星的温度进行了测定[1]，结果表明，这些恒星极有可能全部属于“白矮星”类型，其平均密度比地球上已知的任何物质都大。他对大约 20 颗这类恒星进行观测，测得的温度在 30,000℃到 100,000℃之间，而它们的光度却相对较低。对于一个典型的行星状星云核，我们可以假设照相星等为 12.5 等时的视差为 0.002 角秒，这使得这类恒星的绝对星等为 +4，大约比典型的银河系内 *O* 型星暗 8 个星等。诚然，在这个范围里，随着温度的升高发光效率会降低，但是在半径恒定时，温度升高一定伴随着亮度成比例地增大，而即使在瑞利–金斯黑体辐射区域的最远端，亮度的增大也至少与温度的一次方成正比。因此这些恒星的半径一定非常小。

赞斯特拉根据测定的恒星亮度与星云亮度之间的差别得到了这些数值，而现在人们认为星云的发光是由恒星辐射（舒曼区激发）产生的，并且全部被转化成长波辐射。因此星云给我们提供了由目视星等到热星等的改正方法。如果不是完全转化，那么恒星温度要比计算得到的温度高。改正值介于 2 个星等到 7 或 7.5 个星等之间，平均情况在 5 个星等左右，因此温度应当大于 55,000℃。这样，恒星的热星等应当比太阳亮大约 5.9 个星等，而表面温度是太阳的 10 倍。这就意味着其半径为太阳的 1/43，因而即使它的质量等于太阳的质量，其密度也要大于 100,000 g/cm^3。

然而，这些恒星的质量当然会更大。就一个典型的行星状星云而言，在距离核心 6 个角秒处，其自转速度沿视线方向的速度分量可取为 5 km/s，而视差为 0.002

pressure, or if the axis is inclined to the line of sight, the figure comes out even greater. This may be set off against the allowance for the mass of the nebular envelope. The data indicate, therefore, a mean density of 10^6 or 10^7 gm./c.c., which is greater than has even been suggested for any other bodies, and points conclusively to a degenerate state of matter.

These bodies appear to be at the upper end of a sequence of "white" dwarfs, as may be seen from the following summary:

	Mv	Spectrum.
Planetary nuclei	+4	*O*
o Ceti *B*	6	*B8e*
o^2 Eridani *B*	11	*A0*
Sirius *B*	10	*A5*
van Maanen's star	14.5	*F*
Wolf 489	13	*G*

It looks as if this sequence were roughly parallel to the main sequence, probably separated from it by a sparsely populated band, and with considerable scattering within it; at least, the general trend is evident, and the reason why no red stars of the "collapsed" type have been discovered is of course obvious.

There are grave difficulties in the assumption that the source of energy within these white dwarfs is of the same nature as that within giants and main sequence stars, so that it seems worth while to point out that they have no very obvious need for a subatomic source at all. Just before degeneracy sets in, the internal temperatures must be of the order of at least 50 times those of a main sequence star built on the "diffuse" model, or fully 10^9 degrees; Milne's calculation for the companion of Sirius[3] indicates a central temperature not exceeding (but apparently approaching) 3×10^9 degrees. The rate of radiation of such a star is only about 1/100 of that of the sun, or say one calorie per gram in 60 years; taking the specific heat as 3 (when the gas is still on the edge of degeneracy) we see that there is an internal store of heat sufficient for radiation at the white dwarf rate for something of the order of 10^{11} years. Of course, the very fact that the star is about to become degenerate means that not nearly all the kinetic energy of the nuclei and electrons will actually be available for radiation; much of it must remain as zero-point energy permanently in the star. But against this we must set the further energy to be obtained from such contraction as is still possible; at these small radii this energy is large. On the whole, then, the life of a white dwarf comes out, without any subatomic sources, entirely comparable with that of any star that derives its energy from the transmutation of hydrogen into heavier elements.

If the available subatomic energy of a main sequence star is exhaustible, the period spent in the main sequence will be followed by one of gravitational contraction; in the early stages this will be rapid, because the radius is large, and because at least half the energy gained will go into heating the interior of the star; in the late stages, the internal temperature will actually be falling, gravitational contraction will be very effective (if degeneracy is not too marked), and the rate of radiation will have become small. The

角秒，由此推出其质量是太阳的 80 倍 [2]。如果气体部分地由辐射压力支撑，或其自转轴与视线方向倾斜，其质量数值将更大。这就可以作为补充从而平衡星云外包层的质量。这个数值还表明，其平均密度为 10^6 或 10^7 g/cm³ 之间，因此远远大于任何其他物体的密度，可以认为这里的物质处在简并态。

这些天体出现在“白”矮星星序的上端，参见如下的汇总结果：

	目视星等	光谱型
星云核	+4	*O*
鲸鱼座 *o* 星	6	*B8e*
波江座 o^2 星	11	*A0*
天狼星伴星	10	*A5*
范玛宁星	14.5	*F*
豺狼座489号星	13	*G*

粗看起来这个星序与主星序大致平行，似乎通过一个稀少族带和主星序分开，并且内部有明显的弥散。至少这个总趋势是显而易见的，同时为何没有发现“塌陷”的红矮星的原因也是明显的。

很难假设这些白矮星的能源与巨星和主序星的能源是一样的，因此，值得指出的是它们并没有非常明确地表现出对亚原子能源的需要。在进入简并态以前其内部温度的数量级至少是按“扩散”模型计算的主序星温度的 50 倍，可达 10^9 度。米耳恩对天狼星伴星 [3] 的计算表明，其中心温度没有超过（但明显接近）3×10^9 度。其辐射率仅是太阳的 1/100，或者说在 60 年中每克仅产生 1 卡路里热能。假设比热是 3（当气体仍处在简并态的边缘）我们可以看到如果以天狼星伴星的辐射率计算，其内部的储能足以辐射约 10^{11} 年。当然，白矮星即将变为简并态表明，并非原子核和电子的所有动能都将可以用来辐射，大部分动能会以零点能量的形式永久地保留在白矮星内。与此相反，我们也必须假定仍有可能从这类收缩过程中获取更多的能量，因为收缩到如此小的体积时这种能量也是很大的。总之在没有亚原子提供能量的情况下，白矮星的寿命完全可以与任何从氢转变成重元素的过程中获得能量的恒星的寿命相当。

如果主序星恒星可被利用的亚原子能被耗尽，那么在主星序阶段之后将会出现一个引力收缩阶段。在最初阶段这个过程进行得很快，因为其半径很大，而且至少有一半的能量被用来加热中心部分。到晚期恒星内部温度将下降，此时引力收缩效应将非常明显（如果简并还不太显著），而辐射率将变小，因此演化晚期要比早期缓

fairly late stages will therefore be passed through very much more slowly than the early ones, and there will be a marked statistical concentration of the stars at radii perhaps two or three times the minimum figures calculated by Milne for a fully degenerate star. The minimum values are roughly $M^{-1/3}/80$ times the sun's radius, for a star of mass M times that of the sun, and where the masses are known the radii are all of about the anticipated size. Particularly high densities are clearly to be expected for the *O*-stars. In van Maanen's star, with a radius of 0.007 times the sun, it is tempting to suppose we have a massive star in a very late stage indeed; if Milne's formula is applicable, the mass must be at least 8 times the sun, which would mean an Einstein shift corresponding to at least 700 km./sec. This could easily be tested by observation.

It is worth remarking that in a century or less the true radial velocity of a star of such large proper motion and parallax as this could be found from the second order term in the proper motion; the relativity effect, even if much smaller than 700 km./sec., could then be fairly definitely determined.

However that may be, we have now fairly good evidence that stars of all masses can "die"; this does not prove that transmutation rather than annihilation of matter is the source of stellar energy, but it clearly favours it. For Milne's theory leads to the conclusion that a mainly or nearly degenerate star will have a central temperature that may be much less than, but cannot be greater than $3.9\times10^9 M^{5/6}$, and this seems inadequate to stimulate a source of energy that was not active in the main sequence, especially if all stars have dense hot cores, as Milne believes. Thus the very latest stages of degeneracy, where the specific heat is small and contraction difficult, must be run through rapidly, and the total duration of the white dwarf stage probably is not much greater than we have already calculated. The probable relative abundance of white dwarfs is then somewhat difficult to reconcile with a time-scale of 10^{13} or 10^{14} years, but fits well with the transmutation theory time-scale.

The view that the nuclei of planetary nebulae are "white dwarfs" has, we now find, already been propounded by Jeans in "The Universe Around Us", pp. 309–311. That we had both failed to notice this can be excused, if at all, only by the great popularity of the publication in which the theory was announced, but we sincerely regret the oversight. In the application of this result we differ from Jeans, since he tentatively placed these stars at the beginning of stellar evolution, while we place them, with the other white dwarfs, at the end.

(**127**, 661-662; 1931)

H. N. Russell: Princeton University, Mar. 31.
R. d'E. Atkinson: Rutgers University, Mar. 31.

References:

1. *Zeit. f. Astrophysik*, 2, 1 (1931).

2. Cf. Russell, Dugan, and Stewart, *Astronomy*, 835.

3. *Mon. Not. R.A.S.*, 91, 39 (1930).

慢得多。当恒星半径达到米耳恩计算的简并最小半径的 2~3 倍时，将出现明显的统计富集现象。对于质量为太阳质量的 M 倍的恒星，其半径最小值约为太阳的 $M^{-1/3}/80$ 倍。只要已知质量，半径大小就可以预测出来。显然我们可以预期 O 型星具有特别高的密度。对半径为太阳 0.007 倍的范玛宁星，我们倾向于把它看作是演化到很晚阶段的大质量恒星。假定可以使用米耳恩公式，其质量至少应是太阳的 8 倍，这说明对应的爱因斯坦红移大约是 700 km/s，这很容易用观测来检验。

值得一提的是，在近一个世纪的时间里，一个如此大自行和视差的恒星，其视向速度可以由自行的二次项而求得。即使是远小于 700 km/s 的相对论效应也能很确切地测定出来。

无论如何，我们已有很好的证据证明所有质量的恒星都会“死亡”，虽然这并不能证明恒星能源是由物质转换产生的而不是由湮灭产生的，但却明显地对此有利。由米耳恩的理论可以推出：接近简并态的恒星的中心温度可能会远远小于，但不可能大于 $3.9\times10^{9} M^{5/6}$, 因此不足以激活在主星序没有被激活的产能机制，尤其是在如米耳恩认为的所有恒星均具有致密而热的核心的情况下。因此在比热很小而收缩困难的简并最后阶段，演化进行得非常快，处于白矮星阶段的总时间很可能不会大大高于我们已经计算出来的数值。因此白矮星的相对丰度很难用 10^{13} 或 10^{14} 年的时间尺度来估算，但却与转换理论的时间尺度相符。

我们现在发现，行星状星云核是“白矮星”的观点已经被金斯在《我们周围的宇宙》一书的第 309 页 ~ 311 页中提出。我们俩以前没有提及这一点也许可以得到大家的原谅，因为发布这个理论的文章很多，但我们也为自己的疏忽而感到遗憾。不过在运用这个理论时，我们与金斯的观点不一致，因为他试探性地认为这类恒星是处在恒星演化的初始阶段，而我们认为这些恒星以及其他白矮星都处在演化的最后阶段。

（曹惠来 翻译；蒋世仰 审稿）

Atomic Synthesis and Stellar Energy

R. d'E. Atkinson

Editor's Note

The source of the energy that keeps stars shining was a great mystery in the 1930s. The Welsh astronomer Robert d'Escourt Atkinson here puts forward the view that it arises from the radioactive decay of relatively heavy elements in the interiors of stars. His views were proved mistaken the following year when Hans Bethe published an account of how hydrogen atoms can fuse directly to make helium atoms by simple nuclear processes in which the existence of the neutron (also discovered in 1932) is essential.

SOME time ago F. G. Houtermans and the present writer investigated the possibility of synthesis of elements, in stellar interiors, by the wave mechanics process of penetration of nuclei by protons.[1] The theory was not strictly correct, and various modifications have been proposed since, of which the theory of Wilson[2] is perhaps the most important; all theories, however, lead to a probability of proton penetration having the same exponential dependence on both the temperature, T, and the atomic number, Z. The importance of this factor far outweighs that of the multiplicative forefactor which alone is different in the different theories, and it seems therefore desirable to discuss somewhat more fully the consequences of the assumption that any of these theories will give the right order of magnitude for the temperature at which synthesis will occur in large amounts. The effect of the exponential is roughly to make the synthesis probability vary as T^{20}, or some comparable power, and thus even a change of 1,000 in the fore-factor does not seriously affect T. The investigation is being discussed fully in the *Astrophysical Journal*, but in view of the interest of the subject at present, and also of the comparative unfamiliarity of the line of attack, a short summary may both appeal to a wider audience and prepare the way for the more detailed treatment.

Direct synthesis of helium from hydrogen is clearly a very unsatisfactory process, but we do not need to assume its existence at all; as in the above paper, we assume that helium is produced entirely indirectly, by the spontaneous disintegration of unstable nuclei that must first themselves be formed. In addition to the known radioactive elements, Gamow's theory of nuclear stability[3] now indicates that we may expect a large number of lighter elements to be unstable if they were to be formed, and in fact we rely mainly on these. For example, if above argon the incorporation of electrons, which we clearly must suppose can occur, is somewhat difficult (and the existence of apparently permanent non-radioactive *isobars* seems to show that it can be extremely difficult), nuclei such as Fe^{52}, Ni^{56}, and Zn^{60} may be formed; according to Gamow, the last of these, and quite possibly all three, should be unstable. After emitting one α-particle, they would again collect four protons and two electrons in such order and at such intervals as they could, combine them into a fresh α-particle, and re-emit this one also in due course. At various points in this cycle they

原子合成与恒星能量

罗伯特·阿特金森

编者按

在 20 世纪 30 年代，人们完全不清楚维持恒星不断发光的能量来源于哪里。威尔士天文学家罗伯特·阿特金森在本文中提出这种能量来源于较重元素在恒星内部的放射性衰变过程。一年后，他的观点被证明是错误的，因为汉斯·贝特发表了一篇关于氢原子可以通过简单的核反应直接聚变成氦原子的报告，当然这个结论只有在人们了解到中子的存在（也是在 1932 年发现的）之后才能得出。

豪特曼斯和本文作者曾经研究过在恒星内部，质子经由波动力学过程进入核内而引起元素合成的可能性[1]。理论并非严格正确，后来人们又提出了各种修正意见，这其中威尔逊的理论[2]应该是最为重要的；然而，所有理论一致认为质子渗透概率与温度 T 和原子序数 Z 有相同的指数函数关系。这一因素的重要性远远超过倍增因子，而后者也是不同理论中的唯一不同之处。所以，有必要更充分地讨论一下由“任何理论都能给出在大量合成时温度的正确数量级”这一假设所导致的结论。指数效应使合成概率大致按 T^{20}（或别的某个可与之比拟的幂律）变化。所以即使前面的因子改变了 1,000 倍也不会严重影响 T。这项研究在《天体物理杂志》上有完整的讨论，但考虑到目前人们对此课题很感兴趣，却又不太熟悉这个理论，所以一个简短的概述或许会吸引更多的读者，也能为更细致的探讨做准备。

直接由氢合成氦显然是一个非常不能令人满意的过程，但我们根本不需要假设这种过程的存在；如上一篇文章中所述，我们可以假设所有氦都是通过不稳定核的自发衰变产生的，但需要首先生成不稳定核。除已知的放射性元素外，伽莫夫的核稳定性理论[3]表明：我们现在可能要寄希望于为数众多的轻元素（如果它们能够形成的话）是不稳定的，事实上我们会主要依赖这些轻元素。例如，氩以上原子对电子的吸收是有一定困难的，虽然我们确信上述过程可以发生（而明显稳定的非放射性**同量异位素**的存在似乎表明这个过程可能极为困难），同时可能会形成类似 Fe^{52}、Ni^{56} 和 Zn^{60} 这样的核；按伽莫夫的说法，最后一种，或很可能所有三种，应该不会是稳定的。在释放掉一个 α 粒子之后，它们会再次按某种顺序和间隔收集四个质子和两个电子，并将其组合为一个新的 α 粒子，而后再次按既有程序释放掉。在这个周期的不同点，它们辐

would also have to emit a total of just as much energy as would be set free by the direct synthesis of a helium nucleus; this energy can then be used to maintain the star's radiation. This method of evading the well-known difficulty of the 6-body collision as a source of helium obviously opens up important avenues; it is, however, not at all necessary to regard helium as the final product. In fact, since small Z-values favour synthesis, all the helium formed will be rapidly built up again. The energy developed is roughly the same, however, *per proton consumed*, whatever the products.

Since no other theory proposes a lower temperature for an energy source than ours turns out to do, we may take it that stars will at any rate contract until this process becomes operative; they will then be unable to contract further, since even a small contraction will enormously stimulate the energy development and force them to expand again. Milne's theory seems in this way to be ruled out until such time as the hydrogen supply near the centre has run low.

Rosseland has shown that when there is no great excess of hydrogen, electrostatic forces will tend to drive it from the centre, and the centre is in any case the only place where it is being consumed. Thus in any star a time will arrive when the disappearance of hydrogen near the centre prevents the generation of energy there altogether; the star must then condense towards the degenerate state. It is known that the energy of the white dwarfs may be entirely gravitational if their lifetimes may be supposed to be only of the order of 10^{11} years; in addition to explaining why stars intermediate between white dwarfs and the main sequence seem to be scarce, this theory explains why heavy white dwarfs should be the commonest[4] (the minimum radius is smaller and the square of the mass larger, so that they have very much more gravitational energy available).

So long, however, as hydrogen is present, synthesis should continue. Since, now, the star's mass remains practically constant, its energy generation must remain moderately so, and it is easily seen that this involves an approximate constancy of the helium supply, but with an ever-present possibility of adding a little to it. This is the fundamental condition for stellar stability, and determines the central temperature. For a star to keep control, it must slowly decrease its central temperature, since the number of helium sources is being added to; thus, after an initial contraction to start the process, stars spend probably the greater part of their lives *expanding*. During most of their lifetime the expansion is very slow, and the central temperature at any one mass is almost determinate; this accounts for the main sequence.

The actual value for the central temperature cannot, however, be what this simple theory would indicate. If iron is to be synthesised, a temperature of perhaps 200 million degrees would be necessary, and at this temperature all the light elements would be so readily converted to heavier ones that they could not become abundant at all. With a constant helium supply we can use the ordinary equilibrium law of radioactivity theory, namely, that the amounts of the various elements should be directly as their "average lives"; the

射出的总能量正好等于直接合成一个氦核所释放的能量，而这些能量就可以用来维持恒星的持续辐射。这种方法避开了把六体碰撞作为氦核来源所遇到的著名困难，从而明显地打开了解决问题的重要思路。但也没有必要把氦核看成是最终产物，事实上，因为小的原子序数 Z 更有利于合成，因此所有已形成的氦核会迅速地再次增大。而不论产物是什么，释放出的能量，即**每个质子所消耗的能量**，都是大致相同的。

因为所有其他理论要求的能量源温度都比我们的理论高，所以我们可以认为恒星无论如何都会收缩，直到这种过程开始为止；之后它们将不能继续收缩，因为即使是一个小的收缩也会极大地促进能量的释放，并使得它们再次膨胀。如此看来，在靠近中心的氢供给变低之前，米耳恩的理论就可以被排除了。

罗斯兰德已经阐述过，当没有大量氢过剩的时候，静电力会倾向于将氢从中心推出，而中心是氢得以消耗的唯一地方。所以对于任何恒星都必然会出现靠近中心处的氢不再存在的时候，这时能量产生被完全阻止，这样恒星也必然会收缩到简并态。人们知道，如果可以认为白矮星的寿命约为 10^{11} 年的话，它们的能量可能会全部来自引力，因此这个理论除了能解释为什么介于白矮星和主星序间的恒星看起来很少外，还解释了为什么大质量的白矮星应当是最常见的[4]（最小半径值越小，质量平方越大，以至于它们具有更多的引力能可以使用）。

然而，只要还存在氢，合成就应该继续。因为既然恒星的质量实际上是保持恒定的，那么它产生的能量也会保持在一个适度的值，很容易看出这关系到基本恒定的氦供给，但再增加一点的可能性是经常存在的。这是恒星稳定的基本条件，也决定了恒星中心的温度。一个恒星要想维持可控，它就必须慢慢地降低自己的中心温度，因为氦源的数量在增加；所以在经过了最初的收缩以启动这一过程之后，恒星用于**膨胀**的时间可能占去了它生命中的一大半。在它们生命的绝大部分时间里，膨胀进行得都很缓慢，并且对于任何给定的质量，中心温度几乎都是确定的，这也解释了主星序的存在。

然而，实际的中心温度值不可能由这种简单的理论给出。如果要合成铁，温度可能需要达到 2 亿度，在这种温度下，所有的轻元素都会非常容易转化成重元素，以至于它们不可能大量存在。在氦供给恒定的条件下，我们可以使用普通的放射性平衡律，即，不同元素的数量应该就相当于它们的“平均寿命”，不管我们面对的是合成过程还是分解过程都不会影响这个定律的有效性。另外，由于原子序数 Z 以指

fact that we are dealing with synthesis and not disintegration does not affect the validity of the principle. Since the atomic number, Z, affects the synthesis probabilities, that is, the average lives, exponentially, all light elements ought to be very scarce; oxygen is, however, more abundant than anything except hydrogen, and a number of other elements near it are also at least as plentiful as iron.

In fact, it is easily seen that the most abundant element of all in a star must have an average life (until further synthesis) comparable with the past life of the star itself; otherwise some heavier element would be more abundant. If now we assume the central temperature is so low that oxygen is as long-lived as this, we find that it and the lighter elements are nevertheless abundant enough for their synthesis alone to supply enough energy, and that the actual temperature is about 16 million degrees in the sun. This is in agreement with the figure obtained on Eddington's theory for a polytrope of index about three and constitution rather above 50 percent hydrogen by weight.

At this temperature there is, however, no synthesis of iron, and thus no further supply of helium, and the process will soon exhaust itself. The difficulty may be overcome by an arbitrary assumption, and it has been found possible to make one that accounts at the same time for the permanence and actual position of the main sequence, and for the relative proportions of all the elements in main sequence stars. The assumption is that there is a second synthesis process, which is more probable than the first when $Z>8$ and has a probability that increases somewhat with increasing Z; it must depend about as extremely on the temperature as the other process. Even if this assumption is wrong, it is probably valuable to have the theory investigated in detail; in point of fact, it does at least lead to a number of correct results.

Oxygen will now certainly be the longest lived element, and the products of synthesis will "pile up" at and near this value of Z. It is, as a matter of fact, desirable to keep them from adding to the iron group which is supplying the helium, for a constant supply is wanted. The iron group itself must, however, be abundant enough to be in "equilibrium" with the lightest elements, since it must produce as many α-particles per second as there are helium-lithium syntheses. It is found that the situation which will develop involves a marked minimum between the oxygen and the iron maxima and a marked fall after, say, zinc, in very good qualitative agreement with observation.[5] A consideration of Gamow's theory leads us to expect in addition a maximum among and below the lightest rare earths, and possibly one in the lead region; both are found. Practically the entire range of elements thus shows a qualitative agreement with what the theory requires.

The same will be true of any main sequence star; but since the age of the star at a point when it is, say, half hydrogen is much greater for small stars than for large ones, and the density is also larger in small stars, the central temperature must be smaller in them if oxygen is still to have a long enough lifetime. This is satisfactory, for Jeans's modification of Eddington's theory does involve a polytropic index varying systematically with the mass in about the right sort of way to produce this effect. Jeans's modification results from using

数方式影响着合成概率，或者说是平均寿命，所以所有的轻元素都应该非常稀少，但是氧含量比除去氢之外的所有轻元素都要多，而且邻近氧的另外一些元素也至少和铁元素一样丰富。

事实上，很容易看出，恒星中最丰富的元素一定具有与恒星目前的年龄相当的平均寿命（直到后面的合成），否则一些重元素将会更多。如果现在我们假设中心温度很低，以至于氧的寿命就这么长，那么我们就会发现它和更轻的元素仍然会非常丰富以至于仅由它们的合成就可以提供足够的能量，而太阳的实际温度约为 1,600 万度。这与从爱丁顿理论中得到的多方指数约为 3、氢的重量百分比明显超过 50% 的数字相符。

然而，在这个温度下不存在铁的合成，也就没有进一步的氦供给，并且该过程会迅速耗尽。这种困难可能会因一个任意的假设而得以克服，我们可以选择一个能同时解决主星序的持久和实际位置问题，又能给出主序星中所有元素的相对比例的假设。这个假设认为存在另一个合成过程：这个过程在 $Z>8$ 时比第一个过程更有可能发生，且发生第二个过程的可能性会随着 Z 的增大而增加；它肯定与另一个过程一样与温度有很大的关系。即使这个假设是错误的，仔细地研究这个理论可能也很有价值；实际上，它至少能给出一些正确的结果。

现在，氧必定是最长寿的元素，合成的产物会在这个 Z 值附近“堆积”。事实上，阻止它们增加到提供给氦的铁族元素中是可以理解的，因为需要的是恒定的供给。然而，铁族元素本身必须足够丰富以便与最轻的元素形成“平衡”，因为它每秒必须产生与氦–锂合成一样多的 α 粒子。这个论断将导致在氧和铁的最大值之间存在一个显著的最小值以及在锌之后出现一个明显的下降，这和观测结果在定性上是非常相符的 [5]。根据伽莫夫的理论，我们认为除了在最轻的稀土元素及以下的区域会出现一个最大值以外，很可能在铅区还有一个极大值，并且这两者都被发现了。实际上从定性角度上看，元素的整个范围都和理论要求的一致。

同样的结论将适用于所有主序星；但当恒星的年龄处于某一点，比如只剩一半氢的时候，质量小的恒星的存在时间远远大于质量大的恒星，小恒星的密度也比大恒星高。如果氧此时仍然有足够长的寿命，则小质量恒星的中心温度一定会相对低一些。这是令人满意的结论，因为金斯对爱丁顿理论的修正的确包含了一个多方指数，它以一种合适的方式有规律地随质量变化从而产生了这种效应。金斯的修正来

the theoretical Kramers value for the absorption coefficient, and the main reason why it has not been more generally adopted (by followers of Eddington) than it has is probably that the absolute values obtained by Kramers' theory did not seem to fit the facts, When, however, we adopt Russell's high value for the hydrogen content of stars the discrepancy disappears.

The vast majority of the stars are thus accounted for. The main sequence consists of stars built on the Eddington-Jeans model, with central temperatures fairly sharply defined at any one mass and rising very slowly with the mass, and with a constitution very similar to that actually observed; the central temperature seems to be about 16–20 million degrees in the sun. The white dwarfs consist of a roughly parallel band of stars built nearly on the Milne degenerate model, with central temperatures up to about 3×10^9 degrees; they should have about the same constitution except for a shortage of hydrogen at the centre, and will be fainter for a given mass.

In addition to these two main classes we may account for the low density giants. These have a comparatively low central temperature and can only obtain their energy (*a*) if they have a large amount of free helium or unstable atoms already present, or (*b*) if some very light element can also be unstable. In case (*a*) they will not be able to live very long; but if they are very heavy their total lifetimes will be short anyway, and their life in this state may be a fairly large fraction of the total. The wave mechanics formula used shows that a star of mass 30 suns, if 10 percent of it were helium and 80 percent hydrogen, could develop enough energy for an absolute magnitude of −6 even at the density of an $M5$ supergiant, with a central temperature of only 4 million degrees. Such stars will, however, be "overstable", that is, they will be liable to develop pulsations; these are well known to be a common feature of very massive red stars. If they are not very massive, only a small fraction of their lifetimes can be spent in this state, and they would, in addition, have a very large colour index indeed; we should thus scarcely expect to see any. In case (*b*) a long life would be possible, but as synthesis would now certainly result in an immediate increase in the amount of the unstable element, the stars would have to change their central temperatures over a fairly large range during their lifetimes. The unstable element is assumed to be the isotope Be^8, which exists in very small quantities on the earth and probably has in fact a mass defect (referred to helium as a unit) of very nearly zero. Many beryls contain a large and otherwise unexplained amount of helium, and when the idea of the instability of Be^8 was first proposed, Lord Rayleigh at once pointed out the significance of this fact.[6] The "Hertzsprung gap" and its prolongation between the Cepheids and the B stars may be shown to follow if Be^8 has a long life, and its presence on the earth guarantees this. A long life is also in harmony with the Be^8/Be^9 ratio and the He/Be ratio.

A number of other observations may readily be fitted into the theory. We may mention in particular the absence of low density stars at medium and small masses, the occurrence of R and N types among giants but not among dwarfs, the existence of binaries in which the brighter star is the cooler and less dense, and the fact that the brightest stars in clusters are usually all red or all blue.

源于对吸收系数采用理论的克拉莫值，而这为什么没有被后人（爱丁顿的追随者们）更广泛地采纳，主要原因可能是因为由克拉莫理论得到的绝对值看起来与事实不符。然而，当我们采用罗素对于恒星中氢含量的较高的估计值时，这种不符合也就消失了。

绝大多数恒星都可以用这种方式进行解释。主星序中的恒星符合爱丁顿–金斯模型，这些恒星的中心温度在任一质量处都有非常明确的定义，并且中心温度随质量增加的速度很慢，它们的组成与实际观测结果非常相似，太阳的中心温度约为1,600万到2,000万度。白矮星由大体平行的恒星带组成，这些恒星基本上符合米耳恩的简并模型，其中心温度高至30亿度；它们与主序星应该有大体相同的组成，但在中心处氢元素较少，并且在质量一定时也会更暗一些。

除了这两个主要的类别之外，我们还知道有低密度的巨星。它们的中心温度相对较低，并且只能在两种情况下获得能量：(*a*) 如果它们有大量的自由氦或是业已存在的不稳定原子，(*b*) 如果一些非常轻的元素也是不稳定的。在情形 (*a*) 中，它们的寿命将不会很长，即使它们非常重，其总寿命仍然会很短，而处于这种状态下的时间长度或许要占总寿命中相当大的一部分。我们使用的波动力学公式显示，一个质量为太阳质量30倍的恒星，如果它的10%是氦，80%是氢，那么即使它的密度只相当于一个*M*5超巨星，中心温度只有400万度，它也足以释放出相当于绝对星等为–6的能量。然而，这样的恒星会"过于稳定"，即它们将有可能发展出脉动，这是质量非常大的红巨星广为人知的常见特性。如果它们的质量不是很大，停留在这种状态下的时间仅是整个寿命中的很小一部分，它们还将具有非常大的色指数，我们也因此很难看到这样的恒星。在情形 (*b*) 中，恒星可能会具有很长的寿命，但是因为合成肯定会导致这种不稳定元素的数量即刻增长，所以在一个相当长的寿命阶段内，恒星将不得不改变它们的中心温度。人们猜测这种不稳定的元素是同位素Be^8，它在地球上只有很小的量，可能会有非常接近于零的质量缺损（以氦核为单位）。许多绿柱石中含有大量的氦，除此之外无法解释为什么有这么多氦，当Be^8的不稳定性刚刚被提出时，瑞利勋爵立刻指出这一发现的重要性[6]。如果Be^8有很长的寿命，也许可以证明介于造父变星和*B*型星的"赫兹伯伦空隙"和它的延长线将是必然的结果，而Be^8在地球上的存在保证了这一假设的成立。另外，较长的寿命也与Be^8/Be^9比以及He/Be比相符合。

一些其他的观测结果也很容易纳入这个理论中来。我们需要特别提到的是，不存在低密度的中小质量恒星，*R*和*N*型星在巨星而不是矮星中出现，在一些双星系统中较亮的星会是密度和温度较低的星，并且星团中最亮的恒星通常全部是红的或者全部是蓝的。

The arguments that have been urged in support of the "long time scale" (10^{13}–10^{14} years) may all be met with some plausibility. In particular the well-known theory of Jeans for the eccentricities of binaries, and similar "kinetic theory" arguments can all be reconciled with the "short time scale" (10^{11} years) if the galaxy is expanding as fast as the universe in general is; this expansion (for the universe as a whole) seems to be demanded by the general theory of relativity.

It thus appears that as a result of the wave mechanics on one hand, and the general theory of relativity on the other, the universe may have developed its present complexity of stars and of atoms from an initial state consisting of a fairly dense, nearly uniform, nearly stationary mass of cold hydrogen. This comparatively simple beginning constitutes at least a pleasant ornament, if not an actual support, for our theory. It must, however, be admitted that there are still some serious difficulties; those that have been noticed are discussed in the full account which will shortly appear.

(**128**, 194-196; 1931)

R. d'E. Atkinson: Rutgers University.

References:

1. *Zeits. f. Physik*, **54**, 656 (1929).

2. *Mon. Not. R. A. S.*, **91**, 283 (1931).

3. *Proc. Roy. Soc.*, **126**, 632 (1930).

4. Russell and Atkinson, *Nature*, 661 (May 2, 1931).

5. Russell, *Astr. Jour.*, **72**, 11 (1929).

6. *Nature*, **123**, 607 (1929).

强烈支持“长时间尺度”（$10^{13} \sim 10^{14}$ 年）的论证也许都是貌似正确的。尤其是如果星系膨胀得和整个宇宙一样快的话，金斯关于双星偏心率的著名理论以及与之相似的“动力学”理论都可以与“短时间尺度”（10^{11} 年）相符合；而这种膨胀（对作为一个整体的宇宙来说）看起来正是广义相对论所要求的。

波动力学和广义相对论都认为，宇宙似乎是从一个相当致密、近乎均匀、近乎稳定的冷氢物质发展到现在的恒星以及原子所具有的复杂程度。这种相对简单的开端如果没有在事实上支持我们的理论，至少也是一个令人愉悦的点缀。然而，必须承认还有一些重要的问题有待于解决，已经被注意到的一些问题将会在很快就要出版的完整报告中进行讨论。

（汪浩 翻译；蒋世仰 审稿）

The Internal Temperature of White Dwarf Stars

E. A. Milne

Editor's Note

With the development of quantum mechanics, astronomers began to incorporate some of its tenets into their study of the structure of stars. A prediction of quantum mechanics is that, in certain circumstances, particles and atoms will form into "degenerate" states, in which the particles all occupy a single quantum state and which are extremely dense. This happens inside white dwarf stars, formed by the collapse of ordinary stars near to the end of their life cycle, which have a Sun-like mass within an Earth-like volume. This paper by E. Arthur Milne, attempting to clarify the internal temperature of white dwarfs, refers to Subrahmanyan Chandrasekhar, who first predicted the existence of black holes and who deduced the maximum possible mass for white dwarfs.

IT has recently been discovered by S. Chandrasekhar,[1] B. Swirles,[2] and R. C. Majumdar,[3] independently, that the opacity of a degenerate gas is very small compared with what would be computed for a classical gas at the same density and temperature, the ratio being an inverse power of Sommerfeld's degeneracy-criterion parameter. This discovery seriously affects estimates of the internal temperatures in white dwarf stars. It has previously been held that interiors of the white dwarf stars are amongst the hottest of stellar interiors; for example, Russell and Atkinson[4] remark that their internal temperatures must be of the order of 50 times those of a main sequence star built on the "diffuse" model. Again, Jeans[5] says "it appears that the central temperatures of the white dwarfs must be enormously high, while those of giant stars of large radius must be comparatively low". This has given rise to the paradox that the coolest stellar interiors appeared to be the best generators of stellar energy, the hottest the worst. To quote Jeans[6] again, "… many of the hottest and densest stars are entirely put to shame in the matter of radiation by very cool stars of low density, such as Antares and Betelgeuse".

If, however, the opacity in the interior of a white dwarf is very low, the temperature gradient in the interior must be very small. In the limit of zero opacity (assuming also small conductivity) the temperature-distribution is isothermal. The degenerate core is therefore a mass at an approximately uniform temperature, and the value of this temperature is determined purely by the observed mass M and luminosity L and the intrinsic opacity κ_1 of the gaseous envelope which surrounds the core. On the "generalised standard model", in which the energy-sources are uniformly distributed and the opacity takes a constant value κ_1 in the gaseous envelope, the temperature T' of the approximately isothermal degenerate core is, in the standard notation:

$$T' = \frac{(R/\mu)^{\frac{5}{3}}}{\left(\frac{1}{3}a\right)^{\frac{2}{3}} K}\left(\frac{\kappa_1 L}{4\pi c G M - \kappa_1 L}\right)^{\frac{2}{3}} \qquad (1)$$

白矮星的内部温度

爱德华·阿瑟·米耳恩

编者按

随着量子理论的发展，天文学家们开始把一些量子力学的原理融入他们对恒星构造的研究中。量子力学的一个预言是：在特定条件下，粒子和原子将形成“简并”态。在这种状态中，所有粒子都占据着一个简单的量子态，并且处于这种状态下的恒星将具有很高的密度。白矮星的内部就会出现上述情况，白矮星是由普通恒星在其生命周期的末期塌陷形成的，它和地球的体积差不多，但却具有如太阳那么大的质量。这篇由阿瑟·米耳恩撰写的文章试图对白矮星的内部温度作出解释。文中还提到了第一个预言黑洞存在的苏布拉马尼扬·钱德拉塞卡，是他推测出了白矮星的最大可能质量。

钱德拉塞卡[1]、斯怀尔斯[2]和马宗达[3]最近各自独立地发现：与同等密度和温度的经典气体相比，简并气体的不透明度是非常小的，其比值是索默菲尔德简并判据参数的倒数。这一发现对估算白矮星的内部温度影响重大。过去人们一直把白矮星的内部温度看作是恒星中最高的；例如，罗素和阿特金森[4]就认为白矮星的内部温度50倍于建立在“弥散”模型基础上的主序星。而且，金斯[5]也说过：“看起来白矮星的中心温度非常高，而那些大半径巨星的中心温度则会相对较低一些。”这就产生了一个悖论，温度最低的恒星内部是最好的恒星能量产生器，最热的反而最差。金斯[6]还说过：“……很多最热和最致密的恒星在辐射上远远比不上那些具有较低密度的低温恒星，比如心宿二和参宿四。”

然而，如果一个白矮星内部的不透明度很低，那么它内部的温度梯度也一定非常小。在不透明度为零的极限情况下(假设传导性也很小)，温度分布是等温的。因此，简并核就是一团温度近似恒定的物质，而这个温度的值完全取决于观测质量 M、光度 L 和核心周围气体包络层的内禀不透明度 κ_1。在“一般标准模型”中，各能量源是均匀分布的，气体包络层的不透明度是常数 κ_1，则温度近似恒定的简并核的温度 T' 用标准符号表示如下：

$$T' = \frac{(R/\mu)^{\frac{5}{3}}}{\left(\frac{1}{3}a\right)^{\frac{2}{3}}K}\left(\frac{\kappa_1 L}{4\pi cGM - \kappa_1 L}\right)^{\frac{2}{3}} \qquad (1)$$

For the observed mass and luminosity of the Companion of Sirius, T' is $0.34 \times 10^6 \kappa_1^{\frac{2}{3}}$ degrees, or, even if we adopt the high value $\kappa_1 = 300$ for the gaseous envelope, the value of T' is only 15 million degrees. For smaller values of the envelope-opacity it will be still smaller. For an almost completely degenerate star the internal temperature is determined by the *photospheric* opacity in the thin gaseous envelope.

According to my conclusion that all stars contain a degenerate zone surrounded by a gaseous envelope, formula (1) applies to all stars. Stars with a high value of L/M have small, incompletely degenerate cores, in which the temperature gradients though small are larger than in completely degenerate cores. Formula (1) still gives the interfacial temperature between core and envelope, and is thus a lower limit to the central temperature. It follows that stars with large internal generation of energy, that is, large values of L/M, have very hot central cores. Such stars will not be built on the standard model, but the effect of concentrating the energy sources to the centre, keeping other parameters constant, is only to increase the central temperature. Jeans's paradox, therefore, completely disappears; the best generators of energy have the hottest cores, and this applies to stars of all types, from white dwarfs to giants.

This result is quite obvious physically. The gaseous envelope acts simply as a blanket the role of which is to keep the core warm. A high energy-generator surrounds itself with a thick blanket, which keeps it very warm; a low energy-generator with a thin blanket. The actual value of the temperature attained in the core depends naturally on the intrinsic opacity of the blanketing envelope.

The above considerations illustrate, by a particular example, my contention that we cannot discuss the internal state of a star without discussing the opacity of its outer layers. In the case of a completely degenerate white dwarf, to ignore the effect of the photospheric opacity would be to obtain an utterly false estimate of the internal temperature.

The above results were communicated to Section A of the British Association on Sept. 29, 1931, at the discussion on the evolution of the universe, but did not appear in the printed accounts. I may add that the full theory of the "generalised standard model", now fairly completely worked out, affords possible explanations of many of the observed characteristics of the stars in general, including some of those summed up in the "Russell diagram", the approximate "mass-luminosity" law for non-dense stars, the occurrence of pulsating stars and Novae, and the possible existence of several types of configurations for large M and L (O-type, giant M-type, N-type, etc.) stars.

(**128**, 999; 1931)

E. A. Milne: Wadham College, Oxford, Nov. 17.

代入天狼星伴星的观测质量和光度值后，可得 T' 为 $0.34\times10^{6}\kappa_1^{\frac{2}{3}}$ 度，就算我们给气体包络层的不透明度取一个较高的数值，比如令 $\kappa_1=300$，T' 的值也只能达到 1,500 万度。如果气体包络层的不透明度低于 300，则 T' 还会更低一些。对于一个几乎完全简并的恒星而言，其内部温度是由气体包络薄层中的**光球**不透明度决定的。

我认为所有恒星都包含一个由气体层围绕的简并区域，根据这个结论，公式 (1) 可以适用于所有恒星。*L*/*M* 值较高的恒星具有小的、不完全的简并核，其温度梯度虽然小，但要比完全简并核的温度梯度大。公式 (1) 还给出了核与包络层界面处的温度，这个温度是中心温度的下限。因此内部能产生大量能量的恒星，也就是 *L*/*M* 值很大的恒星，其中心核的温度相当高。这样的恒星不可能建立在标准模型之上，但这种将能源积聚到中心的效应，只会提高中心的温度，而不会使其他参数发生变化。因此，金斯的悖论完全消失了；产能最大的恒星有最热的核，这对所有类型的恒星都适用，不管是白矮星还是巨星。

这个结果的物理意义也很明显。气体包络层就像毯子一样起到了给核保温的作用。一个高效能量产生器周围包着一条使它保持高温的厚毯子；而一个低效能量产生器周围包着一条薄毯子。所以核心温度的实际值当然会取决于包络层的内禀不透明度。

上述结果通过一个特别的例子说明了我的观点，即在未了解一个恒星外层的不透明度之前，我们无法讨论它的内部状态。对于一颗完全简并的白矮星来说，忽略光球层的不透明效应将使我们对其内部温度的估计完全不可信。

我已经在 1931 年 9 月 29 日讨论宇宙演化时将以上结果提交给英国科学促进会的 A 分会（译者注：数理分会），但在后来的书面记录中没有看到。我要补充说明的是：“一般标准模型”的完整理论现在已经圆满地确定下来了，它能够对许多观测到的恒星特性给出可能的解释，包括一些从“罗素图”中总结出来的规律、非致密恒星的“质量—光度”定律、脉动变星和新星的出现以及高质量高光度恒星可能存在的几种构型（*O* 型，巨 *M* 型，*N* 型等）。

（伍岳 翻译；蒋世仰 审稿）

References:

1. Chandrasekhar, S., *Proc. Roy. Soc.*, **133**, A, 241 (Sept. 1931).
2. Swirles, B., *Monthly Notices, R.A.S.*, 861 (June 1931).
3. Majumdar, R. C., *Astr. Nach.*, No. 5809 (Aug. 1931).
4. Russell, H. N., and Atkinson, R. d'E., *Nature*, 661 (May 2, 1931).
5. Jeans, J. H., *Astronomy and Cosmogony*, 139.
6. Jeans, J. H., *Astronomy and Cosmogony*, 125.

The Expanding Universe

Editor's Note

This editorial describes a recent lecture of James Jeans on developments in cosmology. Most puzzling were attempts to reconcile the apparent rapid expansion of the universe—suggested by the progressive redshift detected in more distant galaxies—with evidence on the past duration of the universe, putting its age at millions of millions of years. One possibility, Jeans suggested, is that the expansion seen today may not have continued uniformly into the past, but may have begun only after an earlier quiescent period. Jeans noted that recent calculations by Arthur Eddington, attempting to explain the expansion in theoretical terms, were rather speculative. But as Eddington intimated, the expansion really was predicted by the theory of general relativity.

IN his Ludwig Mond lecture, delivered at Manchester on May 9, on the subject of "The Expanding Universe", Sir James Jeans began with a review of the system of galaxies, our knowledge of which has been greatly extended by recent work with the 100-inch reflector at Mt. Wilson. From these results, he concludes that some two million nebulae lie within a distance of 140,000,000 light-years—a sphere of observation which bears the same ratio to the whole of space as the Isle of Man to the whole surface of the earth. Reference was then made to the conclusions of Friedmann and Lemaître that the equilibrium of such a universe would be unstable, and if expansion started it would continue. Sir James conjectures that the initial impulse which started the expansion may have arisen in the process of the condensation of the primeval chaotic gases into nebulae. Spectrograms of the distant galaxies indicate such an expansion, the rate of recession being about 105 miles per second at a distance of a million light-years, and increasing in the same proportion as the distance, so that it attains the amount of 15,000 miles per second for the most distant nebula yet measured.

Allusion was then made to the difficulty of reconciling this rapid recession with a past duration of the universe extending to millions of millions of years. Sir James has himself given strong reasons in favour of such a past duration, but he now admits that it may be necessary to abandon it. There are, however, some alternatives; there might have been a long period before the recession got fairly started; or the spectral shift that appears to indicate recession may be due to some other cause. Allusion was made to Sir Arthur Eddington's attempt to evaluate the cosmical constant, and so obtain a theoretical value for the rate of expansion; he obtained a value quite close to the observed rate. Sir James noted that this result, while intensely interesting as linking up the largest and the smallest objects of observation, is still a matter of controversy, and cannot be accepted as certain.

(**129**, 787-788; 1932)

膨胀的宇宙

编者按

这篇社论描述了詹姆斯·金斯最近就宇宙学的发展状况所作的演讲。越遥远的星系探测到的红移越大证明了宇宙在快速膨胀，另有证据表明宇宙已经存在了成千上万亿年，调和二者之间的关系是最令人费解的事。金斯提出，一种可能是：宇宙过去并不一直都像现在所观测到的那样膨胀，膨胀可能只是在宇宙形成之初的某段静止期后才开始的。金斯认为阿瑟·爱丁顿最近对宇宙大爆炸的理论计算只不过是一种推测。但正如爱丁顿所述，由广义相对论确实可以预言宇宙的膨胀。

詹姆斯·金斯爵士5月9日在曼彻斯特的路德维格·蒙德讲座上发表了题为“膨胀的宇宙”的演讲，他首先评述了星系系统，指出最近用威尔逊山上的100英寸反射式望远镜进行的观测工作极大地拓展了我们对星系系统的认识。根据这些结果他推断：在距离地球140,000,000光年范围内大约存在着200万个星云——这个观测范围相对于整个宇宙空间的比例相当于马恩岛相对于整个地球表面面积的比例。然后，他引用了弗里德曼和勒梅特的结论，即宇宙的平衡是不稳定的，膨胀一旦开始就会一直持续下去。詹姆斯爵士推测使宇宙开始膨胀的最初动力可能起源于原始混沌气体凝缩为星云的过程。遥远星系的光谱证实了这种膨胀的存在，100万光年处星系的退行速度约为105英里/秒，随着距离的增加，退行速度也会以相同的比例增加，目前能够观察到的最远的星云的退行速度可以达到15,000英里/秒。

于是就出现了这样一个困难：难道宇宙在过去的成千上万亿年中能一直保持如此快的退行速度吗？詹姆斯爵士本人曾经给出强有力的证据证明膨胀的长期存在，但他现在承认这种观点也许必须被放弃。然而，也存在着另外的可能：如在真正开始退行之前可能已经经过了很长一段时间；或者看似证明退行的谱线移动也许是由其他原因引起的。他提到阿瑟·爱丁顿爵士曾试图计算过宇宙常数，并由此得到了宇宙膨胀速率的理论值；爱丁顿的计算值与观测到的速率非常接近。而詹姆斯爵士认为：虽然这个结果能非常有趣地将观测中的最大物体和最小物体联系起来，但这一结果仍然是有争议的，不能作为确定的事实接受。

（史春晖 翻译；邓祖淦 审稿）

White Dwarf Stars

Editor's Note

This account of a lecture by English astronomer E. Arthur Milne shows that the superdense stars that now preoccupy many astronomers were already recognised in the 1930s. Indeed, Milne suggests that the first of these, a companion to Sirius, was inferred by the nineteenth-century German astronomer Friedrich Bessel, and observed in 1862. This "Sirius *B*" was later deduced to have a tremendous density, as well as being as source of dim but bluish-white light. It was the first so-called "white dwarf" star, whose physical composition could be explained using the new relativistic quantum theory. This blend of quantum, relativistic and stellar theories is now central to our understanding of dense cosmic bodies such as neutrons stars and black holes.

THE Halley Lecture delivered at the University of Oxford by Prof. E. A. Milne on May 19 was on the subject of the "White Dwarf Stars". He said that the discovery by Halley of the proper motion of some of the fixed stars led to a remarkable succession of researches in pure astronomy, in modern physics, and in cosmogony generally. The proper motion of one of Halley's stars, Sirius, was found by Bessel not to be uniform, but to contain a periodic element of about fifty years. This led him to suggest that Sirius was in reality double, consisting of a pair of stars, one much fainter than the other. In 1862 a faint star, Sirius *B*, was actually seen by Alvan Clark close to the place that had been theoretically assigned to the supposed companion. In 1915, W. S. Adams at Mount Wilson Observatory succeeded in obtaining a photograph of the spectrum of Sirius *B*, which led to the unexpected conclusion that the density of Sirius *B* was of the order of one ton to the cubic inch. It was shown by Eddington that this surprising density was not physically improbable, and further, that in the light of Einstein's general theory of relativity, the relative displacement of the lines of the spectrum of Sirius *A* and Sirius *B* could be estimated. The measurement when actually carried out by Adams in 1925 gave a result so near that of Eddington's estimate that the computed small radius and high density of Sirius *B* may now be accepted with confidence. A few other stars besides Sirius *B* are known in which low luminosity and abnormal blueness are combined with high density; these are known as "white dwarfs". They are all within five parsecs of the sun, but there is no reason to suppose that this is an abnormal region of space. Consideration of the phenomenon of nova-outbursts and the study of the nuclei of planetary nebulae lead to the conclusion that the list of dense objects can be largely extended. The physical state of matter at these high densities has been elucidated by R. H. Fowler in the light of the researches of Fermi and Dirac. The existence of white dwarf stars shows that it is possible for any gas to exist in either of two states or phases, the "perfect" or the "degenerate" phase; the dense state being identified with that of the second phase of a gas. It is suggested that, as foreshadowed by Bessel in regard to Tycho's nova of 1572, the system of Sirius may owe its origin to the nova phenomenon of the original Sirius; two companions

白矮星

编者按

这篇文章通过报道英国天文学家阿瑟·米耳恩的一次演讲说明现在许多天文学家潜心研究的超密恒星早在20世纪30年代就已经被发现了。米耳恩指出：第一个超密恒星是天狼星的一颗伴星，19世纪德国天文学家弗里德里希·贝塞尔推出了它的存在，这颗星于1862年被发现。人们后来发现“天狼B星”具有很高的密度，它是一颗暗淡的星，发出蓝白色的光。这颗星是最先被人们称为“白矮星”的恒星，其物理组成可以用相对论性量子理论来解释。现在这种融合量子力学、相对论和恒星理论的新理论在我们认识和理解像中子星和黑洞这样的致密宇宙天体时是必不可少的。

5月19日，米耳恩教授在牛津大学的哈雷讲座上发表了关于“白矮星”的演讲。他说，自哈雷发现了一些恒星的自行现象以后，人们便开始在纯天文学、现代物理学以及天体演化学方面展开一系列重大的后续研究。天狼星是其中一颗哈雷认为有自行的恒星，贝塞尔发现天狼星的自行并不是均匀的，但包含一个大约50年的周期。据此，贝塞尔提出，天狼星实际上是双星，即由一对恒星组成，其中一颗比另一颗暗很多。1862年，阿尔万·克拉克在理论给出的伴星位置附近发现了一颗暗淡的恒星，即天狼*B*星。1915年，亚当斯在威尔逊山天文台成功地得到了天狼*B*星的光谱图，由这张图推出了一个意外的结果：天狼*B*星的密度约为1吨/立方英寸。爱丁顿认为，自然界存在这么高密度的天体也不是不可能，而且由爱因斯坦的广义相对论可以估算出在天狼*A*星和*B*星光谱图中的谱线相对位移。1925年，亚当斯进行了相关的测量，测量结果非常接近于爱丁顿的估算值，以至于人们开始相信天狼*B*星真的具有计算得到的半径小和密度高的特点。天狼*B*星和其他一些恒星在具有较低光度和显示出反常蓝色的同时，也具有很高的密度；这些恒星被称为“白矮星”。它们与太阳的距离都在5秒差距以内，但不能据此就认为这一区域是宇宙中的反常区域。根据对新星爆发现象和行星状星云核的研究，我们可以认为宇宙中可能还存在着很多像这样的致密天体。福勒在费米和狄拉克研究的基础上阐明了物质在如此高密度下的物理状态。白矮星的存在表明，任何气体都可能以两种状态或相的形式存在，即“理想态”和“简并态”；致密状态就属于气体的第二种状态。这说明，正如贝塞尔在第谷1572年发现了新星之后所作的预言那样，天狼星系统可能会起源于原始天狼星的新星现象，从而产生了两颗伴星，其中一颗不断膨胀，而另一颗仍然保持致密。贝塞尔认为这些与人类探索宇宙物质组成相关的现象将会受到人们的关注，这一预

resulting, of which one re-expanded and the other remained dense. Bessel's anticipation of the interest of these phenomena in relation to our knowledge of the physical constitution of the universe has been amply justified by the course of events.

(**129**, 803; 1932)

言已经被事实充分地印证了。

（金世超 翻译；蒋世仰 审稿）

Internal Temperature of Stars

G. Gamow and L. Landau

Editor's Note

In the early 1930s astronomers were still at a loss to understand the high internal temperatures of stars, since they had yet to comprehend fully the process of nuclear fusion. George Gamow was a Russian physicist who emigrated to the United States in the late 1930s and there first articulated the "Big Bang" theory of how the universe began. Here he and his compatriot Lev Landau, one of the leading scientists of the Soviet era, try to constrain internal stellar temperatures by observing that lithium should be consumed by nuclear reaction with hydrogen before it can reach the star's surface, if the temperature inside exceeds several millions of degrees.

IT may be of interest to notice that the investigation of the process of thermal transformation of light elements in stars[1] enables us to check the upper limit for the temperature of internal regions. In fact, so far as lithium is present, for example, on the star surface, it is natural to accept that it is in equilibrium with the lithium content in the internal regions of the star near the stellar nucleus, where the production of different elements takes place. On its way from the stellar nucleus through the hot regions of the star, lithium atoms will be partly destroyed by thermal collisions with hydrogen atoms ($Li^7 + H^1 \rightarrow 2He^4$) and will not reach the surface at all if the temperature of the internal regions is too high.

For the rate of the reaction in question we have:

$$\omega \sim \int \pi \left(\frac{h}{mv}\right)^2 \cdot e^{-2\pi Ze^2/hv} \cdot v \cdot N \cdot 4\pi v^2 \left(\frac{m}{2\pi kT}\right)^{3/2} \cdot e^{-mv^2/2kT}\, dv \tag{1}$$

where Z is the atomic number of the element, v and N the velocity and density of protons, and T the absolute temperature.

Calculating the integral we obtain:

$$\omega \sim Nh^{5/3} \frac{(4\pi Ze^2)^{1/3}}{m^{4/3}(kT)^{2/3}}\, e^{-3/2(m/kT)^{1/3}(2\pi Ze^2/h)^{2/3}} \tag{1^1}$$

On the other hand, in the time $1/\omega$ a lithium atom will travel through the distance

$$l \sim \sqrt{D/\omega} \tag{2}$$

where the diffusion coefficient D is given by the expression:

$$D \sim (kT)^{1/2}/N^1 \sigma M^{1/2} m^{1/2} \tag{2^1}$$

恒星的内部温度

乔治·伽莫夫，列夫·朗道

编者按

20 世纪 30 年代早期的天文学家们仍然不能理解恒星内部的温度为什么会那么高，因为他们那时还没有完全了解核聚变反应。乔治·伽莫夫本是一位俄罗斯的物理学家，他于30年代后期移居美国，并在美国首先明确地阐明了关于宇宙起源的“大爆炸”理论。在这篇文章中，伽莫夫和他的同胞，苏联时代最有名的科学家之一列夫·朗道一起提出：如果恒星的内部温度超过了几百万度，那么锂就应该因不断与氢发生核反应而在到达恒星表面前被完全消耗掉，所以恒星的温度不可能有那么高。

我们注意到一个很有意思的事情：对恒星内部轻元素热转变过程的研究[1]使我们能够验证恒星内部区域的温度上限。事实上，如果以恒星表面处的锂为例，我们会很自然地认为它的含量应该与星核附近的锂含量平衡，我们知道星核处是生成各种不同元素的地方。锂原子在从星核处穿越恒星热区的过程中会因同氢原子发生热碰撞（$Li^7+H^1 \rightarrow 2He^4$）而被部分摧毁，如果内部区域温度太高的话，锂原子将无法到达恒星表面。

上述过程的反应速率为：

$$\omega \sim \int \pi\left(\frac{h}{mv}\right)^2 \cdot e^{-2\pi Ze^2/hv} \cdot v \cdot N \cdot 4\pi v^2\left(\frac{m}{2\pi kT}\right)^{3/2} \cdot e^{-mv^2/2kT}\,dv \tag{1}$$

这里 Z 是元素的原子数，v 是速度，N 是质子密度，T 是绝对温度。

将上式积分后我们得到：

$$\omega \sim Nh^{5/3}\,\frac{(4\pi Ze^2)^{1/3}}{m^{4/3}(kT)^{2/3}}\,e^{-3/2(m/kT)^{1/3}(2\pi Ze^2/h)^{2/3}} \tag{1'}$$

另一方面，在 $1/\omega$ 的时间里，一个锂原子将穿过的距离是：

$$l \sim \sqrt{D/\omega} \tag{2}$$

其中的扩散系数 D 可以表示为：

$$D \sim (kT)^{1/2}/N^1\sigma M^{1/2}m^{1/2} \tag{2'}$$

Here N^1 is the total number of atoms in a cubic centimeter, σ the cross-section of collision, and M the atomic weight of the atoms in question.

Using (1^1) and (2^1), we obtain from (2):

$$l \sim \frac{m^{5/12}(kT)^{7/12}}{\sqrt{NN^1}\,\sigma^{1/2}h^{5/6}M^{1/4}(4\pi Ze^2)^{1/6}}\,e^{3/4(m/kT)^{1/3}(2\pi Ze^2/h)^{2/3}} \quad (3)$$

Accepting $N \sim N^1 \sim 10^{24}$ and $\sigma \sim 10^{-18}$ cm.^{2}, we obtain for lithium (Z = 3) the following numbers:

T (C°)	10^6	5×10^6	10^7	5×10^7	10^8
l (cm.)	10^{13}	10^6	10^4	10	1

From this table the conclusion is reached that either lithium is present on the star surface only occasionally or that no regions with temperatures of more than several millions of degrees can exist in the interior of a star.

(**132**, 567; 1933)

G. Gamow and L. Landau: Ksoochia Basa, Khibini, Aug. 10.

Reference:

1. Atkinson and Houtermans, *Z. Phys.*, 54, 656 (1929).

这里 N^1 为 1 立方厘米体积内的总原子数，σ 是碰撞的横截面积，M 是与该过程相关的原子的原子量。

把 (1^1) 和 (2^1) 代入 (2)，我们得到：

$$l \sim \frac{m^{5/12}(kT)^{7/12}}{\sqrt{NN^1}\,\sigma^{1/2}h^{5/6}M^{1/4}(4\pi Ze^2)^{1/6}}\,e^{3/4(m/kT)^{1/3}(2\pi Ze^2/h)^{2/3}} \tag{3}$$

考虑到 $N \sim N^1 \sim 10^{24}$ 和 $\sigma \sim 10^{-18}$ cm^2，对于锂 ($Z = 3$) 我们可以得到下面的数字：

T (℃)	10^6	5×10^6	10^7	5×10^7	10^8
l (cm)	10^{13}	10^6	10^4	10	1

根据这个表我们可以得出以下结论：要么锂元素在恒星表面的存在量很少，要么在一个恒星内部不可能存在温度超过几百万度的高温区。

（魏韧 翻译；蒋世仰 审稿）

Atomic Transmutation and the Temperatures of Stars

A. S. Eddington

Editor's Note

Here English astronomer Arthur Eddington challenges a recent claim in *Nature*, by Soviet physicists George Gamow and Lev Landau, that the internal temperature of stars can be gauged by whether or not lithium can be detected at their surfaces. Gamow and Landau argued that if the temperature exceeds several million degrees, lithium would be destroyed by reaction with hydrogen before it could reach the surface. Not so, says Eddington, because the lithium would be carried not by slow diffusion but by more rapid circulation currents induced by the star's rotation. That would make its presence consistent with the temperatures (up to 10–20 million degrees) that astronomers had estimated.

THE letter of Gamow and Landau[1] suggests that an upper limit to the internal temperature of a star can be obtained by considering the disintegration of lithium. Investigations of this kind will probably be of great importance in the future development of astrophysics, but the actual proposal of Gamow and Landau rests on an assumption which is scarcely likely to be true. They postulate that any lithium found at the surface must have been carried there by diffusion from the central region, where it is presumed to have been created. Diffusion in a star is an exceedingly slow process, the time of relaxation being of the order 10^{13} years[2]. It would make small progress during the maximum age of the giant stars. But there is a process of mixing which is likely to operate much faster, namely, the circulating currents in meridian planes indirectly caused by the rotation of the star. The order of magnitude is indicated in an example treated by the writer in which the speed of the vertical current was found to be 60 metres a year[3]. The example was chosen with the view of giving an upper limit to the amount of this circulation; but, allowing for slower currents in an average star, the lithium will be brought to the surface far more quickly in this way than by diffusion.

It is difficult to see how any consistent theory of distribution could be given if diffusion alone were operating. If there is time for lithium produced at the centre to reach the surface, there is time for the heavy elements to have disappeared from the surface by downward diffusion; or if it is supposed that they, like lithium, were created at the centre, there is no mechanism by which they could ever reach the surface. In the steady distribution towards which diffusion is slowly tending, there should not be a single atom of lead in the outer half of the volume of the star.

The existence of these circulating currents will raise the upper limits given by Gamow and Landau. Since the disintegration is sensitive to changes of temperature, the increase may

原子嬗变和恒星温度

阿瑟·爱丁顿

编者按

在这篇文章中，英国天文学家阿瑟·爱丁顿对苏联物理学家乔治·伽莫夫和列夫·朗道最近发表在《自然》杂志中的观点提出质疑，他们认为在恒星表面是否能检测到锂元素可以作为判断其内部温度高低的标准。伽莫夫和朗道声称：如果温度超过了几百万度，锂在到达恒星表面之前就会因与氢发生反应而被完全破坏。爱丁顿说，并非如此，因为锂不是被缓慢的扩散过程而是被更快的由恒星自转产生的环流所带动的。这样锂在恒星表面的存在与天文学家推算出的恒星内部温度（高达10~20百万度）之间就不再矛盾了。

伽莫夫和朗道在快讯[1]中提出：恒星内部温度的上限可以通过对锂元素衰变的考虑得到。这一研究对天体物理学未来的发展很可能是至关重要的，但伽莫夫和朗道就这一问题的提议是建立在一个几乎不可能成立的假设基础上的。他们假设在恒星表面发现的所有锂元素都是在中心区域产生并通过扩散而来的，而实际上锂有可能是在恒星表面产生。恒星内部的扩散过程非常缓慢，其弛豫时间的数量级可达10^{13}年[2]。仅在年龄最大的巨星中会出现少许快一点的扩散，但是还可能有一个过程会使混合过程速度加快，那就是间接由恒星自转引起的在子午面上的环流。作者通过对一个实例的分析得出垂直流的速度约为60米/年[3]。选择这个例子的目的在于由它可以得出环流效应的上限；但就算在大多数恒星中环流速度会慢一些，锂元素以这种方式到达表面仍要远快于通过扩散的方式。

如果只有扩散这一个过程在起作用，我们就很难给出一个自洽的分布理论。如果在恒星中心产生的锂有足够的时间扩散到恒星表面，那么所有的重元素就会通过向下的扩散过程从表面消失；如果假设重元素像锂一样产生于恒星的中心，那么我们将无法说明它们是如何到达恒星表面的。如果认为扩散是一个缓慢进行的稳态分布过程，则在恒星的靠外的一半体积内一个铅原子也不可能找到。

环流效应的存在将会提高伽莫夫和朗道给出的恒星内部温度的上限。由于衰变过程对温度变化非常敏感，所以增加值不会太大；但这也许足以使我们安然接受通

not be very large; but it may well be sufficient to remove any difficulty in accepting the temperatures of the order 10^7–2×10^7 found by astronomical methods, whilst negativing any suggestion of considerably higher temperatures.

(**132**, 639; 1933)

A. S. Eddington: Observatory, Cambridge, Oct. 9.

References:

1. *Nature*, **132**, 567 (Oct. 7, 1933).
2. *Internal Constitution of the Stars*, §§195—196.
3. *Monthly Notices R.A.S.*, **90**, 54 (1929).

过天文方法得到的恒星内部温度的数量级 10^7~2×10^7，同时也否定了认为恒星内部温度显著高于这个值的提议。

（魏韧 翻译；邓祖淦 审稿）

Atomic Transmutation and Stellar Temperatures

T. E. Sterne

Editor's Note

George Gamow and Lev Landau in the Soviet Union had recently argued that the astronomical observations suggesting stars have internal temperatures of tens of millions of degrees should be inconsistent with the presence of lithium on their surfaces, since this should be burnt up in the interior in nuclear reactions with hydrogen. Arthur Eddington then wrote to *Nature* to dispute the idea. Here Theodore Eugene Sterne of Harvard College Observatory weighs in, saying that Eddington's objection invokes conditions under which stars would be unstable, but that lithium might actually be formed as well as destroyed en route to the stellar surface, if temperatures were even higher than Eddington supposed.

GAMOW and Landau[1] suggest either that lithium of mass 7 can be present only occasionally on a star's surface, or that no regions with temperatures of more than several millions of degrees can exist in the interior of a star; their argument is that at higher temperatures lithium could not find its way by diffusion from "the internal regions of the star, where the production of different elements takes place" to the surface, before being disintegrated. Eddington[2] has replied by noticing that the presence of ascending currents may decrease the time required for the ascent of the lithium, so as to remove the difficulty in accepting the central temperatures of the order of 2×10^7 found for his models, "whilst negativing any suggestion of considerably higher temperatures".

Most will agree with Eddington that temperatures of only some few millions of degrees are too low for the liberation of sufficient energy; his central temperatures are about the lowest which will yield the correct rates of liberation. But it appears from our present knowledge of disintegrations that an Eddington star would be violently over-stable, unless there were some other important source of energy than transmutations. Stellar matter at his central temperatures would behave not merely like gunpowder, but like gunpowder at just so high a temperature as to be deteriorating steadily, with any decrease in temperature stopping its liberation of energy, and any increase causing it to explode! Eddington shows[3] that one of his stars will be over-stable if the rate ε of liberation of energy increases more rapidly than about T^3, unless there is a delay of the order of months or years between an increase in T and the resulting change in ε. The most important contribution to the total energy liberated by transmutations comes from the disappearance of hydrogen because of its large packing fraction, and the rate at which the speed of disappearance of hydrogen increases with the energy of the collisions can be calculated by Gamow's theory[4] of the nucleus. Except for a constant factor, the calculated speeds appear to be in satisfactory agreement with the observed speeds[5]; the factor does not particularly concern us because we are interested in the exponent, s say, of T for that temperature at which ε

原子嬗变和恒星温度

西奥多·尤金·斯特恩

编者按

苏联的乔治·伽莫夫和列夫·朗道最近提出：从天文观测推断恒星的内部温度高达几千万度，这与恒星表面存在锂元素的事实相矛盾，因为锂在恒星内部会与氢发生核反应从而被消耗掉。阿瑟·爱丁顿在《自然》杂志上撰文反对这个观点。在这里，哈佛大学天文台的西奥多·尤金·斯特恩也加入了讨论，他说爱丁顿的反对意见是以恒星处于不稳态为条件的，不过当温度高于爱丁顿所设想的温度时，也许可以认为锂在向恒星表面扩散的途中，损耗和生成的过程在同步进行着。

伽莫夫和朗道[1]认为，或者质量数为7的锂元素只能偶尔在恒星表面存在，或者恒星内部根本不存在温度超过几百万度的区域。他们的理由是当温度更高时，锂在衰变之前恐怕无法从“合成各种元素的恒星内部”扩散到恒星表面。爱丁顿[2]回复道：由恒星自转引起的上升流或许会减少锂元素在上升过程中所用的时间，所以可以排除接受他模型中认为恒星中心温度为 2×10^7 量级的困难，“同时也否定了认为恒星内部温度显著高于这个值的提议”。

大多数人都认同爱丁顿的观点，即认为几百万度的温度对于恒星释放充足的能量来说是太低了；爱丁顿模型的中心区温度大约是能保证恒星以适当速率释放能量的最低温度。但根据我们目前对衰变过程的认识，在除嬗变外没有其他重要能量来源的情况下，爱丁顿模型中的恒星将是极端超稳定的。恒星物质在他假定的中心温度下表现得不仅像火药，而且还是在这样的高温下会不断质变的火药，如果温度稍有下降，它就会停止能量的释放，而如果温度略有提高它将发生爆炸！爱丁顿证实[3]，如果能量释放速度 ε 比 T^3 增加得更快，那么他的恒星中的一个将会是超稳定的，除非在 T 的增加和由此引起的 ε 的改变之间有数月或数年的延迟。在嬗变过程释放的总能量中有很大一部分来自氢元素的消耗，因为氢元素所占的比例较大，氢元素消失的速率随碰撞能量的增加而增大，这一速率可以根据伽莫夫的核理论[4]计算出来。计算出来的速率在乘上一个常数因子后似乎同观测到的结果吻合得很好[5]；这个因子和我们的关系不大，因为我们感兴趣的是 T 的指数因子 s，在某一特定温度下我们要考虑 s 取什么样的值才能使 ε 的数量级与 L/M 相符。在温度为 T 时，

is of the right order of magnitude to agree with L/M. One can calculate s, considering the statistical distribution over all energies of collision at a temperature T, and it is found[6] that s lies between 9 and 30. There is no delay, and an Eddington star with ε varying like T^{15} would be violently over-stable. These figures refer to the disintegration of lithium; s is increased, and matters are made considerably worse, if elements other than lithium are being disintegrated. The possibility that there may be another important source of sub-atomic energy, "annihilation", cannot be disproved, but there is not the least experimental evidence for the occurrence of any kind of annihilation that could supply useful energy to a star. The creation and disappearance of positive electrons would serve merely to increase the specific heat of the material, while at Eddington's temperatures even this increase would probably be trivial.

It is difficult to see how more than traces of elements like lithium could be formed at temperatures no higher than Eddington's, but if the temperatures are considerably higher than his, then the lithium can be made[7] as well as disintegrated, and by the aid of ascending currents some of it could perhaps appear on the surface. Since it would not be subjected to disintegration *alone* throughout the trip to the surface, for a time the abundance might even increase. If elements are being made as well as being disintegrated, the difficulty of over-stability is avoided, for there is no longer an ε which increases rapidly with T, but merely an ε which depends upon the rate of loss of energy by radiation into space.

There is still another way out of the difficulty raised by Gamow and Landau, and that is that lithium may have been present from the beginning in the star's atmosphere, while diffusion and currents may not yet have carried all of it into the far interior where transmutations occur. This is consistent with Eddington's calculations[8], for the vertical current of 60 metres a year which he found was an upper limit which applied to the neighbourhood of the surface only; at a place where ε and the mean value of ε interior to this are nearly equal (as presumably they are in regions where transmutations occur frequently) the vertical velocity by Eddington's calculations is considerably less. In this case the internal temperatures could well be as high as Eddington's, or higher. The considerations of over-stability suggest the higher temperatures.

(**132**, 893; 1933)

T. E. Sterne: Harvard College Observatory, Cambridge, Mass., Nov. 6.

References:

1. *Nature*, **132**, 567 (Oct. 7, 1933).

2. *Nature*, **132**, 639 (Oct. 21, 1933).

3. *Internal Constitution of the Stars*, § 136.

4. *Z. Phys.*, **52**, 510 (1928).

5. Lawrence, *et al.*, *Phys. Rev.*, **42**, 150 (1932); Henderson, *Phys. Rev.*, **43**, 98 (1933).

6. A paper of the author's on this and allied topics is published in the *Mon. Not. R.A.S.*, **93**, No. 9 (Oct. 1933).

7. A paper of the author's on the equilibrium of transmutations is in the *Mon. Not. R.A.S.*, **93**, No. 9 (Oct. 1933).

8. *Mon. Not. R.A.S.*, **90**, 54 (1929).

我们可以通过所有碰撞能量的统计分布计算出 s 值，结果发现这个指数因子的范围在 9 到 30 之间 [6]。如果没有延迟，ε 按照 T^{15} 变化的爱丁顿恒星将会是极端超稳定的。这些数字来源于锂元素的衰变；如果除锂之外还有其他元素发生衰变，s 会增加，于是问题将变得更糟。我们不能排除可能存在另外一种重要的亚原子源，"湮灭"，但是现在还没有丝毫实验上的证据可以证明某种形式的湮灭能为恒星提供可用的能量。正电子的产生和湮灭只会提高物质的比热，而在爱丁顿所说的温度下即使是这样的增长也是微不足道的。

很难想象在不高于爱丁顿温度的情况下，恒星还能生成少量像锂这样的元素，但如果温度显著地高于爱丁顿的温度，则锂可能在产生 [7] 的同时也在分解，而且在上升流的作用下，其中一部分锂也许能到达恒星表面。由于锂在朝恒星表面运动的过程中不会**只**在分解，在一段时间内锂的丰度可能还会增加。如果元素在产生的同时也在衰变，那么超稳定性的问题就可以迎刃而解了，因为 ε 不再随 T 的升高而迅速增长，它只取决于向太空辐射造成的能量损失率。

还有一种方法可以解决伽莫夫和朗道提出的难题，锂可能从一开始就存在于恒星的大气中，而扩散和气流也许并没有把所有的锂元素都带到恒星很深的内部，也就是嬗变过程发生的地方。这同爱丁顿的计算结果是一致的 [8]，他所说的 60 米 / 年的垂直流只不过是一个上限，仅仅适用于恒星的表面附近；在 ε 和 ε 的内部平均值大致相等的地方（这个区域也被认为是嬗变过程频繁发生的区域），由爱丁顿计算出的垂直速度大大低于 60 米 / 年 [8]。如果事实果真如此，则恒星的内部温度可以等于或高于爱丁顿的温度。而超稳定性的存在需要有更高的温度。

（魏韧 翻译；邓祖淦 审稿）

Planetary Photography*

V. M. Slipher

Editor's Note

Vesto Slipher was director of the Lowell Observatory in Flagstaff, Arizona, when he wrote this review of planetary imaging carried out there. The observatory was founded by American astronomer Percival Lowell in 1894, and in subsequent decades it offered some of the clearest direct views of the planets. Slipher's description of Mars, thought to have a substantial atmosphere of oxygen and water vapour, makes it clear why many regarded the seasonal colour changes as being due to vegetation. Lowell himself believed there were even signs of intelligent life. Slipher has been under-rated as an astronomer, having understood the recession-induced redshift of galaxies before Edwin Hubble, and overseeing the observations that led to the discovery of Pluto in 1930.

THE Lowell Observatory was founded in 1894, by the late Percival Lowell, who maintained and directed it during his lifetime and endowed it by his will, that it might permanently continue astronomical research and in particular that of the planets. For nearly four decades now, it has been occupied with planetary investigations. It is situated at Flagstaff, Arizona, because, of the numerous places he had tested, it was here that Lowell found the conditions best for planetary studies. The major instruments of the Observatory are: (1) 24-inch aperture Clark refractor of 32 feet focus, (2) 42-inch Clark reflecting telescope, (3) a new 13-inch photographic telescope, (4) 15-inch Petitdidier reflector, and in addition several smaller instruments, together with a number of spectrographs, special cameras for photographing the planets, radiometric apparatus for use with the 42-inch reflector, for measuring the heat of the planets, and such laboratory equipment as is needed in the work carried on.

During the first decade, the work at the Observatory was mainly visual observations of the planets, then it was extended to include their spectrographic study, and during the second decade direct photography of the planets was added and has been continued since, giving a permanent record of them to the present time. During the past decade, their heat measurement has also been made a regular part of the observational programme. In short, whenever it has been possible to apply new means, they have been made use of in order that the planets be studied from every possible point of view.

During the early years of the Observatory, Lowell was able to observe Mercury and to confirm Schiaparelli's conclusion that the planet constantly keeps its same face to the

* From a discourse entitled "Planet Studies at the Lowell observatory", delivered at the Royal Institution on Friday, May 19.

行星的照相分析*

斯里弗

编者按

当维斯托·斯里弗写下这篇关于洛威尔天文台行星照相的综述时，他正是这个天文台的台长。该天文台位于亚利桑那州弗拉格斯塔夫市，于1894年由美国天文学家珀西瓦尔·洛威尔创建，并在之后的几十年中提供了一些对于行星最清晰的直接观测结果。斯里弗认为火星的大气中含有氧气和水蒸气，他对于火星的描述明确地解释了火星上为何会有如植被存在而出现的季节性颜色变化。洛威尔本人甚至认为那里有智慧生物存在的迹象。作为一个天文学家，斯里弗是超前的，因为他先于埃德温·哈勃理解了退行运动导致的星系光谱红移，并指导了促使1930年发现冥王星的那些观测。

洛威尔天文台由已故的珀西瓦尔·洛威尔于1894年创建，洛威尔先生生前一直领导并维护该天文台，后来根据他的遗嘱该天文台被捐赠出来以用于永久性地开展天文学特别是行星方面的研究。在过去将近四十年间，该天文台一直致力于行星方面的研究。它位于亚利桑那州的弗拉格斯塔夫市，因为在洛威尔先生曾经勘查过的众多地方中，该地是最适于进行行星研究的。该天文台的主要设备包括：(1) 24英寸口径、32英尺焦距的克拉克折射式望远镜，(2) 42英寸口径克拉克反射式望远镜，(3) 一架新型的13英寸照相望远镜，(4) 15英寸珀蒂迪迪埃反射式望远镜，此外还包括一些小型仪器，如摄谱仪、用于拍摄行星的特殊相机、用于测量恒星热量的辐射测量仪（与42英寸折射式望远镜一起使用），以及其他一些工作所需的实验室设备。

在最初的十年间，洛威尔天文台的主要工作是目视观测行星，后来扩展到光谱研究。在第二个十年间，开始对行星进行直接照相并且一直持续到现在，对这些年的行星活动进行了持续的记录。在最近的十年间，热辐射测量也已成为该天文台进行观测的例行程序之一。简而言之，无论何时只要有可能使用新方法，这里的天文学家就会加以利用，以便从每个可能的角度去研究行星。

在该天文台成立初期，洛威尔对水星进行了观测，并证实了斯基亚帕雷利的结论，即水星总是保持同一个面朝向太阳，如同月亮总是同一个面朝向地球一样。因此，

* 基于5月19日星期五在英国皇家科学研究院发表的一篇演讲，题为“在洛威尔天文台的行星研究”。

Sun, as our Moon does to the Earth. Thus its small mass and the intense heating by the Sun long since dissipated its atmosphere. Venus proved more difficult, and with very faint surface markings, its length of day was left somewhat uncertain, while from all considerations it appeared that this planet also keeps the same face constantly toward the Sun, for even the spectrograph showed no evidence of a day shorter than a few weeks. Spectral studies of Venus have failed to give any evidence of an Earth-like atmosphere, no bands of oxygen or water being found, although it might have been expected that Venus would be the planet most like the Earth.

From this non-committal and veiled planet we pass to the best observed of all, Mars, which has long attracted wide interest. Martian seasonal change shows itself clearly in the polar caps, which alternately increase and decrease, and in the blue-green markings which darken in the growing season and pale again as winter approaches, the great ochreish expanses, changing little from winter to summer, except as influenced by light spots and clouds. The shrinking of the polar cap with summer's coming is to be seen in Fig. 1, where are shown five photographs of the same face of the planet showing particularly the upper hemisphere, but made at Martian seasonal dates. With the contraction of the cap the shaded areas darken and enlarge, as may readily be seen in the photographs.

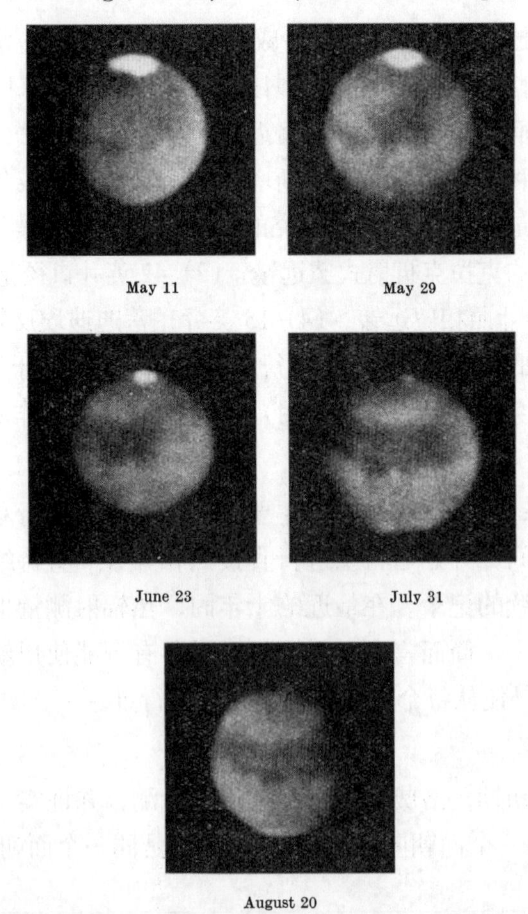

Fig. 1. Photographs of Mars showing the shrinking of the polar cap and the growth of dark areas.

由于水星自身较小的质量以及太阳长期对其强烈的加热，其大气已经消失殆尽。由于表面特征模糊，金星上每日的长度存在某种程度的不确定性。要证明金星始终是同一个面朝向地球是很困难的。但是综合各方面来看，金星还是很可能保持同一个面朝向太阳的，因为光谱研究也没有证据表明金星上的一天会短于地球上数周。尽管人们曾认为金星是最接近地球的行星，但在金星的光谱研究中，我们并没有发现其大气与地球大气类似的证据，它的大气光谱中并不含有氧或水的谱带。

让我们把目光从这个不合作且隐藏起来的星球上移开，下面看一下目前为止获得了最好的观测的行星——火星，天文学家一直对火星观测有着浓厚的兴趣。火星的季节性变化可以清楚地反映在极冠的交替消长变化中；也会反映在蓝绿斑的变化中，当生长季节到来时，火星上的蓝绿斑会变深，而当冬季到来时其又会变淡；而火星上赭红色的宽阔区域除非受到光线和云层的影响，否则从冬季到夏季几乎没有什么变化。夏季来临时极冠收缩的情况如图 1 所示，图中五幅照片显示的是火星的同一个面，以上半球为主，其中标出的时间为火星日期。随着极冠的收缩，阴暗区域变暗并且面积扩大，这点在图中非常明显。

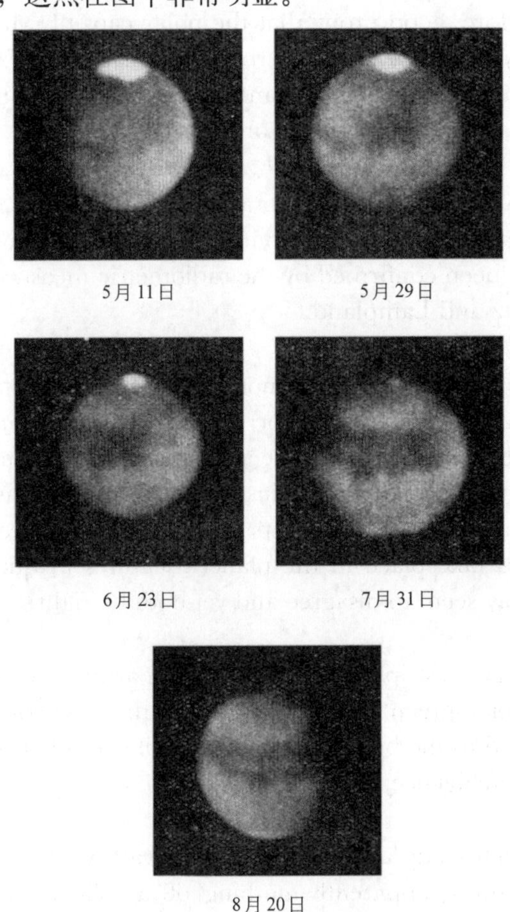

图 1. 火星表面的照片显示了极冠的收缩和阴暗区域的增加

Dark rifts appear in the melting caps, always at precisely the same time and the same places each Martian year, which clearly prove the caps to be deposits on the planet's surface. Irregularities of the surface must cause this patchy melting of the caps to be repeated always with most punctual harmony to the Martian calendar. Such features of the melting caps are to be seen in Fig. 1. The melting cap is bordered by a dark collar, and is more disposed to be regular in outline than the forming cap, which is irregular in outline and indefinite, and to begin with is erratic storm clouds only. An autumn cap appears at the opposite side of the planet to the polar cap.

The behaviour of the caps means that Mars has an atmosphere, for that is the only vehicle which does such transportation of substance. Occasionally, when Mars is so placed that we look a little into its night sky, we see on it a bright streak of light due to a cloud high in its atmosphere, catching the sunlight, while the surface is dark beneath it. Such allow us to measure their height above the Martian surface, and a fine one in 1903 was fully 15 miles high, whereas clouds are rarely more than 5 miles above the Earth. Hence Mars must have quite a considerable atmosphere, and the spectrograph at Flagstaff showed it to contain water and oxygen, but no strange substances. Thus it closely resembles that of the Earth, but is less dense, because the Martian surface gravity is only three-eighths of ours. There is, therefore, good proof that the polar caps of Mars are snow. Long ago someone suggested they might be frozen carbonic acid gas, but Faraday himself showed experimentally the conditions of pressure and temperature required to solidify this gas; conditions which we are sure cannot prevail on Mars.

Lowell, some years ago, deduced the temperature of Mars from a full evaluation of the factors involved, such as albedo, the behaviour of the caps, etc., and arrived at a value of 48 °F. This has recently been confirmed by the radiometric measurements made at Lowell Observatory by Coblentz and Lampland.

While there is room for difference of opinion as to the interpretation of the canals of Mars, their existence as true markings on the planet has been clearly established, for they have been photographed and have been seen by nearly all skilled observers who have observed the planet carefully with powerful instruments. The Lick astronomers Schaeberle, Campbell and Hussey of the early observers, and Trumpler more recently, all drew the canals. Because changes take place in the planet's features in quite short time intervals sometimes, observers may seem to disagree and yet both be right.

Lowell regarded the canals as strips of vegetation along artificially produced water courses, for they, like the larger blue-green areas, darken when the time comes for seasonal growth in vegetation; and this led to the belief that vegetable life, and hence also probably animal life of some degree of intelligence, exist on Mars.

Jupiter has received much study at the Lowell Observatory. What we see on Jupiter are mostly atmospheric features, apparently nothing of a solid surface appearing. Usually

在每个火星年中，火星上极冠融化时出现的黑色裂缝总是准确出现在相同的时间和相同的地方，这就明确地证明了极冠的确是存在于火星表面的。火星表面的不规则必然引起极冠不均匀的融化，并且按照火星日历每年准时地重复。这些极冠融化的特征可以参见图1。消融中的极冠边界上嵌有黑边，并且相对于形成中的极冠它的轮廓要更规则，而形成中的极冠轮廓既不规则也不清楚，且总是起源于无规律的风暴云。秋季的极冠出现在与这个极冠位置相对的火星的另一端。

极冠的变化特征表明火星存在着一种大气，因为大气是物质输运的唯一途径。偶尔，当火星的位置合适到我们可以看到一点火星上的夜空时，我们可以发现一条明亮的光带，这是由于火星高层大气中的云受太阳光照射形成的，而云层下面的火星表面则是暗黑的。我们可以据此测量出火星云距离火星表面的高度，1903年的一个测量结果显示这些云的高度有15英里，而地球上云的高度很少超过5英里。因此，火星的大气应该是相当可观的，同时在弗拉格斯塔夫进行的光谱观测表明火星大气中含有水和氧气，但没有什么特别的物质。因此，火星的大气与地球大气非常的相似，但后者较为稠密，因为火星表面的引力只是地球的八分之三。也有很好的证据表明，火星极冠是由雪构成的。很久以前，有人认为极冠可能是冰冻的碳酸气体，但是法拉第用实验给出了这种气体固化时所需的温度和压力条件，而我们确信这种条件在火星上并不普遍存在。

若干年前，洛威尔通过全面测定星体反照率以及极冠的变化特征等各种影响因素，从而对火星的温度进行了推算，得出的温度数值为48°F。最近洛威尔天文台的科布伦茨和兰普朗德通过辐射测量确认了这一温度值。

尽管对于火星运河的解释存在着不同的见解，但作为火星的特征，它们的存在已得到确认，因为这些运河已经被拍成照片而且几乎所有曾经通过强大的仪器对火星进行过仔细观察的训练有素的天文观测者都曾经看到过它们。利克天文台早期的观测者舍贝勒、坎贝尔、赫西，以及最近的特朗普勒，都曾绘制过这些运河。但是由于这些火星的特征有时可能会在很短的时间内发生变化，所以不同天文学家的观测结果看起来不太一致，但他们都是正确的。

洛威尔认为这些运河是沿人工开凿的水渠生长的植被带，因为当植物生长季节来临时它们会变深，这和火星上较大片的蓝绿色区域的变化一致。这就使人相信，火星上存在着植物，同时也很可能存在具有某种程度智慧的动物。

洛威尔天文台对木星也进行了大量的研究。我们所看到的木星主要是其大气的特征，几乎看不到其固态表面上的任何特征。由于木星的转动速度很快，在观测条

so much detail is present that the visual observer, owing to the planet's rapid rotation, has difficulty in recording properly in drawings and notes all he is able to see under good observing conditions. In these circumstances the aid of photography has been very important, and a photographic record of the planet, as complete as possible, has been kept at Flagstaff since 1905. Fig. 2 indicates the nature of the Jupiter markings and gives some idea of their rapid and sometimes extensive changes, which give some hint of the very great activity present on the planet.

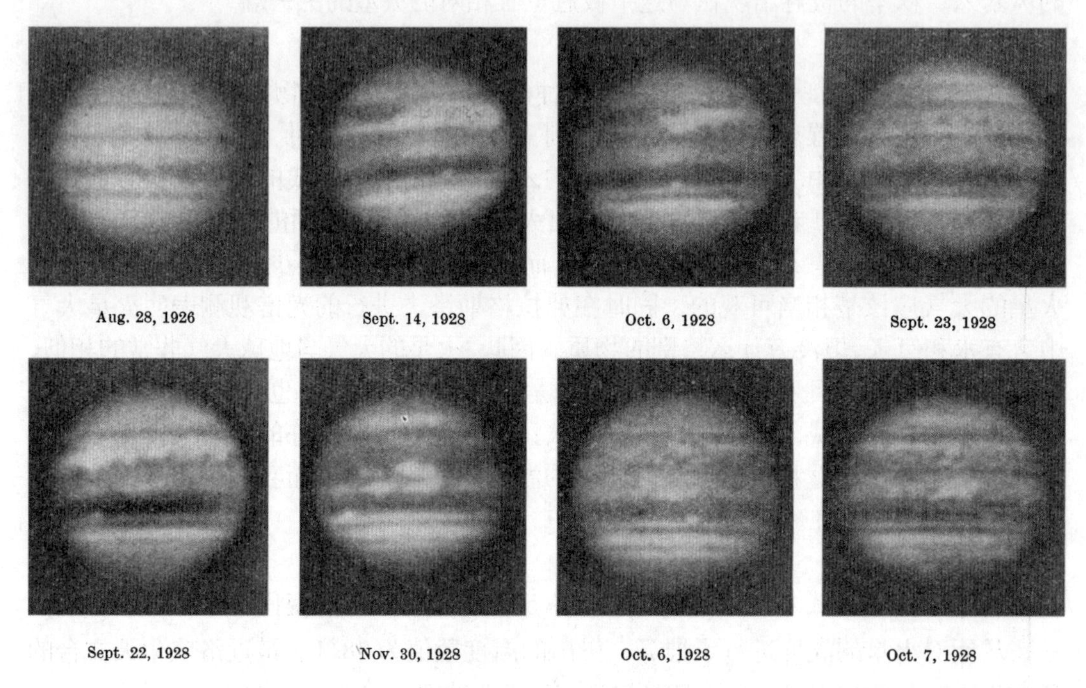

Fig. 2. Photographs of Jupiter.

Spectrum analysis of the light of Jupiter has revealed a great number of dark bands in the red and infra-red, due to the selective absorption of its atmosphere. Most of these are yet unidentified, but ammonia is present, and possibly also methane gas. The most remarkable quality of the planet's atmosphere is its rapidly increasing absorption into the longest wave-lengths, which must affect the radiation in a decided manner.

Saturn has been regularly observed at Flagstaff, visually, photographically, and spectrographically. Lowell studied theoretically the planet's law of mass distribution, the polar flattening and relation of satellites to divisions in the ring system, leading to new results. Photographs of the planet and rings in light of different colours show some surprising changes, sometimes from year to year. It was found in 1921, when the Earth and Sun were very near the plane of Saturn's rings, that, contrary to previous belief, the rings could always be seen, and that the rings caused two dark lines across Saturn's ball, one the shadow of the rings and the other the rings themselves as seen dark against Saturn (Fig. 3).

件较好时，会有太多的细节呈现于眼前以至于目视天文观测者难以绘制和记录下他的所见。在这种情况下，我们借助于照相技术就显得非常重要了；弗拉格斯塔夫天文台保存了自 1905 年以来几乎最为齐全的图像资料。图 2 显示了木星斑纹的特征，可以使我们对于木星急剧的、有时大范围的变化有一些了解，这些变化表明木星表面是非常活跃的。

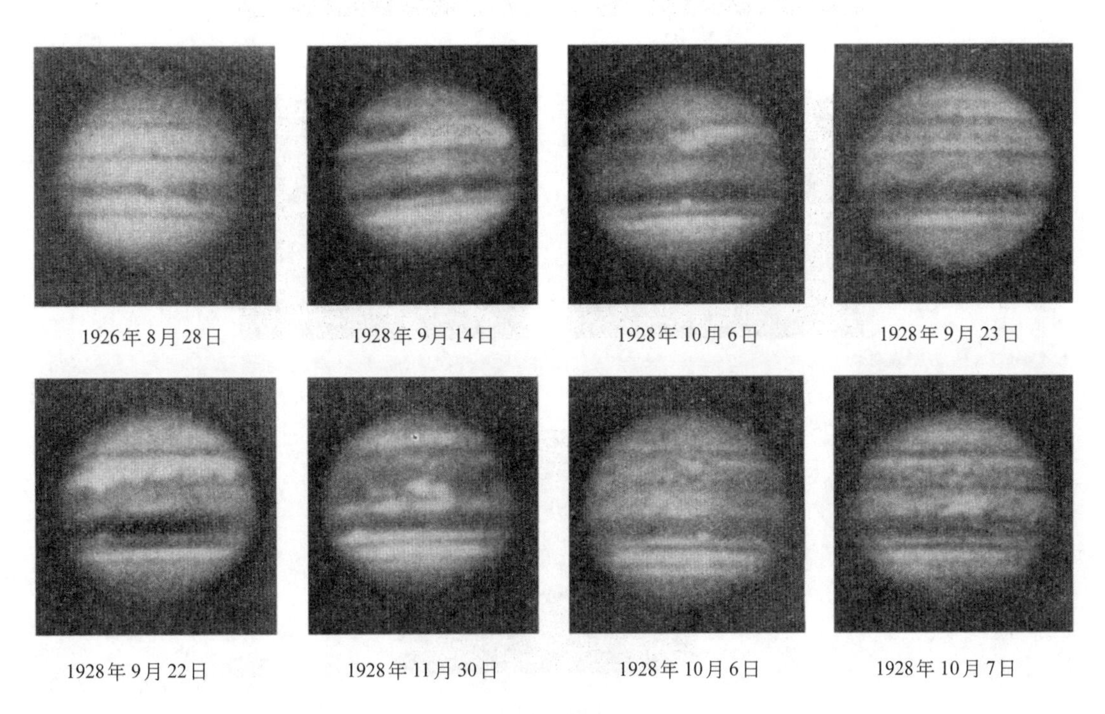

图 2. 木星的照片

对来自木星的光进行的光谱分析显示，在光谱红端和红外端存在着大量的吸收带，这是由于木星大气的选择性吸收造成的。这些吸收带中的大多数还未确认，但可以确定有氨气存在，可能还存在甲烷。木星大气最显著的特征就是，在波长最长处吸收量迅速增加，这对于木星的辐射有着决定性的影响。

在弗拉格斯塔夫，也采用目视、照相以及光谱的方式对土星进行了常规观测。洛威尔从理论上研究了土星质量分布规律、两极扁平化以及卫星与土星环中缝隙的关系，并给出了一些新的结果。土星及土星环的不同颜色的照片会显示出一些奇异的变化，有时不同年份也会有所不同。1921 年，观测发现当地球和太阳非常靠近土星环平面时，仍然可以看到这些土星环，这与先前的观点是相反的；这时，这些土星环在土星的球体上呈现为两道黑线，其中一道是环的阴影，另一道是环自身，这是因为环的颜色比土星深（图 3）。

Fig. 3. Photographs of Saturn.

Spectrum analysis of Saturn's light shows much the same absorption bands as were found for Jupiter (except that those of ammonia are weaker in Saturn), so their atmospheres are much alike. The rings show no atmosphere, but are meteoric. The fact that the cloud belts of Saturn are so much weaker than those of Jupiter is doubtless due to the former having a very great seasonal disturbance owing to its highly tipped axis. This factor is practically absent from Jupiter, and so allows its clouds to form and continue strongly belted parallel to the equator, whereas for Saturn the seasonal disturbance tends to destroy such belts.

While Uranus and Neptune are each more than sixty times the volume of the Earth, their great distances, nineteen and thirty times our distance from the Sun, give them only tiny discs even in the largest telescopes, and markings on them are very difficult of observation. Hence to get the rotation of Uranus the spectrograph was employed; it showed the planet's day to be 10.7 hours, and the rotation to be in the direction in which the satellites revolve.

1911年11月4日　　1916年2月11日

1921年4月22日　　1922年5月15日

1929年9月2日

图3. 土星的照片

对来自土星的光进行的光谱分析表明，它的吸收带几乎与木星完全相同（只不过土星中的氨气吸收带比较弱），由此可见这两颗行星的大气非常相似。土星环没有大气而仅仅是流星体。土星的云带比木星的云带弱得多，这肯定是由于前者高度倾斜的自转轴会导致非常大的季节性扰动。对木星而言这个影响因素是不存在的，这就允许了木星中云的形成，并且持续的呈带状平行分布于赤道附近。在土星上，季节性的扰动往往会破坏这样的云带。

天王星和海王星的体积均大于地球的60倍，但由于它们与地球的距离分别是日地距离的19倍和30倍，因此，即使我们借助于最大的望远镜，它们看起来仍只是很小的圆盘，因而难以观测其表面细节特征。因此引进了摄谱仪对天王星的自转进行研究；结果显示天王星上的一天时间是10.7个小时，它的自转方向与其卫星的转动方向一致。

The spectrum analysis of these two planets has also taught us much as to their atmospheres. They bear resemblance to those of Jupiter and Saturn, but show much more intense and numerous absorption bands, the strongest of which are present in the two latter planets. This atmospheric band system is much more intense in Neptune than in Uranus; in short, the bands increase from Jupiter to Uranus and again from the latter to Neptune, somewhat with the distance of the planet from the Sun.

Fig. 4 shows the spectra of these four planets compared with that of the Moon, and gives a good idea of the manner in which the absorption bands increase from Jupiter to Neptune. It is of interest to note that the ammonia band clearly evident in Jupiter, a little way to the left of C, is weak in Saturn, Uranus and also in Neptune.

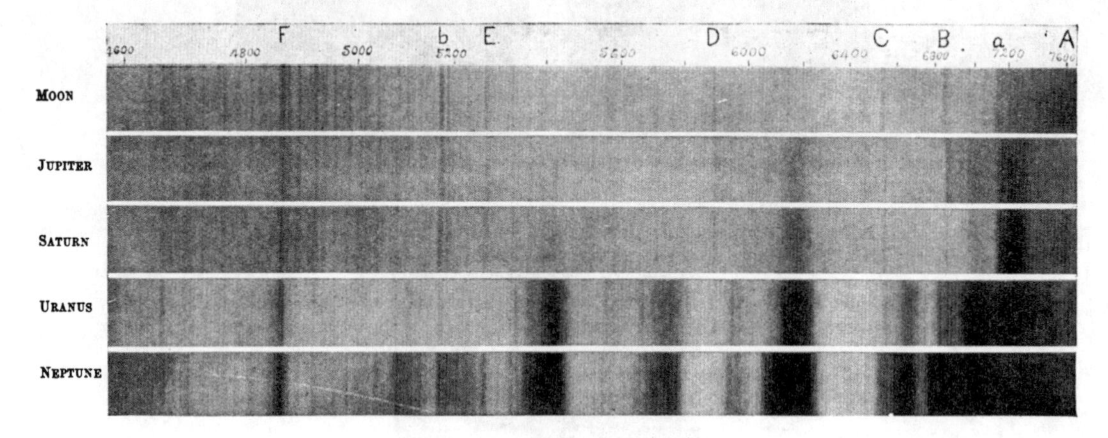

Fig. 4. Spectra of planets and the Moon.

This study of the planets at the Lowell Observatory, in addition to many results not given here relative to the several planets, has much emphasised the differences of the two main groups of planets: Earth, Venus, Mercury and Mars, and the giant group—Jupiter, Saturn, Uranus and Neptune. The first group are comparable with the Earth in size, in density, in energy they receive from the Sun and in atmospheres, so far as they show any at all. The other group are much larger bodies, but of much lower densities, and have a very different type of atmosphere, while the solar energy they receive is much less than the Earth's share—ranging from 1/26 for Jupiter to 1/900 for Neptune. But these studies indicate that these planets may be much more effectively utilising this small energy gift from the Sun than does the nearer group of planets, for their atmospheres, as their spectra show, are as blankets retaining energy of the longer heat-waves, and may let little or none pass out in the heat spectrum available to observers on the Earth.

These studies further direct attention to that important break between the two groups of planets between Mars and Jupiter, and emphasise the need of its further study, and perhaps from theoretical grounds as well, for when we know what has happened to produce the asteroids and cause this vast change in the planetary bodies, we shall better understand the past of the solar system.

(**133**, 10-13; 1934)

对这两颗行星的光谱分析可以使我们了解到很多关于它们大气的信息。它们的光谱与木星和土星相似，但是却拥有更加密集且数目更多的吸收带，其中最强的吸收带就出现在了天王星和海王星的光谱中。相比而言，海王星大气吸收带比天王星更加密集；简言之，吸收带从木星到天王星再到海王星逐渐增加，这在一定程度上与行星距离太阳的远近有关。

将上述四颗行星的光谱与月球的光谱进行比较，如图 4 所示，从中可以清楚地看到吸收带数目从木星到海王星递增的特点。值得注意的是，木星光谱中氨气的吸收带非常明显，其位于 C 波段的左侧附近，但是在土星、天王星和海王星中氨气的光谱则较弱。

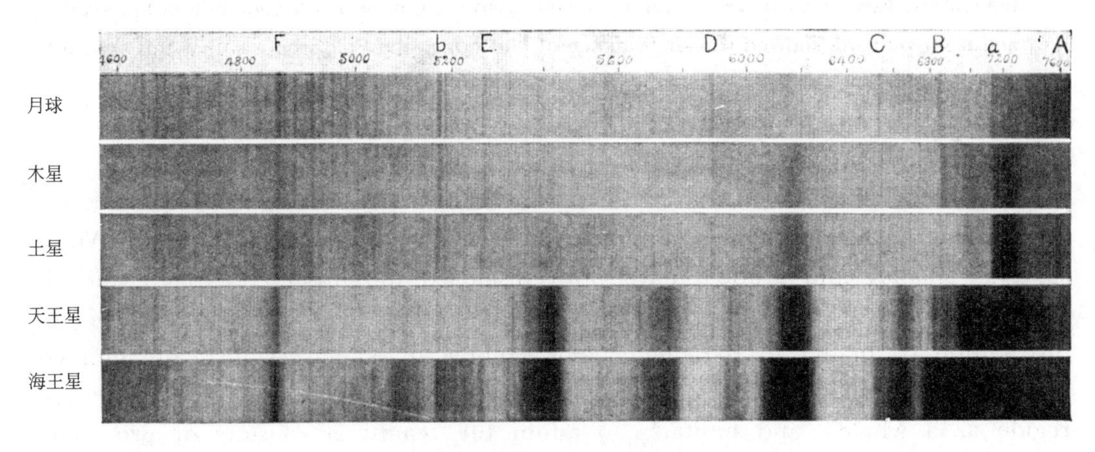

图 4. 行星与月球的光谱图

洛威尔天文台根据以上研究，以及这里未给出的其他一些相关观测结果，从而清楚地表明了两组行星的差别：一组是地球、金星、水星和火星，另一组是木星、土星、天王星和海王星。第一组行星在尺寸、密度、大气以及吸收的太阳能量方面均与地球相近，而后一组天体具有非常大的体积，但是密度较低，其大气构成与前一组显著不同，吸收的太阳能量比地球少得多——其中木星最多，其吸收的能量是地球的 1/26，海王星最少，其吸收的能量是地球的 1/900。研究表明，这些巨行星可能会比第一组距离太阳更近的行星更加有效地利用所获得的有限的太阳能，因为它们的大气可能只让很少一部分或者不让具有热效应的光谱波段透出，因而地球上的观测者很难观察到它们，正如光谱所显示的，在热长波长段显示的能量基本为空白。

这些研究使我们的注意力进一步集中到以火星和木星为界的两组行星的巨大差别上；同时也强调了要对此进行深入研究的必要，也许还要进行基础理论方面的研究，因为只有当我们最终弄清楚小行星是如何产生的，以及行星之间为什么存在这样巨大的差异，我们才有可能对整个太阳系的形成有更好的理解。

（金世超 翻译；蒋世仰 审稿）

Apparent Clustering of Galaxies

B. J. Bok

Editor's Note

Recent surveys of the placement of distant galaxies made it possible to begin describing their distribution. Here Bart Bok of the Harvard Observatory reviews evidence that galaxies are clustered, not distributed at random. This is particularly evident in statistical analyses of the number of galaxies brighter than a certain threshold within equal-area regions of each pole of the Galaxy. Two recent surveys of faint galaxies, from the Californian Mount Wilson Observatory and from Harvard, showed similar evidence of clustering, and Bok proposes that this should be considered a general and significant feature of our universe.

A considerable amount of material on the distribution of external galaxies has become available through the publication of the Harvard and Mount Wilson surveys. Shapley and Hubble have both discussed the observed irregularities in the distribution of these galaxies. Shapley emphasises the non-uniformity of the distribution of matter in the metagalaxy. Hubble finds that "statistically uniform distribution of nebulae appears to be a general characteristic of the observable region as a whole", and hesitates to admit the reality of clusters or groups of galaxies with the exception of the few that are readily recognised as such. Statistical analysis of the available material is now possible; and as the comparison between the observed distribution curves, corrected for the effect of dispersion in the limiting magnitudes, and the theoretical frequency curves, computed on the assumption of random distribution, has yielded some rather definite results, it seems worth while to communicate them in advance of publication in more detail.

The Shapley-Ames catalogue of galaxies brighter than the thirteenth magnitude[1] exhibits conspicuous deviations from a random distribution. Both galactic polar caps were divided into a number of equal areas (well-known clusters being excluded), and the number of galaxies was counted in each area. The observed frequency curve had a much larger dispersion than the theoretical curve, computed on the assumption of random distribution. The accompanying table shows conclusively that the irregularities in the distribution cannot have been caused by galactic or extragalactic absorption.

星系的视成团

博克

编者按

近来对于遥远星系位置的巡天观测使得我们可以开始描述它们的分布。本文中，哈佛天文台的巴特·博克评述了星系是成团而非随机分布的证据。特别是对银河系的两极方向相同面积区域内亮于某一定阈值的星系数的统计分析更是证明了这一点。近来分别由加州威尔逊山天文台和哈佛大学天文台所做的暗星系巡天观测显示出了类似的成团性证据，博克提出这应是宇宙普遍且重要的特征。

根据哈佛大学天文台和威尔逊山天文台所发表的巡天观测数据，人们可以得到大量关于河外星系分布的资料。沙普利和哈勃都曾就这些观测到的星系讨论过其分布的不规则性。沙普利强调了物质在总星系中分布的不均匀性。而哈勃则发现“星云统计上的均匀分布似乎是整个可观测区域的一个普遍特征”，但是除了少数容易识别者之外，他不愿意承认存在星系群或星系团的这一事实。现在已经能够对这些可用的资料进行统计分析；通过对极限星等弥散效应进行修正后得到的观测分布曲线和由随机分布的假设计算得到的理论频率曲线进行比较，我们已经得到了一些较为确定的结果，在更多细节发表之前，有必要先将这些结果公布出来。

亮于 13 星等的沙普利－艾姆斯星系表[1]中的星系显示出与随机分布存在明显的偏离。将银河系的两个极冠区平均分成若干区域（在去除了那些我们所熟知的星团后），并对每一个区域中的星系进行计数。观测到的频率曲线的弥散远远大于由随机分布假设计算得到的理论曲线的弥散。附表清楚地表明，分布中的不规则性不可能是由银河系或河外星系的吸收所造成的。

North Galactic Polar Cap.

No. of galaxies (Shapley-Ames)	$\overline{\log N}$ (Hubble)	No. of galaxies (Shapley-Ames)	$\overline{\log N}$ (Hubble)
$\frac{1}{2}$	1.92	15	1.79
$1\frac{1}{2}$	1.99	17	1.85
4	1.86	$18\frac{1}{2}$	1.87
$5\frac{1}{2}$	1.90	22	1.96
$6\frac{1}{2}$	1.88	24	1.95
10	1.87	26	1.87
$10\frac{1}{2}$	1.88	$29\frac{1}{2}$	1.86
$12\frac{1}{2}$	1.87	31	1.88
14	1.83	$31\frac{1}{2}$	1.86
$14\frac{1}{2}$	1.95	36	1.94
$14\frac{1}{2}$	1.93		

The first column of this table gives the number of galaxies counted for one of the areas in the Shapley–Ames catalogue. The centres of 9–13 survey fields used by Hubble in his study of the distribution of faint galaxies (down to mag. 19.5) fall within the limits of each area, and the second column of the table contains the mean value of log N for these faint galaxies. The absence of any progression in the values of $\overline{\log N}$ shows that the deviations from random distribution are due to a real clustering of galaxies and are not caused by the absorption of light in space.

Both the Mount Wilson[2] and Harvard[3] surveys of faint galaxies show evidence of clustering. The diagram (Fig. 1) gives a comparison between Hubble's observed distribution curve (dots), corrected for a dispersion of ±0.15 mag. in the limiting magnitude of the Mount Wilson plates, and the theoretical curve (crosses) computed on the assumption that the galaxies are distributed at random.

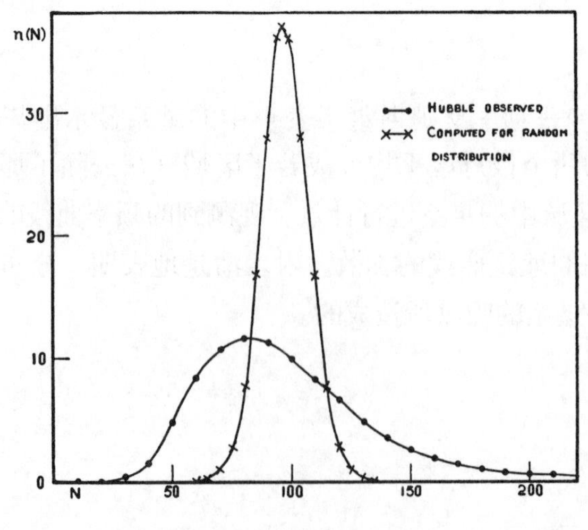

Fig. 1

银河系北极冠区

星系数（沙普利 – 艾姆斯）	$\overline{\log N}$ (哈勃)	星系数（沙普利 – 艾姆斯）	$\overline{\log N}$ (哈勃)
$\frac{1}{2}$	1.92	15	1.79
$1\frac{1}{2}$	1.99	17	1.85
4	1.86	$18\frac{1}{2}$	1.87
$5\frac{1}{2}$	1.90	22	1.96
$6\frac{1}{2}$	1.88	24	1.95
10	1.87	26	1.87
$10\frac{1}{2}$	1.88	$29\frac{1}{2}$	1.86
$12\frac{1}{2}$	1.87	31	1.88
14	1.83	$31\frac{1}{2}$	1.86
$14\frac{1}{2}$	1.95	36	1.94
$14\frac{1}{2}$	1.93		

此表格的第一列给出了沙普利 – 艾姆斯星表中每个区域中的星系数。哈勃在关于暗星系（暗至 19.5 星等）分布的研究中所使用的 9~13 巡天区域的中心，落在了每个区域的界限以内，而表格的第二列是这些暗星系数目的对数 $\log N$ 的平均值。$\overline{\log N}$ 的值的没有任何规律可循，这说明随机分布中的偏离是由实际中的星系成团造成的，而不是由光线在空间中被吸收所致。

威尔逊山天文台 [2] 和哈佛大学天文台 [3] 对暗星系的巡天观测都显示出星系成团的证据。图中（图 1）对哈勃观测得到的分布曲线（圆点表示）与理论曲线（叉号表示）进行了比较，其中哈勃观测的分布曲线是对威尔逊山天文台底板的极限星等进行 ±0.15 星等的弥散修正后得到的，而理论曲线则是在假设星系随机分布的基础上进行计算得到的。

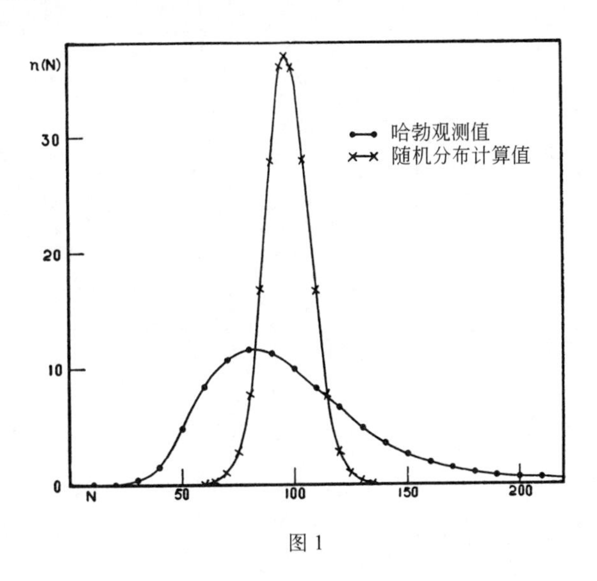

图 1

Similar deviations from random distribution are found in the Harvard material. The observed frequency curve in log N has, for the north galactic polar cap, a dispersion of ±0.25, and as the maximum value of the error dispersion amounts to only ±0.15 (most probable value ±0.09), the true dispersion must be of the order of ±0.20 in log N. The dispersion computed theoretically for random distribution is not larger than ±0.03 in log N. For the south galactic polar cap the discrepancy is even greater. We should in addition consider Shapley's elegant and definite proof for the presence of clustering in nine regions[4].

We can scarcely escape the conclusion that a widespread tendency towards clustering among galaxies is one of the chief characteristics of our universe.

(**133**, 578; 1934)

Bart J. Bok: Harvard Observatory, Cambridge, Mass, Jan. 27.

References:

1. *Harv. Ann.*, **88**, No. 2; 1932.
2. *Astrophys. J.*, 79, 8; 1934.
3. *Harv. Bull.*, 889; 1932. Harvard Reprint 90; 1933.
4. *Harv. Bull.*, 890; 1932.

在哈佛大学天文台的资料中也同样发现了相对于随机分布的偏离。银河系北极冠观测到的频率曲线，对 $\log N$ 有 ±0.25 的弥散度，而误差的最大弥散值只有 ±0.15（最可几值为 ±0.09），因此 $\log N$ 的真实的弥散必定在 ±0.20 量级。而理论计算得到的随机分布的 $\log N$ 的弥散则不会超过 ±0.03。而在银河系的南极冠区这个差异则更加明显。另外我们还应该考虑沙普利对 9 个天区中星系成团的存在所做的简练而明确的证明[4]。

我们很难回避这样一个结论，即星系成团的普遍趋势是宇宙的一个主要特征。

（周杰 翻译；邓祖淦 审稿）

Solar Magnetism

G. E. Hale

Editor's Note

On the occasion of Dutch physicist Pieter Zeeman's 70th birthday, American astronomer George Hale surveys applications of Zeeman's methods to the study of solar magnetism. These methods—based on how a magnetic field influences atomic energy levels, and hence the frequency of light an atom emits—had recently been used to document a 23-year cycle of sunspot activity, and also to show the existence of huge vortices of flowing material surrounding sunspots. The technique had also established that the Sun as a whole is also a large magnet. It is almost certain, Hale concludes, that these fields are all caused by charged particles in some kind of rotational motion.

THE recent celebration of Prof. Zeeman's seventieth birthday offers a favourable opportunity to describe current applications of his powerful method of research to the study of solar magnetism. Our latest results include the completion of the first observed 23-year magnetic cycle of sunspots and the conclusion of a long investigation of the Sun's general magnetic field, made for the purpose of checking beyond question the original measurements begun in 1912.

Zeeman Effect in Sunspots

As explained eleven years ago in *Nature*[1], I was led in 1908 to the discovery of magnetic fields in sunspots by a hypothesis based upon the results of two series of studies, begun at the Kenwood Observatory in 1890, and continued at the Yerkes and Mount Wilson Observatories. The first of these related to the nature of various phenomena of the solar atmosphere revealed by spectrographs and spectroheliographs. The hydrogen flocculi, as first shown by the H_α line at Mount Wilson in 1908, indicated the existence of immense vortices surrounding sunspots, and suggested that electrically charged particles might be whirled within the spots in such a way as to produce appreciable electric currents. Such currents would set up magnetic fields, possibly of sufficient strength to be detected by a powerful spectroscope. Zeeman had shown how the spectrum lines of luminous metallic vapours between the poles of a magnet are widened or split into several components, polarised in distinctive ways. Meanwhile our studies of sunspot spectra, supplementing those made with less powerful spectrographs by Young, had reached a point where many lines on our photographs were not only widened but also separated into apparent doublets or triplets. These had previously been regarded as reversed lines, due to the superposition of two vaporous layers of different temperature and density. Such reversals actually exist in certain cases, notably in the lines of hydrogen and calcium. Thus, the true understanding of the sunspot spectrum had been obscured.

太阳磁场

海耳

编者按

在荷兰物理学家彼得·塞曼 70 岁寿辰的庆典上，美国天文学家乔治·海耳综述了塞曼的方法在研究太阳磁场方面的各种应用。这些基于磁场如何影响原子的能级进而影响原子发光频率的方法近期被用于记录以 23 年为周期的太阳黑子活动，同时这种方法也表明在太阳黑子附近存在流动物质的巨大的涡旋。这一技术同时表明整个太阳是一个大磁体。海耳得出的结论认为，几乎可以肯定太阳磁场完全是由带电粒子的某种旋转运动造成的。

最近塞曼教授的 70 岁生日庆典为我们提供了一个很好的机会，对他提出的有效研究方法在太阳磁场研究方面的最新应用进行讨论。我们最新的应用结果包括首次完成的对太阳黑子 23 年磁周期的观测以及通过对太阳普遍磁场的长期观测得到的结论，这些观测是为了检验始于 1912 年的对太阳磁场的最初观测，1912 年的观测结果曾被人们认为是无可争辩的。

太阳黑子的塞曼效应

正如 11 年前在《自然》杂志上的[1]解释一样，1908 年，基于之前的两项研究成果，我做了一个假设，并由此发现了太阳黑子的磁场，这两项研究工作均于 1890 年在肯伍德天文台开始，后来在叶凯士天文台和威尔逊山天文台继续进行。其中第一项工作是关于太阳大气中一些现象的本质的研究，通过使用光谱仪和太阳单色光照相仪，我们对此有了一定的了解。1908 年，根据在威尔逊山天文台观测到的 H_α 线，人们第一次发现了太阳上的氢谱斑，这一现象意味着在太阳黑子附近存在巨大的涡旋，从而暗示了带电粒子在太阳黑子内部做回旋运动，并因此产生了可被测量的电流。而这些电流可以产生一个强度很大的磁场，足以被一台功能强大的分光镜观测到。塞曼已经为我们展示了在两个磁极之间的发光金属蒸汽的谱线是如何被展宽或分裂成不同部分并发生明显的极化现象的。同时，将我们对太阳黑子的光谱研究和扬用功能稍弱的光谱仪对太阳黑子所做的研究综合起来，可以得到这样一个结论：照片中的很多谱线不仅仅被展宽了，而且明显地分裂成了双线或三线。这种由温度和密度不同的两层蒸汽气体重叠所造成的谱线曾被前人称为反向线。这样的反向线在某些情况下确实是存在的，而在氢元素和钙元素的谱线中尤为明显。所以说，关于太阳光谱的正确解释在当时还不是十分明朗。

In the hope of disentangling the question, a new attack on sunspots was begun. Aided by the 60-foot tower telescope on Mount Wilson, equipped with a 30-foot grating spectrograph and suitable polariscopic apparatus, it was easy to test my hypothesis. The presence of magnetic fields was readily established in all the sunspots observed, and the polariscopic phenomena of the sunspot lines, varying as the solar rotation changed the angle between the lines of force and the line of sight, was quickly found to harmonise with Zeeman's laboratory results on the spectra of vapours. My solar work was greatly facilitated by experiments made in our own laboratory by King, provided with a Du Bois magnet and all the essential equipment.

Magnetic Polarity of Sunspots

The sunspot spectrum contains many thousands of lines, and its complete investigation is an extensive task. After a sufficient number of these lines had been examined in order to establish the existence, strength and general character of the magnetic fields, another phase of the problem was attacked.

Speaking broadly, sunspots in the northern hemisphere of the Sun were found to be opposite in polarity to those in the southern hemisphere. But occasional apparent exceptions indicated the need for a more careful analysis. The earliest drawings of sunspots, made by Galileo and Scheiner, suggest their complex character. They often appear at first as single spots, but soon develop into groups, frequently containing many components, large and small. No observer could fail to detect, however, a remarkable tendency of spots to occur in pairs, consisting of large spots with small companions, or of two groups of small spots. Here was an interesting chance for polarity tests, which showed that such pairs are almost invariably bipolar: that is, they consist of two spots or groups having opposite magnetic poles. The smaller spots that frequently cluster about the preceding (western) and following (eastern) major spots usually agree in polarity with the larger spots they accompany, though this is not an invariable rule.

From such characteristics a scheme of magnetic classification developed, which has been used ever since on Mount Wilson in recording the magnetic phenomena of thousands of spots examined with the 150-foot tower telescope and the 75-foot spectrograph. This long task, in which Nicholson, Ellerman, Joy and many others have taken part, has now covered more than two of the well-known sunspot cycles of approximately eleven years duration.

Law of Sunspot Polarity

It is well known that the first spots of each of these 11-year frequency cycles break out in comparatively high latitudes some time before the last of the spots of the previous cycle disappear near the equator. From 1908, the spots of the then existing cycle continued to show the same polarity, opposite in the two hemispheres, while slowly decreasing in mean latitude. Not long before the minimum of solar activity in 1912, the forerunners of the next 11-year cycle began to appear. To our surprise, their polarity was opposite to

人们怀着解决这个问题的期望迎来了又一轮太阳黑子的爆发。借助于威尔逊山天文台 60 英尺塔状望远镜、30 英尺光栅光谱仪以及合适的偏振光仪，很容易检验出我的假说是否正确。通过观测,我们确定在所有观测到的太阳黑子中都存在磁场。此外，当太阳旋转使磁力线方向和视线方向的夹角发生变化时，观测到的偏振光的偏振方向也随之改变，而且人们很快发现了这个变化关系与塞曼在实验室蒸汽谱线中发现的变化规律是相符的。我们实验室的金所做的实验大大推动了我在太阳磁场方面的研究工作，他在这个实验中使用了杜波依斯磁铁和其他相关的必要设备。

太阳黑子磁场的极性

太阳黑子的光谱中包含有成千上万条谱线，全面研究这些谱线所需的工作量很大。在对大部分的谱线进行了研究之后，我们确定了磁场的存在，并得到了磁场的强度和一般性特征。之后，我们又开始着手解决这个研究中的另一个部分了。

概括地说，我们发现太阳北半球上和南半球上的太阳黑子的磁场具有相反的极性。但是偶尔也会有例外出现，这表明对此现象我们还需要更加细致的分析。最早由伽利略和沙伊纳绘制的太阳黑子图显示了太阳黑子的复杂特性。它们起初通常是单个出现的，然后发展为一群，常常包含大大小小的不同部分。然而，黑子都有明显的成对现象，所有的观测者都注意到了这个现象，成对的现象可能表现为在大黑子的附近伴随有小的黑子，或者是两群小的黑子在一起。这些黑子对让人们有机会去检测磁场极性，人们发现这些黑子对总是具有不同的极性：即成对出现的两个或两群黑子的磁场极性总是相反的。较小的黑子经常聚集在前面（西部）和后面（东部）的大黑子的周围，并且与相伴的较大黑子呈现出相同的磁性，尽管这种模式并不是一成不变的。

人们从这些磁场的特性出发对磁场进行了分类，这种分类自从在威尔逊山天文台上被用来记录成千上万个黑子的磁现象之后，就一直被沿用至今，当时人们是通过 150 英尺塔状望远镜和 75 英尺光谱仪来观测黑子的磁现象的。这个工作已经开展了长达两个太阳黑子周期之久，众所周知，每个周期大约为 11 年，尼克尔森、埃勒曼、乔伊和很多其他科学家都曾参与了这项工作。

太阳黑子磁场极性的规律

众所周知，在每个 11 年的太阳活动周期开始时爆发的第一个黑子，总是在上一个周期的最后一个黑子在赤道附近消失之前出现在纬度相对稍高的地方。在 1908 年之后出现的这个太阳活动周期中，黑子的磁场极性一直保持不变，而在南北两个半球上黑子的磁极则是相反的，同时黑子的平均纬度慢慢变小。而在太阳活动极小年到来之前的 1912 年，就已经出现了下一个 11 年活动周期的先兆。令我们惊讶的是，

that of the spots of the preceding cycle. Moreover, the succeeding spots of the new cycle, which overlapped for a time the remnants of the old cycle in lower latitudes, retained the same reversed polarities for approximately eleven years. Then another frequency cycle commenced, with another reversal of polarity. Thus the complete magnetic cycle, bringing back spots of the same polarity as those first observed, occupies some twenty-two or twenty-three years, and comprises two frequency cycles. The northern and southern hemispheres represent this novel effect with opposite signs. The diagram shown in Fig. 1 summarises the changes of latitudes and polarities during the period 1908–35.

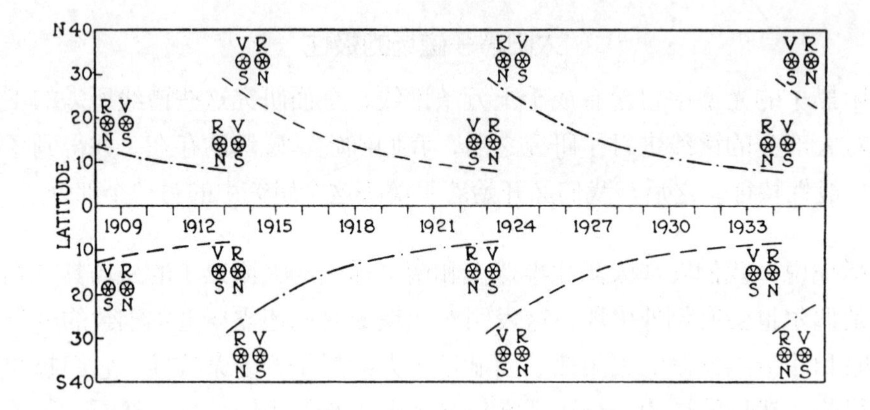

Fig. 1. Law of sunspot polarity. The curves represent the approximate variation in mean latitude and the corresponding magnetic polarities of sunspot groups observed at Mount Wilson form June 1908 until January 1935. The preceding spot is shown on the right.

A more detailed examination of the observations, many of which have been published in the *Astrophysical Journal* and the *Publications of the Astronomical Society of the Pacific*, would suffice to show that occasional exceptional phenomena complicate the explanation of these changes. About ninety-seven percent of consistent results, however, obviously point toward some general solution, applicable to the Sun and countless other stars, but still remaining in the form of the empirical law illustrated in Fig. 1.

General Magnetic Field of the Sun

Soon after the detection of strong magnetic fields in sunspots, I began to wonder whether the Sun as a whole might possess a general magnetic field. There was no very promising theoretical ground for such speculation, but the magnetic field of the Earth, with poles not far removed from the poles of rotation, was at least suggestive. Schuster had queried in 1891: "Is every rotating body a magnet?" and the structure of the solar corona resembles that of a magnetic field*. Thus while it was a far cry from the solid Earth to the vaporous Sun, it seemed worth while to undertake a trial.

The first attempts, made with the 60-foot tower telescope, were fruitless. In 1912, with the

* In the present brief statement no attempt is made to enumerate other speculations and theories.

在新一轮的周期中出现的黑子的磁场极性与前一个周期中的黑子的磁场极性完全相反。而且，在新的周期中最先出现的那些黑子与上一个周期中残存的黑子曾一度在低纬度发生重合，并且二者在约 11 年的时间里一直保持相反的极性。因此，具有另一种反向极性的太阳活动周期被提了出来。这使太阳黑子变回到与前一次观测时具有同极性的一个完整的磁场活动周期变成了大约是 22 或 23 年，即由两个 11 年的太阳活动周期组成。虽然太阳的南北半球中黑子磁场的极性相反，但都出现了这种奇特的现象。我们在图 1 中总结了 1908~1935 年黑子的纬度和磁场极性的变化情况。

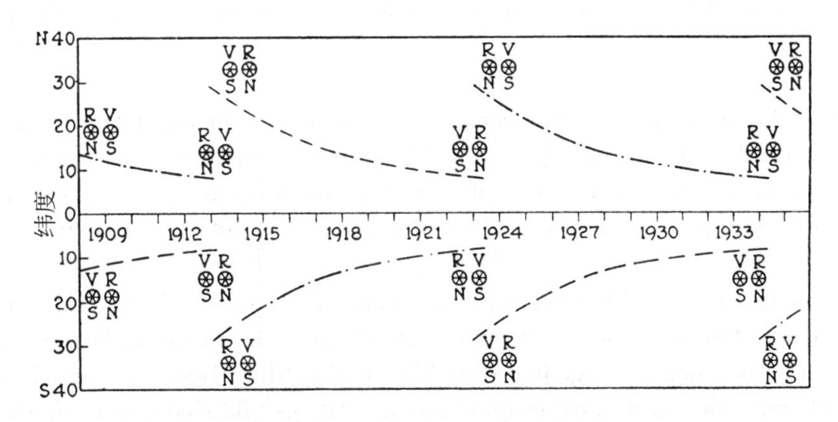

图 1. 太阳黑子磁场极性的规律。这些曲线近似地表示了黑子群平均纬度的变化和相应的磁场极性。这些黑子群是 1908 年 6 月至 1935 年 1 月在威尔逊山上观测到的。前导黑子出现在右边。

《天体物理学杂志》和《太平洋天文学会会刊》上刊登了大部分关于太阳磁场的观测结果，而对所有的观测结果进行仔细研究后我们可以发现：偶尔出现的小概率现象增加了人们解释此种变化关系的难度。虽然 97% 的数据都一致明确指向了那些普遍的结论，该结论适用于太阳和其他无数的恒星，但是这种结论仍只是一种如图 1 所示的经验规律。

太阳的普遍磁场

在探测了太阳黑子中的强磁场后，我开始好奇整个太阳是否拥有一个整体的磁场。虽然我的这个猜想没有任何可靠的理论基础，但是地球磁场的磁极与转动极相差不远这一事实使我产生了这种联想。舒斯特曾在 1891 年提出这样的疑问："是不是所有的旋转物体都带有磁场？"太阳日冕层的结构与磁场的结构很像 *。因此，虽然固态的地球与气态的太阳大不相同，但是这样的猜想还是很值得我们去尝试检验一下的。

我用 60 英尺的塔状望远镜进行了第一次的观测尝试，结果一无所获。1912 年，

* 目前简单的声明论述中没有对其他猜想和理论进行列举。

completion of the 150-foot tower telescope and 75-foot spectrograph, a better opportunity offered itself. Obviously no such widening and complete splitting of lines as had been found in sunspots existed in regions away from spots. But by using a purely differential method, comprising a compound quarter-wave plate overlying a long Nicol prism on the slit, it seemed barely possible that minute magnetic displacements of suitable lines might be detected by measurements on successive strips. Assuming the magnetic poles of the Sun to correspond with the poles of rotation, and mounting the quarter-wave strips with their principal sections alternately at angles of +45° and −45° with the slit of the spectrograph, the displacements on odd and even strips should attain maximum values on the central meridian at about 45° north and south latitude, and decrease to zero at the poles and equator.

In the search for such minute displacements, every precaution was taken to obviate any possible bias on the part of the measurers. Thus the quarter-wave plates were frequently inverted, and the measurer was never allowed to know in advance in which position they stood, nor the hemisphere or latitude of the photographs under measurement.

Great difficulties attended this investigation, which was continued for several years. Many members of the Mount Wilson staff joined me in the task, including Seares, Anderson, Ellerman and van Maanen, together with Miss Lasby, Miss Richmond and Miss Felker, while check measures and tests were made by Adams, Babcock and others. Several different types of measuring machines were employed, and every possible means of avoiding personal or instrumental errors was adopted. The results, fully described in a series of papers in the *Astrophysical Journal*, seemed to leave no room for doubt regarding the existence, polarity and approximate strength of a weak general magnetic field of the Sun, having poles lying within a few degrees of the Sun's poles of rotation.

The fact remains, as pointed out in these papers, that some of the many measurers engaged in the work could not detect the general field, though the results of all those who succeeded were in agreement regarding its polarity and order of magnitude. The difficulties of measurement can be appreciated only by those who have endeavoured to detect such minute displacements of lines, rendered broad and diffuse by the great dispersion employed. The observed displacements ranged from zero near the equator and poles to maximum values of 0.001 A. at mid-latitudes.

Several years ago I renewed this investigation with the coelostat telescope (equivalent focal length 150 feet) and the 75-foot spectrograph of my solar laboratory in Pasadena. Having made a new series of photographs of the same dispersion as those previously taken on Mount Wilson, I endeavoured to measure them by several different instruments, including a Zeiss microphotometer and a tipping plate micrometer of the form used in our earlier work. The results showed little, if any, general magnetic field, and finally I thought it advisable to undertake new check measures of the plates made more than twenty years

当 150 英尺的塔状望远镜和 75 英尺的光谱仪的建成并投入使用后，我们刚好又遇到了一个极好的观测时机。显然在黑子以外的地方人们没有观测到谱线的展宽和完全的分裂。但若使用完全不同的方法，如把一个 1/4 波片和一个长的尼科尔棱镜重叠装在狭缝上，通过测量连续的条纹还是勉强有可能会观测到磁场在相应谱线上引起的微小位移。若假设太阳的磁极与旋转极是重合的，那么在装上一个与主截面成正 45 度角、与光谱仪狭缝成负 45 度角的 1/4 波片后，奇偶条纹之间的位移应该会在中心子午线南北纬 45 度左右的地方达到最大值，而在两极和赤道减小到为零。

为了测出这种微小的位移，我们采取了很多预防措施来避免任何可能在测量仪器上出现的偏差。因此我们要经常翻转 1/4 波片，并且在测量前不能预先知道其所处的位置，或照片上被测量区域所在的半球和纬度。

这项研究持续了几年，期间也遇到了很多困难。很多在威尔逊山天文台的工作人员都加入了我的这项工作，其中包括西尔斯、安德森、埃勒曼和范玛宁，还有拉斯比、里士满和费尔克三位女士。测试工作和测量结果的检验是由亚当斯、巴布科克等人来完成的。我们使用了几种不同类型的测量仪器，并采用了所有可能的办法来避免由人为和仪器所造成的误差。在《天体物理学杂志》上发表的一系列文章详细描述了我们的观测结果，这些结果明确阐述了太阳弱普遍磁场的存在、极性以及大概的强度，结果还表明太阳的磁极与其转动极的夹角只在几度的范围内。

正如我们在上述文章中所指出的，虽然所有成功的观测结果在太阳磁场极性和磁场强度的量级上都是吻合一致的，但事实上这项研究工作使用的很多测量仪器并不能探测到太阳普遍磁场的相关信息。只有那些致力于观测谱线微小位移的人们才能深刻地体会到测量过程中的困难，因为太阳磁场引起相应谱线的微小位移在使用的显著色散条件下变得很宽而散漫。所观测到的微小位移的变化范围从两极和赤道附近的零到中纬度附近的最大值 0.001 埃。

几年前，我利用帕萨迪纳太阳实验室中的（等效焦距为 150 英尺）定天望远镜和 75 英尺的光谱仪重新更新了这些研究数据。我用这些仪器观测得到了一系列新的照片，这些照片与之前在威尔逊山天文台观测到的照片呈现出一样的色散关系，我努力尝试用各种各样不同的仪器来测量这些照片，这些仪器包括蔡司显微光度计和一台形态上与我们先前使用过的仪器类似的有转动盘的千分尺。这些结果几乎没有显示出有关太阳普遍磁场的信息，即便有也是极少量的，所以我认为有必要对二十多年前在威尔逊山天文台观测得到的照片底片进行重新的检验测量工作。之后的测

ago on Mount Wilson. More difficulties were encountered, and most of the experienced measurers who had overcome them before were no longer available. However, others kindly enlisted, and a study lasting four years has at last yielded a sufficient number of independent confirmations to satisfy us of the validity of our former conclusions.

As before, some of the measurers have been unable to detect the general magnetic field shown by our old plates. On the contrary, Dr. John Strong with an improved Zeiss microphotometer, used visually, and Dr. R. M. Langer with the original tipping plate micrometer, have found systematic average displacements of the same sign. Mr. J. Evershed, who very kindly volunteered to assist, obtained excellent confirmatory measures of our original plates at his observatory in England, using his own admirable method of measurement. Within the last few weeks [date of communication, September 14, 1935], Dr. Langer has obtained two more unmistakable confirmations with the aid of a new type of combined measuring, recording and computing machine, built in my laboratory after a design due chiefly to himself. The chief advantages of this machine are its speed of operation, permitting a very large number of measures to be made in a short time, and its complete freedom from any possibility of bias.

Taken altogether, the evidence is overwhelmingly in favour of the existence, polarity and order of magnitude of the general magnetic field of the Sun given in our original papers. There is thus far no evidence of change of polarity at sunspot minima. As for any possible changes of intensity, every single determination is necessarily based upon thousands of measures, and hence represents mean values for many points on the Sun. Thus much time may elapse before the question of variability can be settled.

The striking magnetic phenomena of sunspots, and the evidence we have offered that the entire Sun is a magnet, would seem to have important bearings on the problems of terrestrial magnetism and the fundamental nature of magnetism itself. It is difficult to avoid the belief that the strong magnetic fields of the spots, and the much weaker general magnetic fields of the Sun and the Earth, arise from the same general cause, namely, the rotation of bodies carrying electrically charged particles. Many different hypotheses based upon this view have been tested, but there is much room for further work. While this will naturally deal at first with the simplest general assumptions, a detailed study of such anomalous phenomena as are presented by about three percent of all sunspots should not be overlooked.

(**136**, 703-705; 1935)

George E. Hale: Mount Wilson Observatory, Pasadena, California.

Reference:

1. "Sun-spots as Magnets and the Periodic Reversal of their Polarity", *Nature*, 113, 105, Jan. 19, 1924.

量工作中，我们遇到了更多的困难，而且大部分曾经克服过这些困难的经验丰富的工作人员也都不能再加入这项工作。不过，经过另外征集到的一些工作人员的努力，以及四年的研究最终得出了大量独立的证据，这些足以使我们相信以前的结论是有效的。

像前面提到的那样，有些测量仪器在老的照片底片上测不到太阳的普遍磁场。与之相反，约翰·斯特朗博士利用改良过的蔡司显微光度计、兰格博士利用早期的带转动盘的千分尺，通过目视都发现了由这一现象造成的系统平均位移。埃费希德先生非常主动且热心地帮助了我们，他在英国的天文台中，用他自己发明的绝妙的测量方法对我们最初的照片底片进行了测量，并得到了极好的确定的测量结果。在过去的几周内 [1935 年 9 月 14 日，通信日期]，兰格博士借助于一种新型的复合仪器又获得了另外两组确凿无误的数据核实了上述结果，这种仪器兼容了测量、记录和计算三种功能，主要由兰格博士自己设计并在我们实验室制成。这台仪器的主要优点在于它的运行速度，它可以在短时间内进行大量的测量，并可以避免产生任何可能的偏差。

综上所述，在我们最初的文章中已经提供了充分确凿的证据来证实太阳普遍磁场的存在，并测得了它的极性和强度量级。即便如此，我们手头上还是没有任何关于磁场极性在太阳黑子极小年会改变的证据。由于我们做出的每一个判断都是基于成千上万次测量的结果，因此它反映的是太阳上很多点的平均值。因此，要彻底弄清太阳上磁场强度的变化还需要更多的时间。

在解决地球磁场和磁场本身基本性质的问题上，太阳黑子显著的磁现象和我们提供的整个太阳是一个磁体的证据似乎有很重要的意义。人们很难回避这样的想法，那就是太阳黑子的强磁场和相比之下弱得多的太阳和地球上的整体磁场是由相同的原因造成的，即载有带电粒子的物体的旋转。基于这种观点的很多不同假说都已被检验过了，但是进一步研究的空间仍然很大。当然，我们的工作会最先涉及一些最简单的普遍假设，但是也不应该放弃对那些由 3% 的太阳黑子所引起的反常现象的深入研究。

（史春晖 翻译；蒋世仰 审稿）

The Expanding Universe and the Origin of the Great Nebulae*

J. H. Jeans

Editor's Note

The astronomer James Jeans here criticises recent speculations of George Gamow and Edward Teller on the origins of galaxies. Gamow and Teller started from an estimate of the mass density of the universe, and reasoned that for the condensation of galaxies ever to have taken place, the average distance between galaxies must have once been 1,000 times smaller than today, a ratio curiously similar to that of the distance between galaxies to their typical diameters. Jeans rejects this argument as following from poor estimates of these quantities, and instead offers his own view: that current physics allows for the possibility of galaxy formation under a wide range of conditions, indeed so wide that nothing can be ruled out as impossible.

PROF. G. Gamow and Dr. E. Teller have propounded[1] an interesting view of the genesis of the nebulae, but I doubt if either their arguments or their calculations can survive criticism. In brief, they think that the average density of matter in the present universe is about 10^{-30}; that matter of this density, spread uniformly as a gas through space, could only condense into the present nebulae if the gas-molecules were at a temperature of "several million degrees, which seems to be very unlikely"; that for condensation to have taken place in the past, inter-nebular distances must have been less than now by a factor of about 1,000. "Since the present average distance between the nebulae is just about 1,000 times larger than their diameters, we conclude ..." and so on. Their whole cosmogony is based on the supposed equality of these two factors of 1,000. I believe both to be wrongly calculated.

The average internebular distance is about 2,000,000 light-years[2], but the average diameter of a nebula[3] is certainly more than 2,000 light-years. The true ratio here is probably nearer to 100 than to 1,000.

Recent investigations by Sinclair-Smith, Holmberg and others suggest that Hubble's estimate of 10^{-30} for mean density (which was anyhow only given as a lower limit), is emphatically on the low side. The average nebula appears to have a mass of at least 10^{11} Suns. This gives a mean density, for nebular matter alone, of 2.5×10^{-29}, which reduces the needed contraction from 1,000:1 to 40:1. But this is not all; in calculating the criterion for the formation of condensations, the authors neglect the gravitational field of all uncondensed matter outside the incipient condensation. My own calculations[4] indicate

* Translator's Note: "The Great Nebulae" in this paper refers to galaxies, which are different from the modern notion of nebulae as gas clouds.

膨胀的宇宙和大星云*的起源

金斯

编者按

本文中天文学家詹姆斯·金斯评论了近期乔治·伽莫夫和爱德华·特勒对星系起源的推测。伽莫夫和特勒从对宇宙中物质密度的估算开始，继而推测如果星系的塌缩曾经发生过，那么过去星系间的平均距离应该是今天的1/1,000，这个比值非常接近于星系间的距离与它们的典型直径的比值。本文中，金斯认为这些估算数据的准确性太差从而否定了这个论据，并进一步提出了他自己的观点：根据现在的物理学，星系可以在一个极为广泛的条件下形成，甚至没有什么条件是不可能的。

乔治·伽莫夫教授和爱德华·特勒博士提出了[1]一个关于星云诞生的有趣观点，不过我怀疑他们的论据以及他们的计算能否经得起推敲。简而言之，他们认为在今天的宇宙中，物质的平均密度大约是 10^{-30}；而只有当这些气体分子处于“看似不可能的数百万度的”温度时，像气体一样均匀散布于空间中的这一密度的物质才能刚好塌缩成今天的星云；而由于过去曾发生了这种塌缩，因此过去星云间的距离应该是今天的1/1,000。“因为今天星云之间的平均距离正好比它们的直径大差不多1,000倍，所以我们可以得出……”等等。实际上，他们的整个宇宙演化论都是基于假定的这两个1,000因子是相等的。但我认为这两个因子都算错了。

星云间的平均距离大约是2,000,000光年[2]，但是星云的平均直径[3]必定大于2,000光年。因此两者之间真实的比值应该更接近100而不是1,000。

辛克莱·史密斯、霍姆伯格以及其他人的最新研究认为，哈勃估算的平均密度值 10^{-30}（只是作为一个下限而言）无疑是偏低的。一般星云的质量至少是太阳质量的 10^{11} 倍。如果仅考虑星云物质，则给出的平均密度为 2.5×10^{-29}，这把所需的塌缩比从1,000∶1降到了40∶1。不过这还不够全面；在计算塌缩产生的条件时，作者忽略了初始塌缩区域之外所有未塌缩物质的引力场。我自己的计算[4]表明，这会进一

* 译者注：这里的“大星云”指的是现在通常所说的星系，和现在通常说的星云的概念不同。

that this introduces a further wrong factor of $8\pi^2/9$, or about 9.

When all this is put straight, I find that the present universe, as it now stands, could be formed by the condensation of a gas with a "thermal" velocity of about 20 km. a second. To obtain this velocity, we need not postulate gas molecules (!) "at temperatures of several million degrees"; we can get it from hydrogen atoms at 19,000°C., or from free electrons at −260°C., or from a mixture of atoms and electrons at any temperature we choose between −260°C. and 19,000°C. The range is so wide that nothing can be ruled out as impossible.

The range of universes which could be formed by gas-condensation would appear, then, to include the actual present universe. If so, we cannot form any estimate of the epoch, or any picture of the manner, in which our universe was formed, from the considerations proposed by Gamow and Teller.

I believe other objections can be brought against their arguments without being hypercritical. For example, in the expanding universe of the theory of relativity, material particles gain kinetic energy during the process of expansion. But this gain must not be included in the equation of energy; if it is, there is no longer conservation of energy. This makes me think that the authors' arguments about velocity of escape are invalid.

(**143**, 158-159; 1939)

J. H. Jeans: Cleveland Lodge, Dorking.

References:

1. *Nature*, 143, 116 (1939).

2. Hubble, "The Realm of the Nebulae", 189.

3. *Loc. cit.*, 178.

4. "Astronomy and Cosmogony" (second edit.), 348.

步引入一个 $8\pi^2/9$ 或者说大约为 9 的错误因子。

当所有这些都被纠正之后我发现，今天的宇宙的状态可以由“热运动”速度大约为 20 千米 / 秒的气体塌缩而成。为得到这个速度，我们不需要假设气体分子（!）“处于数百万度的温度”；我们可以从 19,000℃的氢原子，或者 −260℃的自由电子，或者处于我们选择的 −260℃到 19,000℃之间的任意温度的原子和电子的混合物中得到这个速度。这个范围如此之广，几乎在任何条件下都可以实现。

于是，由气体塌缩形成的宇宙范围就似乎包括了今天实际存在的宇宙。如果是这样，我们就不能根据伽莫夫和特勒提出的理论得到任何对演化时期的估算或者宇宙形成方式的图像。

我相信对他们的论据再提出一些反对意见还是很容易的。例如，相对论认为，在不断膨胀的宇宙中，物质粒子在膨胀过程中获得动能。但是，所获得的这份动能不包含在能量方程中；否则的话，就不再有能量守恒。这使我认为，作者关于逃逸速度的论点的推理是错误的。

（钱磊 翻译；何香涛 审稿）

Interpretation of the Red-shifts of the Light from Extra-galactic Nebulae

M. E. J. G. de Bray

Editor's Note

Edwin Hubble, the distinguished American astronomer, was the first to point out that the galaxies identified in the sky as groups of stars held together by their mutual gravitational attraction also appeared to be receding from the Earth (or our own galaxy) with velocities that may be a substantial fraction of the velocity of light. This letter shows that the new facts did not disconcert those who believed in a secular decline of the velocity of light with the mere passage of time.

IN his most recent contribution on this subject[1] Hubble mentions a red-shift of +0.231 A. This corresponds to a velocity of recession of approximately 43,000 miles per second, nearly a quarter of the velocity of light! The conviction grows upon one that these fantastic velocities are not real.

Alternative interpretations of the red-shift have been proposed, but have failed to provide an acceptable explanation. No one has taken account of the fact that a list of the available determinations of the velocity of light shows a steady decrease of velocity with time, following a linear law, $c = 299{,}900 - 3.855\ T$, for the epoch 1900[2] (see Table 1).

Table 1

T	c
1902.4	299,901 ± 84 km./sec.
1924.6	299,802 ± 30
1926.0	299,796 ± 4
1928.5	299,778 ± 20
1932.1	299,774 ± 11
1937.7	299,764 ± 15
1938.5	299,761 ± ?

This may be due to a "fatigue" phenomenon: there are in Nature no inexhaustible stores of energy, no examples of perpetual motion. There are tremendous stores of energy and the ensuing motion may continue without perceptible alteration for periods almost beyond reckoning, but neither lasts indefinitely. The energy of a photon is not considerable and an explanation of the slowing-down may be sought in a decrease of frequency as the light journeys through space. It may be sought in the fact that the velocity of propagation of a vibration depends on the physical properties (gravitational, electrical, magnetic, atomic, thermal and probably others) of the space through which it is propagated, and that these properties are gradually altering. Whatever its explanation, the slowing-down is an observed fact.

河外星云光谱红移的解释

德布雷

编者按

美国著名天文学家埃德温·哈勃第一个指出，星系——那些在太空中由于恒星自身之间的引力作用而聚集在一起的星群——似乎也在以与光速相比相当可观的速度渐渐远离地球（或者说是银河系）。本文表明，对于那些相信仅仅随着时间的消逝光速就会持续减少的人们，新得到的观测事实并没有使他们感到惶恐不安。

哈勃在其红移课题[1]的最新研究中提到了河外星云光谱存在 +0.231 埃的红移量。这意味着该河外星云的退行速度大约为每秒 43,000 英里，这几乎是光速的 1/4！但是人们逐渐相信，如此巨大的速度并不是真实的。

人们已经提出若干种对红移的解释，但都没有达到令人满意的程度。因为这些解释中都没有考虑以下事实，即：一系列对光速的测量表明，光速随着时间流逝而持续减小，在 1900 年之后，其满足线性关系，c=299,900 − 3.855T[2]（参见表 1）。

表 1

T	c
1902.4	299,901 ± 84 千米 / 秒
1924.6	299,802 ± 30
1926.0	299,796 ± 4
1928.5	299,778 ± 20
1932.1	299,774 ± 11
1937.7	299,764 ± 15
1938.5	299,761 ± ?

这可能是一种“疲劳”现象，即：自然界中存储的能量并不是无穷无尽的，也没有永恒运动的实例存在。尽管目前自然界中存在着巨大的能量，并且在我们无法想象的很长的时间内运动会不断持续，这段时间内并不会发生任何可以察觉的变化，但这些运动并不会无限地持续。尽管一个光子的能量微不足道，但根据光在空间的传播过程中频率的减小，或许可以找到光速减慢的原因。我们借助以下事实加以理解，即：光振动的传播速度取决于它所通过空间的物理属性（包括引力、电、磁、原子、热和其他可能的作用），而这些属性是在逐渐变化的。但无论哪一种解释是正确的，光速变慢都是不争的事实。

Besides this, a satisfactory interpretation of the red-shifts must conform to the following further observed facts:

(1) Strömberg, by measuring the aberration of light from nebulae of the Ursa Major I cluster, has shown that the velocity of the light from this cluster is the same as that of light from laboratory sources[3]. Light from all sources, near and far, reach therefore the eye with the same velocity, c.

(2) Planck's constant h is constant, for Adams and Humason[3] have shown that the relation $E\lambda=ch$ (1) holds good for sources of light near and far, by comparing a grating and a prism spectrogram of the emission spectrum of the nucleus of *N.G.C.* 4151.

(3) The wave-length is constant during the light's journey: this assumption is accepted as correct when measuring radial velocities.

(4) The frequency n of a newly generated photon of energy E is the same for all sources of light near and far, since (1) may be written $E=hn$, and it is admitted that the energy of a photon generated by an atom of the same element under the same conditions is the same in all sources near and far. For example, the energy of the photon giving the K line of Ca^+ has the same energy whether generated in a remote nebula or on the Earth.

It follows from the above that light leaves a remote nebula with a velocity $c_0>c$, and a wave-length λ_0 given by the relation $c_0=n\lambda_0$ (2). During the light's journey, both c_0 and n decrease until, when the light reaches the Earth, they have become c and n' respectively, with $c=n'\lambda_0$ (3) and n' is less than n. Since $c_0>c$, from (2), $\lambda_0>\lambda$, the corresponding wave-length of light from laboratory sources, and $\lambda_0=\lambda+d\lambda$, which is the observed red-shift.

In Table 2, m_c is the luminosity of the fifth brightest nebula in each cluster (the clusters are so similar that this gives a precise measurement of distance) and $d\lambda/\lambda$ is the observed red-shift[1]. R is the distance in light years, from the relation $\log r=0.2\ m_c+4.803$; λ_0 is the initial wave-length of the K line of Ca^+, calculated from the red-shift, taking its Earth value as $\lambda=3933$ A.—it is the wave-length at both ends of the light's journey; n' is the frequency of the light from the nebula when it reaches the Earth, and c_0 is the initial velocity of the light. The nebulae are in order of increasing distances. The data relating to Ursa Major II are inconclusive: they have been derived from a single spectrogram of a single nebula only.

此外，一种令人满意的关于红移的解释还需要与以下几条进一步的观测事实相吻合：

(1) 斯特伦贝里测量了大熊座 I 星团中星云的光行差，结果表明来自该星团的光速与实验室光速相同[3]。因此，可以认为远近不同的所有光源发出的光到达人眼时的速度是一样的，均为 c。

(2) 亚当斯和赫马森[3]分别用光栅和棱镜测量了 *N.G.C.*4151 核心区域的光谱，对比结果表明，无论光源远近，关系式 $E\lambda=ch$ (1) 都可以很好地满足。这说明普朗克常数 h 确实是一个常数。

(3) 在测量视向速度时，我们认为此假设是正确的，即：光在传播过程中波长保持恒定。

(4) 由于 (1) 式也可以写为 $E=hn$，那么无论光源远近，新产生的能量为 E 的光子的频率 n 都是相同的。这相当于承认，无论远近，同种元素的原子在相同条件下产生的光子能量都是相同的。例如，Ca^+ 的 K 线对应的光子，无论产生于地球还是遥远的星云中，能量都是一样的。

根据上述可以得出以下推论：光离开遥远星云时的速度 $c_0>c$，其波长 λ_0 可以由关系式 $c_0=n\lambda_0$ (2) 得到。光在传播的过程中，光速 c_0 和频率 n 都不断减小，直至到达地球，此时二者分别变为 c 和 n'，但是它们仍然满足关系 $c=n'\lambda_0$ (3)，其中 $n'<n$。由于 $c_0>c$，根据 (2) 式可得 $\lambda_0>\lambda$，其中，λ 为实验室光源的波长。该不等式也可以写为 $\lambda_0=\lambda+d\lambda$，这就是我们观测到的红移现象。

在表 2 中，m_c 表示每个星团中亮度排名第五的星云的光度（这些星团非常相似，所以可以用于精确的距离测量），$d\lambda/\lambda$ 表示观测到的红移[1]。R 表示以光年为单位的距离，由下面的关系式可确定 R：$\log r = 0.2\, m_c+4.803$。λ_0 是 Ca^+ 的 K 线的初始波长，可以由观测到的红移量计算得到，地球上的光源发出的相应谱线的波长取为 λ=3,933 埃，我们认为光发出时和被接收到时波长都是该值。n' 表示来自星云的光到达地球时的频率，而 c_0 表示光的初始速度。表 2 中的星云按距离由近及远排列。大熊座 II 的数据是由单个星云的单一光谱推出来的，所以是不确定的。

Table 2

Cluster	m_c	$d\lambda/\lambda$	r light-years	λ_0 of K line of Ca^+	n'	c_0
—	—	—	—	(λ=3,933 A.)	(n=762.2×10^12^)	(c=299,774 km./sec.)
Virgo	10.49	0.0041A.	8×10^6^	3,949 A.	759.1×10^12^	300,990
Pegasus	12.88	0.0127	24	3,983	752.6	303,585
Perseus	13.48	0.0174	31.6	4,001	749.1	305,030
Coma	14.23	0.0245	44.6	4,029	744.0	307,090
U. Ma. I	16.12	0.0517	106.4	4,137	724.6	315,320
Leo	16.33	0.0653	117	4,190	715.5	319,360
Cor. Bor.	16.54	0.0707	129	4,211	711.9	320,960
(U. Ma. II	17.73	0.1403	223.4	4,485	668.4	341,850)
Boötes	17.89	0.1307	210.4	4,447	674.1	338,950

The slowing down of the frequency and of the velocity both follow linear laws: $n'(\times 10^{12})=762.2-0.4T$; and c_0(Km./sec.)=299,774–173T, where T is in millions of years.

Accordingly, 240 millions years ago, light travelled with a velocity approximately 13 percent greater than now, and the wave-length of the K line of Ca^+ was 4,447 A.

Should the above be confirmed, our measurements of stellar radial velocities will require a correction depending on the distance.

(**144**, 285; 1939)

M. E. J. Gheury de Bray: First Avenue House, High Holborn, London, W. C. 1, June 29.

References:

1. Hubble, "The Observational Approach to Cosmology".

2. *Nature*, **120**, 602 (1927). A later list of observations appears in *Isis*, **25** (2), No. 70 (Sept. 1927).

3. Reports Mount Wilson Obs., 1935–1936.

表 2

星云	m_c	$d\lambda/\lambda$	r 光年	λ_0 为 Ca^+ 的 K 线的初始波长	n'	c_0
—	—	—	—	(λ=3,933 埃)	(n=762.2×10^{12})	(c=299,774 千米 / 秒)
室女座	10.49	0.0041 埃	8×10^6	3,949 埃	759.1×10^{12}	300,990
飞马座	12.88	0.0127	24	3,983	752.6	303,585
英仙座	13.48	0.0174	31.6	4,001	749.1	305,030
后发座	14.23	0.0245	44.6	4,029	744.0	307,090
大熊座 I	16.12	0.0517	106.4	4,137	724.6	315,320
狮子座	16.33	0.0653	117	4,190	715.5	319,360
北冕座	16.54	0.0707	129	4,211	711.9	320,960
(大熊座 II	17.73	0.1403	223.4	4,485	668.4	341,850)
牧夫座	17.89	0.1307	210.4	4,447	674.1	338,950

由表可知，频率和速度的减小都遵循以下的线性规律：$n'(\times 10^{12})$=762.2−0.4 T；c_0 (千米 / 秒)=299,774−173T，其中 T 的单位为百万年。

因此，2.4 亿年前，光传播的速度比现在的值大约高 13%，那时 Ca^+ 的 K 线的波长为 4,447 埃。

如果以上理论得到证实的话，那么我们测量的恒星视向速度都需要根据距离的大小予以修正。

(金世超 翻译；何香涛 审稿)

The Crab Nebula: a Probable Super-nova

Editor's Note

In 1932, Subrahmanyan Chandrasekhar had proposed that massive stars which had exhausted their energy source would collapse and then, if sufficiently massive, would explode in a thermonuclear explosion called a supernova. The Crab Nebula has been recognised as an unusual object in the sky—a ring of light—and by the beginning of the twentieth century it was known to be expanding. By a remarkable sequence of inferences in which ancient Chinese records played an important part, the object has now been recognised to be the remnant of a supernova, as defined by Chandrasekhar, that exploded in 1054.

THE Crab Nebula in Taurus, which has the distinction of being No. 1 in Messier's Catalogue, is of peculiar interest, as is shown by articles published respectively in *L'Astronomie* (August 1939) and in the *Telescope* (September 1939). These articles summarize the results of several technical papers published on this subject during the past few years. It seems likely that the Crab Nebula was first recorded by John Bevis, an English physician, in 1731. It was rediscovered in 1758 by the French astronomer, Charles Messier, and was later the subject of careful scrutiny by Sir William Herschel. It was observed with the great reflector at Parsonstown, Ireland, by Lord Rosse, whose drawing of the nebula, published in 1844, probably suggested its name. It remained, however, for astronomical photography to show, first in 1892 by Isaac Roberts and later by the American astronomers (Keeler, Curtis and Ritchey), its peculiar filamentary structure and afterwards to provide data for measuring its linear rate of expansion of about 0.18″ per annum.

Using this value and the present angular dimensions of the nebula and extrapolating backwards, an interval of about 800 years is obtained for the nebula to expand from a point of origin. A similar time interval (900 years) was obtained by Hubble. Spectra of the Crab Nebula, first obtained in 1913–1915 at the Lowell Observatory, showed a bowing of the emission lines, when the slit of the spectrograph crossed the whole extent of the nebula (major axis 6′). Interpreting this feature as a differential Doppler effect due to the approaching nearer side of a shell of gas and the receding further side, Mayall in 1937 derived a velocity of expansion of about 1,300 km./sec. from Lick spectrograms. Assuming a constant rate of expansion, he concluded from the available data that the epoch of the outburst was about A.D. 1100. Meanwhile, Lundmark had pointed out that the Crab Nebula was near the position of the bright object recorded in Chinese and Japanese annals as having been seen for six months in A.D. 1054. In 1934, a translation by Y. Iba of the Japanese records gave the position of the object as near the star ζ Tauri and

蟹状星云可能是一颗超新星

编者按

1932年，苏布拉马尼扬·钱德拉塞卡提出，已耗尽能量的大质量恒星会发生塌缩，而且如果质量足够大，塌缩的恒星将会发生热核爆炸，从而形成所谓的超新星。人们已经认识到蟹状星云是太空中一个独特的天体——一个光环，并且早在20世纪初就知道它在不断地膨胀。通过一系列惊人的推断（其中中国古代的相关记载起了很大的作用）表明该天体是超新星爆发后留下的遗迹。按照钱德拉塞卡的观点，这颗超新星爆发于1054年。

在梅西耶星云表中金牛座的蟹状星云编号为No.1，人们对它特别感兴趣。关于这一星云的介绍可以分别参考《天文学》（1939年8月）和《望远镜》（1939年9月）中的文章。这些文章总结了在过去的几年中人们对这一天体的研究成果。1731年，英国内科医生约翰·贝维斯记载了蟹状星云，这很可能是关于该星云的最早记录。后来，法国天文学家夏尔·梅西耶于1758年再次发现该星云。紧接着，威廉·赫歇尔爵士对这一星云又进行了仔细的研究。罗斯勋爵在爱尔兰的帕森斯城也使用大反射望远镜对这一星云进行了观测，他所绘制的蟹状星云发表于1844年，这很可能就是其星名的由来。然而，直到1892年艾萨克·罗伯茨对该星云进行了拍摄，以及后来美国天文学家（基勒、柯蒂斯和里奇）所进行的类似工作，人们才发现该星云具有独特的细丝状结构，之后从这些相关的数据中人们还测出了其膨胀的线速度为每年0.18″。

根据蟹状星云这一膨胀速度和目前观测到的张角大小，通过反推可以求得蟹状星云从原点向外膨胀至今约有800年。哈勃也曾得到类似的结果（为900年）。洛威尔天文台于1913~1915年首次测量了蟹状星云的光谱；当摄谱仪的狭缝横跨整个星云（长轴为6′）时，观测到了许多弓状的发射谱线。这些特征被解释为由于靠近观测者一侧的星云气体壳的逼近和远端气体壳的远离造成的差分多普勒效应，梅奥尔于1937年根据利克天文台的光谱图推导出该星云的膨胀速度约为1,300千米/秒。他假定该膨胀速率一直不变，从而由现有的数据计算出该星云大约爆发于公元1100年。同时，伦德马克指出，中国和日本的史书中记载，公元1054年曾观测到持续了6个月（编者注：北宋至和元年）的特亮的客星（超新星），其位置与蟹伏星云的位置相近。1934年，伊巴翻译了日本古书中的相关记录，发现这一天体的位置靠近金

its brightness as equalling that of Jupiter. By combining the apparent linear expansion in seconds of arc per annum with the absolute expansion in km./sec., the order of distance was derived as 1,500 parsecs, equivalent to nearly 5,000 light years. Using this distance and the apparent magnitude of Jupiter (–2.2 m.), the absolute magnitude of –13.1 is obtained for the nova, which must have been at least one hundred times as bright as an ordinary nova.

In *Contributions* from the Mount Wilson Observatory No. 600, W. Baade assembles the evidence for the existence of two classes of novae, common novae and super-novae, which differ in luminosity by a factor of about 10,000. Typical of the former class is the nova which appeared in the Andromeda nebula in 1885 and reached a maximum apparent visual magnitude of 7.2, equivalent to an absolute magnitude of –15.0. To this recently recognized class of super-novae, so the evidence suggests, the nova of 1054 may have belonged, and the expanding shell of gas originating with the cosmic explosion is still visible as the Crab Nebula.

(**144**, 874; 1939)

牛座 ζ 星，其亮度相当于木星的亮度。综合上面已经得到蟹状星云的视膨胀角速度（单位：角秒 / 年）和绝对膨胀线速度（单位：千米 / 秒），我们可以计算出蟹状星云与地球的距离为 1,500 秒差距，约等于 5,000 光年。利用这一距离和木星的视星等（–2.2 等），我们可以得到这颗新星的绝对星等为 –13.1，该亮度至少是普通新星的 100 倍。

在第 600 期的《威尔逊山天文台研究通讯》上，巴德根据收集到的观测数据将新星分为两类：普通新星和超新星，后者的光度大约是前者的 10,000 倍。1885 年出现在仙女座星云中的新星就属于典型的普通新星，其最大视目视星等为 7.2，相当于绝对星等 –15.0。至于最近才发现的这类新星，即超新星，有关证据表明公元 1054 年观测到的那颗新星就属于这一类，并且这类新星中源于宇宙爆炸时期的气体壳层膨胀现象，至今仍然可以在蟹状星云中看到。

（金世超 翻译；何香涛 审稿）

The Evolution of the Stars

F. Hoyle and R. A. Lyttleton

Editor's Note

Astronomers Fred Hoyle and Raymond Lyttleton review how recent advances in nuclear theory were fundamentally changing astronomers' understanding of stellar evolution. It was now possible, as George Gamow had recently argued, to probe the nuclear chemistry likely to be important at the densities and temperatures prevailing in stars. But Hoyle and Lyttleton think Gamow went rather too far in suggesting most problems of stellar evolution were now solved. In Gamow's view, for example, all red giant stars must be of very recent origin, which seemed most unlikely. Hoyle and Lyttleton were correct: physicists then understood only the rudiments of nuclear chemistry, and knew nothing of nuclear fusion, a key process driving all stellar activity and evolution.

PROF. G. Gamow has recently discussed in *Nature*[1] the consequences of recent developments in nuclear theory on the problem of stellar evolution. In the light of the exact knowledge that is now available of a large number of nuclear reactions, it is possible to decide which processes rise to importance at the densities and temperatures prevailing in the stars, and on this basis trustworthy estimates have been given for the rate of liberation of subatomic energy. These results, which are the outcome of laboratory investigations, furnish the mathematical theory of internal constitution of the stars with a new equation that enables the luminosity of a star to be calculated by direct methods. The information so obtained has been utilized to attempt to resolve the many paradoxes and discrepancies encountered in discussing the general problem of stellar evolution. All this recent work has been authoritatively summarized by Gamow with great clarity and understanding in the article referred to above. It is therefore with some surprise that we find that Prof. Gamow concludes his article with the impression that these new developments practically solve the problems of stellar evolution. This seems to us to be so far from being the case that some further discussion of the claims of nuclear theory as the main factor in stellar evolution would be desirable.

In the first place, it should be noticed that the application of nuclear theory in its present state inevitably leads to a result at least as embarrassing as any of the questions that it might possibly resolve. For the theory maintains that no synthesis of atomic nuclei from hydrogen is possible within the stars except for the very light elements. This would imply that the stars can no longer be regarded as the building place of the heavy elements, which must have formed before they became part of the star—if indeed they ever were formed. Now although such a conclusion does not itself constitute a logical contradiction, it seems to us to present such overwhelming difficulty that it is much more reasonable to conclude that the basis of nuclear theory is in need of revision rather than that the heavy

恒星的演化

霍伊尔，利特尔顿

编者按

天文学家弗雷德·霍伊尔和雷蒙德·利特尔顿回顾了核物理理论的近期进展是如何从根本上改变天文学家对恒星演化的理解的。如乔治·伽莫夫近期所指出的，现在有可能证明核化学在恒星当时的密度和温度下是有重要作用的。伽莫夫认为恒星演化的大多数问题现在都已经解决了，但霍伊尔和利特尔顿对此并不认同。例如，伽莫夫认为所有红巨星一定是在新近才产生的，但这似乎是最不可能的事情。霍伊尔和利特尔顿是对的：那时的大多数物理学家只知道核化学方面的初级知识，但对核聚变这一驱动恒星活动和演化的关键过程却一无所知。

乔治·伽莫夫教授最近在《自然》[1] 杂志上讨论了核物理理论的最新进展在恒星演化问题上所取得的结果。根据现有可用大量核反应得到的准确知识，人们有可能确定在恒星当时密度和温度下哪些过程具有重要的作用，并且在此基础上，人们得到了对亚原子能量释放率的可靠估计。这些实验研究得到的结果为计算恒星内部结构的数学理论提供了一个新的方程，由此可以采用直接的方法计算恒星的光度。我们利用所获得的信息尝试解决恒星演化的普遍问题中遇到的许多矛盾和差异。在上面所引用的那篇文章中，伽莫夫以十分清晰易懂的方式对近来的所有工作进行了权威性的总结。令人有些惊讶的是伽莫夫教授在他的文章中得出的结论给人这样的印象：这些新的进展从实际上解决了恒星演化的多种问题。而在我们看来情况远非如此，核物理理论在恒星演化中作为主要因素的论断还需要进一步讨论。

首先，我们应该注意到目前核物理理论的应用和它有可能解决的任何问题一样，会不可避免地导致一种同样令人为难的结果。因为该理论认为，在恒星内部氢仅能合成一些非常轻的元素，而不能合成其他的原子核。这意味着恒星不能再被当成是重元素产生的地方，如果这些重元素确实曾经产生过，那么它们必须在变成恒星的一部分之前就已经产生了。现在尽管这个结论自身在逻辑上并不矛盾，但是目前在我们看来它却造成了非常巨大的困难，所以更为合理的结论是：核物理理论的基础所需要的是修正而非重元素不是由合成产生的。另一方面，许多研究者似乎已经认

elements were not formed by synthesis. On the other hand, many investigators seem to have accepted the former result as satisfactory, and in particular Gamow has proceeded to make it the basis of a theory of the red giant stars.

Secondly, Prof. Gamow apparently regards the conclusion that various classes of stars should be of totally different ages as a natural one. Thus the fact that the present theory leads to a life-time for certain massive stars of order 10^{-3} the life-time of the Sun is not regarded as a difficulty at all; indeed it is merely supposed that this is the case and the theory remains unquestioned. In point of fact, it is an essential part of the theory as proposed by Gamow that all red giant stars are considered as of very recent formation, since the presence of lithium, etc., is required to enable them to radiate with their supposed low internal temperatures.

Now if all the stars could be regarded as single autonomous bodies, it would be difficult to dispute the validity of these views by direct means. But it so happens that the frequent occurrence of binary systems enables a simple test of the theory to be made, for in the case of binary stars both components must be of comparable age. It is immediately clear that the well-known difficulty concerning the relative emission per unit mass of the components of doublestars must remain in any theory that appeals only to the internal properties of the stars, although by making very artificial assumptions in Gamow's theory some of these discrepancies might be avoided. For example, it would require as a general result that the less massive components of binary systems form with a hydrogen content differing systematically from that of their companions. In certain cases this would lead to even more dubious initial conditions: thus, whilst Sirius must have formed with high hydrogen content, it would have to be assumed (according to the more generally accepted theory of degenerate matter) that the companion formed almost solely from heavy elements for it to have been practically exhausted during the whole of its existence. Such a solution of this difficulty could scarcely be regarded as satisfactory even if there were no other objection to the theory described by Gamow.

It should be particularly noticed that the foregoing suggestion assumes that the components have not evolved by fission, for this latter process (even if it were dynamically satisfactory) would clearly lead to two stars of closely similar compositions. We have been able to show, however, that to produce close binary systems, periods of order 5×10^{10} years are necessary[2]. Thus the existence of spectroscopic binary systems in which one component is a red giant or any highly luminous star presents an immediate contradiction of the theory given by Gamow. Moreover, there seems to be a general tendency for the mass of binary systems to increase with decreasing separation.

While on the subject of fission, it is perhaps only proper to point out here that although the mathematical investigations of the development of rotating fluid masses clearly demonstrate that binary stars cannot be generated by fission[3], many astronomers do not yet seem to have realized the physical significance of the mathematical results. As a consequence of this there are still many who "believe" in the fission theory. But as will

同前一个结论是令人满意的，特别是伽莫夫，他已着手把它作为了红巨星理论的基础。

其次，伽莫夫教授显然认为各种类型的恒星有完全不同的年龄是很自然的事情。因此，由现有理论得出的某些大质量恒星的寿命仅仅是太阳寿命的 10^{-3} 倍的结论完全没有被认为是一件困难的事情；事实上大家认为这就是事实且现有理论是毋庸置疑的。其实正如伽莫夫所提出的，所有红巨星形成于最近这样一个观点是现有理论里必不可少的一部分，这是因为需要锂等元素的存在，以使其可以在假定的内部低温中发出辐射。

现在，如果所有恒星都可以被看作是独立的天体，那么可能很难用直接的方法对这些观点的正确性提出质疑。但幸亏多次发现的双星系统使我们可以对该理论进行简单的检验，因为在此情况下，两颗成员星应该有同等的年龄。尽管伽莫夫理论中人为的假设避免了某些矛盾，但很显然，对于任何单纯只研究恒星内部性质的理论来说，著名的关于双星中成员星的单位质量相对辐射的困难仍然存在。例如，作为一个一般性结果人们可能会假设双星系统中质量较小的成员星在形成的时候氢含量系统性地和它们另外的成员星不同。在某些情况下，这可能会让初始条件变得更加可疑：所以，虽然天狼星在形成时氢含量可能很高，但我们不得不假设（考虑到简并态物质的理论已被更为广泛地接受了）其成员星几乎只由重元素组成，以使其在存在的整个期间耗尽主要的核燃料。即使对于伽莫夫提出的理论没有其他反对的理由，但是对于上述困难，这样的解决方案也是绝不可能令人满意的。

应该特别注意，上述理论假设了成员星不是通过分裂演化而来的，因为分裂演化(尽管在动力学上是合理的)会产生两颗组分相似的恒星。然而，我们已经能够证明，必须有 5×10^{10} 年量级的时间 [2] 才能产生密近双星系统。于是分光双星中，若其中一颗成员星是红巨星或者任一颗高亮度的星，就必然同伽莫夫所提出的理论相矛盾。另外，似乎存在一个普遍趋势，即双星系统的质量随间距的减小而增加。

在恒星分裂的问题上，或许在这里指出最为合适，即尽管转动流体演化的数学研究结果清楚地证明了双星不可能由分裂产生 [3]，但是许多天文学家似乎还并没有意识到这些数学结果在物理方面上的重要性。因此仍然还有很多人“相信”分裂理论。但是正如人们将要看到的，即使求助于分裂理论也不能挽救伽莫夫的红巨星理论，

be seen, even an appeal to fission could not save Gamow's theory of the red giant stars, in addition to which it would raise afresh the difficulty of the relative emissions of the components in binary stars. Even if some process of break-up of a single star led to a binary system, it is evident that the components must have similar chemical compositions, while in close binary systems consisting of a red giant star and a class *B* star, Gamow's theory would require the red star to be the less massive component on account of the mass luminosity relation. But observation shows that in such pairs the giant star tends to be the more massive component. This is the case, for example, in the three stars υ Sagittarius, *V V* Cephei and ζ Aurigae. Thus it seems that no matter from what angle we approach the questions raised by the observed properties of binary systems, the paradoxes already recognized by astronomers must remain in one form or another.

However, quite apart from the foregoing objections to the constructive portion of Gamow's article, it is very noticeable that no reference is made to the wide class of dynamical features that is associated with the stars. This, of course, is the direct result of attending only to the internal physical properties of the stars; but the dynamical features we have in mind are altogether too marked to remain unaccounted for in a satisfactory theory. Thus such questions as the formation of individual stars, and of binary and multiple systems, together with the general increase of mass with decreasing separation, and the observed approximation to equipartition of energy among the stars seem to present the real key to any theory of stellar evolution. An internal theory can give no explanation for the correlation between peculiar velocity and spectral class or the observed tendency for massive stars to lie in the galactic plane, features that must be related to the previous history of the stars.

It has been customary during recent years for investigators on stellar evolution to devote attention to internal constitution with little or no regard for the dynamical features. It appears that Prof. Gamow has followed essentially in this tradition and therefore confined his article to the modifications effected by the introduction of modern nuclear theory. Thus, in dealing with the properties of variable stars, no attempt is made to account for the three distinct periodicity groups comprised by stars of periods of order half a day, four days and 300 days. These variables also show a marked preference as regards spectral class, the first being largely of classes *B* and *A*, the second of *F* and *G* and the third of class *M*. Moreover, the two short-period groups exhibit a most remarkable property in that none of them, out of more than two hundred available examples, possesses a close companion, whereas about one star in five of normal stars of similar spectral classes does possess a close companion. On the other hand, long-period variables appear to possess a normal complement of companions. The first and third types are stars of moderate luminosity and show no pronounced galactic concentration, whereas the variables of intermediate period, the Cepheids, are strongly concentrated to the galactic plane and are among the most luminous known stars. Thus it is clear that very remarkable dynamical properties are intimately connected with even the different types of variability, and therefore that purely internal considerations are most unlikely to prove capable of elucidating the nature of the connexion.

此外双星成员星的相对辐射问题将再次出现。就算单星分裂的某一阶段会产生双星系统，但可以肯定的是成员星之间仍应该有相似的化学组成，对于一个由一颗红巨星和一颗 B 型星组成的密近双星系统，伽莫夫的理论认为基于质光关系，红巨星的质量相对较小。但观测结果表明在这种双星系统中，红巨星的质量往往相对较大。例如在人马座 υ、仙王座 VV、御夫座 ζ 这三个双星系统中，情况均是如此。这样看来，似乎无论我们从什么角度来处理双星系统观测性质上的问题，这个已被天文学家意识到的矛盾总会以某种形式存在。

然而，除了前面对伽莫夫文章理论构建部分提出的异议外，很明显，还没有文献提到与恒星相关的广泛动力学特征。当然，这是只关注恒星内部物理性质的直接结果；但是，总的来说我们考虑的动力学性质是个很重要的问题，为此任何一个理论要令人信服都应该对此给出说明。于是，诸如单个恒星、双星以及聚星系统的形成，双星质量普遍随间距的减小而增加，以及观测到类似在恒星之间近似均分能量，类似这样的问题似乎是所有恒星演化理论的真正关键。一个恒星内部结构的理论不能解释本动速度和光谱类型之间的关系，也不能解释为什么观测到的大质量恒星都倾向位于银道面上，而这些特征必然和恒星早先的历史有关。

近年来，致力于恒星演化研究的科学家们都习惯性地只关注恒星内部组成而很少或者不关心其动力学特征。伽莫夫教授似乎也基本上遵循了这个传统，因此他的文章局限于引入现代核物理理论加以改进。因此，在处理变星性质的时候，他没有试图解释由周期为半天、四天和 300 天量级的恒星组成的三个不同的周期组。这些不同周期的变星也明确显示出了与光谱型相关的顺序，第一类主要是 B 和 A 型星，第二类是 F 型星和 G 型星，而第三类是 M 型星。此外，两个短周期的组还显示出一个最显著的特性：在已有的两百多个例子中，没有一个有密近的伴星。而在具有类似光谱型的一般双星中，每 5 对就有一个具有密近的伴星。另一方面，长周期的变星有正常的伴星。第一类和第三类中等亮度的恒星并没有显著的向银道面聚拢，而中等周期的变星，如造父变星，却都位于已知最明亮的恒星之中并且都高度聚集在银道面上。这清晰地表示，明显的动力学性质甚至与不同类型的光变有着紧密的联系，因此只考虑恒星的内部结构是不大可能诠释这些联系的本质的。

From these and many other dynamical qualities associated with various types of stars, it appears to us that Prof. Gamow has over-estimated the importance of nuclear theory in the problem of stellar evolution. Indeed, in our opinion nuclear physics has very little to add to the results already conjectured by astrophysicists, and can merely serve to confirm these conjectures, a typical instance being the mass-luminosity relation itself. Finally, we wish to point out that although the present article consists largely of criticism, we have discussed elsewhere a number of the questions raised[2], and it has been found that purely dynamical considerations may be sufficient to provide a natural explanation of many of the difficulties mentioned in this article.

(**144**, 1019-1020; 1939)

F. Hoyle and R. A. Lyttleton: St. John's College, Cambridge.

References:

1. *Nature*, **144**, 575, 620 (Sept. 30 and Oct. 7, 1939).

2. *Proc. Camb. Phil. Soc.*, (4), **35** (1939).

3. *Mon. Not. Roy. Astr. Soc.*, **98**, 646 (1938).

考虑到上述这些情况以及其他与各种类型恒星相联系的动力学性质，在我们看来伽莫夫教授高估了核物理理论在恒星演化问题中的重要性。事实上，我们认为核物理对天体物理学家所推测出的结果几乎没有什么补充，只能用其来验证这些推断，一个典型的例子是质光关系本身。最后，我们想指出，尽管本文主要是批评，但是我们已经在另外一篇文章中对所提出的许多问题[2]进行了讨论，并且我们发现从纯粹的动力学角度考虑可能足以为本文中提到的这些困难提供一个合理的解释。

（钱磊 翻译；何香涛 审稿）

Interpretation of Nebular Red-shifts

K. R. Popper

Editor's Note

The chief interest of this paper is the name of the author—Karl R. Popper. He was born in New Zealand, trained as a physicist and in the 1940s took a lively interest in controversies about the expansion of the Universe. Popper eventually became a philosopher, based at the London School of Economics, and introduced into science the idea that the only valuable scientific theories are those that can be falsified.

IN a recent communication[1], M. E. J. Gheury de Bray discusses the possibility of explaining the red-shifts by assuming that the velocity of light (c_T) is constant throughout the universe at any given time but decreases with time (T).

This hypothesis implies the assumption (*a*) that the atomic frequencies remain constant throughout space and time and may therefore be used as clocks (atomic clock "*AC*") for time measurements; (*b*) that measuring rods, other than light years (which contract with a decrease of c_T), can be found. For such measuring rods we may now choose the distances of *the material points* (for example, the nebulae) in the universe, as according to the hypothesis the universe neither expands nor contracts. Such measuring rods we call "material rods" ("*MR*").

It shall be shown here that the proffered hypothesis, based on the measuring system *AC* + *MR*, is one of three alternative ways of formulating the hypothesis of the expanding universe.

The following clocks and measuring rods can be taken as basis of a measuring system for cosmological purposes:

(*a*) Clocks	(*b*) Measuring rods
AC (atomic clocks); *or*	*MR* (material rods); *or*
LC (light-clocks: the time taken by light over a given distance—*c* is here assumed as constant).	*LR* (light-rods, for example, light years: the distance covered by light in a given time—*c* is here assumed as constant)

Only the following three combinations of these clocks and measuring rods can be taken as basis of a measuring system: *AC* + *MR*, which is the basis of Mr. Gheury de Bray's hypothesis; *AC* + *LR*, a basis which leads to the expansion theory of the red-shifts, and *LC* + *MR*, which leads to what Milne[2] calls "dynamical time scale", and to the "speeding up" theory of the red-shifts.

对星云红移的解释

波普尔

编者按

这篇文章最有意思的是作者本人——卡尔·波普尔。他出生于新西兰并被培养成为一名物理学家，20 世纪 40 年代他对有关宇宙膨胀的争论产生了极大的兴趣。但波普尔最终成为了伦敦经济学院的一名哲学家，并把只有有价值的科学理论才是可证伪的这一理念引入到科学中。

在最近的通信文章中[1]，谷瑞·德布雷讨论了一种对引力红移的解释，该解释假设：对于一个给定的时刻，宇宙中各处光速（c_T）都是常数，但会随时间（T）的流逝而减小。

该假说包含以下假设：(*a*) 原子的频率保持恒定，不随空间和时间变化，因此可以作为测量时间的钟（即原子钟，缩写为“*AC*”）；(*b*) 除了随着 c_T 减小而缩短的光年外，还存在其他的测量标尺。根据宇宙既不扩张也不收缩的假说，我们可以选择宇宙中**物质点**（比如星云）之间的距离作为这种测量标尺。我们称这种测量标尺为“物质标尺”（缩写为“*MR*”）。

我们将在本文中证明，基于 *AC*+*MR* 测量系统提出的假设是构建宇宙膨胀假说的三种可选方法之一。

下文中的钟和测量标尺可以作为宇宙学测量的基准：

(*a*) 钟	(*b*) 测量标尺
AC（原子钟）；**或者** *LC*（光钟：光传播一段给定距离所需的时间——*c* 在这里被假设为常数）。	*MR*（物质标尺）；**或者** *LR*（光标尺，例如光年：光在一段给定时间内传播的距离——*c* 在这里被假设为常数）。

只有以下三种钟和测量标尺的组合才能作为测量系统的基准：*AC*+*MR*，这是谷瑞·德布雷先生假说的基础；*AC*+*LR*，这是红移膨胀理论的基础；*LC*+*MR*，这构成了米耳恩[2]的“动力学时标”理论以及红移“加速”理论的基础。

($AC + LR$). The theory of the expanding universe states that the distances between material points, measured in LR, increase; in other words, it maintains an increase of MR against LR. This leads to the recession of distant nebulae and to the Doppler effect. The identity of the frequencies of characteristic spectral lines throughout the universe (atomic clocks) is thereby assumed; the measuring system of the theory is thus the $AC + LR$ -system. In tracing back the recessive movements it is found that 1.86×10^9 years ago the size of the universe was zero. Thus an absolute scale of time T (based on the atomic clock) is assumed in which the unit is the (atomic) year and the *present value* of T is $T_P = 1.86\times10^9$. (This "age of the world" is confirmed by other measurements based on atomic radioactivity clocks.)

($AC + MR$). The above statement that MR increases if measured in LR is equivalent to saying that LR decreases if measured in MR. If MR is chosen and the clocks are not changed, the system $AC + MR$ is adopted. The decrease of LR means a decrease of the velocity of light (measured in MR). This permits an explanation of the red-shift of old light, since c_T was greater at the time $T-\Delta T$ when it was emitted and therefore $\lambda_{T-\Delta T}$ is longer than λ_T.

($LC + MR$). There is a third way to express the same facts. If we allow the light a longer time for its voyage, by defining our clocks in such a manner that they are slowing down as compared with the atomic clocks, then the decrease of c can be made to disappear. In other words, we adopt a new time scale τ in which the atomic frequencies throughout the universe are speeding up; this explains the red-shift of old light.

The three alternative ways of expressing things agree in regard to the observable effects they describe. All predict the red-shift of old (distant) light. All therefore also predict a violet shift of characteristic spectral lines in the course of time to come (in $AC + LR$, that is, in the expansion theory, this has to be explained by an expansion of the spectroscope's grating). The three theories are logically equivalent, and therefore do not describe alternative *facts*, but the same facts in alternative *languages*. (To ask whether "in reality" the universe expands, or c decreases, or the frequencies speed up, is not more legitimate than, when prices of goods fall throughout the economic system, to ask whether "in reality" the value of money has increased or the value of the goods has decreased.) Nevertheless, the $AC + MR$-language seems to offer a particularly simple mathematical treatment and, furthermore, an *observational approach* to Milne's "dynamical" time scale. It shall therefore be briefly examined.

In $AC + MR$, where characteristic atomic frequencies are assumed constant and c_T depends in some way upon T,

$$(d\lambda/\lambda)_{T+\Delta T} = (\Delta c_T/T)_{T+\Delta T}; \tag{1}$$

on the other hand, the observational law of red-shifts (velocity-distance relation) can be written:

$$(d\lambda/\lambda)_{T_P-\Delta T} = (\Delta T/T)_P, \tag{2}$$

($AC+LR$)。宇宙膨胀理论认为，用 LR 测量得到的物质点之间距离是不断增加的；换言之，相对于 LR 测量结果，MR 测量结果有所增加。这种膨胀将导致远距离星云的退行以及多普勒效应。因此假设宇宙各处的特征谱线频率（即原子钟）是相同的；则对应的测量系自然是 $AC+LR$ 系。根据反演退行运动可知，1.86×10^9 年前宇宙的尺寸为 0。因此，我们假定时间 T（基于原子钟）的绝对标度以（原子）年为单位，T 的**当时值**为 $T_P = 1.86 \times 10^9$。（基于原子放射钟的其他测量方法也可证实这一“世界年龄”。）

($AC+MR$)。上文中提到，如果用 LR 测量会发现 MR 值增加了，这相当于说如果用 MR 测量会发现 LR 值减小。如果保持原子钟 AC 不变，而用 MR 代替 LR，那么就构成了另一种测量系统 $AC+MR$。LR 的减小意味着光速的减小（用 MR 测量）。这就可解释远距离光的红移现象了，因为光速 c_T 在光发出的时刻 $T-\Delta T$ 时较大，所以 $\lambda_{T-\Delta T}$ 大于 λ_T。

($LC+MR$)。本文中第三种描述相同的物理事实的方式。如果我们定义一种新的相对原子钟变慢的钟，这样就可以使光传播一段相同的距离但时间变长，于是光速 c 就不会减少了。换言之，我们采用一种新的时间标度 τ，在这种时标中，宇宙各处的原子频率都会加速；这同时也解释了远距离光的红移现象。

以上三种可选的描述方式与观测到的效应相符。三者均预言了远距离光的红移以及特征谱线随着时间到来的过程中的紫移现象（在 $AC+LR$ 系统中，即宇宙膨胀理论中，这必须用分光镜光栅的膨胀来解释）。这三种理论在逻辑上是等价的，因此它们并不是用于描述不同的物理**事实**，而是用不同的**语言**描述相同的事实。（若要问“实际上”是宇宙在膨胀，还是光速 c 在减小，还是频率在增加，这无异于在经济系统中物价下降时询问“实际上”是钱的价值在增加还是商品的价值在降低，疑问本身就缺乏合理性。）然而，$AC+MR$ 方法似乎可以提供一种十分简单的数学处理办法，并且可以为米耳恩的“动力学”时标提供**观测方法**。下面我们就对这方法进行简要的检验。

在 $AC+MR$ 系统中，假定原子的特征频率为常数，而 c_T 以某种规律随着 T 的变化而变化，

$$(d\lambda/\lambda)_{T+\Delta T} = (\Delta c_T/T)_{T+\Delta T} \tag{1}$$

另一方面，红移的观测定律（速度－距离关系）可以用下式表达：

$$(d\lambda/\lambda)_{T_P-\Delta T} = (\Delta T/T)_P \tag{2}$$

if the "apparent distance" in light years, calculated on observed luminosities, is identified with ΔT. From (1) and (2) we get

$$(\Delta c_T/\Delta T)_P = (dc_T/dT)_P = -(c_T/T)_P, \tag{3}$$

Generalizing this and integrating we get the law of the decrease of c:

$$c_T = c_P T_P/T. \tag{4}$$

If we proceed from $AC + MR$, to $LC + MR$, we have to introduce a time scale τ so that $c_\tau = c_P =$ constant. In order to get the general formula for $d\tau/dT$, we make use of (4), which can now be written as

$$d\tau/dT = c_T/c_\tau = c_T / c_P = T_P/T; \tag{5}$$

integrating and choosing $\tau_P = 0$ (Milne chooses $\tau_P = T_P$), we arrive at

$$\tau = T_P \log (T_\tau/T_P), \tag{6}$$

which is essentially Milne's formula $(A)^2$; the index "τ" in "T_τ" indicates the value of T at the instant τ of the τ-scale. Equivalent to (6) is $T_\tau = T_P e^{\tau/T_P}$. From this and (5) we get the law of the "speeding up" of characteristic frequencies when measured in LC+MR ; that is, on the τ-scale:

$$\nu_\tau = \nu_P dT_\tau/d\tau = \nu_P e^{\tau/T_P}, \tag{7}$$

and from this and $c_\tau = c_P$ we get, corresponding to (2), the law of red and violet shifts in the form

$$(d\lambda/\lambda)_{\tau+\Delta\tau} = -\Delta e^{\tau/T_P}/e^{\tau/T_P} = -\Delta T_\tau/T_\tau. \tag{8}$$

In the above deduction we have identified the "apparent distance" (calculated on observed luminosities, that is, energy densities ρ, and not corrected for departures) with ΔT. Thus our assumption regarding ρ is that it measures the square of the *time* ΔT the light travels, which in AC+MR cannot be identified with the ("real") *distance* r; for as the light was quicker when it was emitted, $r>\Delta T$. If the usual assumption that ρ measures not $(\Delta T)^2$ but r^2 is upheld, our deductions would have to be altered and we would neither get (4) nor its equivalent (6), that is, Milne's scale of τ. Vice versa, if our assumption regarding ρ is upheld and if we consequently arrive in $LC + MR$ at Milne's τ-scale, in which $r = \Delta\tau$, then neither r^2 nor $(\Delta\tau)^2$ can be taken as being measured by ρ. Whether this behaviour of ρ in τ can be deduced *a priori* in Milne's theory I do not know. But available *empirical* data seem to speak at least not against our assumption that $(\Delta T)^2$— which is smaller than r^2 —is a measure of ρ, for "present evidence points to observed luminosities (ρ) decreasing

如果根据观测的光度计算出“视距离”（以光年为单位）等同于 ΔT，那么根据公式（1）和（2），我们可以得到：

$$(\Delta c_T/\Delta T)_P = (dc_T/dT)_P = -(c_T/T)_P \tag{3}$$

将上式进行推广并积分，我们可以得出 c 的递减定律，公式如下：

$$c_T = c_P T_P/T \tag{4}$$

为了将 $AC+MR$ 系统转化为 $LC+MR$ 系统，我们必须引入时标 τ，这样可得 $c_\tau = c_P =$ 常数。为了得到 $d\tau/dT$ 的一般公式，我们可以将（4）式变形如下：

$$d\tau/dT = c_T/c_\tau = c_T / c_P = T_P/T \tag{5}$$

对上式积分并取 $\tau_P = 0$（米耳恩选取 $\tau_P = T_P$），我们可以得到下式：

$$\tau = T_P \log(T_\tau/T_P) \tag{6}$$

该式本质上就是米耳恩的公式（A）[2]；其中，T_τ 的下标 τ 表示在 τ 时标下 τ 时刻的 T 值。（6）式可以变形为 $T_\tau = T_P e^{\tau/T_P}$。根据此式和（5）式，我们可以用 $LC+MR$ 测量得到特征频率的“加速”定律；在 τ 时标中，可以表达为：

$$\nu_\tau = \nu_P dT_\tau/d\tau = \nu_P e^{\tau/T_P} \tag{7}$$

根据此式和 $c_\tau = c_P$，并结合（2）式，红移和紫移规律可以表达为以下形式：

$$(d\lambda/\lambda)_{\tau+\Delta\tau} = -\Delta e^{\tau/T_P}/e^{\tau/T_P} = -\Delta T_\tau/T_\tau \tag{8}$$

在上述推导中，我们假定“视距离”（视距离是根据观测光度，即能量密度 ρ 计算得到的，而且没有经过偏差校正）等同于 ΔT。因此，我们假设密度 ρ 与光传播一段距离所用的**时间** ΔT 的平方是相关的，在 $AC+MR$ 的测量系统中这种视距离与（“真实”）**距离** r 是不能等同的。因为这个系统中光刚发射时的速度要更大些，所以 $r > \Delta T$。如果采用通常的假设，即密度 ρ 并不符合 $(\Delta T)^2$ 的规律，而符合 r^2 的规律，那么我们就不得不修改我们的推导，这样的话，我们既不会得到（4）式也不会得到（6）式（即米耳恩 τ 时标）。反之亦然，如果我们关于密度 ρ 的假设成立，而且在 $LC+MR$ 系统中米耳恩时标 τ，即 $r = \Delta\tau$ 成立的话，那么 r^2 和 $(\Delta\tau)^2$ 便都不能用密度 ρ 来测量。我尚不清楚在米耳恩的理论中是否可以**先验地**推导出时标 τ 中 ρ 的规律。但是目前的**经验**数据至少没有与我们的假设相矛盾，在我们的假设中 ΔT^2（小

with distance *not even quite as rapidly* as we should expect. ..."[3].

(**145**, 69-70; 1940)

K. R. Popper: Canterbury University College, Christchurch, New Zealand, Nov. 27.

References:

1. *Nature*, 144, 285 (1939).
2. Milne, E. A., *Proc. Roy. Soc.*, A, 158, 327.
3. Schroedinger, E., *Nature*, 144. 593(1939).

于 r^2）是 ρ 的一种量度，因为“目前的证据显示，尽管观测到的光度（ρ）随距离减小，但**并没有**预期的**那样迅速**……”[3]。

（金世超 翻译；何香涛 审稿）

A Quantum Theory of the Origin of the Solar System

J. B. S. Haldane

Editor's Note

The British biologist J. B. S. Haldane here proposes a provocative hypothesis for the origin of the Solar System. The belief that it formed out of a rotating nebular disk has, he suggests, fallen into difficulty. Perhaps instead the planets were ejected from the Sun sometime in the distant past by the impact of photons of extraordinarily high energy. The energy required would (via $E=mc^2$) be equivalent to a mass of roughly 10^{19} tons. Haldane argues that such energies could have arisen in the early universe. Nearby stars may also have been hit with photons, and may thus also have planets. Haldane's speculations were soon forgotten, but inspired others to ponder the nature of the very early universe.

THE hypothesis of Kant and Laplace that the solar system originated by a gradual process from the contraction of a rotating nebula has become more and more improbable as the theory of such a process was investigated (cf. Jeans[1]). As a consequence, catastrophic theories of its origin have been put forward. In these theories another star, or even two stars, passed close to the Sun, or collided with it. In this article, which lays no claim to do more than open the discussion of possibilities, I suggest a quite different catastrophic origin, namely, a quantum transaction or perhaps a series of such transactions. I shall try to show that on Milne's[2] cosmological theory, this is a plausible hypothesis, and further that certain other cosmological problems are made less difficult if it is accepted.

According to Milne's cosmology, the universe can be represented in two distinct ways. On the kinematical representation, time t has a finite past t_0 of about 2×10^9 years or 6.3×10^{16} sec. Space is Euclidean, but every observer on a "fundamental particle" has his own private space. The infinite assemblage of fundamental particles, identified with the nuclei of galaxies, is contained in a finite sphere of radius ct, expanding with the velocity of light. An observer on any particle judges himself to be at the centre of this sphere, with the others receding from him. The different private spaces are related by the Lorentz–Larmor transformation. On the dynamical representation the time $\tau=t_0(\log t-\log t_0+1)$ has an infinite past, and the fundamental particles are at rest in a public hyperbolic space. The radii of planetary and atomic orbits are constant, as are the periods of planets and electrons, whereas in kinematic time and space the orbital radii and angular momentum increase with t.

太阳系起源的量子理论

霍尔丹

编者按

本文中，英国生物学家霍尔丹就太阳系的起源提出了一个富有争议的假设。他认为，太阳系是由一个旋转的星云盘形成的理论已经陷入了困境。真实的情况可能是在很久以前的某个时期，极端高能光子的冲撞导致行星被太阳喷射出来，所需的能量（根据 $E=mc^2$）相当于约 10^{19} 吨的质量。霍尔丹证明这样高的能量在宇宙早期是有可能存在的。近邻的恒星也会被光子所碰撞过，因而也可能有行星。霍尔丹的猜想很快就被人们遗忘了，但却引起了其他研究者们更深入地思考早期宇宙的性质。

康德和拉普拉斯的假说认为太阳系起源于旋转星云收缩的渐变过程，人们在对这个过程的理论进行研究之后，发现其越来越不可能（参见金斯的书[1]）。结果有人提出了太阳系起源的灾变理论。这些理论认为，另外一颗恒星，甚至两颗恒星，近距离地经过太阳，或者与其相撞。在本文里，我将集中对各种可能性展开讨论，并提出一种完全不同的灾变起源，即量子转换或者一系列类似的转换。我将尝试根据米耳恩[2]的宇宙学理论对此进行证明，这是一个合理的假说，而且如果它被接受的话，某些宇宙学问题就会变得不那么难以理解。

根据米耳恩的宇宙论，宇宙可以以两种截然不同的方式来表示。从运动学角度来看，时间 t 存在有限的过去，大约为 2×10^9 年或者 6.3×10^{16} 秒，用 t_0 表示。空间是欧几里德空间，但是每个处在“基本粒子”上的观察者都有其自己的专有空间。无限个这种被认为是星系核的基本粒子的集合，被包容在一个半径为 ct、以光速膨胀的有限体积的球内。处在任何粒子上的观察者均认为自己就处于这个球的中心，其他粒子则在不断远离他退行。不同的专有空间可由洛伦兹–拉莫尔变换联系起来。而在动力学表象中，时间 $\tau=t_0(\log t-\log t_0+1)$ 有无限长的过去，基本粒子在一个公共的双曲空间中处于静止。行星和原子的轨道半径都是一定的，行星和电子的周期也是恒定的，然而在运动学表象的时间和空间中，行星和原子的轨道半径及角动量都随着时间 t 的增加而增加。

One difficulty of the collision or encounter theory is the extreme rarity of such events. On some versions of the expanding universe theory such encounters were more probable in the remote past, when the stars were densely packed. But in Milne's cosmology an encounter was never more probable in a given stretch of dynamical time than it is now. It could be argued that as the dynamical past is infinite, an encounter is certain. However, it is no part of Milne's hypothesis that the stars have always existed.

Milne has not yet succeeded in deducing quantum mechanics from his few and simple postulates. His mechanics are in fact mainly classical. However, he has considered the behaviour of photons. The quantum parameter h, defined as E/v, where E is the energy radiated in a transition, and v its frequency, is invariant on the kinematical time-scale, for the red-shift of the distant galaxies is explained by the Doppler effect due to their recession; and the energy radiated in an atomic transition is invariant on either scale.

The main difficulty to be overcome in any theory of the origin of the solar system is this. The total angular momentum of the system is about 3.3×10^{50} erg-seconds. This is conserved on the dynamical scale. Unless most of the mass of the Sun is concentrated in a very dense core, all this angular momentum could be present in the Sun, due to its rotation, without its showing any more tendency to burst than does Jupiter at the present time. Hence some external source of energy must be postulated before it could emit the matter which condensed into the planets. The source of this energy has usually been supposed to be a star. I suggest that it may have been a photon.

The mass of the Sun is about 2.0×10^{33} gm., that of the planets about 0.00134 of this value; the solar radius 7×10^{10} cm.; and the gravitational constant 6.66×10^{-8}. The mechanical energy of the solar system is almost wholly given by the work required to lift the planets to their present orbits against solar gravitation. This again is almost equal to the work required to lift them to infinity, namely, $\gamma mM/R$, where γ is the gravitational constant, m and M the masses of the planets and Sun, and R the solar radius. The kinetic energy and the energy of the fall from infinity to the present orbits involve corrections of the order of $\gamma mM/r$, where r is the radius of a planetary orbit. Since for Jupiter $r = 1{,}100R$, these can be neglected.

$$\frac{\gamma mM}{R} = 5 \times 10^{45} \text{ ergs.}$$

Now at first sight a photon of this energy (and therefore of mass 6×10^{19} tons) appears a ridiculous conception. It would have, on the kinematical scale, a frequency of 8×10^{71} sec.$^{-1}$, and a wave-length of 4×10^{-62} cm. But now consider the conditions at time t, when t was very small. The radius of the universe was ct. It could not accommodate radiation of a wave-length greater than ct, and the past would be too short for such radiation to have accomplished even a single oscillation. At any time t there is a minimal possible size of photon, the frequency of which is of the order of t^{-1}. Probably it is a good deal less. This is borne out by the following consideration.

碰撞理论遇到的一个困难是发生这样的事件的概率极端稀少。根据某些膨胀宇宙理论，这种碰撞更有可能发生在非常久远的过去，因为当时恒星非常密集。但是在米耳恩的宇宙论里，在给定的过去某段动力学时间里发生一次碰撞的概率绝不比现在同样的一段时间内发生的概率大。也许可以争辩说，由于动力学的过去是无限的，所以碰撞肯定会发生。但是，米耳恩的假说中并没有假设恒星总是存在的。

米耳恩还没有成功地从其少数而简单的假设中推导出量子力学。实际上他的力学理论主要还是经典力学。然而，他还是考虑到了光子的行为。量子参数 h 被定义为 E/v，其在运动学时标下是不变的，E 是一次跃迁过程中辐射出的光子能量，v 是光子的频率，这是因为遥远星系的红移现象被解释为星系退行产生的多普勒效应，而一次原子跃迁所辐射出的能量在任何一种尺度下都是不变的。

所有太阳系起源的理论中都要克服一个主要困难。太阳系的总角动量大约为 3.3×10^{50} 尔格 · 秒。这在动力学时标内是守恒的，由于太阳在转动而且并没有比今天的木星有更明显的爆发倾向，所以除非太阳的大部分质量集中在极其致密的核心上，否则太阳系总的角动量将主要是太阳的角动量。因此，必须假设存在某些外在的能量来源，由此太阳才能抛射出之后凝聚成行星的物质。这个外在能量的来源通常被假定为一个恒星。而我认为它可能是一个光子。

太阳质量大约是 2.0×10^{33} 克，而行星质量大约是它的 0.134%；太阳的半径是 7×10^{10} 厘米；引力常数是 6.66×10^{-8}。太阳系的机械能几乎完全等于克服太阳的引力从而将太阳系中的行星送到它们现有的轨道所需做的功。也几乎等于将太阳系中的这些行星送到无穷远所做的功，即 $\gamma mM/R$，其中 γ 是引力常数，m 和 M 分别是行星和太阳的质量，而 R 是太阳的半径。对于行星的动能以及行星从无限远运动到现在的轨道上的能量，我们需要量级为 $\gamma mM/r$ 的修正，其中 r 是行星轨道的半径。由于对木星来说，$r = 1{,}100R$, 因此这个修正项可以忽略不计。

$$\frac{\gamma mM}{R} = 5\times10^{45}\ \text{尔格}$$

尽管乍看起来一个光子具有这个能量（质量为 6×10^{19} 吨）是非常荒唐的想法。从运动学的尺度来看，其频率将会是 8×10^{71} 次每秒，其波长仅有 4×10^{-62} 厘米。但是，如果我们考虑时间 t 很小的情况。当时宇宙的半径为 ct，因此它容不下一个波长比 ct 更长的辐射，而由于过去的时间过短以至于这种辐射无法完成哪怕一次振荡。在任何一个时间 t 都有一个最小的光子的可能尺度，其频率为 t^{-1} 的量级。或许还要小得多。这可由下述考虑来证明。

The mean lives of excited atoms liable to radiate light of visible frequency always appear to exceed 10^{-8} sec., though shorter lives are associated with higher frequencies. Thus out of a group of excited atoms existing from the beginning of kinematic time, only a minority would have radiated before $t = 10^{-8}$ sec., when the universe had a radius of 3 metres, or about 10^7 wave-lengths. At a time of the order of $t = 10^{-15}$ sec. there could, on any hypothesis, have been extremely little visible radiation, as it could not have been produced by the ordinary radiation processes. This argument suggests that radiation of frequency less than t^{-1} is impossible, while radiation with a frequency less than about $10^6 t^{-1}$ is produced, if at all, with some difficulty.

We can conclude, then, that at $t = 10^{-72}$ sec. the minimum photon corresponding to a completed oscillation would have had an energy of about 6.5×10^{45} ergs. So if there was any radiation at all at this time, it was sufficiently hard to lift the planets out of the Sun, if the Sun absorbed it. Its contribution of momentum would of course have been negligible. If some planetary matter was shot out of the solar system, and if some hydrogen was lost even from the major planets, the energy required must be multiplied by a small factor. If the radius of the Sun was larger it must be divided by a small factor. But we can conclude that the earliest date for the formation of the solar system is about $t = 10^{-72}$, or $\tau=-4.1\times10^{11}$ years, that is to say, the Earth cannot have revolved round the Sun much more than 4×10^{11} times. An error of 5 in the exponent of t would alter τ by about 5 percent.

At a time about $t = 10^{-75}$ sec. the minimal photon, which on the dynamical scale had a period of 2×10^9 years at any date, would have had an energy and frequency about 1,000 times greater than a planet-generating photon. If absorbed by a star of solar dimensions it would have been sufficient to split it into a pair the distance of which was large compared with their radii. In such a case the parent star could not have contained enough angular momentum to allow its progeny to move in circular orbits. Otherwise it would previously have broken up by centrifugal action. Hence the orbits of distant binaries would be expected to be very eccentric, as in fact they generally are. On this hypothesis the more widely separated binary stars were formed about 1–2×10^{10} years earlier than the solar system, on the dynamical scale, in agreement with the arguments based on gravitation, which ascribe to them an age of the order of 10^{11} years.

To return to the solar system, it may be asked whether it was formed by the absorption of a single photon, or of several in succession. The analogy with an atom, now less striking than at the time of Bohr's original theory, suggests the former hypothesis, but the latter must also be considered. The formation of the solar system would appear to have been in principle unobservable, since any radiation with which it could have been observed would either have passed through it unaltered or destroyed it. However, the correspondence principle can be applied to events of this character. The primitive Sun, containing the angular momentum of the whole solar system, had a period of rotation of the order of a day on the dynamical scale, or somewhat more if it was larger than at present on this scale, while Jupiter has a period of revolution of about twelve years. When the correspondence principle is applied to an atom, we find that the frequency of the absorbed radiation lies

发射可见光的受激原子的平均寿命大致都超过 10^{-8} 秒，不过发射更高频率辐射的受激原子寿命则相对较短。因此，在运动学时间一开始就存在的一组受激发原子中，只有少数会在 $t = 10^{-8}$ 秒之前辐射，这时宇宙的半径是 3 米，即相当于原子波长的 10^7 倍。由于可见辐射不能由通常的辐射过程产生，所以在 $t = 10^{-15}$ 秒量级的时间内，无论如何都只可能有极少的可见辐射。这个论证说明频率小于 t^{-1} 的辐射是不可能产生的，而如果真的能发生频率小于 $10^6 t^{-1}$ 的辐射，那也是相当困难的。

由此我们可以得出结论，在 $t = 10^{-72}$ 秒时，一个完整振荡的最小光子能量大约为 6.5×10^{45} 尔格。因此，如果这个时间内真的存在任何辐射，且被太阳吸收那么这个能量就足以使行星从太阳中爆发出来。当然，其动量的贡献在此已被忽略。如果某些行星物质被抛射出太阳系，并且甚至连某些大行星也丢失掉一些氢时，所需能量就必须乘以一个小的因子了。如果太阳的半径再大些的话，那么它就必须再除以一个小的因子。但是我们可以得出结论，太阳系最初形成的时间大约是 $t = 10^{-72}$, 或者 $\tau = -4.1\times10^{11}$ 年，这也就是说，地球围绕太阳旋转的圈数不可能超过 4×10^{11}。t 的指数上 5 的误差将会使 τ 变化约 5%。

时间约为 $t = 10^{-75}$ 秒时，在任何日期动力学标度上一个周期为 2×10^9 年的最小光子，其质量和频率都比可产生行星的光子高大约 1,000 倍。如果被一个太阳尺度的恒星吸收，它将足以将这颗恒星分裂成一对相距大于各自半径的星体。在这种情况下，母星将没有足够的角动量使其子量沿圆轨道运动。否则，它之前就会由于离心作用而分裂开来了。因此可预期，远距离的双星轨道将是有很大偏心率的，正如它们通常那样。由这个假说可得，距离较大的双星的形成比太阳系早了约 1×10^{10}~2×10^{10} 年，在动力学标度下，这与基于引力的考虑而得到的结果 10^{11} 年相符合。

再回到太阳系，人们或许会问，它是由于单个光子的吸收，还是由于几个连续光子的吸收而形成的。由于现在已经远没有在玻尔理论诞生之初时那么惊人了，所以与原子的相似显示了前一个可能是正确的，但是后一种可能性也必须予以考虑。太阳系的形成原则上似乎是不可观测的，因为任何能够对之进行观测的辐射要么是通过它后没有变化，要么就会毁坏它。然而，对应原理却能够被应用到具有这种特征的事件上。在动力学标度上，包含整个太阳系角动量的原初太阳，其自转周期是一天的量级，而如果它比现在大点的话，这个周期还会稍长点，相对应的木星公转的周期大概是 12 年。当对应原则被应用于一个原子时，我们发现被吸收辐射的频率处于该原子的初态和终态的频率之间。如果行星的形成也是如此的话，那么，在动力

between those of the atom in its initial and final states. If this was so for the formation of planets, the period of the photon required to produce the solar system (or Jupiter alone) is of the order of a year on the dynamical scale, so its frequency was about 2×10^9 times that of the minimum photon, and the epoch of origin was, on the t scale, about 2×10^9 times that calculated above as a minimum. Alternatively, we might argue as follows, The planet-making photon was a train of electro-magnetic waves. A train with a suitable period would set up electro-magnetic oscillations in the Sun, which might lead to the ejection of one or more planets. Given the size and physical state of the Sun, the period would be calculable. It would probably be rather shorter than that calculated above on the correspondence principle. In either case a photon would be most likely to be absorbed if it approached in the direction of the solar axis.

Since $\nu = 10^{72}$, and if T be the corresponding period on the dynamical scale, while t is the epoch of formation of the solar system, $\nu = t_0/t\mathrm{T}$; hence if T is about a year, $t = 2\times10^{-63}$ sec. roughly, whence $\tau = -3.7\times10^{11}$ years. If, on the other hand, the Sun absorbed a number of photons (say 9 in all, in order to form the major planets with Pluto and the parent of the asteroids) the value of ν for Jupiter would be only slightly less, but that for Mercury would be about 10^{66}, while the values of T would not differ among themselves so much. In this case the origins of the various planets were strung out over a period of about 4×10^{10} years of dynamical time, while the larger satellites of the outer planets (but probably not those of the Earth and Mars) could have been generated by photons absorbed by these planets at a still later date.

We must now consider the probable state of matter at this time. There could, of course, have been no radiation from atoms, nuclei, or electrons; and it is fairly clear that all matter was fully ionized, since any atomic systems would be ionized by thermal collisions, and free electrons would be unable to enter quantized orbits by emitting radiation. Thus stars formed by gravitational condensation could only lose the energy liberated in this process by emitting matter. Their radii would be those at which protons and electrons were just lost. Thus the solar radius on the dynamical scale might well have been ten times its present value. If so, the energy of the postulated photon must be diminished by a factor of 10, which would only decrease the dynamical date—τ by 4.6×10^9 years. The planets would, however, lose a good deal of matter immediately on formation, so that their original mass was greater than at present. This would give a correction in the opposite direction, while tidal friction would give a smaller correction.

The planets remained gaseous for a very long stretch of dynamical time. About $t=10^{-10}$ sec., loss of energy by radiation became appreciable, and by $t=10^{-4}$ sec., or $\tau= -10^{11}$ years it was in full swing. By about $t=10^{10}$ sec. or earlier, the planets had liquefied, and by $t=10^{13}$ sec. or $\tau= -1.5\times10^{10}$ years, the stars had contracted to normal stellar dimensions. These contractions were probably responsible for the origin of many close binary systems, of the Moon, and perhaps of the asteroids. During more than 3×10^{11} dynamical years the planets were gaseous. I suggest that during this period most of them acquired days equal to their years, while the Sun rotated in a period which was some sort of average

学尺度上，产生太阳系（或者仅仅是木星）所需的光子周期具有一年的量级，因此它的频率大概是最小光子的 2×10^{9} 倍，而在 t 尺度上，其起源的年代大约是上面计算所得到的最小值的 2×10^{9} 倍。或者也可以这么论证，产生行星的光子是一系列电磁波。具有合适周期的一系列电磁波会在太阳中产生电磁振荡，并由此导致一个或者多个行星的喷射。给定太阳的大小和物理状态，是可以计算出周期的。结果有可能比上面根据对应原则所计算出的结果略短。在任何一种情况下，如果光子沿太阳轴方向接近的话，它将很可能被吸收掉。

当频率 $\nu=10^{72}$，如果 T 是动力学标度上相应的周期，而 t 是太阳系形成的时间，则 $\nu=t_0/t\mathrm{T}$；因此，如果 T 大约是 1 年，大致取 $t=2\times10^{-63}$ 秒，则 $\tau=-3.7\times10^{11}$ 年。另一方面，如果太阳吸收了一些光子（例如说，为了产生包括冥王星在内的大行星以及小行星的母星，总共吸收了 9 个光子），木星的 ν 值将只会稍小一点，但水星的 ν 值将是 10^{66}，虽然它们各自的 T 值将不会有很大的差别。在这个情况下，各个行星大约是在动力学时标的 4×10^{10} 年时间里接连产生的，而外行星的较大卫星（很可能不包含地球和火星的卫星）可能是在此后的时期内由这些行星吸收光子所产生的。

我们现在必须考虑此时物质的可能状态。当然，原子、原子核和电子已经不再发出辐射了；并且很显然，所有物质都被完全电离了，这是因为任何原子系统都会通过热碰撞而被电离，而自由电子又不能够通过发出辐射而进入量子轨道。因此引力塌缩形成的恒星在这一过程中产生的能量只能通过抛射物质来释放。它们的半径恰好就是质子和电子会被丢失的半径大小。因此，在动力学标度上，太阳系的半径将正好是现在半径的 10 倍。 如果是这样的话，假设的光子的能量必须除以一个因子 10，而这只会使动力学表象的时间 τ 减少 4.6×10^{9} 年。然而，这会使行星在形成时立即丢失大量物质，因此它们原初的质量比现在的更大。这将会导出与潮汐摩擦力修正相反作用的一个修正，而潮汐摩擦力给出的是一个更小的修正值。

行星在一段很长的动力学时间里都为气态，大约到 $t=10^{-10}$ 秒时，其辐射所损失的能量开始变得可观，而到 $t=10^{-4}$ 秒或者 $\tau=-10^{11}$ 年为止其达到最活跃的状态。大约到 $t=10^{10}$ 秒或者更早，行星已经液化，到 $t=10^{13}$ 秒或者 $\tau=-1.5\times10^{14}$ 年时，恒星就已收缩到通常的恒星尺度了。在这个收缩过程中很可能导致很多密近双星系统、月球、可能还包括小行星的形成。在多于 3×10^{11} 的动力学年里，行星是气态的。我认为，在这个时期中，其中大多数行星的自转周期等于公转周期，而太阳的自转周期则相当于行星公转周期的某种平均值。行星收缩时，动力学标度下的角动量是

of the planetary years. On contraction, angular momentum on the dynamical scale was conserved, and the days therefore shortened to their present lengths on the dynamical scale, except in so far as they were lengthened by the ejection of satellites and by later tidal friction. This would involve contractions of the radii by factors varying between about 20 and 100. The exceptions may be said to prove the rule. Uranus has a retrograde relation. Its satellites revolve at a high inclination to the ecliptic, and that of Neptune has a retrograde motion. It would seem that tidal friction did not complete its work on the outer planets. The other cases of retrograde satellites are probably better explained by capture.

Energy is generally thought to be liberated in stars by the breakdown of unstable nuclei generated by thermal nuclear collisions. At present the rate of liberation is limited by the number of effective collisions, and is thus roughly constant in dynamical time. In the remote past nuclear breakdown was the limiting factor; so the Sun's radiation per dynamical year gradually rose to its present level, and has been fairly steady through geological time. Since through most of the history of the stars and planets in dynamical time nuclei of all kinds were effectively stable, but thermal collisions occurred, and moreover through a long dynamical period the minimum photons were capable of providing the energy for nuclear synthesis, it is suggested that the heavy elements, including the radioactive ones, were built up from hydrogen between the formation of the stars and the effective beginning of their thermal radiation.

If the solar system was generated by nine or more photon absorptions, most of the stars in our neighbourhood must have absorbed several photons, and produced planets. If it only absorbed one, the frequency of long-period binaries suggests that events of this type were not rare, so that our galaxy may include some hundreds of millions of planetary systems. If so, the field of biology is probably wider than has been suggested.

The galaxies have masses of the order of 10^{45} gm. This is the mass of a photon of period 10^{-92} sec., that is, of the minimum photon at $t=10^{-92}$ sec. Even if the galaxies were originally particles of matter as closely packed as atomic nuclei, and therefore of rather less than the size of the Sun, the energies needed to disrupt them into gas were considerably less than that of such a photon. Hence if the galaxies originated by the absorption of radiation, in which case some of Milne's "fundamental particles" may still exist in a compact form, or even if their whole mass arose from radiation, they cannot date from before $t = 10^{-92}$ sec., or $\tau = -5\times10^{11}$ years. Thus the long time-scale of about 10^{12} years deduced from a study of gravitational interactions of stars, which are naturally measured in dynamical time, appears as a consequence of Milne's theory.

The above arguments must be regarded as the attempt of a layman to deduce some of the consequences implicit in Milne's cosmology, consequences which he had partly envisaged when he wrote in 1936 that "all dynamical theories of the origin of the solar system may require drastic revision". I have doubtless missed other consequences as important as any which I may have elicited. Even if my hypothesis is found to be logically coherent, it may well prove, when fully developed, to be as untenable as Laplace's nebular theory. In

守恒的，因此它们的昼夜长度就这样缩短到动力学标度下现在的长度，只是在一定程度上受到抛射出卫星和后来的潮汐摩擦影响而使行星周期略微有所加长。这涉及的半径收缩因子大约在 20~100 之间。例外的情况也可以证明这个规律。天王星存在一个逆行的关系，其卫星转动轨道与黄道平面存在很大的倾角，而海王星的卫星也有逆行。看来，潮汐摩擦在外行星上并没有完成其作用。而将其他逆行的卫星解释或是被俘获而来的似乎更好一些。

人们通常认为，恒星中释放的能量的来源是热核碰撞产生的不稳定核的分裂。现阶段的能量释放率受有效碰撞数所限制，因此在动力学时间上几乎是常量。在遥远的过去，核碎裂的确是限制因素；所以太阳辐射在一个个动力学年后逐渐上升直到现在的水平，并在整个地质时间上都保持相当稳定。因为从动力学时标上看，在大部分恒星和行星的演化历史中，各种核子实际上都是稳定的，但由于热碰撞的发生，加之经过动力学时标下很长的一段时间后最小光子已能够提供核合成的能量，因此可以认为，包括放射性元素在内的比较重的元素，是在恒星形成之后和它们热辐射真正开始之前的这段时间中由氢所形成的。

如果太阳系是通过吸收九个或者更多的光子而产生的，那么我们周围的大部分恒星也必定吸收了一些光子并由此产生了行星。如果恒星只吸收一个光子，长周期双星的发现频次表明这种情况并不罕见，因此，我们的星系可能包括上亿个行星系统。如果真是这样的话，生物学的领域很可能比一般认为的要更为广阔。

星系质量的数量级为 10^{45} 克，相当于一个周期为 10^{-92} 秒的光子的质量，这也就是在 $t = 10^{-92}$ 秒时极小光子的质量。即使星系最初是像原子核内粒子那样紧密地聚在一起的物质粒子，因此其大小远小于太阳，要使它们瓦解变成气体所需的能量要比这样一个光子的能量小很多。因此如果星系起源于吸收辐射，在这种情况下米耳恩的某些“基本粒子”仍然可能以致密的形式存在，或者即使它们所有的质量都是由辐射产生，它们也不可能起源于 $t = 10^{-92}$ 秒或者 $\tau = -5 \times 10^{11}$ 年以前。因此，从恒星之间的引力相互作用研究中可以导出星系起源这个漫长的时间尺度是 10^{12} 年，基于动力学时间可以很自然地得到这一测量结果，而且似乎也是米耳恩理论的一个结论。

上面的论证应该被看作是一个外行人为了导出暗含于米耳恩宇宙论中某些结果所做的尝试，米耳恩在 1936 年写道，“所有关于太阳系起源的动力学理论都将需要做彻底的修改”，由此可见当时他就已经在一定程度上预见到了这些结果。我一定还忽略了某些和我所提到的结论同样重要的结论。即便我的假设在逻辑上被证明是通顺的，但当相关的理论完全发展起来后，可能事实会证明我的理论就如拉普拉斯的

particular, the secular stability of non-radiating ionized gaseous spheres and the relation of the uncertainty principle to the scale of time will require investigation. Above all, the details of the postulated process were in principle unobservable, and it will therefore be hard to test the proposed theory as rigorously as others have been tested in the past. This is a serious defect, since the value of a scientific theory increases with the number of ways in which it can be tested. But much of current physical theory has the same defect.

I have not suggested an origin for the postulated photon or photons. To do so would involve either a further step in a possibly infinite regress or the assumption that they were primordial constituents of the universe. They might, for example, have been generated by the acceleration of large charges during the origin of the galaxies. It may be asked what is their present state, if any of them have not been wholly or mainly converted into kinetic energy. The energy of a photon is invariant on the kinematic scale appropriate to the particle emitting it; but since a particle absorbing it is moving away from its source, its frequency and energy are lowered by the Doppler effect, and on the kinematical scale appropriate to such a particle, both vary as t^{-1}, where t is the epoch of absorption. Thus the postulated planet-making photons are now trains of electromagnetic waves of a period of the order of a year, and much too small to be observable in practice. The mass of matter at any time is thus the fraction of the mass at an earlier time which has not been degraded by the Doppler effect, and at a sufficiently early date most of the mass of the universe, or all of it, may have been radiation rather than matter.

In conclusion, I wish to thank Prof. Milne for his encouragement, and for elucidating several details of his cosmology in letters; and to emphasize that if the theory here sketched has any value at all, it will only prove its value by serving as a basis for exact calculations by persons better versed than myself in physics and astronomy.

(**155**, 133-135; 1945)

J. B. S. Haldane: F.R.S., University College, London.

References:

1. "Problems of Cosmogony and Stellar Dynamics" (Cambridge, 1919). "Astronomy and Cosmogony" (Cambridge, 1928).
2. "Relativity, Gravitation, and World Structure" (Oxford, 1935). *Proc. Roy. Soc.*, A, **154**, 22 (1936); **156**, 62 (1936); **158**, 324 (1937); **159**, 171, 526 (1937); **160**, 1, 24(1937); **165**, 313, 333 (1937). *Phil. Mag.*, **34**, 73 (1943).

星云理论那样站不住脚。特别是，对无辐射的电离气态球的长期稳定性以及不确定性原理与时间标度的关系还需进一步的研究。最重要的是，所假设过程中的细节在原则上是不可观测的，因此，对所提出的理论难以像过去的理论那样进行严格的检验。这是一个致命的缺陷，因为科学理论的价值随着可检验方式的数量的增加而增加。不过，当今很多物理理论都有同样的缺陷。

我并未给出所假设的一个或多个光子的起源。如果这样做的话势必要倒退到无限远的时间之初，或者是假定它们是宇宙的原初组成物。例如，它们可能产生自星系起源之初的大量电荷的加速运动。可能有人会问，如果它们中的任何一种还没有完全或者大部分转化成动能的话，它们现在是什么状态。在粒子发射出光子的情况下，光子的能量在运动学标度下是不变的；但是由于一个吸收光子的粒子远离发出光子的源，根据多普勒效应，其频率和能量都降低了，而在适合此粒子的运动学标度下，则都与 t^{-1} 成正比，其中 t 是吸收发生的时期。因此假设产生行星的光子这时就是一系列周期在年的数量级上的电磁波，这对实际观测来说太小了因而不能被观测。因此，任何时刻物质质量都是早先还没有因多普勒效应而减少时的物质质量的一部分，并且在足够早期时，宇宙的大部分质量，甚至全部质量都可能是辐射而非物质。

最后，我想对米耳恩教授的鼓励，以及他在来信中对他宇宙论的一些具体内容所做的解释表示感谢；我还想强调，如果本文所概述的理论有任何价值的话，它的价值仅限于作为比我更精通物理学和天文学的人进行严格计算时的一个基础。

（李忠伟 翻译；邓祖淦 审稿）

A Quantum Theory of the Origin of the Solar System

E. A. Milne

Editor's Note

Edward Arthur Milne here responds to a highly speculative idea by J. B. S. Haldane about the origin of the Solar System. Haldane noted that, owing to the expansion of the universe, photons in the remote past may have had extremely high energies. Some may have been so energetic that their absorption by matter of any sort may have been sufficient to create planets and perhaps even entire galaxies. Milne supports and adds to this notion. These speculations are in themselves of little interest today. But Milne does recognize that physical conditions in the early universe would be extreme, involving enormous energy densities utterly beyond those observable today—and that quantum theory might be needed to describe that situation.

PROF. Haldane's idea as developed in the foregoing article seems to me to be fundamentally important. As all may not be familiar with the details of kinematic cosmology, and as readers may have difficulty in keeping pace with the rapier-like speed of Prof. Haldane's mind, I beg to be allowed to traverse some of the same ground in more pedestrian fashion.

To begin with, a word of explanation: I first announced my ideas on the two time-scales at the Blackpool meeting of the British Association, in a discussion on the origin of the solar system; but the consequences of the ideas were so bizarre that I felt it to be absolutely necessary to develop the formal and philosophical aspects of the theory in full detail before proceeding to the more speculative consequences. This programme I carried out in a series of papers published by the Royal Society during 1936–38, and, though hindered by war-work, in *Philosophy* (1941), in addresses before the London Mathematical Society (1939), the Royal Society of Edinburgh (1943), the Royal Astronomical Society (1944) and in a series of papers in the *Phil. Mag.*(1943). I am at present wrestling with the difficult problem of the conservation of linear momentum for gravitating bodies in the expanding universe, and I do not wish to be hustled. However, in *Proc. Roy. Soc.*, A, **165**, 354(1938), discussing the role of the correspondence principle on the two time-scales, I wrote: "It is not a fanciful speculation to see in the interplay of radiation keeping t-time with matter obeying the classical laws of mechanics on the τ-scale a phenomenon giving rise to the possibility of change in the universe *in time*, and so an origin for the action of evolution in both the inorganic and organic universes". A possible mode of that interplay has now been pointed out by Haldane.

I have long been aware that all theories of the origin of the solar system require drastic

太阳系起源的量子理论

米耳恩

编者按

爱德华·阿瑟·米耳恩在这篇文章中回应了霍尔丹提出的关于太阳系起源的纯粹猜测性的想法。霍尔丹认为，由于宇宙的膨胀，光子在遥远的过去可能具有极高的能量。有的光子可能已经有非常高的能量，以至于无论何种物质对它们的吸收都足以创造出行星，甚至可能是整个星系。米耳恩支持这个想法，并丰富了该内容。在今天看来，这些推测本身几乎没有价值。但是米耳恩确实认识到宇宙早期的物理条件应该是极端的——包含着巨大的完全超出今天能观测到的能量密度，而要描述那种情况可能需要用到量子理论。

在我看来，霍尔丹教授在前面的文章中提出的观念非常重要。考虑到可能并不是所有人都熟悉运动宇宙学的细节，而且有些读者可能也很难跟上霍尔丹教授灵敏的思维，因此，请允许我在这里以更加通俗的方式对某些相同的话题加以讨论。

首先需要说明一点：事实上，在英国科学促进会的布莱克浦会议上，在讨论太阳系的起源时，我首次提出对于两种时间标度的观点；但是这一观点所导致的结论过于奇异，因而我觉得在进一步得出更加奇怪的结论之前，绝对有必要详细推导出这一理论的具体形式及其哲学内涵。在进行这项计划的过程中，我于 1936~1938 年间完成并由皇家学会发表了一系列文章，尽管由于战争工作受到阻碍，但我的文章仍然在《哲学》（1941 年）上发表了，我在伦敦数学学会（1939 年）、爱丁堡皇家学会（1943 年）、皇家天文学会（1944 年）都进行过这方面的讲演，此外我也在《哲学杂志》上发表过一系列文章（1943 年）。目前，我正在努力解决膨胀宇宙中受引力作用天体线动量守恒的难题，但我不希望被人催促。然而，在 1938 年《皇家学会学报》（A 辑，第 165 卷，第 354 页）的一篇文章中，当探讨相应原理对于两种时间标度所起的作用时，我写道："对于遵循经典力学定律下时间标度为 τ 的物质与持续时间 t 的辐射，它们之间的相互作用并不是异想天开的猜测；这种现象使得宇宙及时演化成为可能，也是无机与有机宇宙世界演化活动的根源。"现在霍尔丹指出了这种相互作用的一种可能的模式。

很久之前我就意识到，由于在时间接近 t=0，即太阳系刚刚诞生时，动力学和

re-consideration in the light of the fact that at times of the order of $t = 0$, when the solar system was born, dynamical and optical conditions were very different. Haldane works with equal facility in either time-scale; but it must be remembered that the τ-scale is a concession to our Newtonian predilections, that it has in its description a constant t_0 (the present age of the system on the t-scale) which has nothing to do with *phenomena*; it has to do only with the language by which we describe the phenomena. Phenomena themselves are best studied through the t-scale, and in this scale the precise value of t at the epoch studied is all-important.

In Haldane's calculation of the order of magnitude of the energy required to be communicated to the Sun to form the solar system of planets, he uses the formula $\gamma mM/R$, with the present values of γ and R. It might be objected that on my theory $\gamma \propto t$, and that therefore the required energy was then much smaller. The answer is that $R \propto t$ also, that energy is a "time-invariant", and that Haldane's calculation is accordingly correct. On his data, the value of $\gamma mM/R$ is 5×10^{45} ergs, as he says.

Previous speculators on the early history of the universe had always argued that since the universe is expanding, collisions must have been then more frequent, forgetting that lengths of material objects (that is, radii) would have then been much smaller. By translating to the τ-scale (stationary universe) we see that collisions would be just as frequent, or as infrequent, as now. The new contribution which Haldane makes is that the optical situation would be entirely different. At epoch t, when the radius of the expanding universe was ct, there cannot well have been photons of wave-length exceeding ct. The inequality $l<ct$ implies for the frequency n the relation $n=c/l>1/t$. (Here l and n are measured on the t-scale.) Working again on the t-scale, the inequality $\Delta E=h_0n>h_0/t$ gives the minimum permissible photon energy. Taking $h_0=6.55\times10^{-27}$, at epoch $t=10^{-72}$ sec., we get $\Delta E > 6.5\times10^{45}$ ergs, so that such photons as were then possible would have sufficient energy to disrupt the Sun and form a solar system.

There is no difficulty as to where the photons could come from. For according to kinematic relativity the mass (actual) and energy of the universe are infinite; and light must be present. Hence it must be, at small t, of enormous frequency and energy. The state of material atoms would be one of complete ionization; and the history of any photon would be one of successive degradations of frequency by interaction with matter, until at the present epoch light is *mostly* as we know it. This degradation of the individual photons due to interacting with matter must be distinguished from their constancy of frequency in time (t-scale) as they are propagated through empty space.

The epoch at which a photon ΔE was not less than 6.5×10^{45} ergs was, on the t-scale, 10^{-72} sec. The τ-measure of this epoch was $\tau=t_0\log(t/t_0)+t_0$. The "time ago" at which it occurred is τ_0-t, where τ_0, the present epoch on the τ-scale, is equal to t_0. This gives

光学条件与现在非常不同，所以关于太阳系起源的所有理论都需要进行全盘地重新考虑。霍尔丹对于这两种时间标度采用了相同的处理方法；但必须记住的是，这里的时间标度 τ 是对牛顿学说的偏好做出了一定的妥协，因为在它的描述中含有常数 t_0（时间标度 t 中系统的当前年龄），该常数与**现象**本身并无任何关联，只与我们用来描述现象的语言有关。现象本身最好是通过时间标度 t 来加以研究，因为在这样的时标下，研究时期内 t 的精确值是十分重要的。

霍尔丹在计算形成由行星组成的太阳系所需要传送给太阳的能量大小的量级时，采用了公式 $\gamma mM/R$，其中 γ 和 R 均采用了当前值。可能有人反对在我的理论中所采用的 $\gamma \propto t$，因为这样需要的能量就会小得多。对于这一疑问的解答是，也存在 $R \propto t$，这样能量就是一个“不随时间变化的量”，所以霍尔丹的计算是正确的。根据他的数据，如他所说 $\gamma mM/R$ 值为 5×10^{45} 尔格。

以前探讨宇宙早期历史的研究者们总是认为，既然宇宙在膨胀，碰撞在早期的宇宙中一定更加频繁，然而他们忘记了物质实体在那时的尺度（即半径）也比现在小很多。转换到 τ 时间标度中（定态宇宙），我们就会发现碰撞的频繁程度与现在大体相同。霍尔丹的新贡献在于，他认为宇宙早期与现在的光学环境应当完全不同。在时间 t，当膨胀的宇宙的半径为 ct 时，不可能存在波长超过 ct 的光子。根据不等式 $l<ct$ 可知，对于频率 n 而言，存在 $n=c/l>1/t$。（这里的 l 和 n 均在时间标度 t 中测量。）仍然在时间标度 t 中，不等式 $\Delta E=h_0n>h_0/t$ 给出了许可的最小光子能量。取 $h_0=6.55\times10^{-27}$，时间 $t=10^{-72}$ 秒，我们可以得到 $\Delta E>6.5\times10^{45}$ 尔格，因此这样的光子可以提供足够的能量使最初的太阳瓦解，从而形成太阳系。

解释这些光子的来源并不困难。根据相对论运动学，宇宙的（实际）质量和能量是无限的，而光是肯定存在的。因此在 t 很小时，光子必然具有极高的频率和能量。物质原子应处于一种完全电离状态，任何光子的演化史都将是通过与物质不断作用，频率持续降低的过程，直至现在，光就成为了我们**通常**所了解的样子。但我们必须区分出单个光子由于与物质作用而导致的能量衰减以及这些光子在真空中传播过程中它们的频率对于时间（时间标度 t）的恒定性。

当一个光子的能量 ΔE 不小于 6.5×10^{45} 尔格时，在时间标度 t 中是 10^{-72} 秒。这一时间在时间标度 τ 中的测量为 $\tau=t_0\log(t/t_0)+t_0$。它在“时间以前”发生是用 τ_0-t 表示，其中 τ_0 是现在的时间标度 τ，它就等于 t_0。由此，我们得到

$$\begin{aligned}\tau_0 - \tau = t_0 - \tau &= t_0 \log_e(t_0/t) \\ &= 6.3 \times 10^{16} \times 2.3 \times \log_{10}(6.3 \times 10^{16}/10^{-72}) \text{ sec.} \\ &= 6.3 \times 10^{16} \times 2.3 \times 88.8 \text{ sec.} = 4.1 \times 10^{11} \text{ yr.,}\end{aligned}$$

in agreement with Haldane. This is of the order of the "long" time-scale estimated by gravitational methods, that is, on the τ-scale.

Haldane's fundamental idea (pressing it to its limit) may be stated in the form that, just as the epoch $t = 0$ is a singularity in the mechanical t-history of the universe—an epoch at which the density was infinite—so the epoch $t = 0$ is a singularity in the optical history of the universe, namely, an epoch at which the frequency of radiation was infinite, because the wave-length had to be zero. Actually we can only make significant statements about the radiation for *small* epochs t, when the frequency would on the whole be very large. A spectrum would soon come into existence, by the absorption and backward emission (or backward scattering) of radiation by the naturally receding particles, with resulting degradation of frequencies by the cumulative Doppler effects. But some of the original high-frequency radiation would traverse space unscathed, and, in spite of the inevitable Doppler effect at the terrestrial receiving end, a small fraction of this would retain a still very high frequency, and might be the origin of the undulatory component of the present cosmic rays.

I think it would be wise, in this preliminary discussion of Haldane's idea, not to go into details as to how a primordial photon of huge energy could disrupt a star. It is sufficient to dwell on the remarkable result that Haldane has deduced from kinematic relativity, namely, that at very early epochs in the history of the universe, such photons as there were must have possessed enormous energies.

(**155**, 135-136; 1945)

E. A. Milne: F.R.S., Wadham College, Oxford.

$$\tau_0 - \tau = t_0 - \tau = t_0 \log_e(t_0/t)$$
$$= 6.3 \times 10^{16} \times 2.3 \times \log_{10}(6.3 \times 10^{16}/10^{-72}) \text{ 秒}$$
$$= 6.3 \times 10^{16} \times 2.3 \times 88.8 \text{秒} = 4.1 \times 10^{11} \text{年}$$

该值与霍尔丹的计算结果相一致。这正好是由引力方法估算出的“长”时标（即时间标度 τ）的量级。

霍尔丹的基本思想（推到其极致）可以通过以下形式说明：正如时刻 $t = 0$ 是宇宙力学 t 历史中的奇点（那时密度为无穷大）一样，时刻 $t = 0$ 也是宇宙光学史中的奇点，即在该时刻辐射频率无穷大，因为波长必须为0。实际上，我们只能对那些小的时间 t 中的辐射进行有意义的阐述，此时频率总的说来会非常高。随后由于处于自然退行的粒子对辐射的吸收和逆向发射（或逆向散射），很快就会形成一个光谱，同时，逐渐积累的多普勒效应会导致频率的降低。然而，一些原始的高频辐射将不受干扰地在空间传播，尽管到达地球时不可避免地会存在多普勒效应，但是仍然会有一小部分辐射仍保留着很高的频率，这可能正是现在的宇宙线中波动组分的来源。

我认为在对霍尔丹想法进行的这一初步探讨中，未对有极高能量的原始光子是如何瓦解一颗恒星的问题进行细节的讨论是非常明智的。详细地论述霍尔丹通过运动相对论推导出的非凡的结果，即在宇宙演化早期存在的光子一定具有极高的能量就已经足够了。

（金世超 翻译；邓祖淦 审稿）

Chinese Astronomical Clockwork

J. Needham *et al.*

Editor's Note

Joseph Needham was born in 1900, the son of a leading physician in London. He studied medicine and biochemistry at the University of London but in the mid-1930s he encountered three young people visiting his laboratory from China and became deeply interested in the history of Chinese science. One of the visitors was Lu Gwei-djen, the daughter of a Chinese pharmacist, who taught him classical Chinese. Needham and Lu spent the period 1942–46 in China where he was director of the Sino-British Science Cooperation Office. Here he formed a working relationship with Wang Ling, a distinguished historian who remained a close collaborator. The main product of his Chinese work was a series of 23 volumes entitled *Science and Civilisation in China*. Needham died in 1995. This article is presumably one of the those collected for his mammoth work.

IT is generally agreed that the invention of the mechanical clock was one of the most important turning-points in the history of science and technology. According to the view accepted until recently[1-4], the problem of slowing down the rotation of a wheel to keep a constant speed continuously in time with the apparent diurnal rotation of the heavens was first solved in Europe in the early fourteenth century A.D. Trains of gearing were then combined with the verge and foliot escapement and powered by a falling weight. Recent research has shown, however, that these first mechanical time-keepers were not so much an innovation as has been supposed[5]. They descended, in fact, from a long series of complicated astronomical "clocks", planetary models, mechanically-rotated star-maps and similar devices designed primarily for exhibition and demonstration rather than accurate time-keeping. Although such devices are of the greatest interest as the earliest complex scientific machines, it has not hitherto been possible to adduce more than a few fragmentary remains, and literary descriptions tantalizingly incomplete, which lack sufficient detail for clear understanding of the mechanical principles involved. But the examination of certain medieval Chinese texts, the relevance of which has not been realized, has now permitted us to establish the existence of a long tradition of astronomical clock-making in China between the seventh and the fourteenth centuries A.D.

The key text is the "Hsin I Hsiang Fa Yao" [New Design for a (Mechanized) Armillary (Sphere) and (Celestial) Globe], written by Su Sung in A.D. 1090, the appropriate sections of which we have fully translated. This describes in great detail an astronomical clock of large size (Figs. 1 and 2) powered not by a falling weight, but by a scoop-wheel using water or mercury. Besides the rotation of the sphere and globe by trains of gear-wheels, the clock embodied elaborate time-keeping jack-work. The escapement consisted of a weigh-bridge

中国的天文钟

李约瑟等

编者按

李约瑟出生于1900年，是伦敦一个首席医生的儿子。他在伦敦大学学习医学和生物化学，然而在20世纪30年代中期，他遇到了三个参观他实验室的中国年轻人，开始对中国科学的历史产生了浓厚的兴趣。其中一个来访者鲁桂珍是一位中国药师的女儿，教他学习文言文。1942年～1946年间李约瑟和鲁桂珍在中国工作，李约瑟时任中英科学合作馆馆长。在这里他结识了出色的历史学家王铃，并保持密切的合作关系。他在中国工作的主要成果是出版了一套23卷的、名为《中国科学技术史》的书。1995年李约瑟逝世。这篇文章可能是他当时庞大工作中的一部分。

人们通常将机械钟的发明看作科技史上最重要的转折点之一。按照最近才被人们广为接受的观点[1-4]，在公元14世纪早期欧洲人首先解决了机械钟转轮变慢的问题，从而使钟表保持恒定的速率，与大自然的日夜变化相一致。他们将一系列传动装置与带有摆轮心轴与原始平衡摆的擒纵装置结合起来，然后用落锤提供动力。然而，最近的研究表明，这些最早的计时器并非如之前认为的那样富有创新性[5]。事实上，这些钟表起源于一系列复杂的天文“钟”、行星模型、机械转动星图以及类似的装置，这些设计最初是为了演示而不是为了精确计时。尽管这些装置作为最早的复杂的科学机械具有重要的意义，但是迄今只能引证少量破碎的遗物以及残缺到令人着急的文字描述，要想清楚全面地理解其中的机械原理，还缺乏足够的资料。然而，阅读那些人们尚未意识到存在关联的中世纪的中国史料，使我们承认在公元7世纪到14世纪中国已经存在着悠久的天文钟制造传统。

这方面的主要书籍是公元1090年苏颂撰写的《新仪象法要》（对浑天仪和天球仪的新设计），我们已将相关的章节全部翻译。这本书详细描述了一种尺寸很大的天文钟（图1和图2），它不是由落锤，而是由水或水银驱动的水斗轮提供动力。除了一系列传动转轮装置带动浑天仪和天球仪的球体旋转外，该钟具有精确的守时机制。擒纵装置包括一个衡桥，用于在水斗充满前托住水斗防止其下降；还包括一个撞击

which prevented the fall of each scoop until full, and a trip-lever and parallel linkage system which checked the rotation of the wheel at another point. An anti-recoil device was also built in. The basic principle involved is thus more like the anchor escapement of the late seventeenth century than the verge and foliot type, although the time-keeping is, of course, governed mainly by the flow of water rather than by the escapement action itself. This type of effect is therefore the "missing link" between the time-keeping properties of a steady flow of liquid and those of mechanically produced oscillations. So complete is the description, which has yielded more than a hundred and fifty technical terms of eleventh-century mechanics, that it has been possible to prepare detailed working drawings of the clock (Figs. 3 and 4).

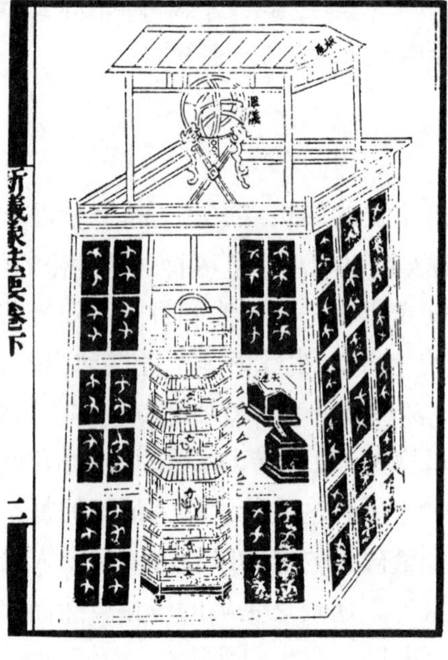

Fig. 1

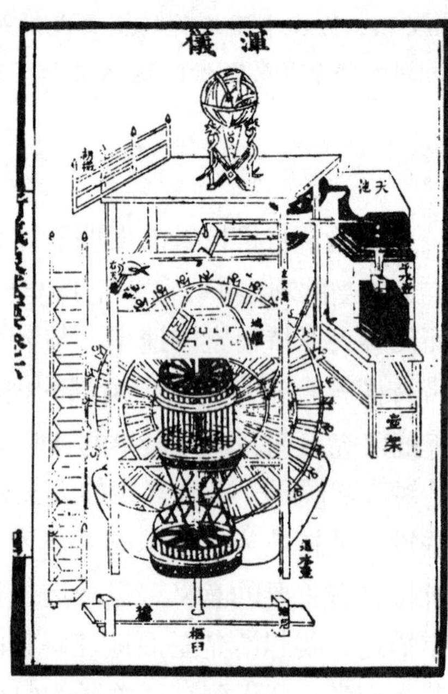

Fig. 2

Fig. 1. Astronomical clock of A.D. 1088 from Su Sung's "Hsin I Hsiang Fa Yao", ch. 3, p. 2*a*. External appearance. The armillary sphere on the platform above, the celestial globe in the upper chamber of the tower, and the pagoda with the time-announcing jacks below this. On the right, the housing removed to show the water-tanks. Estimated total height about 30 ft.

Fig. 2. The same as Fig. 1. General view of the works. Vertical shaft with jack-wheels in the foreground, behind this the driving-wheel with its scoops. Water-tanks on the right, part of the anti-recoil device on the left at the top. Escapement not shown in this diagram, though the device above the wheel is probably intended for part of it

杆和平行连杆系统，用于在另一点检查转轮的转动。还装有一个抗反冲装置。因此，整个驱动系统的基本原理更加类似于 17 世纪晚期的锚状擒纵装置而非使用摆轮心轴和原始平衡摆的类型，虽然守时主要是由水流而不是擒纵装置本身驱动的。因此，在稳定液体流守时性能和机械振荡守时性能之间，这种效应类型是二者之间“缺失的一环”。史料对于这一机械钟装置的记录非常完备，其中包含了 11 世纪的 150 多个机械部件，使得我们能够给出详细的时钟机械工作示意图（图 3 和图 4）。

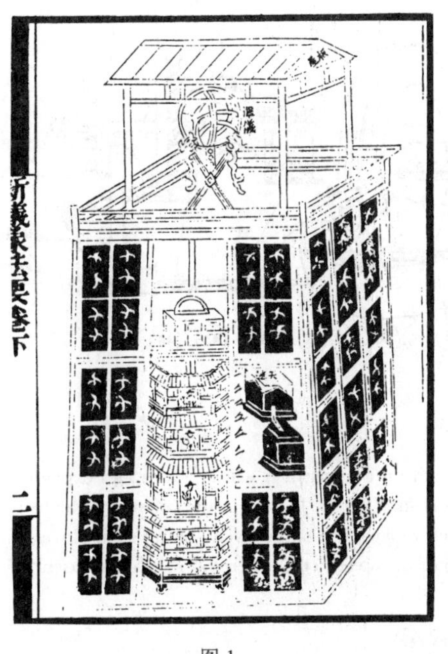

图 1

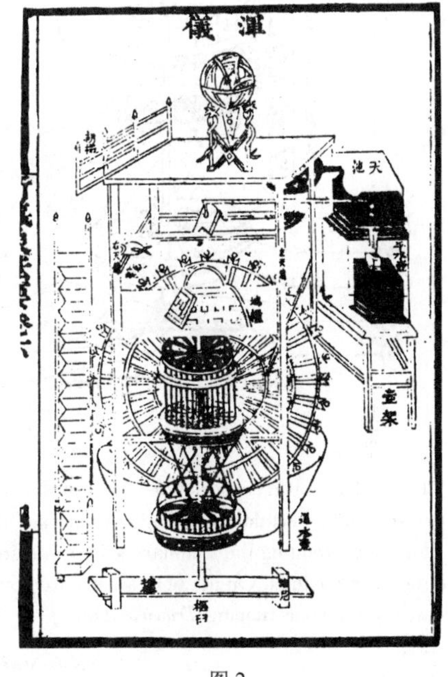

图 2

图 1. 公元 1088 年苏颂在《新仪象法要》中记载的天文钟的外观图（第 3 章第 2*a* 页）。浑天仪位于最上面的平台上，天球仪位于塔内上层，具有报时功能的支架塔位于塔内下层。图中右侧去除了房屋外层，以显示出水箱。整个装置的高度大约为 30 英尺。

图 2. 同图 1 一样给出了装置的内部结构纵览。装置前面装有支架轮和立轴，后面是驱动轮及其水车。右侧装有水箱，左上方露出部分抗反冲装置。图中未给出擒纵装置，但转动轮上方的装置可能是擒纵装置的一部分。

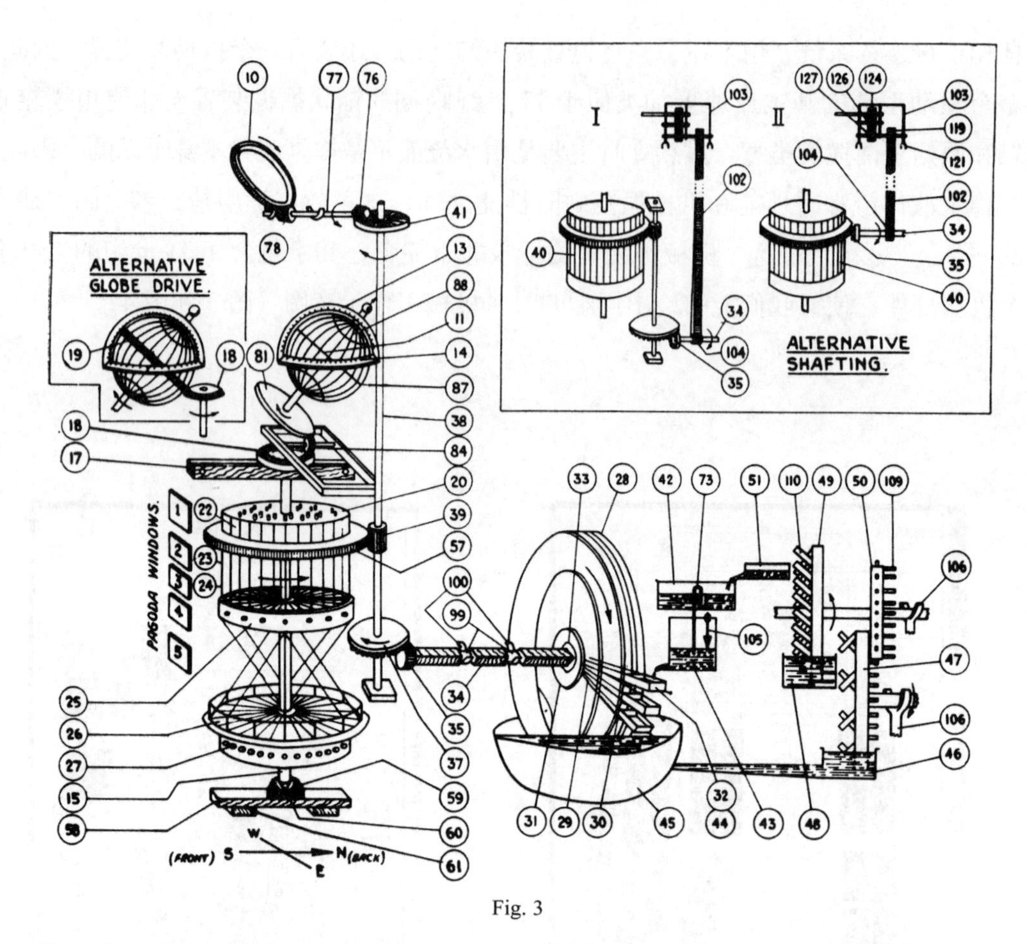

Fig. 3

Fig. 3. General reconstruction. The numbers indicate technical terms which will be given and explained in the full publication. Gearing at the top of the jack-wheel shaft on the left drives the celestial globe. The long shaft rotating the armillary sphere was in later models replaced by successively shorter chain drives, as shown in the inset. On the right are the two norlas which raised the water to the uppermost reservoir, sufficient for twenty-four hours running. (*Drawn by Angel*)

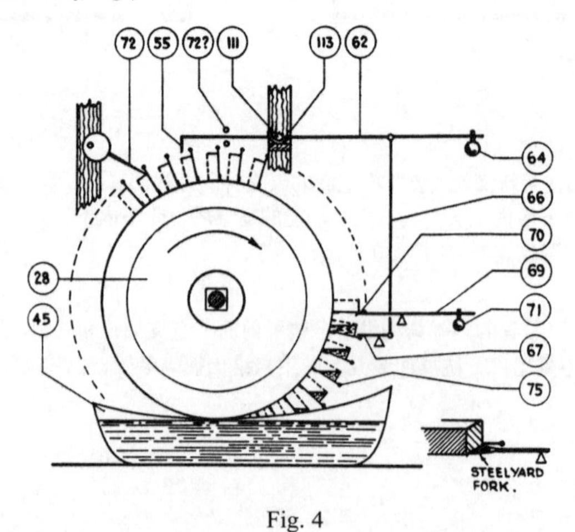

Fig. 4

Fig. 4. Diagram of the escapement, with its weigh-bridge and parallel linkwork. The clepsydra flow delivered into the scoop at 70. (*Drawn by Angel*)

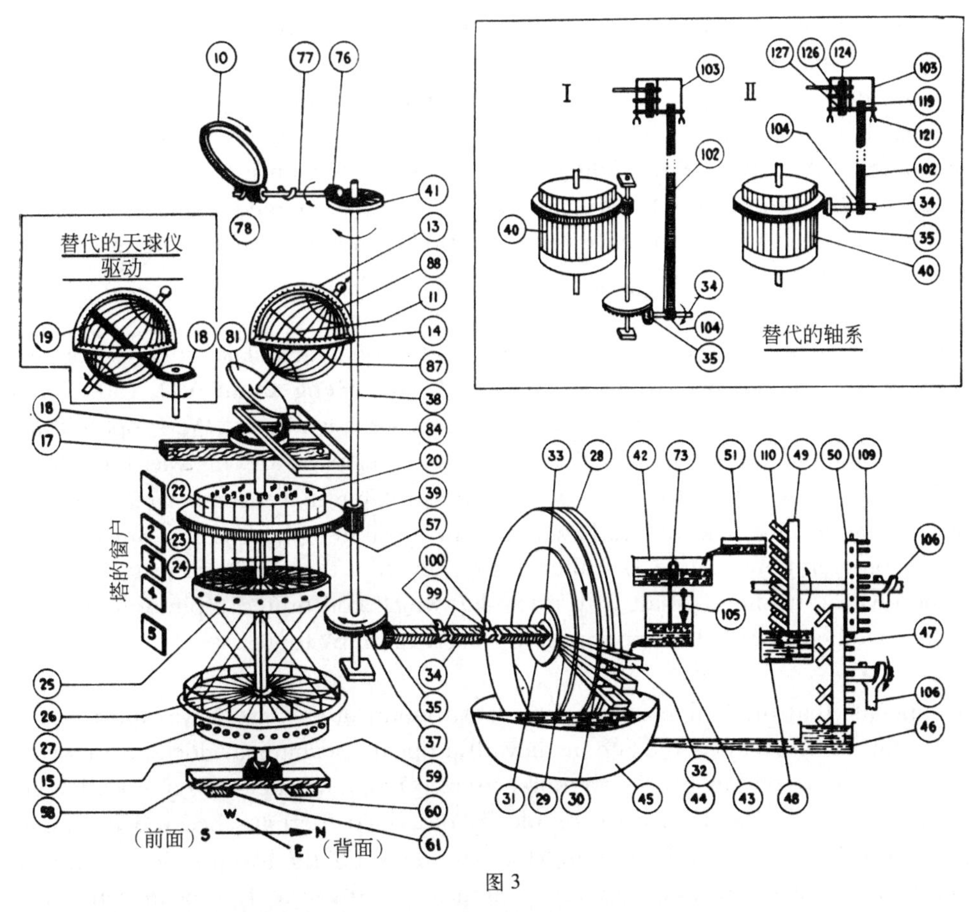

图 3

图 3. 中国天文钟结构的复原重建图。将在以后完整的出版物中给出并解释图中数字代表的部件。图中左侧位于支架轮轴顶部的传动装置用于驱动天球仪。如插入框图所示，用于转动浑天仪的长杆在后续的模型中被更短的链式传动替代了。图中右侧是两个轮轴，用于将水汲取到最高处的储水槽，从而保证系统 24 小时连续运转。（本图由安杰尔绘制）

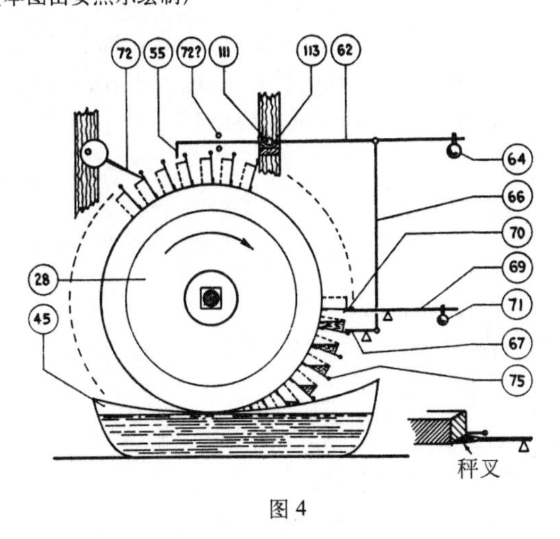

图 4

图 4. 擒纵装置示意图，包括衡桥装置和平行连杆系统。漏壶的水流从标号 70 的位置流入水斗。（本图由安杰尔绘制）

The full understanding of this text has enabled us to interpret many descriptions of other clocks contained partly in the dynastic histories and partly in other sources, some from books now lost, but preserved in the "Yü Hai" encyclopaedia of A.D. 1267. Thus an important astronomical clock driven by mercury was built by Chang Ssu-Hsün, a Szechuanese, in A.D. 979. The tradition seems to start with an instrument completed in A.D. 725 by the Tantric monk, I-Hsing, and an engineer named Liang Ling-Tsan, the description of which would not have been comprehensible without thorough prior study of Su Sung's text.

Earlier texts describe celestial globes or demonstrational armillary spheres rotated by clepsydra water. These range from the work of Chang Hêng about A.D. 130 to that of Kêng Hsün about 590, but evidence of any escapement is absent. We supposed at first that these employed only the sinking float of a large clepsydra. This was the system of the Hellenistic anaphoric clock with its rotating astrolabic dial[6], and perhaps also of the famous striking water-clocks of the Byzantine[7] and Arabic[8] culture-areas. But textual and historical considerations incline us rather to the view that the clepsydra water dripped on to a scoop-wheel, turning a shaft with a trip-lug which constituted a pinion of one. This acted on a toothed ring on the apparatus, moving it tooth by tooth.

It thus appears that the Chinese tradition of astronomical clockwork was more nearly in the direct line of ancestry of the late medieval European mechanical clocks. Moreover, the detailed description of this previously unrecognized type of water-driven clock has made it possible to find similar devices incompletely described (perhaps because incompletely known) in Indian[9], Arabic and Hispano-Moorish[10] texts. Of the transmission of influences little can as yet be said, though there are indications that the European centuries just preceding the fourteenth knew devices with water-powered and mechanically checked driving-wheels[11]. This would suggest that the time of transmission was rather that of "the Crusades" (as in the case of the wind-mill) than that of the Marco Polo.

All the texts now translated, with commentary and discussion, will, it is hoped, be published as a special monograph by the Antiquarian Horological Society, to which the results of the investigation have been communicated.

(**177**, 600-602; 1956)

Joseph Needham: F.R.S., Caius College, Cambridge.
Wang Ling: Trinity College, Cambridge.
Derek J. Price: Christ's College, Cambridge.

References:

1. Beckmann, J., "A History of Inventions, Discoveries, etc.", vol. 1, pp. 340 ff. (Bohn, London, 1846).

2. Usher, A. P., "A History of Mechanical Inventions", pp. 191 ff., 304 ff., 2nd edit. (Harvard Univ. Press, Cambridge, Mass., 1954).

3. Frémont, C., "Origine de l'Horloge à Poids" (Études Expérimentales de Technologie Industrielle, No. 47; Paris, 1915).

对苏颂一文的全面理解使我们得以理解部分王朝史书上或其他书上记载的对其他各种钟的描述，虽然其中一些书目已失传，但其内容在《玉海》百科全书（成书于公元 1267 年）中保留了下来。公元 979 年，一位名叫张思训的四川人建造了一座用水银驱动的天文钟。公元 725 年，密宗高僧一行和一位名叫梁令瓒的工程师制造了一个钟，这似乎是制钟历史的开始。如果没有对苏颂一文深入的研究，就不会对这些描述有充分的理解。

更早的史料记载了用漏壶中的水驱动的天球仪或演示用的浑天仪。从大约公元 130 年张衡到大约公元 590 年耿询都进行了这方面的工作，但是史料中没有任何擒纵装置的内容。起初我们推测，这些装置只使用了大漏壶中的浮子结构。就类似于装有转动拨号星盘的希腊式浮子升降钟[6]，也可能类似拜占庭[7]和阿拉伯地区[8]著名的水钟。但是考虑到原文记载和具体历史条件，我们倾向于以下观点：漏壶中的水滴到斗式水车上，触动附有凸缘开关的传动轴，从而驱动系统相互啮合转动。

综上所述，中国古代的天文钟可能就是欧洲中世纪晚期机械钟的直系祖先。此外，对这种以前未获认可、靠水驱动的天文钟的详细描述，使我们可能在印度[9]、阿拉伯和西班牙摩尔[10]的史料中也会找到描述不完整的（可能是由于我们不完全知晓）类似装置。虽然欧洲直到 14 世纪才掌握了水驱动和机械驱动轮装置[11]，但是这种传播的影响迄今还很难说。这表明，中国时钟制造术传到欧洲的时间是在“十字军”东征的时候（和风车一样），而不是马可波罗时代。

目前我们已经翻译完毕所有相关的史料，添加了注释和讨论，并与古钟协会进行了沟通，希望他们将其作为一本特别专著出版。

（金世超 翻译；沈志侠 审稿）

4. Howgrave-Graham, R. P., "Some Clocks and Jacks, with Notes on the History of Horology", *Archaeologia*, 77, 257 (1927). Baillie, G. H., "Watches" (Methuen, London, 1929).

5. Price, D. J., "Clockwork before the Clock", *Horological J.*, 97, 810 (1955); and **98**, 31 (1956).

6. Vitruvius, IX. 8. Cf. Drachmann, A. G., "The Plane Astrolabe and the Anaphoric Clock", *Centaurus*, **3**, 183 (1954).

7. Diels, H., "Über die von Prokop beschriebene Kunstuhr von Gaza; mit einem Anhang enthaltende Text und Übersetzung d. *ekphrasis horologiou* des Prokopios von Gaza", *Abhandlungen d. preuss. Akad. Wiss.* (Phil.-Hist. Kl.), No. 7 (1917).

8. Wiedemann, E., and Hauser, F., "Über die Uhren im Bereich der islamischen Kultur", *Nova Acia* (*Abhandlungen d. K. Leop.-Carol. deutschen Akad. d. Naturforsch.*) Halle, **100**, No. 5 (1915).

9. Burgess, E., "The Sürya-Siddhânta, a. Textbook of Hindu Astronomy", pp. 282, 298, 305 ff., edit. Phanindralal Gangooly (University, Calcutta, 1935).

10. Rico y Sinobas, M., "Libros del Saber de Astronomia del Rey D. Alfonso X de Castilla" (Aguado, Madrid, 1864).

11. Drover, C. B., "A Mediaeval Monastic Water-Clock", *Antiquarian Horology* (Dec. 1954).

A Test of a New Type of Stellar Interferometer on Sirius

R. Hanbury-Brown and R. Q. Twiss

Editor's Note

Robert Hanbury-Brown and Richard Twiss had recently demonstrated that photons in two separate but coherent beams of light tend to arrive together at two distinct detectors. Here they show how this effect could be applied in stellar interferometry, which measures the size (the angular diameter, or angle subtended by lines to each edge) of stars. Testing the technique for the star Sirius, they eventually found a signal indicating an angular size for the star of 0.0063 seconds of arc. This was the first measurement of the angular diameter of Sirius, and showed how the quantum nature of light could be put to advantage in extremely sensitive interferometry.

WE have recently described[1] a laboratory experiment which established that the time of arrival of photons in coherent beams of light is correlated, and we pointed out that this phenomenon might be utilized in an interferometer to measure the apparent angular diameter of bright visual stars.

The astronomical value of such an instrument, which might be called an "intensity" interferometer, lies in its great potential resolving power, the maximum usable base-line being governed by the limitations of electronic rather than of optical technique. In particular, it should be possible to use it with base-lines of hundreds, if not thousands of feet, which are needed to resolve even the nearest of the *W*-, *O*- and *B*-type stars. It is for these stars that the measurements would be of particular interest since the theoretical estimates of their diameters are the most uncertain.

The first test of the new technique was made on Sirius (α Canis Majoris A), since this was the only star bright enough to give a workable signal-to-noise ratio with our preliminary equipment.

The basic equipment of the interferometer is shown schematically in Fig. 1. It consisted of two mirrors M_1, M_2, which focused light on to the cathodes of the photomultipliers P_1, P_2 and which were guided manually on to the star by means of an optical sight mounted on a remote-control column. The intensity fluctuations in the anode currents of the photomultipliers were amplified over the band 5–45 Mc./s., which excluded the scintillation frequencies, and a suitable delay was inserted into one or other of the amplifiers to compensate for the difference in the time of arrival of the light from the star at the two mirrors. The outputs from these amplifiers were multiplied together in a linear mixer and, after further amplification in a system where special precautions were taken to eliminate the effects of drift; the average value of the product was recorded

一种新型恒星光干涉法对天狼星的测试

汉伯里–布朗，特威斯

编者按

最近，罗伯特·汉伯里–布朗和理查德·特威斯演示了两个分离的但是在相干光束中的光子能够同时到达两个不同的探测器。在本文中他们介绍了如何将这一原理运用到恒星干涉法中来测量恒星的大小（即角直径，或沿视线指向恒星边缘所形成的张角）。他们运用该技术测试天狼星，最终发现信号显示该恒星的角大小为0.0063角秒。这是首次测量天狼星的角直径，也显示了光的量子性质是如何发挥高敏感干涉测量法的优势的。

我们最近记述的实验[1]证明，相干光束中光子的到达时间是相关的。我们指出，这个现象可以应用在干涉仪测量可见亮星的视角直径上。

这种被称为“强度”干涉仪的设备在天文学方面的意义在于其超强的潜在分辨率，其最大可用基线受电子学技术而不是受光学技术的限制。尤其是用几百英尺而非几千英尺的基线时，这种仪器甚至可以分辨最接近的 W 型、O 型和 B 型星。正是由于这类恒星直径的理论值具有很大的不确定性，所以我们特别关注这种测量方法。

由于天狼星（大犬座α星A）是唯一一颗明亮到足以用我们早期实验设备就能给出可用信噪比的恒星，所以利用该项新技术对天狼星进行了首次测试。

干涉仪的基本装置如图1所示。其中，两面反射镜 M_1、M_2 将光束汇聚到光电倍增管 P_1、P_2 的阴极上。这两面反射镜通过人工操作安装在远程控制栏上的光学瞄准器来对准所选恒星。在光电倍增管阳极电流中，在5兆周/秒~45兆周/秒的波段内强度起伏（不包括闪烁频率）被放大，并向其中任何一个放大器内置入恰当的延迟时间，来补偿恒星的光到达两面反射镜时所产生的时间差。在一个采取了特殊预防措施以消除漂移影响的系统中，信号被进一步放大；其后在一个线性混频器中将放大器的输出叠加在一起。结果的平均值记录在一个积分电动机的转数计上。该测量读数直观地给出了光束到达两面反射镜产生的强度起伏的相关系数。然而，读数的大小依

on the revolution counter of an integrating motor. The readings of this counter gave a direct measure of the correlation between the intensity fluctuations in the light received at the two mirrors; however, the magnitude of the readings depended upon the gain of the equipment, and for this reason the r.m.s. value of the fluctuations at the input to the correlation motor was also recorded by a second motor. Since the readings of both revolution counters depend in the same manner upon the gain, it was possible to eliminate the effects of changes in amplification by expressing all results as the ratio of the integrated correlation to the r.m.s. fluctuations, or uncertainty in the final value. The same procedure was also followed in the laboratory experiment described in a previous communication[1].

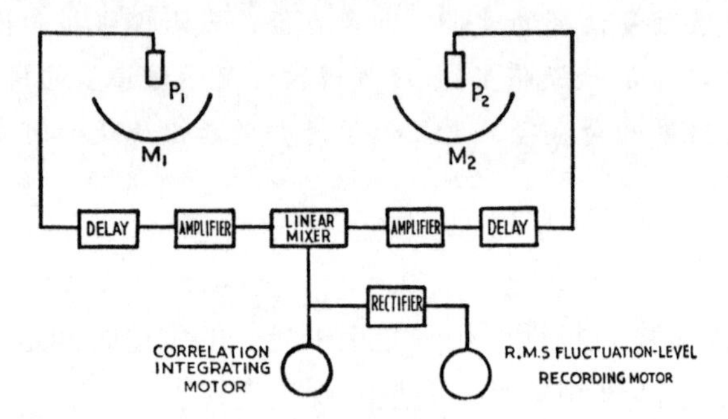

Fig. 1. Simplified diagram of the apparatus

There is no necessity in an "intensity" interferometer to form a good optical image of the star. It is essential only that the mirrors should focus the light from the star on to a small area, so that the photocathodes may be stopped down by diaphragms to the point where the background light from the night sky is relatively insignificant. In the present case, the two mirrors were the reflectors of two standard searchlights, 156 cm. in diameter and 65 cm. in focal length, which focused the light into an area 8 mm. in diameter. However, for observations of Sirius, the circular diaphragms limiting the cathode areas of the photomultipliers (*R.C.A.* type 6342) were made as large as possible, namely, 2.5 cm. in diameter, thereby reducing the precision with which the mirrors had to be guided.

The first series of observations was made with the shortest possible base-line. The searchlights were placed north and south, 6.1 metres apart, and observations were made while Sirius was within 2 hr. of transit. Since the experiments were all carried out at Jodrell Bank, lat. 53° 14′ N., the elevation of the star varied between $15\frac{1}{2}°$ and 20°, and the average length of the base-line projected normal to the star was 2.5 metres; at this short distance Sirius should not be appreciably resolved.

Throughout the observations the average d.c. current in each photomultiplier was recorded every 5 min., together with the readings of the revolution counters on both the integrating motors. The small contributions to the photomultiplier currents due to the night-sky background were measured at the beginning and end of each run. The gains of the photomultipliers were also measured and were found to remain practically constant over periods of several hours.

赖于设备增益，因此，均方根起伏值在输入相关电动机时也被记录在了另一个电动机中。因为两个转数计的读数以相同的方式依赖于增益，所以将所有结果表示为积分相关值与均方根起伏值（即最终值的不确定性）的比值，可能可以消除在信号放大过程中变化的影响。与以前的通讯[1]中描述的实验室实验步骤相同。

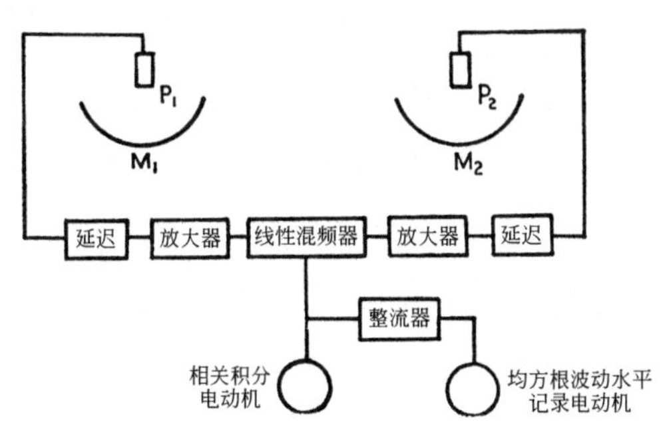

图 1. 仪器结构简图

用“强度”干涉仪进行测量时不必形成一张很好的恒星光学图像；只有当反射镜将星光聚集到很小的区域内时才是必要的，可用光阑使光电阴极缩小光圈至夜空的背景光相对而言可忽略的地步。在此情况中，两面反射镜是两个标准探照灯的反射面，直径为 156 厘米，焦距为 65 厘米，能够将光束聚集在一个直径为 8 毫米的区域内。然而，在观测天狼星时，限制光电倍增管（*R.C.A.* 6342 型）阴极区域的圆形光阑的直径可大至 2.5 厘米，因此降低了控制反射镜所要达到的精确度。

第一批观测采用了尽可能最短的基线。探照灯分别放置在相距 6.1 米的南北两侧，在天狼星位于中天的两个小时内进行了观测。因为实验全部是在北纬 53° 14′的焦德雷尔班克天文台进行的，所以天狼星的地平高度在 15.5° 和 20° 之间变化，垂直于星体的基线投影的平均长度是 2.5 米，在如此近的距离内不会明显地分辨出天狼星。

在观测的整个过程中，每隔 5 分钟就记录一次每个光电倍增管的平均直流电流以及两个积分电机上转数计的读数。在每次观测的开始和结束时，测量夜空的背景光对光电倍增管电流的微小的影响。我们也测量了光电倍增管的增益，结果发现实际上它在几个小时内保持恒定。

In order to ensure that any correlation observed was not due to internal drifts in the equipment, or to coupling between the photomultipliers or amplifier systems, dummy runs of several hours duration were made before and after every observation; for these runs the photomultiplier in each mirror was illuminated by a small lamp mounted inside a detachable cap over the photocathode. In no case was any significant correlation observed.

In this initial stage of the experiment, observations were attempted on every night in the first and last quarters of the Moon in the months of November and December 1955; the period around the full moon was avoided because the background light was then too high. During these months a total observation time of 5 hr. 45 min. was obtained, an approximately equal period being lost due to failure of the searchlight control equipment. The experimental value for the integrated correlation $C(d)$ at the end of the observations is given in the line 3 of Table 1. The value of $C(d)$ is the ratio of the change in the reading of the counter on the correlation motor to the associated r.m.s. uncertainty in this reading.

Table 1. Comparison between Theoretical and Observed Correlation

1. Base-line in metres	2.5 (N.S.)	5.54 (E.W.)	7.27 (E.W.)	9.20 (E.W.)
2. Observing time (min.)	345	285	280	170
3. Observed ratio of integrated correlation to r.m.s. deviation: $C(d)$	+8.50	+3.59	+2.65	+0.83
4. Theoretical ratio of integrated correlation to r.m.s. deviation, assuming star has an angular diameter of 0.0063″: $C(d)$	+9.35	+4.11	+2.89	+1.67
5. Theoretical ratio of integrated correlation to r.m.s. deviation, assuming star is a point source : $C(o)$	+10.15	+5.63	+5.06	+4.40
6. Theoretical normalized correlation coefficient for star of diameter 0.0063″: $\Gamma^2(d)$	0.92	0.73	0.57	0.38
7. Observed normalized correlation coefficient with associated probable errors : $\Gamma^2(d)$	0.84±0.07	0.64±0.12	0.52±0.13	0.19±0.15

In the second stage of the experiment the spacing between the mirrors was increased and observations were carried out with east-west base-lines of 5.6, 7.3 and 9.2 metres. These measurements were made on all possible nights during the period January–March 1956, and a total observing time of $12\frac{1}{4}$ hr. was obtained. The observed values of the integrated correlation $C(d)$ are shown in line 3 of Table 1.

As a final check that there was no significant contribution to the observed correlation from any other source of light in the sky, such as the Čerenkov component from cosmic rays[2], a series of observations was made with the mirrors close together and exposed to the night sky alone. No significant correlation was observed over a period of several hours.

The results have been used to derive an experimental value for the apparent angular diameter of Sirius. The four measured values of $C(d)$ were compared with theoretical values for uniformly illuminated disks of different angular sizes, and the best fit to the observations was found by minimizing the sum of the squares of the residuals weighted

为了确保任何观测得到的相关系数不是由于仪器的内部漂移或者光电倍增管和放大器系统之间的耦合作用所致，在每次观测前和观测后都要进行几个小时的试运行。运行时，通过光电阴极上方活盖内的一个小灯照射每个反射镜的光电倍增管。在任何情况下都没有观测到任何显著的相关性。

在该实验初期阶段，我们试图在 1955 年的 11 月及 12 月的上弦月和下弦月的每个夜晚都进行观测。满月那段时间没有观测，因为夜空背景光过亮。在这两个月里观测时间总计达 5 小时 45 分，由于探照灯控制系统的失灵而浪费的观测时间与其大致相同。观测结束后综合相关系数 $C(d)$ 的实验值列在表 1 的第 3 行中。$C(d)$ 是相关电机的计数器读数与该读数的均方根不确定性之间的比值。

表 1. 理论相关系数和观测得到的相关系数之间的比较

1．基线长度（单位：米）	2.5（南北向）	5.54（东西向）	7.27（东西向）	9.20（东西向）
2．观测时间（单位：分钟）	345	285	280	170
3．综合相关系数与均方根偏差的观测比：$C(d)$	+8.50	+3.59	+2.65	+0.83
4．假设恒星角直径为 0.0063″ 时，综合相关系数与均方根偏差的理论比：$C(d)$	+9.35	+4.11	+2.89	+1.67
5．假设恒星是点光源时，综合相关系数与均方根偏差的理论比：$C(o)$	+10.15	+5.63	+5.06	+4.40
6．角直径为 0.0063″ 恒星的理论归一化相关系数：$\Gamma^2(d)$	0.92	0.73	0.57	0.38
7．观测得到的理论归一化相关系数及其相应可能误差：$\Gamma^2(d)$	0.84 ± 0.07	0.64 ± 0.12	0.52 ± 0.13	0.19 ± 0.15

在该实验的第二阶段，增加两面反射镜之间的间距，在东西向基线为 5.6 米、7.3 米和 9.2 米时进行观测。1956 年 1 月到 3 月期间在所有可观测夜晚进行观测，总计观测时间达 12.25 小时。观测得到的综合相关系数 $C(d)$ 列在表 1 第 3 行中。

将天空中其他光源（如宇宙线的切伦科夫辐射 [2]）对观测得到的相关系数没有显著的影响作为最后的检验，我们进行了一系列观测，将反射镜紧靠在一起并只对着夜空单独曝光，在连续的几个小时时间段并没有观测到显著的相关性。

现已用这些结果来推导天狼星视角直径的实验值。将 $C(d)$ 的 4 个观测值与从不同角大小的均匀照射圆盘得到的理论值进行比较，可得当各点的观测误差加权平方和最小时，与观测结果拟合得最好。在比较时，假设圆盘的角直径和 $C(o)$（零基线时的相关系数）的值都是未知的，并且考虑了每点的不同光通量和观测时间。因此，

by the observational error at each point. In making this comparison, both the angular diameter of the disk and the value of $C(o)$, the correlation at zero base-line, were assumed to be unknown, and account was taken of the different light flux and observing time for each point. Thus the final experimental value for the diameter depends only on the relative values of $C(d)$ at the different base-lines, and rests on the assumption that these relative values are independent of systematic errors in the equipment or in the method of computing $C(d)$ for the models. The best fit to the observations was given by a disk of angular diameter 0.0068″ with a probable error of ±0.0005″.

The angular diameter of Sirius, which is a star of spectral type *A*1 and photovisual magnitude −1.43, has never been measured directly; but if we assume that the star radiates like a uniform disk and that the effective black body temperature[3,4] and bolometric correction are 10,300°K. and −0.60, respectively, it can be shown that the apparent angular diameter is 0.0063″, a result not likely to be in error by more than 10 percent. (In this calculation the effective temperature, bolometric magnitude and apparent angular diameter of the Sun were taken as 5,785°K., −26.95, and 1,919″, respectively.) Thus it follows that the experimental value for the angular diameter given above does not differ significantly from the value predicted from astrophysical theory.

A detailed comparison of the absolute values of the observed correlation with those expected theoretically has also been made, and the results are given in Table 1 and in Fig. 2. In making this comparison, it is convenient to define a normalized correlation coefficient $\Gamma^2(d)$, which is independent of observing time, light flux and the characteristics of the equipment, where $\Gamma^2(d)=C(d)/C(o)$ and $C(d)$ is the correlation with a base-line of length d, and $C(o)$ is the correlation which would be observed with zero base-line under the same conditions of light flux and observing time. The theoretical values of $\Gamma^2(d)$ for a uniformly illuminated disk of diameter 0.0063″ are shown in line 6 of Table 1. For monochromatic radiation it is simple to evaluate $\Gamma^2(d)$, since it can be shown[5] that it is proportional to the square of the Fourier transform of the intensity distribution across the equivalent strip source; however, in the present case, where the light band-width is large, the values of $\Gamma^2(d)$ were calculated by numerical integration.

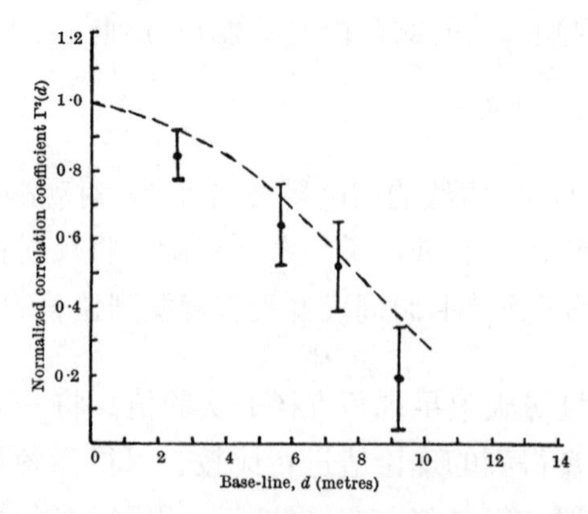

Fig. 2. Comparison between the values of the normalized correlation coefficient $\Gamma^2(d)$ observed from Sirius and the theoretical values for a star of angular diameter 0.0063″. The errors shown are the probable errors of the observations

基于这些相关的数值与设备的系统误差无关或与计算 $C(d)$ 模型的计算方法也无关的假设，直径最终的实验值只取决于不同基线下 $C(d)$ 的相对值。与观测值拟合的最好的是角直径为 0.0068″的圆盘，误差可能为 ±0.0005″。

天狼星的光谱型是 *A*1 型，仿视星等是 –1.43。至今没有直接测出其角直径；但是如果我们假设该恒星像一均匀圆盘一样辐射，且有效黑体温度 [3,4] 和热星等改正分别是 10,300 K 和 –0.60，则可得视角直径为 0.0063″，该结果的误差不太可能超过 10%。（在该计算过程中，太阳的有效温度、热星等和角直径取值分别为 5,785 K、–26.95 和 1,919″。）因此可以看出，以上给出的角直径的实验值和由天体物理学理论预测的数值并没有很大的差异。

观测得到的相关系数的绝对值和理论上得到的预期值之间的详细比较见表 1 和图 2。在比较时，定义一个归一化相关系数 $\Gamma^2(d)$ 是方便的，$\Gamma^2(d)=C(d)/C(o)$。该系数与观测时间、光通量和设备的特性无关。其中，$C(d)$ 是基线长度为 d 时的相关系数，$C(o)$ 是在光通量和观测时间相同的条件下基线为零时观测得到的相关系数。表 1 的第 6 行列出了一个角直径为 0.0063″均匀辐射圆盘的 $\Gamma^2(d)$ 理论值。因为已知 $\Gamma^2(d)$ 与等价条形光源强度分布的傅立叶变换的平方成比例 [5]，所以用单色辐射估计 $\Gamma^2(d)$ 是很简单的。然而，真实的情况是光束的带宽很宽，所以 $\Gamma^2(d)$ 值由数值积分得到。

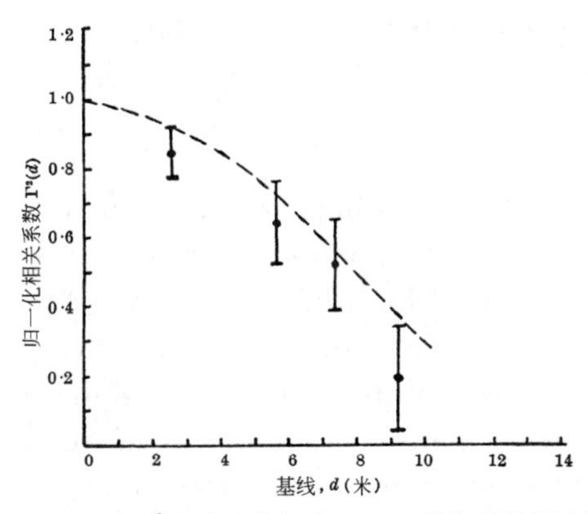

图 2. 观测天狼星所得归一化相关系数 $\Gamma^2(d)$ 和角直径为 0.0063″的恒星的理论值之间的比较。图中所示误差为观测中可能产生的误差。

The theoretical values of $C(o)$, given in line 5, were calculated for the conditions of light flux and observing time appropriate to each base-line by means of equations (1) and (2) of our previous communication[1] (though in the present experiment the r.m.s. fluctuations were smaller by a factor $1/\sqrt{2}$ than the value given in the previous paper, which refers to an alternative electronic technique). The most important quantities in this calculation are the gains and output currents of the photomultipliers and the band-widths of the amplifiers; but it is also necessary to make a small correction for the combined spectral characteristics of the photocathodes, the atmospheric attenuation, the star and the mirrors. Finally, in line 4 of Table 1 the theoretical values of the correlation $C(d)$ are shown; they were calculated from the theoretical values of $C(o)$ and $\Gamma^2(d)$ by means of the relation given above.

The correlation observed at the shortest base-line (2.5m.) can be used as a rough test of the effects of atmospheric scintillation on the equipment, since the corresponding theoretical value depends only on well-known quantities and is almost independent of the angular diameter of the star. Throughout the observations Sirius was seen to be scintillating violently, although the corresponding fluctuations in the d.c. anode currents of the photomultiplier tubes, which were smoothed with a time constant of about 0.1 sec., were only of the order of ± 10 percent, as might be expected with mirrors large by normal telescope standards. Nevertheless, the observed correlation $C(d)=+8.50$ does not differ significantly from the calculated value of +9.35, and it follows that it cannot be greatly affected by scintillation.

The experimental values of $C(d)$ obtained at the four base-lines may be compared with the corresponding theoretical values $C(d)$ by means of lines 3 and 4 of Table 1. However, it is more convenient, since these values depend upon the different values of observing time and light flux at each base-line, to normalize the observed values of $C(d)$ by the corresponding values of $C(o)$, so as to give the normalized correlation coefficients $\Gamma^2(d)$ shown in line 7. In Fig. 2 these experimental values of $\Gamma^2(d)$ are shown together with their probable errors, and may be compared with the broken curve, which gives the theoretical values for a uniform disk of 0.0063″. It can be seen that both the relative and absolute values of $\Gamma^2(d)$ are in reasonable agreement with theory, and that within the rather wide limits of this preliminary test there is no significant difference between the correlation predicted and observed.

In assessing the potentialities of the technique described here, it is important to note that, although the measurements took five months to complete, the visibility was so poor that the total observing time was only 18 hr., while in this limited period additional absorption of 0.25–0.75 magnitudes due to haze or thin cloud was often present. If the observations had been made at a latitude where Sirius transits close to the zenith, the improved signal-to-noise ratio, due to decreased atmospheric absorption, would have made it possible to obtain the same data in a total observing time of about four hours.

表 1 中第 5 行给出 $C(o)$ 的理论值是通过对应于各基线长度的光通量和观测时间，运用我们以前研究[1]中给出的式（1）和式（2）计算出来的（尽管现在实验得到的均方根起伏是以前文章中涉及其他电子技术给出值的 $1/\sqrt{2}$）。在这一计算中最重要的量是光电倍增器的输出电流和增益以及放大器的带宽；但是对光阴极、大气衰减、恒星和反射镜的联合光谱特征做一些微小的修正也是很必要的。最后，在表 1 第 4 行中给出了 $C(d)$ 的理论值，这些值是用 $C(o)$ 和 $\Gamma^2(d)$ 的理论值通过上文给出的相互关系计算得到的。

由于相对应的理论值只和已知量有关而与恒星的角直径几乎无关，在最短基线（2.5 米）情况下观测得到的相关系数可以作为大气闪烁对该设备影响的一个粗略测试。整个观测过程中，天狼星看起来闪烁得厉害，尽管在光电倍增管阳极的直流电中（时间常数约为 0.1 秒时电流平稳），在反射镜比正常望远镜大的情况下，预期相应的起伏只有 ±10% 的量级。然而，观测得到的相关系数 $C(d)$=+8.50 和计算值 +9.35 相差不大，这表明闪烁对观测结果没有产生很大影响。

对表 1 中第 3 行、第 4 行 4 种不同基线情况下得到的相关系数 $C(d)$ 的实验值和相对应的理论值进行比较。然而，由于这些数据和在每一基线处不同的观测时间值和光通量值有关，所以更为方便的是用相对应的值 $C(o)$ 对观测值 $C(d)$ 进行归一化，以便得到表 1 第 7 行给出的归一化相关系数 $\Gamma^2(d)$。在图 2 中，给出了 $\Gamma^2(d)$ 的实验值及其各自的可能误差，并可以和角直径为 0.0063″ 均匀圆盘理论值的虚线进行比较。我们可以看出，$\Gamma^2(d)$ 的相对值及绝对值都与理论吻合，而且在这种初步的测试的较宽限制下，理论预测和观测分别得到的相关系数并没有显著差异。

在评估本文所描述的这种技术的潜能时，值得我们注意的是尽管该测量工作耗费了 5 个月的时间才完成，但能见度太低以至于总观测时间只有 18 小时，同时在这一有限的时期内，由于常常有雾或薄云造成了 0.25~0.75 星等的额外吸收。如果在天狼星的中天靠近天顶的纬度地区进行观测，由于减少了大气吸收的影响，信噪比提高，这样也许在总共 4 个小时的观测时间内得到等量的数据信息。

Thus, despite their tentative nature, the results of this preliminary test show definitely that a practical stellar interferometer could be designed on the principles described above. Admittedly such an instrument would require the use of large mirrors. Judging from the results of this test experiment, where the peak quantum efficiency of the phototubes was about 16 percent and the overall bandwidth of the amplifiers was about 38 Mc./s., one would need mirrors at least 3 metres in diameter to measure a star, near the zenith, with an apparent photographic magnitude +1.5. Mirrors of at least 6 metres in diameter would be required to measure stars of mag. +3, and an increase in size would also be needed for stars at low elevation because of atmospheric absorption. However, the optical properties of such mirrors need be no better than those of searchlight reflectors, and their diameters could be decreased if the overall band-width of the photomultipliers and the electronic apparatus could be increased, or if photocathodes with higher quantum efficiencies become available. It must also be noted that the technique of using two mirrors, as described here, would probably be restricted to stars of spectral type earlier than *G*, since cooler stars of adequate apparent magnitude would be partially resolved by the individual mirrors.

The results of the present experiment also confirm the theoretical prediction[6] that an "intensity" interferometer should be substantially unaffected by atmospheric scintillation. This expectation is also supported by experience with a radio "intensity" interferometer[5,7,8] which proved to be virtually independent of ionospheric scintillation. It is also to be expected that the technique should be capable of giving an extremely high resolving power. Without further experience it is impossible to estimate the maximum practical length of the base-line; however, it is to be expected that the resolving power could be at least one hundred times greater than the highest value so far employed in astronomy, and that almost any star of sufficient apparent magnitude could be resolved.

We thank the Director of Jodrell Bank for making available the necessary facilities, the Superintendent of the Services Electronics Research Laboratory for the loan of much of the equipment, and Dr. J. G. Davies for his assistance with setting up the searchlights. One of us (R. Q. T.) wishes to thank the Admiralty for permission to submit this communication for publication.

(**178**, 1046-1048; 1956)

R. Hanbury-Brown: Jodrell Bank Experimental Station, University of Manchester.
R. Q. Twiss: Services Electronics Research Laboratory, Baldock.

References:

1. Hanbury Brown, R., and Twiss, R. Q., *Nature*, 177, 27 (1956).
2. Galbraith, W., and Jelley, J. V., *J. Atmos. Terr. Phys.*, **6**, 250 (1955).
3. Kuiper, G. P., *Astrophys. J.*, **88**, 429 (1938).
4. Keenan, P. C., and Morgan, W. W., "Astrophysics", edit. J. A. Hynek (McGraw-Hill, New York, 1951).
5. Hanbury Brown, R., and Twiss, R. Q., *Phil. Mag.*, **45**, 663 (1954).
6. Twiss, R. Q., and Hanbury Brown, R. (in preparation).
7. Hanbury Brown, R., Jennison, R. C., and Das Gupta, M. K., *Nature*, **170**, 1061 (1952).
8. Jennison, R. C., and Das Gupta, M. K., *Phil. Mag.*, 1, viii, 55 (1956).

因此，尽管这一初步测试具有一定的尝试性，但是其结果确实表明：可以用上述原理设计一个实际的恒星干涉仪，但是我们需要承认，这样的观测需要更大的反射镜。根据本次测试实验结果，当光电管的峰值量子效率约为 16%、放大器的总带宽约为 38 兆周 / 秒时，如果所选星体位于天顶、视照相星等为 +1.5，则至少需要直径为 3 米的反射镜。如果所测量星是 +3 等星，则需要至少直径为 6 米的反射镜。由于大气吸收的作用，观测星体的地平高度越低，所需反射镜的直径也就越大。然而，这些反射镜的光学性质只需要和探照灯反射面一样即可。而且，如果光电倍增管和其他电子器件的总带宽增加或者光电阴极的量子效率进一步提高并可用，那么反射镜的直径可以相应减小。需要指出的是，上述两面反射镜的观测技术仅仅适用于光谱型早于 G 型恒星的恒星。因为有足够视星等的较冷恒星可以用单面反射镜部分分辨。

现阶段的实验结果也证明了理论上预期的结果 [6]，即这种“强度”干涉仪基本上不会受到大气闪烁的影响。用无线电“强度”干涉仪的实验结果 [5,7,8] 也支持这一预期结果：实际上无线电干涉仪的观测和电离层的闪烁无关。人们还希望这种技术可以达到更高的分辨率。没有进一步的实验就想要预估出基线实际最大长度是不可能的；然而，人们预期该技术的分辨率至少能提高到目前天文学领域内分辨率最高值的 100 倍。这样一来，就都能够分辨几乎所有有足够视星等的恒星。

我们感谢焦德雷尔班克天文台的主管提供可用的必要仪器，电子服务研究实验室的负责人借给我们大部分的实验设备，以及戴维斯博士在探照灯放置上给予的帮助。我们中的特威斯非常感谢英国海军部批准此内容发表。

（冯翀 翻译；蒋世仰 审稿）

Investigation of the Radio Source 3C 273 by the Method of Lunar Occultations

C. Hazard *et al.*

Editor's Note

Radio astronomy was just getting started in the 1950s and 1960s. Before the development of the interferometer, positions of radio sources were best determined by "lunar occultations" when the Moon passed in front of the object. The position of the peculiar radio source 3C 273 is determined by such a method in this paper by Cyril Hazard and coworkers, leading to the identification of 3C 273 as the first known quasi-stellar object or quasar, a large black hole surrounded by gas.

THE observation of lunar occultations provides the most accurate method of determining the positions of the localized radio sources, being capable of yielding a positional accuracy of the order of 1 sec of arc. It has been shown by Hazard[1] that the observations also provide diameter information down to a limit of the same order. For the sources of small angular size the diameter information is obtained from the observed diffraction effects at the Moon's limb which may be considered to act as a straight diffracting edge.

The method has so far been applied only to a study of the radio source 3*C* 212 the position of which was determined to an accuracy of about 3 sec of arc[1,2]. However, 3*C* 212 is a source of comparatively small flux density and although the diffraction effects at the Moon's limb were clearly visible the signal-to-noise ratio was inadequate to study the pattern in detail and hence to realize the full potentialities of the method. Here we describe the observation of a series of occultations of the intense radio source 3*C* 273 in which detailed diffraction effects have been recorded for the first time permitting the position to be determined to an accuracy of better than 1″ and enabling a detailed examination to be made of the brightness distribution across the source.

The observations were carried out using the 210-ft. steerable telescope at Parkes, the method of observation being to direct the telescope to the position of the source and then to record the received power with the telescope in automatic motion following the source. Three occultations of the source have been observed, on April 15, at 410 Mc/s, on August 5 at 136 Mc/s and 410 Mc/s, and on October 26 at 410 Mc/s and 1,420 Mc/s, although in October and April only the immersion and emersion respectively were visible using the Parkes instrument. The 410 Mc/s receiver was a double-sided band receiver, the two channels, each of width 10 Mc/s, being centred on 400 Mc/s and 420 Mc/s, while the 136 Mc/s and 1,420 Mc/s receivers each had a single pass band 1.5 Mc/s and 10 Mc/s wide respectively.

用月掩的方法研究射电源3C 273

哈泽德等

编者按

射电天文学在20世纪50年代和60年代才刚刚起步。在干涉仪发明之前，利用月球从天体前经过时所产生的“月掩效应”是确定射电源位置的最佳方法。本文中西里尔·哈泽德及其同事们利用该方法测定了3C 273这个特殊射电源的位置，从而证认了3C 273为第一个已知的类星体——由气体包围着的一个巨大的黑洞。

对月掩的观测为确定射电源位置提供了最为准确的方法，由此能得到的位置准确度达1角秒的量级。哈泽德[1]已指出，观测提供的直径信息达到了相同量级的最低限度。对于具有小的角大小的射电源，通过从月亮的边缘处观测到的衍射效应也可获取直径的信息，其中月亮的边缘可视为一个直的衍射边缘。

这种方法目前只用于射电源3C 212的研究中，其位置测定的准确度约为3角秒[1,2]。然而，3C 212是一个流量密度相当小的射电源，虽然在月亮边缘处的衍射效应清晰可见，但其信噪比还无法满足详细研究其模式的需要，因此还无法体现出该方法的全部潜力。本文中我们描述了对强射电源3C 273一系列掩源现象的观测，之前为了使该射电源位置测量精度优于1″，已详细记录了衍射效应，并且对整个射电源的亮度分布进行了详细检测。

这次观测用的是位于帕克斯的一台210英尺的可跟踪望远镜，观测的方法是首先将望远镜对准射电源的方向，之后在望远镜自动跟踪射电源时记录所接收到的功率。目前已观测到射电源的三次掩源，其时间和频率分别为：4月15日频率为410兆周/秒，8月5日频率为136兆周/秒和410兆周/秒，以及10月26日频率为410兆周/秒和1,420兆周/秒，尽管在10月和4月用帕克斯望远镜分别只能观测到掩始和复现。其中，接收到410兆周/秒频率的是一个双边波段双通道接收器，两个通道带宽均为10兆周/秒，其中心频率分别为400兆周/秒和420兆周/秒，而接收到136兆周/秒和1,420兆周/秒频率的是两个单通道接收器，其带宽分别为1.5兆周/秒和10兆周/秒。

The record of April 15, although of interest as it represents the first observation of detailed diffraction fringes during a lunar occultation, is disturbed by a gradient in the received power and is not suitable for accurate position and diameter measurements. Therefore, attention will be confined to the occultation curves recorded in August and October and which are reproduced in Fig. 1. It is immediately obvious from these records that 3*C* 273 is a double source orientated in such a way that whereas the two components passed successively behind the Moon at both immersions, they reappeared almost simultaneously. The prominent diffraction fringes show that the angular sizes of these components must be considerably smaller than 10″, which is the order of size of a Fresnel zone at the Moon's limb.

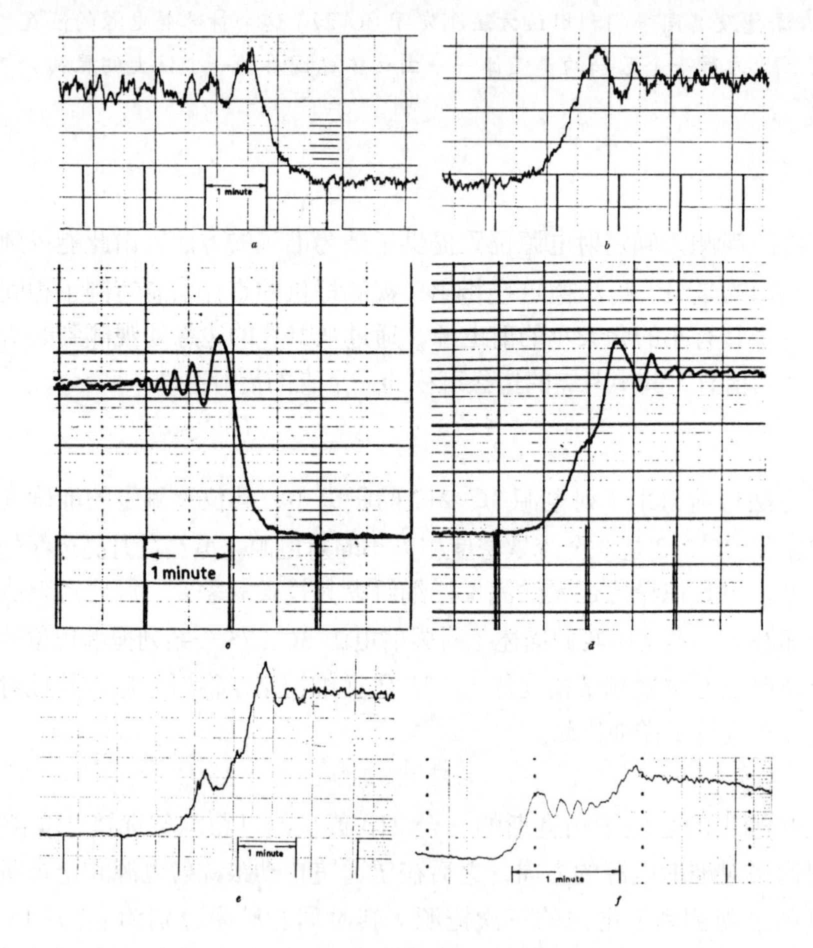

Fig. 1. Facsimiles of records showing occultations on August 5 and October 26, 1962, at different frequencies. (*a*) Emersion of August 5, 1962, at 136 Mc/s; (*b*) immersion of August 5, 1962, at 136 Mc/s; (*c*) emersion of August 5, 1962, at 410 Mc/s; (*d*) immersion of August 5, 1962, at 410 Mc/s; (*e*) immersion of October 26, 1962, at 410 Mc/s; (*f*) immersion of October 26, 1962, at 1,420 Mc/s. Abscissae, U. T.; ordinates, flux density

尽管4月15日的记录由于展示了在月掩期间对详细的衍射条纹的首次观测而非常令人感兴趣，然而却受到了接收功率中梯度的干扰，因此并不适用于对位置和直径的准确测量。因此，注意力应集中在8月和10月所记录的掩源曲线上，如图1所示。根据这些记录可以很快发现3*C* 273是一组双射电源，因为它们以下列方式出现，即两个子源相继通过月亮后面两个掩始处，并几乎同时复现。这组显著的衍射条纹表明，两个子源的角大小必然远小于10″，而10″则是月球边缘处菲涅耳区大小的量级。

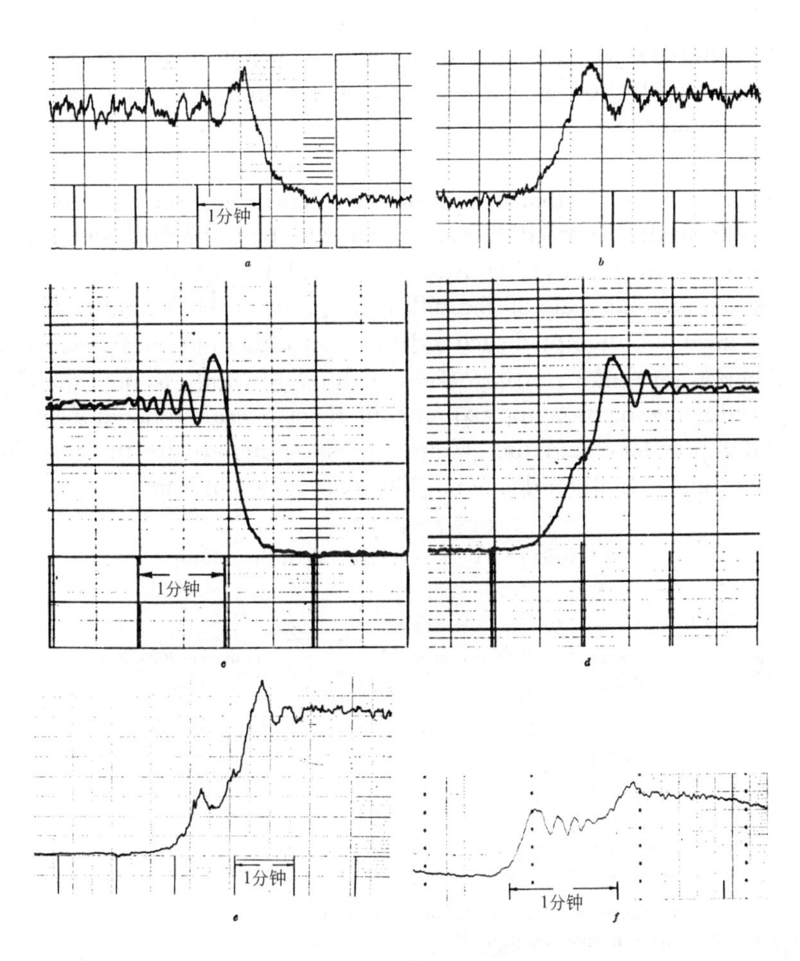

图1. 1962年8月5日和10月26日在不同频率时掩源记录的复制图。(*a*)1962年8月5日频率为136兆周／秒时的复现；(*b*)1962年8月5日频率为136 兆周／秒时的掩始；(*c*)1962年8月5日频率为410兆周／秒时的复现；(*d*)1962年8月5日频率为410 兆周／秒时的掩始；(*e*)1962年10月26日频率为410兆周／秒时的掩始；(*f*)1962年10月26日频率为1,420 兆周／秒时的掩始。横坐标为世界时；纵坐标为流量密度。

The most interesting feature of Figs. 1(*e*) and 1(*f*) is the change in the ratio of the flux densities of the two components with frequency. The ratio of the flux density of the south preceding source (component *A*) to that of the north following source (component *B*) is 1:0.45 at 410 Mc/s and 1:1.4 at 1,420 Mc/s, indicating a striking difference in the spectra of the two components. If it be assumed that the flux densities[3] of 3*C* 273 at 410 Mc/s and 1,420 Mc/s are 60 and 35 Wm^{-2} $(c/s)^{-1}$ and that over this frequency-range the spectrum of each component may be represented by $S \propto f^n$, then the above ratios correspond to spectral indices for components *A* and *B* of −0.9 and 0.0 respectively. The spectral index of *A* is a representative value for a Class II radio source; but the flat spectrum of *B* is most unusual, no measurements of a comparable spectrum having yet been published. If the spectral indices were assumed constant down to 136 Mc/s then at this frequency component *A* must contribute almost 90 percent of the total emission, a conclusion which is confirmed by a comparison of the times of immersion at 136 Mc/s and 410 Mc/s on August 5.

It has been shown by Scheuer[4] that it is possible to recover the true brightness distribution across the source from the observed diffraction pattern, the resolution being subject only to limitations imposed by the receiver bandwidth and the finite signal to noise ratio and being independent of the angular scale of the diffraction pattern. However, in this preliminary investigation we have not attempted such a detailed investigation but based the analysis on the calculated curves for uniform strip sources of different widths as published by Hazard[1]. As a first step in the investigation approximate diameters were estimated from the intensity of the first diffraction lobe and the results corresponding to the three position angles defined by the occultations and indicated in Fig. 2. are given in Table 1.

Table 1. Effective Width of Equivalent Strip Source

(Sec. of arc)

Frequency Mc/s	Component *A* Position angle			Component *B* Position angle		
	106°	313°	84°	105°	314°	83°
136	6.4	6.4	—	—	—	—
410	3.1	4.2† 2(6)*	4.2	3.1	3.0†	2.7
1,420	—	—	2.9	—	—	2.1 0.5(7)*

* Estimated from an analysis of the whole diffraction pattern.

† Component *B* assumed to have width of 3″.

图 1(*e*) 和 1(*f*) 最令人感兴趣的特征是两个子源的流量密度之比随频率的变化。当频率为 410 兆周 / 秒时，南方的前导源 (子源 *A*) 与北方的后继源 (子源 *B*) 的流量密度之比为 1:0.45，而当频率为 1,420 兆周 / 秒时则为 1:1.4，这说明两个子源在光谱上的明显差异。假设当频率为 410 兆周 / 秒和 1,420 兆周 / 秒时，3*C*273 的流量密度 [3] 分别是 60 瓦 · 米$^{-2}$(周 / 秒)$^{-1}$ 和 35 瓦 · 米$^{-2}$(周 / 秒)$^{-1}$，并且在此频率范围内，每个子源的频谱可由 $S \propto f^n$ 表示，则上述比值相应于子源 *A* 和 *B* 的谱指数分别为 –0.9 和 0.0。*A* 的谱指数是 II 类射电源的典型值，但 *B* 的平谱是最特别的，目前尚未发表过类似的频谱的测量结果。如果假定谱指数在频率降至 136 兆周 / 秒时仍为常数，则在此频率下子源 *A* 几乎贡献总发射的 90%，通过比较 8 月 5 日当频率为 136 兆周 / 秒和 410 兆周 / 秒时的掩始时间可以证实这一结论。

朔伊尔 [4] 曾指出，可以通过观测到的衍射图案来重现整个射电源的实际亮度分布，分辨率的大小仅仅受到接收器带宽和有限信噪比的限制，并且和衍射图案的角尺度无关。然而，在我们这篇初步的研究中尚未进行如此详细的研究，而是建立在对哈泽德 [1] 发表的为具有不同宽度的均匀条带状射电源的计算曲线的分析之上。作为研究中的第一步，通过第一衍射波瓣的强度估算出近似直径，与图 2 中由月掩确定的三个位置角相对应的结果在表 1 中给出。

表 1. 等效条带源的有效宽度

(角秒)

频率 兆周 / 秒	子源 *A* 位置角			子源 *B* 位置角		
	106°	313°	84°	105°	314°	83°
136	6.4	6.4	—	—	—	—
410	3.1	4.2† 2(6)*	4.2	3.1	3.0†	2.7
1,420	—	—	2.9	—	—	2.1 0.5(7)*

* 由整个衍射图案估算得到。

† 假设子源 *B* 的宽度是 3″。

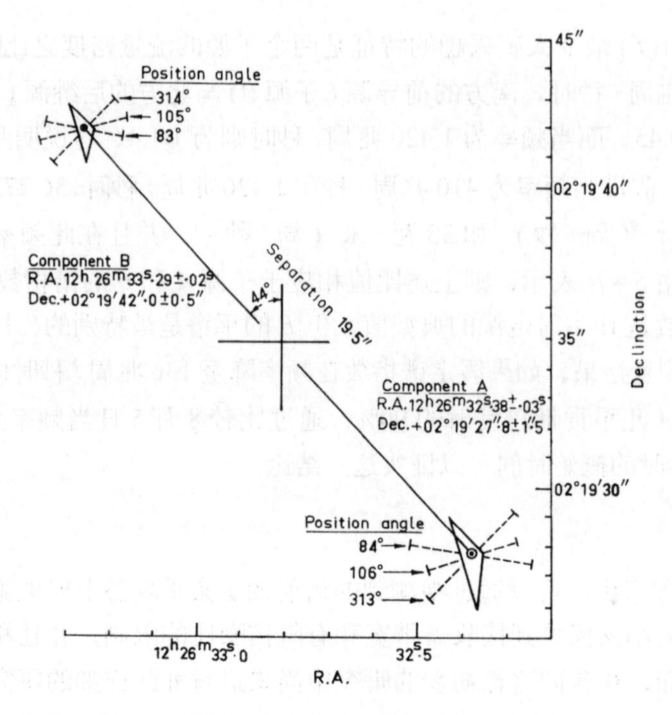

Fig. 2. Diagram of the radio source 3*C* 273. The sides of the full line triangles represent the positions of the limb of the Moon at the times of occultation. The broken lines represent the widths of the equivalent strip source as measured at 410 Mc/s for each of three position angles indicated

As already indicated here, the 136-Mc/s measurements refer only to component *A* and hence no diameter measurements are available for *B* at this frequency. The 410-Mc/s observations of the August occultation are the most difficult to interpret owing to the components having both comparable flux density and small separation relative to the angular size of the first Fresnel zone. At immersion the widths were estimated by using a process of curve fitting to reproduce Fig. 1(*d*); at emersion (position angle 313°) the diameter of component *B* was assumed to be 3″ as indicated by the estimates at position angles 105° and 83°. The individual measurements at each frequency are reasonably consistent but there is a striking variation of the angular size of component *A* with frequency and evidence of a similar variation for component *B*. As at the time of the August occultation the angular separation of the Sun and the source was about 50° and hence coronal scattering of the type observed by Slee[5] at 85 Mc/s is not likely to be significant, this variation in size suggests that the model of two uniform strip sources is inadequate.

Therefore, a more detailed analysis was made of the intensity distributions of the lobe patterns given in Figs. 1(*c*) and 1(*f*), and it was found that in neither case can the pattern be fitted to that for a uniform strip source or a source with a gaussian brightness distribution. The 1,420-Mc/s observations of component *B* can be explained, however, by assuming that this source consists of a central bright core about 0.5″ wide contributing about 80 percent of the total flux embedded in a halo of equivalent width of about

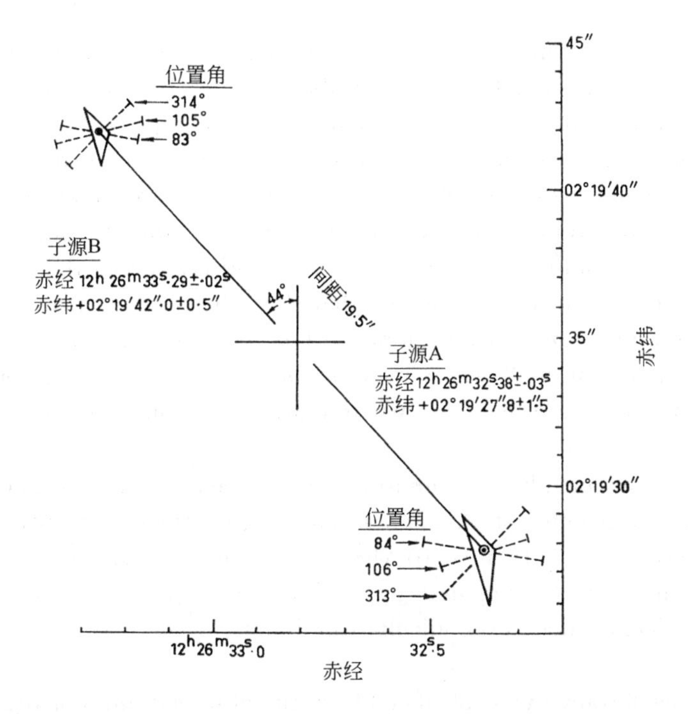

图2. 射电源3*C* 273的示意图。实线三角形的边表示掩源时月球边缘的位置。虚线表示等效条带源的宽度，测量是频率为410兆周/秒时对三个指定位置角分别进行的。

如本文已指出的，136兆周/秒的测量仅仅针对子源*A*，因此在这个频率上并无对子源*B*可用的直径测量。8月掩源的410兆周/秒的观测是最难以解释的，因为两个子源的流量密度相近，并且其间距相对于第一菲涅耳区的角大小来说很小。掩始时的宽度是用曲线拟合图1(*d*)来估算的；在复现时（位置角313°），根据位置角为105°和83°时所估算的，子源*B*的直径假定为3″。每个频率的独立测量结果是相当一致的，但子源*A*的角大小随频率有显著变化，有证据显示子源*B*也存在类似的情况。如在8月掩源发生期间，太阳与射电源的角间距约为50°，因此由斯利[5]在频率为85兆周/秒时观测到的这种类型的日冕散射似乎并不显著，这类角大小的变化表明，用这种两个均匀条带状的模型来描述射电源是不恰当的。

因此，我们对图1(*c*)和图1(*f*)中波瓣条纹的强度分布进行了更细致的分析，发现在任何情况下，都不能将图案拟合为均匀条带源或具有高斯亮度分布的源。然而，只要假定这个源是由宽度约为0.5″的中心亮核组成，并且贡献了相当于一个等效宽度约为7″的晕圈所包含的总流量的80%左右，就可以解释子源*B*在频率为1,420兆周/秒

7″. Fig. 1(*b*), where component *A* predominates, suggests that this source has a similar structure but with a core of effective width about 2″ at 410 Mc/s and a halo of width 6″. It therefore seems that the overall extent of both components are comparable but that the emission is more highly concentrated to the nucleus in *B* than in *A*. The close agreement between the halo size of *A* and its effective diameter at 136 Mc/s suggests that the observed variation of effective size with frequency may be due to a difference in the spectra of the halo and central regions. This would imply that the spectrum becomes steeper in the outer regions of the sources, that is, in the regions of lower emissivity. It is of interest that the integrated spectral indices of the two components show an analogous effect. Thus the spectrum of *B*, where most of the emission arises in a source about 0.5″ wide, is markedly flatter than that of *A*, where it arises in a source about 2″ wide.

The analysis is not sufficiently accurate to reach any reliable conclusions on the ellipticity of the individual components of 3*C* 273, but allowing for the uncertainty in the estimated widths and position angle 314°, the 410-Mc/s observations indicate that both components may be elliptical with *A* elongated approximately along the axis joining the two components and *B* elongated perpendicular to this axis.

The position of each source was calculated from the observed times of disappearance and reappearance, which were estimated from the calculated flux density at the edge of the geometrical shadow and, where possible, from the positions of the diffraction lobes; these times are given in Table 2. In estimating the values of T_D^A and T_R^A from the 136-Mc/s records a small correction was applied for the effects of component *B*, this correction being estimated by comparison with the 410-Mc/s records. The corresponding times for *B* were estimated from the 410-Mc/s observations using the estimated position of component *A* and the known flux density ratio of the two components. For each component the times and associated errors given in Table 2 define three strips in each of which the source should lie; the centre lines of these strips represent the limb of the Moon at the time of observation and define in each case a triangular-shaped area. In principle, the position of the source lies in the area common to the three associated strips but it was found that for each component, and in particular for component *A*, that the size of the triangles defined by the Moon's limb was larger than would be expected from the estimated timing errors. This suggests that errors in the positions of the Moon's limb are more important than the estimated timing errors, and possibly that the effective position of the source varies slightly with frequency. The position of each source was therefore assumed to be given by the centre of the circle inscribed in the triangle defined by the Moon's limb at the relevant times. Dr. W. Nicholson of H.M. Nautical Almanac Office has kindly carried out these calculations and the estimated positions are as follows:

Component *A*	R.A.	12h 26m 32.38s ± 0.03s
(Epoch 1950)	Decl.	02° 19′ 27.8″ ± 1.5″
Component *B*	R.A.	12h 26m 33.29s ± 0.02s
(Epoch 1950)	Decl.	02° 19′ 42.0″ ± 0.5″

的观测结果。图 1(*b*) 中的子源 *A* 是占主要地位的，这表明该源具有类似的结构，但在 410 兆周 / 秒处有一个有效宽度约为 2″的核，宽度约为 6″的晕。因此两个子源的总体宽度相近，但是 *B* 的中心区的发射比 *A* 更高度集中。子源 *A* 的晕大小与它在 136 兆周 / 秒的有效直径的一致性表明，观测到的有效大小随频率的变化可能是由晕和中心区域在光谱上的差异导致的。这表明光谱在射电源的外部区域或者说在发射率较小的区域变得陡峭。非常有趣的是，两个子源的累积光谱指数显示出了一种类似的效应。子源 *B* 大部分的发射来自一个宽度约为 0.5″的源，其光谱比大部分发射来自宽度约为 2″的源的子源 *A* 的光谱要平缓得多。

这些分析尚未精确到足以得出任何关于 3*C* 273 各个子源椭率的可靠结论，但即便考虑到估算宽度及位置角 314°时的不确定性，410 兆周 / 秒的观测也表明两个子源可能都是椭圆状的，其中 *A* 的长轴方向大约是沿着连接两个子源的轴线，而 *B* 的长轴方向则垂直于该轴线。

每个源的位置是根据观测到的消失时间及重现时间而计算出来的，消失和重现是通过几何阴影边缘的流量密度以及可能从衍射波瓣的位置估算的；表 2 中给出了这些时间。在根据 136 兆周 / 秒的记录估算 T_D^A 和 T_R^A 值的过程中，对子源 *B* 的作用做了一项小的修正，这项修正是通过与 410 兆周 / 秒记录的对比估算出来的。子源 *B* 的相应时间是通过 410 兆周 / 秒观测采用子源 *A* 的估计位置及两个子源的已知流量密度比来估算的。对每个子源而言，表 2 中给出的时间和相关误差确定了源应该所在的三个条带；这些条带的中线代表了观测时月球的边缘，并且确定了在每种情况下的三角形阴影区。原则上，源的位置应位于三个相关条带共同的区域中，但发现对每个子源而言，尤其是对子源 *A*，由月球边缘所确定的三角形面积大于根据估算时间误差而预期的值。这表明月球边缘的位置误差比估算的时间误差更重要，并且源的有效位置可能随频率稍有变化。因此，每个源的位置可能是由在相关时间内月球边缘所确定的三角形中内切圆的中心所给出的。航海天文历编制局的尼克尔森博士已做过计算，估算的位置如下：

子源 *A*	赤经	12h 26m 32.38s ± 0.03s
(历元 1950)	赤纬	02° 19′ 27.8″ ± 1.5″
子源 *B*	赤经	12h 26m 33.29s ± 0.02s
(历元 1950)	赤纬	02° 19′ 42.0″ ± 0.5″

Table 2. Observed Occultation Times of the Two Components of *3C* 273

	Component *A* (U.T.)	Component *B* (U.T.)
Time of disappearance August 5, 1962	07h 46m 00s ± 1s	07h 46m 27.2s ± 0.5s
Time of reappearance August 5, 1962	09h 05m 45.5s ± 1s	09h 05m 45.7s ± 1.5s
Time of disappearance October 26, 1962	02h 55m 09.0s ± 1s	02h 56m 01.5s ± 0.4s

The average positions of the two sources given here represent the most accurate determination yet made of the position of a radio source. The quoted errors were estimated from the size of the triangles defined by the Moon's limb at the times of disappearance and reappearance, for the method is not subject to uncertainties introduced by refraction in the Earth's ionosphere or troposphere and is also free from the effects of confusion. A comparison of the times of disappearance and reappearance at different frequencies indicates that there is also no significant source of error due to refraction in either the solar corona or a possible lunar ionosphere; any refraction appears to be less than 0.3″ even at 136 Mc/s. This may be compared with the upper limit of 2″ at 237 Mc/s and 13″ at 81 Mc/s as estimated by Hazard[1] and Elsmore[6] respectively, and allows a new limit to be set to the density of the lunar ionosphere. Thus, from his observations at 81.5 Mc/s, Elsmore has set an upper limit to the electron density of 10^3 cm^{-3}; and it follows that the present measurements set a limit of about 10^2 cm^{-3}. Similarly, Buckingham[7] has estimated that at 50 Mc/s a ray passing at 50° to the Sun would be deviated by 1″ if the electron density in the solar corona at the Earth's distance from the Sun is 100 cm^{-3}. The present observations at 136 Mc/s and 410 Mc/s on August 5 indicate that at 50 Mc/s the deviation is less than 2″ at this angle, setting an upper limit to the electron density of about 200 cm^{-3}, which may be compared with an upper limit of 120 cm^{-3}, set by Blackwell and Ingham[8] from observations of the zodiacal light.

In a preliminary examination of a print from a 200″ plate it was noted that the position of component *B* agreed closely with that of a thirteenth magnitude star. We understand that the investigations by Drs. A. Sandage and M. Schmidt of the Mount Wilson and Palomar Observatories have revealed that this star and an associated nebulosity is very probably the source of the radio emission.

We thank Mr. J. G. Bolton for his interest in this work and his assistance, with that of the staff at Parkes, in ensuring the success of these observations. We also thank Dr. W. Nicholson, who calculated the positions of the sources, for his valuable co-operation and interest in the occultation programme. One of us (C. H.) thanks Dr. E. G. Bowen for his invitation to continue occultation work at Parkes as a guest observer from the Narrabri Observatory of the School of Physics of the University of Sydney.

(**197**, 1037-1039; 1963)

表 2. 3*C* 273 两个子源的观测掩源的时间

	子源 *A*（世界时）	子源 *B*（世界时）
消失的时间 1962 年 8 月 5 日	07h 46m 00s ± 1s	07h 46m 27.2s ± 0.5s
重现的时间 1962 年 8 月 5 日	09h 05m 45.5s ± 1s	09h 05m 45.7s ± 1.5s
消失的时间 1962 年 10 月 26 日	02h 55m 09.0s ± 1s	02h 56m 01.5s ± 0.4s

此处给出的两个源的平均位置是迄今对射电源位置的最准确的测定。其中误差是根据由消失和重现时月球边缘所确定的三角形的大小进行估算的，因为该方法并不受地球电离层或对流层的折射所引入的不确定性的影响，也没有干扰效应。对不同频率时消失和重现时刻的比较表明，并没有由日冕或可能的月球电离层的折射所引起的重要误差来源；即使频率为 136 兆周 / 秒时折射也小于 0.3″。这与分别由哈泽德 [1] 和埃尔斯莫尔 [6] 估算的当频率为 237 兆周 / 秒时的 2″ 和 81 兆周 / 秒时的 13″ 的上限相一致，并允许对月球电离层的密度取一个新的极限。因此，根据埃尔斯莫尔在频率为 81.5 兆周 / 秒时的观测，他取的电子密度的上限为 10^3 厘米$^{-3}$；而目前的测量取的极限约为 10^2 厘米$^{-3}$。与此类似，白金汉 [7] 假设在频率为 50 兆周 / 秒时有离开太阳 50° 的射线通过，如果在地球距离太阳的距离为 1 个天文单位那么远的地带日冕中的电子密度是 100 厘米$^{-3}$ 时，所估算射线的偏离将为 1″。8 月 5 日频率为 136 兆周 / 秒时和 410 兆周 / 秒时的观测表明，当频率为 50 兆周 / 秒时的角度偏离小于 2″，所取的电子密度的上限约为 200 厘米$^{-3}$，这与由布莱克韦尔和英厄姆 [8] 根据黄道光的观测所得的 120 厘米$^{-3}$ 的上限一致。

在对 200″ 的望远镜照相底片翻印片的初步检验中，注意到子源 *B* 的位置与一颗十三等的恒星的位置非常接近。我们了解到，威尔逊山天文台和帕洛玛山天文台的桑德奇博士和施密特博士的研究已揭示了这颗恒星及有关的星云状物质很可能就是该射电辐射的源。

我们特别感谢博尔顿先生对这项工作的关注以及他的帮助，并感谢帕克斯的同事为成功进行这些观测所做出的努力。我们还要感谢尼克尔森博士对月掩项目的兴趣和非常有价值的合作以及对射电源位置所进行的计算。我们其中的一员，来自悉尼大学物理学院纳拉布里天文台的哈泽德要感谢鲍恩博士的邀请，使得他能继续以客座观察员的身份在帕克斯的月掩项目中工作。

（沈乃澂 翻译；蒋世仰 审稿）

C. Hazard, M. B. Mackey and A. J. Shimmins: C.S.I.R.O. Division of Radiophysics, University Grounds, Sydney.

References:

1. Hazard, C., *Mon. Not. Roy. Astro. Soc.*, **134**, 27 (1962).
2. Hazard, C., *Nature*, **191**, 58 (1961) .
3. Bolton, J. G., Gardner, F. F., and Mackey, M. B. (unpublished results).
4. Scheuer, P. A. G., *Austral. J. Phys.*, **15**, 333 (1962).
5. Slee, O. B., *Mon. Not. Roy. Astro. Soc.*, **123**, 223 (1961).
6. Elsmore, B., *Phil. Mag.*, 2, 1040 (1957).
7. Buckingham, M. J., *Nature*, **193**, 538 (1962).
8. Blackwell, D. E., and Ingham, M. F., *Mon. Not. Roy. Astro. Soc.*, **122**, 129 (1961).

3C 273: A Star-like Object with Large Red-shift

M. Schmidt

Editor's Note

Here astronomer Maarten Schmidt uses the position of the unusual radio source 3C 273 determined by Cyril Hazard and colleagues to establish that it is a star-like object with a small jet apparently emanating from it, within 1 arcsecond of the radio sources. Schmidt also reports an optical spectrum, which can be explained only if the source were at a redshift of 0.158, making it very distant. This constituted the discovery of the first quasar. We now know that quasars are massive black holes surrounded by gas. As the gas falls into the black hole it becomes very hot, and the light emitted can swamp the glow of the galaxy at whose centre the quasar lies.

THE only objects seen on a 200-in. plate near the positions of the components of the radio source *3C* 273 reported by Hazard, Mackey and Shimmins in the preceding article are a star of about thirteenth magnitude and a faint wisp or jet. The jet has a width of 1″–2″ and extends away from the star in position angle 43°. It is not visible within 11″ from the star and ends abruptly at 20″ from the star. The position of the star, kindly furnished by Dr. T. A. Matthews, is R.A.12h 26m 33.35s ±0.04s, Decl. +2° 19′42.0″+0.5″ (1950), or 1″ east of component *B* of the radio source. The end of the jet is 1″ east of component *A*. The close correlation between the radio structure and the star with the jet is suggestive and intriguing.

Spectra of the star were taken with the prime-focus spectrograph at the 200-in. telescope with dispersions of 400 and 190 Å per mm. They show a number of broad emission features on a rather blue continuum. The most prominent features, which have widths around 50 Å, are, in order of strength, at 5,632, 3,239, 5,792, 5,032 Å. These and other weaker emission bands are listed in the first column of Table 1. For three faint bands with widths of 100–200 Å the total range of wave-length is indicated.

The only explanation found for the spectrum involves a considerable red-shift. A red-shift $\Delta\lambda/\lambda_0$ of 0.158 allows identification of four emission bands as Balmer lines, as indicated in Table1. Their relative strengths are in agreement with this explanation. Other identifications based on the above red-shift involve the Mg II lines around 2,798 Å, thus far only found in emission in the solar chromosphere, and a forbidden line of [O III] at 5,007 Å. On this basis another [O III] line is expected at 4,959 Å with a strength one-third of that of the line at 5,007 Å. Its detectability in the spectrum would be marginal. A weak emission band suspected at 5,705 Å, or 4,927 Å reduced for red-shift, does not fit the wave-length. No explanation is offered for the three very wide emission bands.

3C 273：一个具有很大红移的类星体

施密特

编者按

本文中，天文学家马丁·施密特利用西里尔·哈泽德及其同事们测定的特殊射电源 3C 273 的位置确认了 3C 273 是处于该射电源 1 角秒范围内的一个看来发射着一个小喷流的类似恒星天体。他还获得了这个源的光谱，只有当该射电源处于红移 0.158 时才能解释此光谱。从而使得该源被认为是一个距离非常遥远的源，这就是第一个类星体的发现。现在我们知道，类星体其实是由气体包围着的大质量黑洞。当气体落入黑洞时它会变得非常炽热，它所发出的光线能够掩盖星系的光芒，类星体处于该星系的中心。

在之前的文章中哈泽德、麦基和西敏报道了射电源 3C 273 子源的位置，在 200 英寸望远镜的底片上该射电源附近唯一能看到的天体是一颗亮度约为 13 等的星和一缕暗淡的亮条或喷流。喷流的宽度为 1″~2″，从恒星沿 43° 位置角向外延伸。在从恒星开始 11″ 的范围内不可见，并在位于恒星的 20″ 处突然结束。由马修斯博士友好提供的星的位置是赤经 12h 26m 33.35s ± 0.04s，赤纬 +2° 19′ 42.0″ ± 0.5″（1950 年）或射电源子源 *B* 以东 1″。喷流末端位于子源 *A* 以东 1″。射电结构与具有喷流的恒星之间的密切关系是非常具有启发性和耐人寻味的。

恒星的光谱取自 200 英寸望远镜主焦点处的摄谱仪，其色散为每毫米 400 埃和 190 埃。它们在偏蓝色连续谱上表现出大量的宽发射特征。这些光谱宽度约 50 埃，最突出的特征是按强度排序为 5,632 埃、3,239 埃、5,792 埃、5,032 埃。表 1 第 1 列中给出了这些及其他较弱的发射带。对于宽度为 100 埃 ~ 200 埃的 3 个较暗的波段则给出了其总的波长范围。

对这种光谱仅有的解释涉及一个相当大的红移。如表 1 所示，一个红移值 $\Delta\lambda/\lambda_0$ 为 0.158 的红移可证认出 4 个发射带为巴耳末谱线。它们的相应强度与这种解释是吻合的。基于上述红移的其他证认还包括 2,798 埃附近的迄今为止仅在太阳色球的发射中发现的 MgII 线以及在 5,007 埃处的 [O III] 禁线。在此基础上可以推测在 4,959 埃处还应该有一条 [O III] 线，其强度是 5,007 埃处线的 1/3。该谱线在光谱中刚好能被检测到。一个疑似在波长 5,705 埃或是由 4,927 埃红移后的微弱的发射带与其波长并不相符。对于其中 3 个非常宽的发射带，目前还没有给出任何解释。

Table1. Wave-lengths and identifications

λ	$\lambda/1.158$	λ_0	
3,239	2,797	2,798	Mg II
4,595	3,968	3,970	Hε
4,753	4,104	4,102	Hδ
5,032	4,345	4,340	Hγ
5,200–5,415	4,490–4,675		
5,632	4,864	4,861	Hβ
5,792	5,002	5,007	[O III]
6,005–6,190	5,186–5,345		
6,400–6,510	5,527–5,622		

It thus appears that six emission bands with widths around 50 Å can be explained with a red-shift of 0.158. The differences between the observed and the expected wave-lengths amount to 6 Å at the most and can be entirely understood in terms of the uncertainty of the measured wave-lengths. The present explanation is supported by observations of the infra-red spectrum communicated by Oke in a following article, and by the spectrum of another star-like object associated with the radio source *3C* 48 discussed by Greenstein and Matthews in another communication.

The unprecedented identification of the spectrum of an apparently stellar object in terms of a large red-shift suggests either of the two following explanations.

(1) The stellar object is a star with a large gravitational red-shift. Its radius would then be of the order of 10 km. Preliminary considerations show that it would be extremely difficult, if not impossible, to account for the occurrence of permitted lines and a forbidden line with the same red-shift, and with widths of only 1 or 2 percent of the wave-length.

(2) The stellar object is the nuclear region of a galaxy with a cosmological red-shift of 0.158, corresponding to an apparent velocity of 47,400 km/sec. The distance would be around 500 megaparsecs, and the diameter of the nuclear region would have to be less than 1 kiloparsec. This nuclear region would be about 100 times brighter optically than the luminous galaxies which have been identified with radio sources thus far. If the optical jet and component *A* of the radio source are associated with the galaxy, they would be at a distance of 50 kiloparsecs, implying a time-scale in excess of 10^5 years. The total energy radiated in the optical range at constant luminosity would be of the order of 10^{59} ergs.

Only the detection of an irrefutable proper motion or parallax would definitively establish *3C* 273 as an object within our Galaxy. At the present time, however, the explanation in terms of an extragalactic origin seems most direct and least objectionable.

表 1. 波长和证认

λ	λ/1.158	λ_0	
3,239	2,797	2,798	Mg Ⅱ
4,595	3,968	3,970	Hε
4,753	4,104	4,102	Hδ
5,032	4,345	4,340	Hγ
5,200~5,415	4,490~4,675		
5,632	4,864	4,861	Hβ
5,792	5,002	5,007	[O Ⅲ]
6,005~6,190	5,186~5,345		
6,400~6,510	5,527~5,622		

因此，看起来可以用 0.158 的红移来解释宽度约为 50 埃的 6 个发射带。观测到的波长和预期的波长最大差值为 6 埃，而这完全可以认为是由波长测量的不确定度引起的。随后由奥凯发表的一篇关于红外光谱的观测结果，以及另一篇由格林斯坦和马修斯发表的旨在讨论与射电源 3*C* 48 相关的另一个类星体光谱的通讯都支持上面的这种解释。

这一史无前例的对一个具有很大红移的类星体的证认暗示了下面两种可能的解释。

(1) 该类星体是一颗具有很大引力红移的恒星，它的半径的量级约为 10 千米。初步研究结果表明，考虑到在相同红移下容许谱线和禁线的出现，并且宽度仅为波长的 1% 或 2%，这种情况即便不是不可能，其存在也是极其困难的。

(2) 类星体是宇宙学红移为 0.158 的星系的核心区域，其相应的视速度为 47,400 千米 / 秒。其距离约为 500 兆秒差距，而核心区域的直径应小于 1 千秒差距。这个核心区域的亮度在光学上是迄今为止所有通过射电源证认的亮星系的亮度的 100 倍。如果射电源的光学喷流和子源 *A* 与星系相关，那么它们的距离应为 50 千秒差距，这意味着其时间尺度超过 10^5 年。在光度恒定的情况下，其光学波段内辐射的总能量量级应为 10^{59} 尔格。

只有确定无误地检测到其自行或视差才能最终肯定 3*C* 273 为银河系中的天体。然而，目前来看，将其视为起源于河外的天体似乎是最直接、最不会引起争议的。

I thank Dr. T. A. Matthews, who directed my attention to the radio source, and Drs. Greenstein and Oke for valuable discussions.

(**197**, 1040; 1963)

M. Schmidt: Mount Wilson and Palomar Observatories, Carnegie Institution of Washington, California Institute of Technology, Pasadena.

非常感谢马修斯博士将我的注意力引向射电源，并感谢格林斯坦博士和奥凯博士与我进行了有益的讨论。

（沈乃澂 翻译；邓祖淦 审稿）

Possible Anti-matter Content of the Tunguska Meteor of 1908

C. Cowan *et al.*

Editor's Note

The explosion that occurred over southern Siberia on 30 June 1908 is the most energetic meteoritic event of modern times (see also NTLRS Volume II). This paper addresses the question whether some fraction of the meteor responsible may have consisted of "anti-matter", whose annihilation by the constituents of the atmosphere would account for the large amount of energy released. A particular interest of this paper is that one of the authors was William F. Libby of the University of California at Los Angeles, who was then a member of the US Atomic Energy Commission and who had made a special study of the newly discovered radioactive isotope carbon-14: the use of carbon-14 had made it possible to use tree-rings for the more accurate determination of the age of ancient specimens of wood. The modern view is that the Tunguska event was within the limits to be expected of meteoritic phenomena.

PERHAPS the most spectacular meteor fall to be observed in modern times occurred on June 30, 1908, at $0^h\ 17^m\ 11^s$ U.T. in the basin of the Podkanemaia Tunguska River, Siberia (60°55′ N, 101°57′ E), some 500 miles to the north of Lake Baykal[1]. It was seen in a sunlit cloudless sky over an area of about 1,500 km in diameter and was described as the flight and explosion of a blindingly bright bolide which "made even the light of the Sun appear dark". The fall was accompanied by exceptionally violent radiation and shock phenomena. Although seismic, meteorological and geomagnetic field disturbances were registered at points around the world at the time, and descriptive accounts of the phenomena accompanying the fall were collected from witnesses during the years following, the first inspection of the place of fall was not made until 1927. No trace of a crater was found, though great damage of the forest was still evident due to thermal and blast effects.

Various hypotheses have been advanced to explain the massive phenomena as having been caused by a large meteor, a small comet or a nuclear explosion. All are argued cogently, for and against. Estimates of the total yield of energy, made from the records of the disturbances already mentioned and from deductions based on blast and thermal damage at the site of the fall, agree with one another quite well and place the yield at something in excess of 10^{23} ergs, probably about 10^{24} ergs. If this were the result of a nuclear explosion of some sort, then the yield of neutrons into the atmosphere with the consequent formation of carbon-14 from atmospheric nitrogen should be detectable by analysis of plant material which was growing at that time. It seemed worth while, therefore, to make such an analysis, and the growth rings of a tree were chosen for this purpose.

1908年通古斯陨石可能含有的反物质

考恩等

编者按

1908年6月30日，在西伯利亚南部上空发生了一次近代以来能量最大的陨石事件(也可参考《〈自然〉百年科学经典》第二卷)。本文主要讨论了陨石的残余部分中是否存在"反物质"成分的问题，这些成分与大气组成物质发生湮灭并释放出大量能量。特别值得注意的是，本文作者中有一位当时担任美国原子能委员会委员，美国洛杉矶加州大学的威廉·利比，他对新近发现的放射性同位素碳-14做了专门研究：碳-14的应用使得通过树木的年轮来准确测定古代树木标本的年龄成为可能。现代的观点是，通古斯陨石事件在预期的陨石现象的界限之内。

1908年6月30日世界时0点17分11秒，在距西伯利亚贝加尔湖以北500英里的石泉通古斯河流域（北纬60°55′，东经101°57′），发生了可能是现代观测到的最壮观的陨石坠落事件[1]。在晴朗天空直径约1,500千米的范围内，飞过一个能使人致盲的明亮火球并发生了爆炸，其光芒"甚至使太阳光都黯然失色"。与陨石坠落相伴而来的还有极其强烈的辐射和冲击波现象。尽管在该事件发生的同时世界各地都记录了地震、气象和地磁场方面的扰动，而且在随后几年也从目击者那里收集了对陨石下落时伴随现象的描述，但对陨石坠落地带进行的首次考察直到1927年才展开。虽然由热气流和爆炸效应对森林造成的巨大破坏依然明显，但并没有发现陨石坑的痕迹。

为了解释由大流星、小彗星或核爆炸引起的规模巨大的现象，人们已提出了各种假设。无论是支持意见还是反对意见都很有说服力。通过之前提到的扰动记录以及基于对陨石坠落地带爆炸和热破坏的推论，所得到的对产生总能量的估计是相互符合的，且产量超过 10^{23} 尔格，可能约为 10^{24} 尔格。如果这是某种核爆炸的结果，则爆炸产生的中子进入大气后，会与大气中的氮形成碳-14，能在当时生长的植物物质中分析检测得到。因此，做这种分析看来是具有价值的，并为此选择树木年轮来进行分析。

Phenomena of the Fall

Before reporting the present work it may be of interest to repeat some of the accounts of the effects of the Tunguska meteor. The general area of the fall is composed of taiga with peat bogs and forest and is (fortunately) very sparsely populated. One eye-witness, S. B. Semenov, a farmer at Vanovara some 60 km away, told L. A. Kulik, who investigated the meteor first in 1927 (ref. 2), that he was sitting on the steps outside his house around 8 a.m., facing north, when a fiery explosion occurred which emitted so much heat that he could not stand it: "My shirt was almost burnt on my body". However, the fireball did not last long. He just managed to lower his eyes. When he looked again, the fireball had disappeared. At the same time, an explosion threw him off the steps for several feet, leaving him briefly unconscious. After regaining his senses, a tremendous sound occurred, shaking all the houses, breaking the glass in the windows, and damaging his barn considerably.

Another observer[3], P. P. Kosolopov, a farmer and neighbour of S. B. Semenov, was working on the outside of his house, when suddenly he felt his ears being burnt. He covered them with his hands and ran into his house after asking Semenov if he had seen anything, on which Semenov answered that he too had been burnt. Inside the house, suddenly earth started falling from the ceiling and a piece from his large stove flew out. The Windows broke and he heard thunder disappearing to the north. Then he ran outside, but could not see anything.

A Tungus, Liuchetken, told Kulik on April 16, 1927, that his relative, Vassili Ilich, had some 500 reindeer in the area of the fall and many "storage places". With the exception of several dozen tame deer, the rest were grazing in that area. "The fire came by and destroyed the forest, the reindeer and all other animals". Then several Tungus went to investigate and found the burnt remains of several deer; the rest had completely disappeared. Everything was burnt in Vassili Ilich's storage including his clothing. His silverware and samovars (tin?) were molten. Only some large buckets were left intact.

According to Krinov[1] the dazzling fireball moved within a few seconds from the south-east to north-west leaving a trail of dust. Flames and a cloud of smoke were seen over the area of the fall. Visible phenomena were observed from a distance as great as 700 km, and loud explosions were heard after the passage of the fireball at distances up to 1,000 km.

The first inspection of the site was carried out by Kulik in 1927 (ref. 2). Trees were blown down over an area with a radius of 30–40 km. Exposed trees were uprooted with their roots pointing toward the center of the explosion in a radial manner. Additional expeditions by the Academy of Sciences of the U.S.S.R. were sent in 1928 and 1929–30. The center of the explosion area was found to have been ravaged by fire and searing could be traced to a radius of 15–18 km from the center of the explosion. Numerous holes with a diameter from several to several tens of metres had been found in the first expedition of 1927; however, subsequent work including excavations up to 34 m depth did not yield

陨石坠落现象

在陈述现阶段的工作之前，可以先来回顾一下有关通古斯陨石造成的影响的报告。陨石坠落的位置一般是泥炭沼泽地和森林，（值得庆幸的是）这里人迹罕至。住在距离陨石坠落大约 60 千米的法诺伐拉的一位目击者，名叫谢苗诺夫的农夫告诉曾在 1927 年首先研究陨石的库利克 (参考文献 2)，在上午大约 8 点他正坐在屋外的台阶上面朝北方，发生剧烈爆炸时所释放出的巨大热量使他几乎承受不了："我的衬衫几乎在我身上着火了"。然而，火球并没有持续很久。他只是眨了下眼，当他再往前看时，火球已经消失。同时，爆炸将他从台阶上抛到几英尺外，使他陷入了短暂的昏迷。在恢复知觉后，只听一声巨响，所有的房屋都为之震动，窗上的玻璃都碎裂了，他的谷仓也遭到了严重的破坏。

另一位目击者[3]是农夫科索洛波夫，他是谢苗诺夫的邻居，当时正在屋外工作，他突然感到耳朵在燃烧。他询问谢苗诺夫是否看见了什么，谢苗诺夫说他也烧伤了，之后他便用双手捂住耳朵冲进屋里。此时屋里突然开始从天花板上掉土，炉子也有一部分碎裂飞了出去。窗户破了，接着他听见雷鸣般的声音渐渐消失在北方。然后他跑了出去，但是什么都看不见。

1927 年 4 月 16 日一位名叫柳切特坎的通古斯人告诉库利克，他的亲戚瓦西里 · 伊里奇在陨石坠落的地区大约有 500 头驯鹿和许多"贮料棚"。除几十头驯服的鹿以外，其余的正在这个区域吃草。"这场突如其来的大火烧毁了森林、驯鹿和其他动物。"然后几位通古斯人又对该地区进行了调查，发现了一些被烧死的鹿的残骸；其余的动物完全消失了。瓦西里 · 伊里奇贮料棚里的所有东西包括他的衣服也都被烧着了。他的银器和水壶(也许是马口铁做成的)也熔融了。只有几个大水桶还完好无缺地立在那儿。

根据克里诺夫[1]所说，在短短几秒的时间内耀眼的火球就从东南移动到西北方向，只留下了一些烟尘的痕迹。那些产生的火焰和烟气在整个陨石坠落地区都看得见。光学可见范围高达 700 千米，而巨大的爆炸声在火团经过的 1,000 千米距离处都听得见。

库利克在 1927 年对这片地区进行首次调查（参考文献 2)。半径 30 千米 ~ 40 千米范围内的树木被刮倒。被连根拔起的树的树根呈辐射状指向爆炸发生的中心。另外的调查是在 1928 年和 1929 年 ~ 1930 年间由苏联科学院组织进行的。他们发现爆炸的中心区域已经被大火烧得面目全非，并且距爆炸中心 15 千米 ~ 18 千米范围内也被烧焦。在 1927 年第一次调查时发现了许多直径在几米到几十米之间不等的洞，然而在后来包括深达 34 米的挖掘工作中都未再发现任何陨石类物质。库利克后来将

any meteoric material. These holes were explained later by Kulik as natural formations[4]. During 1938–39, an aerial survey was conducted over the devastated area to assess more completely the extent of the destruction.

The fall of the meteor resulted in a seismic wave recorded on the Zöllner-Repsold pendulums of the Irkutsk Magnetic and Meteorological Observatory[1]. Subsequent analyses for the epicentre of the earthquake coincide with the location of the fall and also established the accurate time of the event.

In addition, several observatories in Russia and Europe recorded the barometric waves caused in the atmosphere by the meteor. The seismic and barometric effects have been discussed in detail by Krinov[1], Fesenkov[5], and Whipple[6]. The Tunguska meteor also caused a definite disturbance of the Earth's magnetic field as registered at the Irkutsk Observatory and others around the world. The disturbances were similar to those recorded following nuclear explosions in the atmosphere.

After the fall of the meteor, the nights were exceptionally bright everywhere in Europe and Western Siberia. As far south as the Caucasus, newspapers could be read at midnight without artificial light. The brightness slowly diminished and disappeared after a duration of two months[7].

In comparison, if the fall had occurred in the United States over, say, Chicago, visible phenomena would have been noticed as far away as Pittsburgh, Pennsylvania, Nashville, Tennessee, and Kansas City, Missouri. The thunder would have been heard in Washington, D.C., Atlanta, Georgia, Tulsa, and in North Dakota.

The Meteorite Hypothesis

The results of the first investigations of the Tunguska site led to the belief that a meteorite of very large initial mass penetrated the Earth's atmosphere and hit the surface, destroying itself in a violent explosion[8]. This explanation sought to account for the absence of meteoritic debris in the fall area. Since a crater was never found, it was assumed that one might have been formed in a layer of permanently frozen soil which lost its form rapidly and could no longer be distinguished after the first summer.

In the analysis of Fesenkov[5] all evidence points to a retrograde orbit around the Sun with considerable inclination of its orbit to the ecliptic. This is atypical for meteorites derived from asteroidal disintegration. Another interpretation of the motion of the meteor is that it moved parallel to the Earth at much lower speed, in which case its relative speed had to be very low. In view of the great energy released, the explanation of a retrograde orbit associated with high relative speed is to be preferred over a slow relative speed which is difficult to reconcile with the effects of the meteor, such as burning the area, etc.

这些洞解释为天然形成[4]。在 1938 年 ~1939 年间，又对被毁坏区域进行了一次空中探测，以更加全面地估计破坏程度。

伊尔库兹克地磁与气象台[1]的佐尔诺摆记录到了陨石坠落造成的地震波。对数据进行进一步分析得到的震中位置与陨石坠落的地点相一致，而且也确定了该事件发生的具体时间。

另外，在俄国和欧洲的许多台站都观测到了陨石在大气中造成的气压波。这些地震波和气压波的影响已经由克里诺夫[1]、费先科夫[5]和惠普尔[6]详细论述。正如伊尔库兹克台站及世界各地的台站记录到的，通古斯陨石也对地球的磁场造成了巨大的扰动。这种扰动与在大气中发生核爆炸后记录到的扰动类似。

陨石坠落后，欧洲和西伯利亚西部各处的多个夜晚都格外明亮。南至高加索山脉人们都可以在午夜不用人工照明就阅读报纸。这种亮光逐渐减弱，在持续了将近两个月[7]后才彻底消失。

相对而言，如果陨石坠落在美国的芝加哥，那么在远至宾夕法尼亚州匹兹堡、田纳西州纳什维尔、密苏里州堪萨斯城都可以看见光学现象。在华盛顿、亚特兰大、佐治亚州、塔尔萨和北达科他州均可以听见这种雷鸣般的声音。

陨石假设

对通古斯地区第一次调查的结果使人们相信，这是一颗原本具有巨大质量的陨石，它穿过地球大气层然后撞击地面，并且在一次剧烈的爆炸[8]中完全销毁了。这种理论试图解释在陨石坠落区域没有发现陨石残骸的原因。由于没有发现任何一个陨石坑痕迹，所以人们猜想有可能是在永久冻土层中曾经存在陨石坑，但很快就变形了，因此在第一个夏天之后就不能再辨认。

在费先科夫[5]的分析中，所有的证据都表明存在一个与黄道夹角很大、逆向围绕太阳公转的轨道。这是来源于小行星解体后的陨石的非典型轨道。关于陨石运动的另一种解释是以较低的速度相对于地球平行方向运动，但这种相对速度必须非常小。考虑到释放出的巨大能量，较高相对速度的逆行轨道的解释要优于低速轨道的解释，因为后者很难与陨石产生的一些影响相一致，比如烧焦的地区等。

The Cometary Hypothesis

This hypothesis was proposed by A. S. Astapovich and independently by F. J. W. Whipple in 1930 (ref. 6). The evidence in favour of a cometary nature of the meteor is the motion of the meteorite opposite to that of the Earth and the resulting high velocity of an estimated 60 km/sec[5] which yielded on impact the calculated 10^{23} ergs. Since F. L. Whipple's comet mode[8] consists of a conglomerate of frozen ices such as methane, water and ammonia interspersed with solid mineral matter, the meteor or small comet appears likely to have exploded above the Earth's surface without leaving significant traces of matter on the ground. Based on the observations of the Potsdam Geodetical Institute which permit the velocity determination of the shock wave propagated through the atmosphere, the speed of 318 m/sec measured corresponds to an atmospheric height of 5–6 km, which is the altitude of the main explosions of the meteor[5,6].

Further evidence favouring a small comet is the unusual luminescence of the night sky immediately after the fall over Siberia, Russia, and Western Europe, but not the United States or in the southern hemisphere[5]. Evidently the dust tail was directed away from the Sun, as expected for comets, and extended in a north-westerly direction at the moment the main body hit. The dissipation of this tail resulted in the night sky being brighter initially by about 50–100 times the normal value, but 10^4 times less than daylight.

Abbot in California found that approximately from the middle of July, or 2 weeks after the explosion, until the second half of August 1908, the coefficient of transparency of the atmosphere was noticeably depressed[9]. Fesenkov suggested that this was caused by the loss of vast amounts of material from the meteor during its flight through the atmosphere, possibly of the order of several million tons of matter[10].

It appears unusual, however, that such a comet was not observed on its collision course with the Earth, as it should have been seen unless it approached from a direction with very small angular distance from the Sun. Fesenkov estimated the size of the cometary nucleus as about several hundred metres[5], which is perhaps only one order of magnitude below that of well-known comets seen at great distances.

The Nuclear Reaction Hypothesis

In an article by F. Y. Zigel discussing the results of A. V. Zolotov's expeditions of the past three years, the events of the Tunguska fall have been re-examined[11]. The velocity of the meteor has always been required to be large in order to account for the release of 10^{23} ergs on impact. This can be determined from the ratio between the amplitudes of the ballistic wave caused by the velocity of the body in the atmosphere and the blast wave caused by the explosion of the body itself.

Zolotov selected trees which had remained standing and on which traces of the effects of both waves remained. Apparently, the ballistic wave arrived from the west and broke only very slender branches, whereas the blast wave from the north broke large tree branches.

彗星假设

1930年阿斯塔波维奇和惠普尔独立地提出了这个假设(参考文献6)。支持陨石是彗星的证据是，陨石是对着地球运动的，并且根据其释放出来的10^{23}尔格的能量可以得到其速度高达约60千米/秒[5]。因为惠普尔的彗星模型[8]包含诸如甲烷、水和氨与固态矿物质混杂在一起冻结成冰状的聚合物，这种流星或小彗星可以在地球表面上方爆炸,并在地面上不留明显的物质痕迹。基于波茨坦大地测量研究所的观测，通过测定大气传播的冲击波的速度，测量的速度达318米/秒，相应地大气高度为5千米~6千米，这是流星的主要爆炸高度[5,6]。

进一步支持小彗星的证据是，在西伯利亚、俄罗斯和西欧范围内陨石坠落后，夜空立即非常明亮，但在美国或南半球并非如此[5]。显然，正如对彗星所预期的，尘尾是背向太阳的，主体此刻在西北方向扩散。夜空中这种尘尾的耗尽产生的亮度最初约是正常值的50~100倍，但其是日光的万分之一。

在加州阿博特发现,大约从7月中旬（即在爆炸后两周）到1908年8月下半月，大气的透明度明显下降[9]。费先科夫提出，这是由于陨石落下前在大气中飞行损失大量物质所引起的，可能有几百万吨量级的物质损失[10]。

然而，奇怪的是，在与地球碰撞的过程并没有观察到这种彗星，这本应该被观察到的，除非这种彗星来自的方向与太阳的夹角很小。费先科夫估计，彗星核的大小约是几百米[5]，这可能比在远距离看到的熟识的彗星仅低一个量级。

核反应假设

在席格尔讨论佐洛托夫的考察队过去三年研究结果的论文中，对通古斯陨石事件已重新做了研究[11]。为了能在碰撞时释放10^{23}尔格的能量，通常要求陨石的速度很大。这可以根据在大气中由物体速度产生的弹道波振幅和物体本身爆炸产生的冲击波的比例得出。

佐洛托夫挑选了依然竖立并保留着两种波作用痕迹的树木。显然，从西方到达的弹道波只破坏了细长的树枝，而从北方来的冲击波却破坏了更大的树枝。根据这些结果，他计算出了在主要爆炸前的弹道波，事实上与冲击波相比是较小的。通

From these results he calculated that the ballistic wave before the main explosion was, in fact, of minor size as compared with the blast wave. Eye-witnesses of the Tunguska fall recalled that the flight of the meteor was dimmer than the Sun, corresponding to a velocity in the atmosphere of less than 4 km/sec. If the velocity was more than an order of magnitude lower than this, then the explosion could not have possessed the required energy for the explosion.

These considerations led to the question whether or not a massive chemical or nuclear release of energy occurred at the final break-up of the meteor. The nature of an explosion can be determined by the distribution of the energy released, one factor being the amount of radiant energy emitted.

At 17–18 km from the epicenter, Zolotov found trees which had been subjected to a thermal flash and had started to burn. A natural forest fire was ruled out for the area. In order to start a fire in a living tree, about 60–100 cal/cm^2 of incident thermal radiation is required. By calculation the radiant energy of the explosion was found to be 1.5×10^{23} ergs. Other energy-yield estimates for different locations placed the thermal energy of the explosion between 1.1 and 2.8×10^{23} ergs.

Since the estimated yield of thermal energy is so close to the estimate of the total explosive energy, Zolotov favours a nuclear rather than a chemical explosion.

The Chemical Radical Reaction Hypothesis

In an examination of the records of the fall, the radiation flash stands out among the others discussed by different authors during the past decades. Specifically, the remarks by Semenov and Kosolopov of experiencing burning sensations, and the melting of Vassili Ilich's metal ware, appear to confirm the emission of considerable amounts of thermal radiation by the explosion.

Very large chemical high-energy explosions can create sufficiently intense shock-waves in air which, in turn, will radiate thermal energy, perhaps sufficient to account for the fire-setting in the taiga. From the examination of nuclear explosions, the phenomena accompanying the release of large amounts of energy in air are well known[12]. In the case of a nuclear explosion and a fraction of a second after the detonation, a high-pressure, intensely hot and luminous shock front forms and moves outwards from the fireball.

While the dissipation of kinetic energy in the Tunguska explosion probably accounts for the major portion of energy released, the reaction with air of vast amounts of chemical high-energy species such as the radicals observed on comets can be an additional source of energy. For high meteor velocities, the relative contribution of chemical energy to the final explosive break-up will be small, but for a low-velocity body it may be significant. Theoretical considerations place the output of energy of a system using the recombination energy of chemical radicals midway between that of conventional chemical propellants and nuclear reactions in energy released/unit mass.

古斯陨石的目击者回忆起陨石的运动轨迹比太阳要模糊，相当于在大气中以小于4千米/秒的速度运动。如果速度比这个估计值大不到一个量级，则爆炸将无法拥有爆炸时所需要的能量。

这些考虑产生了一个问题：在陨石最终解体时，是否发生了大规模的化学或核能量的释放。爆炸的性质可以根据释放能量的分布来确定，其中一个因素是发射出的辐射能量。

在距震中17千米~18千米处，佐洛托夫发现了受辐射闪光影响而燃烧的树木。在此区域外的森林自然起火。为了在一棵活树上点着火，需要的热辐射约为60卡/厘米^2~100卡/厘米^2。通过计算爆炸的辐射能量，发现其值为1.5×10^{23}尔格。估计在不同位置爆炸所产生能量数值在1.1×10^{23}尔格~2.8×10^{23}尔格之间。

由于估计产生的热能与估计总的爆炸能量非常接近，因此佐洛托夫认为这是核反应，而不是化学爆炸。

自由基的化学反应假说

在陨石坠落记录的确认中，过去几十年内不同的作者对辐射闪光进行了很多讨论。尤其是谢苗诺夫和科索洛波夫感受到的烧伤感，以及瓦西里·爱里斯的金属制品的融化，似乎证明了爆炸产生的热辐射有相当大的能量。

很多化学高能爆炸可以在空气中制造出足够强的冲击波，由此而来的辐射热能也许足以使针叶树林地带着火。根据核爆炸检验，其爆炸过程会伴随着在空气中释放大量能量，这已被人所熟知[12]。在核爆炸的情况下以及爆炸声之后的几分之一秒内，会形成一团高压、强热和亮闪光的冲击波波前，并从火球向外移动。

通古斯陨石事件中耗损的动能也许是所释放能量的主要部分，在彗星上大量化学高能的与空气会发生反应的化学物质，例如在彗星上观测到的自由基可能是附加能源。对于高速陨石而言，化学能对于最终爆破能的相对贡献将是很小的，但对于低速物体而言，这可能是很重要的。利用单位质量释放能量介于传统的化学燃料和核反应能之间的化学基的重组，理论考虑了系统的能量输出。

A very large chemical radical explosion of the meteor would account for many of the observed phenomena. Our very limited knowledge of the actual concentration of radicals on comets, their exact nature and the mechanism of radical reactions make a quantitative calculation of the release of energy by such a model very difficult, especially in context with the uncertainties of the exact orbit of the Tunguska meteor[13].

The Anti-matter Hypothesis

Discounting any but purely natural phenomena, it becomes difficult to construct a model for either a fission or fusion chain-reaction which would produce the effects observed. For the former, an almost-critical mass of fissionable material might be conceived which became tamped on entering the atmosphere. The tamping would have to be such, however, as to take material far beyond criticality in a very short time to prevent its disassembly with low yield. The multi-megaton yield observed, however, coupled with the very low efficiency known for the best of such devices, would require a large initial mass—well above the critical mass of normal density uranium or plutonium. Thus, super-criticality obtained by tamping alone could scarcely be credited as the mechanism. On the other hand, to obtain it by increasing the density of a sub-critical mass by compression seems equally unlikely, for this must be a result of the mechanical forces generated by penetration into the atmosphere.

To obtain the effects from a fusion reaction, a sufficient amount of deuterium, and possibly tritium, must be contained in a compressed state and heated to several million deg. C. It must then be maintained in that state so that self-heating can carry the reaction the explosion stage. Again, it is difficult to conceive of a model for such a mechanism which is attained merely by entry into the atmosphere of the Earth.

In searching for other natural means by which a large nuclear energy yield might be obtained, we are unable to find one other than the annihilation of charge-conjugate ("anti-") matter with the gases of the atmosphere. Several objections immediately arise to this hypothesis, all different from those raised above. No mechanical extremes are required of the model, however.

The first objection is that no evidence is known for the existence of anti-matter in the gross state. Other than as anti-particles produced by high-energy interactions of ordinary matter with itself or with electromagnetic radiation, no anti-matter has been observed. This is understandable in the environment of the Earth, and so one must look to astronomy for such evidence. The complete symmetry between the two charge-conjugate states of matter, however, makes an astronomical test of an isolated, distant object difficult.

The second problem arises in considering the flight of an "anti-rock" through the atmosphere. If the rock is approximately spherical with diameter d cm and of density ρ g/cm^3, then it might penetrate a distance of $d\rho$ g/cm^2 into an absorbing medium before being consumed. The minimum distance through the atmosphere is about 10^3 g/cm^2.

陨石的大规模化学基爆炸将会解释许多观察到的现象。我们对彗星上自由基的实际浓度及其准确的性质和自由基反应机制不够了解，使采用这种模型对释放能量作定量计算还相当困难，尤其是考虑到通古斯陨石轨道的不确定性[13]。

反物质假设

除了这些完全的自然现象之外，很难再建立出能产生已观测效果的裂变或聚变链式反应模型。对裂变反应而言，可以假定根据其中原有的已接近临界的大量裂变物质在进入大气层时被急剧填实。然而，这种填实是指在极短的时间内要使裂变物质密度急速上升到远大于临界值，以防止它爆裂飞散而导致能量释放过低。考虑人们熟知的这种情况下核爆炸效率极低，要获得所观测到的几百万吨量级的爆炸能量，对于通常的铀密度，发生这种情况将需要一个很大的初始裂变物质质量。因此，这种由填实过程获得超临界的假设不足以成为令人信服的机制。另一方面，通过在穿过大气时所产生的机械力作用使亚临界质量的裂变物质密度压缩到超临界的机制假设，似乎同样也不大可能。

要得到聚变反应效应，必须有足够量的氘和可能的氚处于压缩态，并加热到几百万摄氏度。并且还必须保持这种状态，使其自身产热足以将反应导入爆炸阶段。另外，基于这种仅仅是在其进入地球大气的机制，该模型还是很难被接受的。

在研究其他可以产生巨大核能量的自然方法中，除电荷共轭（“反”）物质与大气中气体的湮灭外，我们找不到其他方法。立刻出现了对此假说的若干反对意见，都与上述方法不同。然而，这种模型却没有设定一个呆板的限制。

第一种反对意见是，在所有领域中都没发现反物质存在的证据。除了通过普通物质之间及其与电磁辐射的高能相互作用产生反粒子之外，尚未观测到反物质。这在地球环境是可以理解的，因此人们将寻找相应证据的范围扩大到了天文学领域。然而，由于两个共轭电荷物质态之间的完全对称性，使得进行孤立的、远距离的天文实验变得很困难。

在考虑“反物质性岩石”穿过大气飞行时出现了第二个问题。如果岩石近似为球体，直径为 d 厘米，密度为 ρ 克 / 厘米^3，则其在烧毁前要穿透一段 $d\rho$ 克 / 厘米^2 的距离进入吸收介质。穿过大气层所要经过的最短距离是大约 10^3 克 / 厘米^2。因此，

Thus, if the density is of the order 10 g/cm^3, then the diameter of the rock is of the order 100 cm. The number of nucleons in such an object is approximately $\frac{1}{2}Ad^3\rho$, or about 3×10^{30}, and the yield of energy would be of the order 10^{27} ergs, rather than 10^{23} as observed. The fact that the bolide did not reach the surface of the Earth is ignored in this estimate, and is off-set by the additional distance due to the inclined trajectory of the object. The discrepancy factor is, nevertheless, quite large. In addition, the flight of the bolide would have exhibited its largest yield somewhere toward the middle of its path, rather than towards its end—it would have thinned-down and died out.

A second look at the process tempers these conclusions, however. The exceedingly strong radiation shock accompanied by heating of the air ahead of the bolide, in addition to the pressure of electrons and other particles ejected in the forward direction by the annihilation reaction of complex nuclei with other, different complex nuclei, would rarefy the atmosphere ahead and greatly increase the range. A carefully calculated model for such a process may be in order. It could be that only a small fraction of the bolide could annihilate in flight, but that it remains essentially solid until it reaches a point where it is travelling slowly deep in the atmosphere. Here, continued annihilation might heat it to the gaseous stage and dissemble it explosively, resulting in a final annihilation as the gases mixed with atmospheric gases. In any event, the process seems far too complex to dismiss on the basis of a rapid estimate.

Of the three models for a nuclear explosion, we choose the annihilation model as a basis for an estimate of the amount of carbon-14 produced. We must first estimate the number of neutrons produced/nucleon annihilated.

Annihilation of Anti-rock in the Atmosphere

We have, of course, no information concerning the state or the chemical composition of the supposed anti-matter comprising the bolide. Assuming it to be molecular compounds similar to those of ordinary meteorites, we ignore annihilation of the electrons, for these would produce a small fraction of the yield and would form no neutrons in the process.

The simplest case of nucleon annihilation is that of $p\bar{p}$. Even in this instance, the annihilation is not limited to *S* states, and the process becomes complex due to the various possible angular momentum states in the initial system and various charge states in the final system. The final system may contain pairs of kaons and various numbers of positive, negative and neutral pions. A measure of the number of charged particles emitted in the annihilation of $p\bar{p}$ is given by Horwitz *et al.*[14] as an experimentally obtained histogram extending from zero to seven prongs/event, with a flat maximum in the region of 3–4 prongs. The high average multiplicity greatly complicates the situation because of the many possible quantum numbers in the final state. Refinements[15] in an estimate by taking into account $p\bar{n}$, $n\bar{n}$ and $n\bar{p}$ are obviated by the realization that we may be dealing here with reactions between complex nuclei and between fragments of such nuclei as they become broken by partial annihilation. Let us take four charged pions, on the average, as the basis for proceeding, two positive and two negative/nucleon pair annihilated.

如果密度为10克/厘米^{3}数量级，则岩石的直径为100厘米数量级。在这类物体中核的数量近似为$\frac{1}{2}Ad^3\rho$，约为3×10^{30}，产生的能量将为10^{27}尔格，而不是观测到的10^{23}尔格。在上述估计中忽略了陨石并未到达地球表面的事实，而且用天体轨道倾斜增加的距离进一步偏置了。无论如何这两种情况的差异巨大。此外，陨石的最大状态并非是在最后出现的，而是在飞行的过程中出现的，只有这样陨石才能慢慢变小并消失。

然而，再次考虑这个过程缓解了上述结论的矛盾。除了各种复杂的核湮灭过程中在前进方向上喷出的电子及其他粒子压力外，先于陨石到达的超强的辐射冲击波及伴随着的热气团也提前使空气变得稀薄，并且很大程度上扩大了射程。在此模型下对这个过程进行精确的计算也许的确是合情合理的。可能只有一小部分陨石在飞行过程中烧毁了，但是直到其缓慢地运动到大气深层时仍然保留着一部分必要的固体形态物质。至此，进一步的燃烧会将其加热至气态，并以爆炸的形式发生，导致最后是陨石气体物质和大气气体混合后燃烧。在任何情况下，根据快速的估计，这一过程显得过于复杂而不考虑。

在核爆炸的三种模型中，我们选择湮灭模型作为估计产生碳-14含量的基础。我们必须首先估计产生的中子/湮灭的核子的数量。

大气中反物质岩石的湮灭假设

当然，对于被认为是组成火球的反物质，我们还没有得到其组成或化学组分的相关信息。假设它是类似于普通陨石的分子化合物，由于电子湮灭过程中只产生一小部分物质，并且此过程中没有中子产生，所以可以忽略掉电子的湮灭。

核子湮灭的最简单情况是$p\bar{p}$湮灭。即使在这种情况下，湮灭并不只限于S态，在初始系统中各种可能的角动量，以及在最终系统中的各种带电状态，会使过程变得复杂。最终系统可能会含有一对K介子以及各种数量的正、负和中性的π介子。在$p\bar{p}$湮灭中发射出的多种带电粒子个数的测量方法是由霍维兹等人[14]给出的，在实验中每个事件都有0~7个射线径迹得到的直方图，其中在3~4个射线径迹时有一个平缓的极大值。由于在末态中有许多可能的量子数，因此高平均值的多重性使状态极大地复杂化了。当我们意识到我们在处理复杂原子核之间和原子核因部分湮灭而破碎成的碎片之间的反应，则这种将$p\bar{n}$、$n\bar{n}$和$n\bar{p}$考虑进去所进行的估算精确[15]就可以排除了。让我们取4个带电的π介子作为平均意义上的基础过程，每对湮灭的核子分别是2个正粒子和2个负粒子。

The positive pions will decay in the atmosphere, but the negative ones will, in general, be captured by oxygen, nitrogen and carbon nuclei. In view of the overall uncertainty in this estimate, we will assume that all of the negative pions are absorbed at rest by nuclei. A simplified picture of the process, obtained from the measurement of prongs produced in stars in nuclear emulsions from negative pion absorption, is that of the 140 MeV rest energy of the pion gained by the nucleus, 40 MeV is lost by fast neutron emission at the time of absorption and 100 MeV is then lost by boiling-off of neutrons and charged particles in an evaporative process. Taking the mean energy[16] of the prompt neutrons as 12 MeV and their binding energy as 8 MeV, the mean number of prompt neutrons is 2. Assuming that the probability for then boiling off a neutron is the same as that for a proton, and weighting the probabilities obtained from prong counts accordingly, we find that two more neutrons are produced from light nuclei. Thus, four neutrons are produced per pion absorbed, or eight neutrons per nucleon pair annihilated. In view of the great uncertainties in this estimate, we take the number to be 8 ± 4. Thus, for a total energy yield of 10^{24} ergs by nucleon–antinucleon annihilation and a yield of about 3×10^{-3} ergs/nucleon pair, about $(2.7\pm1.4)\times10^{27}$ neutrons would be released to the atmosphere.

Effect on Atmospheric Radiocarbon Content

We may make some estimates of the effects of releasing neutrons in amounts such as this in the following way: Assume that every neutron produced is absorbed in the reaction $^{14}N(n,p)^{14}C$, and that the radiocarbon so produced is rapidly oxidized to carbon dioxide in the atmosphere. Thus $(2.7\pm1.4)\times10^{27}$ molecules of radio-CO_2 mix with the atmospheric gases. Taking the total mass of the atmosphere as 5.3×10^{21} g, and the mean carbon dioxide content of the air as 0.030 volume percent (though this varies geographically and seasonally), we readily calculate the atmospheric carbon dioxide to contain 6.6×10^{17} g carbon as carbon dioxide. Taking the decay constant of carbon-14 as 2.3×10^{-10} m^{-1}, our new radiocarbon should exhibit 9.4×10^{-1} d m^{-1} g^{-1} of atmospheric carbon. As the specific activity of atmospheric carbon is $13.56\pm$d m^{-1} g^{-1}, this represents an increase of some 7 percent in the radiocarbon activity[17].

In making this estimate, we have taken the radiocarbon to be uniformly distributed in the atmosphere after both vertical mixing and mixing between the northern and southern hemispheres, and have neglected absorption in the ocean and biosphere. Thus the result is approximate.

An alternative basis for an estimate of the yield of radiocarbon by an anti-matter Tunguska explosion is provided by the data on the yield of this isotope by the testing of nuclear explosives in the atmosphere. By September 1961, the equivalent of 70 MT (1 MT, megaton TNT equivalent is 4×10^{22} ergs) of fission and fusion nuclear explosive was released in air bursts and about 100 MT in surface tests[18]. The specific radiocarbon-level taken up by plants at that time[19] was about 25 percent above the natural cosmic-ray level of radiocarbon. We may estimate an upper limit to the anti-matter in the Tunguska meteor in the following way:

正的π介子将在大气中衰变，但是负的π介子通常被氧核、氮核和碳核俘获。考虑到该估计中的各种不确定性，我们将假定所有的负的π介子都是在静止时被核吸收的。通过测量核乳胶负的π介子吸收产生的星状径迹，得到的简化过程是：原子核得到的π介子的140 兆电子伏静止能，在吸收时由于快中子发射损失40 兆电子伏，而蒸发中子和带电粒子损失100 兆电子伏。取瞬发中子的平均能量[16]为12 兆电子伏，其结合能为8 兆电子伏，则瞬发中子的平均数为2。假定随后蒸发放射1个中子的概率与放射1个质子的概率相同，并且通过相应的计数方式可以得到概率的加权，我们发现从轻核中会多产生2个中子。因此，每吸收1个π介子产生4个中子，或每湮灭1对核子产生8个中子。由于估计中有很大的不确定性，我们取的数为 8 ± 4。因此，由核子–反核子湮灭产生的总能量是 10^{24} 尔格，每个中子对产生的能量约 3×10^{-3} 尔格，可得到约 $(2.7\pm1.4)\times10^{27}$ 个中子将在大气中被释放。

对大气中放射性碳含量的影响

我们可以用以下方法对释放中子产生的总体效果进行估计：假定产生的每个中子是被反应 $^{14}N(n,p)^{14}C$ 吸收的，同时产生的放射性碳在大气中被氧化变成二氧化碳。因此，有 $(2.7\pm1.4)\times10^{27}$ 个放射性-CO_2 分子与大气的气体混合。取大气的总质量为 5.3×10^{21} 克，空气中二氧化碳平均含量为总体积的0.03%(虽然这随地点和季节会有些变化)，我们容易计算大气中的二氧化碳含有 6.6×10^{17} 克碳。取碳-14的衰变常数为每分钟 2.3×10^{-10}，则新的放射性碳将以 $9.4\times10^{-1}d$ 分钟$^{-1}\cdot$克$^{-1}$的含量存在于大气中。当大气中碳的比放射性为 $13.56\pm d$ 分钟$^{-1}\cdot$克$^{-1}$，这表示放射性碳的活度增加了7%[17]。

在进行这项估计时，我们假定在大气的垂直方向及南、北半球都很好的混合后放射性碳在大气中均匀分布，并且忽略海洋和生物对大气的吸收作用。因此这是近似值。

根据测试这次核爆炸后大气中产生的放射性碳同位素的数据，可以得到另外一种关于通古斯陨石事件的反物质估计理论基础。直至1961年9月，由核裂变和聚变在大气中爆炸反应释放出的能量相当于70 MT当量(1 MT，一百万吨TNT炸药相当于 4×10^{22} 尔格)，在地表内的试验中得到的结果约为100 MT[18]。当时从植物中得到的这种放射性碳水平约在自然宇宙射线水平的25%以上。我们对通古斯陨石的反物质上限可做估计：

Taking the full 70 MT of air bursts and one-half of the 100 MT of surface bursts as effective for producing radiocarbon, we have $\frac{70 + 50}{25}$ or 5 MT of fission or fusion fired in the atmosphere producing a 1 percent rise in radiocarbon activity.

If, now, the known damage parameters of the Tunguska explosion are used as input data for the Nuclear Bomb Effects Computer, a value of about 30 MT (10^{24} ergs) energy yield is obtained (supplement to publication cited as ref. 12), which at 2 BeV (3×10^{-3}ergs) per nucleon pair consumed and 8±4 neutrons yield gives a total neutron yield as shown above, of $(2.7\pm1.4)\times10^{27}$ neutrons. Since the meteor disintegrated in the atmosphere, this would be expected to give $(2.7\pm1.4)\times10^{24}$ carbon-14 atoms. Therefore, if the Tunguska explosion had been due to anti-matter, it should have behaved like 35 MT of fission or fusion fired at the same latitude (say the U.S.S.R. test site at Novaya Zemlya, 74°N, 150°E) and, using our experience with bomb test carbon-14 as a basis of comparison, we can estimate the possible anti-matter content of the 1908 Siberian meteorite.

Radiocarbon Analysis

A section of a 300-year-old Douglas fir (*Pseudotsuga taxifolia*), the "Hitchcock" tree, which fell in the winter of 1951 in an unsurveyed area (35°15′N, 111°45′W) of the Santa Catalina Mountains about 30 miles from Tucson, Arizona, was provided by the Laboratory for Tree-Ring Research of the University of Arizona, Tucson. About 20 g of wood was stripped from each ring for the interval of 1870–1936, and the radiocarbon contents of the rings of each fifth year were measured, excepting for the years around 1908. Table 1 contains the results expressed as percentage deviations from the international standard reference level of 1890 (0.95 of the count-rate of the National Bureau of Standards oxalic acid). Column IV contains carbon-13 mass spectrometric corrections in per mil deviation from the Chicago PDB standard. (The mass spectrometric analyses were provided with the help of R. McIver and W. Sackett of the Jersey Production Research Co., Tulsa, Oklahoma.) The percentage deviations in carbon-14, corrected by these figures for isotopic fractionation[20], are contained in column V, according to $\left(\frac{1 + \delta^{14}C}{1 + 2\delta^{13}C} - 1\right) \times 100$.

Finally, the last column contains the results corrected again for the effects of dilution of atmospheric carbon dioxide by the burning of industrial fossil fuels[20]. We have used Fergusson's[21] values for this correction (the Suess effect).

Additional tree-ring samples were measured from an oak tree (samples provided by L. Wood, Inst. Geophysics, Univ. Calif., Los Angeles) cut in 1964 near Los Angeles (in the Simi Valley, 34°12′ N, 118°48′ W). They are given as UCLA-776, 778, 779.

取空气中爆炸的全部 70 MT 和表面爆炸 100 MT 的一半作为产生放射性碳的有效值，要使放射性碳活度产生 1% 的增长，大气中裂变或聚变燃烧产生的能量约需要 (70+50)/25 或 5 MT。

如果将现已知的通古斯陨石事件的破坏参数用作为核爆炸效应计算机的输入数据，得出约产生 30 MT 能量的值（10^{24} 尔格）(补充部分见参考文献 12)，其中每对中子消耗 2 吉电子伏（3×10^{-3} 尔格），从上文给出的 8±4 个中子产量也可得到总中子产量为 $(2.7\pm1.4)\times10^{27}$ 个中子。由于陨石是在大气中解体的，所以估计给出了 $(2.7\pm1.4)\times10^{24}$ 个碳-14 原子。因此，如果通古斯陨石爆炸是由反物质引发的，那么这将等同于在相同的纬度处（例如苏联的试验地点为新地岛，北纬 74°，东经 150°）引发 35 MT 裂变或聚变的行为，根据核试验的经验将碳 -14 作为比较的基础，可以估计得出 1908 年西伯利亚陨石可能含有的反物质量。

放射性碳的分析

图森的亚利桑那大学树木年轮研究实验室提供了一棵 300 年树龄的道格拉斯冷杉的截面，这棵“龙爪”树是 1951 年冬在离亚利桑那的图森约 30 英里的圣卡塔利娜山未勘察区域（北纬 35° 15′，西经 111° 45′）倒下的。研究中选取了 1870 年～1936 年间每个年轮里约 20 克的木材，对除了 1908 年外，每隔 5 年年轮内的放射性碳含量进行了测量。表 1 列出了与 1890 年的国际标准参考基准（美国国家标准局草酸的计数速率 0.95）对比后的百分比偏差结果。第 IV 列给出了碳 -13 的质谱修正与芝加哥 PDB 标准每密耳的偏差值。（在俄克拉何马州塔尔萨的马克依夫和萨基特的帮助下由泽西产品研究公司提供了质谱分析。）在第 V 列内给出的是根据 $\left(\frac{1+\delta^{14}C}{1+2\delta^{13}C}-1\right)\times100$ 利用同位素分馏数据计算对碳 -14 中的百分比偏差做出的修正[20]。最后一列给出了因工业矿物燃料燃烧而对大气中的二氧化碳造成稀释的修正结果[20]。我们在这项修正中采用了弗森格[21]的数值（修斯效应）。

1964年，利用在洛杉矶附近（在半山谷中，北纬34° 12′，西经118°48′）取得的橡树又进行了年轮样品测量（样品由美国加州大学地球物理研究所的伍德提供）。它们分别用UCLA-776、778、779标记。

Table 1. Radiocarbon Content of Tree-Rings around 1908[22]

I	II	III	IV	V	VI
Sample No.	Year	% $\delta^{14}C$ uncorrected	Per mil $\delta^{13}C$	% $\delta^{14}C$ corrected for isotopic fractionation	% $\delta^{14}C$ corrected for Suess effect
UCLA-769	1873	0	−22.3	0	+0.05
UCLA-768	1878	−0.72	−23.0	−0.75	−0.67
UCLA-767	1883	−0.31	−22.9	−0.32	−0.22
UCLA-766	1888	−1.64	−22.2	−1.69	−1.59
UCLA-765	1893	−3.75	—	—	−3.60
UCLA-782	1894	−1.26	—	—	−1.11
UCLA-763	1898	−0.48	−22.9	−0.50	−0.30
UCLA-760	1903	−0.28	−23.1	−0.29	−0.02
UCLA-761	1908	−1.07	—	—	−0.72
UCLA-778	1908	−0.96	—	—	−0.61
UCLA-774	1909	+0.26	−22.6	+0.25	+0.60
UCLA-776	1909	+0.17	−24.8	+0.16	+0.51
UCLA-780	1910	−0.70	−22.2	−0.73	−0.38
UCLA-779	1910	−1.50	−24.5	−1.55	−1.20
UCLA-762	1913	−0.81	−22.6	−0.84	−0.45
UCLA-764	1918	−1.20	−22.4	−1.24	−0.69
UCLA-770	1923	−0.63	−23.0	−0.66	+0.04
UCLA-771	1928	−2.40	−22.4	−2.45	−1.58
UCLA-772	1933	−1.50	−22.0	−1.55	−0.27

The results of columns V and VI are plotted in Figs. 1 and 2, respectively.

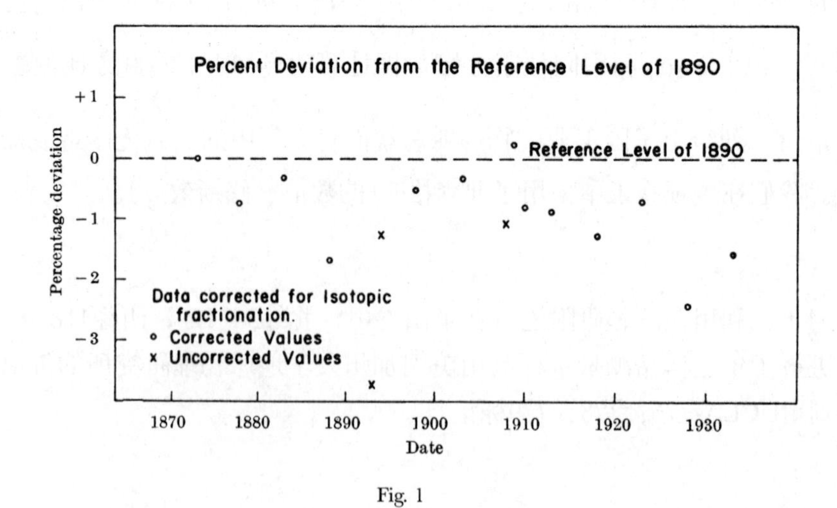

Fig. 1

表1. 1908年左右树木–年轮的放射性碳的含量[22]

I	II	III	IV	V	VI
样品号	年代	% 未修正的$\delta^{14}C$	每密耳$\delta^{13}C$	%同位素分馏修正后的$\delta^{14}C$	%修斯效应修正后的$\delta^{14}C$
UCLA-769	1873	0	–22.3	0	+0.05
UCLA-768	1878	–0.72	–23.0	–0.75	–0.67
UCLA-767	1883	–0.31	–22.9	–0.32	–0.22
UCLA-766	1888	–1.64	–22.2	–1.69	–1.59
UCLA-765	1893	–3.75	—	—	–3.60
UCLA-782	1894	–1.26	—	—	–1.11
UCLA-763	1898	–0.48	–22.9	–0.50	–0.30
UCLA-760	1903	–0.28	–23.1	–0.29	–0.02
UCLA-761	1908	–1.07	—	—	–0.72
UCLA-778	1908	–0.96	—	—	–0.61
UCLA-774	1909	+0.26	–22.6	+0.25	+0.60
UCLA-776	1909	+0.17	–24.8	+0.16	+0.51
UCLA-780	1910	–0.70	–22.2	–0.73	–0.38
UCLA-779	1910	–1.50	–24.5	–1.55	–1.20
UCLA-762	1913	–0.81	–22.6	–0.84	–0.45
UCLA-764	1918	–1.20	–22.4	–1.24	–0.69
UCLA-770	1923	–0.63	–23.0	–0.66	+0.04
UCLA-771	1928	–2.40	–22.4	–2.45	–1.58
UCLA-772	1933	–1.50	–22.0	–1.55	–0.27

第 V 列和第 VI 列中的结果分别在图 1 和图 2 中绘出。

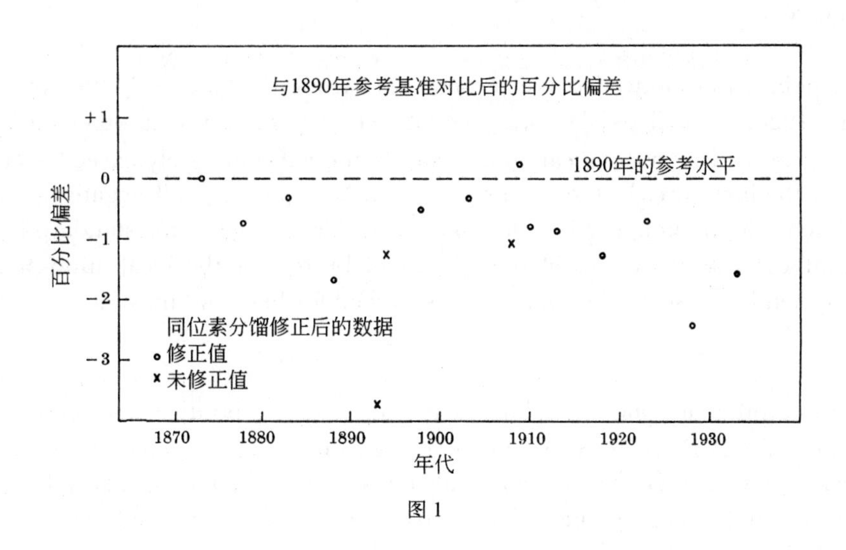

图 1

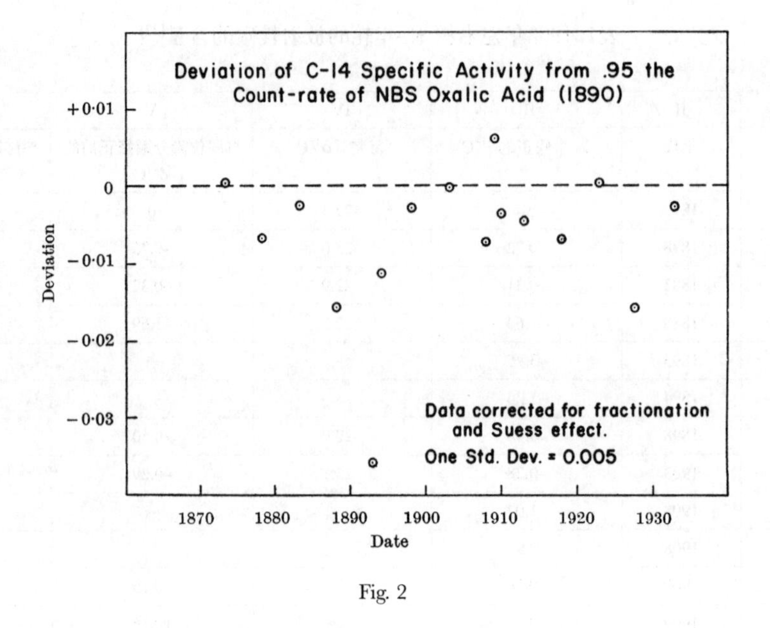

Fig. 2

As some 90,000 counts were taken on each sample, the standard deviation in each value is of the order 0.005 of that value. Experience has shown that the equipment is sufficiently stable, so the statistical uncertainty is the principal one.

Discussion

Inspection of Table 1 yields some interesting points: of all the numbers in columns III and V, only those values for the year 1909 exceed the reference-level. In column V, two others also exceed the standard, but by relatively small amounts. When a mean value is calculated for the points in a forty-year span around 1909, the latter exceeds this value by about 1 percent.

A second point to be noticed is the presence of strong fluctuations in the years around 1893 and 1928, as well as the presence of other, lesser ones at other times. These fluctuations are typical and appear to be real[23-26], though they rarely exceed 2 percent, as reported in the literature. In the results presented here, they are all negative with respect to the reference-level, though this is due to the arbitrary choice of the standard level. They are, evidently, due to variations in the carbon-14 burden of the local atmosphere. Such fluctuations tend to obscure the small effect searched for here and make its value the more uncertain.

At least three other instances are known in which strong positive deviations appear to occur[24]. They are A.D. 1687 (+2.65 percent), 1297 B.C. (+2.23 percent), and 1925 B.C. (+2.34 percent), where the deviations are taken with respect to the average values obtained from 39 oak samples ranging in age from 110 to 203 years prior to 1960. When compared with the deviation of the oxalic acid standard, however, which was +4.99±1.06 percent with respect to the oak average, these deviations are also negative.

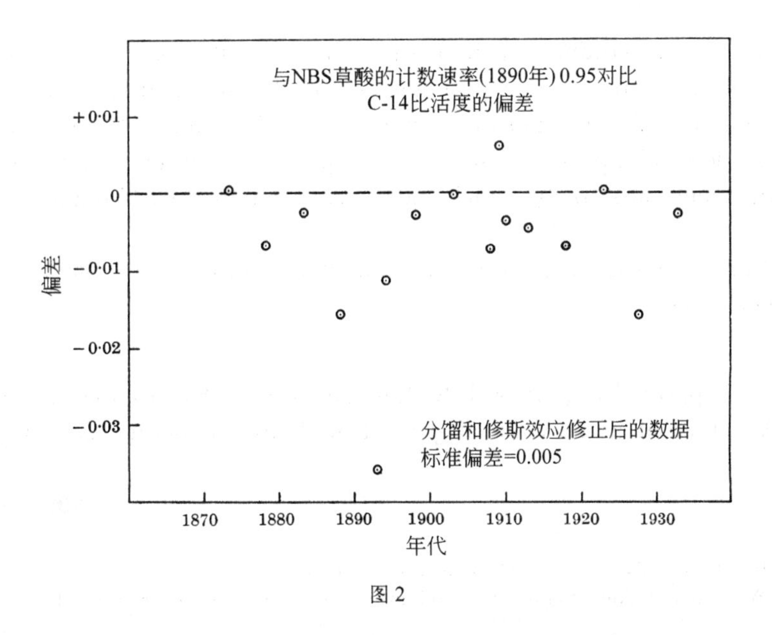

图 2

每个样品约测量 90,000 个计数，每个数值的标准偏差为该值的 0.005 量级。经验表明，仪器是足够稳定的，所以主要因素是统计数据的不确定度。

讨　论

观察表 1 数据可以发现以下有趣的特点：在第 III 和第 V 列的所有数中，仅 1909 年的数值超出参考标准水平。在第 V 列中，另两个值也超出了标准，但只超出了很小的量。当对 1909 年前后的 40 年进行平均值计算时发现，后者约超出了这个平均值 1%。

第二点值得注意的是，在 1893 年和 1928 年附近数据出现了很大的波动，在其他时段也出现了相对较小的波动情况。尽管正如在文献中提到的，这些扰动很少超出 2% 的范围，但这些的确是典型的，并且是真实存在的 [23-26]。在本文列出的结果中，由于标准水平的选择是任意的，所以此处这些值相对于参考标准水平值都是负的。显然，这些是由于局部大气中碳-14 含量的变化而引起的。这种扰动给我们研究更小尺度的作用增加了难度，并且也使其数值的不确定度更高。

至少已经知道了其他三个发生明显正偏移的例子 [24]。即公元 1687 年 (+2.65%)、公元前 1297 年 (+2.23%) 和公元前 1925 年 (+2.34%)，其中的偏移量是相对于 1960 年前在 110 年～203 年范围内的 39 个橡树样品的平均值给出的。然而，这些相对于橡树平均值的偏差为 +4.99 ± 1.06%，在与草酸标准值的偏差进行比较时，这些偏差也是负的。

Although there are uncertainties in both the estimate of the expected radiocarbon yield on the basis of the anti-matter hypothesis for the Tunguska meteor and in any extra radiocarbon burden of the atmosphere in the years following 1908 as reflected in this work, the data do yield a positive result. They appear to set an upper limit of 1/7 for the fraction of the meteorite's energy which could have been due to anti-matter.

We thank Rainer Berger, Mrs. Gera Freeman, Bette Davis and Emilio Cueto for their advice and assistance.

This work was supported in part by U.S. National Science Foundation grant *GP*-1893 to one of us (W. F. L.) and by the Walter F. Joyce Foundation to another (C. L. C.).

(**206**, 861-865; 1965)

Clyde Cowan: Department of Physics, Catholic University of America.
C. R. Atluri and W. F. Libby: Institute of Geophysics and Department of Chemistry, University of California, Los Angeles.

References:

1. Krinov, E. L., *The Solar System*, edit. by Middlehurst, B. M., and Kuiper, G. P., 4, 208 (Univ. Chicago Press, 1963).
2. Kulik, L. A., *Akademiia Nauk S.S.S.R., Co. R. (Doklady)*, No. 23, 399 (1927).
3. Told to Kulik, L. A., on March 30, 1927. See ref. 2.
4. Kulik, L. A., *Trans. Lomonosov Inst. Akad. Sci.* (U.S.S.R.), 2, 73 (1932)
5. Fesenkov, V. G., *Astronomicheskii Zhurnal*, **38**, No. 4, 577 (1961).
6. Whipple, F. J. W., *Quart. J. Roy. Meteorol. Soc.*, **56**, 287 (1930); Astapovich is credited for putting forward this suggestion first by Krinov, E. L., in his *Principles of Meteoritics* (Pergamon Press).
7. Shenrock, A. M., *Monthly Bull. Nikolaev Main Phys., Obs.*, No. 6 (1908). Rudnev, D. D., *St. Petersburg Univ. Student Sci. Trans.*, **1**, 69 (1909). Aposlolov, L., *Mirovedenie*, No. 3, 281 (1926). Whipple, F. J. W., ref. 6.
8. Whipple, F. L., *Astrophys. J.*, **111**, 375 (1950).
9. Fesenkov, V. G., *Meteoritika*, **6** (1949).
10. Fesenkov, V. G., *Meteoritika*, **6**, 8 (1949).
11. Zigel, F. Y., *Znaniye-Sila*, No. 12, 24 (1961).
12. Glasstone, S. (edit.): *The Effect of Nuclear Weapons*, prepared by the U.S. Dept. Defense and published by the U.S. Atomic Energy Comm., Washington (1962). McMillan, W. G. (personal communication).
13. Berger, R. (personal communication).
14. Horwitz, N., Miller, D., Murray, J., and Tripp, R., *Phys. Rev.*, **115**, 474 (1959).
15. See, for example, *Nuclear Interactions*, by DeBenedetti, S. (Wiley and Sons, New York, 1964).
16. Thorndike, A. M., *Handbook of Physics*, edit. by Condon and Odishaw (McGraw-Hill, New York, 1958).
17. Karlen, I., Olsson, I., and Karlberg, P., *Arkiv Geofysik*, 4, 465 (1964).
18. Libby, W. F., *Proc. U.S. Nat. Acad. Sci.*, **45**, 959 (1959).
19. Fergusson, G. J., *J. Geophys. Res.*, **68**, 3933 (1963).
20. Suess, H. E., *Science*, **122**, 416 (1955).
21. Fergusson, G. J., *Proc. Roy. Soc.*, A, **243**, 561 (1958).
22. Berger, R., Fergusson, G. J., and Libby, W. F., *Radiocarbon*, 7, 1975 (in the press).
23. Whitaker, W. W., Valastro, jun., S., and Williams, M., *J. Geophys. Res.*, **64**, 1023 (1959).
24. Suess, H. E., *Science*, **122**, 415 (1955).
25. Ralph, E. K., and Stuckenrath, R., *Nature*, **188**, 185 (1960).
26. Damon, P. E., Long, A., and Sigalove, J., *Radiocarbon*, **5**, 283 (1963).

虽然正如在本文中所提到的，对在通古斯陨石的反物质假设基础上的放射性碳含量估计，以及 1908 年后大气中任何额外的放射性碳的含量估计，这两个值都存在不确定度，但确实得到了一个正值的结果。这表示陨石中由反物质产生的部分能量所占比例的上限可能在 1/7。

我们感谢雷纳 · 伯杰、弗里曼 · 格拉夫人、贝特 · 戴维斯和埃米利奥 · 奎托的建议及帮助。

本工作一部分由美国国家科学基金会（基金 *GP*-1893）向我组人员（利比）提供资助，另一部分由沃尔特 · 乔伊斯基金会向另一组员（考恩）提供资助。

（沈乃澂 翻译；尚仁成 审稿）

A Radar Determination of the Rotation of the Planet Mercury

G. H. Pettengill and R. B. Dyce

Editor's Note

Remarkably, Mercury's rotation rate was determined only in 1965, in this paper by Gordon Pettengill and Rolf Dyce. Mercury had been expected to be "tidally locked" with one face pointing always to the Sun, just as the Moon is tidally locked to Earth. But it turned that its rotation rate was two-thirds of its orbital period.

DURING the recent inferior conjunction of the planet Mercury in April, 1965, radar observations were obtained by the Arecibo Ionospheric Observatory in Puerto Rico (operated by Cornell University with the support of the Advanced Research Projects Agency under a Research Contract with the Air Force Office of Scientific Research). The system operated at a frequency of 430 Mc/s, with an antenna gain of 56 dB and a transmitted power of 2 MW. The resulting sensitivity was sufficient to obtain significant echoes not only from the nearest part of the planetary disk but also from more distant regions, removed by up to 0.06 of the planet's radius. By using short transmitted pulses of 500 μsec duration, it was possible to isolate the echo power from these more distant regions, and to carry out a Fourier analysis of their spectral composition.

Since the source of the delayed echoes can quite reliably be associated with a known area of the planetary surface, the magnitude of the apparent planetary rotation can be inferred from the measured spectral dispersion through a simple geometrical relationship. The apparent rotation is the vector sum of an intrinsic rotation and a contribution arising from the relative motion of the observer and the target planet. Since the latter is quite accurately calculable from the known orbital motions of the Earth and Mercury and the known rotation of the Earth, a constraint is set on the allowable vector magnitude and position assigned to the intrinsic planetary rotation.

By carrying out observations spread over a period of time it is possible to solve for both the magnitude and direction of the planetary rotation. In the present series of observations, data have been obtained for April 6, 10, 12 and 25, 1965. On most of these days it was possible to check the results by comparing the inferred angular rotation obtained from data at various delays measured with respect to the earliest (and strongest) echo component. From this comparison a degree of confidence could be established, and an estimate obtained of the measurement error.

The data were used to compute a most likely value of intrinsic planetary rotation with

雷达测定水星的自转

佩滕吉尔，戴斯

编者按

引人注目的是直到 1965 年，戈登·佩滕吉尔和罗尔夫·戴斯才在本文中给出水星自转速率。预期中的水星是“潮汐锁定”的，有一面始终朝向太阳，就像月球是被地球潮汐锁定的一样。但实际上其自转速率是轨道周期的三分之二。

在 1965 年 4 月水星最近一次的下合期间，位于波多黎各的阿雷西博电离层天文台（它由康奈尔大学在高等研究计划局的支持下，在空军科学研究办公室的一个研究合同框架下运行）对其进行了雷达观测。该系统的工作频率为 430 兆周 / 秒，天线增益为 56 分贝，发射功率为 2 兆瓦。该系统的灵敏程度不仅可以接收到来自紧靠行星盘最近部分的显著回波信号，还能把观测区扩大到行星半径的 0.06 倍处。利用时长为 500 微秒的短发射脉冲，可以分离出更远区域的回波功率，并利用傅里叶分析研究其频谱组成。

由于延迟回波的反射源与行星表面已知区域的关系相当可靠，故行星视自转速率的大小可以通过一个简单的几何关系从测得的光谱色散中得到。视自转是行星本身的自转与观测者和目标行星间相对运动的矢量和。由于后者可以根据已知的地球和水星轨道运动情况及地球自转率很精确地计算出来，因此也就可以对行星本身的自转速率的矢量大小和方向给出一个约束值。

通过一段时期的连续观测，就有可能求解出行星自转速度的大小和方向。目前的观测数据序列是 1965 年 4 月 6 日、10 日、12 日和 25 日得到的。在大部分时间中，可以通过对比分析由测得的各种延迟数据所推算出的角自转速率与最早（同时也是最强）的回波组分来检验结果。通过这些比较可以确定置信度，并估算测量误差。

利用麻省理工学院林肯实验室欧文·夏皮罗博士所开发的程序，这些数据可以

a procedure developed by Dr. Irwin Shapiro of the Lincoln Laboratory, Massachusetts Institute of Technology. Fig. 1 shows the measurements together with the best fit curve. The curve for a retrograde rotation which would be permitted on the basis of the data of April 6, 10 and 12 alone is also included, as is the behaviour that would be expected on the assumption of rotation which is synchronous with the orbital period. As shown in Fig. 1, the rotation is direct with a sidereal period of 59 ± 5 days. The direction of the pole is not well-determined from these limited data, but is approximately normal to the planetary orbit.

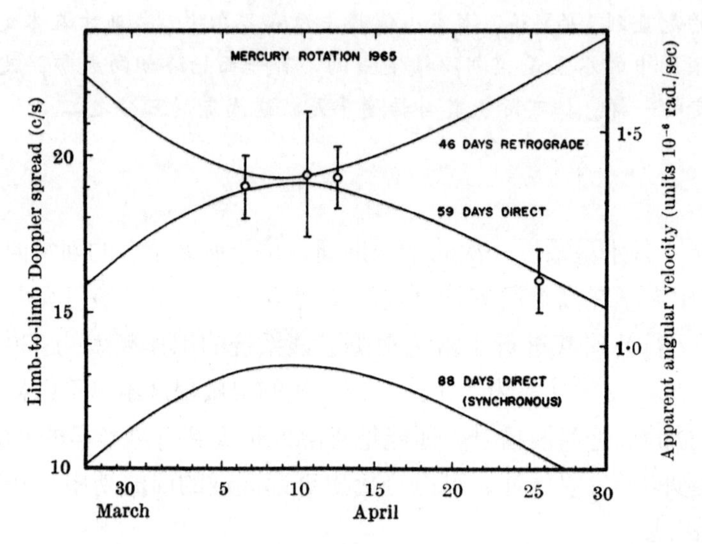

Fig. 1. Plot of the apparent rotational angular velocity of the planet Mercury versus date for several values of rotation during the inferior conjunction of April 1965. The values inferred from the measurements are shown with their estimated errors

The finding of a value for the rotational period of Mercury which differs from the orbital period is unexpected and has interesting theoretical implications. It indicates either that the planet has not been in its present orbit for the full period of geological time or that the tidal forces acting to slow the initial rotation have not been correctly treated previously, as suggested in the following communication.

We thank the staff of the Arecibo Ionospheric Observatory for their assistance in carrying out the measurements.

(**206**, 1240; 1965)

G. H. Pettengill and R. B. Dyce: Cornell-Sydney University Astronomy Center, Arecibo Ionospheric Observatory, Arecibo, Puerto Rico.

用来计算出水星本身自转速率的最大可能值。图 1 给出了观测值以及最佳拟合曲线。在假定自转与轨道周期同步的情况下，逆行自转是允许存在的，图中也包括了基于 4 月 6 日、10 日和 12 日的数据所绘制的逆行自转曲线。如图 1 所示，自转与 59±5 天的恒星周期是相一致的。尽管无法从这些有限的数据中准确测定极轴的方向，但其大致是与行星轨道面相垂直的。

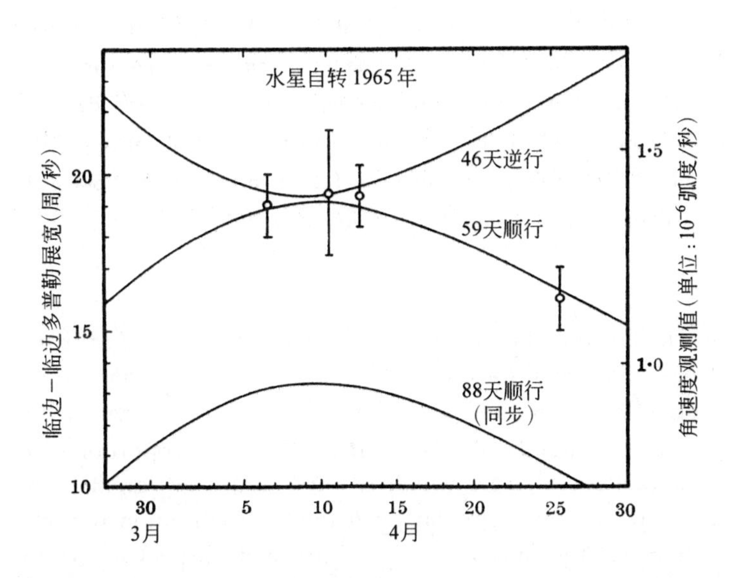

图 1. 在 1965 年 4 月下合期间，根据在不同时间所得自转值绘出了水星视自转角速度图。图中还给出了测量估计值及其误差。

对水星自转周期值和轨道周期值不一致的发现是出人意料的，这也激发了人们对其理论含义的兴趣。如在后面的报道中所指出的，它表明或者在整个地质历史时期水星并非一直处于现在的轨道上，或者之前我们并未正确对待潮汐力对减慢初始转动速率影响。

在此我们感谢阿雷西博电离层天文台的工作人员在测量过程中所给予的协助。

（钱磊 翻译；肖伟科 审稿）

Rotation of the Planet Mercury

S. J. Peale and T. Gold

Editor's Note

Here Stan Peale and Thomas Gold show that because of Mercury's large eccentricity, its rotation rate determined by Pettengill and Dyce is actually is to be expected, contrary to the previously prevailing view that the planet would be "tidally locked" with the same face always pointing sunward. They also are able to infer that Mercury is not permanently rigid, which they suggest might be partially explained by its high surface temperature. Relatively little is known about Mercury even now. The Messenger mission was launched to study the planet in 2004, and after three flybys, it will enter orbit around Mercury in March 2011.

SOLAR tidal friction must be an intense effect for Mercury, and it must be expected that the planet's spin would have relaxed from any original value to one that is under the control of this effect in a time short compared with the age of the solar system. The retarding torque exerted by the Sun on a planet is proportional to $1/r^6$ (where r is the distance Sun-planet), a factor which is some 300 times greater for Mercury than for the Earth. For a planet on a circular orbit the final condition would then be one of synchronous rotation like the motion of the Moon with respect to the Earth. Mercury's motion around the Sun takes 88 days and for synchronous rotation the sidereal period would thus be 88 days also. The observed value of 59 ± 5 days differs markedly from this (see preceding communication).

For a planet with substantial orbital eccentricity the condition is different, however, and synchronous rotation with the orbital period need not then be expected. With the $1/r^6$ dependence the tidal torque at perihelion will exceed that at other times, and the angular velocity of the planet will thus settle at a value greater than the mean orbital angular velocity, but not quite as great as the orbital angular velocity at perihelion. For Mercury, where the eccentricity is 0.2, 2π times the reciprocal of the orbital angular velocity at perihelion is 56.6 days. A spin with a sidereal period lying between 56 and 88 days thus must be expected.

A more precise calculation may be made based on the consideration that the final angular velocity of the planet will be such that the time average of the tidal torque around the orbit is zero. No further change in the planet's spin will then occur except on the much longer time scale on which other orbital elements can be influenced by tidal friction, effects which probably are unimportant in the age of the solar system. The precise calculation of the angular velocity to fit this condition can be made without a quantitative description of the dissipation properties of the planet, but involving certain assumptions. The tidal phase lag has to be assumed to be a small angle only, and one may make the

水星的自转

皮尔，戈尔德

编者按

在本文中，斯坦·皮尔和托马斯·戈尔德表明，由于水星具有较大的偏心率，佩滕吉尔和戴斯测定的水星自转速率实际上是在预料之中的，这与先前盛行的观点相反，即该行星应该是“潮汐锁定”的，有一面始终朝向太阳。他们也推断，水星不是刚性不变的，他们暗示这其中的部分原因或许是其表面的高温造成的。但即使现在我们对水星也知之甚少。用来对水星进行研究的信使号于2004年发射，在三次飞掠过水星之后它将在2011年3月进入环绕水星的轨道。

太阳的潮汐摩擦作用对水星产生强烈的影响，并且可以预期在与太阳系年龄相比更短的时间内，水星自转速率已经在潮汐作用下从某个初始值减少到了现在的数值。太阳作用于行星的黏滞力矩正比于 $1/r^6$（r 是太阳和行星间的距离），对水星来说该系数大约是地球的300倍。位于圆轨道上的行星，其最终的状态就是类似月球相对于地球的同步绕转。水星绕太阳转动一圈需要88天，并且由于同步绕转，其恒星周期也应该是88天。但观测到的 59 ± 5 天的结果与该值明显不符(见前一篇报道)。

然而对于轨道偏心率较大的行星来说情况会不一样，并且同步绕转与轨道周期并不需要一致。由于系数 $1/r^6$ 的关系，近日点的潮汐力矩将超过其他时候的值，于是行星的角速度也将出现比平均轨道角速度大的值，但也没有近日点的轨道角速度大。对于水星，其偏心率是0.2，2π 乘以近日点轨道角速度的倒数是56.6天。因此预期的自转恒星周期一定是在56天和88天之间。

更精确的计算基于以下考虑，行星的最终角速度应该满足沿轨道的潮汐力矩的时间平均等于零。这时行星的自转就不会再变化了，除非是在更长的时标上，其他轨道根数受到了潮汐摩擦的影响，但这些影响在太阳系年龄的时标下是不太重要的。在这种条件下对角速度进行精确计算并不需要对行星的耗散特性进行定量描述，但是会涉及某些特定假设。必须假设潮汐相位滞后只是一个小角度，才能利用各种依赖幅值和频率的 Q 函数方法进行计算（$1/Q$ 是比耗散函数[1]）。如果假设 Q 不依赖

calculation with Q being dependent on amplitude and frequency in a variety of ways ($1/Q$ is the specific dissipation function[1]). If Q is assumed to be independent of amplitude and inversely proportional to frequency, then the condition will yield a period of 71 days. If Q is assumed independent of amplitude and of frequency the result is 65 days. Any additional effect of amplitude dependence would go in the sense of increasing the dissipation near perihelion where the amplitudes are greatest and therefore decreasing the period still further. It thus seems likely that Mercury is indeed in its final state of spin, and that its present rotation therefore reflects very accurately certain characteristics of the dissipation process.

The condition discussed here is based on the supposition that the solar torque exerted on the tidal bulge exceeds that exerted on any permanent deformation from axial symmetry. In the converse case a period of 88 days for the rotation would indeed result. This may imply that Mercury has not much permanent rigidity. The high surface temperature may be partly responsible for this. The solar system has thus provided us with an example of each of the two final states of rotation that tidal friction can bring about: the Moon, which has locked into the synchronous rotation, and Mercury, which has come to the rotation that is enforced in the absence of any permanent asymmetry of the body.

(**206**, 1240-1241; 1965)

S. J. Peale and T. Gold: Cornell-Sydney University Astronomy Center, Cornell University, Ithaca, New York.

Reference:

1. MacDonald, G. F. J., *Rev. Geophys.*, **2**, No. 3 (1964).

于幅值并且与频率成反比，那么得到的周期是 71 天。如果假设 Q 不依赖于幅值和频率，结果是 65 天。任何其他幅值相关的效应都应该是增加近日点附近（那里的幅值最大）的耗散，并进一步减小自转周期。因此很有可能水星确实是处于它的最终自转状态，它现在的自转非常准确地反映了耗散过程的某些特征。

这里讨论的条件基于以下假设：太阳作用于潮汐隆起的力矩超过任何偏离轴对称的永久形变的效果。在相反的情况下，88 天的自转周期也确实存在。这也可能暗示了水星并不能永久保持刚性。它表面的高温可能是其中的部分原因。可见太阳系为我们分别提供了由潮汐摩擦导致的两个不同的最终转动状态：月球，已经锁定到同步绕转；水星，由于其本身缺少恒定的非对称性而处于强迫性的自转状态中。

（钱磊 翻译；肖伟科 审稿）

A Physical Basis for Life Detection Experiments

J. E. Lovelock

Editor's Note

James Lovelock is best known for his Gaia hypothesis, which postulates that the Earth regulates its climate with feedback processes much like the homeostatic mechanisms that maintain constant conditions in living organisms. Lovelock developed these ideas as an independent scientist, funded from the proceeds of inventions such as instruments for detecting small amounts of trace gases in the atmosphere. In the mid-1960s Lovelock was engaged in discussions at NASA about experiments to search for life on the surface of Mars, to be conducted by lander spacecraft. Here Lovelock sidesteps difficult questions about the unknown chemical basis of extraterrestrial life by proposing a general physical argument rooted in the way life will create a departure from thermodynamic equilibrium in its environment.

THE design of an efficient and unequivocal experiment in extra-terrestrial life detection should take into account: (1) A definition of life stated in terms favourable for its recognition. (2) A description of the past and present environment of the planet to be sampled.

As yet, there is no formal physical statement to describe life from which an exclusive definition for experimental purposes could be drawn. Moreover no comprehensive description is available of the atmospheric as well as the surface physical and chemical environment of any of the planetary bodies.

It is not surprising, in view of the vast expense of space-probe experiments and of the formidable uncertainties already stated here, that the proposed experiments in life detection all ask the cautious geocentric question: "Is there life as we know it?" Most certainly it is difficult to envisage in detail an alien biochemistry; it would seem pointless and very uneconomic to send a space probe to detect a speculative life-form.

It is the object of this article to show that we are not necessarily limited to experiments based on the recognition of a specific life-form, either Earth-like or alien. Also, that it is possible, by accepting a limited phenomenological definition of life, to design simple experiments from the general recognition of life phenomena, including that with which we are familiar. The application of this approach to experiments in life detection is the basis of the discussion which follows.

Recognition of Life

It is a relatively simple matter to distinguish between living and inorganic matter on

生命探测实验的物理基础

洛夫洛克

编者按

令詹姆斯·洛夫洛克最为出名的是他提出的盖亚假说，它假设地球通过反馈过程对其气候进行调节，这非常类似于生物有机体内保持恒定状态的稳态机制。洛夫洛克是作为一名独立科学家提出以上想法的，他的经费来自他发明诸如检测大气中痕量气体的仪器获得的收入。20 世纪 60 年代中期，洛夫洛克曾参与过美国国家航空航天局关于利用登陆航天器在火星表面进行生命搜索实验的讨论。本文中洛夫洛克以生命会在其生存环境中使热力学平衡发生偏离为基础提出一个普遍的物理论据，绕开了由于对地外生命的化学基础未知而产生的一系列难题。

为了对地球外的生命进行有效而明确的检测，实验设计应考虑以下两点：(1) 从便于识别生命出发确定生命的定义；(2) 对被调查行星的过去和现在环境的描述。

到目前为止，尚无正式的物理表述可以给出生命唯一可用的定义。此外，对任何行星的大气、表面物理和化学环境都还没有一个全面的描述可供使用。

考虑到航天探测器实验的巨大花费以及本文已经阐述过的巨大的不确定性，人们就不会奇怪，为何对所有提出的生命探测实验都会谨慎地问一个以地球为中心的问题：“那里的生命和我们所知的相同吗？”详细地设想一种外星生物化学绝对是困难的；因此发射航天探测器去寻找一种推测的生命形式没有意义，也不划算。

本文的目的就是要说明，我们不必将实验的基础局限于识别某种特别生命形式，无论其是类似地球上的生命形式，还是外星球的生命形式。此外，我们也可以接受对生命有限的唯象定义，并根据普遍承认的，包括一些我们所熟知的生命现象设计一些简单的实验，这是可能的。将这种方法应用到生命探测的实验中是下面讨论的基础。

生命的识别

尽管还没有用生化术语对生命做出的正式定义，但用生化实验来区分地球上

Earth by biochemical experiments even though no formal definition of life in biochemical terms exists. Experience suggests, for example, that a system capable of converting water, atmospheric nitrogen and carbon dioxide into protein, using light as a source of energy, is unlikely to be inorganic. This approach for recognition of life by phenomenology is the basis of the experiments in detection of life so far proposed. Its weakness lies not in the lack of a formal definition but in the assumption that all life has a common biochemical ancestry.

It is also possible to distinguish living from inorganic matter by physical experiments. For example, an examination of the motion of a salmon swimming upstream suggests a degree of purpose inconsistent with a random inorganic process. The physical approach to recognition of life is no more rigorous, at this stage, than is the biochemical one; it is, however, universal in application and not subject to the local constraints which may have set the biochemical pattern of life on Earth.

Past discussions of the physical basis of life[1-3] reach an agreed classification as follows: "Life is one member of the class of phenomena which are open or continuous reaction systems able to decrease their entropy at the expense of substances or energy taken in from the environment and subsequently rejected in a degraded form". This classification is broad and includes also phenomena such as flames, vortex motion and many others. Life differs from the other phenomena so classified in its singularity, persistence, and in the size of the entropy decrease associated with it. Vortices appear spontaneously but soon vanish; the entropy decrease associated with the formation of a vortex is small compared with energy flux. Life does not easily form, but persists indefinitely and vastly modifies its environment. The spontaneous generation of life, according to recent calculations from quantum mechanics[4,5], is extremely improbable. This is relevant to the present discussion through the implication that wherever life exists its biochemical form will be strongly determined by the initiating event. This in turn could vary with the planetary environment at the time of initiation.

On the basis of the physical phenomenology already mentioned, a planet bearing life is distinguishable from a sterile one as follows: (1) The omnipresence of intense orderliness and of structures and of events utterly improbable on a basis of thermodynamic equilibrium. (2) Extreme departures from an inorganic steady-state equilibrium of chemical potential.

This orderliness and chemical disequilibrium would to a diminished but still recognizable extent be expected to penetrate into the planetary surface and its past history as fossils and as rocks of biological origin.

Experiments for Detection of Life

The distinguishing features of a life-bearing planet, described here, suggest the following simple experiments in detection of life:

生命体和无机物质相对来讲是比较简单的事情。例如经验表明，一个能够利用光作为能源，将水、大气中的氮和二氧化碳转换成蛋白质的系统，就不太可能是无机的。这种通过现象表征来识别生命的方法是至今为止提出的探测生命实验的基础。它的缺陷并不在于缺乏正式的定义，而在于假定所有生命的生物化学源头都相同。

也可以用物理实验来区分有生命的和无机的物质。例如，对鲑鱼逆流而上的游动的调查显示出一定程度的目的性，与随机的无机过程并不一致。现阶段识别生命的物理研究方法与生化研究方法相比并不是很严格；然而，它在应用中是普适的，并不受局部条件的限制，这种限制可能已经限定了地球上生命的生化模式。

过去对生命的物理基础[1-3]的讨论达成了如下普遍认同的分类："生命是一类开放的或连续反应系统中的一员，它可以消耗从外界环境中摄取的物质或能量来使自己的熵减少，随后以降解后的形式排出"。这个分类覆盖面很广，也包括了诸如火焰、涡旋运动及其他许多现象。生命与上述分类中其他现象不同还表现为它的奇特性、持久性以及伴随生命过程的熵的减少程度。涡旋自发形成，但很快就会消失；与能流相比涡旋形成时伴随的熵的减少是很小的。生命不容易形成，但一旦形成存在期限就无限定，并在很大程度上改变其环境。近来根据量子力学的计算结果表明[4,5]，生命的自发产生几乎是不可能的。与此相关的是现今的讨论暗示了无论生命在哪里存在，它的生化形式很大程度上受初始事件所决定。并且这又可能因初始时行星的环境条件的不同而变化。

在已经提到的物理现象学的基础上，一个存在生命的行星与一个不毛之地的区别在于：(1) 高度有序性及基于热力学平衡几近不可能的结构和事件的普遍存在。(2) 完全偏离无机物化学势的稳态平衡。

这种有序性和化学不平衡虽然微小，但仍有希望以能被识别的程度穿透到行星的表面，并成为生物起源的历史化石和基础。

生命探测的实验

基于上述维持有生命的行星所具有的显著特性，我们提出了下列简单的探测生命的实验：

(*A*) *Search for order.* (1) Order in chemical structures and sequences of structure. A simple gas chromatograph or a combined gas chromatograph-mass spectrometer instrument would seek ordered molecular sequences as well as chemical identities.

(2) Order in molecular weight distributions. Polymers of biological origin have sharply defined molecular weights, polymers of inorganic origin do not. A simple apparatus to seek ordered molecular weight distributions in soil has not yet been proposed but seems worthy of consideration.

(3) Looking and listening for order. A simple microphone is already proposed for other (meteorological) purposes on future planetary probes; this could also listen for ordered sequences of sound the presence of which would be strongly indicative of life. At the present stage of technical development a visual search is probably too complex; it is nevertheless the most rapid and effective method of life recognition in terms of orderliness outside the bounds of random assembly.

(*B*) *Search for non-equilibrium.* (1) Chemical disequilibrium sought by a differential thermal analysis (DTA) apparatus. Two equal samples of the planetary surface would be heated in a DTA apparatus: one sample in the atmosphere of the planet, the other in an inert gas, such as argon. An exotherm on the differential signal between the two samples would indicate a reaction between the surface and its atmosphere, a condition most unlikely to be encountered where there is chemical equilibrium as in the absence of life. It should be noted that this method would recognize reoxidizing life on a planet with a reducing atmosphere. This experiment could with advantage and economy be combined with, for example, the gas chromatography mass spectrometry experiment (*A*1) where it is necessary to heat the sample for vaporization and pyrolysis.

(2) Atmospheric analysis. Search for the presence of compounds in the planet's atmosphere which are incompatible on a long-term basis. For example, oxygen and hydrocarbons co-exist in the Earth's atmosphere.

(3) Physical non-equilibrium. A simplified visual search apparatus programmed to recognize objects in non-random motion. A more complex assembly could recognize objects in metastable equilibrium with the gravitational field of the planet. Much of the plant life on Earth falls into this category.

Experiments *A*1, *B*1 and *B*2 are the most promising for the development of practical instruments. Indeed, the gas chromatography-mass spectrometry combination experiment and the DTA experiment already proposed for planetary probes[7] are, with minor modifications, capable of recognizing the ordered sequences and chemical disequilibrium discussed earlier. Experiment *B*2, atmospheric analysis, is simple and practical as well as important in the general problem of detection of life. A detailed and accurate knowledge of the composition of the planetary atmosphere can directly indicate the presence of life in terms of chemical disequilibrium; such knowledge also is complementary to the

(*A*) **搜寻有序性** (1) 化学结构和结构序列的有序性。一台简单的气相色谱仪或气相色谱–质谱联用仪就能检测出有序的分子序列和其化学对应物。

(2) 分子量分布的有序性。生物学来源的聚合物具有明确的分子量，而无机物起源的聚合物并非如此。尚未提出在土壤中探测分子量分布有序性的简便设备，但似乎值得考虑。

(3) 对有序性的“看”和“听”。已经有人提出将用于其他目的（气象学）的简单话筒应用在未来的行星探测器上。我们也可以用其听取能够强烈显示出生命存在的有序序列的声音。对于现阶段的技术发展水平，可视化搜索可能太复杂了；而从无序组织中依据有序性来识别生命却是最快和最有效的方法。

(*B*) **搜寻非平衡性** (1) 用差热分析（DTA）装置搜寻化学的不平衡性。在DTA 装置中对行星表面上两份相同的样品加热，一份样品在行星大气中，另一份样品置于惰性气体中（例如在氩气中）。两份样品放热曲线的不同信号即可说明地表和大气之间发生反应。在没有生命的情况下是存在化学平衡的，这时不会产生上述的差异。值得注意的是这种方法还可以用来识别存在于还原性行星大气中的再氧化生命。本实验还可以和气相色谱质谱联用实验结合起来（*A*1），使得分析更加有效和经济，但这时需要加热样品使其汽化并产生热分解。

(2) 大气分析。在行星的大气中寻找是否有从长期看来不相容的化合物共存。比如在地球大气中氧和碳氢化合物的共存。

(3) 物理不平衡。一种简化过的可视化搜索设备，经设计可识别出非随机运动的物体。更加复杂的组装仪器可以识别出在行星重力场下处于亚稳平衡态的物体。地球上的大部分植物都属于这一类。

由于发展了实用的仪器，实验 *A*1、*B*1 和 *B*2 是最有希望的。实际上，对行星探测器已提出的气相色谱–质谱联用实验和 DTA 实验[7]只需较小的改动便可识别出有序序列和前面提到的化学不平衡。大气分析实验 *B*2 简单实用，在解决生命探测的一般问题中很重要。依据化学不平衡，对行星大气组分详细和准确的了解能够直接证明生命的存在；这些了解对其他生命探测实验的理解以及后续实验的规划也都是有帮助的。即使在生物很丰富的地球上，也存在很多区域，诸如被新雪所覆盖的区域，

understanding of other life detection experiments and to the planning of subsequent experiments. Even on Earth where life is abundant there are many regions, such as those covered by fresh snow, where a surface sample might be unrewarding in the search for life. The atmospheric composition is largely independent of the site of sampling and provides an averaged value representative of the steady state of chemical potential for the whole planetary surface.

Fig. 1 shows the abundance of hydrocarbons of carbon number between 11 and 33 for abiotic hydrocarbons of the Fischer-Tropsch process[8] and hydrocarbons of biological origin, wool wax[9]. Poisson distributions around the predominant hydrocarbon numbers are shown as solid lines. The inorganic hydrocarbons fit closely the expected Poisson distribution for a state of chemical equilibrium. By contrast the biological hydrocarbons show large departures in the distribution of their abundance from this equilibrium state; also, especially for the higher molecular weight alkanes, a two-carbon ordered sequence is well established.

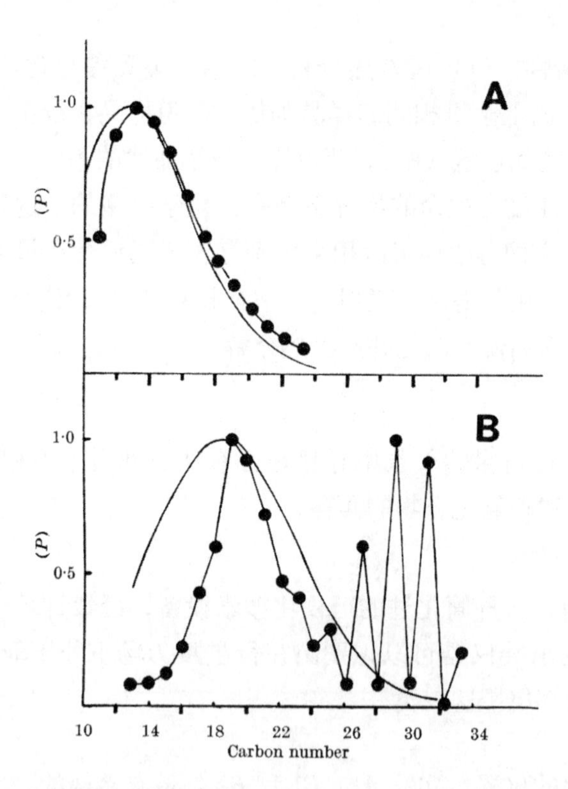

Fig. 1. The abundance of *n*-alkanes from an inorganic source (*A*), Fischer–Tropsch hydrocarbons, and from a biological source (*B*), wool wax. The observed abundances (●——●) are compared with normalized Poisson distributions (——) around the preponderant alkanes

In a similar manner with experiment *B*1 the disequilibrium associated with life can be demonstrated. A few mg of soil heated in a DTA apparatus in air shows a large exotherm when compared with a similar reference sample heated in argon. The combustion of even

取自其中的表面样品可能找不到生命存在的迹象。大气组分与取样的地点大体无关，并且可以提供一个可以代表整个行星表面的化学势稳恒态的平均值。

图 1 分别显示了来自费–托合成过程的非生物源的烃[8]和来自生物源羊毛蜡的烃[9]中碳原子数目在 11 到 33 之间的烃的丰度。实线表示的是在含量最多的烃周围的泊松分布。无机烃和化学平衡态所预期的泊松分布很好地吻合了。相反，生物源的烃丰度分布却和这个平衡态有较大的偏离；尤其对于具有较高分子量的烷烃，建立了双碳的有序序列。

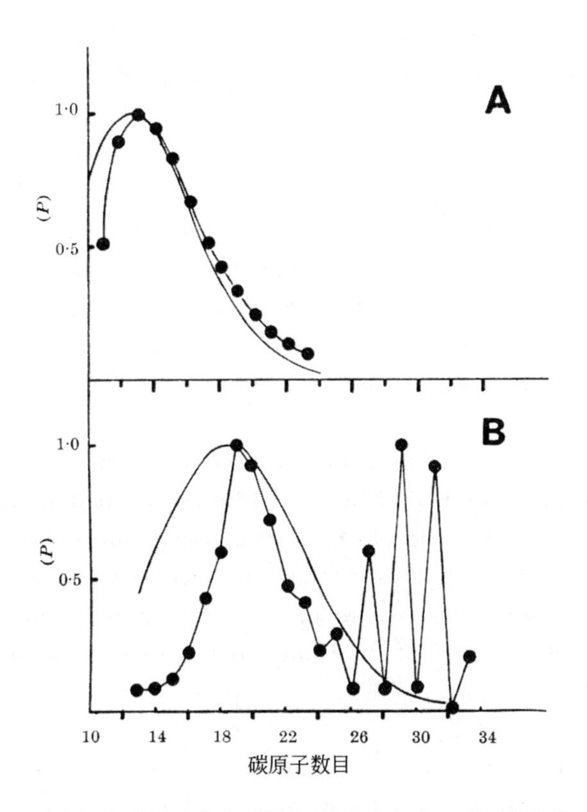

图 1. 来自费–托烃的无机源（*A*）以及来自生物源羊毛蜡（*B*）的 *n*- 烷烃的丰度。在主要烷烃附近，观察到的丰度（●——●）与标准泊松分布（——）的比较。

通过与实验 *B*1 类似的方法，还可以证明与生命有关的不平衡。加热放置在空气中的 DTA 装置中的几毫克土壤，当与在氩中加热的类似的参考样品相比时，有更

a few micrograms of organic matter in these circumstances is capable of generating a detectable signal.

Detection of Life on Mars

Ordinarily one does not look for fish in a desert, nor for cacti on an ice cap. Should we, therefore, look for microorganisms of Earth-like habits on Mars, or should we rather ask one or more of the general questions discussed here? The answer to this must depend on the history of Mars, past and present.

The following is the only sure information so far available on Mars. It is dry. The atmosphere is thin and contains no more than a trace of oxygen. The flux of solar radiation at the surface, although less than on Earth, is also less filtered and may include an appreciable content of energetic radiation; in particular, short wave-length ultra-violet. The temperature range includes periods above zero centigrade. Finally, but less certain, is the possibility[6] that oxides of nitrogen are present in appreciable amounts.

If these conditions are representative of Mars in the past as well as now, there seems no reason to assume that life, if present at all, can resemble that on Earth. However, it is possible that Mars was once Earth-like (primeval non-living Earth) and has changed physically to its present state; or less likely that the present state of Mars, like that of the Earth, is a consequence of biological change. For success a geocentric biochemical experiment must assume that Mars was once Earth-like and that life is still surviving in a highly adapted form yet still recognizable to the experiment.

If Mars is as it always was or has been changed to its present state by biological action, then life, if there now, would be very different from that we know. Could we conceive of living systems in liquid N_2O_4 as an ionizing solvent? Could they use hard ultra-violet as a source of energy? Is cellular life necessary in a dry environment, or did cell membranes evolve on Earth to offset the overwhelming effects of dilution in the primeval seas? What sort of Martian biochemistry could have generated the present atmosphere?

Answers to these questions are important in the design of experiments to detect particular life-forms; thus, with a growth experiment, or a biochemical experiment, the strength, composition and conditions of incubation of the medium are of vital importance. This information, however, is not needed in the general detection of life. In view of what is not known of conditions on Mars, the physicochemical experiments in life recognition such as experiments *A*1 and *B*1 and *B*2 seem more worth considering for early probe experiments. These simple experiments do not require a prior knowledge of the planetary environment and are not limited to Earth biochemistry.

大的放热。在这种环境下即便有几微克的有机物燃烧，也可以检测到信号。

探测在火星上的生命

通常，人们不会在沙漠中寻找鱼，也不会在冰盖上寻找仙人掌。那么我们应该在火星上寻找具有类似地球上习性的微生物吗？或者我们更应该问一些在本文中提到的一个或多个一般性的问题？这些问题的答案必然取决于火星的历史，包括过去和现在。

以下是至今所能获得的关于火星的有限确切信息。它很干燥，大气稀薄，仅含有痕量的氧气。火星表面的太阳辐射量虽然要比地球表面的低，但相对来讲被过滤掉的量也较少，因此可能含有相当的高能辐射（尤其是短波紫外线辐射）。温度范围有一部分时间在零摄氏度以上。最后还有一点不太确定的是火星上可能[6]存在相当数量的氮的氧化物。

如果上述信息可以代表火星的过去和现在，那么似乎没有理由假设火星上的生命（如果存在的话）和地球上的是类似的。但有可能火星曾经是类似地球的（原初无生命的地球），只是后来才按照自然规律演变成了现在的状态。或者还有一种比较小的可能性是火星现在的状态，正如地球一样，是由于生物演变的结果。为了成功进行一个以地球为中心的生化实验，必须先要假设火星曾经是类似地球的，并且现在生命仍以一种高度适应的形式存活下来，而且这种形式仍可以为我们的实验所识别。

如果火星一直以来就是现在的模样，或者通过生物活动而变成了现在的样子，那么火星上的生命（如果现在还有的话）将与我们所知道的生命非常不同。我们可以构想以液态 N_2O_4 为离子化溶剂的生命系统吗？它们能否将硬紫外线作为能量的来源呢？细胞生物是否必须要在干燥环境下生存，或者是否在地球上细胞膜得到了进化以防止被远古时期的海水过度的稀释？什么样的火星生化反应才能产生现在的火星大气呢？

以上问题的解答对于我们在设计检测某种特别形式的生命存在的实验时具有很重要的意义。因而，对于发育实验或生化实验，孵化环境的强度、成分和条件是至关重要的。但这些信息对于一般的生命检测是不必要的。鉴于火星上还有一些我们不知道的情况，生命识别的 *A*1、*B*1 和 *B*2 这三个物理化学实验作为早期的探测器上的实验更值得我们优先予以考虑。这些简单的实验并不要求我们对行星环境有预先的了解，并且也不局限于地球上的生物化学。

I thank A. Zlatkis and P. G. Simmonds of the University of Houston for their advice and for conducting and providing me with the results of the differential thermal analysis experiments suggested in the discussion here. I also thank G. Hobby and G. Mamikunian of the Jet Propulsion Laboratory, California Institute of Technology, Pasadena, California, for their advice.

This work was supported by a grant from the National Aeronautics and Space Administration (*NSG* 199-62, J. E. Lovelock).

(**207**, 568-570; 1965)

J. E. Lovelock: Bowerchalke, Nr. Salisbury, Wiltshire.

References:

1. Schrödinger, E., *What is Life?* (Camb. Univ. Press, 1944).
2. Bernal, J. D., *The Physical Basis of Life* (Routledge and Kegan Paul, London, 1951).
3. Denbigh, K. G., *The Thermodynamics of the Steady State* (Methuen and Co., London, 1951).
4. Wigner, E. P., *The Logic of Personal Knowledge* (Routledge and Kegan Paul, London, 1961).
5. Landsberg, P. T., *Nature*, **203**, 928(1964).
6. Kiess, C. C., Karrer, S., and Kiess, H. K., *Publ. Astro. Soc. Pacific*, **72**, 256 (1960).
7. Lipsky, S. R., and Lovelock, J. E., *Recommendations report.* Submitted to Donald Easter, Planetary Atmospheres Section, NASA Headquarters, Washington, D.C. (April 1964).
8. Meinschein, W. G., *Space Sci. Revs.*, **2**, 665 (1963).
9. Mold, J. D., *et al.*, *Biochemistry*, **3**, 1293 (1964).

我要感谢休斯敦大学的兹拉特基斯和西蒙兹对本文中的差热分析实验给出建议、进行操作并将结果提供给我。我还要感谢来自加利福尼亚州帕萨迪纳市的加州理工学院喷气推进实验室的霍比和马米库尼安的建议。

本项工作受来自美国国家航空航天局的基金（*NSG*199-62，洛夫洛克）支持。

（沈乃澂 翻译；邓祖淦 审稿）

Spectral Data from the Cosmic X-ray Sources in Scorpius and near the Galactic Centre

R. Giacconi *et al.*

Editor's Note

Riccardo Giacconi was an Italian working at the Massachusetts Institute of Technology (MIT) when he developed the X-ray telescope—a device based on the formation of images of extraterrestrial X-ray sources involving the scattering of X-rays incident on the shaped mirror at glancing angles. Using such a telescope launched by a rocket, Giacconi discovered the existence of X-ray sources in 1962. In this paper, Giacconi and his colleagues described the discovery of further X-ray sources, many of them near the centre of our Galaxy. The culmination of this work was the launch of the Einstein satellite by the United States in 1978. Giacconi left MIT to become director of the European Southern Observatory in Chile (with an administrative and research base in Munich, Germany).

DURING two rocket flights conducted from White Sands Missile Range on August 28,1964 (flight I), and October 26, 1964 (flight II), we have obtained information regarding the spectral composition of the X-ray source in Scorpius (*ScoX*-1) and those along the Galactic equator in the vicinity of the Galactic centre. The results were obtained from two separate detectors and extend over the spectral region 1–25 keV. We find that the radiation from *ScoX*-1 extends from 15 keV to at least 2.5 keV, and that the spectral distribution is not consistent with black-body radiation. The radiation from the region along the Galactic centre extends to about 25 keV. The latter region apparently contains several distinct X-ray sources as reported by Giacconi *et al.*[1], Bowyer *et al.*[2] and Fisher *et al.*[3].

Since the discovery of cosmic X-ray sources by Giacconi *et al.*[4] in 1962, only gross information regarding the spectral composition of the radiation has been reported. In the same paper an effective wave-length near 3 Å was reported based on atmospheric attenuation and the difference in counting rate in two separate Geiger counters. Giacconi *et al.*[5] afterwards reported that the spectrum was consistent with a black-body temperature of about 10^7 °K. Bowyer *et al.*[6] attempted to measure the spectral composition of radiation from the Crab Nebula between 1 and 10 keV by making observations in two independent Geiger counters. Later, they published[2] a critical evaluation of their experiment which showed that one of the counters did not function as expected. In the same paper they presented results on the atmospheric attenuation of X-radiation from the Scorpius source which was consistent with a black-body temperature of $2–3\times10^6$ °K. In another experiment

天蝎座和银河系中心附近的宇宙 X 射线源光谱数据

贾科尼等

编者按

里卡尔多·贾科尼是一位在麻省理工学院工作的意大利人，期间，他研制了 X 射线望远镜，这是基于地球外 X 射线源在按不同掠射角入射到特定形状的镜面上时所产生的散射成像的一种装置。1962 年，用这种由火箭发射到太空的望远镜，贾科尼发现了 X 射线源。本文中，贾科尼及其同事们描述了发现更多 X 射线源的过程，其中很多射线源靠近我们银河系的中心。1978 年，美国爱因斯坦号人造卫星的发射标志着这项工作达到了顶峰。贾科尼离开麻省理工学院后，成为位于智利的欧洲南方天文台的主管（在德国慕尼黑拥有行政部门和科研基地）。

白沙导弹靶场分别于 1964 年 8 月 28 日（第一次飞行）和 1964 年 10 月 26 日（第二次飞行）进行了两次火箭飞行实验，对天蝎座（*ScoX*-1）和银道面上银河系中心附近的 X 射线源进行了探测，获得了大量光谱信息。这些结果由两个独立的探测器获取，能谱范围为 1 千电子伏 ~ 25 千电子伏。结果表明，天蝎座 *X*-1 的辐射能谱范围为 2.5 千电子伏 ~ 15 千电子伏，但其光谱分布与黑体辐射谱并不一致；来自银河系中心区域的辐射能谱延伸到大约 25 千电子伏，这一区域明显包括若干不同的 X 射线源，相关报道可以参见贾科尼等人 [1]、鲍耶等人 [2] 和费希尔等人 [3] 的研究。

自从贾科尼等人于 1962 年发现宇宙 X 射线源以来 [4]，关于该辐射谱组成的报道均较为粗略。贾科尼的小组在同一篇文章中还基于大气衰减和两个单独的盖革计数器的计数率差异而报道了有效波长位于 3 埃附近的辐射。随后，他们进一步报道 [5]，该辐射谱与温度约为 10^7 K 的黑体相符。鲍耶等人 [6] 则尝试使用两个独立盖革计数器，在 1 千电子伏 ~ 10 千电子伏范围内观察蟹状星云的辐射谱组成。不久，他们在发表的论文中 [2] 对自己的实验进行了关键性的评估，发现其中的一个计数器并没有按照预期进行工作；在同一篇文章中他们给出了天蝎座射电源 X 射线辐射的大气衰减数据，其与温度为 2×10^6 K~3×10^6 K 的黑体相符。在另一项实验中，克拉克 [7] 则使用气球搭载设备，在 20 千电子伏 ~ 60 千电子伏范围内，对蟹状星云的 X 射线

X-radiation from the Crab has been observed between 20 and 60 keV by Clark[7] using balloon-borne instrumentation.

Discussions of the source mechanism for the generation of the observed X-rays has centred on three possibilities: (1) black-body radiation from a neutron star; (2) radiation from an optically thin hot plasma; (3) synchrotron radiation from high-energy electrons in a weak magnetic field. Specific literature citations appear in refs. 5 and 7 cited here and in a review article by Giacconi and Gursky[8].

We are reporting here spectral data obtained by placing absorbers in front of the Geiger counters during flight I and by performing a pulse height analysis of the output of a sodium iodide (NaI) scintillation counter during flight II. The several sources along the Galactic equator could not be resolved with these counters and all contribute to the spectral data from that region. Other results from these rocket flights have been reported by Giacconi *et al.*[9], which demonstrated the separation between *ScoX*-1 and the sources near the Galactic centre, and by Oda *et al.*[10], which presented the results of the measurement of the angular diameter of *ScoX*-1.

The Geiger counter detector used during flight I of this experiment was a bank of twelve individual argon-filled counters with beryllium windows of 9.0 mg/cm^2 thickness and a total sensitive area of about 70 cm^2. The efficiency, calculated as the product of the absorption in the gas filling and the transmission of the window, lies above 10 percent between 17 keV and 1.2 keV and peaks at between 3 and 4 keV. This computed efficiency was checked experimentally by exposing counters to the beam from a tungsten target, windowless X-ray tube operated from 1.8 to 10 kV. The observed counting rates were in agreement over the entire range of voltage with those predicted using thick target yield curves and the computed counter efficiency.

The Geiger counters were equipped with rectangular collimators which limited the field of view to 15° full width at half maximum (FWHM) in the direction of rotation of the rocket and 20° FWHM in the direction of the rocket long axis. The counter bank was mounted so that its axis of maximum sensitivity made an angle of 80° with respect to the long axis. Two filters, *F*1 of 7.04 mg/cm^2 of beryllium and *F*2 of 1.72 mg/cm^2 of mylar, were placed in front of the detector and were sequentially removed during the flight to allow measurements for approximately equal time-intervals of the X-ray fluxes with *F*1 and *F*2 in place, *F*1 alone in place, and no filter.

The motion of the rocket during this flight (flight I) consisted of a rapid rotation with a period of 0.55 sec about the long axis plus a slow precession of the rotation axis with a period of 83 sec along a cone of a small opening angle. The X-ray source regions were observed during each rotation but, because of the low counting rates, data from successive rotations were added in order to obtain sufficient statistical precision.

辐射进行了观测。

目前，围绕着所观察到的 X 射线源的产生机制主要存在三种可能性：(1) 中子星产生的黑体辐射；(2) 光学薄热等离子体产生的辐射；(3) 弱磁场中高能电子引起的同步辐射。这方面具体的相关文献可见本文中引用的参考文献 5 和参考文献 7，以及贾科尼与古尔斯基的综述文章[8]。

在第一次飞行中，我们将吸收体放置在盖革计数器前；在第二次飞行中，我们对碘化钠闪烁计数器的输出进行了脉冲高度分析，并将在本文报道这两次探测的光谱数据。根据这些计数器并不能分辨出银道附近的各个辐射源，但是这些辐射源对该区域的光谱数据都有贡献。关于这两次火箭飞行探测，贾科尼等人[9]先前也曾报道过一些其他结果，验证了天蝎座 *X*-1 与银河系中心附近的辐射源是分离的；同时小田稔等人[10]给出了天蝎座 *X*-1 角直径的测量结果。

第一次飞行实验所使用的盖革计数器包括一组 12 个单独的充氩计数器，其窗口材料为铍，厚度为 9.0 毫克 / 厘米^2，总传感面积约为 70 厘米^2。计数器的效率用氩气的吸收率乘以窗口的透射率计算，在 1.2 千电子伏 ~17 千电子伏之间一般高于 10%，而且峰值在 3 千电子伏~4 千电子伏之间。这种计数器的理论计算效率可以通过实验进行检验，方法是将计数器暴露在钨靶的射线中，无窗 X 射线管工作电压为 1.8 千伏 ~10 千伏。在整个电压范围内，观测到的计数率与使用厚靶产额曲线和计数效率计算值所得到的预测值相符。

实验中使用的盖革计数器配备了矩形准直器，从而可以限制其视场，在火箭的旋转方向上，半高全宽为 15°；在火箭的长轴方向上，半高全宽为 20°。氩气计数器采用最大灵敏度方向与火箭长轴方向呈 80° 夹角的方式安装。探测器前面安装有两个滤光片 *F*1 和 *F*2，*F*1 的窗口材料为 7.04 毫克 / 厘米^2 厚的铍，*F*2 的窗口材料为 1.72 毫克 / 厘米^2 厚的聚酯薄膜。在飞行过程中，两滤光片先后被去掉，这样就可以分别测量同时存在 *F*1 和 *F*2、只有 *F*1 以及没有滤光片三种条件下的 X 射线流量，三次测量的持续时间间隔大体相同。

在第一次飞行中，火箭的运动包括周期为 0.55 秒绕长轴的快速转动，以及周期为 83 秒的沿小孔径角圆锥面的缓慢进动。在每次转动过程中，X 射线源区域都会被观测一次；但是，由于计数率较低，所以我们将连续转动获取的数据叠加在一起，以便获得足够的统计精确度。

During each rotation the detector swept out a band on the celestial sphere the width of which, 20°, equalled that of the collimator. The precession results in an approximately sinusoidal variation of the angle of closest approach of a given celestial object to the centre of this band. During each precession cycle the centre of the band came within 2° of *ScoX*-1 and moved along the Galactic equator between $l_{II} = 2°$ to $l_{II} = 27°$.

Three full precession cycles were recorded during the rocket flight, and since a filter change occurred almost in phase with the precession we have essentially a complete precession cycle of data for each of the three filter conditions. The data from the precession cycle with no filter in place have been the subject of a previous paper[9]. The counting rates with background subtracted observed with the three filter conditions when the detector crossed *ScoX*-1 and crossed the region along the Galactic equator are listed in Table 1. The counting rates represent an average over the same fraction of the precession cycle for the three cases. Listed also is the attenuation, defined as the ratio of the counting rates for the two filter combinations compared with no filter.

Table 1. Counting Rate observed in the Two-source Regions with and without Filters

Filter condition	Source region	Counting rate measured c.p.m	Attenuation relative to no filter
No filter	*ScoX*-1	620±20	1
*F*1	*ScoX*-1	440±18	0.71±0.04
*F*1+*F*2	*ScoX*-1	350±17	0.56±0.03
No filter	Gal cen	350±18	1
*F*1	Gal cen	240±18	0.70±0.06
*F*1+*F*2	Gal cen	210±15	0.62±0.05
Typical background rate		30 c.p.m	

The significance of the attenuation factor can be partially understood from the following arguments. Photons at 2.5 keV will give the observed attenuation of about 0.7 per filter. In order to obtain the same attenuation factor in the case of a distribution of energies, the number of lower energy photons present, which are absorbed more strongly, must be compensated for by high-energy photons which are absorbed less strongly. Thus, there must be comparable numbers of photons above and below 2.5 keV. The magnitude of the attenuation depends on the true spectral distribution and can be calculated from the relation:

$$\text{Attenuation} = \int \varphi(\lambda) e^{-\mu(\lambda)x} \varepsilon(\lambda) d\lambda \Big/ \int \varphi(\lambda) \varepsilon(\lambda) d\lambda \tag{1}$$

where $\varphi(\lambda)$ is an assumed spectrum in terms of number of photons per unit wave-length interval, $\varepsilon(\lambda)$ is the counter efficiency, $\mu(\lambda)$ is the linear absorption coefficient of the filter and x is the thickness of the filter. We performed this calculation for three assumed spectra, namely, (1) a power law spectrum of the form:

$$\varphi(\lambda) = A\lambda^{-(\alpha+1)} \tag{2}$$

对于每一次转动，探测器在天球上都扫过 20° 的区域，这与矩形准直器角宽度一致。指定天体与扫描带中心最接近点的角度近似成正弦变化，这是由进动引起的。在每个进动周期内，扫描带中心都会进入与天蝎座 *X*-1 呈 2° 的范围内，同时在 $l_{II}=2°$ 到 $l_{II}=27°$ 之间沿银道移动。

本次火箭飞行中，我们记录了三个完整的进动周期，并且由于滤光片的转换几乎与进动同相，所以基本上对于三种滤光片情况下的每一种我们都得到了完整的进动周期数据。对于没有滤光片的情况，我们在先前的一篇文章中已经分析过这方面的进动周期数据[9]。当探测器扫过天蝎座 *X*-1 和银道区域，在以上三种滤光片情况下观测到的计数率参见表 1，该计数率已经减去了背景计数率。这里的计数率是指在三种情况下进动周期某一小段相同时间内的平均值。表 1 也给出了滤光片的衰减，定义为两滤光片条件下的计数率与无滤光片条件下的计数率的比值。

表 1. 不同滤光片条件下两个辐射源区域的计数率

滤光片条件	辐射源区域	测量计数率（每分钟计数）	相对没有滤光片的衰减
无滤光片	天蝎座 *X*-1	620 ± 20	1
*F*1	天蝎座 *X*-1	440 ± 18	0.71 ± 0.04
*F*1+*F*2	天蝎座 *X*-1	350 ± 17	0.56 ± 0.03
无滤光片	银河系中心	350 ± 18	1
*F*1	银河系中心	240 ± 18	0.70 ± 0.06
*F*1+*F*2	银河系中心	210 ± 15	0.62 ± 0.05
典型背景计数率	30（每分钟计数）		

衰减因子的重要性可以由以下分析看出。能量为 2.5 千电子伏的光子穿过每个滤光片的衰减大约为 0.7。滤光片对低能光子的吸收较强，对高能光子的吸收较弱，因此就能量分布而言，为了获得相同的衰减因子，需要对前者进行补偿。所以，在高于和低于 2.5 千电子伏的范围内，必然存在数目相近的光子数目。衰减的具体数值取决于实际的光谱分布，并可以通过以下关系计算：

$$衰减=\int \varphi(\lambda)e^{-\mu(\lambda)x}\varepsilon(\lambda)d\lambda \Big/ \int \varphi(\lambda)\varepsilon(\lambda)d\lambda \quad (1)$$

其中 $\varphi(\lambda)$ 是给定的光谱分布，为单位波长间隔内的光子数，$\varepsilon(\lambda)$ 为计数器效率，$\mu(\lambda)$ 为滤光片的线性吸收系数，x 为滤光片的厚度。下面对以下三种给定的光谱分布进行计算。也就是，（1）幂律谱形式：

$$\varphi(\lambda)=A\lambda^{-(\alpha+1)} \quad (2)$$

Equation (2) is equivalent to a distribution of the form $\nu^{-\alpha}$ (ν = frequency) when the spectrum is expressed in terms of power per unit frequency interval.

(2) An exponential spectrum of the form:

$$\varphi(\lambda) = (A/\lambda)\exp(-hc/\lambda kT) \tag{3}$$

Except for the weak energy dependence of the Gaunt factor this spectral distribution describes free-free emission by electrons having a Maxwellian distribution of energy.

(3) A thermal spectrum of the form:

$$\varphi(\lambda) = A\lambda^{-4}[\exp(hc/\lambda kT)-1] \tag{4}$$

The quantity A that appears in these three distribution laws is a constant that can be determined from the relation:

$$A = N/\int\varphi(\lambda)\varepsilon(\lambda)d\lambda \tag{5}$$

where N is the observed counting rate.

For each of the distribution laws and for each filter condition, equation (1) was evaluated by numerical integration for a series of values of α in the case of the power law spectrum and of T in the case of exponential and thermal spectra. The observed attenuations for *ScoX*-1 are obtained with $\alpha = -1.1\pm0.3$ for a power law spectrum, with $T = (3.8\pm1.8)\times10^7$ °K for an exponential spectrum, and $T = (9.1\pm0.9)\times10^6$ °K for a blackbody spectrum. The listed uncertainties result from the statistical fluctuations expected from the total accumulated counts. Within the precision of the measurement, the spectrum from the two-source regions is the same over the interval of wave-lengths to which the Geiger counter is sensitive. Furthermore, it is not possible to decide between the three spectral types from these data alone.

Additional spectral results come from the observation during flight II of the same two-source regions by an NaI(T1) crystal with area of 38.5 cm^2 and a thickness of 1 mm covered by 6.9 mg/cm^2 of aluminium. The crystal was viewed by a 7188 CBS photomultiplier and the combination had a measured resolution of 48 percent FWHM at 22 keV. The detector field of view was limited to 5°×25° FWHM and the detector was mounted with its axis at 60° from the long axis of the rocket. During the flight this detector scanned substantially the same source regions in the sky as did the Geiger counters in flight I, and the rocket motion was comparable with that of flight I.

事实上，如果光谱分布表达为每单位频率间隔功率的形式，式(2)就等价于 $\nu^{-\alpha}$（ν 表示频率）分布。

（2）指数谱形式：

$$\varphi(\lambda) = (A/\lambda)\exp(-hc/\lambda kT) \tag{3}$$

除了冈特因子的弱能量依存性之外，这种光谱分布描述了具有麦克斯韦能量分布的电子的自由–自由发射。

（3）热辐射谱形式：

$$\varphi(\lambda) = A\lambda^{-4}[\exp(hc/\lambda kT)-1] \tag{4}$$

以上三种分布律中的系数 A 为一常数，其大小可以由下式决定：

$$A = N/\int\varphi(\lambda)\varepsilon(\lambda)\mathrm{d}\lambda \tag{5}$$

其中 N 为观测到的计数率。

对于每一种分布律和每一种滤光片条件，在幂律谱情况下取不同 α 值，以及在指数谱和热辐射谱情况下取不同 T 值，从而对式(1)中进行数值积分求解。观测到的天蝎座 X-1 的衰减，对于幂律谱，对应的 $\alpha = -1.1 \pm 0.3$；对于指数谱，对应的温度 $T =（3.8 \pm 1.8）\times 10^7$ K；对于黑体辐射谱，对应的温度为 $T =（9.1 \pm 0.9）\times 10^6$ K。以上给出的不确定度是由累积计数中的统计涨落造成的。在测量精度范围内，两个X 射线源区域的辐射谱在盖革计数器敏感的波长区间上是相同的。此外，仅仅通过这些数据不太可能决定光谱分布的具体形式。

另外，在第二次飞行实验中，我们也对以上两个辐射源区域进行了观测并获取了光谱数据。这次探测中的探测器使用了面积为 38.5 厘米^2 的碘化钠（Tl）晶体，其厚度为 1 毫米，镀有 6.9 毫克 / 厘米^2 的铝。该晶体由一个 7188 CBS 光电倍增管观测，该组合的分辨率在 22 千电子伏处为半高全宽的 48%。探测器的半高全宽视场为 5°×25°，探测器的固定轴与火箭长轴方向夹角为 60°。在飞行过程中，探测器基本上是持续地扫描天空中相同的辐射源区域，这和第一次飞行中的盖革计数器扫描的类似，而且火箭的运动也和第一次飞行类似。

The results from this detector consisted of those photomultiplier pulses above a threshold equivalent to about 8 keV. These were stretched to 1 ms with their pulse height preserved and telemetered in real time from the rocket. Pulses from successive spins of the rocket were summed as a function of both pulse height and rocket azimuth. The azimuthal distribution of all telemetered pulses is shown in Fig. 1. Sources are apparent at azimuths of 2° and 26° which correspond respectively to traversal of the regions containing *ScoX*-1 and the Galactic equator. We have assumed that the same X-ray sources are responsible for the radiation detected in the NaI and Geiger counter detection systems.

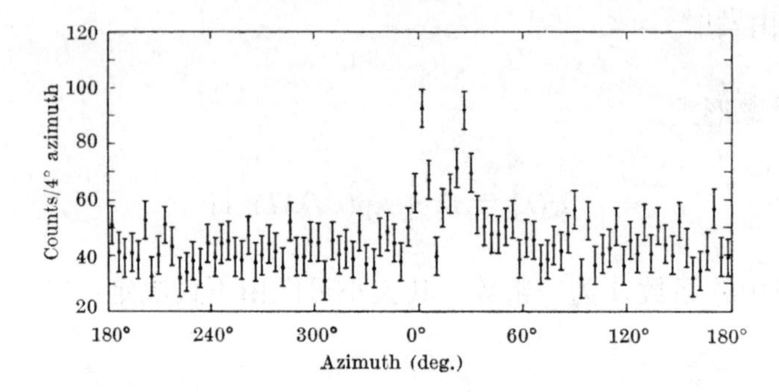

Fig. 1. Azimuthal distribution of counts observed in NaI detector

The differential pulse height distributions of the data within the source regions are shown with background subtracted in Fig. 2, as well as the background spectrum itself. The data below 10 keV show the cut-off resulting from the electronic threshold. Above 10 keV the distribution of pulse heights observed in the two-source region is markedly different from the background distribution. The distribution observed from *ScoX*-1 is consistent with a cut-off in the spectrum at about 15 keV, whereas the radiation from the Galactic equator region extends to higher energies and there is evidence for a possible peak at 20 keV from the same region. The observation of this peak must be regarded as only tentative because of the low statistical precision of the results; however, this feature cannot be an instrumental effect since it does not appear in either of the other two spectra.

The ambiguity in the choice of spectral distribution law that arises from the Geiger counter data alone can be partially resolved by considering the NaI data as well. The counting rates observed for the two-source regions by the two detectors are listed in Table 2. The ratio of counting rates for *ScoX*-1 is (14±3):1, which can be obtained for either a power law or exponential spectral distribution with the parameters determined from the Geiger counter data. To fit a black-body distribution, however, requires a temperature of $(17\pm1)\times10^{6}$ °K which cannot be reconciled with the Geiger counter data. The black-body temperature consistent with the Geiger counter data predicts a ratio of about 200:1. This analysis applied to the ratio of counting rates observed in the Galactic counter region yields the same general results; namely, that the ratio can be fitted with either an exponential or power law spectrum but that a black-body spectrum yields an insignificant

探测器的数据与那些能量强度均大于阈值约为 8 千电子伏的光电倍增管脉冲相等。在 1 毫秒的时间长度上保持这些脉冲的高度不变，并实时地从火箭传回地面。然后将火箭连续旋转过程中探测到的脉冲进行加和，最后结果可以作为脉冲高度和火箭方位角的函数。图 1 给出了所有遥测脉冲的方位角分布。显然辐射源位于 2° 和 26° 方位角，分别对应天蝎座 X-1 和银道区域。事实上，在以上分析中，我们已经假设碘化钠和盖革计数器探测系统探测到的辐射来自相同的 X 射线源。

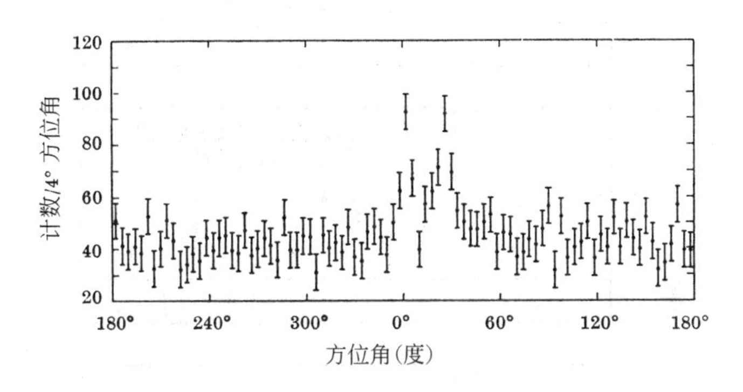

图 1. 碘化钠探测器计数的方位角分布

图 2 给出了辐射源区域内不同的脉冲高度分布和背景辐射谱，前者已经扣除了背景计数。由图可知，10 千电子伏之下的数据存在能谱的截止现象，该现象是由电子的阈值引起的。在高于 10 千电子伏范围内，观测到的两个辐射源的脉冲高度分布与背景分布非常不同。所观测到的天蝎座 X-1 的分布与在 15 千电子伏左右存在截止的光谱一致，而银道区域的辐射则延伸到更高的能量范围，但同样在该区域 20 千电子伏附近可能存在另一个峰值。由于结果统计精度较低，20 千电子伏附近的峰尚属于推测；但该峰不可能是仪器引起的，因为它并没有在另外两个能谱中出现。

仅使用盖革计数器数据而难以确定采用何种光谱分布律的问题，可以通过同时参考碘化钠数据得到部分解决。两探测器对两个辐射源观测得到的计数率如表 2 所示。天蝎座 X-1 的计数率的比值为（14 ± 3）:1，该结果由幂律或指数谱分布得到，两种分布的具体参数可以由盖革计数器数据确定。为了拟合黑体辐射分布，需要满足特征温度为（17 ± 1）$\times 10^6$ K，然而，该值与盖革计数器数据不符。另一方面，与盖革计数器数据相符的黑体温度给出的计数率比值约为 200:1。将以上分析应用于银河系区域计数率比值也会得到相似的结果，即指数谱或幂律谱均可以拟合计数率比值，但是黑体辐射谱给出的碘化钠探测器流量与实际观测值相比几乎可以忽略。

flux in the NaI detector compared with what is observed. It is possible that a series of black-body sources gives rise to the observed radiation which, however, implies that one of the sources has a temperature of several times 10^7 °K.

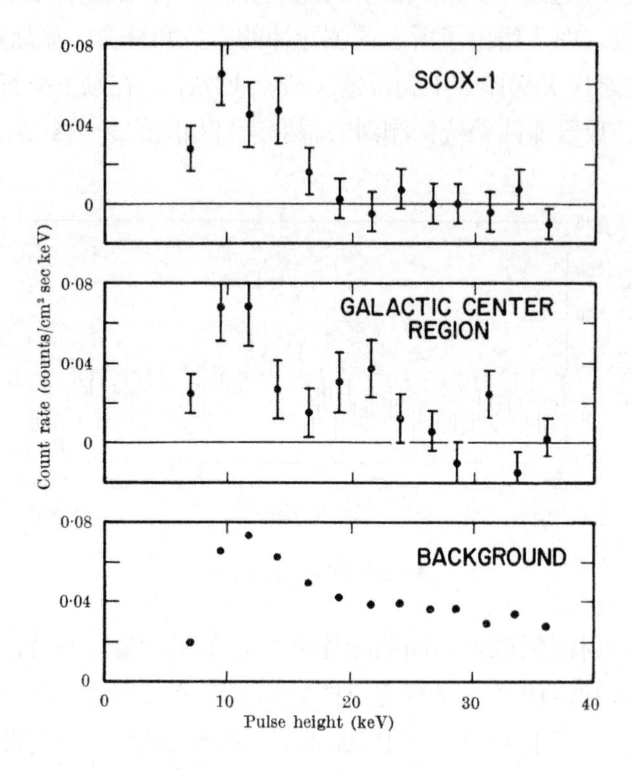

Fig. 2. Pulse height distribution of counting rates observed in NaI detector

Table 2. Counting Rates and Absolute Fluxes Observed for the Two–source Regions

	Counting rate counts/cm^2 · sec	Integrated power ergs/cm^2 · sec	Integrated flux photons/cm^2 · sec
ScoX-1			
1–10 keV	16.8±0.6	(1.61±0.4)×10^{-7}	32±6
> 8 keV	1.2±0.3	(3.3±0.8)×10^{-8}	1.8±0.4
Galactic equator			
1–10 keV	4.3±0.2	(0.4±0.1)×10^{-7}	8±2
> 8 keV	0.67±0.15	(1.9±0.4)×10^{-8}	0.8±0.2

Numbers listed for 1–10 keV are derived from Geiger counters and those listed for >8 keV are derived from the NaI detector. The listed errors are based on the count statistics and in the case of the integrated power and flux in the 1–10 keV range include the effect of the uncertainty of the spectral index

In the case of *ScoX*-1, if one wishes to fit a thermal spectrum of the order of 10^7 °K as is consistent with the Geiger counter data alone plus a high-energy tail to account for the NaI results, the power in the tail must be the order of 25 percent of that in the thermal portion of the spectrum and to fit a lower temperature requires an even more substantial non-thermal tail. Thus, if a neutron star is postulated as the X-ray source, a large fraction of the observed flux cannot arise from the black-body radiation from that object.

原因可能是，若干黑体辐射源都对观测值有贡献，但是其中一个具有数倍于 10^7 K 的温度。

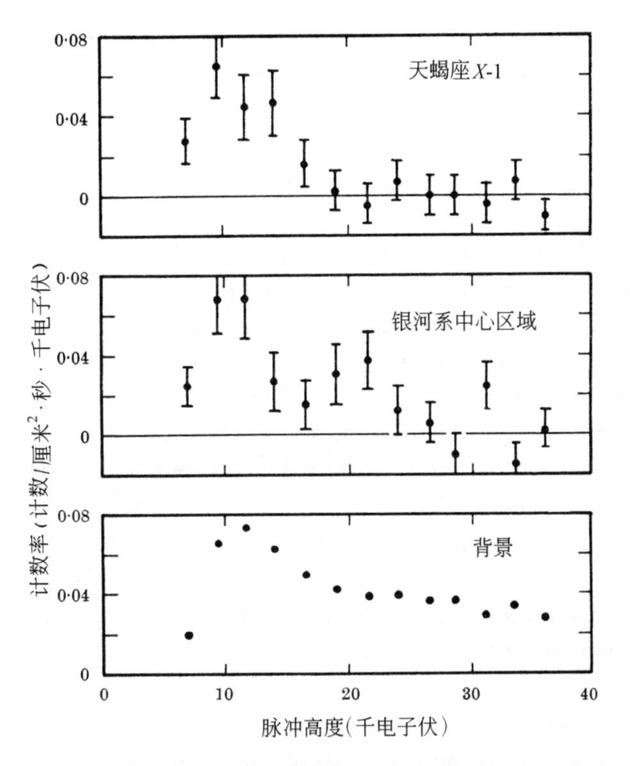

图 2. 碘化钠探测器观测到的计数率的脉冲高度分布

表 2. 对于两个辐射源区域观测得到计数率和绝对流量

	计数率 个 / 厘米²· 秒	积分功率 尔格 / 厘米²· 秒	积分流量 个光子 / 厘米²· 秒
天蝎座 *X*-1			
1 千电子伏 ~10 千电子伏	16.8 ± 0.6	$(1.61 \pm 0.4) \times 10^{-7}$	32 ± 6
> 8 千电子伏	1.2 ± 0.3	$(3.3 \pm 0.8) \times 10^{-8}$	1.8 ± 0.4
银道			
1 千电子伏 ~10 千电子伏	4.3 ± 0.2	$(0.4 \pm 0.1) \times 10^{-7}$	8 ± 2
> 8 千电子伏	0.67 ± 0.15	$(1.9 \pm 0.4) \times 10^{-8}$	0.8 ± 0.2

表中给出的1千电子伏~10千电子伏能量范围内的数据由盖革计数器获得，大于8千电子伏的数据来自碘化钠探测器。表中给出的误差来源于计数统计；对于积分功率和流量两栏，1 千电子伏 ~10 千电子伏范围内的误差也考虑了光谱指数的不确定度。

对于天蝎座 *X*-1，如果希望拟合量级为 10^7 K 的热辐射谱，使其与盖革计数器数据一致，并符合碘化钠探测结果中的高能拖尾，那么拖尾部分的能量必须大约是光谱热辐射部分的 25%。如果希望以较低的温度拟合，则要求存在一个更大的非热辐射拖尾。因此，如果假定 X 射线源是中子星的话，那么观测流量中的大部分不可能来自中子星的黑体辐射。

More generally the absence of a substantial thermal contribution to the spectrum indicates that the observed X-rays are being generated in regions that are optically thin. Whether the X-rays are generated in a hot plasma or by the synchrotron process cannot be determined on the basis of these data except for noting that, in a hot plasma line, emission and edge discontinuities can arise, either of which could account for the peak at 20 keV for which there is some evidence from the NaI detector. The synchrotron process cannot give rise to such features.

Table 2 lists the counting rate, the integrated power and photon flux observed in the two-source regions by the two detectors. For *ScoX*-1 the counting rates are corrected for the shadowing of the collimators and result from the fact that the detector axis comes only within 2° of the source. No such correction was made in the case of the sources near the Galactic centre. In the 1–10 keV region the integrated power and photon flux were obtained by assuming a power law distribution with the spectral index obtained from the filter attenuation of the Geiger counter results. The uncertainty in the spectral index yields the uncertainties in the integral quantity. In the >8 keV region the integrated power and photon fluxes were obtained by summing over the NaI pulse-height distribution.

In summary, the results presented here indicate the following:

(1) The ratio of the counting rate in the Geiger counters to that in the NaI detector is not consistent with a single thermal spectrum.

(2) The NaI detector does not detect significant radiation from *ScoX*-1 beyond about 15 keV, whereas radiation is observed from the Galactic centre region to beyond 20 keV. The minimum detectable flux density at 20 keV (defined as equivalent to 1σ above background) is about 3×10^{-27} ergs/cm^2·sec(c/s). Clark[7] measures a flux of $(2.4\pm0.6)\times10^{-27}$ ergs/cm^2·sec (c/s) over the range of 20–40 keV from the Crab Nebula. The power from *ScoX*-1 at these high energies is thus considerably below what is emanating from the Crab even though at lower energies (1–10 keV) *ScoX*-1 is the brighter object as reported by Friedman[2].

(3) For both source regions the attenuation of the counting rate in the Geiger counters by the filters indicates that the X-ray flux below 2.5 keV is comparable with that above 2.5 keV. This result does not agree with that reported by Fisher[3] that little or no radiation is present below 4 keV from these sources.

The work reported here was sponsored by the U.S. National Aeronautics and Space Administration, Office of Space Sciences, under contract *NASw*-898.

(**207**, 572-575; 1965)

一般而言，辐射谱中缺乏大量的热辐射表明观测到的 X 射线源是产生自光学薄区域的。并不能仅仅根据以上数据来判断 X 射线源究竟是产生于热等离子体还是同步辐射过程，我们需要注意到：热等离子体谱线存在的发射和边缘非连续性均可能引起 20 千电子伏峰（碘化钠探测器显示出的一些证据）的形成，而同步辐射过程并不能解释这一峰值。

表 2 给出了两探测器探测到的两个辐射源区域的计数率、积分功率和光子通量。对于天蝎座 *X*-1，根据计数率仅来自探测器轴与辐射源呈 2° 夹角的范围内，对准直器的遮蔽效应进行了校准。但对于银河系中心附近的辐射源并没有进行这样的校准。通过假定幂律分布，可以求得 1 千电子伏 ~10 千电子伏范围内的积分功率与光子通量，其中光谱指数可以由盖革计数器的滤光片衰减推出。但是，光谱指数的不确定性会引起积分数值的不确定性。在大于 8 千电子伏的范围内，通过在碘化钠脉冲高度分布上进行加和，可以求得积分功率和光子通量。

综上所述，通过本文给出的结果可以得出以下推论：

（1）盖革计数器与碘化钠探测器二者计数率的比值与单一热辐射谱并不符合。

（2）对于天蝎座 *X*-1，碘化钠探测器并没有探测到超过 15 千电子伏的显著辐射；而观测到的银河系中心区域的辐射延伸范围超过 20 千电子伏。位于 20 千电子伏的最小可探测流量密度（其值定义为比背景高 1σ）大约为 3×10^{-27} 尔格 / 厘米^2·秒（周 / 秒）。克拉克[7]对于蟹状星云在 20 千电子伏 ~40 千电子伏范围内测量，得到的流量为 $(2.4\pm0.6)\times10^{-27}$ 尔格 / 厘米^2 · 秒（周 / 秒）。在这么高的能量范围内，天蝎座 *X*-1 的辐射功率明显低于蟹状星云的辐射功率。但是，弗里德曼[2]曾报道过天蝎座 *X*-1 在较低能谱范围内（1 千电子伏 ~10 千电子伏）是相对较亮的天体。

（3）对于本文中的两个辐射区域，由滤光片引起的盖革计数器计数率的衰减表明，低于 2.5 千电子伏的 X 射线流量与高于 2.5 千电子伏的 X 射线流量相当。这一结果与费希尔[3]的报道并不一致，后者认为这些辐射源在低于 4 千电子伏的区域几乎没有辐射。

本研究得到了美国国家航空航天局空间科学办公室的资助，项目合同号为 *NASw*-898。

（金世超 翻译；蒋世仰 审稿）

R. Giacconi, H. Gursky and J. R. Waters: American Science and Engineering, Inc., Cambridge, Massachusetts.

References:

1. Clark, G., Garmire, G., Oda, M., Wada, M., Giacconi, R., Gursky, H., and Waters, J., *Nature* (this issue, p.584). Reported also by Giacconi, E., *et al.*, at Sec. Conf. Relativistic Astrophysics, Austin, Texas (December 1964).
2. Bowyer, S., Byram, E. T., Chubb, T. A., and Friedman, H., *Science*, 147, 394 (1965). Reported also by Friedman, H., at Sec. Conf. Relativistic Astrophysics, Austin, Texas (December 1964).
3. Fisher, P. C., Johnson, H. M., Jordan, W. C., Meyerott, A. J., and Acton, L. W. (submitted to *Astrophys. J.*).
4. Giacconi, R., Gursky, H., Paolini, F. R., and Rossi, B., *Phys. Rev. Letters*, **9**, 439 (1962).
5. Giacconi, R., Gursky, H., Paolini, F. R., and Rossi, B., *Proc. Fifth Intern. Space Sci. Symp.* (to be published).
6. Bowyer, S., Byram, E. T., Chubb, T. A., and Friedman, H., *Science*, **146**, 912 (1964).
7. Clark, G., *Phys. Rev. Letters*, 14, 91 (1965).
8. Giacconi, R., and Gursky, H., *Space Sci. Rev.*, **4**, 151 (1965).
9. Giacconi, R., Gursky, H., Waters, J., Clark, G., and Rossi, B., *Nature*, **204**, 981 (1964).
10. Oda, M., Clark, G., Garmire, G., Wada, M., Giacconi, R., Gursky, H., and Waters, J., *Nature*, **205**, 554 (1965).

Positions of Three Cosmic X–ray Sources in Scorpio and Sagittarius

G. Clark *et al.*

Editor's Note

The 1960s saw the opening up of a new part of the electromagnetic spectrum in which signals from star-like objects could be detected. Riccardo Giacconi, an Italian national working at the Massachusetts Institute of Technology, played a crucial part in the development of what is now called X-ray astronomy by developing the mirrors by means of which X-rays can be focused as if they were radiation from optically visible stars. (The technique is to reflect the X-rays from glass or metal surfaces at very small angles.) In this paper, Giacconi and his colleagues described the detection of three previously unknown X-rays stars based on evidence collected during a rocket flight from White Sands in New Mexico. In the remainder of the decade, Giacconi and his colleagues built satellites able to monitor the whole sky for X-ray sources. It is interesting that two of Giacconi's collaborators in this study, M. Oda and M. Wada, both returned to Japan to important positions in Japanese science—Oda as director of the national space research enterprise and Wada as a molecular biologist using techniques borrowed from physics.

WE have determined the positions of three cosmic X-ray sources in the constellations of Scorpio and Sagittarius within uncertainty areas of 1.2–3 square degrees. Positions for these sources were reported earlier by us at the Austin Conference on Relativistic Astrophysics on the basis of a preliminary analysis of the same data[1]. One of these sources is *ScoX*-1, which was first detected by Giacconi *et al.* in 1962 (ref. 2). (Our designations specify the constellation and the order of discovery, for example, *ScoX*-2 for second X-ray source in Scorpio.) The other two, *SgrX*-1 and *ScoX*-2, lie near the galactic plane in the complex of sources the existence of which was reported in a previous publication[3]. We have also located a probable fourth source to within 1° of a segment of a great circle that is almost parallel to and about 4° below the galactic equator. The Kepler supernova remnant *SN*1604 was scanned and no evidence of an X-ray source at its position was found.

Table 1. Positions (Epoch 1950.0) of Three X-ray Sources

Source	Right ascension		Declination	
ScoX-1	16h 12m		−15.6°	
	or 16h 19m		−14.0°	
ScoX-2	16h 50m	(16h 50m)	−39.6°	(−41°)
SgrX-1	17h 44m	(17h 37m)	−23.2°	(−24°)

The preliminary values of the position co-ordinates given earlier in ref. 1 are shown in parentheses

We obtained the data with a rocket that was launched from White Sands, New Mexico,

天蝎座和人马座中三个 X 射线源的位置

克拉克等

编者按

从 20 世纪 60 年代开始，电磁波谱增加了一个新的作用即能检测到类星体的信号。在现今称为 X 射线天文学的开创中，在麻省理工学院工作的意大利人里卡尔多·贾科尼扮演了重要角色，他开发了可聚焦 X 射线的镜面，就好像这些 X 射线来自光学可见星体的辐射一样。（这项技术是在玻璃或者金属表面，以非常小的角度反射 X 射线。）在本文中，贾科尼和他的同事们基于在新墨西哥州的白沙导弹靶场发射的火箭在飞行过程中收集的资料，描述了探测到的三颗以前未知的 X 射线星体。在这十年贾科尼和他的同事们还建造了能够监测整个天空 X 射线源的卫星。有趣的是，在这次研究中贾科尼的两位合作者小田稔和和田后来都回到了日本，并在日本科学界担任重要职务——小田稔担任国家空间研究院的主任，和田成为了一名借鉴物理学技术做研究的分子生物学家。

我们已经确定了位于天蝎座和人马座中的三个宇宙 X 射线源的位置，面积不确定度为 1.2 平方度 ~ 3 平方度范围。早前在奥斯汀的相对论天体物理学会议上，我们也曾基于相同数据的初步分析报道过这些射线源的位置 [1]。其中一个射线源是天蝎座 *X*-1（*ScoX*-1），由贾科尼等人于 1962 年首先探测到（参考文献 2）。（我们主要根据星座和发现顺序对射线源进行命名，比如，天蝎座 *X*-2（*ScoX*-2）表示在天蝎座中发现的第二个 X 射线源。）另外两个射线源人马座 *X*-1（*SgrX*-1）和天蝎座 *X*-2 位于银道面附近，先前的一篇论文曾报道过该区域具有一系列的辐射源 [3]。同时我们也找出了可能的第四个 X 射线源，它位于近似平行于银道但比银道低约 4° 的大圆中的一段（1° 之内）。此外，对开普勒超新星遗迹 *SN*1604 也进行了扫描，但没有发现存在 X 射线源的证据。

表 1. 三个 X 射线源的位置（历年 1950.0）

辐射源	赤经	赤纬
天蝎座 *X*-1	16h 12m 或者 16h 19m	–15.6° –14.0°
天蝎座 *X*-2	16h 50m (16h 50m)	–39.6° (–41°)
人马座 *X*-1	17h 44m (17h 37m)	–23.2° (–24°)

先前参考文献 1 给出的位置坐标的初步值标示在括号里。

1964 年 10 月 26 日在恒星时 20h 20m 的时候，一枚用于探测 X 射线源的火箭

on October 26, 1964, at a sidereal time of 20h 20m, and reached an apogee of 224 km. Above the atmosphere the rocket spun around its long axis with an almost constant angular velocity of approximately 806 deg/sec. The spin axis precessed with an angular velocity of 5.51 deg/sec around a circular cone the apex angle and spatial orientation of which did not vary by more than 0.3° during the 260 sec when useful X-ray data were obtained.

The effects of the rocket motion on the observations are most easily visualized in the rocket frame of reference which we define to be a co-ordinate system fixed in the rocket with its z-axis parallel to the spin axis. During each spin any given point on the celestial sphere moved through 360° of azimuth in the rocket frame at an almost constant angle of elevation with respect to the rocket equator. In addition, the precession caused the elevation angle of an object to vary periodically with an amplitude equal to one-half of the apex angle of the precession cone. Thus an object was observed repeatedly in successive spins whenever it moved within the elevation range of a given detector's field of view.

In describing our analysis, we shall refer to the precession frame of reference which we define to be a co-ordinate system the z-axis of which is parallel to the axis of the precession cone, and the x-axis of which is in the direction of the vector product $\hat{z}_C \times \hat{z}_P$, where $\hat{z}_C$ and $\hat{z}_P$ are unit vectors parallel to the z-axes of the celestial and precession co-ordinate systems, respectively. We define the spin azimuth of a direction to be its azimuthal angle measured around the spin axis from the direction of $\hat{z}_P \times \hat{z}_S$, where $\hat{z}_S$ is a unit vector in the direction of the spin axis. We define the bearing of a direction to be the sum of its spin azimuth and the precession azimuth of the spin axis. During one complete precession cycle the spin azimuth of a given direction fixed in space decreases non-uniformly by 360°. In the same period the precession azimuth of the spin axis increases uniformly by 360°. Therefore, the bearing of a fixed direction varies about a mean value which, as one can easily see, differs by 90° from the azimuth of the fixed direction in the precession frame. The amplitude of this variation is determined by the angle of the precession cone and the elevation of the direction with respect to the precession equator.

The rocket carried a variety of instruments, among which were several X-ray detectors and two star sensors. The data consisted essentially of the occurrence times of pulses from these detectors and sensors. Four of the X-ray detectors were banks of Geiger tubes, each having sensitive areas of 70 cm^2, beryllium windows 9 mg cm^{-2} thick, and argon fillings giving a gas thickness of 5.4 mg cm^{-2}. Three of the banks had slat collimators that gave rectangular fields of view with full widths of 3.0° in the narrow dimension, and 30° (detectors *GH*0 and *GH*20) or 40° (detector *GV*10) in the wide dimension. The narrow dimensions of detectors *GH*0 and *GH*20 were perpendicular to the rocket equator, while that of detector *GV*10 was parallel. The fourth bank, detector *GMC*0, had a modulation collimator[4] the response function of which, in the direction perpendicular to the rocket equator, was a saw-tooth function with maxima at rocket elevation angles given by the formula $1.4° \pm \arctan(0.0417n)$, where n is an integer. In the direction parallel to the rocket equator the

在新墨西哥州的白沙导弹靶场发射升空，火箭到达 224 千米的远地点，我们由此火箭获得了数据。穿过大气层之后，火箭以大约 806 度 / 秒的固定角速度围绕其长轴旋转，同时该旋转轴以 5.51 度 / 秒的角速度围绕圆锥进动；其中在采集 X 射线有用数据的 260 秒内，该圆锥顶角和空间方向变化不超过 0.3°。

我们将火箭参考系定义为固定于火箭之上且 z 轴平行于火箭旋转轴的坐标系；那么，火箭运动对于观测的影响就可以很容易在该参考系中观测到。对于每一次旋转，天球上的任意给定点在火箭坐标系中都会移动 360° 方位角，但该点与火箭赤道面的仰角几乎不变。此外，火箭的进动可以使得天体的仰角周期性变化，其幅度等于进动圆锥顶角的一半。因此，在连续旋转过程中，无论它何时进入探测器视场的仰角范围内，都可以重复的观测到同一天体。

为了便于分析，下面我们定义进动参考系，该坐标系的 z 轴平行于进动圆锥的轴，x 轴方向与 $\hat{z}_C \times \hat{z}_P$ 矢量积的方向一致，其中 $\hat{z}_C$、$\hat{z}_P$ 分别表示平行于天球坐标系的 z 轴和进动坐标系 z 轴的单位矢量。我们将自旋方位定义为自旋轴相对 $\hat{z}_P \times \hat{z}_S$ 矢量积方向的方位角，其中 $\hat{z}_S$ 代表自旋轴方向的单位矢量。我们将特定的方位定义为自旋方位与自旋轴进动方位之和。在一个完整的进动周期内，在给定的空间方向的固定的自旋方位将非均匀的减小 360°；同样在这个周期中，自旋轴的进动方位将均匀增加 360°。因此，特定的方位会围绕某一平均值变化，显然，该平均值与进动参考系中固定方向的方位相差 90°。围绕平均值变化的幅度取决于进动圆锥的角度和该方向相对进动赤道面的仰角。

在这次飞行试验中，火箭携带了各种仪器，其中包括若干 X 射线探测器与两个星敏感器。需要采集的数据主要包括这些探测器和敏感器记录下的脉冲发生次数。四个 X 射线探测器各由一组盖革管组成，每个管子的灵敏面积为 70 厘米^2，铍窗的厚度为 9 毫克 · 厘米$^{-2}$，氩气的厚度为 5.4 毫克 · 厘米$^{-2}$。其中三组盖革管含有狭缝准直器，准直器矩形视场的窄方向全宽为 3.0°，宽方向上全宽为 30°（探测器 *GH*0 和 *GH*20）或 40°（探测器 *GV*10）。探测器 *GH*0 与 *GH*20 的窄方向是与火箭赤道平面相垂直的，而探测器 *GV*10 的窄方向平行于该平面。第四组盖革管，即探测器 *GMC*0，包括一个调制式准直器 [4]，其响应函数在垂直于火箭赤道平面的方向上是锯齿函数；当火箭仰角为 1.4° ± arctan(0.0417n)，n 为整数，该函数取值最大。在平

response function of detector *GMC*0 was triangular with a base width of 15°. The star sensors, *SS*0 and *SS*30, had effective fields of view of 1.0°×5.2°, with their long dimensions perpendicular to the rocket equator. In our detector designations the number following the letter code specifies the nominal elevation in degrees of the centre of its field of view above the rocket equator. We determined the actual orientations of the detectors and star sensors in the rocket frame of reference to accuracies of ±0.3° in elevation and azimuth by optical measurements on the ground before and after the flight.

We solved the problem of determining the orientation of the rocket in the celestial frame of reference at any given instant in the following way. We first made use of the fact that *ScoX*-1 was within the elevation range of the modulation collimator throughout the precession cycle. We could therefore determine the amplitude of its variation in elevation from the observed modulation caused by the motion of *ScoX*-1 back and forth over the saw-tooth response function. This amplitude, which is equal to the full apex angle of the precession cone, was found to be 5.1±0.1°. (This angle could also be determined from the star sensor data, though not with as high accuracy as from the modulation collimator data.) From the relative times of the signals from the star sensor *SS*0, we identified seven stars between the second and fourth magnitude as the sources of the signals. We then determined the orientation of the precession cone axis and the phase of the precession motion which minimized the mean square difference between the predicted and observed times when these seven stars entered and left the elevation range of the star sensor during the precession cycle. We found that the precession axis lay in the direction with right ascension 302.2°±0.5° and declination 38.2°±0.3°. From these data we could calculate the celestial orientation of the spin axis at any given instant. Analysis of the nearly periodic pulses from *SS*0 produced by various stars repeatedly crossing the field of view showed that the spin frequency increased approximately linearly with time by about 0.2 percent during the useful period of the flight. Thus the bearing of the axis of any given detector, which for constant precession and spin frequencies would be a linear function of time, could be expressed by a quadratic function of time. The average deviation between the computed and the actual bearing was less than ±0.5° over the useful observation period.

To determine the relative bearings of the X-ray sources and reference stars, we plotted the numbers of counts per bearing interval from each of the detectors as a function of the bearing of the detector's axis, reduced modulo 360°. The data accumulated over four precession periods are shown in Fig. 1. The main peak of the distribution for detector *GV*10 in Fig. 1*a* is due to *ScoX*-1. The difference between its average bearing and that of the star β-Cetus, observed by the star sensor *SS*0, is 239.8°±0.3°. A segment of the great circle through the precession axis direction with this azimuth relative to β-Cetus is marked *A* in Fig. 3. We conclude that *ScoX*-1 lies within 0.3° of this segment.

行于火箭赤道平面的方向上，探测器 *GMC*0 的响应函数是三角函数，其底宽为 15°。星敏感器 *SS*0 和 *SS*30 的有效视场为 1.0° × 5.2°，其长轴方向垂直于火箭赤道平面。在我们的探测器命名中，字母代码后面紧跟数字，这个数字表示视场中心相对火箭赤道的标定仰角，以度为单位。通过火箭飞行前后的地面光学测量，我们可以确定在火箭参考系中，探测器与星敏感器的仰角和方位角实际指向的准确度为 ±0.3°。

我们将用下面的方法阐述如何在任意时刻确定天球参考系中的火箭方向。我们首先注意到以下事实：在整个进动周期内，天蝎座 *X*-1 始终位于调制式准直器的仰角范围内。而天蝎座 *X*-1 伴随锯齿响应函数的往返运动会造成调制现象，这样，我们就可以通过观测调制确定仰角变化的幅度。这一幅度等于进动圆锥的整个顶角大小，具体大小为 5.1±0.1°。（这一角度也可以由星敏感器数据得到，但不如调制式准直器数据得出的结果准确度高。）根据星敏感器 *SS*0 捕获信号的相对次数，我们可以识别出信号源为七颗恒星，星等位于 2 等和 4 等之间。然后，我们可以确定进动圆锥轴的指向和进动运动的相位，这需要满足在进动周期内，以上七颗恒星进入和离开星敏感器仰角范围的预计次数和观测次数的均方差最小。这样我们发现，进动轴方向位于赤经 302.2° ±0.5°，赤纬 38.2° ±0.3°。通过这些数据，我们可以计算出任意时刻自旋轴的天球方向。反复穿越视场的各个恒星在敏感器 *SS*0 中都会产生近似周期性脉冲，对这些脉冲分析可知，在整个有效的飞行期间，自旋频率大约随时间线性增加了 0.2%。因此，任意给定的探测器轴的方位可以表达为时间的二次函数，因为对于恒定进动和自旋频率它是时间的线性函数。在有效观测时间内，方位计算值和实际值之间的平均差值小于 ±0.5°。

为了确定 X 射线源和参考星的相对方位，我们需要绘制出探测器分布函数曲线，如图 1 所示，纵轴为探测器每方位间隔内计数，横轴为探测器轴对 360° 取模后的方位角，需要注意的是图 1 中数据是由四个进动周期累积得到的。图 1*a* 中是探测器 *GV*10 的分布，其中主要的峰值是由天蝎座 *X*-1 引起的。通过星敏感器 *SS*0 观测，其平均方位角与恒星 β–鲸鱼座相差 239.8° ±0.3°。图 3 中标记出了通过进动轴方向的大圆周上的一段，用 *A* 表示，相对 β–鲸鱼座的方位也为 239.8° ±0.3°。我们可以得出结论，天蝎座 *X*-1 位于该段上 0.3° 以内。

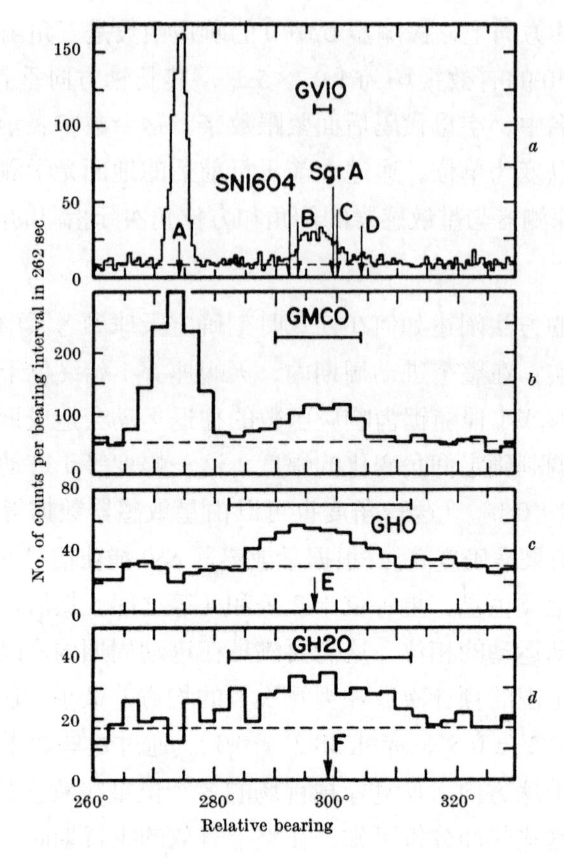

Fig. 1. Bearing distributions of the counts from the four detectors in the vicinity of *ScoX*-1 and the galactic centre complex. The full widths of the fields of view are indicated under the label of each detector. The background levels are indicated by the dashed lines

The second, broader peak in Fig. 1*a* is due to the complex of sources near the galactic centre. From the fact that the ratio of the amplitude to the area of the peak is less than one-half times that of the *ScoX*-1 peak we conclude that it contains more than two sources. Arcs *B* and *C* in Fig. 3 are loci of positions the average bearings of which are equal to the values marked *B* and *C* in Fig. 1*a*. They are segments of great circles drawn through the precession axis, and they bound the region within which these sources lie. The Kepler's supernova remnant *SN*1604 lies 2° outside this region, and the radio source *SgrA* at the galactic centre is just barely within it. The separate peak, the average bearing of which is marked *D* in Fig. 1*a*, is 3.5σ above the background. It is probably caused by a separate source that lies somewhere along the great circle segment marked *D* in Fig. 3.

The bearing distribution for the *GMC*0 detector is shown in Fig. 1*b*. It has separate peaks due to *ScoX*-1 and the galactic centre complex, but it has a low angular resolution and does not improve the bearing determinations already make with the *GV*10 data.

Figs. 1*c* and 1*d* show the bearing distributions for the two detectors *GH*0 and *GH*20 the fields of view of which were narrow slits parallel to the rocket equator. Both show the

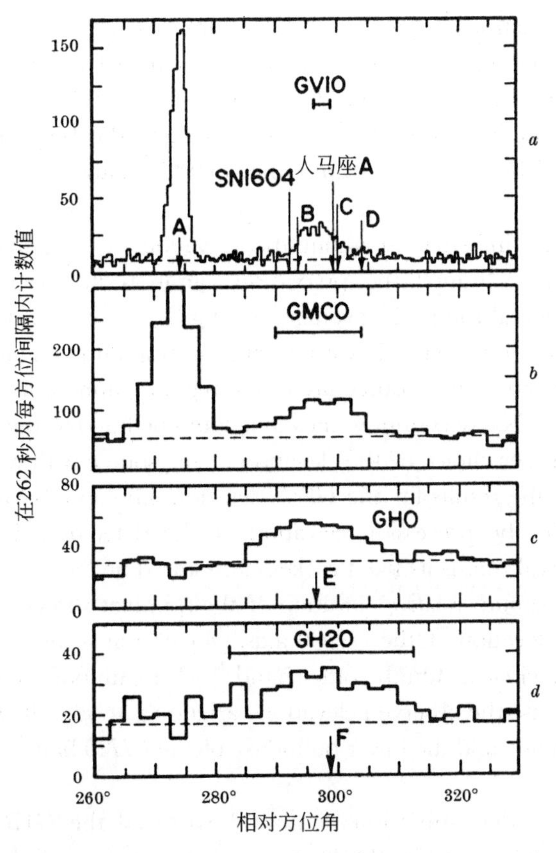

图 1. 四个探测器对天蝎座 X-1 附近和银心复合辐射源探测计数的方位分布。其中，探测器命名中的数字代表视场的全宽，图中虚线表示本底水平。

图 1*a* 中第二个较宽的峰是由银河系中心附近的若干辐射源形成的。基于振幅与峰面积的比值小于天蝎座 X-1 峰一半的事实，因此我们可以得出结论：宽峰对应的辐射源不止两个。图 3 中的 *B* 弧、*C* 弧表示位置轨迹，平均方位角分别等于图 1*a* 中的 *B*、*C* 标记值，它们都是经过进动轴的大圆周上的一段，所围绕的区域正好是辐射源所在的区域。开普勒超新星遗迹 *SN*1604 位于该区域外 2°，而位于银河系中心的射电源人马座 *A* 正好位于这一区域。图 1*a* 中另一个平均方位角的独立峰用 *D* 标记出，比背景高 3.5σ。该峰可能由另外一个独立辐射源引起，该源位于图 3 中 *D* 段上的某处。

图 1*b* 给出了探测器 *GMC*0 的方位分布。该图中的两个独立峰分别源自天蝎座 *X*-1 和银河系中心复合辐射源，但是角分辨率较低，且并不能进一步提高由 *GV*10 数据给出的方位精度。

图 1*c* 和图 1*d* 分别给出了探测器 *GH*0 和 *GH*20 的方位分布，二者的视场均是平

presence of sources which are a part of the complex between circles *B* and *C*, and which lie within the two 8-degree wide elevation bands scanned by the detectors during the precession cycle. These bands are bounded by the small circles marked *GH*0 and *GH*20 in Fig. 3. It is apparent from the absence of a peak at the bearing of *ScoX*-1 in either distribution that *ScoX*-1 does not lie within either of these bands.

To obtain more precise information about the elevation of a source with respect to the precession equator we examined the precession variation in the rates of counts in an appropriate bearing interval round the average bearing of the source. At any fixed bearing the precession elevation of a detector axis varied with an amplitude equal to one-half the apex angle of the precession cone. If a source lay in the region scanned, this variation in elevation caused a variation in the X-ray counting rates that could be fitted to the known response curve of the detector by a proper choice of the elevation of the source with respect to the precession equator. Fig. 2*a* shows the results for the *GV*10 detector. The abscissa is the relative elevation which we define to be the precession elevation of the detector axis at the centre of the indicated bearing interval minus its fixed rocket elevation. The fact that the counting rate was a maximum at the maximum relative elevation of the spin axis shows that *ScoX*-1 is above the maximum precession elevation of the *GMC*0 axis, which limit is indicated by the dashed line that crosses the bearing circle *A*. On the other hand, *ScoX*-1 cannot lie above the band bounded by the *GH*20 lines because the observed elevation variation is too small. Therefore, *ScoX*-1 must lie between the dashed line and the lower boundary of the *GH*20 band.

To refine the elevation determination of *ScoX*-1 we used the *GMC*0 data the elevation variation of which is shown in Fig. 2*b*. From the completeness of the modulation, as well as that of the modulation observed in a previous experiment with a similar but higher resolution collimator, it was possible to conclude that the angular diameter of *ScoX*-1 is less than 7 arc min (ref. 5). One sees here that *ScoX*-1 lines on one of the maxima of the response function when the precession elevation of the *GMC*0 axis is -2.3°, -0.2° or $+1.8^{\circ}$. The solid lines crossing the bearing circle *A* show the positions of the two maxima which fall between the limits previously determined. We are not able to choose between the two intersections on the basis of the data from this experiment. The boxes around each of the two intersections are 0.6° wide by 1.0° long, and they indicate the estimated errors in the bearing and elevation determinations, respectively.

Figs. 2*c* and 2*d* show the elevation variation of the counts from the sources observed by the *GH*0 and *GH*20 detectors. In both cases the sources appear to lie about 0.3° below the midpoint of the elevation scan as indicated by the line crossing the bearing circles *D* and *E*. We determined these latter bearing circles using only the data from the parts of the precession cycle when the sources gave the highest counting rates. As before, the boxes around each intersection indicate the estimated errors in the position determinations.

The two alternative positions we find for *ScoX*-1 are both within the uncertainty circle of 2° radius around the position published by the *NRL* group[6]. The lower of the two is within 0.5° of the position given by Fisher *et al.*[7].

行于火箭赤道平面的狭缝。两个探测器都表明，存在的辐射源是圆周 *B* 和 *C* 之间复合源的一部分，且在一进动周期内分别位于两个 8 度宽的仰角扫描带内。同时这两个带区位于用 *GH*0 和 *GH*20 标记的小圆周之间，如图 3 所示。显然，图 1*c* 和图 1*d* 中并未出现天蝎座 *X*-1 对应的峰，因此天蝎座 *X*-1 并不位于以上两个扫描带内。

为了使辐射源相对进动赤道平面的仰角信息更加精确，我们在辐射源平均方位附近提取合适的方位区间，然后在该区间检查了进动过程中的计数率变化。对于任意给定的方位，探测器轴的进动仰角的变化幅度等于进动圆锥顶角的一半。如果某个辐射源位于扫描区域内，仰角变化会引起 X 射线计数率的变化；通过合理选取辐射源相对于进动赤道平面的仰角，X 射线计数速率可以用已知的探测器响应曲线进行拟合。图 2*a* 给出了探测器 *GV*10 的探测结果。图中横坐标是相对仰角，我们将其定义为指定方位区间中心的探测器轴的进动仰角减去固定的火箭仰角。当自旋轴相对仰角最大时，计数率达到最大。这一事实表明天蝎座 *X*-1 位于 *GMC*0 轴最大进动仰角之上，这一限制对应于图 3 中与方位圆周 *A* 交叉的虚线。另一方面，天蝎座 *X*-1 不可能位于 *GH*20 线包围的带上，因为观测到的仰角变化过小。因此，天蝎座 *X*-1 一定位于虚线和 *GH*20 扫描带下边界之间。

为了进一步改善天蝎座 *X*-1 的仰角精度，我们采用 *GMC*0 的数据，图 2*b* 给出了这些数据的仰角变化。根据本次实验以及以前一次实验（使用了一个相似的但分辨率更高的准直器）观测到的完整调制，也许可以得出结论：天蝎座 *X*-1 的角直径小于 7 角分（参考文献 5）。这样，我们可以发现，当 *GMC*0 轴的进动仰角为 -2.3°、-0.2° 或 $+1.8^{\circ}$ 的时候，天蝎座 *X*-1 位于响应函数某个极大值上。与方位圆周 *A* 交叉的实线给出了两个极大值的位置，正好落在以前确定的极限之间。但根据实验数据，我们尚不能确定究竟是哪个点。位于两个交叉点附近的方框，长宽分别为 1.0°、0.6°，分别表示仰角和方向角的估计误差。

图 2*c* 和图 2*d* 给出了探测器 *GH*0 和 *GH*20 对辐射源观测计数随仰角的变化。在这两种情况下，辐射源看起来大概都位于仰角中点以下 0.3°，其中仰角可以由与方位圆周 *D*、*E* 交叉线得到。当辐射源产生最大的计数率时，我们可以只采用进动周期的部分数据，确定该方位圆。如前所述，每个交叉点附近的方框代表位置的估计误差。

我们发现，天蝎座 *X*-1 存在两个可以选择的位置，均位于美国海军研究实验室研究小组确定的位置附近半径为 2° 的不确定圆内[6]。两个位置中较低的那个距离费希尔等人[7]确定的位置小于 0.5°。

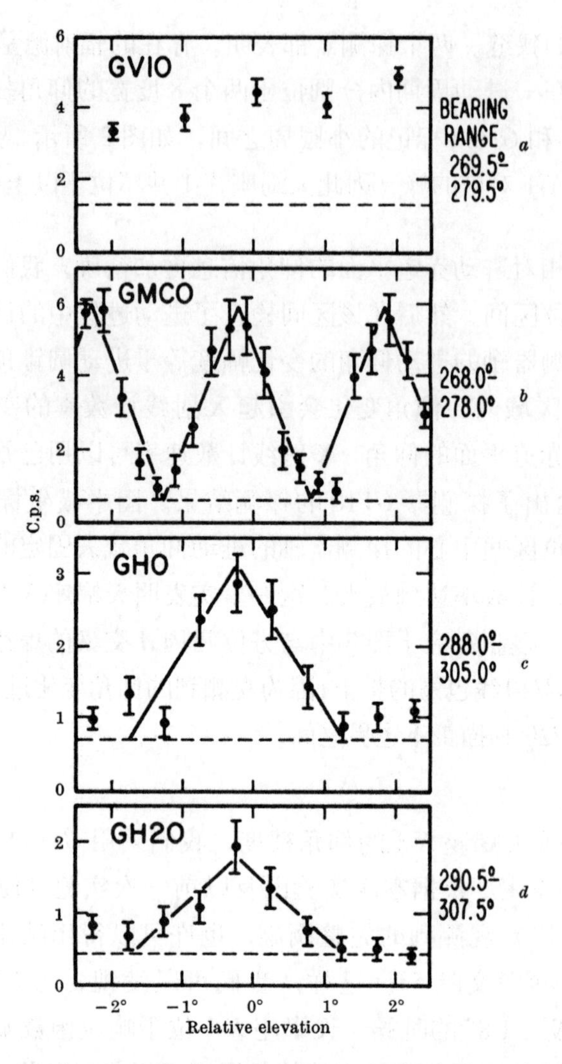

Fig. 2. Variation of the counting rates with the relative elevation of the detector axis at the midpoint of the indicated bearing interval. The background levels are indicated by the dashed lines

SgrX-1 is 5° away from the Kepler supernova remnant *SN*1604. The absence of a significant peak at the average bearing of *SN*1604 in the *GV*10 data in Fig. 1*a* shows that any X-ray emission from this remnant must be several times smaller than that of the nearby sources. Therefore, our observations do not support the identification of an X-ray source with *SN*1604, as suggested by the *NRL* group[6] on the basis of data obtained with an instrument of lesser angular resolution.

ScoX-2 lies more than 5° away from the position of the nearest source reported by the *NRL* group[6].

Two of the circular arcs along which Fisher *et al.*[7] have located sources pass within 1° of our locations for *SgrX*-1 and *ScoX*-2, respectively.

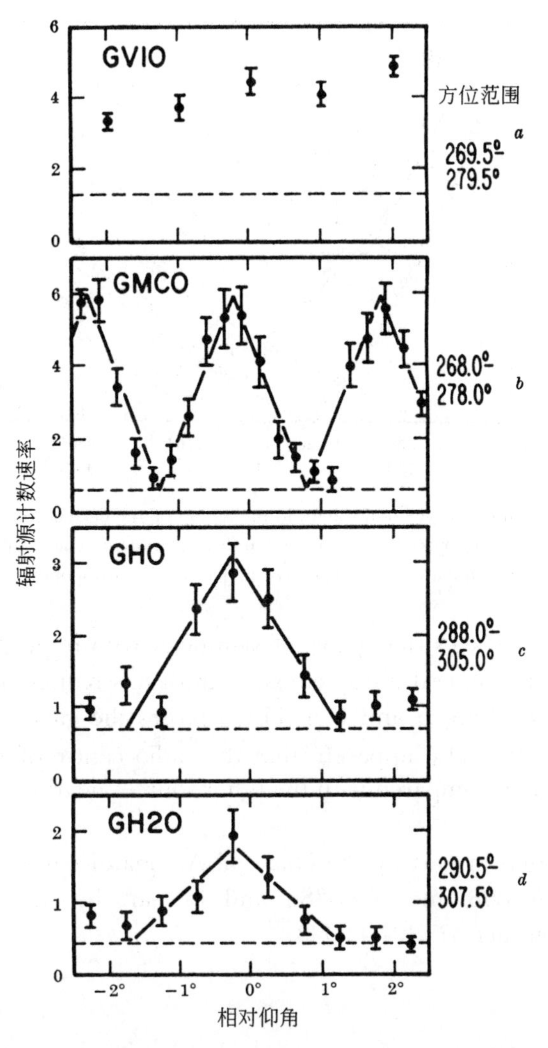

图 2. 计数率随探测器轴相对仰角的变化，其中仰角是方位间隔中心点对应的值，图中虚线表示本底。

人马座 *X*-1 与开普勒超新星遗迹 *SN*1604 的距离为 5°。如图 1*a* 给出了探测器 *GV*10 的数据，在 *SN*1604 平均方位附近缺乏明显的峰值；由此图可知来自遗迹的 X 射线辐射比附近的辐射源小若干倍。因此，我们的观测并不能确认是否在 *SN*1604 中存在 X 射线源；基于使用角分辨率较低的仪器获得的数据，美国海军研究实验室研究小组也表明了这一点 [6]。

天蝎座 *X*-2 与美国海军研究实验室研究小组报道的最近辐射源位置相差 5° [6]。

费希尔等 [7] 定位辐射源的两条圆弧位于我们对人马座 *X*-1、天蝎座 *X*-2 定位的距离小于 1° 范围内。

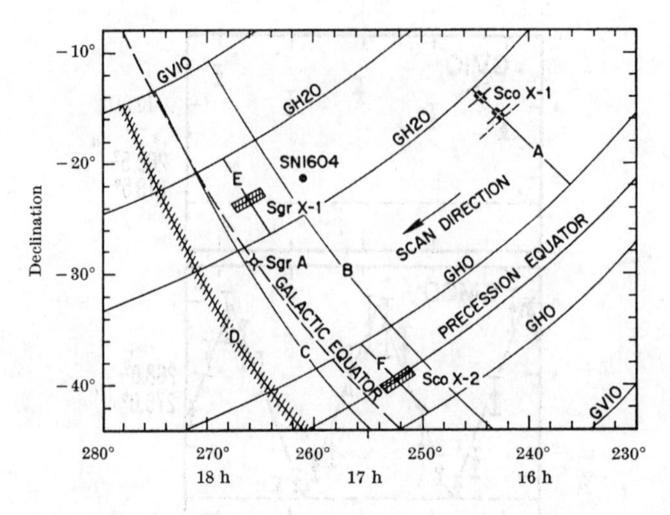

Fig. 3. Mercator projection of the celestial sphere in the region of the galactic centre. The cross-hatched areas show the locations of the sources as determined in this experiment. The small circles labelled *GV*10, *GH*0, etc., are the extreme boundaries of the regions scanned by the corresponding detectors.

Finally, we note that there is a striking lack of symmetry with respect to the galactic centre in the distribution of the general X-ray emission from this region. The centre line of the emission region between lines *B* and *C* in Fig. 3 passes the galactic centre at a galactic latitude of $b^{\mathrm{II}} = +2.5$. It is also apparent that the radio centre of the galaxy, *SgrA*, is at most a weak X-ray source compared with the other sources nearby.

This work was supported in part by the National Aeronautics and Space Administration under contracts *NASw*-898 and *NsG*-386 and in part by the U.S. Atomic Energy Commission under contract *AT* (30-1) 2098.

(**207**, 584-587; 1965)

G. Clark, G. Garmire, M. Oda and M. Wada: Laboratory for Nuclear Science and Department of Physics, Massachusetts Institute of Technology.

R. Giacconi, H. Gursky and J. R. Waters: American Science and Engineering, Inc.

References:

1. Giacconi, R., *Proc. Austin Conf. Relativistic Astrophysics*, Dec. 1964 (to be published).
2. Giacconi, R., Gursky, H., Paolini, F., and Rossi, B., *Phys. Rev. Letters*, 9, 439 (1962).
3. Giacconi, R., Gursky, H., Waters, J. R., Clark, G., and Rossi, B., *Nature*, **204**, 981 (1964).
4. Oda, M., *App. Optics*, 4, 143 (1965).
5. Oda, M., Clark, G., Garmire, G., Wada, M., Giacconi, R., Gursky, H., and Waters, J., *Nature*, **205**, 554 (1965).
6. Bowyer, S., Byram, E. T., Chubb, T. A., and Friedman, H., *Science*, 147, 394 (1965).
7. Fisher, P. C., Johnson, H. M., Jordan, W. C., Meyerott, A. J., and Acton, L. W. (preprint).

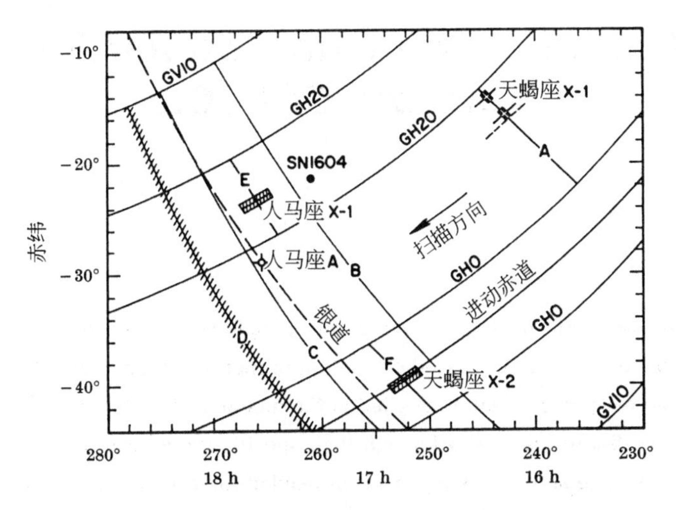

图 3. 在银河系中心区域天球的墨卡托投影。交叉影线区域给出了实验中确定的辐射源的位置。标记为 *GV*10、*GH*0 等的小圆，表示相应探测器扫描区域的极限边界。

最后，需要指出的是，银河系中心区域一般 X 射线辐射分布相对银河系中心明显缺乏对称性。如图 3 所示，辐射区域的中心线位于 *B*、*C* 两线之间，以银纬 $b^{\mathrm{II}} = +2.5$ 穿过银河系中心。另一点值得注意的是，与附近其他辐射源相比，银河系中心的人马座 *A* 是一个比较弱的 X 射线源。

本工作部分由美国国家航空航天局资助，合约序号为 *NASw*-898 和 *NsG*-386；同时也受到美国原子能委员会的部分资助，合约序号为 *AT*(30-1)2098。

（金世超 翻译；尚仁成 审稿）

A Model of the Quasi-stellar Radio Variable *CTA* 102

M. J. Rees and D. W. Sciama

Editor's Note

In 1965 quasars were still newly discovered, and people were struggling to understand how to generate both the brightness and periodic variation of their emission. Here Martin Rees and Denis Sciama report an early interpretation applied to the quasar *CTA* 102, which crucially contains the idea of a rotating disk of material. Although their specific model was soon superseded, it served to outline the size scale and energy mechanisms needed for a better understanding.

ACCORDING to Sholomitsky[1], the quasi-stellar[2] radio source *CTA* 102 (ref. 3) has a variable flux density at 32.5 cm, the period being about 100 days. Sholomitsky takes this to mean that the source cannot be larger than ~0.1 parsec, which is the distance light travels in one period. Since its angular diameter is not less than about 0.01 sec[4,5], he concludes that it must be closer than 2 Mpc, and is possibly inside our own Galaxy. However, Schmidt[6] has recently announced that the optical object identified with *CTA* 102 has a red shift $z=\delta\lambda/\lambda$ of 1.037, and so is probably at a distance comparable with the radius of the universe (~3,000 Mpc).

Although it is by no means certain that the observed variations originate in the source itself, we wish to propose a model which assumes this, and is consistent with the red shift observations.

The model is illustrated in Fig. 1. The radio emission is produced in a spheroidal shell the axis of symmetry of which is approximately along the line of sight. (Shell models for radio sources have been discussed by several authors[7], and spheroidal shells in particular by Layzer[8].) The main part of the emission comes from the region *ADB*, and its spectrum is taken to have a peak at about 300 Mc/s (Fig. 2). The variable part is assumed to come from the disk-like region *ACB*, which is pulsating (perhaps as a result of an explosion occurring at O, the effect of which may reach all parts of the disk at about the same time). When this region is compressed, the magnetic field strength will rise, the individual electrons will be accelerated by the betatron mechanism, and the radiated power will be greatly enhanced. When radiating at its maximum, it is required to emit ~ 25 percent of the total flux observed at 1,000 Mc/s in order to account for the observed variations (that is, its flux density must be $\sim 2\times 10^{-26}$ W/m^2/(c/s)). Its spectrum is taken to be as shown in Fig. 2. The total spectrum then agrees with the observed spectrum of *CTA* 102 (ref. 9).

类星体射电变星 *CTA* 102 模型

瑞斯，夏玛

编者按

1965 年类星体才刚刚被发现，在当时人们致力于对其同时产生的亮度及周期性变化的辐射机制进行研究。本文中马丁·瑞斯和丹尼斯·夏玛给出了适用于类星体 *CTA* 102 的早期解释，他们的观点中最关键的就是提出了存在旋转的盘状物质。尽管他们给出的具体模型很快就被更合适的模型取代了，但是他们的工作有助于给出类星体的大致尺寸并且进一步帮助人们更好地理解能源机制。

根据肖洛米斯基的论文[1]，类星体[2]射电源 *CTA* 102（参考文献 3）在波长 32.5 厘米附近存在流量密度变化，周期大约为 100 天。因此，肖洛米斯基认为，该辐射源的尺寸不可能大于 ~0.1 秒差距，而这恰好是光在一个周期内传播的距离。由于该射电源的角直径不小于 0.01 角秒[4,5]，所以他得出以下结论：*CTA* 102 的距离要小于 2 兆秒差距，很可能就位于银河系内。然而施密特最近的研究表明[6]，认证出的 *CTA* 102 的光学天体存在 $z=\delta\lambda/\lambda=1.037$ 的红移，所以该辐射源的距离可能相当于宇宙半径（~3,000 兆秒差距）。

尽管不能确定上述观测到的变化是否源于辐射源本身，但在我们提出的模型中将仍然假设这种变化源自辐射源，并尽量使模型符合红移观测。

图 1 给出了该模型的示意图。射电辐射产生于图中球状壳，球状壳的对称轴基本上是沿着视线方向的。（一些研究者已经讨论过射电源的壳模型[7]，而雷泽尔也特别提到了球状壳[8]。）大部分辐射主要来自 *ADB* 区域，其光谱在约 300 兆周 / 秒附近有一个峰（如图 2）。假设辐射变化源自 *ACB* 盘状区域，该区域处于脉动状态（这也许是因为 O 点发生爆炸，其影响可以在大致相同的时间到达盘状区域的各部分）。当该区域被压缩时，磁场强度就会增加，单个的电子基于电子回旋加速机制被加速，这样辐射功率就会大大增强。当辐射达到最大时，辐射源在 1,000 兆周 / 秒的辐射需要达到总流量的约 25%，这才能够解释观测到的变化（即其流量密度必须为 $\sim 2\times10^{-26}$ 瓦 / 米^{2}/（周 / 秒））。辐射源模型的光谱如图 2 所示，该总光谱与观测到的 *CTA* 102 光谱较符合（参考文献 9）。

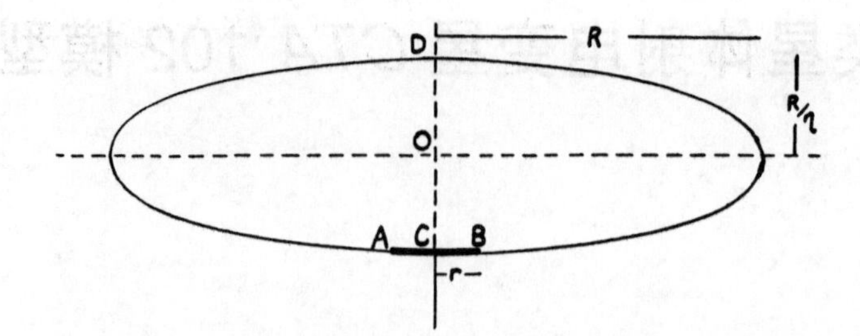

Fig. 1. Proposed model of *CTA* 102

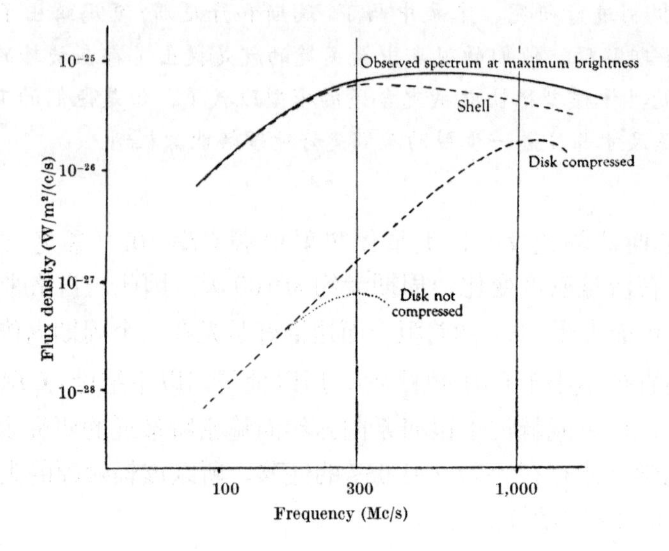

Fig. 2

The disk will have its minimum size consistent with the required flux if it is opaque at (proper) frequencies up to ~1,000(1+z) Mc/s when compressed. Furthermore, if its emission is to vary with a (proper) period of 100/(1+z) days, its thickness cannot exceed $3\times10^{17}/(1+z)$ cm. The electrons which radiate at frequencies around 1,000(1+z) Mc/s have energies of $\sim2\times10^{-5}(1+z)^{\frac{1}{2}}H^{-\frac{1}{2}}$ ergs, each electron producing $\sim2.16\times10^{-22}H$ ergs/sec/(c/s) (where H is in gauss). If synchrotron self-absorption is occurring, the power radiated from the surface of the disk at this frequency is $\sim3\times10^{-8}(1+z)^{\frac{5}{2}}H^{-\frac{1}{2}}$ ergs/sec/(c/s)/cm². The number density of these electrons is therefore $\sim5\times10^{-4}(1+z)^{\frac{7}{2}}H^{-\frac{3}{2}}$ per c.c., and their energy density $\sim10^{-8}(1+z)^{4}H^{-2}$ ergs/c.c. Allowing for the fact that they only contribute a few percent of the particle energy density, and assuming that the total particle energy is comparable with the magnetic energy, we conclude that the magnetic field when the disk is compressed is $\sim5\times10^{-2}(1+z)$ gauss. A threefold increase in the field strength will probably be sufficient to produce the required increase of about 20 in luminosity at 1,000(1+z) Mc/s (though the exact factor depends on the energy spectrum of the electrons). The pulsations will be sufficiently rapid if the Alfvén speed ~c, and this will be true if the particle density of the ambient gas does not exceed ~1 per c.c.

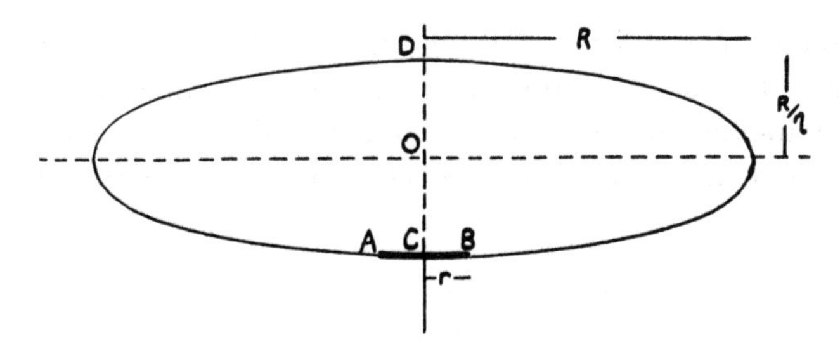

图 1. 提出的 *CTA* 102 模型

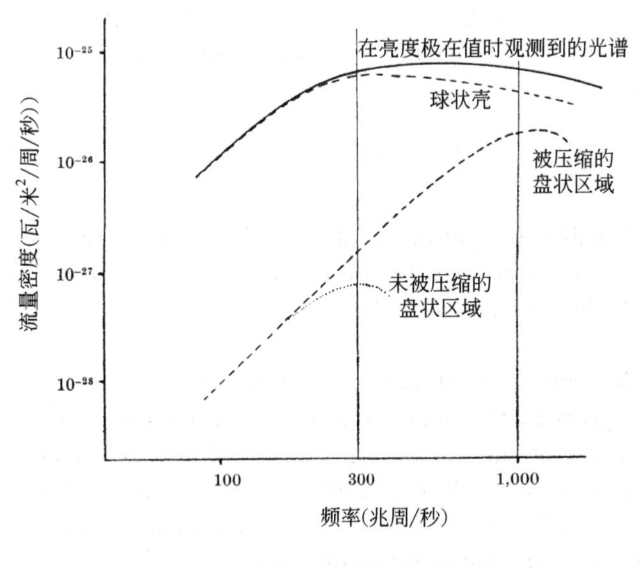

图 2

如果压缩过程中在直到 ~1,000(1+z) 兆周 / 秒的（固有）频率范围内，盘状区域都是不透明的，那么该区域不但具有最小尺寸，而且也符合流量的要求。此外，如果其辐射变化的（固有）周期为 100/(1+z) 天，那么它的厚度不可能超过 $3\times10^{17}/(1+z)$ 厘米。辐射频率约为 1,000(1+z) 兆周 / 秒附近的电子，能量为 $\sim2\times10^{-5}\,(1+z)^{\frac{1}{2}}H^{-\frac{1}{2}}$ 尔格，其中每个电子产生的功率为 $\sim2.16\times10^{-22}H$ 尔格 / 秒 /（周 / 秒）（H 单位为高斯）。如果发生了同步加速自吸收现象，则盘状区域表面在该频率的辐射功率为 $\sim3\times10^{-8}(1+z)^{\frac{5}{2}}H^{-\frac{1}{2}}$ 尔格 / 秒 /（周 / 秒）/ 厘米^2。因此，这些电子数密度是 $\sim5\times10^{-4}(1+z)^{\frac{7}{2}}H^{-\frac{3}{2}}$ 个 / 厘米^3，电子的能量密度为 $\sim10^{-8}(1+z)^4H^{-2}$ 尔格 / 厘米^3。考虑到电子只贡献了一部分粒子能量密度，并假设粒子总能量可与磁能相比的事实，我们就可以得出结论：在盘状区域压缩的情况下，磁场为 $\sim5\times10^{-2}(1+z)$ 高斯。场强增加到 3 倍将可能足以使频率 1,000(1+z) 兆周 / 秒的光度增加到大约 20 倍（尽管准确的因数取决于电子的能量谱）。如果阿尔文速度接近于 c，则这种脉动变化就会足够的迅速；然而，这只在周围气体粒子密度不超过 ~1 个 / 厘米^3 的条件下成立。

The angular diameter of the disk when the magnetic field has the foregoing value is[4,5] $\sim 3\times 10^{-3}(1+z)^{\frac{1}{2}}$ sec. (This assumes that the electrons are radiating incoherently. If they were coherent the angular diameter of the source could be much smaller, which might permit its linear diameter to be ~0.1 parsec, consistently with a large red shift.) It follows that:

$$r \simeq 22.5\frac{z}{(1+z)^{\frac{1}{2}}} \text{ parsecs}$$

in the steady state cosmology. (In the Einstein–de Sitter model this value must be decreased by a factor ~2 if $z\sim 1$.) Estimating the average magnetic field over the whole shell as $10^{-2}(1+z)$ gauss, we deduce from the occurrence of self-absorption below 300 Mc/s that, in the steady state model:

$$R \simeq 225\frac{z}{(1+z)^{\frac{1}{2}}} \text{ parsecs}$$

This estimate of R enables the parameter η to be determined, for since the disk must not deviate from the tangent plane at C by more than $\sim 0.1/(1+z)$ parsec, it follows from the geometry that $R \simeq 2{,}500z^2/\eta$ and so $\eta \simeq 11z(1+z)^{\frac{1}{2}}$.

If $z\sim 1$ the dimensions are: $R\sim 160$ parsecs, $r\sim 16$ parsecs, and $\eta\sim 15$. The lifetime of the electrons is 50-100 years in the fluctuating region, and rather longer in the rest of the source. The total energy of the source is at least $\sim 10^{57}$ ergs, of which $\sim 10^{54}$ ergs is in the disk. The probability that the shell should be oriented so that the disk points towards us is about 3×10^{-5}, which implies that very few radio sources can conform to our model. This model enables us to make the following predictions:

(1) The amplitude of the variations depends on the frequency ν of observation as follows:

$$\begin{array}{ll} \sim 0 & 0 < \nu < 300 \\ \sim 0.013\left\{\left(\frac{\nu}{300}\right)^{2.5} - 1\right\} & 300 < \nu < 1{,}000 \\ \sim 0.25 & \nu > 1{,}000 \end{array}$$

where ν is in megacycles. Thus the observations of Caswell and Wills[10], who found no variations at 178 Mc/s, are not necessarily inconsistent with Sholomitsky's observations.

(2) For $300 < \nu < 1{,}000$ the intensity reaches a maximum in a time $t(\nu)$ days, say, ($t < 50$), and then remains constant for a time $(100 - 2t)$ days (that is, while the disk is opaque to frequencies between ν and 1,000 Mc/s). The rise-time $t(\nu)$ is an increasing function of ν, as illustrated in Fig. 3.

当磁场达到上述值的时候，盘状区域的角直径为 $\sim 3\times 10^{-3}(1+z)^{\frac{1}{2}}$ 角秒[4,5]。（这里假设电子进行非相干辐射。如果电子是相干的话，辐射源的角直径会小很多，对应于较大的红移量线直径可能为 ~0.1 秒差距。）基于稳恒态宇宙学理论，角直径可以由以下公式求得：

$$r \doteq 22.5\frac{z}{(1+z)^{\frac{1}{2}}} \text{ 秒差距}$$

（在爱因斯坦－德西特模型中，如果 $z\sim 1$，则上式需要除以因数 ~2。）在整个壳上，以平均磁场 $10^{-2}(1+z)$ 高斯为估计值，从 300 兆周 / 秒以下自吸收的发生率可以推导出以下适用于稳恒态模型的公式：

$$R \doteq 225\frac{z}{(1+z)^{\frac{1}{2}}} \text{ 秒差距}$$

既然盘状区域偏离 C 点切面不会超出 $\sim 0.1/(1+z)$ 秒差距，那么该区域几何结构满足 $R \doteq 2{,}500z^2/\eta$，对比公式可以由 R 确定参数 $\eta \approx 11z(1+z)^{\frac{1}{2}}$。

如果 $z\sim 1$，则代入上式可得，$R\sim 160$ 秒差距，$r\sim 16$ 秒差距，$\eta\sim 15$。电子在辐射脉动区域的寿命为 50 年 ~ 100 年，在该辐射源的其他区域会更长一些。辐射源的总能量至少为 $\sim 10^{57}$ 尔格，其中盘状区域的能量为 $\sim 10^{54}$ 尔格。球状壳的盘状区域具有方向性且指向我们的概率大约为 3×10^{-5}，这意味着符合我们模型的辐射源很少。我们根据该模型进行如下预测：

（1）流量幅度变化取决于观测频率 v，规律如下：

$$\begin{array}{ll} \sim 0 & 0 < v < 300 \\ \sim 0.013\left\{\left(\dfrac{v}{300}\right)^{2.5} - 1\right\} & 300 < v < 1{,}000 \\ \sim 0.25 & v > 1{,}000 \end{array}$$

其中 v 的单位为兆周。因此，尽管卡斯韦尔与威尔斯[10]在 178 兆周 / 秒观测不到变化，但这并不一定与肖洛米斯基的观测结果相矛盾。

（2）在 $300 < v < 1{,}000$ 频率范围内，流量强度在第 $t(v)$ 天（$t < 50$）达到最大，然后在（$100 - 2t$）天内该强度保持不变（换言之，此时盘状区域频率在 v 到 1,000 兆周 / 秒范围内不再透明）。如图 3 所示，强度上升时间 $t(v)$ 是关于 v 的增函数。

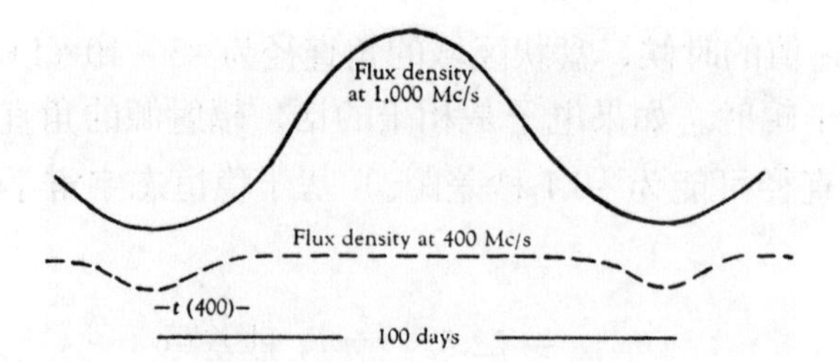

Fig. 3. Time variation of flux density

(3) The times at which the intensity is a minimum should be the same at all frequencies $\nu > 300$ Mc/s, unless there is appreciable dispersion.

However, as regards (3), even if there is negligible dispersion in the source, there may be appreciable dispersion produced by ionized intergalactic gas[11]. At 400 Mc/s, for example, the resulting delay may be as large as 2 h. At this frequency our model predicts an increase in flux of more than 1 percent in about 10 days following the minimum, so that an intergalactic delay might be detectable. Moreover, there may be, superposed on the main variation, an additional small amplitude variation of much shorter time-scale than 50 days, which might then be detectable when the main variation is at a minimum. It would therefore appear to be worth while to develop the sophisticated techniques necessary to detect the possible dispersion, and so to test the hypothesis that there is a significant ionized gas in intergalactic space, and perhaps even to determine the scale-factor of the universe if other radio variables are discovered[11].

We thank Profs. F. T. Haddock, A. Sandage and M. Schmidt for their advice. This work was begun while one of us (D. W. S.) was visiting the Department of Physics and Astronomy, University of Maryland, under National Aeronautics and Space Administration grant *NsG* 5860. He is grateful to Profs. H. Laster and G. Westerhout for their hospitality.

Note added in proof. It has been reported by W. A. Dent (*Science*, **148**, 1458; 1965) that the quasi-stellar source 3*C* 273 (and possibly 3*C* 279 and 3*C* 345 as well) is variable at 8,000 Mc/s, and that again there is a discrepancy between light-size, red shift and angular diameter. However, the discrepancy is much less than for *CTA* 102, and a special geometry of the type considered here might be a possible explanation without leading to such an unfavourable probability factor.

(**207**, 738-740; 1965)

M. J. Rees and D. W. Sciama: Department of Applied Mathematics and Theoretical Physics, University of Cambridge.

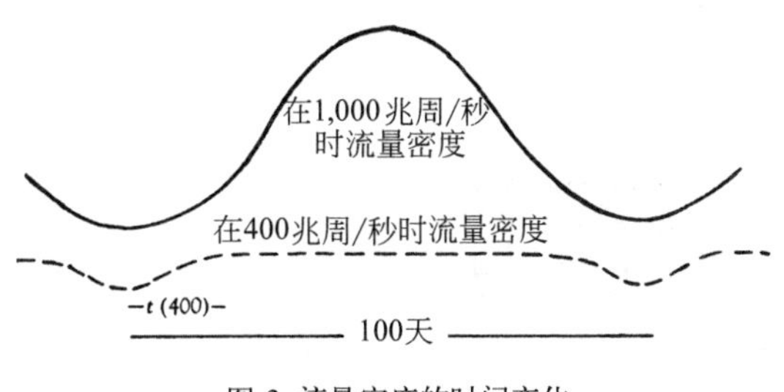

图 3. 流量密度的时间变化

（3）除非存在明显的色散，否则对于 v > 300 兆周 / 秒的频率，辐射强度达到最小的时间是一致的。

针对上面第（3）点，即使辐射源中的色散可以忽略，电离的星系际气体也可能产生明显的色散 [11]。比如，在频率为 400 兆周 / 秒时，造成的延时大约为 2 小时。在该频率下，我们的模型预测到当流量达到最小值后约 10 天其增加值超过了 1%，因此星系际间的延时是可以探测到的。此外，在主要变化之上也可能叠加有小幅度变化，其时间尺度小于 50 天；这样，当主要变化达到最小时，就可以探测到这种小幅度变化。基于以上分析，看起来有必要发展更加精密的技术，用于探测可能存在的色散，并且可以验证星系间存在电离气体的假设；如果可以发现其他射电变星的话 [11]，甚至也许可以确定宇宙的尺度因子。

我们非常感谢哈多克教授、桑德奇教授、施密特教授提供有益的建议。这项工作起始于本文作者之一夏玛访问马里兰大学天文物理系期间，得到了美国国家航空航天局资助，合约序号为 *NsG* 5860。最后也非常感谢拉斯特教授和韦斯特豪特教授的盛情接待。

附加说明　登特（《科学》，第 148 卷，第 1458 页；1965 年）曾报道过，类星体射电源 3*C* 273（可能也包括 3*C* 279 和 3*C* 345）在 8,000 兆周 / 秒是变化的，并且与 *CTA* 102 类似，其光学尺度、红移以及角直径之间并不一致。然而，这一差异远小于 *CTA* 102 的差异，在不引起这种不利因素的条件下，本文中所指出的这种特殊几何模型将成为一种可能的解释。

（金世超 翻译；于涌 审稿）

References:

1. Sholomitsky, G. B., *I. A. U. Information Bulletin on Variable Stars*, No. 83 (Feb. 27, 1965).
2. Sandage, A., and Wyndham, J. B., *Astrophys. J.*, **141**, 328 (1965).
3. Harris, D. E., and Roberts, J. A., *Publ. Astro. Soc. Pacific*, 72, 237 (1960).
4. Slish, V. I., *Nature*, **199**, 682 (1963).
5. Williams, P. J. S., *Nature*, **200**, 56 (1963).
6. Schmidt, M., *Astrophys. J.*, **141**, 1295 (1965).
7. Menon, T. K., *Nature*, **206**, 810 (1965).
8. Layzer, D., *Astrophys. J.*, **141**, 837 (1965).
9. Conway, R. G., Kellermann, K. I., and Long, R. J., *Mon. Not. Roy. Astro. Soc.*, **125**, 261 (1963).
10. Caswell, J. L., and Wills, D., *Nature*, **206**, 1241 (1965).
11. Haddock, F. T., and Sciama, D. W., *Phys. Rev. Letters*, **14**, 1007 (1965).

Recent Developments in Cosmology*

F. Hoyle

Editor's Note

Fred Hoyle was one of the most versatile and productive of astronomers in the first decades after the Second World War. When this paper was published, he was the director of the Institute of Theoretical Astronomy at Cambridge (then being constructed). This article is the text of a talk largely concerned with the implications of the newly discovered class of celestial objects called quasi-stellar objects or quasars. That Hoyle did not win a Nobel Prize for his seminal work on the formation of elements in stars is now widely considered a grave injustice.

I shall start from the observed shift of the spectrum lines of galaxies interpreted in terms of the expansion of the universe. The red shift implies that distances between galaxies, measured with an imaginary ruler for example, increase with time. An immediate question is whether the universe was denser in the past than it is today. If so, how much denser was the universe?

A definitive answer to this critical problem could be obtained in principle by observing the state of the universe in the past. This is possible because of the finite speed of light. We do not observe a galaxy as it is now but as it was at the moment the light started on its journey. In the case of very distant galaxies the light started several billion years ago, so we have direct evidence of the state of affairs several billion years ago. All that need be done, again in principle, is to observe the density of galaxies as it was a few billion years ago and to compare it with the density in our immediate neighbourhood. This would be a direct way of settling the density problem. This indeed was the way in which Hubble tried to settle it more than thirty years ago. He failed to do so because of the extreme difficulty of making a fair count of distant galaxies. The tendency is to count the brighter galaxies but to miss the fainter ones. Since Hubble's attempt nobody has had the hardihood to make a direct assault on the problem by attempting to count galaxies.

Ryle and his associates have counted radio sources instead of galaxies, and here the result has turned out to be much more clear-cut, at any rate so far as the counting process itself is concerned. The difficulty with radio sources is that we are still far from sure exactly what it is that is being counted. Developments in the past two years, the discovery of the quasi-stellar sources in particular, have shown the situation to be more complex than it was at first thought to be. The indication of the radio counts is that the universe was more dense in the past than it is today. However, further knowledge is needed concerning the

---

* Substance of an Evening Discourse delivered on September 6 at the Annual Meeting in Cambridge of the British Association for the Advancement of Science.

宇宙学的新进展*

霍伊尔

编者按

弗雷德·霍伊尔是在二战后最初十年里天文学界最博学、最多产的科学家之一。本文发表时，他是剑桥大学理论天文研究所（当时还正在建设中）的所长。本文是一篇演讲的文字稿，主要是探讨关于新发现的一类被称作类星体的天体的宇宙学含义。霍伊尔没有凭借在恒星内元素形成方面的开创性工作而获得诺贝尔奖，现在人们普遍认为这是很不公正的。

我将从观测到的星系谱线红移被解释为宇宙膨胀开始。假设我们用一把想象的尺子去测量，红移暗示着星系间距离会随着时间而增加。紧接着会产生的问题就是，过去的宇宙是不是比现在的密度要大。如果是的话，密度要大多少？

原则上，通过观测宇宙过去的状态就可以给出这个关键问题的确切答案。因为光速的有限性使得这种观测成为可能。我们看到的不是这个星系当前的状态，而是其在光发出时刻的状态。在观测远距离星系的情况下，我们有可能看到的是其数十亿年前发出的光，所以我们可以直接观测到宇宙数十亿年前的状态。同样从原则上说，我们需要做的仅仅是观测几十亿年前的星系的密度并和我们近邻星系的密度比较。这将是解决密度问题的最直接的方法。三十多年前哈勃就曾试图这样做。但由于对远距离星系做准确的计数极其困难而未能成功。观测愈远愈倾向记录更亮的星系而丢失较暗的星系。从哈勃尝试至今，再没有人试图直接计数星系以破解这个难题。

赖尔和其合作者们选择对射电源而不是星系进行计数，如果仅就计数过程本身而言，这时结果显得更为清楚。用射电源计数的困难在于我们至今还远未准确地知道我们计数的是什么。过去两年中的发展，尤其是类星射电源的发现使人们认识到真实情况远非我们一开始设想的那么简单。射电源的计数结果暗示了宇宙过去比现在要更致密。但是，在我们对射电源本质做出更为准确的了解前，我们还不能给出

* 于9月6日在剑桥大学召开的英国科学促进会年度会议上发表的晚间演讲的内容。

nature of the radio sources before this conclusion can be regarded as definitive.

If the quasi-stellar objects are truly cosmological a great deal becomes pretty well settled. Imagine an object of fixed intrinsic brightness to be moved to increasingly great distances. Two things will happen. The apparent luminosity of the object will decrease and the red-shift of the spectrum lines will increase. A theoretical relation between these quantities can be determined for any specified cosmological theory. The relation is different in different theories, so that in principle it would be possible to distinguish between one cosmological theory and another if we could experiment in this way with a standard object of fixed brightness. Unfortunately it is impossible to move a single object to increasing distances; so the astronomer must rely on similar objects just happening to lie at different distances. The question then arises of how sure we can be that the objects are really similar to each other. Massive galaxies do seem to be quite remarkably similar, but the theoretical differences we are looking for are rather slight in the case of galaxies. This is because galaxies, even the brightest galaxies, cannot be observed far enough away for the theoretical differences to be more than slight. At greater distances the theoretical differences become much more appreciable, however. The technical problem is that of photographing spectrum lines when the light intensity is very small. What is needed is more light. It is here that the quasi-stellar sources are of critical importance. Accepting for the moment that the red-shift of the spectrum lines in these sources is cosmological in character, the quasi-stellar sources are brighter than the most massive galaxies by about four magnitudes, a factor of about forty.

At present, red-shift measurements are available for about fifteen quasi-stellar objects. The shifts are dimensionless numbers given by dividing the wave-length shift of any spectrum line by the laboratory wave-length of the same line. The result is the same for all lines. The measured values range from quite small values, for example 0.16 for 3*C* 273, up to the enormous value of 2 for the source 3*C* 9. The theoretical differences become quite large for red-shifts as great as this, so that a distinction between different cosmological theories should be straightforward once a sample of the order of a hundred quasi-stellar objects has been obtained. The present indication based on the small sample of fifteen is that the universe has expanded from a state of higher density, although the statistical scatter in the sample is large enough to be comparable to the effects that are being looked for.

For spectral shifts as large as two, the Lyman-α line is displaced from the unobservable ultra-violet into the blue, at about 3,700 Å. It is possible to look for a continuum on the blueward side of Lyman-α. This continuum is subject to absorption by neutral hydrogen atoms in intergalactic space. A very small density of neutral atoms would be sufficient to absorb out the light completely. Schmidt's observation of 3*C* 9 shows the continuum to be present but to be weakened, that is, to be lower than the continuum on the redward side of Lyman-α. The implication is that intergalactic gas, if it exists, must be hot, perhaps above 10^6°K. The weakening of the continuum is rather strange, for in a sensitive situation like this one would expect either the continuum to be essentially unweakened or to be

一个确切的定论。

如果类星体确实是宇宙学的，那么许多问题就能够较好地确定下来。想象一下一个本身亮度固定的天体逐渐离我们远去的情况。这时将出现两个现象：该天体的视亮度会降低，同时观测到谱线红移会增加。这些量之间的理论关系对于任一特定的宇宙学理论可由计算确定。不同的宇宙学理论所对应的关系不同，所以，原则上，如果我们有一个本身亮度固定的天体并用上述实验方法将可以甄别不同的宇宙学理论。不幸的是，不可能将一个天体移到愈来愈远的距离，所以天文学家们只能依靠处于不同距离上的相似的天体。于是问题又变成了我们能在多大程度上肯定这些天体彼此间是类似的。从观测上来看，大质量星系看起来确实是很类似的，但就星系而言，我们所要寻求的理论上的差异相当小。这是由于即便是最亮的星系也无法在理论差异稍稍明显的距离上被观测到。在更遥远的距离尺度上，不同理论的差异将变得明显起来。然而，技术上的困难是当光强度非常小时，不能获得照相光谱谱线数据。需要的是更亮的天体。类星射电源正是在此刻成了具有关键重要性的观测对象。在接受这些类星体光谱谱线的红移具备宇宙学特性之后，可以发现类星射电源的光度比最大星系还要高约 4 个星等，即大约亮 40 倍。

迄今为止，人们已经得到了大约 15 个类星体的红移数据。在这里我们定义红移为任意谱线的波长移动除以实验室中该谱线的波长，是无量纲的量。其结果对所有谱线均是相同的。测量得到的谱线红移量范围从十分小的值如类星体 3*C* 273 的 0.16 到非常大的值如 3*C* 9 的 2。在红移如此巨大的情况下，理论的差异将会变得十分明显，一旦我们得到了上百量级类星体样本，不同宇宙学理论中差异的区分也就简单易行了。根据当前我们得到的 15 个类星体的小样本数据，有迹象表明，宇宙可能是由更高密度的状态膨胀而来，尽管在样本中统计弥散跟观测效应差不多明显。

红移大到 2 时，赖曼–α 线将从不可见的紫外区移动到可见光的蓝色区域，大约 3,700 埃。我们将有望观测到赖曼–α 线蓝端出现的连续谱。该连续谱是由星系际空间的中性氢原子对辐射的吸收造成的。密度非常低的中性氢原子就足以将这些光全部吸收。施密特对 3*C* 9 的观测发现了这样的连续谱的确存在，但却被减弱，即要比赖曼–α 线红端的连续谱强度低。这暗示了，如果存在星系际气体的话，这些气体必然非常热，可能超过 10^6K。连续谱的减弱是一个相当奇怪的结果，在这样的十分敏感的情况下，应当期望连续谱基本不被减弱或者完全消失。**事先**并不认为会发生介

absent. The intermediate case seems *a priori* unlikely, since it depends on a closely defined value of the hydrogen density. It is rather in the nature of a coincidence that the density has this critical value.

These remarks are all subject to a cosmological interpretation of the red-shift of the spectrum lines of the quasi-stellar objects. If much smaller, fainter objects were fired out of our own galaxy, or out of some neighbouring galaxy, with speeds close to light the same red-shifts would be observed. Can we be sure that such a "local" interpretation is wrong?

Even though on a "local" hypothesis the quasi-stellars are much less spectacular objects, the total mass involved in all such objects must be as high as $10^6 M_\odot$, or perhaps even more. For such a quantity of matter ejected at speeds close to light, the kinetic energy must be comparable with the rest mass energy, 10^{60} ergs. This is of a similar order of magnitude to the energy involved in the outburst of a major radio galaxy. The energy in the latter case is also in the form of particle motions. The difference is that the particles in the radio galaxies have been thought of in the past as occupying large volumes, not as being condensed into compact objects. However, we now have to ask whether a radio galaxy may not eject compact pieces as well as diffuse clouds of high-speed particles.

Radio galaxies do not eject their material with an initial isotropy. The typical pattern is of two centres of radio emission on opposite sides of a galaxy, with the two centres and the nucleus of the galaxy more or less collinear. To me personally, this has always suggested that an object in the nucleus of the galaxy separates violently into two pieces with a large relative motion. The collinear property then follows from conservation of momentum. Exactly the same phenomenon is observed in the quasi-stellar source *MH* 14-121, two regions of radio emission on opposite sides of, and collinear with, a centre of optical emission. The possibility must be considered that a cascade process is involved. An initial object in the centre of a galaxy breaks violently into two pieces. Later, each of these pieces breaks into two further pieces; and so on. As the cascade develops, and as the objects spread out from the parent galaxy, an approximation to an isotropic situation would then gradually develop.

From independent evidence it has been suggested that an explosion occurred in the nucleus of our own galaxy about ten million years ago. If the quasi-stellar objects emerged in this explosion, as Terrell has suggested, the brightest of the objects, 3*C* 273, would now be distant about 0.5 million parsecs, about one-thousandth of the cosmological distance. The optical emission, instead of the enormous cosmological value of 10^{46} ergs sec^{-1}, would be 10^{40} ergs sec^{-1}. The burning of some 300 solar masses of hydrogen gives sufficient energy to supply such an output for as long as ten million years. Since the mass of 3*C* 273 could be set as $10^4 M_\odot$ there would seem to be no difficulty in explaining the optical output. The kinetic energy of 3*C* 273 would then be $\sim 10^{56}$ ergs, about one order of magnitude less than the total energy of the galactic explosion, as estimated by Burbidge and Hoyle. The masses of other quasi-stellar objects can be set lower than 3*C* 273, because 3*C* 273 is intrinsically considerably brighter than the others, at any rate for

乎中间的情况，因为它取决于非常确定的氢密度值，密度恰巧处于这种临界值将仅仅会是自然界中的一种巧合。

我们以上的评论都是基于这些类星体谱线红移的宇宙学解释。但如果是从我们自己所处的银河系或者一些近邻星系中以接近光速的速度抛射出一个比较小、比较暗的天体，我们也能够观测到类似的红移。我们能完全肯定这种“局域”解释是错误的呢？

尽管在这种“局域”假说下类星体远不是如此令人惊奇的天体，但在所有这样的天体中，其总质量都高达 10^6 个太阳质量甚至更多。这样大质量的天体要被以接近光速的速度抛射出去，其动能必然与其静止质能，10^{60} 尔格相似。差不多是一个大射电星系爆发的总能量的量级。而后者大多数的能量也是以粒子运动形式出现的。其差异在于射电星系中的粒子被认为占有一个很大的体积，而不是凝聚成一个致密天体。我们现在要问的是，射电星系是不是不可以抛出致密碎片以及弥散的高速粒子云。

射电星系最初并不是各向同性地抛射物质。典型的图样是两个射电辐射中心在星系的两边，两个中心和星系核或多或少是共线的。就我个人而言，我认为这总是表示星系核中的某个天体剧烈地分裂成了具有很大相对速度的两部分。而共线性是动量守恒的结果。完全相同的现象也出现在类星射电源 *MH* 14-121 中：两个射电辐射区域分别在中心光学发射的两边并与其共线。因此必须考虑到发生级联过程的可能性。在星系中心的天体最初剧烈的分裂成两部分。接着这两部分又经历各自的分裂过程，如此继续。当其级联过程发生，并随着天体从母星系中散发出来，一个近似各向同性的状态将逐步形成。

有独立的证据显示在我们星系核中在大约一千万年前曾经有一次剧烈的爆炸。如特雷尔所猜想的那样，如果类星体是在这次爆炸中产生的，那么这些类星体中最亮的 3*C* 273 将大约距离我们 50 万秒差距，而这大约是宇宙学距离的千分之一。它的光学发射能流密度也将从以前宇宙学距离所估算的 10^{46} 尔格 / 秒变成 10^{40} 尔格 / 秒。燃烧大约 300 个太阳质量的氢就足以提供这样的能量来输出一千万年。由于 3*C* 273 的质量可设为 10^4 个太阳质量，所以在解释类星体光能输出方面并不存在困难。根据伯比奇和霍伊尔的估算，3*C* 273 的动能约为 10^{56} 尔格，比银河系爆炸的总能量低一个数量级。由于 3*C* 273 的内禀光度比其他类星体都高（仅对于目前观测到的射电源），所以其他类星体的质量应低于 3*C* 273 的质量。取每个天体的质量为 10^3 个太

the sources so far observed. Setting 10^3 $M_\odot$ as the mass per object, and taking the mean speed as half the velocity of light, the kinetic energy per object is comparable with that of 3*C* 273. The total energy estimated by Burbidge and Hoyle, 10^{57} ergs, would provide only for ~10 objects. It would seem therefore that either Burbidge and Hoyle underestimated the violence of the galactic explosion, or an energy difficulty arises.

The same difficulty does not arise in the case of the galaxy *NGC* 5128. This is a nearby massive elliptical, distant about four million parsec. At least one major outburst is known to have occurred in the nucleus of this galaxy within the past ten million years. The probability must also be considered that both *NGC* 5128 and our galaxy are involved, with our galaxy contributing the comparatively low-speed objects and *NGC* 5128 contributing the high-speed objects.

These questions can undoubtedly be resolved by observation. Observations leading to size estimates for the quasi-stellar objects are coming along with rapidly mounting impetus. The light from 3*C* 273 has been known to be variable in a characteristic time of about ten years. This sets the maximum radius of the optical object at ten light years. In addition, rapid flashes in the light over only a few weeks have been suggested. Largely because the maximum radius would have to be reduced to a tenth of a light-year, or even less, there has been a disposition not to believe this evidence in the case of 3*C* 273. On the cosmological hypothesis, how can one have an optical emission a hundred times brighter than the most luminous galaxies pouring out of an object only a tenth of a light-year in diameter? The issue appears to have been resolved by a recent observation of a doubling of the light of a quasi-stellar source in less than a month.

On the radio side, fluctuations from 3*C* 273 have been found, first by Dent and more recently by Moffet and Maltby. The radio data set stronger constraints than the optical data, particularly for the cosmological theory. The present state of the argument is that the cosmological theory just survives the existing data. Whether it will continue to do so remains to be seen. My judgment of the situation is that survival for the cosmological theory depends on there being a sharp saturation in the accumulation of fluctuation data, that not much more in the way of fluctuations can be tolerated. If we are already near the end of the road, the theory will survive and will then probably turn out the correct theory. But if we are still near the beginning of the road, the prospects for the theory will be slight. I would say we have to do with a fifty-fifty situation.

I would like to turn now to quite different issues; but still bearing on the question I asked at the beginning: Has the universe emerged from a more dense state? The kinds of observation I have discussed so far all relate to great distances. Observations can be made in our own neighbourhood which also bear on the problem. I am now going to describe three such observations, together with the related arguments. The three are utterly different in character, illustrating how wide are the issues in cosmology, and how very many phenomena have to be made to fit into a consistent picture.

阳质量，并取其平均速度为光速的一半，依照这样的估计这些天体将会有与 3*C* 273 大致相同的动能。但是按照伯比奇和霍伊尔估计的总能量值，10^{57} 尔格，那场大爆炸的能量就只能供给大约 10 个这样的天体。所以要么是伯比奇和霍伊尔低估了银河系爆炸的强度，要么就要出现能量困难。

但在星系 *NGC* 5128 中不会出现类似的困难。这是一个距离银河系大约四百万秒差距的近邻大质量椭圆星系。在过去的一千万年中，至少有一次大的爆发在这个星系的星系核中发生。当然，我们也必须考虑这种可能性，即观察到的类星体是银河系与这个星系 *NGC* 5128 共同产生的：银河系产生相对低速的天体，而 *NGC* 5128 产生速度较大的天体。

上面谈到的这些问题毫无疑问可以通过观测解决。随着快速积累的观测资料，观测正导致我们对类星体尺度的估算。观测发现，来自 3*C* 273 的光大约以 10 年为特征时标在变化。这说明该光学天体的最大半径为 10 光年。不仅如此，3*C* 273 的辐射中也发现了存在持续时间仅为数周的光闪。主要是因为最大半径会缩小为 1/10 光年或是更少，现在有一种倾向不相信 3*C* 273 提供的证据。因为按宇宙学假设，怎么可能存在直径只有 1/10 光年的天体，却能够释放出比最亮的星系都还要亮 100 倍的光能？最近有观测发现有个类星射电源在一个月之内增亮了一倍，这看起来解决了上面的问题了。

在射电观测方面，登特率先发现了类星体 3*C* 273 辐射的起伏，这个结果随后也被莫菲特和莫尔特比证实。射电数据提供了比光学数据更强的限制，特别是对宇宙学理论而言。现在论据的状态是宇宙学理论仅仅勉强能够满足现有的数据。该结果是不是仍然能够满足以后的数据仍有待观察。我个人的观点是，宇宙学理论是否能够仍然成立取决于积累的光起伏的数据存在很快的饱和，理论已经难以再承受更快的光变了。如果我们已经接近路的终点，理论将经受住考验而被证明是正确的。但如果我们才踏上路的起点，那么这个理论的未来将会非常渺茫。我想说这两种可能性大概对半分。

现在我想谈谈另一个十分不同的话题，但并不会偏离我在本文开头提出的问题：宇宙是否是从一个更为致密的状态演化而来？直到现在，我谈的观测都是与远距离有关的。我们也可以对我们的附近进行观测而这些观测也与这个问题有关。现在我就要谈谈三个这类的观测和与之相关的讨论。这三项观测的特征是完全不同的，显示了宇宙学问题涵盖领域的广阔性，也说明了一个自恰的宇宙学理论必须要能够解释多少现象。

Recently, Penzias and Wilson have observed a radio background at a wave-length of about 7 cm, which they do not believe to be due to their equipment or to the nearby terrestrial environment. The intensity is between 10 and 100 times greater than can be attributed to radio sources. The suggestion is that the universe has a thermodynamic radiation background corresponding to about 3.5°K. Observations at two other wave-lengths at least are needed to confirm this suggestion. One such observation is now being planned by Dicke at Princeton.

There seems no way in which such a background can be explained in terms of current astrophysical processes. Hence, if we accept the suggestion of Penzias and Wilson, the immediate implication is that the universe must have been different in the past from what it is today. Particularly, a higher density is needed to generate the background.

A similar result follows from the entirely different consideration of the helium-to-hydrogen ratio in stars and gaseous nebulae within our galaxy. Determinations of this ratio range from 0.08 to 0.18; and the ratio seems to be just as high in old stars as in most young stars. The ratio to be expected from current stellar activity is only 0.01. So either activity in the galaxy was much greater in the past, or the helium cannot be explained in terms of production from hydrogen through thermonuclear processes within the galaxy. Failure to observe any stars or any object with a low helium content points to the second of these possibilities.

It is possible to show by detailed calculation that, if matter in the universe has emerged from a state in which the temperature was above 10^{10} °K, the helium-to-hydrogen ratio must be about 0.14, a value which falls in the centre of the observed range. However, no values less than a truly universal value should be found. Two independent determinations for the Sun, one from structure calculations, one from observations of solar cosmic rays, give concordant results close to 0.09. This seems significantly below the expected universal value; but further work is needed to establish whether the discrepancy is real or not.

All the lines of investigation which I have mentioned so far point to an affirmative answer to the initial question: they point to the universal density in the past being higher than it is at present. Yet in every case the argument has been fraught with uncertainty. The probability seems against a negative answer, yet the possibility cannot be excluded. Speaking personally, I believe the case for a negative answer would still be arguable if it were not for the third of my three lines of attack. This again is entirely different in character from the helium/hydrogen ratio and from the microwave observation of Penzias and Wilson. I refer to the problem of the origin of elliptical galaxies. In my view, a consideration of this problem points decisively toward the universe having been very much denser in the past than it is at present.

Galaxies have been broadly classified into two types—ellipticals and spirals. There is an incontrovertible argument to show that spirals must have condensed from a more diffuse form. The spirals are rotating. Their angular momenta prevent them from being

最近，彭齐亚斯和威尔逊发现了一种波长在 7 厘米附近的射电背景辐射，他们坚信这个背景与设备和周围环境无关。因为该背景辐射的强度是由射电源辐射产生强度的 10 到 100 倍。所以这说明我们的宇宙有一个大约 3.5 K 的热力学辐射背景。要证实这个结论，还至少需要其他两个波长的观测数据，普林斯顿的迪克正在准备其中一个观测。

现有天体物理过程似乎无法解释这样的宇宙背景辐射。所以如果我们接受彭齐亚斯和威尔逊的结论，那么直接的推论就是宇宙过去处在与现在不同的状态中。特别要指出的是，要产生这样一个背景辐射就需要密度更高的宇宙。

下面一个相似的结论来自一个完全不同的研究方向：对银河系中恒星和气体星云里氦和氢比例的测量。这个比例测定结果大约是 0.08 到 0.18，并且该值似乎在年老的恒星和大多数年轻的恒星中都是类似的。而根据现有恒星活动强度对这个数值的期望值仅为 0.01。这要么说明过去星系中的活动性远比现在更强，要么说明星系中的氦不能被解释为氢在星系里的热核过程的产物。我们在任何恒星或天体中都没有观测到低的氦含量，所以有可能第二种可能性是正确的。

仔细的计算可以证明，如果宇宙中的物质都是经过 10^{10}K 以上的高温状态产生的，那么氦和氢比例将必然是约为 0.14，该值恰好在观测到的数据范围中心。但是不应该发现比实际的普遍值还要低的结果。对太阳的氦和氢比例现有两种独立的推算方法，一是从太阳的恒星结构进行计算，另一是根据太阳宇宙线的观测，而这两种方法都给出了一致的结果近似 0.09。这个结果看起来似乎明显小于所期望的普遍值，但是，我们还需要做进一步的工作来确认这个差异是否是真实的。

至此我介绍的各种不同方向的研究结果对最初的问题都给出了肯定的答复：过去的宇宙的密度比现在的要更高。但是，每项研究工作的论据均具有不确定性。这些结果看上去似乎不支持否定的答案，但是也不能排除这种可能。就我个人而言，如果没有我下面将要介绍的 3 条线中的第 3 条线索，我也会同意否定答案也许还是可以争辩的。这第 3 条线在性质上与氦 / 氢比例、彭齐亚斯和威尔逊的微波背景辐射完全不同。我称这个问题为椭圆星系起源问题。我认为对这个问题的考虑有力地表明了宇宙的过去比现在要更为致密。

星系可以被大致分为两类：椭圆星系和旋涡星系。毋庸置疑，那些旋涡星系肯定是从某种较为疏散的形态凝聚而来。旋涡星系是在旋转的，它们的角动量阻止其被压缩为更致密的形式。尽管从来没有从观测上认真检验过，扁平的椭圆星系也总

compressed into more compact forms. The flattened ellipticals have always been supposed to be similarly in rotation, but this has never been properly checked by observation. Because the ellipticals were thought to be in rotation it was similarly supposed that they were formed by a condensation process. This, I am now convinced, is wrong. I believe ellipticals to have formed through expansion from a higher density state.

Elliptical galaxies are remarkably amorphous. The star distribution is everywhere smooth. If one measures the surface brightness it is found to behave very nearly as an inverse square law, rising with great steepness toward the centre. The centres possess extremely bright central pips. How sharp these centres really are, how star-like, is impossible to say at the moment, for atmospheric seeing effects smear the central pip into an apparent disk.

Suppose the universe expanded from a much denser state, say 10^{-12} g cm^{-3}, and suppose the gas at the beginning of the expansion was not entirely smooth, suppose there were condensation knots already within it. It appears that such condensation knots can restrain the expanding gas to a degree which can be subject to precise calculation. A knot of $10^9 M_\odot$ can restrain a total mass of $10^{12} M_\odot$ within a region of galactic dimensions. A knot of $10^7 M_\odot$ can restrain a mass of about $5 \cdot 10^{10} M_\odot$. The critical point now emerges, that the surface brightness of the resulting aggregation can be calculated (assuming the material forms into stars) and the calculations yield quite unambiguously a law close to the inverse square, in fact just what is observed.

The point I wish to make is that whereas the steep rise towards the centre is expected, and is predicted, by the expansion picture, this characteristic feature of elliptical galaxies cannot, I believe, be understood at all within the condensation picture.

The clinching factor, it seems to me, is that a condensation knot—a memory of the initial dense state—is to be expected at the centre of every major elliptical galaxy. It is these condensation knots that give rise to the phenomenon of the radio galaxy. These are the massive objects which Fowler and I postulated some three years ago. Questions were asked of us at the time as to how our objects ever managed to form. Difficulties of angular momentum were raised. The answer which can now be given is that the objects never formed, in the sense in which the questions were asked. They are relics of a much higher density phase of the universe. They have been there since the galaxies themselves were formed, and in the sense of the radio astronomer they have been smouldering throughout the lifetimes of the galaxies. They are systems which remain at the very edge of stability. Whenever instabilities occur, violent outbursts serve to restabilize them.

Why is my initial question so important? Why make such a fuss about whether the universe has been in a more dense state? Because the present physical theory suggests that there is no limit to how great the density must have been in the past. I use the word "suggests" because the physical divergence of the density was first demonstrated for a homogeneous and isotropic universe. Divergence also occurs when the isotropic restriction is removed. Does it also occur when homogeneity is removed? I have always believed the

是被假定在做类似的旋转。因为认为椭圆星系是在旋转的，所以可以假设它们也经过了一个类似的凝聚过程而形成。但是现在我认为这是错误的。我相信椭圆星系是从一个更高密度态膨胀形成的。

椭圆星系是显著的无定形的，恒星在其中的分布是处处平滑的。如果测量椭圆星系的表面亮度，会发现某一点的面亮度和其距中心的距离非常接近平方反比定律。在接近中心点处面亮度上升十分陡。中心有一个极亮的中心核，但由于大气视宁度效应，将中心点涂抹成看起来成盘状，这让我们现在无法定出其中心有多锐，多像恒星。

假设宇宙是从一个密度高得多，比如，10^{-12} 克 / 厘米^3 的状态膨胀演化而来，并假定刚开始膨胀的气体并非完全平滑的，其中已经存在一些气体凝聚形成的团块，而这些凝聚团块能够在一定程度上使保持住的气体质量将能够被准确地计算。例如，一个大约有十亿个太阳质量的团块将能够使得大约一万亿个太阳质量的气体被束缚在星系大小的范围内。一个大约有一千万个太阳质量的团块将能够使得大约五百亿个太阳质量的气体被束缚。这样就有一个临界点出现了，那么总的聚合体的表面亮度是可以通过计算（假设这些物质形成恒星）得到的，而且计算产生的结果也会很好地符合平方反比定律，即正如我们观测到的。

而我想指明的一点是，鉴于在趋近中心时面亮度很陡的增加是被膨胀图像所期望和预言的，我相信，椭圆星系的这一特征是上面提到的凝聚模型所不能解释的。

就我看来决定因素是质量凝聚的团块，它们含有原初致密态的记忆，应存在于所有大椭圆星系中心。而也正是这些质量凝聚团块产生了射电星系。这就是大约三年前福勒和我提出的大质量天体。我们当时被问到，我们的这些天体是怎么形成的。角动量困难因而产生。而现在对此问题的答案是它们从来没有在问题所问的意义上形成过。那就是这样的大质量天体不是由某种现象或是天文学过程产生的，而是宇宙早期高密度状态至今留下的遗迹。星系形成时它们已经存在在那里了，而从射电天文学家的角度来看它们是贯穿星系寿命期一直保持在焖烧状态的处于稳定性边沿的系统。一旦不稳定性产生，强烈的爆发会使它们恢复稳定。

为什么我最初的问题这么重要？为什么我们要如此关心宇宙早期是不是比现在更致密？因为现有的物理学理论暗示了过去宇宙密度并不存在任何上限。我之所以用“暗示”，是因为物理密度的发散第一次在均匀各向同性宇宙中显示出来。即使我们去掉宇宙各向同性，发散仍然会发生。那么如果我们也不要求宇宙是均匀的，发散也会出现吗？我个人认为答案仍然是肯定的，但我必须指出，我对这个答案的信

answer to this question was also affirmative, but my belief was based more on the failure of those who maintained the opposite to demonstrate their case than on any positive demonstration on the affirmative side. However, progress on the affirmative side has been made very recently, and opinion has generally moved toward the view that the equations of physics contain a universal singularity.

I have always had a rooted objection to this conclusion. It seems as objectionable to me as if phenomena should be discovered in the laboratory which not only defied present physical laws but which also defied all possible physical laws. On the other hand, I see no objection to supposing that present laws are incomplete, for they are almost surely incomplete. The issue therefore presents itself as to how the physical laws must be modified in order to prevent a universal singularity, in other words how to prevent a collapse of physics.

It was with this background to the problem that several of us suggested, some twenty years ago, that matter might be created continuously. The idea was to keep the universe in a steady-state with creation of matter compensating the effects of expansion. In such a theory the density in the universe would not be higher in the past than it is at present. From the data I have presented here it seems likely that the idea will now have to be discarded, at any rate in the form it has become widely known—the steady-state universe. But let me proceed with the theoretical ideas which have grown out of the notions of twenty years ago, for they may turn out to have a value going beyond the first suggestions.

During the past ten years the struggle has been to invent a form of mathematics operating in the manner customary in physics, namely, starting from an action principle. It was found possible to represent the creation of matter through the introduction of a new field. The manner in which the field was treated was quite normal. What was different from ordinary physics was the motive underlying the investigation, the avoidance of a universal singularity, rather than an experiment in the laboratory. Physicists will introduce a new field at the drop of a hat, if experiments in the laboratory should direct them so; but the physicist is unhappy to do so for any other reason.

Having obtained the mathematical structure of the new field it was found that singularities never occur, quite regardless of whether matter is being created or not. So long as the new field exists there will be no singularity either of the universe or of a local imploding body. In other words, the models available for investigation, the models without singularities, were very much wider than the old steady-state theory. During the past few years it is these other models which have been under investigation. What has turned out?

The simplest case is that in which the new field exists but in which there is no creation of matter. It is then possible to obtain a finite, oscillating universe of the kind that has been sought for so long in the usual theory. The universe alternately expands and contracts. Gravitation causes the reversal from expansion to contraction, while the new field causes the rebound from contraction to expansion.

心更多的是建立在试图给出否定答案的人们的失败经历上的，而不是建立在支持方的正确论述上。尽管如此，最近支持方所取得的新进展和看法使得更多人相信我们现有的物理学方程包含着一个一般性的奇点。

事实上，我对这一结论有着根深蒂固的排斥。在我看来，如果一个在实验室中发现的现象不仅抵触现有的物理学定律，还抵触所有可能的物理学定律，那这将是不可接受的。另一方面，我相信现有的物理学定律是不完备的，事实上它们也必然是不完备的。那么我们现在面对的问题就是如何寻找一个方案来修改物理学定律以便消除一般性奇点的出现，换句话说就是要防止整个物理学的坍塌。

正是在这样的背景下，我们中的几个人在大约二十年前提出宇宙中的物质也许是持续不断产生的。提出这个理论的想法是让不断产生的物质来补偿宇宙的膨胀效应，使宇宙保持稳恒态。在这样的理论中过去的宇宙密度并不比现在高。就今天我在此给出的观测数据来看，这样的理论现在应该被抛弃，无论如何，其广为人知的形式，即稳恒态宇宙应该被抛弃。但是，让我进一步阐述其理论概念，它正是从上面那个二十年以前的理论衍生出来的，因为该理论的价值已经超过了其最初所设想的了。

在过去的十年间，人们一直努力于创造一种物理学所熟悉的数学运算形式，即，从作用量原理出发。人们发现，通过引入某种新场以代表物质的产生是可能的。而对场的处理也是很正常的。但这次与通常的物理学不同的是我们的研究的隐含动机不在于在实验室中进行的实验，而是为了消除一般性奇点。物理学家会立刻引入一个全新的场，如果实验室的实验需要他们这样做，但物理学家却不愿意为了任何其他理由而这么做。

在得到了新场的数学结构后，人们发现奇点确实不会出现，而与物质是否产生也没有什么关系。只要这种新的场存在，无论是宇宙还是在我们周围的爆炸天体都不可能有奇异性存在。换句话说，这是一个可用于探讨无奇异性的模型，而它比从前的稳恒态宇宙模型要广泛得多。它是最近的几年中人们一直在研究的另一个模型。那么研究的结果怎样呢？

最简单的一种情况是，存在这种新的场，但没有物质产生。那么我们可以得到有限而振荡的宇宙。这正是迄今为止很多通常理论所寻求的。宇宙交替地膨胀和收缩。引力使得宇宙从膨胀转为收缩，而新的场使宇宙从收缩转为膨胀。

So far as I am aware, such an oscillating model is in satisfactory agreement with all available data. The model is less dull than it seems at first sight, for it contains the possibility of some carry-through from one cycle to the next. Suppose the universe as we observe it eventually stops expanding. Suppose it falls back to a state of comparatively high density, a state in which stars are evaporated, a state in which even the nuclei of heavy elements are disrupted, a state from which matter emerges with the helium-to-hydrogen ratio I described before, about 0.14. In the state of high density things will not be quite uniform. Because of the existence of galaxies, and of clusters of galaxies, there will inevitably be some departures from uniformity. These departures will form the condensation knots round which a new generation of galaxies will form. Thus the condensation knots of which I spoke at an earlier stage are not merely random perturbations. One generation of galaxies acts as the seeds for the next generation. Magnetic fields will also persist from cycle to cycle.

There are two objections to this model. The new field is without sources. It is introduced *ad hoc*, along with the matter. There is never any coupling between the matter and the field. Then there is the subtle, but I believe the correct, objection that a series of oscillations must eventually damp out. Unless dissipative processes are precisely zero, which seems unlikely, the amplitude of the oscillation will gradually die away and the universe will come to rest in an intermediate static state.

For these reasons it is of interest to examine the models with creation of matter, noticing there is no specification from theory as to what the density must be in the steady-state situation. In the past the density was set empirically, by requiring it to be equal to the present-day density. Perhaps this step was wrong. Perhaps the true steady-state density should be very much higher.

During the past year, Dr. Narlikar and I managed to investigate a possibility which had previously proved too difficult to handle, the case in which there are departures from homogeneity, the case in which there are fluctuations from one region of space to another. To our surprise we found that under certain conditions the creation of matter could fall away in a localized region, and that if it did so the region would break into a series of oscillations of a kind that were closely analogous to the oscillations I have just been describing. In the former case we had oscillations of the whole universe, but of a universe of finite volume and finite mass. In this new case we have oscillations of a finite region of an infinite universe. From the point of view of an observer living in such a region it would be difficult to tell the difference. The oscillations would eventually damp away, but in this second case there would simply be a return to the steady-state condition. In this second case there will be many localized oscillating regions, not merely one. The regions will not in general be in juxtaposition with each other; they will usually be separated like islands in an ocean: and, like islands, they will be of different sizes and the amplitudes of their oscillations will be different.

I have already mentioned the philosophy of the physicist, that the whole of physics is

据我所知，这个振荡模型和现在所有数据都能很好地吻合。这个模型并不像乍看起来那样死板，因为其中包含着从一个循环转入下一个循环的可能性。如果我们现在的这个宇宙最终停止膨胀了，并且返回相对高的密度状态，在这种状态中，恒星将会蒸发，甚至重元素原子核也将瓦解，物质将形成，而氦和氢的比例也如我在上面提到的约为 0.14。在高密度状态，物质分布并非是十分均匀的。因为存在着星系或星系团，所以将不可避免地偏离均匀分布。而这样的不均匀将形成凝聚的团块，新一代的星系将会围绕它们形成。这样，我前面提到的凝聚的团块将不仅仅是来源于随机的扰动。上一代的星系将会成为下一代星系的种子。磁场也将在一次次循环中保存下来。

对于该模型，现在有两方面的反对意见。首先是这个模型中新引入的场是无源的。该场是我们为了此问题特别引入的。这个场与物质也绝不会有任何耦合。另一个微妙的但我相信是正确的问题是，经过一系列的振荡后，这种振荡将会逐渐衰减掉。除非宇宙的耗散过程精确地为零，而这几乎是不可能的。所以随着漫长的振荡，振幅逐渐减小，宇宙将会逐渐停止于一个中间静态。

正因为如此，又由于现在没有特定的由理论给出的明确的稳恒态宇宙的密度值，所以考察一下有物质产生的模型是有意义的。过去我们是经验地设定宇宙过去的密度，设定其大约与现在密度相同的数值。也许这一步是错误的。或许真正的稳恒态宇宙密度应当远高于此。

过去的几年间，纳利卡博士和我一直致力于研究宇宙显著偏离均匀性和各向同性状态的可能性，在此状态下空间从一个区域到另一个区域之间存在起伏，这种状态因为过于复杂而没有被人研究过。我们惊讶地发现，在某些条件下，在一个局部区域的物质产生有可能消失，如果这真的发生了，那么这个特定的区域也会进入一系列振荡的状态中，该状态非常类似于前面提到的宇宙振荡。前者是在宇宙中整体发生振荡，但是一个有限体积和有限质量的宇宙。而新的情况是在无限宇宙中的有限区域内发生的振荡。如果从正好生活在这样的一个区域中的观测者的观点，将无法区别两者的不同。同样，这种局部区域发生的振荡也可能衰减掉，但在第二种情况下振荡一旦衰减了，这片区域将就是回到稳恒态条件。在这第二种情况下，有可能存在非常多的局部振荡区域，而不仅仅只有一处。它们通常并不是并列相连的，而一般是像大海中分离的孤岛一样。它们有着不同的大小，其振荡的幅度也不相同。

我曾经提到过，物理学家的哲学是所有物理均能够在实验室中发现。而我们现

discoverable in the laboratory. What has been discovered is a remarkable mixture of elegance—invariance properties for example—and of ugliness, the fine structure constant being 137 … for example. The properties of matter depend critically on the dimensionless numbers of physics, as well as on the structure of the laws. One can take three views on the dimensionless numbers:

(1) They just happen to have the values we find for them and no explanation of these values will ever be found.

(2) The observed values are necessary to the logical consistency of physics.

(3) The observed values are of non-local origin.

I imagine few will be satisfied with the first of these possibilities. I also imagine most physicists prefer (2). But what if (3) should be correct? Could the curious values we observe for the dimensionless numbers be connected with the particular oscillating and finite region in which we happen to live? If this were so, the universe would be far richer in its possibilities and content than we normally imagine. In other regions the numbers would be different and the gross properties of matter, the science of chemistry for example, would be entirely changed.

(**208**, 111-114; 1965)

F. Hoyle: F. R.S., Professor of Astronomy and Experimental Philosophy, University of Cambridge.

在所发现的是如守恒性质般优雅和如精细结构常数 137 般丑陋的异常混合体。物质的性质不仅取决于物理定律，也十分敏感的依赖于一些无量纲的数。人们对这些无量纲的数可以有三种看法：

(1) 我们获得这些常数值纯粹巧合，并没有什么特别的解释。

(2) 这些观测值是与物理学具有逻辑一致性所必需的。

(3) 这些观测值并非局部起源。

我相信大多数人对第一个观点都不会满意，并且我也相信大多数物理学家会倾向于第二个观点。但是有没有人考虑过如果第三个观点是正确的呢？可不可能这些奇怪的无量纲数值恰好与特定的振荡现象及我们现在所处的宇宙区域有关？如果是这样的话，那么宇宙的可能及内容就可能远比我们想象的要丰富得多。在宇宙的其他区域中，这些数值会不同，物质的总体性质将完全不同，例如化学科学。

（樊彬 翻译；邓祖淦 审稿）

New Limits to the Angular Sizes of Some Quasars

R. L. Adgie *et al.*

Editor's Note

Quasars were first recognized as powerful sources of radio emission a very great distance from our own galaxy. By the mid-1960s, radioastronomers energetically set about gathering information about these mysterious objects. This paper describes the use of two radiotelescopes separated by 134 km (one at Jodrell Bank at the University of Manchester in the north of England and one at the Radar Research Laboratory at Malvern, further south), which enabled the size of some quasars to be deduced using the Technique of interferometry

IN previous experiments at Jodrell Bank the angular sizes of discrete radio sources have been investigated by means of long baseline interferometers[1,2]. The highest resolving power used previously was obtained during observations at a wave-length $\lambda=0.73$ m with telescopes 180,000 wave-lengths apart (134 km). Four quasi-stellar and one unidentified source were found to be unresolved in those observations. Their angular sizes were thus shown to be smaller than 0.4 sec of arc. In a further attempt to resolve these sources another experiment has been carried out using the Mark I 250-ft. radio telescope at Jodrell Bank, and one of the 82-ft. radio telescopes operated by the Royal Radar Establishment, Malvern. The separation of these telescopes is 127 km, in a direction which is close to north-south. This interferometer worked on a wave-length $\lambda=0.21$ m, so that the maximum resolving power was more than three times greater than had been obtained in the previous observations. The effective resolving power changes as sources are observed in different directions, but its maximum value is greater than 600,000 wave-lengths for most sources. With this baseline the output fringe frequency produced as the radio source moves through the lobes of the interferometer pattern varies with hour angle from 0 to as much as 40 c/s. An almost identical frequency was produced continuously by a digital "fringe speed machine" and this was subtracted from the interferometer output, so that the fringe patterns displayed on the chart recorder were normally slower than 0.5 cycles/min. A microwave link system was established between the two observatories, via two repeater stations. A new very-high-frequency phase-locking system was also developed, with equipment at each site.

During these observations, the five sources which had not been resolved previously (3*C* 119, 286, 287, *CTA* 21 and 102) all gave clear fringe patterns over a wide range of hour angles, though for only one source did the amplitude remain approximately constant. These sources are listed in column 1 of Table 1. Column 2 shows their catalogued[8] flux density, *S*, at $\lambda=0.21$ m. The observed fringe amplitudes were normalized by daily observations of the source 3*C* 147, which was found to give clear fringe patterns at all hour angles. Corrections arising from various instrumental effects have been applied to these

类星体角大小的新极限

阿吉等

编者按

类星体最早是作为离我们星系非常遥远的强大射电源而为人所知的。在 20 世纪 60 年代中期，射电天文学家们开始积极着手对这些神秘天体的信息进行收集汇总。本文叙述了两架相隔了 134 千米射电望远镜的观测应用（一架在英国北部曼彻斯特大学的焦德雷尔班克，另一架在莫尔文以南的雷达研究实验室），这使人们能够通过干涉测量技术推断类星体的大小。

在焦德雷尔班克之前的实验中，人们曾用长基线干涉仪对分立射电源的角大小做了研究[1,2]。从前进行的最高分辨率的观测是在波长 λ = 0.73 米处，用相隔 180,000 倍波长（即 134 千米）距离的望远镜观测到的。在这些观测中，发现有 4 个类星体和 1 个未证认源未能被分辨。它们显示的角大小小于 0.4 角秒。为了进一步分辨这些射电源，采用位于焦德雷尔班克的马克 I 250 英尺射电望远镜以及莫尔文的皇家雷达研究所的 82 英尺射电望远镜进行另一项实验。这两个望远镜大致以南北方向排列，间距为 127 千米。其干涉仪的工作波段是 λ = 0.21 米，因此测得的极限分辨率提高为之前实验结果的 3 倍多。尽管有效分辨率会随着观测角度的改变而改变，但对大多数源来说其基线最大值超过 600,000 个波长。在这一基线下，当射电源随观测时角穿越干涉图样的波瓣时，会使输出的条纹频率发生 0 周 / 秒 ~40 周 / 秒的变化。而利用一台数字式的“条纹变速机”可以连续得到非常接近的频率值，然后可将此值从干涉仪的输出结果中剔出，这样在记录仪中显示的载波条纹通常就要低于 0.5 周 / 分。在两个天文台之间已借助两个中继站建立了微波连接系统。现已在每个台站新安装了甚高频锁相系统装置。

在这些观测中，之前尚未分辨的 5 个源（*3C* 119，286，287，*CTA* 21 和 102）在较宽的时角变化范围内均给出了清晰的条纹图样，但这其中也只有一个源的振幅大致保持恒定。这些源列于表 1 中的第 1 列。第 2 列给出了 λ = 0.21 米波段处各编目对应的[8]流量密度 *S*。由于 3*C*147 的源在任何时角情况下都给出了清晰的条纹图样，所以对观测到的条纹振幅都用 3*C* 147 的日观测数据进行了归一化。对这些条纹振幅

normalized values of fringe amplitude. It was found that when the sources were observed at elevations less than 15° these corrections were frequently greater than 20 percent, and such observations have not been used. The corrected values of fringe amplitude have been calibrated by considering the maximum value observed for each source, which is shown in column 3 of Table 1. It was found that these maximum values are in an almost constant ratio to the total flux from each source, as shown in column 4, the mean value of these ratios being 14.1±0.9.

Table 1. Radio Sources Smaller than 0.1 Sec of Arc in at Least One Dimension

Source	Flux S at λ=0.21m (flux units)*	Max, fringe amplitude, A, observed with tracking interferometer of aerial spacing 605,000λ (arbitrary units)	A/S
3*C* 119	8.5±0.8	125±10	14.7±1.8
3*C* 286	16.1±0.4	245±20	15.2±1.5
3*C* 287	7.6±1.5	95±5	12.5±2.5
CTA 21	8.0±0.8	115±10	14.4±2.0
CTA 102	6.6±1.6	95±10	14.4±2.0
			Mean value 14.1±0.9

*One flux unit = 10^{-28} $Wm^{-2}(c/s)^{-1}$

A constant ratio in column 4 would be obtained if the minimum linear dimensions of all these sources were similar, and they were at comparable distances, and were partially resolved to the same extent by effective baselines between 500,000λ and 600,000λ. This is an improbable situation, and it seems more likely, as we assume, that each source was unresolved by the effective baseline at which the maximum value of fringe amplitude was observed. This means that each of these sources is smaller than 0.1 sec of arc in at least one direction. The source *CTA* 102 did not appear to be resolved at any hour angle when its elevation was greater than 15°, so that its angular size is shown to be less than 0.1 sec of arc in all position angles. Each of the other sources was partially resolved at some hour angle, and the corresponding maximum angular dimensions (assuming gaussian source models) were in the range 0.12–0.16 sec of arc, as shown in column 5 of Table 1.

The quasar 3*C* 273 was also observed and was found to give clear fringe patterns of unexpectedly large amplitude with one well-marked minimum near meridian transit. Observations of lunar occultations of this source show that it consists of two components, *A* and *B*, the centres of which are 19.5 sec apart in position angle 044°. If both these components had contributed to the fringe patterns observed at this baseline, a pronounced minimum would have been observed every 10–15 min at most hour angles. As these frequent minima were not observed, it follows that only one component of the source is small enough to give fringe patterns at this baseline, and that it, in turn, probably consists of two parts, which are each smaller that 0.1 sec of arc.

As only one minimum was observed, the angular separation of these parts and the position angle cannot, by inspection, be determined uniquely, though they may be obtained in due course by more detailed analyses. If it is assumed that this position angle is also 044°, the

归一化值的修正考虑到了多种仪器效应。具体观测发现，当射电源的仰角低于 15° 时，往往修正幅度会大于 20%，这种观测结果没有被采用。修正后的条纹振幅值在参考了每个源的观测极大值后进行了标定，这些极大值列于表 1 中的第 3 列。由表可知，如第 4 列所示这些极大值与每个源的总流量呈恒定的比例关系，这些比例的平均大小为 14.1 ± 0.9。

表 1. 角尺度至少在一个方向上小于 0.1 角秒的射电源

源	在 λ=0.21 米处的流量 S（流量单位）*	用天线间距 605,000λ 的追踪干涉仪观测的最大条纹振幅 A（任意单位）	振幅 / 流量
3C 119	8.5 ± 0.8	125 ± 10	14.7 ± 1.8
3C 286	16.1 ± 0.4	245 ± 20	15.2 ± 1.5
3C 287	7.6 ± 1.5	95 ± 5	12.5 ± 2.5
CTA 21	8.0 ± 0.8	115 ± 10	14.4 ± 2.0
CTA 102	6.6 ± 1.6	95 ± 10	14.4 ± 2.0
			平均值 14.1 ± 0.9

*1 流量单位 = 10^{-28} 瓦 · 米$^{-2}$（周 / 秒）$^{-1}$

如果所有这些源都具有相近的最小线尺度，离我们的距离也差不多，并且能利用 500,000λ~600,000λ 范围内的基线进行有效的分辨，在满足所有这些条件时表中第 4 列才会看到一个常值。当然这不大现实，正如我们所猜测的，更有可能的情况是每一个源都未能被有效基线所分辨，观测到的都是条纹幅度的最大值而已。这意味着每个源至少在一个方向上小于 0.1 角秒。当源 *CTA* 102 仰角大于 15° 时，在任何时角处都不能分辨它，因此其角大小在所有方位角上均小于 0.1 角秒。如表 1 的第 5 行所示，其他源在某些时角处仍能部分分辨，相应的角大小极大值（假定是高斯源模型）出现在 0.12 角秒 ~ 0.16 角秒范围内。

类星体 3C 273 也曾被观测到，并被发现它能给出清晰的条纹图样，其出人意料的大振幅在其经过子午线附近时存在一个明显的极小值。该源的月掩星观测表明，它由两个子源 A 和 B 组成，其中心偏离 044° 的方位角 19.5 角秒。如果这两个子源对基线上观测到的条纹图样都有贡献，那么在大多数时角处，每 10~15 分钟将都可以观测到明显的极小值。由于这些频繁的极小值并未被观测到，可知两者之中可能只有一个子源足够小以产生在此基线上观测到的条纹图样；反之，也可能存在两个部分，但每个都小于 0.1 角秒。

尽管经过更进一步的分析也许能得出确定的结果，但由于仅观测到一个最小值，因此根据这些观测值要唯一地确定出这些部分之间的角距离和方位角还是有困难的。

separation is of the order 0.4 sec of arc.

This interpretation of our observations, when calibrated with the factor derived from the sources discussed earlier, shows that the two parts of one component of 3*C* 273 are now giving a flux density not less than 30±2 flux units. As may be seen in Table 2, this is significantly greater than the value of 23.3 flux units derived for the brighter component, *B*, from the ratio given by the occultation observations of 1962.9. The total flux from this source was therefore re-measured recently, with the Mark I telescope at Jodrell Bank, and compared with the radio galaxies 3*C* 348 and 353. These measurements show that, on the assumption that the radio galaxies have remained constant, the value of total flux from 3*C* 273 in 1965.6 was 46±1.5 flux units. This is compatible with the other results if it is assumed that the interferometer observations refer to the component *B*, and that the flux of this component has now increased to 30 flux units, that is, by approximately 30 percent in three years. It is not known which part of component *B* has increased. This conclusion may be compared with the measurements reported by Dent[3] which show that at 8,000 Mc/s, where almost all the radiation comes from component *B*, the flux has increased by 40 percent in 2.5 years.

Table 2. Data on 3*C* 273 at 0.21 M

Total flux density (1962-3)	39.8±2 flux units
Flux ratio component *B* : *A* (1962.9)	1.40
Therefore component *A* was then 16.5 and *B*,	23.3 flux units
Long baseline interferometer (1965.5) Two parts of one component gave	30±2 flux units
Total power measurements (1965.6) *A*+*B* Provisional interpretation (1965.7)	46±1.5 flux units

Component *A* is still 16.6 flux units, and component *B* is now 30 flux units. Flux ratio *B* : *A* now 1.84.

We have summarized in Table 3 our current interpretations, based on these and earlier measurements, of the angular dimensions of the ten quasi-stellar sources the redshifts of which have been published[4,5]. Possible values of the linear dimensions of the radio-emitting regions of these sources are shown in columns 4 and 5. Those in column 4 have been calculated on the hypothesis that the redshifts of these sources arise from the general expansion of the Universe, and so correspond to distances greater than 600 Mpc. The values given in column 5 are calculated on the hypothesis suggested by Hoyle and Burbidge[6], that the emitting regions are at distances of order 10 Mpc, and their redshifts arise from very high intrinsic velocities of recession of the individual objects.

如果假设这个方位角也是 044°，那么各部分之间的角距离则在 0.4 角秒的量级。

在先前讨论源因子的标定时，我们对这类观测的讨论表明，3*C* 273 的一个子源中的两个部分在现阶段辐射出的流量密度不小于 30±2 个流量单位。如将在表 2 中所见的，该值要远大于较亮的子源 *B*，*B* 在 1962.9 掩星观测比值中所得到的结果是 23.3 个流量单位。因此，最近用焦德雷尔班克的马克 I 望远镜重新测量了这个源的总流量，并与射电星系 3*C* 348 和 353 做了比较。这些观测表明，如果假定射电星系保持稳定，那么在 1965.6 射电星系 3*C* 273 的总辐射流量约为 46±1.5 个流量单位。如果假设干涉仪的观测结果是参照子源 *B* 的，而且这个子源的流量在 3 年里大致增加了 30%，即现在增至 30 个流量单位，那么这是与其他结果相一致的。但并不清楚子源 *B* 中哪个区的辐射流量在增加。这个结论可与登特 [3] 给出的测量结果相比较，登特指出在 8,000 兆周 / 秒时，几乎所有的辐射均来自子源 *B*，而其对应的流量在 2.5 年里增大了 40%。

表 2. 3*C* 273 在 0.21 米处的数据

总流量密度（1962-3）	39.8±2 个流量单位
流量比例子源 *B* : *A*（1962.9）	1.40
当时子源 *A* 是 16.5 而 *B* 为	23.3 个流量单位
长基线干涉仪（1965.5） 一个子源的两部分测出	30±2 个流量单位
总功率测量（1965.6）*A*+*B* 临时的解释（1965.7）	46±1.5 个流量单位

A 部分仍为 16.6 个流量单位，*B* 部分现在是 30 个流量单位。流量比 *B*:*A* 现在是 1.84。

我们在表 3 中汇总了当前的结果，这是基于对 10 个已发表红移测量结果 [4,5] 的类星体射电源的本次及以前的角大小的测量而得到的，其中第 4 列和第 5 列中给出了这些源的射电发射区域可能的线尺度大小。假设这些射电源的红移是由宇宙学膨胀造成的，那即可计算得出第 4 列中的数值，相应的离我们的距离大于 600 兆秒差距。第 5 列中给出的数值是根据霍伊尔和伯比奇 [6] 提出的假设计算的，即认为其中发射区距离我们 10 兆秒差距量级，并且其红移是由各个天体非常高的内禀退行速度产生的。

Table 3. Current Interpretations of the Radio Measurements of the Angular and Linear Dimensions of some Quasars

Source	Redshift z	Remarks on individual sources	Approximate linear dimensions (parsecs)	
			If redshifts cosmological*	If sources "local" at about 10 Mpc
3*C* 273	0.158	Source has two components 19.5 sec apart in position angle 044°	Separation of *A* and *B* 32,000	950
		Component B, coincident with optical quasi-stellar object, has two parts, each smaller than 0.1 sec, separation probably 0.4 sec. At $\lambda=0.21$m flux of component B has increased by approx. 30 percent in 2.7 yr	Each part smaller than 170 separation probably 700	5 20
		Component A has dimensions 5×1.5 sec flux probably unchanged	8,500×2,500	240×70
3*C* 48	0.367	At $\lambda=0.73$ m, elliptical, 0.4 by <0.3 sec †‡§	1,100× $\leqslant$ 900	20× $\leqslant$ 15
3*C* 47	0.425	At $\lambda=0.21$ m, this source has two components (ref. 9) Separation 62 sec	$\geqslant$ 180,000	$\geqslant$ 3,000
		Each component $\leqslant$ 10 sec, but little structure $\leqslant$ 2 sec ¶	<30,000 but > 6,000	<500 but >100
3*C* 147	0.545	At $\lambda=0.73$ m, elliptical 0.6 by <0.4 sec †‡§	2,000× $\leqslant$ 1,300	30× $\leqslant$ 20
3*C* 254	0.734	At $\lambda=1.89$ m, ≈6 sec. No structure measurements at any wave-length	≈21,000	≈300
3*C* 286	0.86	At $\lambda=0.21$ m, 0.12 by <0.1 sec ‡	420×<350	6×<5
3*C* 245	1.029	At $\lambda=1.89$ m, >12 sec. No structure measurements at any wave-length ¶	>41,000	>600
CTA 102	1.037	At $\lambda=0.21$ m, <0.1 sec in all position angles	<360	<5
3*C* 287	1.055	At $\lambda=0.21$ m, 0.14× <0.1 sec ‡	500×<360	7.0×<5
3*C* 9	2.012	At $\lambda=1.89$ m, >10 sec. No structure measurements at any wave-length ¶	>32,000	>450

* These linear dimensions have been calculated for model universes (ref. 4) in which the acceleration parameter $q_0=+1$. If $q_0=0$, these values must be multiplied by a factor $(1+0.5z)$ which for these sources lies in the range 1–2.01.

† Results (ref. 7) of detailed analysis of observations at $\lambda=0.73$ m.

‡ None of the available evidence suggests that the source contains fine structure within this elliptical component, but further analyses or observations at even higher resolution could conceivably reveal a more complex structure, within these overall dimensions, as in the case of 3*C* 273 *B*.

§ Fringe patterns corresponding to partial resolution of this source were recorded during the observations at $\lambda=0.21$ m described above. They have not yet been analysed in sufficient detail to improve significantly the earlier interpretation of the angular structure of this source.

¶ A weak source at $\lambda=0.73$ and $\lambda=0.21$ m. Ever if it were unresolved, the signal-to-noise ratio would be poor under optimum conditions, and the fringe pattern might not have been recognized during those observations which were attempted.

(**208**, 275-276; 1965)

R. L. Adgie, H. Gent and O. B. Slee: Royal Radar Establishment, Great Malvern.
A. D. Frost, H. P. Palmer and B. Rowson: University of Manchester, Nuffield Radio Astronomy Laboratories, Jodrell Bank.

References:

1. Allen, L. R., Anderson, B., Conway, R. G., Palmer, H. P., Reddish, V. C., and Rowson, B., *Mon. Not. Roy. Astro. Soc.*, **124**, 477 (1962).

2. Anderson, B., Donaldson, W., Palmer, H. P., and Rowson, B., *Nature*, **205**, 375 (1965).

3. Dent, W. A., *Science*, **148**, 1458 (1965).

表 3. 目前对部分类星体角大小和线尺度的射电观测说明

源	红移 z	对各源注解	近似线尺度（秒差距）	
			假设红移是宇宙学红移*	假设源是“本地的”，距离约10兆秒差距
3*C* 273	0.158	源有两个成分，在方向角 044° 处分开 19.5 角秒	*A* 和 *B* 的距离为 32,000	950
		B 部分，与光学类星体一致，由两部分组成，每个都小于 0.1 角秒，且大约间距 0.4 角秒。在 λ=0.21 米波段上 B 部分在 2.7 年里流量增加了大约 30%	每部分小于 170 大致距离 700	5 20
		A 部分尺度为 5 × 1.5 角秒，其流量大致不变	8,500 × 2,500	240 × 70
3*C* 48	0.367	在 λ=0.73 米，尺度小于 0.3 角秒时椭率为 0.4 †‡§	1,100 × ⩽ 900	20 × ⩽ 15
3*C* 47	0.425	在 λ=0.21 米，源由两部分组成（参考文献 9），相距 62 角秒	⩾180,000	⩾3,000
		每部分均⩽ 10 角秒，但细致结构⩽ 2 角秒 ¶	< 30,000 但 > 6,000	< 500 但 > 100
3*C* 147	0.545	在 λ=0.73 米，小于 0.4 角秒时椭率为 0.6 †‡§	2,000 × ⩽ 1,300	30 × ⩽ 20
3*C* 254	0.734	在 λ=1.89 米，约 6 角秒。任何波长均未测出结构	≈ 21,000	≈ 300
3*C* 286	0.86	在 λ=0.21 米，小于 0.1 角秒时椭率为 0.12 ‡	420 × < 350	6 × < 5
3*C* 245	1.029	在 λ=1.89 米，大于 12 角秒。任何波长均未测出结构 ¶	> 41,000	> 600
CTA 102	1.037	在 λ=0.21 米，在任何方位角观测都小于 0.1 角秒	< 360	< 5
3*C* 287	1.055	在 λ=0.21 米，0.14 × < 0.1 角秒 ‡	500 × < 360	7.0 × < 5
3*C* 9	2.012	在 λ=1.89 米，大于 10 角秒。任何波长均未测出结构 ¶	> 32,000	> 450

* 这些线性尺寸是在加速因子为 q_0 =+1 的宇宙模型（参考文献 4）下进行计算得到的。如果 q_0 = 0，对那些红移在 1~2.01 范围内的源，结果必须乘上因子（1+0.5z）。

† 在 λ=0.73 米处观测值的详细分析结果（参考文献 7）。

‡ 还没有证据表明，在源的这个椭圆部分内包含精细结构，但将来更高分辨率的分析或观测将肯定能在此轮廓内揭示一个更复杂的结构，就如 3*C* 273 *B* 那样。

§ 相对应的该源部分分辨条纹图样是在上文中提到的 λ=0.21 米波段观测得到的。它们还尚未得到足够详细的分析以对该源角结构的早期解释做出重要改进。

¶ 在 λ=0.73 米和 λ=0.21 米处存在一个弱源。它一直都未能被解析，在最佳条件下其信噪比也很小，并且在其他尝试观测中对条纹图样甚至无法认证。

（沈乃澂 翻译；肖伟科 审稿）

4. Schmidt, M., *Astrophys. J.*, **141**, 1295 (1965).
5. Shklovsky, I. S., *Astro, Circ., U.S.S.R.*, No. 250 (1963).
6. Hoyle, F., and Burbidge, G. R. (personal communication).
7. Anderson, B., Ph.D. thesis, Univ. Manch. (1965).
8. Conway, R. G., Kellerman, K. I., and Long, R. J., *Mon. Not. Roy. Astro. Soc.*, **125**, 261 (1963).
9. Ryle, M., Elsmore, B., and Neville, Ann C., *Nature*, **207**, 1024 (1965).

Structure of the Quasi-stellar Radio Source 3C 273 B

M. J. Rees and D. W. Sciama

Editor's Note

The radio source known as 3C 273 B—the designation "3C" refers to the third catalogue of radio sources in the sky compiled by the radioastronomy station at Cambridge—is one of the first to be studied in detail. This paper was based both on radioastronomy measurements and optical observations of the source of the radio waves. Of the two authors, Martin Rees is now Lord Rees of Ludlow and became president of the Royal Society of London in 2005. Dennis Sciama was a theoretical astronomer at Cambridge.

IN this article we show that: (1) the radio variations[1] of 3*C* 273 *B* are consistent with its red shift being cosmological; (2) the variable component of the optical continuum[2] of 3*C* 273 *B* may be partly due to inverse Compton collisions between relativistic electrons and their own synchrotron radiation. In this case the optical variations would be correlated with the radio variations, though with a phase delay.

The problem involved in (1) is to reconcile (*a*) the upper limit on the size of the source derived from the period of the variations[3]; (*b*) the lower limit on the angular diameter of the source derived from the absence of self-absorption down to some particular frequency[4,5] (assuming the synchrotron mechanism and that the electrons are radiating incoherently); (*c*) the distance derived from the red shift of the optical counterpart of the radio source. This problem first arose with the quasi-stellar source *CTA* 102, which Sholomitsky[6] reported to vary with a period of about 100 days at 940 Mc/s. In this case the distance derived from the red shift of 1.037 (ref. 7) is ~1,500 times greater than the upper limit permitted by (*a*) and (*b*). We recently proposed[8] a model of *CTA* 102 which would resolve this discrepancy, but recognized that it required a rather improbable geometry and so was very implausible. Since then Maltby and Moffet[9] have reported that *CTA* 102 did not vary appreciably at 970 Mc/s over a 3-year period ending about 2 years before Sholomitsky's observations began. While it is possible that *CTA* 102 varies sporadically, the more likely inference is that Sholomitsky's observations are in error.

The same problem arises with the quasi-stellar source 3*C* 273 *B*, the flux density of which at 8,000 Mc/s was found by Dent[1] to have increased by about 40 percent in 3 years. This source shows no self-absorption in its spectrum down to at least 200 Mc/s (ref. 10) (in fact the spectrum is flat down to this frequency). For a magnetic field strength of 10^{-3} gauss this implies a lower limit on the angular diameter of 0.03 sec (while the observed[11] angular diameter is 0.5 sec). If the period of the radio variations is comparable with that of the

类星体射电源3*C* 273 *B*的结构

瑞斯，夏玛

编者按

众所周知，3*C* 273 *B* 是最早被详细研究的射电源之一。此处“3*C*”是指在剑桥射电天文学观测站收集整理的射电源星表中的第三表。本文的研究是在对该射电源的射电天文学测量和光学观测的基础上进行的。本文的两位作者之一瑞斯（即现在的瑞斯勋爵）于2005年当选英国皇家学会主席，另一位丹尼斯·夏玛则是剑桥大学理论天文学家。

我们在本文中阐明以下两点：(1) 3*C* 273 *B* 的射电变化与它的宇宙学红移相一致 [1]；(2) 3*C* 273 *B* 的连续光谱中的可变部分（即光变）可能部分是由于相对论电子与其同步辐射之间发生逆康普顿碰撞 [2]。这种情况下，光学波段变化与射电变化相关联，只是存在一定的相位延迟。

上面 (1) 中的内容需要满足：(*a*) 根据变化周期推出的射电源尺寸的上限 [3]；(*b*) 根据不发生自吸收的特定频率推出的射电源角直径的下限 [4,5]（假设存在同步辐射机制，并且电子是非相干辐射的）；(*c*) 通过射电源光学对应体红移推出的距离。这一问题最初出现于类星体射电源 *CTA* 102，据肖洛米斯基报道，该射电源在940兆周/秒处的变化周期大约为100天 [6]，在此由它的红移1.037（参考文献7）计算出来的距离是 (*a*) 和 (*b*) 所允许的上限的近1,500倍。事实上，在最近的研究中，我们已经提出了 *CTA* 102 的一个模型 [8] 以解决上面的分歧，但是该模型构建的不大可能的几何形态却令人难以接受。自那以后，莫尔特比和莫菲特就曾报道 [9]，在肖洛米斯基的观测开始之前两年，*CTA* 102 在970兆周/秒附近大约三年的时间都没有发生明显变化。因此，*CTA* 102 可能只是偶尔变化，而肖洛米斯基的观测很可能存在错误。

类星体射电源 3*C* 273 *B* 也存在相同的问题。登特发现在频率为8,000兆周/秒处，该射电源的流量密度在3年时间里增加了大约40%[1]。该射电源在至少大于200兆周/秒的频谱区域内是不存在自吸收的（参考文献10）（事实上，其频谱在高于200兆周/秒的频率范围内都是很平坦的）。对于大小为 10^{-3} 高斯的磁场强度，其角直径下限为0.03角秒（角直径观测值为0.5角秒 [11]）。如果射电变化的周期与光学波段变化的周期（约12年）相当，那么该射电源的尺寸不可能超过3秒差距，其距

optical variations, that is ~12 years. The size of the radio source cannot exceed 3 pc and its distance cannot exceed 20 Mpc. On the other hand, the red shift of 0.158 (ref. 12) implies a distance of 470 Mpc.

The discrepancy would be resolved if the varying part of the source had an angular diameter $\sim 10^{-3}$ sec. We therefore propose[13]:

Model I. This model consists of a source of angular diameter 0.5 sec and (roughly constant) flux density ~25 flux units down to at least 200 Mc/s, in the centre of which is a varying source of angular diameter 1.3×10^{-3} sec (corresponding to a linear diameter of 3 pc), and at minimum a flux density ~2.5 flux units down to the frequency ν_a at which self-absorption sets in. Such self-absorption would not show up in the observed spectrum of this composite source. For a magnetic field H gauss in the central source, $\nu_a \sim 4{,}000\ H^{1/5}$ Mc/s. For simplicity we assume that the central source varies because of an increase in its relativistic electron flux, while its angular diameter and magnetic field are unchanged. If the total flux density of the source is doubled at high frequencies at maximum, the flux from the central source must increase by a factor ~10. Thus the frequency ν_b at which self-absorption sets in at maximum is given by $(\nu_b/\nu_a)^{2.5} \sim 10$, so $\nu_b \sim 10{,}000\ H^{1/5}$ Mc/s.

We can now predict the time variations that would be observed at various frequencies. For simplicity we take the central source to vary sinusoidally at high frequencies. For $\nu > \nu_b$ no self-absorption occurs, so the observed variation would be sinusoidal. In the intermediate range $\nu_b > \nu > \nu_a$ the flux density would at first increase sinusoidally, but when it reaches the level at which it would be self-absorbed at that frequency the increase would be cut off, and the flux density would remain constant until it decreases again in the second half of the cycle. For $\nu < \nu_a$ the variations would be negligible.

A lower limit on H (the value of which has not yet been specified) can be obtained if we require that the energy density of relativistic particles shall not exceed the magnetic energy density. If only ~1 percent of the total energy is in the electrons, we find that the field strength cannot be much less than 1 gauss. If $H=1$ gauss, ν_a and ν_b have the values ~4,000 and ~10,000 Mc/s respectively, and the expected variations at 8,000 Mc/s are large enough to account for the observations. Our estimates for ν_a and ν_b are only rough, but they indicate what might be expected, and the measurement of their values and of the variations at intermediate frequencies would be a useful test of the model and would provide considerable information about the structure of the source.

It is not known how electrons attain relativistic energies in radio sources. The generation process may occur throughout the volume of the source, or the electrons may be ejected from a massive object at its centre. It is easily seen that the electrons must be produced by a mechanism of the former kind if the source in fact resembles Model I. The reason is that the lifetime of an electron which radiates at, say, 10^4 Mc/s in a field of 1 gauss is only $\sim 5\times10^6$ sec, which is much less than the time taken to cross the source, even for an electron moving at the speed of light. If such an electron were ejected from the centre it

离不会超过 20 兆秒差距。另一方面，大小为 0.158（参考文献 12）的红移量意味着其距离为 470 兆秒差距。

如果射电源可变部分的角直径约为 10^{-3} 角秒，那么上文中的不一致就会得到解决。因此，我们提出了以下模型[13]：

模型 I 在该模型中，射电源角直径为 0.5 角秒，在大于 200 兆周 / 秒的频谱范围内流量密度（基本恒定）约为 25 个流量单位；在该射电源的中心区域，存在一个角直径为 1.3×10^{-3} 角秒的变源（对应线直径为 3 秒差距），在自吸收频率 v_a 内，其最小流量密度约为 2.5 个流量单位。但是，自吸收在该复合源的观测谱中并不会出现。若该中心源磁场强度为 H 高斯，那么频率 v_a 约为 4,000 $H^{1/5}$ 兆周 / 秒。为了简化模型，我们假设中心源由于其相对论电子流量的增加而发生变化，但是其角直径和磁场保持不变。如果该射电源极大时在高频的总流量密度增加一倍，那么中心源的流量则必须增加到近 10 倍。因此，自吸收区最高频率 v_b 可以由 $(v_b/v_a)^{2.5}$ 约为 10 得出，即 v_b 约为 10,000 $H^{1/5}$ 兆周 / 秒。

下面，我们预估各种频率下可能观测到的时变。为了简化，我们假定中心源在高频区域按正弦变化。由于在 $v > v_b$ 的区域不存在自吸收，因此实际观测到的时变应是正弦变化。在频率区间 $v_b > v > v_a$ 范围内，某频率处的流量密度首先按正弦变化增加，当达到自吸收阈值时就会停止增加，并保持恒定直至在后半周期继续降低。$v < v_a$ 范围内的变化可以忽略。

如果要求相对论性粒子的能量密度不能超过磁场能量密度的话，我们就可以得到 H 的下限（这个值尚未指定）。如果电子的能量只占总能量的 1%，我们发现磁场强度不可能远小于 1 高斯。如果 H=1 高斯，那么 v_a、v_b 的大小分别约为 4,000 兆周 / 秒和 10,000 兆周 / 秒，频率 8,000 兆周 / 秒处的预期变化足够大，以解释观测结果。尽管我们对 v_a、v_b 的估计有些粗略，但是在一定程度上仍然可以说明问题。实际测量 v_a、v_b 的频率值，以及测量的该频率区间内的流量变化，对于检验这一模型是非常有用的，同时可以提供关于该射电源结构的重要信息。

电子在射电源中是如何获取相对论能量的，这一问题尚不清楚。这一能量产生过程可能遍布于整个射电源，也可能是电子被射电源中心处的致密天体喷射出。然而，很显然的一点是，如果射电源符合模型 I，电子的能量产生机制必须符合前一种情况。原因在于，在强度为 1 高斯的磁场中，辐射频率为 10^4 兆周 / 秒的电子的寿命仅为 5×10^6 秒左右，远小于它穿过该射电源所需的时间（即使以光速射出）。如果电子由中心喷射出，它将难以到达射电源的边缘，除非射电源对于电子的辐射频率

would not reach the edge, unless it were emitting radiation at a frequency at which the source was opaque. Thus for frequencies exceeding v_a our discussion of the variations will be incorrect. To allow for this we now introduce:

Model II. In this model we assume that the electrons are all generated in a very small region (with diameter < 0.1 pc), which presumably contains a massive object. As in Model I, nearly all the flux at frequencies < ~4,000 Mc/s is assumed to come from a halo, and we only concern ourselves with an intense spherically symmetric core with radius ~1 pc, in which the variable higher frequency radio flux is supposed to originate. We assume that H ~1 gauss throughout this region, and that its radio emission has a flux density $S(\nu)$ which may be variable. For each ν we can determine the minimum radius $R(\nu)$ of a sphere from which $S(\nu)$ could come, if it in fact comes from a region which is opaque at all frequencies up to ν. We find that $R(\nu) \sim 0.3\ (S(\nu))^{1/2}(\nu/10^4)^{1.25}$ parsecs, when S is measured in flux units and ν in Mc/s. This radius can be compared with the distance $r(\nu)$ which an electron radiating at frequencies ~ν would travel in its lifetime, assuming that it moves with speed ~c. This is $\sim 5\times 10^{-2}(\nu/10^4)^{-1/2}$ pc. If $R(\nu) > r(\nu)$, electrons moving outward from the centre cannot radiate freely without producing a radiation field the brightness temperature of which at frequency ν is higher than their kinetic temperature. Therefore they will conserve most of their energy until they reach a distance ~ $R(\nu)$ from the centre, where they can radiate freely.

The available data do not enable us to specify the flux density of the core precisely, but it is consistent with the observations to take $S(\nu)$ to be 10–30 flux units in the frequency range from 5,000 up to ~10^5 Mc/s. $R(\nu)$ is then several times greater than $r(\nu)$ throughout this range, and so we conclude that the radiation at frequency ν comes from the surface of a sphere of radius $R(\nu)$.

If electrons are ejected from the centre at a steady rate, $S(\nu)$ and $R(\nu)$ will adjust themselves to values which depend on the energy spectrum of the input electrons, and the flux of the source will be constant. If, however. The rate of input of electrons alters, $R(\nu)$ (and consequently $S(\nu)$) will change after the lapse of an interval of the order of the time for light signals to travel a distance $R(\nu)$. Thus changes in the behaviour of the massive object will cause changes in the radio flux. The maximum late of increase of flux will be attained if $R(\nu)$ expands with speed ~c. An increase of $S(8\times 10^3)$ from 2.5 to 25 flux units. Which corresponds to an increase in $R(8\times 10^3)$ from ~0.6 to ~1.9 pc, could therefore certainly occur rapidly enough to explain Dent's observations. We can also predict, on the basis of this model, that more rapid variations could occur at higher frequencies. Furthermore, if a sudden change in the rate of injection of particles produces changes in $S(\nu)$ at different frequencies, we would expect to observe the variations at higher frequencies before those at lower (even in the absence of intergalactic dispersion[14]).

We have seen that, if the core emits a significant amount of radiation at, say, 10^5 Mc/s, this radiation must come mainly from a sphere which is opaque up to that frequency. The total energy density of all the radio frequency radiation up to 10^5 Mc/s within it will be

是不透明的。因此，对于大于 ν_a 的频率范围，我们关于流量变化的讨论存在不妥之处，所以我们现在需要引入下面的模型：

模型 II　在该模型中，我们假设所有的电子均产生于一个非常小的区域（直径小于 0.1 秒差距），该区域可能包含了一个大质量天体。正如模型 I 中，假定小于约 4,000 兆周 / 秒的频率范围内的所有流量几乎全部来自一个晕。这样，我们只需关心半径近为 1 秒差距的致密球形对称核，并假定高频可变射电流量就来自该区域。不妨设整个区域内的磁场强度 H 约为 1 高斯，射电辐射的流量密度为可变的 $S(\nu)$。如果辐射流量确实来自某一区域，而该区域对小于 ν 的频率都不透明，那么对于每个频率 ν，我们就可以确定一个最小半径 $R(\nu)$ 的球体，而 $S(\nu)$ 就来自该球体。我们发现，半径 $R(\nu)$ 约为 $0.3(S(\nu))^{1/2}(\nu/10^4)^{1.25}$ 秒差距，其中 S 的单位为流量单位，ν 的单位为兆周 / 秒。即使假设电子的速度近似为光速，半径 $R(\nu)$ 与辐射频率近似 ν 的电子在其寿命内所走的距离 $r(\nu)$ 也是可比较的，该距离约为 $5\times10^{-2}(\nu/10^4)^{-1/2}$ 秒差距。如果 $R(\nu)>r(\nu)$，那么从中心向外运动的电子就不能够自由辐射，除非产生一个在频率 ν 时的亮温度大于动力学温度的辐射场。这样，电子就会一直保存大部分能量，直至接近到达某一距离 $R(\nu)$ 后才能够自由辐射。

目前，我们尚不能通过已有的数据准确确定核心区域的流量密度。但如果将从 5,000 兆周 / 秒到 10^5 兆周 / 秒的频率区间内的流量密度取为 10~30 个流量单位，这些数据就会符合观测结果。那么，在这一区间内，$R(\nu)$ 就会比 $r(\nu)$ 大若干倍。这样，我们可以得出结论，频率为 ν 的辐射主要来自半径为 $R(\nu)$ 的球的表面。

如果电子以稳定速率从中心喷射出，$S(\nu)$ 与 $R(\nu)$ 就会根据输入电子的能谱适当调整到某一取值，该射电源的流量将保持恒定。然而，如果电子的输入速率改变，$R(\nu)$（与对应的 $S(\nu)$）将在某一时间间隔（量级与光穿过距离 $R(\nu)$ 所需时间相当）之后随之改变。因此，大质量天体的行为改变将会导致射电辐射流量的变化。如果 $R(\nu)$ 以近似光速 c 膨胀，则可以求得流量增加的最大速率。$R(8\times10^3)$ 从 0.6 秒差距左右增加到 1.9 秒差距左右，$S(8\times10^3)$ 将随之从 2.5 个流量单位增加到 25 个流量单位，这个增加速率足以解释登特的观测结果。基于这一模型，我们也可以预测，在更高的频率可以出现更快的变化。而且，如果注入粒子速率的突然改变将导致 $S(\nu)$ 在不同频率时的变化，我们可以预期在高频观测到的变化会比低频出现的早（即使不存在星系间散射 [14]）。

在上文中，我们已经分析，如果核心区域在某频率（比如 10^5 兆周 / 秒）发射大量的辐射，这些辐射一定主要来自对小于该值的频率范围不透明的某一球体。那么，该球体内频率小于 10^5 兆周 / 秒的所有射电辐射的总能量密度约为 6×10^{-3} 尔格 / 厘米^3

$\sim 6\times10^{-3}$ ergs/c.c.(compared with $\sim 10^{-6}$ ergs/c.c. within Model I), and this very high value suggests that inverse Compton scattering might be important. The Compton lifetime of an electron of energy $\gamma m_0 c^2$ in this radiation field is $8\times10^9/\gamma$ sec. If $S(10^5)$ ~25 flux units, $R(10^5)$ ~0.1 pc and the time which an electron takes to drift out of this sphere is $\sim 2\times10^7$ sec if its outward velocity $\sim c/2$. But the electrons radiating at 10^5 Mc/s have γ ~300, and so their Compton lifetime is of the same order. The most energetic photons would result from the scattering of 10^5 Mc/s photons by electrons with γ ~300, and their frequency would be[15] $\sim 10^{10}$ Mc/s, which is in the ultra-violet range. Photons of lower energy (including the whole visible range) would be produced, and the spectrum of the scattered radiation would have a low-energy cut-off at a frequency depending on the smallest value of γ represented in the electron energy spectrum. The energy radiated by this process in the visible range may well be greater than that of the synchrotron radio emission. The exact relative importance of the Compton and synchrotron losses is very sensitive to the high-frequency cut-off in the radio spectrum (which we have taken as 10^5 Mc/s) to the value of *H*. to the precise position where the electrons are accelerated. And to their mean rate of outward drift. It would therefore clearly not be worth while to base an exact calculation on this crude model.

The foregoing rough arguments do, however, suggest that at least a part of the visible light emitted by 3*C* 273 *B* may have been produced by Compton scattering. Moreover, such a hypothesis would provide a natural explanation for the observed optical fluctuations of the source, since, if electrons are accelerated in the massive object in irregular bursts, there will be variations in the intensity of the scattered light. Nearly all the scattering occurs in a region of radius ~0.1 parsecs (since not only the radiation energy density, but also the particle density, is much higher near the centre). Therefore, fluctuations with time-scales of the order of months, or even less may occur, and these are in fact observed[16]. The long period variations[2] (~12 years) in the optical luminosity may also arise from periodic variations in the rate of injection electrons, which, as we have already seen, can produce the observed variation at radio frequencies. This model therefore suggests that the radio and the long-period optical variations may be connected.

The increase in the radio flux observed by Dent coincided with a decrease in the optical luminosity, which had the most recent of its 12-yearly maxima in 1962. But according to this model one would expect periodic radio variations to lag (perhaps by several years) behind the variations at centre which cause them, so this fact also accords with our model, and suggests that the radio variations, when they have been observed for longer, will also turn out to have a period ~12 years.

The visible light from 3*C* 273 *B* is unpolarized[17] whereas the radio flux (at ~3,000 Mc/s) is ~3 percent polarized[18]. It will not be possible to find out whether the central part of the source is polarized at radio frequencies until the extent to which the degree of polarization changes when the flux varies has been examined. However, even if the radio flux from the core were polarized (implying large-scale uniformity in the direction of *H*), it would not follow that the scattered visible light must appear polarized, since it is mainly produced

（在模型 I 中，该值约为 10^{-6} 尔格 / 厘米^3），这么高的密度值表明，逆康普顿散射可能起到了很重要的作用。在该辐射场中，能量为 γm_0c^2 的电子的康普顿寿命为 $8\times10^9/\gamma$ 秒。如果流量密度 $S(10^5)$ 取值约为 25 个流量单位，那么 $R(10^5)$ 约为 0.1 秒差距；假定电子向外的运动速率约为 $c/2$，那么电子穿出球体的时间约为 2×10^7 秒。对于辐射频率为 10^5 兆周 / 秒的电子，γ 取值约为 300，则电子的康普顿寿命和电子穿出球体的时间为同一量级。电子（γ 约为 300）对频率为 10^5 兆周 / 秒的光子的散射，将使这些光子获得最大能量，使其频率达到 [15] 约 10^{10} 兆周 / 秒，位于紫外光谱区。同时，散射也会产生较低能量（包括可见光范围）的光子，因此散射辐射谱存在一个较低的截止能量，对应的截止频率取决于电子能量谱中 γ 的最小值。这一过程产生的可见光辐射能量可能远大于同步射电辐射的能量。至于康普顿能量损失和同步辐射能量损失二者的相对重要性，则取决于以下若干因素：射电谱中的高频截止频率（我们取值为 10^5 兆周 / 秒）、磁场强度 H、电子加速的准确位置以及电子向外漂移的平均速率。由于这一模型较为粗糙，我们也无需进行过多的精确计算。

尽管上文论述略显粗糙，但是说明了 3*C* 273 *B* 发射的可见光至少有一部分是由康普顿散射产生的。而且，这一假说为观测到的射电源光变提供了自然合理的解释，即如果电子在不规则爆发的大质量天体中加速，散射光的强度就会发生变化。几乎所有的散射都发生在一个半径约为 0.1 秒差距的区域（因为在靠近中心的区域，辐射能量密度和粒子密度都高得多），所以会发生周期为数月（或更短）的起伏，这也正是实验所观测到的结果 [16]。光学波段的长周期变化 [2]（约 12 年）也可能是由于电子注入速率的周期变化造成，而这也可以引起（正如我们所观测到的）射电频率的变化。所以，这一模型认为射电频段变化和长周期光学变化可能是彼此联系的。

登特观测到的射电流量的增加正好与光学光度的减小相符合，该光学光度值曾在 1962 年达到 12 年以来的最大值。另外，这个模型认为射电的周期性变化落后（可能是几年）于中心区域的变化，而后者恰恰是前者形成的原因，因此该模型与前面的观测事实也一致，这意味着，如果对射电变化进行长期观测，就会发现其周期也近似为 12 年。

3*C* 273 *B* 发射的可见光是非偏振的 [17]，但射电流量（在频率约为 3,000 兆周 / 秒处）存在大约 3% 的偏振 [18]。在我们检测确定流量变化时偏振度的变化达到何种程度之前，我们仍无法弄清射电源中心区域在射电频段是否偏振。然而，即使来自中心区域的射电流量是偏振的（这意味着在磁场强度 H 方向存在大尺度的均匀性），散射的可见光也不一定是偏振的。因为这些可见光主要产生于一个极不透明的区域，

in an extremely opaque region, where the radio-frequency radiation will not be polarized even if the field is uniform.

It has been assumed in the foregoing that H does not change significantly when the density of the relativistic particles alters. This will be a good approximation if the magnetic energy is greater than the particle energy (a strength of 1 gauss is sufficient). The field is presumably "anchored" to the massive object.

It should be emphasized that the Compton losses are significant in this model mainly because of the occurrence of synchrotron self-absorption, which prevents the electrons from radiating away their energy as fast as they otherwise would. Any electrons which are accelerated to high energies ($\gamma>300$, say), and which therefore emit synchrotron radiation at frequencies which can escape freely from the source, will lose their energy in a time much shorter than the Compton lifetime, so they will contribute to the radiation at high radio frequencies ($> 10^5$ Mc/s), rather than in the far ultra-violet. ($> 10^{10}$ Mc/s).

Model III. The first two models have taken no account of the fact that lines are present in the optical spectrum of 3C 273 *B*. According to Greenstein and Schmidt[19], these lines could be produced by gas of density $\sim 10^7$ particles/c.c., temperature $\sim 2\times 10^4$ °K, and mass $\sim 10^5$ $M_\odot$. Though this gas need not radiate all the optical continuum, it must be incorporated in a complete model of the source. The model which we shall now describe indicates one way in which this can be done.

If the gas were in the form of a uniform spherical cloud, its diameter would be ~0.5 pc, and it would be opaque to light of all frequencies because of scattering by free electrons. It would therefore be difficult to account for the observed fluctuations in optical luminosity with a time-scale of a few months, since if the variable flux were emitted by a small region within the cloud its variations would be smeared out and would not be observed. This difficulty would be eased if the density within the cloud were non-uniform, or if it had a filamentary structure[20], but there would remain the problem of feeding energy into the gas to balance its losses.

In an alternative configuration, suggested by Shklovsky[21], the gas is distributed in a thin spherical shell. We shall adopt this suggestion, and show that a model can be constructed on the basis of which the radio variations, and both the long and short period optical variations, can be explained.

If the radius of the shell is taken[21] as 4×10^{18} cm, and its density as $\sim 10^7$ particles/c.c., its thickness will be $\sim 10^{17}$ cm if its total mass is $\sim 10^5$ $M_\odot$. The observed broadening of the spectral lines places an upper limit of ~1,500 km/sec on the velocity of the shell if it is expanding. If there is a mass of $\sim 10^9$ $M_\odot$ at its centre, a field of ~1 gauss would be strong enough to prevent the shell collapsing. Scattering by free electrons would be unimportant at all frequencies. The shell will be transparent to most photons of optical frequency, but will be opaque to radio waves at all frequencies below $\sim 10^5$ Mc/s because of free-free

即使该区域的场是均匀的，射频辐射在该区域也不会发生偏振。

在上述分析中，我们已经假定，当相对论性粒子密度发生变动时，磁场强度 H 并不会明显变化。如果磁能大于粒子的能量（1 高斯的磁场强度足够），上面的假定将是一个很好的近似。因此可以推测，磁场主要“固定于”大质量天体。

需要强调的一点是，在该模型中，只有在出现同步加速自吸收时康普顿耗损才会变得明显，因为自吸收会阻止电子能量快速的辐射。当加速电子达到很高的能量（比如说 $\gamma > 300$）时，就会在那些能够自由穿透射电源的频率上产生同步辐射，之后这些电子会在很短的时间内（比康普顿寿命小得多）损失掉能量。因此，这些高能电子产生高频（$> 10^5$ 兆周 / 秒）辐射，而不是远紫外（$> 10^{10}$ 兆周 / 秒）辐射。

模型 III　上面的两个模型并没有考虑出现在 3*C* 273 *B* 光谱中的光学波段谱线。根据格林斯坦和施密特的研究 [19]，这些谱线可能由密度约为每立方厘米 10^7 个粒子、温度约为 2×10^4 K、质量约为 10^5 个太阳质量的气体产生。尽管这些气体并不辐射所有的光学波段连续谱，但必须将其整合到射电源完整模型中。下面我们将阐述一种整合的思路。

如果这些气体是以均匀球状云的形式存在的话，其直径将为 0.5 秒差距左右。由于自由电子的散射，这些气体对于所有频率的光都将是不透明的。如果变化的流量来自这团气体云中的一小块区域，则这种变化就会变得模糊不清并且难以观察，因此实际观测到的周期为数月的光度起伏会难以解释。如果气体云的密度分布是不均匀的，或具有丝状结构 [20]，则以上困难就会消失。然而，这样就会遇到另一问题，即如何将能量输入气体以平衡其损失。

什克洛夫斯基曾经提出另一种结构 [21]，认为气体分布在一个薄球壳中。下面我们将采纳这一建议，并以射电变化为基础建立一个模型，从而可以同时解释长周期和短周期的光学波段的变化。

不妨令该球壳的半径 [21] 为 4×10^{18} 厘米，密度约为每立方厘米 10^7 个粒子，总质量约为 10^5 个太阳质量，那么球壳的厚度约为 10^{17} 厘米。如果该球壳在膨胀的话，可以通过观测到的谱线展宽计算出膨胀速度的上限近似为 1,500 千米 / 秒。如果其中心质量约为 10^9 个太阳质量，那么约 1 高斯的磁场强度足以支撑球壳。这样，在所有频率上，自由电子的散射变得不再重要。该球壳对于光学波段的大多数光子都是透明的，但是由于自由–自由吸收，对于低于约 10^5 兆周 / 秒的所有射电频段都是不

absorption. The observed radio flux must therefore originate outside the shell.

The gas in the shell would produce the observed line emission, and would radiate at a rate $\sim 10^{46}$ ergs/sec, mainly at visible wave-lengths. Its thermal energy content is insufficient to maintain this rate of energy loss for more than a few weeks, and so it must be absorbing energy at an equal rate from the interior. Since the shell absorbs radio waves, intense but unobserved radio radiation may exist in the spherical region within it, and this suggests the possibility that the shell may be absorbing enough energy in the form of radio waves to compensate for its losses in the visible range. This would require the production of $\sim 10^{46}$ ergs/sec at frequencies below $\sim 10^5$ Mc/s, which is ~ 30 times as great as the observed power radiated in this frequency range. However, such a high rate of energy production is not unreasonable since a sphere within which synchrotron self-absorption were taking place at all frequencies up to 10^5 Mc/s would need to have a radius of only ~ 0.8 pc (little more than half the radius of the shell) to emit $\sim 10^{46}$ ergs/sec. Alternatively the required amount could be produced if the whole interior were opaque up to $\sim 7\times 10^4$ Mc/s. The exact situation within the shell will depend on the energy spectrum of the electrons injected into it, but a sufficiently high rate of energy generation will be achieved if relativistic electrons with $\gamma < 300$ are produced at a rate $\sim 10^{46}$ ergs/sec. We must assume that there is a cut-off in the electron energy spectrum for $\gamma > 300$, since otherwise there would be intense radiation at frequencies above 10^5 Mc/s which would escape through the shell and produce observable radiation with a higher flux-density than is observed.

The observed radiation at frequencies below 10^5 Mc/s (apart from the component which comes from the halo) is, according to this model, emitted outside the shell by electrons which have escaped from the interior. An electron ejected from the massive object at the centre will be absorbing and emitting radiation at the same rate until it approaches the shell. Since the synchrotron lifetime of the electrons (in the absence of self-absorption) is of the order of the time which they take to pass through the shell, most electrons will lose their energy before reaching the exterior. Thus there radiation will not be observed directly, but will simply heat the gas. However, since the power of the observed radio radiation is much less than the power required to heat the gas, the observed radio flux would be produced even if only a few percent of the electrons succeeded in escaping. Despite the high density of the gas in the shell, an electron passing through it has a 90 percent chance of reaching the exterior without losing a significant amount of energy in ionization and collisional losses. If the field ~ 1 gauss extends outside the shell, these electrons will radiate away their energy before they have travelled a further distance ~ 0.1 pc unless self-absorption occurs outside the shell as well. Such self-absorption will occur at sufficiently low frequencies, though the situation is somewhat complicated by the fact that the electrons can radiate energy inwards into the shell.

The production of the visible light is more complicated. Much of it, both in spectral lines and in the continuum, comes from the gas in the shell. However, Compton scattering may also, as in Model II, produce radiation at optical frequencies. The radio radiation

透明的。因此，观测到的射电流量一定来自球壳之外。

球壳中的气体将以约 10^{46} 尔格 / 秒的速率辐射观测到的谱线，且辐射主要集中在可见光波段。由于其热能难以在数周之内一直维持其能量损失速率，因此必须还要以相同速率从内部吸收能量。既然该壳层吸收射电波段，那么其内部的球形区域也可能存在着难以观测的强射电辐射。这表明，球壳很可能依靠吸收射电波段能量来补偿其可见光波段的能量损失。为了维持这种平衡，在低于约 10^5 兆周 / 秒的频段能量生成速率需要达到约 10^{46} 尔格 / 秒，而这一理论值约是该频段实际观测功率值的 30 倍。但是如此高的功率并不是不合理，由于在低于 10^5 兆周 / 秒频率下发生同步辐射自吸收的球体半径仅需要 0.8 秒差距（和球壳的半径一半差不多）来进行约为 10^{46} 尔格 / 秒的放射。另外，如果整个球壳内部对小于约 7×10^4 兆周 / 秒的频率都是不透明的，则可以得到另一数值。球壳内的具体情况依赖于注入该壳层的电子的能量谱，如果 $\gamma<300$ 的相对论电子以约 10^{46} 尔格 / 秒的速率产生，就能达到足够高的能量产生率。我们必须假设对于 $\gamma>300$ 的电子能量谱存在着截止，否则在高于 10^5 兆周 / 秒的频段也会出现强辐射，这些辐射将穿出球壳，产生的流量密度就会高于实际观测值。

根据该模型，实际观测到的低于 10^5 兆周 / 秒频段的辐射（除了其中一部分来自晕）是由内部逃逸出的电子发射的。由中心大质量天体喷射出的电子将以相同的速率吸收和发射辐射，直至这些电子到达球壳。既然电子同步辐射的寿命（不存在自吸收的情况下）与它们穿出球壳的时间同量级，那么大多数电子在到达外部之前将损失掉能量。所以，相应的辐射只是将气体加热，但不能直接观测到。然而，由于可被观测的射电辐射的功率比用于加热气体的功率小得多，那么即使只有一小部分电子成功逃逸，也可以产生出观测到的射电流量。事实上，尽管球壳中的气体密度很高，电子经电离和碰撞穿过球壳到达外部而不明显损失能量的概率仍高达 90%。如果磁场强度（约 1 高斯）延伸至球壳外面，这些电子在运动约 0.1 秒差距的距离之前就已经辐射掉了能量，除非球壳外面也存在自吸收。但这些自吸收一般只发生于足够低的频率。事实上，逃逸出球壳的电子也会向球壳内辐射能量，此时的情况会变得有些复杂。

可见光的产生会更复杂一些。无论是谱线还是连续谱，大部分可见光来自球壳中的气体。然而，正如模型 II 那样，康普顿散射可能也会产生光学波段的辐射。如果同步加速自吸收存在于低于约 7×10^4 兆周 / 秒的频段，那么球壳所包围的射电辐

energy density in the volume enclosed by the shell is $\sim 2\times10^{-3}$ erg/c.c. if synchrotron self-absorption occurs at frequencies up to $\sim 7\times10^{4}$ Mc/s, and the corresponding Compton lifetime of an electron is $2\times10^{10}/\gamma$ sec. The highest energy electrons the synchrotron losses of which are balanced by self-absorption have $\gamma \sim 250$, and their Compton lifetimes are of the same order as the time they would take to travel outwards from the centre to the shell. The Compton losses, therefore, are significant, though their exact magnitude is highly sensitive to the precise model adopted. The frequencies of the scattered photons will, as in Model II, be mainly in the visible and near-ultra-violet, and most photons will escape through the shell, though those the frequencies of which correspond to intense spectral lines may the absorbed. Compton scattering of radio frequency radiation outside the shell will be insignificant.

We now consider how the observed variations in flux can occur in this model. We discuss first the long period variations. If there were fluctuations with a time-scale ~12 years in the rate of production of relativistic particles at the centre, there would be changes in the radiation density in the sphere enclosed by the shell, and a consequent change in the temperature of the gas. The rate of radiation of visible light by the shell would therefore alter. There would also be changes in the rate at which electrons escape through the shell, and consequently in the observed radio flux. Variations with periods less than, say, 5 years would be partly smeared out and so could not have a large amplitude.

The observed short period fluctuations in the optical luminosity have a time-scale of a few months, and obviously cannot be produced in the shell, which has a radius of ~4 light years. In Model II they were attributed to the inverse Compton effect. Since we have shown that light will be produced by Compton scattering in this model, a similar explanation might also be possible here. This would require the scattering to occur in a region smaller than a few light months in radius. However, the scattering will not be concentrated in such a small region as in Model II, where the radiation energy density was high enough for scattering to be significant only within a sphere of radius ~0.1 pc. Nevertheless, if all the electrons are ejected from a massive object, their number density will be higher near the centre, and the amount of scattering will be greater there even if the radiation density is no higher. Consequently, the variations in the component of the optical continuum produced by the Compton effect may be sufficiently rapid to account for the observed variations with time-scales of a few months.

We note that whereas the line strengths will alter with the slow period ~12 years, it would be inconsistent with this model for them to be involved in the rapid fluctuations in optical luminosity. We understand that attempts are now being made to see whether the lines do in fact vary in intensity. This feature of our model will thus soon be tested.

If we could construct a complete model of this kind, we would be able to calculate the expected phase lag between the variations in luminosity at optical frequencies and the related radio variations. However, we lack sufficient information to enable us to do this. To illustrate the complexity of the problem we enumerate some of the factors which would

射能量密度约为 2×10^{-3} 尔格 / 厘米^3，而电子相应的康普顿寿命则为 $2\times10^{10}/\gamma$ 秒。其中，具有最高能量的电子的 γ 值大约为 250，其同步辐射损失由自吸收进行补偿，并且这些电子的康普顿寿命与电子从中心到穿出球壳的时间具有相同的量级。因此，尽管康普顿耗损的具体值与采用的精确模型密切相关，但其损失无疑是很明显的。如模型 II，散射光子的频率主要位于可见光和近紫外区，而且尽管一些较强的谱线对应的频率可能会被吸收，但大多数的光子可以从球壳逃逸出。在球壳外，射频辐射的康普顿散射将变得很少。

下面，我们用模型分析所观测到的流量变化是怎样产生的。我们首先分析长周期变化。如果球心相对论性粒子的产率存在约 12 年时标的起伏，那么壳层包含的球体内的辐射密度就会发生变化，气体的温度也会随之变化。这样，由壳层产生的可见光辐射的速率会因此改变，而同时电子逃逸出球壳的速率以及观测到的射电流量也会发生变化。周期小于 5 年的变动则会部分消减，因此相应的起伏幅度会变小。

观测到的光学波段的短周期光度起伏，时间周期为数月，而球壳的半径大约为 4 光年，因此这种短周期起伏显然不可能是由球壳内部产生的。在模型 II 中，认为这种短周期起伏是由于逆康普顿效应产生的。上文中已经提到，该模型中康普顿散射可以产生光，那么类似的原理也可以在这儿用于解释短周期起伏。这就要求散射必须发生于一个较小的区域（半径小于数光月）。然而，这种散射并不会集中于类似模型 II 那么小的区域（在模型 II 中，辐射能量密度足够高，因此使得散射主要集中在一个半径约为 0.1 秒差距的球形区域）。不过，如果所有的电子都是由一个大质量天体喷射出来的，靠近中心的粒子数密度自然会很高，那么即使辐射密度不是很高，散射光强度也会很大。因此，由康普顿效应产生的光学波段连续谱的变化可能足够快，进而可以解释观测到的周期为数月的变化。

然而，我们注意到，在该模型中谱线的强度将以约 12 年的周期缓慢变化，这与上面光度快速起伏的模型不符。我们知道人们正在进行尝试，以确定观测谱线的强度是否真的存在变化。这样就可以很快验证我们模型的这一特征了。

如果我们能够建立一个类似的完整模型，那么我们就可以通过计算来预测光学波段光度变化与相关的射电变化的相位差。遗憾的是，目前我们仍然缺少足够的信息来这样做。为了说明这一问题的复杂性，我们在下面列举了一些用于决定相位差

determine the magnitude of the lag.

(1) The variations in particle density and radiation energy density propagate outwards with different velocities.

(2) The phase delay of the radio variations will depend on whether self-absorption is taking place outside the shell at the frequency of observation. If it is taking place the observed flux would come from the surface of a sphere with radius greater than 1.5 pc, rather than from just outside the shell, thus increasing the lag of the radio variations. At 8,000 Mc/s, the frequency of Dent's observations, self-absorption would occur if more than ~14 flux units of the observed flux density were coming from the compact source (rather than from the halo). Since the observed variations amount to 10 flux units (and this flux cannot come from the halo), it is likely, though not certain, that self-absorption is in fact occurring.

(3) The sense in which the emission rate of the gas in the shell varies as the additional radio energy falls on is must be known.

(4) The relative intensity of the light produced by the inverse Compton effect and the light emitted by the shell must be known. These two components will have the same period, but may be out of phase with one another. The times at which maxima in the resultant optical luminosity occur will thus depend on the phase lag between the two components, and on the ratio of the amplitude of their variations.

This discussion shows that existing observations do not suffice to determine a well-defined working model of 3*C* 273 *B*. However, we can draw the important conclusion that neither the optical nor the radio variations require the source to be closer to us than the 470 Mpc implied by a cosmological red shift of 0.158. This conclusion considerably weakens the case for the local model[3,22] of quasi-stellar radio sources which places them a few megaparsecs away and in which the observed red shift implies a large velocity relative to their surroundings. More detailed observations of optical variations, and of radio variations as a function of frequency, should enable us to decide whether models of the type described here are appropriate for 3*C* 273 *B*, and for other quasi-stellar sources as well.

Note added in proof. H. Gent and H. P. Palmer reported at the Dublin meeting of the Royal Astronomical Society (September 7, 1965) that they have succeeded in resolving 3*C* 273 *B* at 21 cm with a long base-line interferometer. Their proposed interpretation of their observations is that the source consists of two components separated by ~0.4 sec of arc, the angular diameter of each component being less than ~0.1 sec. This limit on the angular size of the components is still much greater than the angular diameter of the variable core in our model ($\sim 10^{-3}$ sec), and of the lower limit of 3×10^{-2} sec for the halo.

(**208**, 371-374; 1965)

M. J. Rees and D. W. Sciama: Department of Applied Mathematics and Theoretical Physics, University of Cambridge.

大小的因素。

（1）以不同速度向外传播的粒子密度的变化和辐射能量密度的变化。

（2）射电变化的相位滞后取决于在观测频段自吸收是否发生于球壳之外。如果确实发生于球壳外，则观测到的流量来自半径大于 1.5 秒差距的球壳表面，而不是刚好在球壳外，因此也增加了射电变化的相位延迟。在频率 8,000 兆周 / 秒时，即登特观测所使用的频率，如果观测到的流量密度中超过近 14 个流量单位来自致密射电源（而不是来自晕）的话，就会发生自吸收。由于观测到的变化为 10 个流量单位（而这些流量不可能来自晕），因此虽不能肯定，但很可能实际上发生了自吸收。

（3）必须清楚球壳气体的发射速率随额外射电能量进入球壳的变化。

（4）必须清楚逆康普顿效应产生的光以及球壳发射的光的相对强度。二者具有相同的周期，但是可能存在相位差。合成后的光度最大值的发生时刻，取决于二者的相位差以及二者变化幅度的比值。

以上的讨论表明，已有的观测尚不足以确定 3*C* 273 *B* 完备的有效模型。然而，我们可以得出以下重要结论：无论是光学波段变化还是射电波段变化，都不需要射电源与我们的距离小于 470 兆秒差距（由宇宙学红移量 0.158 得出）。这一结论相当程度上削弱了类星体射电源的局域模型 [3,22]，该模型认为，观测到的红移意味着百万秒差距外的射电源与周围存在很大的相对速度。继续全面深入地观测光学波段和射电波段随频率的变化，可以使我们最终确定本文所描述的模型是否适用于 3*C* 273 *B* 以及其他类星体射电源。

附加说明：根据亨特和帕尔默最近在皇家天文学会都柏林会议上的报道（1965 年 9 月 7 日），他们已经使用长基线干涉仪在 21 厘米波段成功分辨 3*C* 273 *B*。他们认为，该射电源结构包含两部分，二者间隔约 0.4 角秒，其中每部分的角直径小于约 0.1 角秒。但是，在我们的模型中，可变内核的角直径约为 10^{-3} 角秒，晕的角直径下限为 3×10^{-2} 角秒，显然比观测值小很多。

（金世超 翻译；蒋世仰 审稿）

References:

1. Dent, W. A., *Science*, **148**, 1458 (1965).
2. Smith, H. J., *Quasi-stellar Sources and Gravitational Collapse*, 221 (Univ. Chicago Press, 1965).
3. Terrell, J., *Science*, **145**, 918 (1964).
4. Slish, V. I., *Nature*, **199**, 682 (1963).
5. Williams, P. J. S., *Nature*, **200**, 56 (1963).
6. Sholomitsky, G. B., *I. A. U. Inf. Sull. Variable Stars*, No. 83 (Feb. 27, 1965).
7. Schmidt, M., *Astrophys. J.*, **141**, 1295 (1965).
8. Rees, M. J., and Sciama, D. W., *Nature*, **207**, 738 (1965).
9. Maltby, P., and Moffet, A. T., *Astrophys. J.*, **142**, 409 (1965).
10. Dent, W. A., and Haddock, F. T., *Quasi-stellar Sources and Gravitational Collapse*, 381 (Univ. Chicago Press, 1965).
11. Hazard, C., Mackey, M. B., and Shimmins, A. J., *Nature*, **197**, 1037 (1963).
12. Schmidt, M., *Nature*, **197**, 1040 (1963).
13. Sciama, D. W., *Proc. Intern. School of Physics "Enrico Fermi"*, *Course* 35, July 12-24, 1965 (to be published).
14. Haddock, F. T., and Sciama, D. W., *Phys. Rev. Letters*, **14**, 1007 (1965).
15. Feenberg, E., and Primakoff, H., *Phys. Rev.*, **73**, 449 (1948).
16. Sandage, A., *Astrophys. J.*, **139**, 416 (1964).
17. Moroz, V. I., and Yesipov, V. R., *I. A. U. Inf. Bull. Variable Stars*, No. 31 (1963).
18. Morris, D., and Berge, G. L., *Astro. J.*, **69**, 641 (1964).
19. Greenstein, J. L., and Schmidt, M., *Astrophys, J.*, **140**, 1 (1964).
20. Schmidt, M., *Proc. Second Texas Conf. Relativistic Astrophysics* (to be published).
21. Shklovsky, I. S., *Soviet Astronomy*, **8**, 638 (1965).
22. Hoyle, F., and Burbidge, G. R. (to be published).

Antimatter and Tree Rings

V. S. Venkatavaradan

Editor's Note

Physicists attempting to explain the vast energy released in the Tunguska meteor strike of 1908 in Siberia had been driven to a radical hypothesis: that part of the meteor may have been made of antimatter. Researchers had indeed found evidence in the area for an altered ratio of the abundance of the isotopes carbon-14 and carbon-12, consistent with significant annihilation of matter and antimatter. Here, however, Indian physicist V. S. Venkatavaradan of the Tata Institute of Fundamental Research in Mumbai notes that the isotope ratios observed vary also with sunspot activity, and that this evidence therefore does not support the antimatter hypothesis. Scientists today believe the Tunguska event was caused by the impact of a large rock meteor.

RECENTLY, Cowan *et al.*[1] discussed the interesting case of the Tunguska meteor—the event and its origin mainly in the context of release of a rather high energy of $\sim 10^{24}$ ergs on its impact. Various theories concerning its origin and the nature of the energy source were discussed (for example, asteroidal origin and energy from impact or nuclear reactions). They have shown that none of the theories can satisfactorily explain the amount of energy released during the impact. The authors have invoked the antimatter hypothesis and, as an experimental verification to this, have calculated the expected increase in the carbon-14/carbon-12 ratios in the atmosphere subsequent to the fall of the meteorite. Considering the total energy release, they obtained a value of 7 percent for the expected increase in activity. Their measurements of the atmospheric carbon-14/carbon-12 ratios, based on annual rings of a 300-yr.-old tree, show a possible increase of 1 percent in the year 1909, leading them to the conclusion that probably 1/7th of the energy release in the Tunguska meteorite impact came from antimatter annihilation.

It is the purpose of this communication to point out that the probability of such an interesting conclusion is unfortunately very much reduced if one considers the nature of secular variations of carbon-14/carbon-12 ratios in the atmosphere. It was Stuiver[2] who first pointed out that there existed a good inverse correlation, for the past 1,300 yr. of record, between the solar activity and carbon-14/carbon-12 ratios in the atmosphere. (In what manner sunspot activity brings about this correlation is, however, not well understood as yet.) If we compare the solar activity and the observed carbon-14/carbon-12 ratio during 1870–1933, within the errors of measurements, we do find a fair anticorrelation between the sunspot activity and the carbon-14 deviations (see Fig. 1), there being some phase differences which are not unexpected because of time delays in interactions of relevance to the carbon-14/carbon-12 ratios, for example, air–biosphere and air–sea exchange.

反物质与树木年轮

文卡塔瓦拉丹

编者按

多年来，物理学家们都在试图解释 1908 年发生在西伯利亚的通古斯陨石撞击所释放出的巨大能量，甚至得出了一个极端的猜测：即一部分通古斯陨石可能是由反物质构成的。研究者们确实在这一地区发现了证据，即同位素碳-14 与碳-12 丰度的比例变化，这符合物质和反物质重要的相互湮灭原理。然而在本文中，孟买塔塔基础研究所的印度物理学家文卡塔瓦拉丹指出，这种同位素比例的变化同样与太阳黑子的活动有关，因此上述证据不足以支持这个反物质的猜测。今天，科学家都认为通古斯事件是由于一块巨大的陨石撞击地球引起的。

最近，考恩等人[1]讨论了有趣的通古斯陨石撞击事件——他们主要根据撞击时释放出的极高能量（约 10^{24} 尔格）对该事件及其起因进行了讨论。讨论内容涉及关于事件起因的各种理论以及能量来源的性质（例如，小行星起源，能量来自撞击还是核反应）。最后他们认为，没有一种理论能够圆满地解释撞击时所释放出的巨大能量。他们提出了反物质的假设，而且作为其假设的实验证据，他们还计算了陨石降落后大气中碳-14 与碳-12 比值增加的预期值。考虑到总的能量释放，他们计算出放射性强度的增加值为 7%。根据一棵树龄为 300 年的老树的年轮所测得的大气中碳-14 与碳-12 的比值，他们指出，在 1909 年该值可能增加了 1%，并因此得出结论：在通古斯陨石撞击中大约 1/7 的能量释放来自反物质的湮灭。

这篇通讯的目的在于指出：假如人们考虑到大气中碳-14 与碳-12 比值长期变化的特征，那么上述这个有趣结论成立的可能性就会大大降低。斯蒂尤艾弗[2]最先指出，在过去 1,300 年的记录中太阳活动与大气中碳-14 和碳-12 比值存在良好的反相关性。（但是，太阳黑子活动何以能产生此种相关性现在还不甚清楚。）假如我们比较一下 1870 年 ~1933 年期间的太阳活动和观测到的碳-14 与碳-12 的比值，在测量误差范围内我们确实可以发现在太阳黑子活动与碳-14 偏离值之间有相当好的反相关（参见图 1），由于一些与碳-14 和碳-12 比值相关的相互作用，例如大气–生物圈和大气–海洋间的物质交换存在时间滞后，所以图中太阳黑子与碳-14 偏离值存在一些相位差也在意料之中。

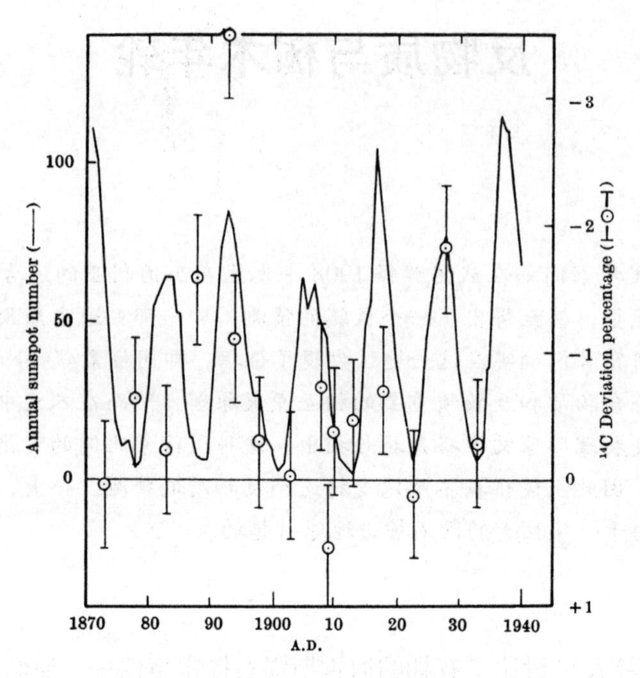

Fig. 1. Measured carbon-14 activity in tree rings (ref. 1) and sunspot activity during the same period

Thus, it is clear that if one takes into account the nature of secular variations of carbon-14/carbon-12 ratios in the atmosphere, it becomes difficult to reach any conclusions which may be of significance to a possible antimatter content of the Tunguska meteorite.

(**208**, 772; 1965)

V. S. Venkatavaradan: Tata Institute of Fundamental Research, Colaba, Bombay-5.

References:

1. Cowan, C., Atlurl, C. R., and Libby, W. F., *Nature*, **206**, 861 (1965).
2. Stuiver, M., *J. Geophys. Res.*, **66**, 273 (1961).

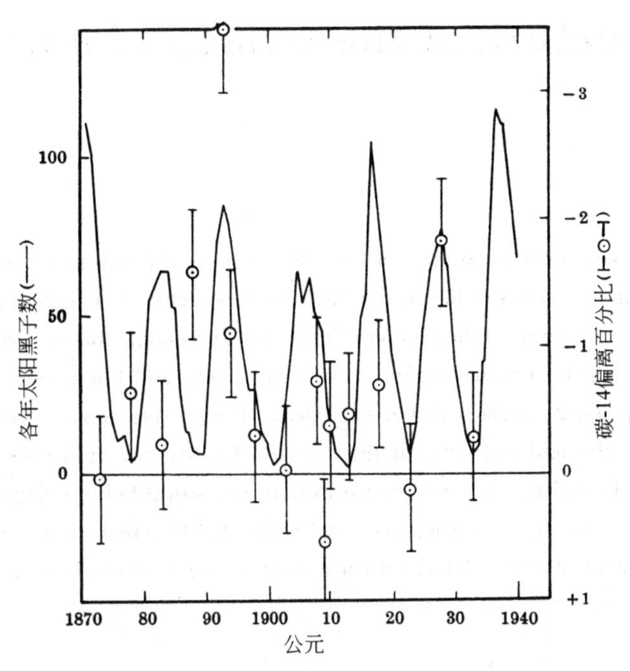

图 1. 树木年轮中所测得的碳-14 放射性强度（参考文献 1）和同时期太阳黑子活动

因此，很明显的是，若考虑到大气中碳-14 与碳-12 比值长期变化的特征，我们就难以得到任何有利于通古斯陨石中可能存在反物质的结论。

（李任伟 翻译；肖伟科 审稿）

Radio Structure of the Galactic Centre Region

D. Downes *et al.*

Editor's Note

Optical astronomers were prevented in the 1960s from constructing a clear view of what the centre of our galaxy consists of because of the confusion between neighbouring bright sources of light and the large amounts of dust present there. Radioastronomy thus became the most effective means of exploring this crucial region of our own galaxy, and this is one of the first attempts to do so. The paper by Dennis Downes at Harvard, who went on to make several important contributions to the understanding of the Galactic Centre, and colleagues, describes a bright source adjacent to Sagittarius *A*, which was at that time suspected of being the galactic nucleus. This presages the discovery of Sagittarius *A in the 1970s, which is awarded that role today. It is widely thought that the Galactic Centre houses a very massive black hole, but this is still unconfirmed.**

RADIO observations have shown that the galactic centre region consists of a number of discrete sources. The brightest of these, Sagittarius *A*, is believed to represent the galactic nucleus. This communication describes a new series of observations of the region, made at frequencies of 8.25 and 15.50 Gc/s with a pencil-beam antenna. The angular resolutions were respectively 4.2 and 2.2 arc min, the latter being the highest pencil-beam resolution so far applied to the galactic centre region. The observations confirm that the microwave spectrum of Sagittarius *A* is non-thermal[1], show that the angular diameter of the source is approximately 3.5 arc min, and demonstrate that, adjacent to Sagittarius *A*, there is an irregular emission region, which is apparently thermal in nature. The relation between radio data concerning the galactic centre and optical information about the centres of nearby normal galaxies is also examined.

The observations were made with the 120-ft. paraboloid antenna at the Haystack field station of Massachusetts Institute of Technology, Lincoln Laboratory. The antenna has a Cassegrain feed system[2]. The receiver at 8.25 Gc/s had a band-width of 0.5 Gc/s, and with an output time constant of 1 sec the minimum detectable signal was of the order of 0.1°K. The receiver at 15.50 Gc/s used two channels centred at 15.25 and 15.75 Gc/s, each having a band-width of 0.5 Gc/s. With an output time constant of 1 sec the minimum detectable signal for this receiver was 0.2°K. The outputs from the receivers were recorded by both analogue and digital equipment. The observations were made by taking drift-scans, at spacings of 2 arc min in declination at 8.25 Gc/s, and at 1 arc min at 15.50 Gc/s. This procedure minimizes the effects of ground radiation scattered or diffracted into the antenna. The scans were taken when the galactic centre region was at meridian transit ±2 h, over which period its elevation ranged from 18 to 13 arc deg above the horizon.

银河系中心区域的射电结构

唐斯等

编者按

由于银河系近邻明亮光源的干扰以及空间中存在大量的尘埃，使得光学天文学家在20世纪60年代难以构建出银河系中心结构的清晰图像。因此射电天文学成为探索我们银河系关键区域最有效的方法，而本文正是最初的尝试之一。哈佛大学的丹尼斯·唐斯对理解银河系中心做出了重要的贡献，他及其同事们描述了邻近人马座*A*的明亮光源，在当时人们猜测人马座*A*是银河系的核心。这预示了现在确认为银河系核心的人马座*A**于20世纪70年代的发现。人们普遍认为银河系中心有一个质量很大的黑洞，但是尚未确认。

射电观测显示银河系中心区域由很多不连续的源组成。其中最亮的是人马座*A*，人们认为它代表了银河系的核心。本文给出了在频率为8.25千兆周/秒和15.50千兆周/秒时用笔束天线对这个区域进行的一系列全新观测结果。其角分辨率分别是4.2角分和2.2角分，后者是目前用于银河系中心区域最高的笔束分辨率。这些观测证实了人马座*A*的微波频谱是非热辐射的[1]，并显示此源的角直径大约是3.5角分，还证实了在人马座*A*附近有一个不规则的、看似具有热辐射性质的辐射区域。本文也研究了银河系中心的射电数据和近邻普通星系中心光学信息之间的关系。

我们用麻省理工学院林肯实验室赫斯塔克野外观测站的120英尺抛物面天线进行观测。这个天线有一个卡赛格林馈源系统[2]。8.25千兆周/秒的接收机带宽是0.5千兆周/秒,输出时间常数1秒时可探测的最弱信号是0.1K量级。15.50千兆周/秒的接收机使用中心位于15.25千兆周/秒和15.75千兆周/秒的两个通道，每个通道带宽为0.5千兆周/秒。输出时间常数1秒时这个接收机可探测的最弱信号是0.2K。模拟以及数字设备记录接收机的输出。采用漂移扫描进行观测，在赤纬上8.25千兆周/秒处的间隔为2角分，15.50千兆周/秒处间隔为1角分。这个方法使散射或衍射进天线的地面辐射效应减到最小。当银河系中心区域位于中天附近±2小时的时候进行扫描,在这段时间中介于视界上的仰角范围为18弧度到13弧度。

Fig. 1*a* shows the radio brightness contours of the galactic centre region at 8.25 Gc/s and Fig. 1*b* shows the contours at 15.50 Gc/s. The bright source is Sagittarius *A*. To the north of Sagittarius *A*, and partially overlapping it, there is a complex emission region, which consists of an irregular curved ridge. This region may also be distinguished on contour maps at other frequencies[3,4]. Beneath these sources, there is a well-known emission region extending over several arc degrees, which is not indicated in the diagrams.

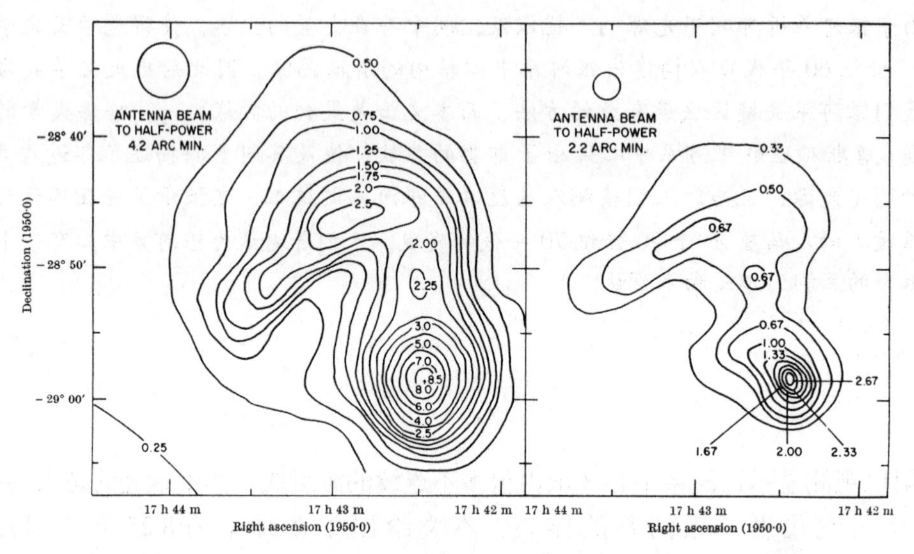

Fig. 1. *a* (Left), galactic centre region at 8.25 Gc/s. Contours represent antenna temperature in °K, corrected by 4 percent for extinction. *b* (Right), galactic centre region at 15.50 Gc/s. Contours represent antenna temperature in °K, corrected by 8 percent for extinction

Estimates of the flux densities of Sagittarius *A* at 8.25 and 15.50 Gc/s are given in Table 1. These values were derived on the assumption that the flux density of *M*87 was 47×10^{-26} M.K.S. units at 8.25 Gc/s and 28×10^{-26} units at 15.50 Gc/s. Beam-width corrections were made for both Sagittarius *A* and *M*87. The flux values for Sagittarius *A* fit closely on the intensity spectral curve of index $\alpha = -0.7$ given earlier by Maxwell and Downes[1], confirming the non-thermal character of the source. The position co-ordinates of Sagittarius *A* given in Table 1 are mean values taken from earlier position measurements of Maxwell and Downes, Hollinger[3] and Broten *et al.*[4].

Table 1. Flux Densities, Angular Diameters, and Positions of Radio Sources in Galactic Centre Region

	Sgr. *A*	Sgr. *B*1	Sgr. *B*2
Flux density ($\times10^{-26}$ M.K.S.)			
at 8.25 Gc/s	150	35	125
at 15.50 Gc/s	100	60	190
Angular diameter(arc min)	4.0×2.5	7×5	17×5
Position (1950.0)			
right ascension	17h 42m 27s	17h 42m 34s	17h 42m 59s
declination	−28°58.5′	−28°51.0′	−28°47.0′

图 1*a* 显示了 8.25 千兆周 / 秒频率下银河系中心区域的射电强度等亮度线图，图 1*b* 显示了 15.50 千兆周 / 秒频率下银河系中心区域的射电强度等亮度线图。其中亮源是人马座 *A*。在人马座 *A* 的北边有一个和它部分重叠并且很复杂的辐射区域，由不规则的弯曲的隆起组成。也可以在其他频率的等亮度线图上辨别出这个区域 [3,4]。在这些源之下有一个延展达几弧度的著名辐射区域，但它在图中没有表示出来。

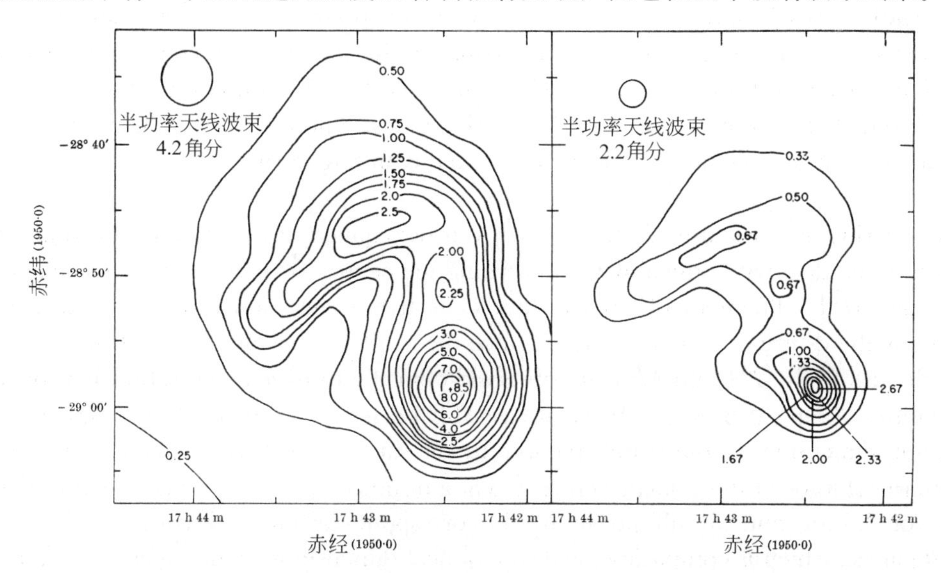

图 1. *a*（左边），8.25 千兆周 / 秒处观测到的银河系中心区域。等亮度线图表示了以 K 为单位的天线温度分布，对消光进行了 4% 的修正。*b*（右边），15.50 千兆周 / 秒处观测到的银河系中心区域。等亮度线图表示了以 K 为单位的天线温度分布，对消光进行了 8% 的修正。

表 1 给出了在 8.25 千兆周 / 秒和 15.50 千兆周 / 秒处对人马座 *A* 流量密度的估计。这些值是在这样的假设下导出的：在 8.25 千兆周 / 秒频率处 *M*87 的流量密度在 M.K.S.（米 · 千克 · 秒）制下为 47×10^{-26} 单位，在 15.50 千兆周 / 秒处是 28×10^{-26} 单位。对人马座 *A* 和 *M*87 都进行了波束宽度的修正。麦克斯韦和唐斯 [1] 早期指出，人马座 *A* 的流量值可以与指数为 $\alpha = -0.7$ 的谱线很好地拟合，这也证实了该源的非热辐射性质。表 1 中给出的人马座 *A* 的位置坐标取自早先麦克斯韦、唐斯、霍林格 [3] 和布拉滕等 [4] 所做位置测量的平均值。

表 1. 银河系中心区域射电源的流量密度、角直径和位置

	人马座 *A*	人马座 *B*1	人马座 *B*2
流量密度 ($\times 10^{-26}$ 米 · 千克 · 秒单位制)			
8.25 千兆周 / 秒处	150	35	125
15.50 千兆周 / 秒处	100	60	190
角直径 (角分)	4.0 × 2.5	7 × 5	17 × 5
位置 (1950.0)			
赤经	17 小时 42 分 27 秒	17 小时 42 分 34 秒	17 小时 42 分 59 秒
赤纬	–28°58.5′	–28°51.0′	–28°47.0′

To estimate flux densities from the irregular emission region north of Sagittarius *A*, we have divided it into two main areas, which we shall refer to as Sagittarius *B*1 and *B*2. The centres and angular diameters of these areas are listed in Table 1, the centres having been measured relative to the position of Sagittarius *A*. On Fig. 1*b* the centres are represented by the closed contours of antenna temperature 0.67°K. Integrated flux densities for Sagittarius *B*1 and *B*2 are given in Table 1. These values, however, should be regarded with caution, particularly at 15.50 Gc/s, since the present survey provides only a few brightness contours for the areas. Comparison of the integrated fluxed from these areas with data taken by Broten *et al.* at 5 Gc/s suggests that the sources are both thermal in nature and that they may therefore be H II regions. There is, however, no evidence to indicate that the sources are physically associated with Sagittarius *A*.

In considering the structure of the centre of our own galaxy, it is instructive to compare the existing radio data with optical data concerning the centres of two nearby normal galaxies, *M*31 and *M*51. (There is, of course, little optical evidence concerning the structure or nature of the centre of our own galaxy, since the centre is heavily obscured by intervening dust clouds.) Both *M*31 and *M*51 are observed optically to have a central nucleus of linear dimensions about 20 parsec[5,6]. In each case this nucleus is embedded in a larger nuclear bulge of dimensions about 1,000 parsec. In the case of our own galaxy, if we believe Sagittarius *A* to be at the galactic centre and at a distance of 10 kiloparsec[7], then its angular width of 3.5 arc min to half power would correspond to linear dimensions of the order of 10 parsec, which is comparable with the optical dimensions of the nuclei of the nearby normal galaxies. Similarly, if the extended source underlying Sagittarius *A* is regarded as radio evidence for a nuclear bulge at the centre of our own galaxy, the radio diameter of about 120 arc min[1] would correspond to linear dimensions of the order of 350 parsec.

Observations of the galactic centre by radio astronomers have now extended over some fifteen years, during which time angular discrimination has increased by a factor of about 30. With each improvement in resolution the central region has generally shown increasing structural complexity. Commencing in 1967, the region will be subject to a series of lunar occultations, and at that time we may look forward to a dramatic increase in the available angular resolution, perhaps by a factor of 100 or more. The detailed structure of this region should then be revealed much more fully.

We thank Mr. A. B. Hull and Dr. S. Weinreb of Lincoln Laboratory, and Mr. R. Rinehart and Mr. J. H. Taylor of Harvard University, for their assistance in making the observations. Lincoln Laboratory is a centre for research and development operated by the Massachusetts Institute of Technology with the support of the U.S. Air Force. The Harvard University part of the programme was supported by the U.S. National Science Foundation.

(**208**, 1189-1190; 1965)

D. Downes and A. Maxwell: Harvard Radio Astronomy Station, Fort Davis, Texas.

M. L. Meeks: Massachusetts Institute of Technology, Lincoln Laboratory, Lexington, Massachusetts.

为估计人马座 *A* 北边不规则辐射区域的流量密度，我们把它分为两个主要区域，我们将其称为人马座 *B*1 和 *B*2。这些区域的中心位置和角直径列于表 1，其中心位置是参照人马座 *A* 的位置测量的。在图 1*b* 中，0.67K 天线温度的闭合等亮度线图描绘出中心位置。表 1 给出了人马座 *B*1 和 *B*2 的流量密度积分。不过，对这些值的采用应该谨慎些，因为目前的巡天只提供了这些区域的几个等亮度线，特别是在 15.50 千兆周 / 秒处。通过比较来自这些区域的积分流量与布拉滕等人在 5 千兆周 / 秒得到的数据，人们发现这两个源实际上都是热辐射的，因此它们有可能是电离氢区。但是没有证据表明这些源和人马座 *A* 有物理上的联系。

在研究我们自己银河系中心的结构时，将已有的射电数据与两个近邻普通星系 *M*31 和 *M*51 中心的光学数据做比较是有意义的。（当然由于银河系中心大量尘埃云遮挡，所以得到的银河系中心结构或性质的光学证据很少。）在光学波段均观测到 *M*31 和 *M*51 具有线尺度大约 20 秒差距的中心核 [5,6]。对于每种情况，这个核都镶嵌于更大尺度的、约 1,000 秒差距的核球中。就我们自己银河系而言，如果我们认为人马座 *A* 是银河系中心并且距离是 10 千秒差距 [7]，那么它半功率处 3.5 角分的角宽度应该对应于 10 秒差距的线尺度，这与近邻普通星系核心的光学尺度差不多。类似地，如果人马座 *A* 背景上的展源被看作银河系存在中心核球的射电观测证据，那么大约 120 角分 [1] 的射电直径应该对应于 350 秒差距的线尺度。

射电天文学家对银河系中心的观测至今已经超过大约 15 年了，在这段时间里角分辨率已提高到原来的 30 倍左右。随着分辨率的每一次提高，中心区域逐渐显示出了不断增加的结构复杂性。从 1967 年开始，这个区域将出现一系列的月掩情况。那时，我们也许可以期待角分辨率大幅度地增加，如 100 倍或者更多，到那时人们可以更加充分地了解这个区域的细致结构。

感谢林肯实验室的赫耳先生和魏因雷布博士，以及哈佛大学的莱茵哈特先生和泰勒先生对观测给予的协助。林肯实验室是由麻省理工学院在美国空军的支持下运行的研发中心。该项目中哈佛大学部分由美国国家科学基金会支持。

（钱磊 翻译；王有刚 审稿）

References:

1. Maxwell, A., and Downes, D., *Nature*, **204**, 865 (1964).

2. Weiss, H. G., *IEEE Spectrum*, **2**, No. 2, 50 (1965).

3. Hollinger, J. P., *Astrophys. J.*, **142**, 609 (1965).

4. Broten, N. W., Cooper, B. F. C., Gardner, F. F., Minnett, H. C., Price, R. M., Tonking, F. G., and Yabsley, D. E., *Austral. J. Phys.*, **18**, 85 (1965).

5. Lallemand, A., Duchesne, M., and Walker, M. F., *Pub. Astron. Soc. Pacific*, 72, 76 (1960).

6. Burbidge, E. M., and Burbidge, G. R., *Astrophys. J.*, **140**, 1445 (1964).

7. *Intern. Astro. Union Inform. Bull.*, No. 11, 11 (1963).

Formation of Hydroxyl Molecules in Interstellar Space

J. L. Symonds

Editor's Note

In the mid-1960s radio astronomers were beginning the process of finding emission from molecules in space. J. L. Symonds here proposes a particular way in which the OH (hydroxyl) radical might be formed, which he crafted to avoid the problem that collisions of two particles seemed very unlikely in the low gas densities of interstellar space. What was missing at the time was the understanding that most of the gas was in the form of molecular hydrogen, and still unseen (only the more dilute atomic hydrogen was known). It would be another five years before molecular hydrogen was found in space, adding the missing part of the puzzle.

DURING the past year, radio astronomers[1,2] have been increasingly interested in microwave signals at frequencies of 1,612, 1,665, 1,667 and 1,720 Mc/s, produced by transitions between four energy levels of the OH molecule in the $^2\Pi_{8/2}$ state. Theoretical and laboratory intensity ratios[3] for the four microwave lines were found to be 1:5:9:1. It was reported[1] recently that anomalous intensity ratios had been measured in the strong OH absorption lines of the radio source Sagittarius *A*. McGee *et al.*[4] have extended their measurements to other radio sources and find emission and absorption at all four frequencies with instances of extremely anomalous intensity ratios.

Emission and absorption lines showing such anomalous intensity distributions indicate unusual populations of energy levels in the OH molecules. The observed distributions are more like those expected of a chemical reaction mechanism, followed by radiative deactivation, rather than those from thermal excitation. Since the gases are tenuous in the regions which produce these signals, the long time-scale between collisions will give the molecules, formed in an excited state, a greater probability of radiative rather than collisional deactivation.

The process of association of two atoms to form a molecule normally has a low probability. Since atomic recombination is known to proceed more rapidly in the presence of a third body, "dust" grains in space have been suggested as a suitable medium. The tenuous nature of the gas would seem to rule out a three-body process, however, and such reactions do not appear likely to produce the observed intensity anomalies. The processes for production and loss of OH molecules by two-body processes, therefore, deserve closer study.

In relation to the known concentrations of oxygen and hydrogen in our galaxy, the

羟基分子在星际空间中的形成

西蒙兹

编者按

20 世纪 60 年代中期，射电天文学家们开始了探寻星际空间分子辐射的历程，西蒙兹在本文中提出了一种可能形成羟基的特殊方法，避免了星际空间中低密度气体环境下两个粒子的碰撞概率过低的问题。那时，人们探测不到分子氢（只能探测到稀薄得多的原子氢），不知道大部分气体是以分子氢的形式存在的。直到 5 年后在星际空间中发现了分子氢，才补上了这一谜题中缺失的部分。

在过去的一年，射电天文学家们[1,2]对频率为 1,612 兆周 / 秒、1,665 兆周 / 秒、1,667 兆周 / 秒和 1,720 兆周 / 秒的微波信号产生了越来越浓厚的兴趣，这些信号都是由处于 $^2\Pi_{8/2}$ 状态的 OH 分子在 4 个能级间跃迁时产生的。4 条微波谱线的强度比的理论值和实验值[3]均为 1∶5∶9∶1。最近有研究称[1]，在观测来自人马座 *A* 射电源的强 OH 吸收谱线时发现了反常的强度比。随后麦吉等人[4]将观测对象扩展到其他射电源，发现不论是发射还是吸收，4 个频率的强度比都出现了极为反常的现象。

具有如此反常强度分布的发射谱线和吸收谱线，意味着 OH 分子的能级布居不同寻常。所观测到的布居似乎更有可能是发生了某种化学反应机制（以及随后的辐射退激发），而不是由热激发引起的。因为产生这些信号的区域中的气体非常稀薄，碰撞之间所具有的长时标将使在激发态形成的分子更有可能发生辐射退激发而不是碰撞退激发。

一般通过两个原子结合而形成一个分子的过程的可能性很低。因为原子复合需要一个第三体作为媒介才能加快进程，太空中的“尘埃”颗粒被认为是一种适当的媒介。但是太空内的气体非常稀薄，发生三体过程似乎是不可能的。而且，即使发生这样的反应也不大可能造成所观测到的反常强度。因此，经过两体过程产生和消耗 OH 分子值得进一步研究。

根据我们所在星系内氧和氢的已知浓度，OH 分子若是完全通过原子间的直接

concentration of OH molecules is such as to make it improbable that they are all formed by the direct collision process between atoms. In the study of gaseous processes in flames by spectroscopic means, OH spectra show evidence of pre-dissociation, made possible by radiationless transitions between states of nearly the same energy where the potential energy curves of the states cross or approach closely. The inverse pre-dissociation process[5] (pre-association) is also possible for forming molecules from a two-body collision. The pre-dissociation of the OH molecule to O and H atoms and its inverse are believed to occur, but with a small probability, because a normally forbidden transition is involved.

It is suggested that other two-body processes should be investigated for an alternative mechanism. Since the atom recombination is low, a mechanism worthy of consideration is the exothermic association of negative oxygen ions, O^-, and positive ions of hydrogen, H^+ (protons).

$$O^- + H^+ = OH + \text{approximately } 12 \text{ eV}$$

The presence of O^- ions seems assured by the strong electron affinity of oxygen (1.45 eV) and the known presence of oxygen atoms in the regions under study. The anomalous emission and absorption has only been found in ionized regions[4,9] where large concentrations of electrons and protons must also exist. The pre-association of O^- and H^+ essentially may be a transition from the coulomb potential energy curve to the $^2\Sigma^+$ or, more probably, the $^2\Pi_{8/2}$ state curve, involving charge transfer. The interaction cross-section should have a maximum value when the relative velocity of the two ions produces kinetic energy close to the differences between the binding energy of the OH molecule and the electron affinity of the oxygen atom, that is , 4.45–1.45 = 3 eV.

In such circumstances, it is possible to form OH molecules in the $^2\Pi_{8/2}$ state with the population of each of the four levels depending greatly on the relative velocity of the O^- and H^+ ions. Whether the molecules emit or absorb energy will depend on the populations of the energy levels and, in bulk, one would not expect to see the theoretical intensity ratios. Situations could arise where the relationships between lines were completely unusual, for example, apparent absorption in some of the four lines and emission in the others. The actual result will depend strongly on the relative velocity distribution of the two ions. Investigation of the intensity ratios may, therefore, give a great deal of information on relative ion velocities.

Since the reaction has a “resonance” character, it will favour relative velocities producing about 3-eV kinetic energy between the ions. If the O^- ion were stationary, the proton velocity would need to be about 25 km/sec, which is certainly in the range found in the ionized regions of the galaxy. The lack of emission or absorption at the four frequencies in galactic regions of lower temperature may be explained in terms of a low cross-section at proton velocities well below 25 km/sec. An accurate calculation of the reaction cross-section and the state of the resulting molecule would be informative.

碰撞来产生，就不可能具有现有的浓度。借助分光镜方法研究火焰中的气体过程时，发现 OH 光谱显示出了预解离的迹象，这使具有几乎相同能量状态间的无辐射跃迁成为可能，这些状态的势能曲线相交或者紧密靠拢。对于以两体碰撞方式形成的分子来说，也有可能发生预解离的逆过程[5]（预结合过程）。OH 分子可以经过预解离过程形成 O 和 H 原子，也可以通过相反过程形成 OH 分子，但概率很低，因为涉及通常条件下的禁戒跃迁。

为了探究是否存在其他机制，我们应当对其他两体化过程进行研究。鉴于原子直接复合的概率很低，值得研究的是带负电荷的氧离子 O^- 与带正电荷的氢阳离子 H^+（质子）结合并放热的机制。

$$O^- + H^+ = OH + \text{约 } 12\ \text{电子伏}$$

已知在所研究区域中存在氧原子，并且氧原子具有强的电子亲和能（1.45 电子伏），这为 O^- 离子的存在提供了可能。目前仅在电离区[4,9]发现了反常的发射和吸收，而该区域也必定存在高浓度的电子和质子。O^- 和 H^+ 离子的预结合过程可能主要是从库仑势能曲线跃迁到 $^2\Sigma^+$ 状态曲线，或者更有可能是到 $^2\Pi_{8/2}$ 状态曲线，其间涉及电荷的转移。当两种离子的相对速度产生的动能接近于 OH 分子结合能与氧原子的电子亲和能之差（即 4.45–1.45 = 3 电子伏）时，碰撞截面达到最大值。

在这种情况下，可能形成处于 $^2\Pi_{8/2}$ 状态下的 OH 分子，其中 4 个能级中每个能级的布居情况主要依赖于各自 O^- 和 H^+ 离子的相对速度。分子究竟是发射能量还是吸收能量取决于能级的布居情况，对于大量粒子而言，将与理论谱线强度比不符。当谱线间的关系完全不同时，就可能出现这种情形，例如 4 条谱线中的某些可能表现为吸收谱线而另外一些则为发射谱线。实际结果主要依赖于两种离子相对速度的分布情况。因此，对强度比值进行研究可以获取大量关于离子相对速度的信息。

由于反应具有“共振”特点，当离子间的相对速度产生大约 3 电子伏动能时是最有利的。如果固定 O^- 离子，则质子的速度大概需要达到 25 千米 / 秒，星系电离区中的粒子速度无疑可以达到这个要求。在星系低温区域 4 个频率处没有发生发射或吸收现象，这可能是由于质子速度明显低于 25 千米 / 秒导致碰撞截面过小而引起的。对反应碰撞截面和产物分子状态进行精确的计算将会获得更多的信息。

The process of association between O^- ions and protons, or for that matter any similar process between other ions, does not appear to have received great attention. The study of recombination rates in flames[6] shows evidence of a non-equilibrium condition in the excitation of molecules and atoms in the flame. Unexplained concentrations and intensity ratios exist which relate to OH molecule formations. These conditions may be the result of ionic recombination to form molecules and not atom-molecule reactions. Evidence of strong negative and positive ion concentrations has been found[7,8], but the result of their presence on the state of excitation of product molecules has not been elucidated.

If the reaction mechanism is as proposed, radiationless transitions will leave the OH molecule with a velocity similar to that of the O^- ion. However, it is not obvious what the "temperature" of the O^- ions will be since their velocity will depend on their mode of formation and their lifetime.

Two effects must be studied. First, the OH line will be shifted by Doppler effects resulting from mass motion relative to the observational point, and the mass motion may not necessarily be related to the motion of the O and H atoms. Secondly, the line broadening may differ from that produced by the "temperature" of the surrounding ionized gas region because the radiationless transitions occurring in pre-association results in some perturbation of the rotational levels in the states involved. Corresponding shifts in line positions and changes in the intensity distributions will occur, with widths also being reduced in the process.

Orientation of the magnetic moment of the O^- ion by a magnetic field should produce a modification of intensity ratios, possibly an alteration of frequency not associated with a mass movement, and certainly polarization effects. Viewing the OH radiation coming from an ionized region, the proposed mechanism suggests that the radiation will be linearly polarized[10] since the movement of hydrogen appears to be radially outward and the axis of rotation of the OH molecules will tend to be in a plane normal to the proton direction.

The known presence of OH molecules must also lead to the assumption that OH^- molecular ions are also present. The electron affinity of the OH molecule is greater than that of the oxygen atom, making possible two reactions of interest. The first is similar to that for the formation of OH and would lead to the formation of the H_2O molecule. The second is the charge transfer process which will result in the production of an atom of hydrogen and an excited OH molecule. Apart from ionizing and other processes which remove OH molecules, it is obvious that there are other modes of OH formation by collision processes between ions, atoms and molecules which must be taken into account.

Finally, the proposed mechanism offers some interesting prospects in relation to the formation of other molecules. Atoms with strong electron affinities are more likely to produce similar reactions with protons. Equally, the rotational states of such molecules will be excited. Atoms of hydrogen, carbon, oxygen, silicon and sulphur would be capable of forming negative ions. Whether the pre-association process is possible is not known for

O^- 离子和质子的结合过程或者其他离子发生的类似过程似乎还没有受到足够的重视。对火焰中复合速率进行研究 [6]，表明在火焰中分子和原子的激发表现出非平衡态的迹象。在 OH 分子形成方面，浓度和强度比值问题仍未得到解释。这可能是离子复合形成分子的结果，而不是原子–分子反应的结果。已证实 [7,8] 存在正、负离子的高度聚集，但是它们的存在对产物分子激发态有何影响，尚未得到阐明。

如果反应机制确如前面所提出的那样，那么无辐射跃迁将使 OH 分子具有类似于 O^- 离子的速度。但是 O^- 离子所具有的“温度”是我们所不知道的，因为它们的初始速度依赖于其形成的方式和寿命。

有两种效应必须加以研究。首先，OH 谱线会因相对于观测点的该团块的整体运动所引起的多普勒效应而变化，而整体运动未必与 O 和 H 原子的运动相关。第二，谱线的致宽机制与由周围电离气体区域的“温度”所产生的 OH 的致宽情况可能是不同的，因为发生在预结合过程中的无辐射跃迁对相关状态中转动能级有些干扰。这些效应会造成相应的谱线位置和强度分布的变化，同时也伴随着谱线宽度的减小。

磁场对 O^- 离子的磁矩定向作用会改变强度比值和产生极化效应，强度比值的改变可能是通过一个与团块整体运动无关的频率变化进行的。观测来自电离区的 OH 辐射时，上述机制意味着辐射将会是线偏振的 [10]，因为氢原子的运动表现为沿径向向外，而 OH 分子的旋转轴则倾向于垂直质子方向的平面内。

已知 OH 分子的存在就势必产生 OH^- 分子离子也存在的假定。OH 分子的电子亲和能比氧原子的电子亲和能大，这使得两个我们感兴趣的反应成为可能。第一个反应类似于形成 OH 分子的反应，只是再形成 H_2O 分子。第二个反应是电荷转移过程，将会产生一个氢原子和一个激发态的 OH 分子。除去离子化和其他一些消除 OH 分子的过程外，很明显，还有另外一些通过离子、原子和分子间碰撞来形成 OH 的方式，这些也是必须加以考虑的。

最后，上述机制为我们提供了一些关于其他分子形成方面的诱人前景。具有高的电子亲和能的原子似乎更有可能与质子发生类似的反应。同样地，这些分子的转动能级将会受到激发。氢原子、碳原子、氧原子、硅原子和硫原子具有形成负离子的能力。对于每一种原子，这种预结合过程是否都会发生还不清楚。因为氮原子无

every case. Since nitrogen does not form a stable negative ion, NH may not be observed. Nevertheless, similar reactions may take place between N^+ and O^- or C^- to give NO and CN molecules.

The formation of H_2 from H^- and H^+ ions may require higher relative velocities, but the cross-section for H^- production from H is four orders of magnitude less than for O^- formation. Such a process may not be observable on this count alone, without invoking the question of the frequency range in which signals may be expected.

In summary, the association mechanism appears to offer some prospect of success in accounting for the anomalous intensity ratios; the emitted radiation may well be linearly polarized if a magnetic field is present; some suggestions can be made as to kinds of molecules that might be formed. More theoretical and experimental work on reactions between negative and positive ions appears to be necessary to establish their importance.

I thank Dr. B. J. Robinson and his colleagues of C.S.I.R.O. Radiophysics Laboratory for their advice and for making available the results of their investigations before publication.

(**208**, 1195-1196; 1965)

References:

1. Gardner, F. F., Robinson, B. J., Bolton, J. G., and van Damme, K. J., *Phys. Rev. Letters*, **13**, 3 (1964).
2. Robinson, B. J., Gardner, F. F., van Damme, K. J., and Bolton, J. G., *Nature*, **202**, 989 (1964).
3. Radford, H. E., *Phys. Rev. Letters*, **18**, 534 (1964).
4. McGee, R. X., Robinson, B. J., Gardner, F. F., and Bolton, J. G., *Nature* (preceding communication).
5. Herzberg, G., *Molecular Spectra and Molecular Structure; Spectra of Diatomic Molecules* (D. Van Nostrand Co., Inc., 1959).
6. Garvin, D., Broida, H. P., and Kostkowski, H. J., *J. Chem. Phys.*, **32**, 880 (1960).
7. King, I. R., *J. Chem. Phys.*, **37**, 74 (1962).
8. Miller, W. J., and Calcote, H. F., *J. Chem. Phys.*, **41**, 4001 (1964).
9. Weaver, H., Williams, D. R. W., Dieter, N. H., and Lum, W. T., *Nature*, **208**, 29 (1965).
10. Weinreb, S., Meeks, M. L., Carter, J. C., Barrett, A. H., and Rogers, A. E. E., *Nature*, **208**, 440 (1965).

法形成稳定的负离子，所以可能无法观测到 NH。但是类似的反应可以在 N^+ 与 O^- 或 C^- 之间发生，分别形成 NO 和 CN 分子。

H^- 与 H^+ 离子结合形成 H_2 可能需要很大的相对速度，但是 H 产生 H^- 的反应截面要比产生 O^- 的反应截面少 4 个数量级。在不知道所产生的谱线可能的频率范围的情况下，这种过程是无法观测到的。

总而言之，结合机制为解释异常的强度比值现象提供了某些可能；如果磁场存在，发射辐射可能会被充分线偏振化；这一机制还可能用于形成其他的分子种类。有必要对正、负离子之间的反应做更多的理论和实验研究，以确定其重要性。

我要感谢鲁滨逊博士及其在澳大利亚联邦科学与工业研究组织放射物理实验室的同事们提出的建议，并感谢他们在发表之前就向我提供了他们的研究结果。

（王耀杨 翻译；沈志侠 审稿）

Stonehenge—An Eclipse Predictor

F. Hoyle

Editor's Note

During the 1960s there was considerable interest in, and debate about, the purpose of the ancient Stonehenge monument in western England. While British astronomer Fred Hoyle was not the first to suggest that Stonehenge was used to predict eclipses, he does demonstrate here how it could more accurately predict them if the "Aubrey circle" represents the ecliptic (the plane of the Solar System, in which the planets orbit the Sun).

THE suggestion that Stonehenge may have been constructed with a serious astronomical purpose has recently received support from Hawkins, who has shown[1] that many alignments of astronomical significance exist between different positions in the structure. Some workers have questioned whether, in an arrangement possessing so many positions, these alignments can be taken to be statistically significant. I have recently reworked all the alignments found by Hawkins. My opinion is that the arrangement is not random. As Hawkins points out, some positions are especially relevant in relation to the geometrical regularities of Stonehenge, and it is these particular positions which show the main alignments. Furthermore, I find these alignments are just the ones that could have served far-reaching astronomical purposes, as I shall show in this article. Thirdly, on more detailed investigation, the apparently small errors, of the order of $\pm 1^\circ$, in the alignments turn out not to be errors at all.

In a second article[2] Hawkins goes on to investigate earlier proposals that Stonehenge may have operated as an eclipse predictor. The period of regression of the lunar nodes, 18.61 years, is of especial importance in the analysis of eclipses. Hawkins notes that a marker stone moved around the circle of fifty-six Aubrey holes at a rate of three holes per year completes a revolution of the circle in 18.67 years. This is close enough to 18.61 years to suggest a connexion between the period of regression of the nodes and the number of Aubrey holes. In this also I agree with Hawkins. I differ from him, however, in the manner in which he supposes the eclipse predictor to have worked. Explicitly, the following objections to his suggestions seem relevant:

(1) The assumption that the Aubrey holes served merely to count cycles of 56 years seems to me to be weak. There is no need to set out fifty-six holes at regular intervals on the circumference of a circle of such a great radius in order to count cycles of fifty-six.

(2) It is difficult to see how it would have been possible to calibrate the counting system proposed by Hawkins. He himself used tables of known eclipses in order to find it. The builders of Stonehenge were not equipped with such *post hoc* tables.

巨石阵——日月食的预报器

霍伊尔

编者按

在20世纪60年代，英格兰西部古老巨石阵的用途是大家非常关注和存在争议的问题。虽然第一个提出巨石阵是用于预测日月食的工具的人并非英国天文学家弗雷德·霍伊尔，但他在这里解释了在“奥布里环”代表黄道面（即太阳系的运行平面，所有行星都在这个平面内绕太阳运动）的前提下如何利用巨石阵更精确地预言日月食。

巨石阵可能是出于一个重要的天文目的而修建的，这一假设最近得到了霍金斯的支持。他指出[1]，在该建筑的不同位置中存在着有天文学意义的准线。一些研究者曾质疑：在一个拥有如此多方位的布局中，这些准线是否应该被认为具有统计学上的显著性。最近我检验了霍金斯发现的所有准线。我认为这个布局不是随机的。正如霍金斯所指出的，有些位置与巨石阵的几何规律之间有特殊的关联，而主要的准线正是在这些特殊位置上发现的。其次，我还发现这些准线恰好就是能长期服务于天文学观测需要的准线，在本文中我会解释这一点。第三，根据更加细致的调查，在这些准线中看似存在的量级为 $\pm 1^{\circ}$ 的小偏差其实根本就算不上偏差。

在第二篇文章中[2]，霍金斯又对早先的一个假说进行了研究，即认为巨石阵可能曾用于预报日月食。月球交点的回归周期为18.61年，这一周期在日月食分析中是非常重要的。霍金斯指出：有一个石标以每年3个洞的速率沿着由56个奥布里洞组成的圆周运动，旋转一周所用的时间恰好是18.67年。这和18.61年非常接近，因而说明月球交点的回归周期与奥布里洞的数量之间是存在相关性的。在这一点上我也同意霍金斯的观点。不过，我与他的分歧之处在于他所说的预测日月食的方式。显然，以下几条对其所持观点的反对意见看起来是合理的：

（1）在我看来，假设奥布里洞仅仅是被用于计算56年的循环未免有点站不住脚。没有必要为了表示出56年的循环，而在这么大半径的圆周上以一定的间距建造56个洞。

（2）很难解释古人是如何校准由霍金斯所提出的计算系统的。为了找到这一系统，他本人使用了已发生过的日月食的记录表。巨石阵的建造者们哪里会有这些在日月食发生之后才统计出的表格。

(3) The predictor gives only a small fraction of all eclipses. It is difficult to see what merit would have accrued to the builders from successful predictions at intervals as far apart as 10 years. What of all the eclipses the system failed to predict?

My suggestion is that the Aubrey circle represents the ecliptic. The situation shown in Fig. 1 corresponds to a moment when the Moon is full. The first point of Aries γ has been arbitrarily placed at hole 14. S is the position of the Sun, the angle $\odot$ is the solar longitude, M is the projection of the Moon on to the ecliptic, N is the ascending node of the lunar orbit, N' the descending node, and the centre C is the position of the observer. As time passes, the points S, M, N and N' move in the senses shown in Fig. 1. S makes one circuit a year. M moves more quickly, with one circuit in a lunar month. One rotation of the line of lunar nodes NN' is accomplished in 18.61 years. In Fig. 1, S and M are at the opposite ends of a diameter because the diagram represents the state of affairs at full Moon.

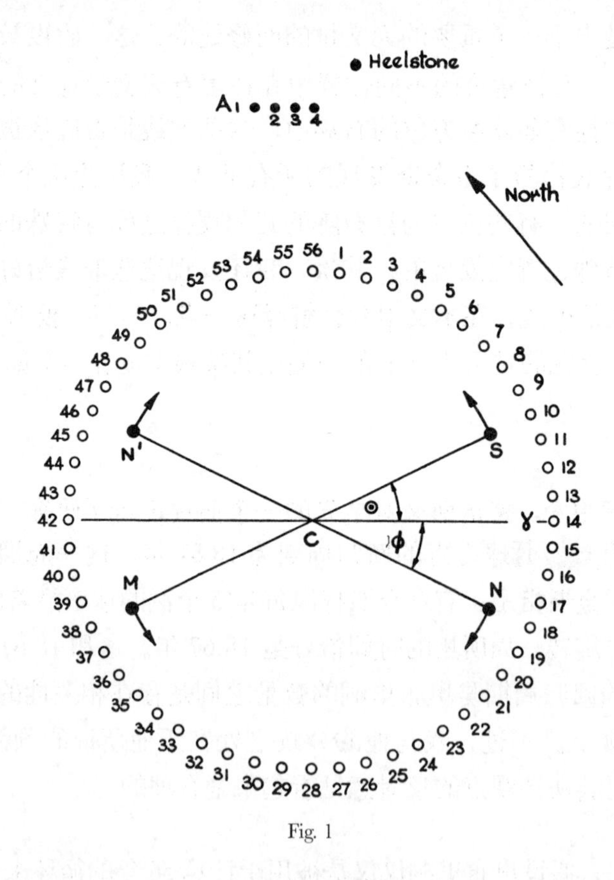

Fig. 1

If the Moon is at N, there is a solar eclipse if the Sun is within roughly $\pm15^\circ$ of N, and a lunar eclipse if the Sun is within $\pm10^\circ$ of N'. Similarly, if the Moon is at N', there will be a solar eclipse if the Sun is within $\pm15^\circ$ of coincidence with the Moon, and a lunar eclipse if it is within roughly $\pm10^\circ$ of the opposite end of the line of lunar nodes. Evidently if we represent S, M, N and N' by markers, and if we know how to move the markers so as to

（3）这种预报方法只预测出了全部日月食中很小的一部分。很难理解这些建造者们会因为成功预言间隔可达 10 年之久的日月食而得到什么好处。怎么解释那么多该系统没有预言出来的日月食呢？

我的观点是奥布里环代表了黄道。图 1 所示的位置对应于满月时的位置。任意取白羊座 γ 作为第一个点放在第 14 号洞处。*S* 是太阳的位置，⊙角代表黄经，*M* 是月球在黄道面上的投影，*N* 是月球轨道的升交点，*N′* 为降交点，中心 *C* 是观测者所在的位置。随着时间的流逝，*S*、*M*、*N* 和 *N′* 点会按图 1 所示的方式运动。*S* 每年转一圈。*M* 运行得会更快一些，一个朔望月循环一周。两个月球交点所连成的直线 *NN′* 旋转一周的时间为 18.61 年。在图 1 中，*S* 和 *M* 位于一条直径的两端是因为这张图代表的是满月时的状态。

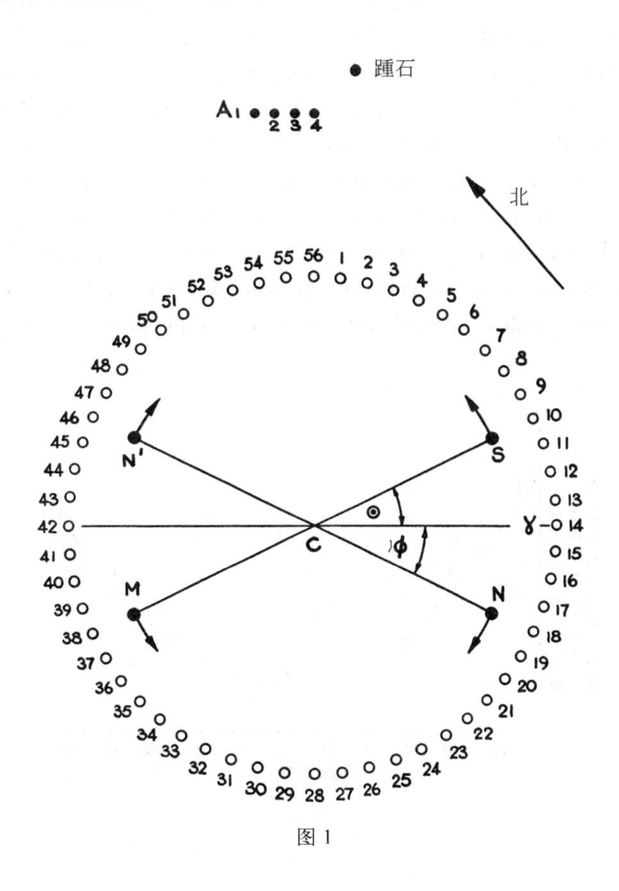

图 1

如果月球位于 *N* 点，那么当太阳在距离 *N* 点大致 ±15°范围之内时就会发生一次日食；而当太阳在距离 *N′* 点 ±10°范围之内时就会发生一次月食。同样，如果月球位于 *N′* 点，那么当太阳在距离月球位置 ±15°范围之内时就会发生一次日食；而当太阳在距离两个月球交点连线的另外一端大致 ±10°范围之内时就会发生一次月

represent the actual motions of the Sun and Moon with adequate accuracy, we can predict almost every eclipse, although roughly half of them will not be visible from the position of the observer. This is a great improvement on the widely scattered eclipses predictable by Hawkins's system. Eclipses can occur as many as seven times in a single year, although this would be an exceptional year.

The prescriptions for moving the markers are as follows: (1) Move *S* anticlockwise two holes every 13 days. (2) Move *M* anticlockwise two holes each day. (3) Move *N* and *N′* clockwise three holes each year.

We can reasonably assume that the builders of Stonehenge knew the approximate number of days in the year, the number of days in the month, and the period of regression of the nodes. The latter follows by observing the azimuth at which the Moon rises above the horizon. If in each lunar month we measure the least value of the azimuth (taken east of north), we find that the "least monthly values" change slowly, because the angle $\phi = NC\gamma$ changes. The behaviour of the "least monthly values" is shown in Fig. 2 for the range $-60^\circ \leq \phi \leq 60^\circ$. (The azimuthal values in Fig. 2 were worked out without including a refraction or a parallax correction. These small effects are irrelevant to the present discussion.) The least monthly values oscillate with the period of ϕ, 18.61 years. By observing the azimuthal cycle, the period of ϕ can be determined with high accuracy by observing many cycles. At Stonehenge sighting alignments exist that would have suited such observations. With the periods of *S*, *M* and *N* known with reasonable accuracy the prescriptions follow immediately as approximate working rules.

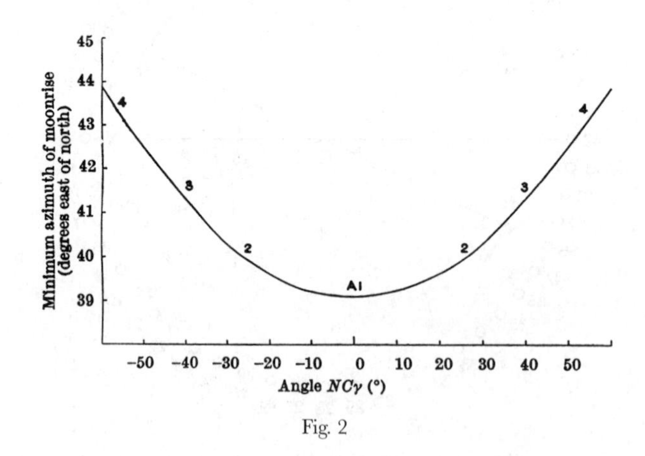

Fig. 2

Suppose an initially correct configuration for *M*, *N* and *S* is known. The prescriptions enable us to predict ahead what the positions of *M*, *N* and *S* are going to be, and thus to foresee coming events—but only for a while, because inaccuracies in our prescriptions will cause the markers to differ more and more from the true positions of the real Moon, Sun, and ascending node. The lunar marker will be the first to deviate seriously—the prescription

食。显然，假设我们用标记标出了 S、M、N 和 N' 的位置，并且假设我们知道如何通过移动这些标记来足够精确地表示太阳和月球的真实移动过程，那么我们就可以预言几乎所有的日月食，尽管有大约一半的日月食在观测者所在的位置上是看不到的。这在很大程度上优于利用霍金斯系统预测非常分散的日月食。在一年之中，日月食的发生次数可多达七次，不过这样的年头是很少见的。

移动这些标记的方法如下所述：(1) 每 13 天将 S 点逆时针移动 2 个洞。(2) 每天将 M 点逆时针移动 2 个洞。(3) 每年将 N 点和 N' 点顺时针移动 3 个洞。

我们可以合理地认为巨石阵的建造者们知道一年中的大致天数、一个月的天数以及月球交点的回归周期。后者可以通过观测月球从地平线升起时的方位角求得。如果在每个朔望月我们都能测量出方位角的最小值（由北向东），我们就会发现“每月最小值”在缓慢变化，这是因为角 $\phi=NC\gamma$ 在不断变化。图 2 中显示出了当 ϕ 的范围处于 $-60°\leq\phi\leq 60°$ 时“每月最小值”的变化情况。（图 2 中的方位角值没有经过折射校正或视差校正。这些较小的效应与现在讨论的内容无关。）每月最小值在 ϕ 的周期——18.61 年内上下波动。通过观测方位角的周期变化，并根据多个周期的观测结果就可以精确地测算出 ϕ 的周期。巨石阵中有一些可用于瞄准的排列很适合进行这样的观测。只要知道具有合理精确度的 S、M 和 N 的周期，就可以马上把上述方法作为大体的工作流程。

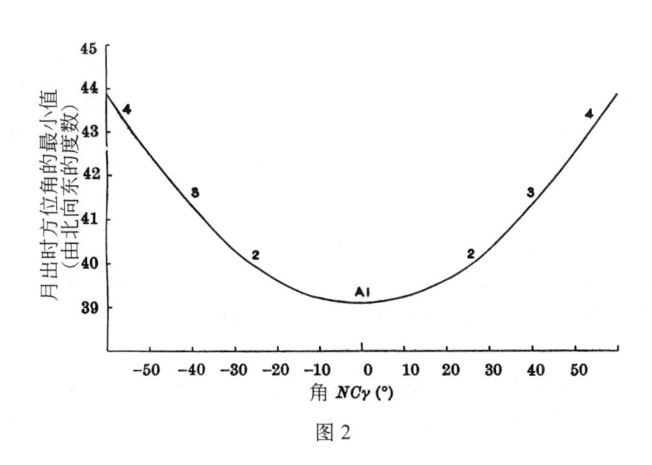

图 2

假设 M、N 和 S 的初始位置是已知的并且是正确的。利用上述方法可以使我们提前预测出 M、N 和 S 未来的位置，并由此预见到将要发生的日月食——但这只能维持很短的时间，因为上述方法中所存在的误差会使标记与真实月球、真实太阳和真实月球轨道升交点的实际位置之间的偏差越来越大。第一个发生显著偏离的将是

gives an orbital period of 28 days instead of 27.32 days. But we can make a correcting adjustment to the M marker twice every month, simply by aligning M opposite S at the time of full Moon, and by placing it coincident with S at new Moon. The prescription for S gives an orbital period of 364 days, which is near enough to the actual period because it is possible to correct the position of S four times every year, by suitable observations made with the midsummer, midwinter, and equinoctial sighting lines that are set up with such remarkable accuracy at Stonehenge.

Stonehenge is also constructed to determine the moment when $\phi=0$, that is, when N should be set at γ. The line C to A 1 of Fig. 1 is the azimuthal direction for the minimum point of Fig. 2. By placing N at γ when the Moon rises farthest to the north, the N marker can be calibrated once every 18.61 years. The prescription implies only a small error over one revolution of N. If N started correctly, it would be out of its true position by only 1° or so at the end of the first cycle. The tolerance for eclipse prediction is about 5°, so that if we were to adjust N every cycle, the predictor would continue to work indefinitely without appreciable inaccuracy. The same method also serves to place N at the beginning.

But now we encounter an apparent difficulty. The minimum of Fig. 2 is very shallow and cannot really be determined in the way I have just described. Angular errors cannot have been less than ±0.25°, and even this error, occurring at the minimum of Fig. 2, is sufficient to produce an error of as much as ±15° in ϕ.

The correct procedure is to determine the moment of the minimum by averaging the two sides of the symmetrical curve, by taking a mean between points 2, for example. The inaccuracy is then reduced to not more than a degree or two—well within the permitted tolerance.

What is needed is to set up sighting directions a little to the east of the most northerly direction. The plan of Stonehenge shows a line of post holes, A 1, 2, 3 and 4, placed regularly and with apparent purpose in exactly the appropriate places.

The same point applies to solsticial measurements of the Sun. In summer the sighting line should be slightly increased in azimuth, in winter it should be slightly decreased.

Hawkins[1] gives two tables in which he includes columns headed "Error Alt.". These altitude errors were calculated on the assumption that the builders of Stonehenge intended to sight exactly the azimuthal extremes. The test of the present ideas is whether the calculated "errors" have the appropriate sign—on the argument given here "errors" should be present and they should have the same sign as the declination. In ten out of twelve values which Hawkins gives in his Table 1 this is so. The direction from C to the Heelstone is one of the two outstanding cases. Here the "error" is zero, suggesting that this special direction

月球的标记M——在该方法中所采用的运行周期为28天而非27.32天。但是我们可以每个月对标记M的位置进行两次校正，只需在满月时把M点调整到S点对面，而在新月时将其摆放到和S点重合即可。S在该方法中的运行周期为364天，这已经非常接近真实的周期值了，因为每年都可以对S的位置进行四次校正，方法是在夏至点、冬至点、春分点和秋分点时刻正确地测量光线的位置，这些都可以在巨石阵中以很高的精度被测定。

巨石阵也可以用于测定$\phi=0$的时刻，即当N被设定在γ点时。图1中从C到$A1$的连线方向即是图2中最小值点的方位角方向。当月球在最靠北的角度升起的时候，需要将N定位于γ点，这样每18.61年就可以对标记N进行一次校准。上述方法要求N在循环一周后只能有很小的偏差。如果N的起始位置是正确的，那么它在环行一周结束时只会与真实位置偏离1°左右。日月食预测所允许的误差在大约5°的范围以内，所以如果我们每循环一周就对N点进行一次校准，那么这个预报器将会在误差很小的情况下无限期地发挥作用。还可以用同样的方法把N的初始位置确定下来。

但是现在我们遇到了一个明显的麻烦。图2中的最小值范围非常宽，用我刚刚描述的方法很难真正确定这个最小值点。角度误差不能小于$\pm 0.25^{\circ}$，而在确定图2中的最小值时即便只存在这么大的误差，也足以使ϕ产生$\pm 15^{\circ}$的误差。

正确的方法是通过在对称曲线的两边取平均来测定最小值的时刻，例如取图中两个点2之间的平均值。利用这种方法就可以把误差减小到不大于1度或2度——远低于误差允许的范围。

现在需要在最北边偏东一点的方向上设置视线方向。在巨石阵的平面图中有一条由柱坑A 1、A 2、A 3和A 4排成的直线，它们的排列很有规律，并且准确地摆放在了适当的位置上，这显然是有意安排的。

可以用同样的方法来测量太阳的二至点。在夏季，视线的方位角会略有增加；而在冬季，它应该略有减少。

霍金斯[1]给出了两张表格，表格中含有标题为“高度误差”的列。计算这些高度误差的前提是假设巨石阵的建造者想要精确地观测出最大方位角。检验上述观点的方法是要看计算出的“误差”是否具有正确的符号——这里给出的论点是应该存在“误差”，并且与偏角的符号一致。对于霍金斯在表1中所给出的12个值，有10个属于这种情况。从C点到那块踵石的方向是两个最突出的偏差点之一。在这里“误

was kept exactly at the direction of midsummer sunrise, perhaps for aesthetic or ritualistic reasons. The other discrepant case is 91→94. Here my own calculation gives only a very small discrepancy, suggesting that this direction was also kept at the appropriate azimuthal extreme.

Negative values of the altitude error correspond to cases where it would be necessary to observe below the horizontal plane, if the objects in question were sighted at their extreme azimuths. This is impossible at Stonehenge because the land slopes gently upward in all directions. Such sighting lines could not have been used at the extremes, a circumstance which also supports this point of view.

It is of interest to look for other ways of calibrating the N marker. A method, which at first sight looks promising, can be found using a special situation in which full Moon happens to occur exactly at an equinox. There is evidence that this method was tried at Stonehenge, but the necessary sighting lines are clearly peripheral to the main structure. Further investigation shows the method to be unworkable, however, because unavoidable errors in judging the exact moment of full Moon produce large errors in the positioning of N. The method is essentially unworkable because the inclination of the lunar orbit is small. Even so, the method may well have caused a furore in its day, as the emphasis it gives to a full Moon at the equinox could have been responsible for the dating of Easter.

An eclipse calibrator can be worked accurately almost by complete numerology, if the observer is aware of a curious near-commensurability. Because S and N move in opposite directions the Sun moves through N more frequently than once a year, in 346.6 days. Nineteen such revolutions is equal to 6,585.8 days, whereas 223 lunations is equal to 6,585.3 days. Thus after 223 lunations the N marker must bear almost exactly the same relation to S that it did before. If the correct relation of N to S is known at any one moment N can be reset every 223 lunations; that is, every 18 years 11 days. The near-commensurability is so good that this system would give satisfactory predictions for more than 500 years. It requires, of course, S to be set in the same way as before. The advantage is that in the case of N it obviates any need for the observational work described above. But without observations the correct initial situation cannot be determined unless the problem is inverted. By using observed eclipses the calibrator could be set up by trial and error. This is probably the method of the Saros used in the Near East. There is no evidence that it was used at Stonehenge. The whole structure of Stonehenge seems to have been dedicated to meticulous observation. The method of Stonehenge would have worked equally well even if the Saros had not existed.

Several interesting cultural points present themselves. Suppose this system was invented by a society with cultural beliefs associated with the Sun and Moon. If the Sun and Moon are given godlike qualities, what shall we say of N? Observation shows that whenever M and S are closely associated with N, eclipses occur. Our gods are temporarily eliminated. Evidently, then, N must be a still more powerful god. But N is unseen. Could this be the origin of the concept of an invisible, all-powerful god, the God of Isaiah? Could it have been the discovery of the

差”为零，说明这个特殊方向与夏至时太阳升起的方向保持精确一致，这也许是出于审美上的需要或者仪式上的要求吧。另一个偏差点是 91 → 94 方向。我通过计算发现这里只有很小的误差，说明这个方向也与最大方位角保持着相当的一致性。

高度误差中的负值对应于需要在地平面以下进行观测的情况，比如在只有这么做才能以最大方位角观察某些天体时。对巨石阵来说这是不可能的，因为在所有方向上地势都是微微向上倾斜的。这样的视线方向不可能用于观测出最大方位角，这一事实再次印证了上述观点。

人们对寻找其他方式校准标记 N 很感兴趣。有一种方法，初看起来前景还不错，但后来发现它用到了一个特殊条件，即要求满月刚好出现在春分点或秋分点。有证据表明，这种方法曾在巨石阵中被尝试过，但视线显然只能位于巨石阵主体结构的外部。然而，进一步的调查表明该方法行不通，这是因为在判断满月的具体时刻时会产生难以避免的误差，而这种误差又会使 N 的定位出现很大的偏差。由于月球轨道的倾角太小，所以这种方法从本质上讲是不可行的。即便如此，该方法很可能在当时曾造成过一时的轰动，因为它要观测在二分点时刻发生的满月，而春分见到的满月可能曾被用于确定复活节的日期。

如果观测者知道一种特殊的近似对等关系，那么几乎完全用数字学就能够准确地进行日月食的校准。因为 S 和 N 向相反的方向运动，所以太阳每次经过 N 所需的时间不到一年，为 346.6 天。19 个这样的循环等于 6,585.8 天，而 223 个朔望月等于 6,585.3 天。因此，在 223 个朔望月之后，标记 N 相对于 S 的位置就会与 223 个月前几乎完全一致。如果能了解到 N 和 S 在任意时刻的正确位置关系，那么每 223 个朔望月，也就是 18 年零 11 天，就可以把 N 的位置重新调整一次。这种近似对等关系非常绝妙，以至于该系统能够在 500 多年的时间里给出令人满意的预测结果。当然，设置 S 点位置的方法应该与以前相同。这种方法的优势在于不需要根据之前描述的观测过程来确定 N 的位置。但没有经过观测就不能确定正确的初始位置，除非这个问题是反过来问的。而利用已经观测到的日月食数据就可以反复地校正这个预报器。这种方法大概就是近东地区所采用的沙罗周期。没有证据能够证明在巨石阵中使用了这一方法。巨石阵的整体结构似乎曾被用于精确的观测。即使不存在沙罗周期，巨石阵的方法也同样可以很好地预报日月食。

以下是几个文化方面的有趣话题。假设巨石阵的建造者来自一个崇尚与日月有关的文化的民族。如果太阳和月亮被赋予了类似神的特性，那么我们应该怎么看待 N？观测结果表明在 M 和 S 非常靠近 N 时总会有日月食现象发生。我们的神就会被暂时抛在一边。那么显然 N 一定是一个更加强有力的神。但 N 是不可见的。这

significance of N that destroyed sun-worship as a religion? Could M, N and S be the origin of the doctrine of the Trinity, the "three-in-one, the one-in-three"? It would indeed be ironic if it turned out that the roots of much of our present-day culture were determined by the lunar node.

(**211**, 454-456; 1966)

Fred Hoyle: University of Cambridge.

References:

1. Hawkins, G. S., *Nature*, **200**, 306 (1963).
2. Hawkins, G. S., *Nature*, **202**, 1258 (1964).

难道就是一个无形而全能的神——以赛亚神概念的起源吗？发现 N 的重要性会不会破坏了宗教中对太阳的崇拜？M、N 和 S 会不会就是三位一体教义，即“三中有一、一中有三”的起源呢？如果能够证明现今文明的根源主要来自月球交点，这还真有点讽刺意味。

（孟洁 翻译；肖伟科 审稿）

Stonehenge—A Neolithic "Observatory"

C. A. Newham

Editor's Note

The astronomical function of the ancient monument of Stonehenge near Salisbury in England is also considered in this contribution by astronomer C. A. ("Peter") Newham. He proposes that the post holes of the stone circle were used to mark the bearing of moonrise, which would be used to help predict eclipses. Archaeologists, however, have remained somewhat sceptical of such speculative interpretations by astronomers.

PROF. G. S. Hawkins[1] has shown how Stonehenge could be regarded as a "computer" to predict the time when eclipses of the Sun and Moon were due, though not always visible at Stonehenge. He ingeniously relates the fifty-six Aubrey holes with a 56-year eclipse cycle. The principle is valid, but there is no evidence to support the idea that Stonehenge was intended to be a "computer" other than that the majority of main features embodied in the "monument" unquestionably have some astronomical connexion. There are, however, a number of "post holes"; however, no satisfactory explanation has so far been put forward to explain their purpose.

An analysis of the position and number of post holes prompts a suggestion as to their purpose, and may well provide the clue to the method by which the builders of Stonehenge acquired elementary knowledge of Moon cycles, possibly including the approximate 56-year eclipse cycle.

Stonehenge post holes. Of the many post holes found in and around Stonehenge, there is one group of about forty holes situated in the "causeway" near the northeast entrance (Fig. 1). The holes seem to radiate from the centre of the Aubrey circle, and lie within a 10-degree arc north of the heelstone or solstice line. They are roughly arranged into six ranks crossing the line of the causeway.

巨石阵——一个新石器时代的“天文台”

纽汉

编者按

在这篇文章中，天文学家纽汉（“彼得”）也在思考英国索尔兹伯里附近的史前纪念碑——巨石阵的天文学功能，他提出石圈中的柱坑可以用于标志月亮升起的方位，这对日月食的预测是有帮助的。然而，考古学家们仍然对这些天文学家的试探性解释持怀疑态度。

霍金斯教授[1]曾解释过如何把巨石阵看作“计算工具”来预言日月食的发生时间，虽然在巨石阵的位置上未必能观测到所有的日月食。他巧妙地将56个奥布里洞与56年的日月食周期联系起来。虽然他提出的原理具有可行性，但是现在还没有证据能证明巨石阵原本是被设计为一种“计算工具”，而不是大部分主体结构无疑与天文学有一定联系的“纪念碑”。然而，这里存在很多“柱坑”；迄今为止还没有令人满意的解释能够说明这些柱坑的用途。

本文根据对柱坑位置和数目的分析得出了一个有关其用途的假设，该假设很可能提供了有关巨石阵建造者如何获取月球周期基础知识方面的线索，其中或许包括约为56年的日月食周期。

巨石阵的柱坑。在巨石阵之内和周围发现的很多柱坑中，有一组柱坑位于“长堤”上靠近东北入口的地方，数量大约有40个（图1）。这些柱坑似乎是从奥布里环的中心延伸出来的，并且分布在从奥布里环中心到踵石或至日点的连线以北10度的范围之内。它们在垂直于长堤的延伸方向上大致排成6行。

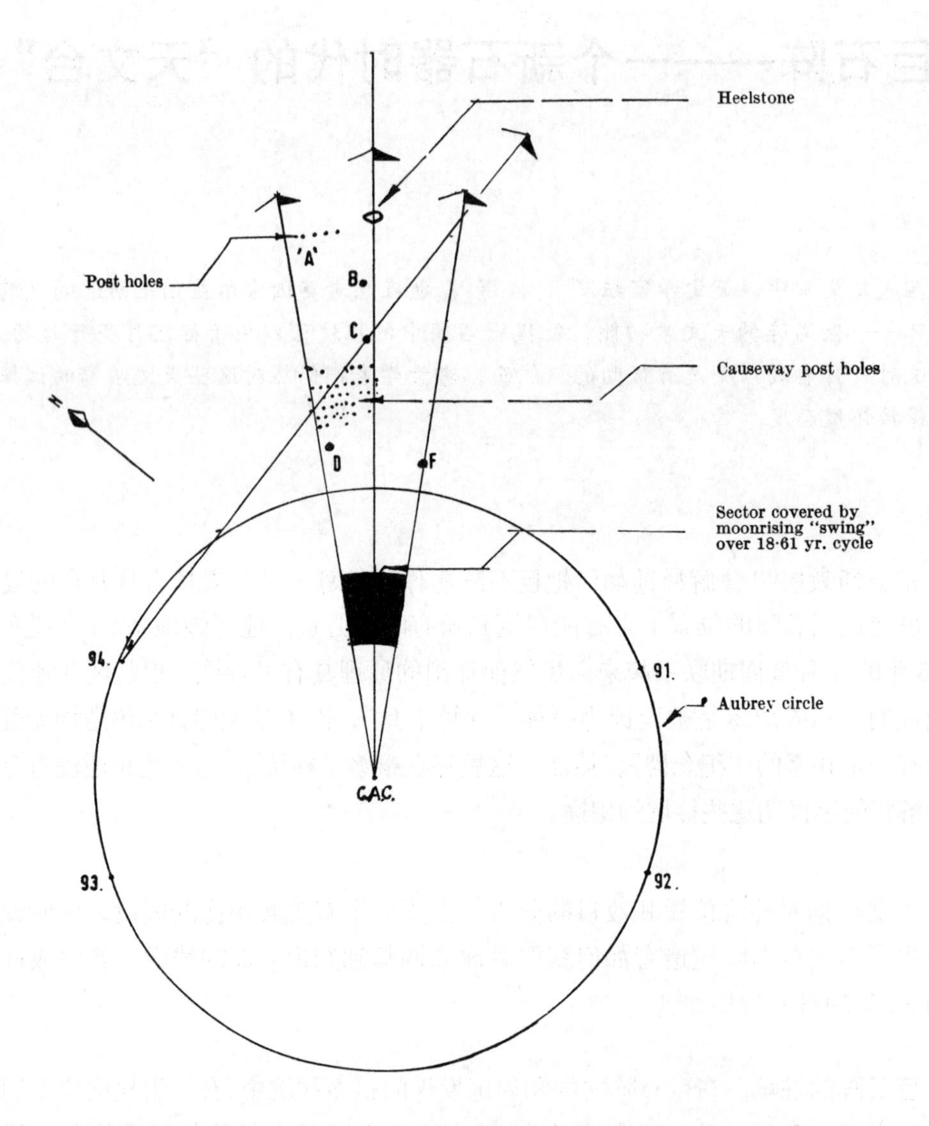

Fig. 1. Schematic diagram showing the position of post holes in relation to other main features

A fairly reliable record of the behaviour of the Moon could be obtained by planting a temporary marker (wooden stake) to align on the point where the winter full-moon appeared above the horizon each year. The indications are such that this was done over a large number of years covering several 18.61-year cycles. Such a period would be sufficient to ascertain a 19-year phase or metonic cycle and possibly the approximate 56-year eclipse cycle as suggested by Hawkins (that is, $3\times18.61=55.83$). It should be appreciated, however, that they would have considerable difficulty in defining this eclipse cycle by their crude methods.

Starting from the premise that the post holes were used for the purpose suggested, the azimuth bearing of each post hole was first ascertained in rank order, numbered 1 to 6 counting in a north-easterly direction.

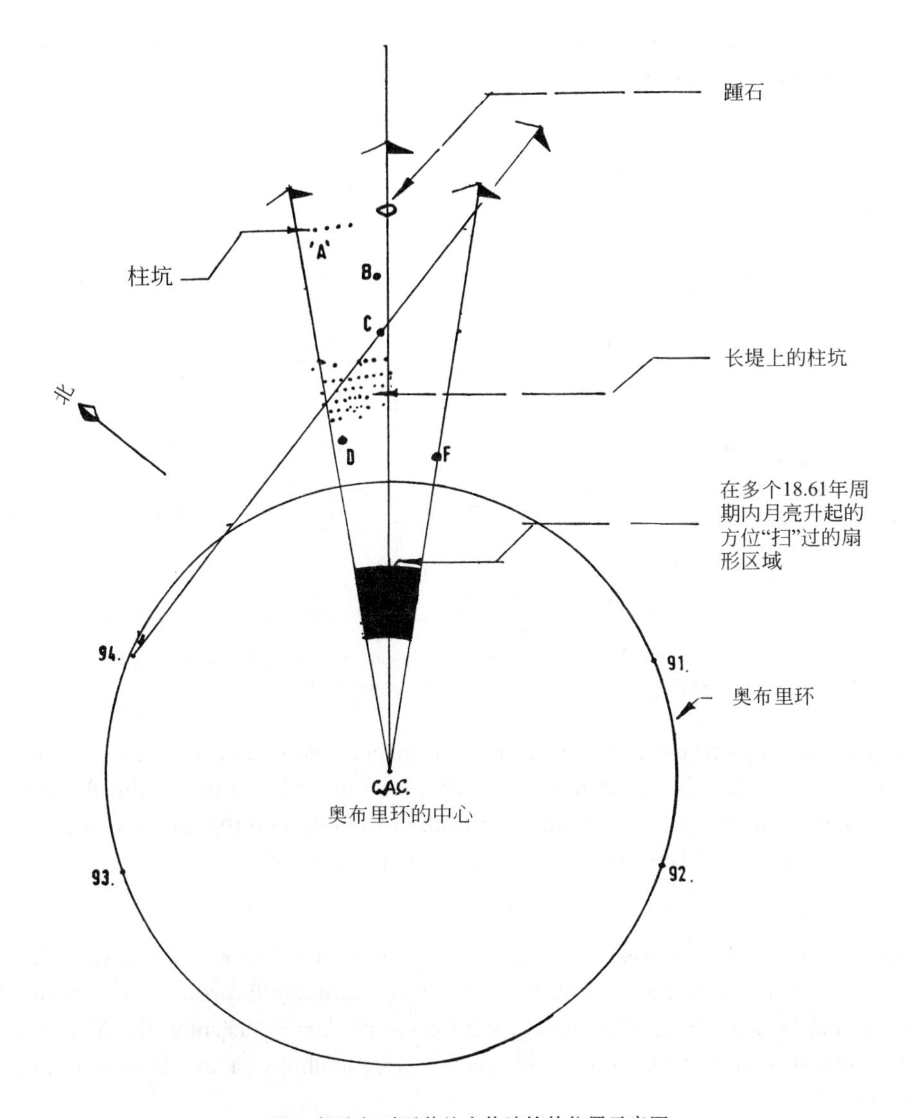

图 1. 柱坑相对于其他主体建筑的位置示意图。

通过在每年冬季满月初升出地平线时的方向点上植入一个临时的标志（木桩）就可以非常可靠地记录月球的活动。一些迹象表明这样的观测记录工作持续了涵盖多个 18.61 年周期的时间。这么长的时间足以使测定者发现 19 年的太阴周期或称默冬章，还可能发现霍金斯所指出的约 56 年的日月食周期（即：$3 \times 18.61 = 55.83$）。然而，可以想象当时人们用如此粗糙的方法来确定这个日月食周期势必会遇到相当大的困难。

假设这些柱坑就是为了达到上述目的而留下的，以此为前提，先将每个柱坑的方位按行的顺序编号，沿着东北方向依次为 1 到 6。

The number of holes attributed to each rank and their azimuth bearings taken from the centre of the Aubrey circle are given in Table 1.

Table 1

Rank No.	1	2	3	4	5	6
Azimuth bearings degrees E. of N.	–	–	(40.4)	(40.7)	–	(40.3)
	41.9	41.9	–	–	41.8	–
	43.1	43.0	43.0	42.9	42.9	42.7
	–	44.1	44.0	43.8	44.0	–
	–	45.3	45.3	45.5	45.6	–
	–	45.6	45.6	–	–	–
	46.3	46.4	46.5	46.3	46.4	46.4
	–	– *47.4	–	–	–	46.7
	–	47.8	47.5	47.2	47.6	47.5
	–	49.3	49.4	49.0	48.9	49.1
	–	49.8	–	–	49.8	–
	–	–	50.4	50.2	50.8	50.3

The figures in parentheses refer to holes beyond the line of full orb but are reached by the line of first gleam of moonrise.
*This hole is situated between ranks No. 2 and No. 3 and cannot be allocated to either rank.

Preliminary calculations reveal that the causeway post holes lie within the arc or sector of the most northerly limits of winter full-moon rise as seen from the centre of the Aubrey circle (C.A.C.); also, the number of holes in any one rank did not exceed the number of risings that could appear north of the heelstone or solstice line in any one cycle.

If individual bearings of moonrise in a sequence of cycles were found to agree with the "hole alignments", any doubt as to their significance would be eliminated; furthermore, dating of the post holes would be feasible. Unfortunately, the necessary data concerning the Moon are not available; to obtain such information would severely tax the ability of the most expert celestial mechanician.

However, computed moonrise bearings covering the period of 2000–1000 B.C. were first used in an attempt to determine whether any similarity existed between general grouping of the post holes and a sequence of moonrisings. The data used were kindly supplied by Prof. G. S. Hawkins. His calculations are based on first-order terms and refer to the full-moon nearest the winter solstice and also require that the instant of moonrise should coincide with the time when the Moon reaches its appropriate maximum declination. This particular feature seldom applies (see Fig. 2).

每一行中柱坑的数量以及它们相对于奥布里环中心的方位角列于表 1。

表 1

行号	1	2	3	4	5	6
方位角度数 由北向东	–	–	(40.4)	(40.7)	–	(40.3)
	41.9	41.9	–	–	41.8	–
	43.1	43.0	43.0	42.9	42.9	42.7
	–	44.1	44.0	43.8	44.0	–
	–	45.3	45.3	45.5	45.6	–
	–	45.6	45.6	–	–	–
	46.3	46.4	46.5	46.3	46.4	46.4
	–	– *47.4	–	–	–	46.7
	–	47.8	47.5	47.2	47.6	47.5
	–	49.3	49.4	49.0	48.9	49.1
	–	49.8	–	–	49.8	–
	–	–	50.4	50.2	50.8	50.3

圆括号中的数字指的是满月线以外的柱坑，但月亮升起时的第一道光线可以照到这些柱坑。
* 这个柱坑位于第 2 行和第 3 行之间，并且无法把它归到这两行的任意一行中去。

初步的计算结果显示：长堤上的这些柱坑是排列在一个圆弧或扇形区域内的，而这个扇形区域对应的是从奥布里环中心（C.A.C）观测冬季满月升起的最靠北的极限；同时，任何一行中柱坑的数目都没有超过在任何一个周期内所能看到的月亮从奥布里环中心与踵石或至日点连线以北升起的次数。

如果在一个周期序列中，每次月亮升起的方位角都与“柱坑的排列”相符，那么所有对这些柱坑重要性的怀疑都将烟消云散；此外，测定这些柱坑的年代也将是可行的。不幸的是，我们没有与月球活动相关的必要数据；即使是最内行的天体力学专家，想要获得这样的信息也要经受得起对其能力的严峻考验。

不过，计算得到的从公元前 2000 年到公元前 1000 年的月亮升起方位角的数据，被首次用于确定柱坑的分组和月出时间序列之间是否存在某种相似性。该数据是由霍金斯教授热心提供的。他的计算基于一阶近似数据，并且参考了最靠近冬至点的满月，还要求月出的同时月球也应当达到其相应的最大赤纬值。这样特殊的条件很少能得到满足（见图 2）。

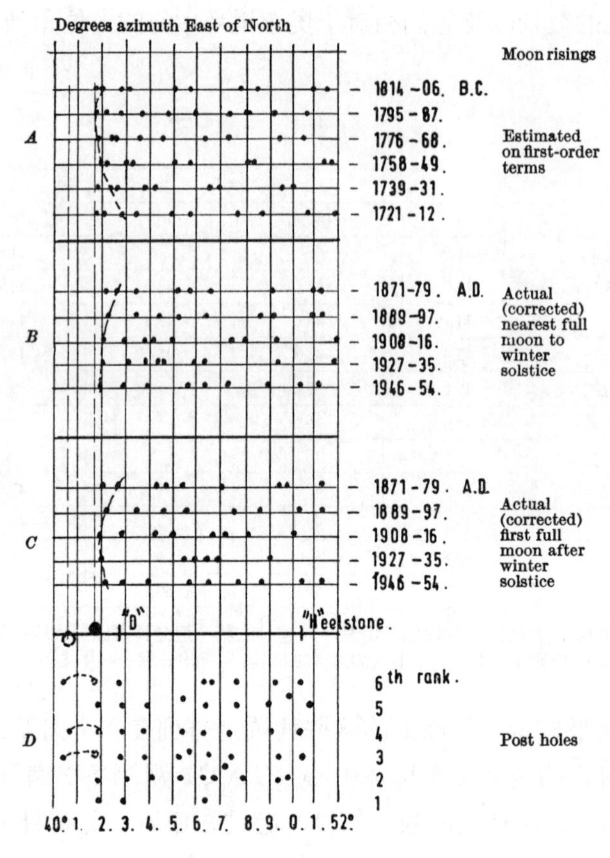

Fig. 2. Comparable azimuth bearings of winter full-moon risings (Full-orb tangent to horizon) and post hole alignments

Group "*A*": Six cycles of moonrisings based on first-order terms as supplied by Hawkins for period 1814–1712 B.C.

Group "*B*": Similar to "*A*" but the "Moons" correspond to the known rising position covering the period A.D. 1871–1954 after correction to allow for the change in the angle to the ecliptic of the Earth–Moon system over the past 3,800 years

Group "*C*": Basically similar to "*B*", but here it applies to the first full-moon after the winter solstice

Group "*D*": Depicts the post holes arranged according to their azimuth bearings

The data connected by broken lines are all separated by a period of 6,940 days or exactly 19 years within a few hours, and differ from other similar Moons in that their rising positions are practically the same. The broken lines in group "*D*" merely indicate the position the three holes would occupy if they were aligned on full orb. The two larger circles immediately above represent the most northerly position of first gleam and full orb of moonrise. The positions of stone "*D*" and the heelstone are indicated on the same line

The pattern of risings that would apply to the early Stonehenge period is more truly represented in groups "*B*" and "*C*". Even so, their arrangements could be subject to slight variations as many individual moons have possible alternative rising positions. This applies in those cases when the Moon reaches "full" during the daytime, for example, on December 12, 1875, the Moon was full at 0745 h (shortly before setting). In similar circumstances, would they regard the critical Moon as being that which rose on the afternoon of December 12, or that of

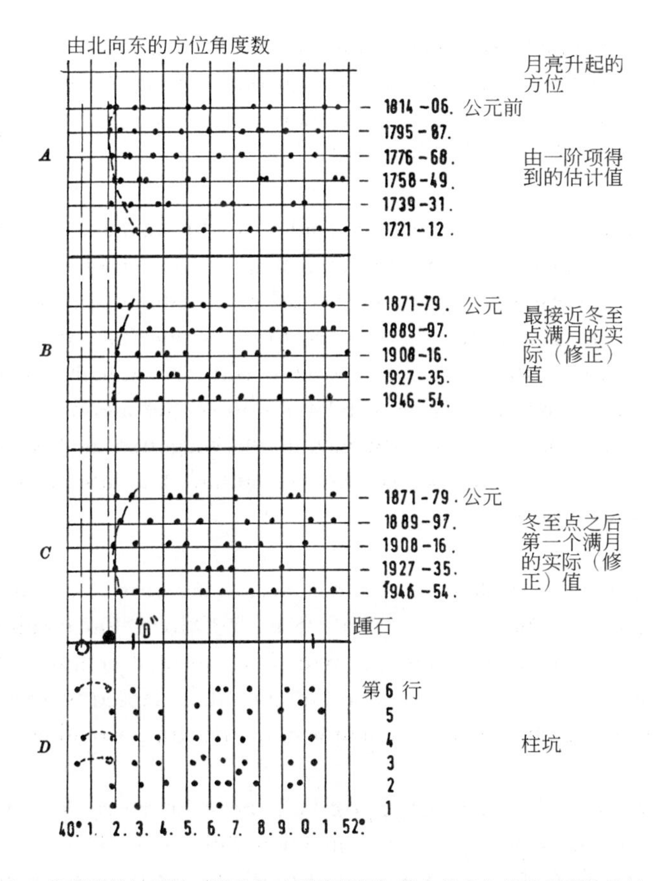

图 2. 冬季满月初升时（满月与地平线相切）的方位角与柱坑排列的比较。

“*A*”组：月亮在 6 个周期中的升起位置，基于霍金斯所提供的公元前 1814 年至公元前 1712 年数据的一阶项。
“*B*”组：与“*A*”组类似，但“月亮”的升起位置是公元 1871 年至公元 1954 年间实际测定的位置，并针对地月系统经过 3,800 年之后与黄道面夹角的变化进行了一定的修正。

“*C*”组：与“*B*”组大体上一致，但在此用到的是冬至点之后的第一次满月。
“*D*”组：描绘了柱坑按照方位角排列的情况。

由虚线连接起来的数据都相隔 6,940 天或者恰好 19 年，误差在几个小时以内，它们与其他类似数据的不同之处在于月亮升起的位置基本相同。“*D*”组中的虚线只是为了说明这三个柱坑应该占据的位置，如果它们在发生满月的时候能排成一条直线的话。在它们上边的两个稍大一点的圆代表了满月初升时第一缕光线最北端的位置。在同一条水平线上还标有“*D*”号石和踵石的位置。

这种适用于巨石阵时代早期的月升模式在“*B*”组和“*C*”组中得到了更真实的体现。即使如此，它们的排列仍有可能会出现一些微小的变化，因为在很多情况下月亮升起的位置可能会有两种选择。这种情况通常出现在当月亮在白天达到“满”月的时候，例如：1875 年 12 月 12 日，月亮于早上 7 点 45 分达到满月（月落前不久）。在类似的情况下，他们会把临界的月亮看作是 12 月 12 日的下午（译者注：原稿可能有误，此处

the following day? If midnight was their datum, the former Moon would apply. On the other hand, it would be the latter if the critical time was on its rising. A similar condition arises in respect of those Moons which are full 14 or 15 days before the solstice and would be followed by the next Moon 14 or 15 days after the solstice.

There is no question that people of early civilizations had to contend with the same difficulty which was partly obviated by choosing the critical Moon as the one which was full after a specified time, such as the day of the winter solstice or equinox. In the second millennium B.C. the Delians regarded the critical Moon as the first one that was full after the winter solstice, and the possibility that such was the case with the Stonehenge people cannot be overlooked. However, the remarkably close relationship of the moonrise sequences and hole patterns must surely be more than a coincidence. If this suggested method of observation had been carried out by the Stonehenge people, they would have had little difficulty in recognizing the 19-year phase or metonic cycle.

It seems fair to assume that these and other post holes served a purpose in obtaining preliminary information of the behaviour of the Sun and Moon. Once satisfactory alignments had been established, markers of a more permanent kind would be installed. If the foregoing assumptions were correct, then the long period of time that elapsed between the inception of the bank, etc., and the building of the first major stone structures would be accounted for.

The position and spacing of the four large post holes near the heelstone indicate a relationship with the causeway holes, especially so in conjunction with stone "*D*" and the heelstone. The size of the holes indicates that the posts were much larger than the causeway posts and, presumably, more permanent. If stone "*B*" were also included, then the seven markers which could be observed would act as a crude "vernier". When correlated with the setting Sun seen in the reverse direction, it would provide a means of defining the time when an eclipse of the Moon was probable.

Whether or not Stonehenge people discovered the 56-year cycle is a matter for conjecture, but their ability to predict pending eclipses after a crude fashion was certainly possible.

It can be shown that a person assisted by nothing more than the crudest equipment in the form of a peg and line and a few stakes could acquire the necessary information. The observer, however, would have to have the necessary tenacity of purpose to undertake systematic observations over a long period of time and a simple method of recording numbers by means of tokens, for example, notches on a stick or pebbles in a bag.

Doubtless, several generations of observers were involved, and if we are to believe that the writings of Diodorus referred to Stonehenge, then the position of supervisor of the "Spherical* Temple" was held by a member of the same family from one generation to another.

---

* Early Greek term for "Astronomical". Heath: History of Greek Mathematics.

似应为 12 月 11 日下午）升起的月亮，还是接下来的一天呢？如果他们的数据以午夜为起止时刻，那么应该以前一天的月亮为准。另一方面，如果临界时间是在上升过程中的月亮，那么就应该以后一天的月亮为准。就至日点前 14 或 15 天发生的满月和至日点之后 14 或 15 天发生的下一个满月来说，情况是相同的。

毫无疑问，在文明的早期，人类不得不去面对同样的难题。通过选择临界时刻的月亮就可以在一定程度上回避这个难题，比如选择某个特定时间，如冬至点或者春 / 秋分点，之后的满月时刻。在公元前两千年，得洛斯人认为临界时刻的月亮是冬至点之后的第一次满月，巨石阵的建造者很可能也注意到了这一点。无论如何，月亮升起方位与柱坑排列之间显然存在的密切联系绝不仅仅是一种巧合。如果巨石阵的建造者们所采用的就是上面提到的观测方法，那么他们就会毫不费力地发现 19 年的周期或称默冬章。

似乎可以合理地假设留下这些以及其他柱坑的目的就是为了获得与太阳和月亮活动有关的初步知识。一旦找到了令人满意的排列方式,就可以建立更为耐久的标志。如果前面的假设是正确的，那么从筑堤等的开工到第一个主体石结构建成之间为什么需要花费那么长时间就可以得到解释了。

踵石附近四个大柱坑的位置和间距表明它们与长堤上的柱坑是有关联的，特别是与“*D*”号石和踵石有关联。从坑的大小来看，踵石附近的四个大柱坑显然比长堤上的柱坑大很多，而且可能更加固定。如果把“*B*”号石也包括在内，那么这七个被用于观测的标志就可以作为一个粗制的“游标尺”。当与从相反方向观察到的落日相联系时，它就可以被作为一种确定月食可能发生时刻的方式了。

关于巨石阵的建造者们是否发现了 56 年的周期这个问题还停留在猜测层面，但利用一种粗糙的方式预测出即将发生的月食是他们绝对可以做到的。

可以证明：一个人仅借助一些最原始的工具，如木钉和线以及几个木桩就可以得到所需的信息。然而，观测者还必须具有长期坚持进行系统观测的毅力，并会用一种简单的方法通过作记号来记录数字，比如在一根小棍上刻痕或者在一个袋子里装鹅卵石。

毫无疑问，这个过程需要好几代观测者才能完成，并且假如我们相信狄奥多在他的著作中所提到的就是巨石阵，那么坐在执掌这个“球形 * 庙宇”宝座上的人必定出自同一个家族，并且一代一代世袭下去。

---

* 在早期希腊语中就是“天文”的意思。希思：希腊数学史。

The strong lunar influence with which Stonehenge must now be associated necessitates revision of hitherto accepted explanations of some salient features. Until new evidence is found pointing to the contrary, it is more logical to conclude:

(*a*) That the small stone (No. 11) in the sarsen circle was intentional, and that the circle represented the 29.5 days of the lunar month.

(*b*) The double circle or spiral[2] of the "*Y*" and "*Z*" holes represented the 59 days of two lunar months. The strong possibility that there were fifty-nine blue stones inside the sarsen circle would provide a more suitable means of representing the same thing.

(*c*) The 19-year phase or metonic cycle was represented by the nineteen blue stones inside the trilithon "horse shoe".

All things considered, including other similar post holes, there seems little doubt that Stonehenge, in its early stages, was a kind of "observatory". It provided a suitable site wherein systematic observations of "Soluna" (Sun and Moon) phenomena were carried out by these neolithic peoples.

(**211**, 456-458; 1966)

C. A. Newham: 5 Sedge Rise, Tadcaster, Yorkshire.

References:

1. Hawkins, G. S., *Nature*, **200**, 306 (1963); **202**, 1258 (1964).
2. Sale, J. L., *The Secrets of Stonehenge* (private publication, 1965).

从现在来看，巨石阵的建造显然与月球活动有很密切的关联，这使我们必须对迄今为止人们所接受的巨石阵主要结构的解释进行修正。在能找到与此相反的新证据之前，我认为以下这些结论更合乎逻辑：

(*a*) 在砂岩漂砾圈中的小石头（11 号石）是被有意放在那里的，这个圈代表的是 29.5 天的朔望月周期。

(*b*) 双圈或螺旋形[2] 的“*Y*”和“*Z*”柱坑代表的是两个朔望月即 59 天。很有可能在砂岩漂砾圈内部曾摆放过 59 块蓝色的石头，这是代表两个朔望月 59 天的一个更为合适的方法。

(*c*) 位于“马蹄”形巨石牌坊内部的 19 块蓝色石头代表了 19 年的周期或称默冬章。

在考虑到包括其他类似柱坑在内的所有因素之后，几乎可以肯定巨石阵在建成之初时就是一座“天文台”。它为新石器时代的古人系统观测与日月有关的现象提供了一个合适的场所。

（孟洁 翻译；肖伟科 审稿）

Observation of a Rapidly Pulsating Radio Source

A. Hewish *et al.*

Editor's Note

A survey of radio emission in the sky revealed a curiously periodic source, as reported here by Anthony Hewish and his colleagues (the actual discovery was made by Jocelyn Bell, who was at that time Hewish's graduate student). The period was determined to be stable to one part in ten million, and the source showed no parallax, so it could not originate within the Solar System. The researchers also found three other sources with similar properties. Hewish speculated that they might arise from radial (in and out) pulsations of white dwarf stars or neutron stars. We now know that they arise from beams on rapidly rotating neutron stars (called pulsars), but the precise mechanism for generating the beams is still controversial. For this discovery, in 1974 Hewish became the first astronomer to be awarded the Nobel Prize in physics.

IN July 1967, a large radio telescope operating at a frequency of 81.5 MHz was brought into use at the Mullard Radio Astronomy Observatory. This instrument was designed to investigate the angular structure of compact radio sources by observing the scintillation caused by the irregular structure of the interplanetary medium[1]. The initial survey includes the whole sky in the declination range $-08° < \delta < 44°$ and this area is scanned once a week. A large fraction of the sky is thus under regular surveillance. Soon after the instrument was brought into operation it was noticed that signals which appeared at first to be weak sporadic interference were repeatedly observed at a fixed declination and right ascension; this result showed that the source could not be terrestrial in origin.

Systematic investigations were started in November and high speed records showed that the signals, when present, consisted of a series of pulses each lasting ~0.3 s and with a repetition period of about 1.337 s which was soon found to be maintained with extreme accuracy. Further observations have shown that the true period is constant to better than 1 part in 10^7 although there is a systematic variation which can be ascribed to the orbital motion of the Earth. The impulsive nature of the recorded signals is caused by the periodic passage of a signal of descending frequency through the 1 MHz pass band of the receiver.

The remarkable nature of these signals at first suggested an origin in terms of man-made transmissions which might arise from deep space probes, planetary radar or the reflexion of terrestrial signals from the Moon. None of these interpretations can, however, be accepted because the absence of any parallax shows that the source lies far outside the solar system. A preliminary search for further pulsating sources has already revealed the presence of three others having remarkably similar properties which suggests that this type of source may be relatively common at a low flux density. A tentative explanation of these unusual sources in terms of the stable oscillations of white dwarf or neutron stars is proposed.

快速脉动射电源的观测

休伊什等

编者按

正如安东尼·休伊什及其同事在本文中所报道的，对天空中射电辐射的探测揭示出了一个奇特周期性源的存在（实际上真正的发现者是乔斯琳·贝尔，那时她还是休伊什的研究生）。经测定，周期稳定在 10^{-7} 的误差内，并且该源没有视差，所以它不可能来自太阳系内部。研究人员还发现了另外三个具有同样特性的射电源。休伊什推测它们可能来自白矮星或中子星的径向（胀缩）脉动。现在我们知道它们是由快速旋转的中子星（被称为脉冲星）所发射的辐射造成的，但人们对产生这种辐射束的确切机制尚存争议。因为上述发现，休伊什在 1974 年成为了第一位获得诺贝尔物理学奖的天文学家。

1967 年 7 月，一台工作频率是 81.5 MHz 的大型射电望远镜在玛拉德射电天文台投入使用。有了这台仪器，研究人员就可以通过观测由行星际介质不规则结构造成的闪烁来研究致密射电源的角结构 [1]。最初的观测范围覆盖了赤纬 δ 为 $-08^\circ < \delta < 44^\circ$ 的全部天区，并且这一区域每周都会被扫描一次。因此大部分天区落在研究人员的定期监测之中。在这台仪器投入运行后不久，研究人员发现：一些起先看上去微弱而零星的干扰信号会在一个固定的赤经和赤纬上被反复观测到。上述结果说明该信号不可能来源于地球。

系统探测从 11 月开始进行。高速记录的结果表明：当这种信号出现的时候，它们总是由一系列脉冲组成，每个脉冲的持续时间约为 0.3 s，而重复周期约为 1.337 s，研究人员很快发现这一周期的稳定性很高。尽管地球的轨道运动可能导致了一定的系统变化，但进一步的观测结果表明，真实周期是一个误差不超过 10^{-7} 的常数。被记录信号的脉冲特性是由一个信号的周期部分产生的，该信号的频率在接收机的整个 1 MHz 通带内递减。

最开始，研究人员认为这些信号的不寻常特性源自一些人造的信号传输装置，比如或许源自深空探测器、行星雷达或者从月球上反射回来的地球信号。但这些解释都不能为人们所接受，因为：没有视差足以说明信号源远在太阳系之外。在对更多的脉动源进行初步探测之后，研究人员发现还有三个源也具有非常类似的性质。这表明此类低流量密度的源在宇宙中可能还是比较常见的。本文提出了一个尝试性的解释，即认为这些不寻常的信号源来自白矮星或者中子星的稳定振荡。

Position and Flux Density

The aerial consists of a rectangular array containing 2,048 full-wave dipoles arranged in sixteen rows of 128 elements. Each row is 470 m long in an E.–W. direction and the N.–S. extent of the array is 45 m. Phase-scanning is employed to direct the reception pattern in declination and four receivers are used so that four different declinations may be observed simultaneously. Phase-switching receivers are employed and the two halves of the aerial are combined as an E.–W. interferometer. Each row of dipole elements is backed by a tilted reflecting screen so that maximum sensitivity is obtained at a declination of approximately +30°, the overall sensitivity being reduced by more than one-half when the beam is scanned to declinations above +90° and below −5°. The beamwidth of the array to half intensity is about $\pm\frac{1}{2}°$ in right ascension and ±3° in declination; the phasing arrangement is designed to produce beams at roughly 3° intervals in declination. The receivers have a bandwidth of 1 MHz centred at frequency of 81.5 MHz and routine recordings are made with a time constant of 0.1 s; the r.m.s. noise fluctuations correspond to a flux density of 0.5×10^{-25} W m^{-2} Hz^{-1}. For detailed studies of the pulsating source a time constant of 0.05 s was usually employed and the signals were displayed on a multi-channel "Rapidgraph" pen recorder with a time constant of 0.03 s. Accurate timing of the pulses was achieved by recording second pips derived from the *MSF* Rugby time transmissions.

A record obtained when the pulsating source was unusually strong is shown in Fig. 1*a*. This clearly displays the regular periodicity and also the characteristic irregular variation of pulse amplitude. On this occasion the largest pulses approached a peak flux density (averaged over the 1 MHz pass band) of 20×10^{-26} W m^{-2} Hz^{-1}, although the mean flux density integrated over one minute only amounted to approximately 1.0×10^{-26} W m^{-2} Hz^{-1}. On a more typical occasion the integrated flux density would be several times smaller than this value. It is therefore not surprising that the source has not been detected in the past, for the integrated flux density falls well below the limit of previous surveys at metre wavelengths.

位置和流量密度

天线中包含着一个由2,048个全波段偶极子组成的矩形阵列，这些偶极子排成16行，每行128个。每一行在该阵列东西方向的长度为470 m，南北方向的宽度为45 m。用位相扫描来确定在赤纬方向的接收模式，使用四个接收机使得可以同时观测四个不同赤纬的目标。利用位相转换接收机，再加上两个半天线阵，就可以组成一台东西方向的干涉仪。每行偶极子单元都被置于一个倾斜的反射屏前面，因而在赤纬+30°附近灵敏度最高。当波束扫描到赤纬+90°以上或者–5°以下时，总灵敏度会降低一半以上。当达到一半强度时，该阵列的波束宽度在赤经和赤纬上分别为 $\pm\frac{1}{2}°$和 ±3° ；而相位匹配被设定为在赤纬方向间隔大约3°产生波束。接收机的带宽是1 MHz，中心频率为81.5 MHz，通常以0.1 s的时间常数进行记录；其中噪声波动的均方根值相当于一个 0.5×10^{-25} W m^{-2} Hz^{-1} 的流量密度。在对脉动源进行更细致的研究时，通常会采用0.05 s的时间常数，并且这些信号在多通道“快速绘图”笔式记录仪上记录时所用的时间常数为0.03 s。通过记录来自英国小镇拉格比的*MSF*（译者注：英国的长波授时编码标准为*MSF*）秒信号可以精确测量脉冲时间。

当脉动源非常强的时候测得的结果见图1*a*。从图中可以清晰地看到规则的周期和脉冲幅度所特有的不规则变化。在这种情况下，尽管对1分钟内数据进行积分得到的平均流量密度只能达到 1.0×10^{-26} W m^{-2} Hz^{-1} 左右，但最大脉冲达到了流量密度（在1 MHz通带内的平均值）的峰值，即 20×10^{-26} W m^{-2} Hz^{-1}。在更典型的情况下，该值会是积分流量密度的数倍。由此可见，以前没有探测到这种源也在情理之中，因为积分流量密度要远小于过去在米波长情况下能够检测到的极限。

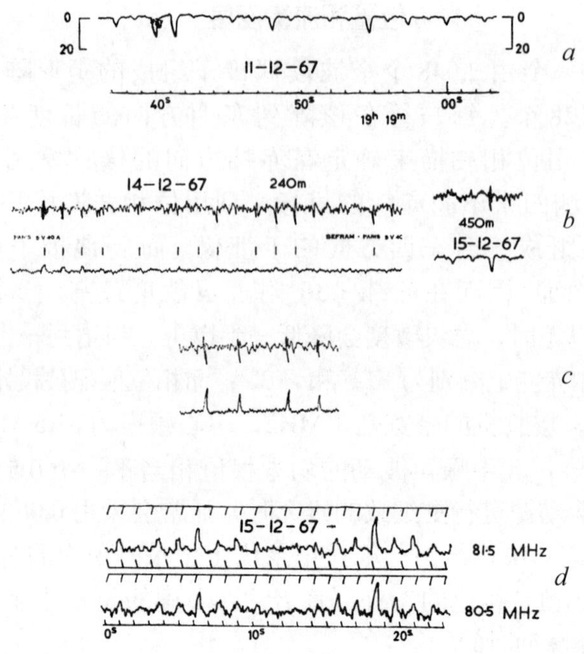

Fig. 1. *a*, A record of the pulsating radio source in strong signal conditions (receiver time constant 0.1 s). Full scale deflexion corresponds to 20×10^{-26} W m^{-2} Hz^{-1}. *b*, Upper trace: records obtained with additional paths (240 m and 450 m) in one side of the interferometer. Lower trace: normal interferometer records. (The pulses are small for l=240 m because they occurred near a null in the interference pattern; this modifies the phase but not the amplitude of the oscillatory response on the upper trace.) *c*, Simulated pulses obtained using a signal generator. *d*, Simultaneous reception of pulses using identical receivers tuned to different frequencies. Pulses at the lower frequency are delayed by about 0.2 s.

The position of the source in right ascension is readily obtained from an accurate measurement of the "crossover" points of the interference pattern on those occasions when the pulses were strong throughout an interval embracing such a point. The collimation error of the instrument was determined from a similar measurement on the neighbouring source 3*C* 409 which transits about 52 min later. On the routine recordings which first revealed the source the reading accuracy was only ±10 s and the earliest record suitable for position measurement was obtained on August 13, 1967. This and all subsequent measurements agree within the error limits. The position in declination is not so well determined and relies on the relative amplitudes of the signals obtained when the reception pattern is centred on declinations of 20°, 23° and 26°. Combining the measurements yields a position

$$\alpha_{1950}=19\text{h } 19\text{m } 38\text{s} \pm 3\text{s}$$

$$\delta_{1950}=22°00' \pm 30'$$

As discussed here, the measurement of the Doppler shift in the observed frequency of the pulses due to the Earth's orbital motion provides an alternative estimate of the declination. Observations throughout one year should yield an accuracy ±1′. The value currently attained from observations during December–January is $\delta=21°58'\pm30'$, a figure consistent with the previous measurement.

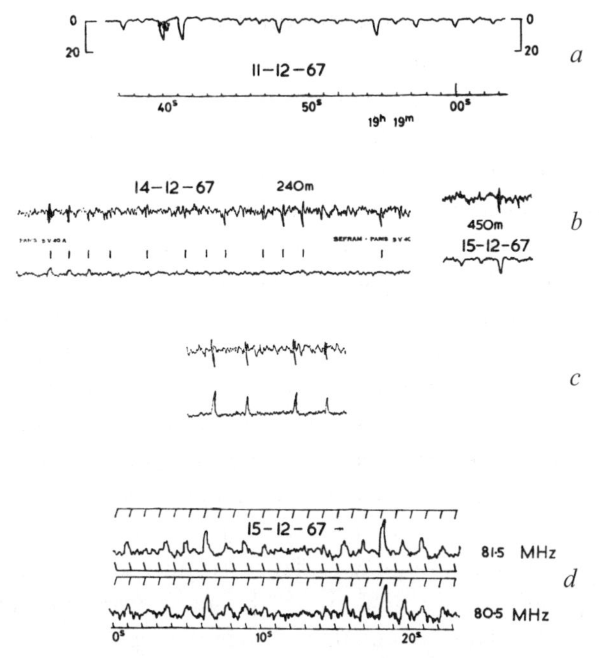

图 1. *a.* 在信号很强的情况下所记录的脉动射电源（接收机的时间常数为 0.1 s）。满刻度的偏转可以达到 20×10^{-26} W m^{-2} Hz^{-1}。*b.* 上半部分曲线：在干涉仪一侧增加了额外路径（240 m 和 450 m）后的记录。下半部分曲线：由正常干涉仪得到的记录。（当 l=240 m 时得到的脉冲很小，因为在干涉图形中这些脉冲都发生在零值附近；通过这种方式只能修正位相，而不会改变上半部分曲线的响应幅度。）*c.* 从信号发生器得到的模拟脉冲。*d.* 使用同样的接收机但在不同频率下得到的同步脉冲记录。频率较低时的脉冲延迟了大约 0.2 s。

当对应于干涉图样中某个“交叉”点的脉冲在整个时间间隔内达到很强时，只要我们对这个点进行精确测定就可以很容易地得到脉动源在赤经方向上的位置。对约 52 min 后出现的邻近源 3*C* 409 进行类似测量即可得到该仪器的瞄准误差。最早用于探测该源的仪器的读数精度通常只有 ±10 s，第一个可用于位置测量的记录是在 1967 年 8 月 13 日得到的。这一次以及之后的所有测量结果都在误差范围内相互吻合。赤纬方向的位置不太好确定，且赤纬的确定依赖于当接收模式以赤纬 20°、23° 和 26° 为中心时所得信号的相对幅度。结合已有的测量结果可以得到具体位置是：

$$\alpha_{1950} = 19\text{h}\ 19\text{m}\ 38\text{s} \pm 3\text{s}$$

$$\delta_{1950} = 22^\circ 00' \pm 30'$$

正如这里所讨论的，由于地球的轨道运动，在脉冲的观测频率上测量多普勒频移可以为我们提供一种新的赤纬估计方法。经过一年的观测，结果的精确度应该能达到 ±1′。目前由 12 月到 1 月的观测数据所得的结果为 δ=21°58′±30′，该值与之前的测量结果是一致的。

Time Variations

It was mentioned earlier that the signals vary considerably in strength from day to day and, typically, they are only present for about 1 min, which may occur quite randomly within the 4 min interval permitted by the reception pattern. In addition, as shown in Fig. 1*a*, the pulse amplitude may vary considerably on a time-scale of seconds. The pulse to pulse variations may possibly be explained in terms of interplanetary scintillation[1], but this cannot account for the minute to minute variation of mean pulse amplitude. Continuous observations over periods of 30 min have been made by tracking the source with an E–W. phased array in a 470 m × 20 m reflector normally used for a lunar occultation programme. The peak pulse amplitude averaged over ten successive pulses for a period of 30 min is shown in Fig. 2*a*. This plot suggests the possibility of periodicities of a few minutes duration, but a correlation analysis yields no significant result. If the signals were linearly polarized, Faraday rotation in the ionosphere might cause the random variations, but the form of the curve does not seem compatible with this mechanism. The day to day variations since the source was first detected are shown in Fig. 2*b*. In this analysis the daily value plotted is the peak flux density of the greatest pulse. Again the variation from day to day is irregular and no systematic changes are clearly evident, although there is a suggestion that the source was significantly weaker during October to November. It therefore appears that, despite the regular occurrence of the pulses, the magnitude of the power emitted exhibits variations over long and short periods.

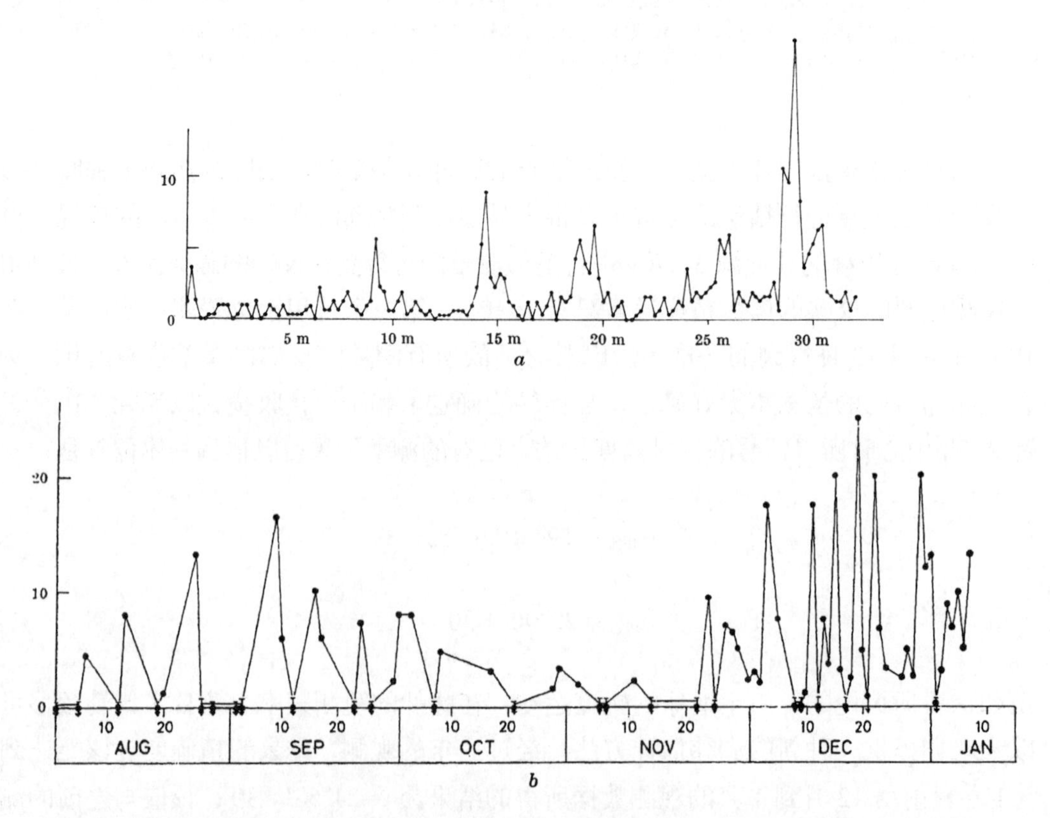

Fig. 2. *a*, The time variation of the smoothed (over ten pulses) pulse amplitude. *b*, Daily variation of peak pulse amplitude. (Ordinates are in units of W m^{-2} Hz^{-1} $\times 10^{-26}$.)

随时间的变化

早先曾提到，这些信号的强度从一天到另一天的变化很大，它们在通常情况下只会在接收模式所允许的 4 min 间隔内随机出现 1 min 左右的时间。另外，正如图 1*a* 所示，脉冲幅度在以秒为单位的时标下有可能会有显著的变化。脉冲与脉冲之间的变化或许可以用行星际闪烁来解释 [1]，但这不能说明平均脉冲的幅度从一分钟到另一分钟的变化。利用在观测月掩星时通常所采用的 470 m × 20 m 反射器组成一个东西方向的相控阵，根据这个相控阵对脉冲源的追踪可以得到时间长达 30 min 的连续观测记录。在图 2*a* 中绘出了时长 30 min 内 10 个连续脉冲的峰值脉冲幅度的平均值。这张图说明有可能存在持续几分钟的周期，但是从相关分析中并没有得到有价值的结论。如果这些信号是线偏振的，则电离层中的法拉第旋转就会产生随机的变化，但是该曲线的形状似乎与这种机制并不相符。该脉动源在首次被发现后的每日变化情况示于图 2*b*。在这张分析图中，每天的数据都是以最大脉冲的峰值流量密度来表示的。另外，虽然该源在 10 月到 11 月期间有明显变弱的迹象，但每天的变化仍是不规则的，而且也没有明显的证据表明存在系统性的变化。因此，尽管这些脉冲的出现是规则的，但是其发射能量的大小在或长或短的时间段内都表现出了一定的波动性。

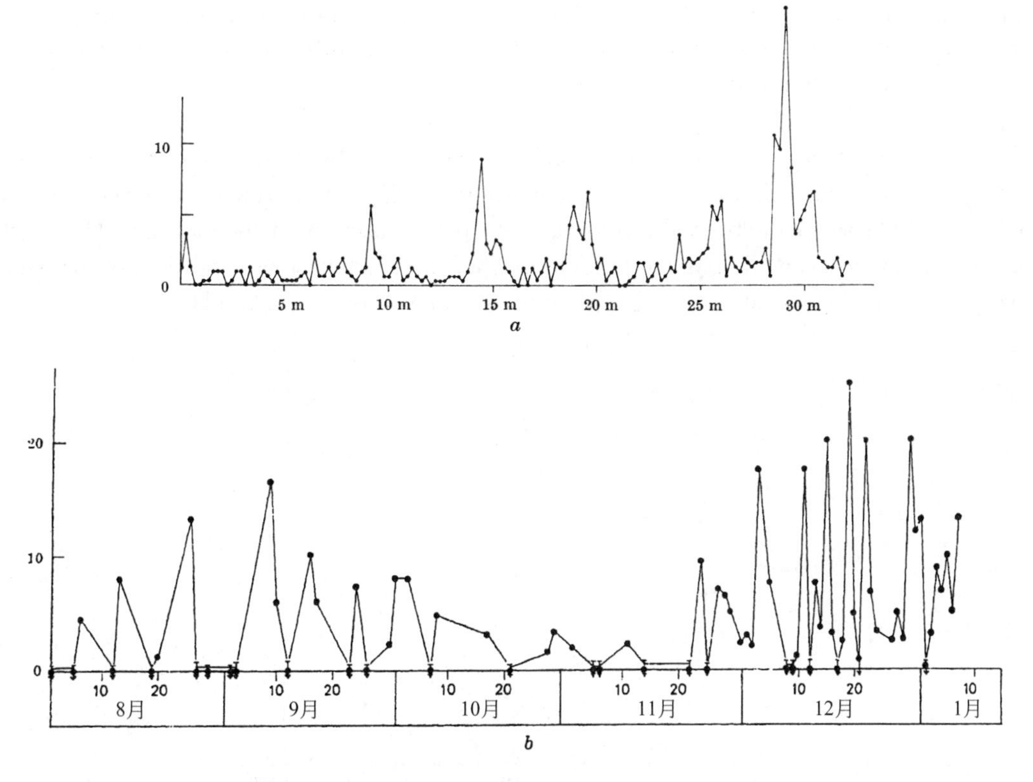

图 2. *a*. 平滑了（10 个脉冲）的脉冲幅度随时间的变化。*b*. 峰值脉冲幅度的逐日变化。（纵坐标的单位是 W m^{-2} Hz^{-1} × 10^{-26}。）

Instantaneous Bandwidth and Frequency Drift

Two different experiments have shown that the pulses are caused by a narrow-band signal of descending frequency sweeping through the 1 MHz band of the receiver. In the first, two identical receivers were used, tuned to frequencies of 80.5 MHz and 81.5 MHz. Fig. 1*d*, which illustrates a record made with this system, shows that the lower frequency pulses are delayed by about 0.2 s. This corresponds to a frequency drift of ~ −5 MHz s^{-1}. In the second method a time delay was introduced into the signals reaching the receiver from one-half of the aerial by incorporating an extra cable of known length l. This cable introduces a phase shift proportional to frequency so that, for a signal the coherence length of which exceeds l, the output of the receiver will oscillate with period

$$t_0 = \frac{c}{l}\left(\frac{\mathrm{d}\nu}{\mathrm{d}t}\right)^{-1}$$

where $\mathrm{d}\nu/\mathrm{d}t$ is the rate of change of signal frequency. Records obtained with l=240 m and 450 m are shown in Fig. 1*b* together with a simultaneous record of the pulses derived from a separate phase-switching receiver operating with equal cables in the usual fashion. Also shown, in Fig. 1*c*, is a simulated record obtained with exactly the same arrangement but using a signal generator, instead of the source, to provide the swept frequency. For observation with l>450 m the periodic oscillations were slowed down to a low frequency by an additional phase shifting device in order to prevent severe attenuation of the output signal by the time constant of the receiver. The rate of change of signal frequency has been deduced from the additional phase shift required and is $\mathrm{d}\nu/\mathrm{d}t$ =−4.9±0.5 MHz s^{-1}. The direction of the frequency drift can be obtained from the phase of the oscillation on the record and is found to be from high to low frequency in agreement with the first result.

The instantaneous bandwidth of the signal may also be obtained from records of the type shown in Fig. 1*b* because the oscillatory response as a function of delay is a measure of the autocorrelation function, and hence of the Fourier transform, of the power spectrum of the radiation. The results of the measurements are displayed in Fig. 3 from which the instantaneous bandwidth of the signal to exp (−1), assuming a Gaussian energy spectrum, is estimated to be 80±20 kHz.

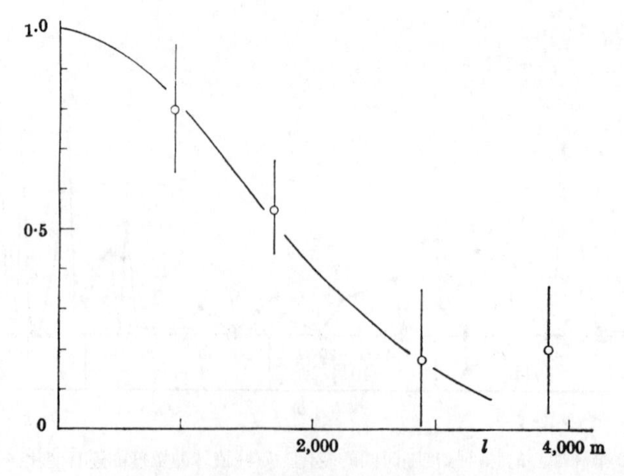

Fig. 3. The response as a function of added path in one side of the interferometer.

瞬时带宽和频率漂移

两个不同的实验都表明，这些脉冲是由一个在接收机的 1 MHz 带宽内频率逐渐下降的窄带信号导致的。在第一个实验中使用了两个同样的接收机，它们的频率分别被设定为 80.5 MHz 和 81.5 MHz。在图 1*d* 中显示的就是用这种系统得到的结果，从图中可以看出频率较低时的脉冲比频率较高时的脉冲延迟了大约 0.2 s，这与 ~−5 MHz s^{-1} 的频率漂移相对应。在第二个实验中，将时间延迟引入从一半天线到达接收机的信号的方法是另外加入一截已知长度为 l 的电缆。由这截电缆导致的相位位移与频率成正比，所以对于相干长度超过 l 的信号，接收机的输出会以 t_0 为周期发生振荡：

$$t_0 = \frac{c}{l}\left(\frac{\mathrm{d}\nu}{\mathrm{d}t}\right)^{-1}$$

其中 $\mathrm{d}\nu/\mathrm{d}t$ 是信号频率的变化率。图 1*b* 给出的是当 l=240 m 和 450 m 时所获得的记录，此外还给出了脉冲同时记录，它是通过一个独立的位相转换接收机用同样的电缆在常规方式下得到的。在图 1*c* 中还有用完全相同的方法得到的模拟记录：其输入信号不是来自脉动源，而是来自一个能提供扫频的信号发生器。当在 l > 450 m 的情况下进行观测时，需要使用一个额外的移相器将周期振荡降至较低的频率，这样就可以避免由接收机时间常数对输出信号造成的严重衰减。由所需的额外相移可以推导出信号频率的变化率，其值为 $\mathrm{d}\nu/\mathrm{d}t = -4.9 \pm 0.5$ MHz s^{-1}。频率漂移的方向可以从记录下来的振动位相推算得到，结果表明从高频到低频的漂移与最初的结论相符。

从图 1*b* 所示类型的记录中还可以得到信号的瞬时带宽。因为作为延迟的函数的振荡响应就是辐射功率谱的自相关函数的量度，也相当于对其进行了傅里叶变换。这些测量结果绘于图 3 中，假设按高斯能谱分布，从图中可估计出降至 exp(−1) 处的信号瞬时带宽为 80 kHz ± 20 kHz。

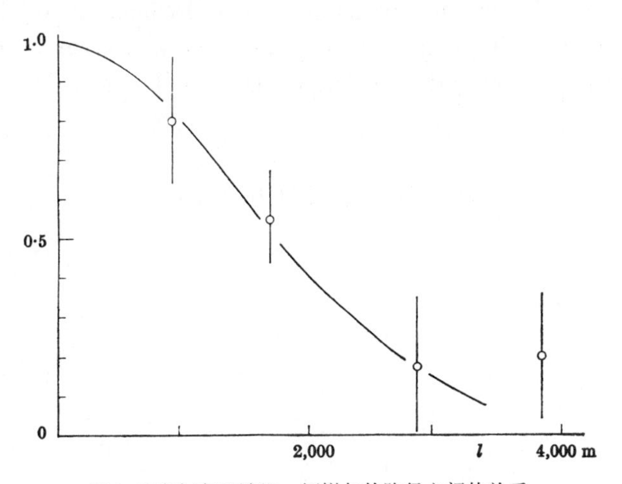

图 3. 响应与在干涉仪一侧增加的路径之间的关系。

Pulse Recurrence Frequency and Doppler Shift

By displaying the pulses and time pips from *MSF* Rugby on the same record the leading edge of a pulse of reasonable size may be timed to an accuracy of about 0.1 s. Observations over a period of 6 h taken with the tracking system mentioned earlier gave the period between pulses as P_{obs}=1.33733±0.00001 s. This represents a mean value centred on December 18, 1967, at 14 h 18 m UT. A study of the systematic shift in the frequency of the pulses was obtained from daily measurements of the time interval T between a standard time and the pulse immediately following it as shown in Fig. 4. The standard time was chosen to be 14 h 01 m 00 s UT on December 11 (corresponding to the centre of the reception pattern) and subsequent standard times were at intervals of 23 h 56 m 04 s (approximately one sidereal day). A plot of the variation of T from day to day is shown in Fig. 4. A constant pulse recurrence frequency would show a linear increase or decrease in T if care was taken to add or subtract one period where necessary. The observations, however, show a marked curvature in the sense of a steadily increasing frequency. If we assume a Doppler shift due to the Earth alone, then the number of pulses received per day is given by

$$N = N_0\left(1 + \frac{v}{c}\cos\varphi \sin\frac{2\pi n}{366.25}\right)$$

where N_0 is the number of pulses emitted per day at the source, v the orbital velocity of the Earth, φ the ecliptic latitude of the source and n an arbitrary day number obtained by putting n=0 on January 17, 1968, when the Earth has zero velocity along the line of sight to the source. This relation is approximate since it assumes a circular orbit for the Earth and the origin n=0 is not exact, but it serves to show that the increase of N observed can be explained by the Earth's motion alone within the accuracy currently attainable. For this purpose it is convenient to estimate the values of n for which $\delta T/\delta n = 0$, corresponding to an exactly integral value of N. These occur at n_1=15.8±0.1 and n_2=28.7±0.1, and since N is increased by exactly one pulse between these dates we have

$$1 = \frac{N_0 v}{c}\cos\varphi\left[\sin\frac{2\pi n_2}{366.25} - \sin\frac{2\pi n_1}{366.25}\right]$$

This yields φ =43°36′±30′ which corresponds to a declination of 21°58′±30′, a value consistent with the declination obtained directly. The true periodicity of the source, making allowance for the Doppler shift and using the integral condition to refine the calculation, is then

$$P_0 = 1.3372795 \pm 0.0000020 \text{ s}$$

脉冲重复频率和多普勒频移

在可以同时显示脉冲和拉格比 *MSF* 时间信号的记录中，对一个正常大小的脉冲前缘的计时也许能达到 0.1 s 左右的精确度。在利用上文中提到的跟踪系统进行长达 6 h 的观测之后可以得到脉冲的周期为 P_{obs}=1.33733 s ± 0.00001 s。这代表了以世界时 1967 年 12 月 18 日 14 时 18 分为中心的平均值。对脉冲频率系统性漂移的研究，可以通过每天测量某个标准时间与紧随其后出现的脉冲之间的时间间隔 T 得到，如图 4 所示。所选的标准时间是世界时 12 月 11 日 14 时 01 分 00 秒（对应于接收模式的中心），随后每隔 23 时 56 分 04 秒（大约为一个恒星日的长度）就会出现下一个标准时间。T 的每日变化示于图 4。如果在必要的时候增加或者减少一个周期，则在脉冲重复频率为常数的前提下就会出现 T 的线性增加或减少。然而，观测结果显示出明显的曲率，说明频率在稳步增长。如果我们假设多普勒频移仅归因于地球，那么每天接收到的脉冲个数可由下式得到：

$$N = N_0\left(1 + \frac{v}{c}\cos\varphi\,\sin\frac{2\pi n}{366.25}\right)$$

其中 N_0 是该源每天发射的脉冲个数，v 是地球的轨道速度，φ 是该源的黄纬，n 是任意指定的天数：在 1968 年 1 月 17 日，即地球沿着源的视向速度为 0 时，令 n=0。这种关系是近似的，因为它假设地球的运行轨道为圆形，而且原点 n=0 也是不精确的。但是由此还是可以证明：在目前能够达到的精度上，观测到的 N 的增加可以仅由地球的运动来解释。为此，可以很方便地估算出当 $\delta T/\delta n = 0$ 时的 n 值，所对应的是一个确切的整数 N。它们会发生在 n_1=15.8 ± 0.1 和 n_2=28.7 ± 0.1 时，而且因为 N 在这些天内刚好增加了一个脉冲，所以：

$$1 = \frac{N_0 v}{c}\cos\varphi\left[\sin\frac{2\pi n_2}{366.25} - \sin\frac{2\pi n_1}{366.25}\right]$$

由上式可得 φ=43°36′± 30′，所对应的赤纬为 21°58′± 30′，这与通过观测直接得到的赤纬值相符。考虑到多普勒频移并使用整数条件对计算进行简化即可得到射电源的真实周期：

$$P_0 = 1.3372795\ \text{s} \pm 0.0000020\ \text{s}$$

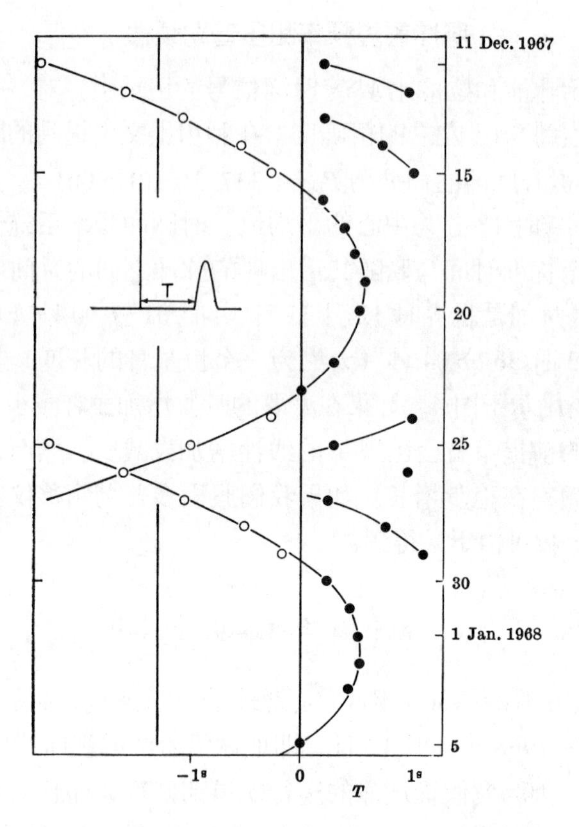

Fig. 4. The day to day variation of pulse arrival time.

By continuing observations of the time of occurrence of the pulses for a year it should be possible to establish the constancy of N_0 to about 1 part in 3×10^8. If N_0 is indeed constant, then the declination of the source may be estimated to an accuracy of $\pm1'$; this result will not be affected by ionospheric refraction.

It is also interesting to note the possibility of detecting a variable Doppler shift caused by the motion of the source itself. Such an effect might arise if the source formed one component of a binary system, or if the signals were associated with a planet in orbit about some parent star. For the present, the systematic increase of N is regular to about 1 part in 2×10^7 so that there is no evidence for an additional orbital motion comparable with that of the Earth.

The Nature of the Radio Source

The lack of any parallax greater than about $2'$ places the source at a distance exceeding 10^3 A.U. The energy emitted by the source during a single pulse, integrated over 1 MHz at 81.5 MHz, therefore reaches a value which must exceed 10^{17} erg if the source radiates isotropically. It is also possible to derive an upper limit to the physical dimension of the source. The small instantaneous bandwidth of the signal (80 kHz) and the rate of sweep (-4.9 MHz s^{-1}) show that the duration of the emission at any given frequency does not exceed 0.016 s. The source size therefore cannot exceed 4.8×10^3 km.

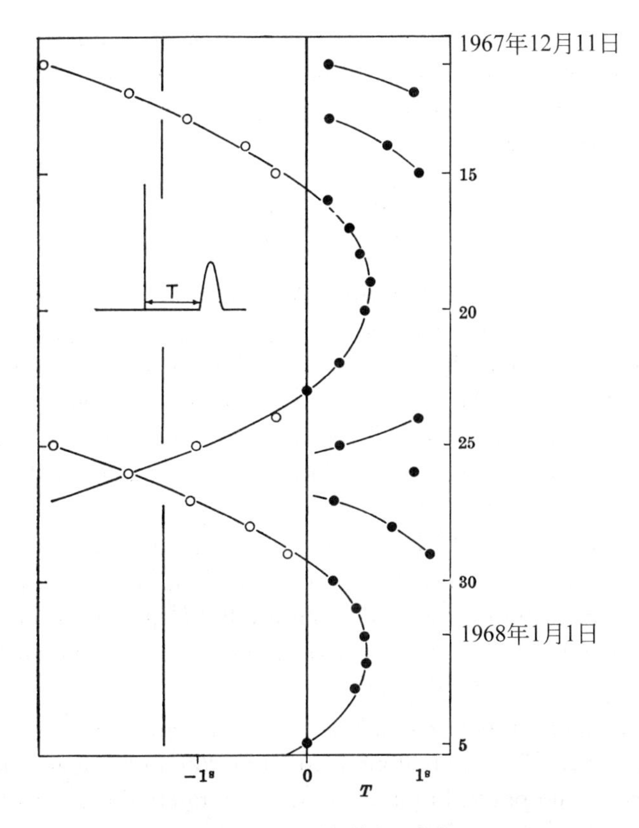

图 4. 脉冲到达时间的每日变化。

根据在一年内连续观测到的脉冲出现时间可以推出 N_0 是一个恒定的数，误差约为 $\frac{1}{3}\times10^{-8}$。如果 N_0 真的是一个常数，那么对该脉动源赤纬的估计值就可以达到 $\pm1'$ 的精度，这一结果是不会受到电离层折射影响的。

另一件值得一提的事是：我们可能会观测到由源自身运动所造成的多普勒频移的变化，这一效应在以下两种情况下或许有出现的可能性：或者脉动源是双星系统中的一个成员，或者该信号与在围绕某颗母恒星的轨道上运行的行星有关。就目前了解的情况而言，N 的系统性增长通常为 $\frac{1}{2}\times10^{-7}$ 量级，所以可以证明：其他的轨道运动所造成的影响都无法与地球的轨道运动相比。

射电源的性质

观测视差不大于 2′ 说明源与地球的距离超过了 10^3 个天文单位。如果该源的辐射是各向同性的，那么它在一个脉冲间隔内释放的能量——在频率为 81.5 MHz 处的 1 MHz 带宽内进行积分——将会超过 10^{17} erg。还可以推算出该源物理尺寸的上限。信号的较窄瞬时带宽（80 kHz）和频率漂移速度（-4.9 MHz s^{-1}）表明，在任意给定频率上的能量释放时间都不会超过 0.016 s。因此，源的大小不可能超过 4.8×10^3 km。

An upper limit to the distance of the source may be derived from the observed rate of frequency sweep since impulsive radiation, whatever its origin, will be dispersed during its passage through the ionized hydrogen in interstellar space. For a uniform plasma the frequency drift caused by dispersion is given by

$$\frac{\mathrm{d}\nu}{\mathrm{d}t} = -\frac{c}{L}\frac{\nu^3}{\nu_p^2}$$

where L is the path and ν_p the plasma frequency. Assuming a mean density of 0.2 electron cm^{-3} the observed frequency drift (–4.9 MHz s^{-1}) corresponds to L~65 parsec. Some frequency dispersion may, of course, arise in the source itself; in this case the dispersion in the interstellar medium must be smaller so that the value of L is an upper limit. While the interstellar electron density in the vicinity of the Sun is not well known, this result is important in showing that the pulsating radio sources so far detected must be local objects on a galactic distance scale.

The positional accuracy so far obtained does not permit any serious attempt at optical identification. The search area, which lies close to the galactic plane, includes two twelfth magnitude stars and a large number of weaker objects. In the absence of further data, only the most tentative suggestion to account for these remarkable sources can be made.

The most significant feature to be accounted for is the extreme regularity of the pulses. This suggests an origin in terms of the pulsation of an entire star, rather than some more localized disturbance in a stellar atmosphere. In this connexion it is interesting to note that it has already been suggested[2,3] that the radial pulsation of neutron stars may play an important part in the history of supernovae and supernova remnants.

A discussion of the normal modes of radial pulsation of compact stars has recently been given by Meltzer and Thorne[4], who calculated the periods for stars with central densities in the range 10^5 to 10^{19} g cm^{-3}. Fig. 4 of their paper indicates two possibilities which might account for the observed periods of the order 1 s. At a density of 10^7 g cm^{-3}, corresponding to a white dwarf star, the fundamental mode reaches a minimum period of about 8 s; at a slightly higher density the period increases again as the system tends towards gravitational collapse to a neutron star. While the fundamental period is not small enough to account for the observations the higher order modes have periods of the correct order of magnitude. If this model is adopted it is difficult to understand why the fundamental period is not dominant; such a period would have readily been detected in the present observations and its absence cannot be ascribed to observational effects. The alternative possibility occurs at a density of 10^{13} g cm^{-3}, corresponding to a neutron star; at this density the fundamental has a period of about 1 s, while for densities in excess of 10^{13} g cm^{-3} the period rapidly decreases to about 10^{-3} s.

If the radiation is to be associated with the radial pulsation of a white dwarf or neutron star there seem to be several mechanisms which could account for the radio emission. It has been suggested that radial pulsation would generate hydromagnetic shock fronts at the stellar surface which might be accompanied by bursts of X-rays and energetic electrons[2,3]. The radiation might then be likened to radio bursts from a solar flare occurring over the entire star during

因为所有的脉冲辐射，不论来源于哪里，都会在穿过星际空间中的电离氢时被色散，所以根据观测到的频率漂移速度就可以推导出该源与地球之间距离的上限。对于均匀的等离子体，由色散导致的频率漂移可由下式给出：

$$\frac{\mathrm{d}\nu}{\mathrm{d}t}=-\frac{c}{L}\frac{\nu^3}{\nu_p^2}$$

其中 L 是路径长度，ν_p 是等离子体的频率。假设平均密度为每 cm^3 有 0.2 个电子，则观测到的频率漂移（$-4.9\ MHz\ s^{-1}$）与 L~65 秒差距相当。当然，某些频散也可能来自源本身，在这种情况下星际物质的色散作用肯定会更小，所以 L 的数值是上限。尽管我们对太阳附近的星际电子密度不甚清楚，但这一结果仍很重要，它表明迄今为止人们探测到的所有脉动射电源都一定来自银河系内的天体。

就现在的定位精度而言，认真尝试进行光学证认是不可能的。现在人们搜索的区域距离银道面很近，在该区域中有两颗 12 等星和大量较暗的天体。由于缺乏更进一步的数据，所以只能对这些奇特的射电源作出可能性最大的解释。

需要解释的一个最显著的特征是这些脉冲的极端规律性。这表明源自一颗恒星整体脉动的可能性要大于恒星大气中的某种更为局部的扰动。就这一点来说，值得一提的是：已经有人提出 [2,3]，中子星的径向脉动可能在超新星和超新星遗迹演化史中起重要作用。

最近，梅尔策和索恩 [4] 对致密星径向脉动的几种标准模式进行了讨论，他们计算了中心密度在 $10^5\ g/cm^3$ 到 $10^{19}\ g/cm^3$ 范围内的恒星所具有的脉动周期。他们论文中的图 4 表明，有以下两种可能性或许可以解释观测到的 1 s 量级的周期。当密度为 $10^7\ g/cm^3$——与白矮星的密度相当时，在基频模式下可以达到的最小周期值约为 8 s；当密度略微增加时，周期也会增加，因为系统有引力塌缩成一颗中子星的趋向。虽然由基频模式得到的周期值太大以至于不能解释观测到的结果，但高阶模式下的周期却具有合适的数量级。如果采纳了上述模型，那么我们就很难解释为什么基频模式下的周期不是主导周期。在目前的观测中应该很容易探测到基频模式下的周期，不能把没有探测到这样的周期归因于观测上的不得力。另一种可能性是：当密度为 $10^{13}\ g/cm^3$——与中子星的密度相当时，基频模式在此密度下的周期值约为 1 s；当密度超过 $10^{13}\ g/cm^3$ 时，周期迅速下降至 10^{-3} s 左右。

如果认为这种辐射与白矮星或中子星的径向脉动有关，那么就会有好几种机制可以解释这种射电发射。有人认为径向脉动将在恒星表面产生磁流体激波阵面，从而可能会伴随着 X 射线和高能电子的爆发 [2,3]。因而或许可以把这种辐射比作是在

each cycle of the oscillation. Such a model would be in fair agreement with the upper limit of $\sim 5\times 10^3$ km for the dimension of the source, which compares with the mean value of 9×10^3 km quoted for white dwarf stars by Greenstein[5]. The energy requirement for this model may be roughly estimated by noting that the total energy emitted in a 1 MHz band by a type Ⅲ solar burst would produce a radio flux of the right order if the source were at a distance of $\sim 10^3$ A.U. If it is assumed that the radio energy may be related to the total flare energy ($\sim 10^{32}$ erg)[6] in the same manner as for a solar flare and supposing that each pulse corresponds to one flare, the required energy would be $\sim 10^{39}$ erg yr^{-1}; at a distance of 65 pc the corresponding value would be $\sim 10^{47}$ erg yr^{-1}. It has been estimated that a neutron star may contain $\sim 10^{51}$ erg in vibrational modes so the energy requirement does not appear unreasonable, although other damping mechanisms are likely to be important when considering the lifetime of the source[4].

The swept frequency characteristic of the radiation is reminiscent of type II and type III solar bursts, but it seems unlikely that it is caused in the same way. For a white dwarf or neutron star the scale height of any atmosphere is small and a travelling disturbance would be expected to produce a much faster frequency drift than is actually observed. As has been mentioned, a more likely possibility is that the impulsive radiation suffers dispersion during its passage through the interstellar medium.

More observational evidence is clearly needed in order to gain a better understanding of this strange new class of radio source. If the suggested origin of the radiation is confirmed further study may be expected to throw valuable light on the behaviour of compact stars and also on the properties of matter at high density.

We thank Professor Sir Martin Ryle, Dr. J. E. Baldwin, Dr. P. A. G. Scheuer and Dr. J. R. Shakeshaft for helpful discussions and the Science Research Council who financed this work. One of us (S. J. B.) thanks the Ministry of Education of Northern Ireland and another (R. A. C.) the SRC for a maintenance award; J. D. H. P. thanks ICI for a research fellowship.

(**217**, 709-713; 1968)

A. Hewish, S. J. Bell, J. D. H. Pilkington, P. F. Scott and R. A. Collins: Mullard Radio Astronomy Observatory, Cavendish Laboratory, University of Cambridge.

Received February 9, 1968.

References:

1. Hewish, A., Scott, P. F., and Wills, D., *Nature*, **203**, 1214 (1964).
2. Cameron, A. G. W., *Nature*, **205**, 787 (1965).
3. Finzi, A., *Phys. Rev. Lett.*, **15**, 599 (1965).
4. Meltzer, D. W., and Thorne, K. S., *Ap. J.*, **145**, 514 (1966).
5. Greenstein, J. L., in *Handbuch der Physik, L.*, 161 (1958).
6. Fichtel, C. E., and McDonald, F. B., in *Annual Review of Astronomy and Astrophysics*, **5**, 351 (1967).

每个振荡周期内从一个贯穿恒星的耀斑发出的射电暴。这样一种模型与认为源的上限尺寸为 ~ 5×10^3 km 的观点吻合得非常好，也与格林斯坦[5]提出的白矮星平均尺寸为 9×10^3 km 的论点相符。可以通过以下方式大致估计出该模型所需的能量：在 1 MHz 带宽内以第三类太阳暴形式发射的总能量将与在假设射电源位于 ~10^3 个天文单位处的前提下得到的射电流量有相同的量级。如果假设射电能量和耀斑总能量(~10^{32} erg)[6]之间的关系与和一个太阳耀斑的关系一样，并假设每个脉冲对应着一个耀斑，那么所需的能量将为 ~10^{39} erg yr^{-1}，而在距离为 65 秒差距处的对应能量值将为 ~10^{47} erg yr^{-1}。有人估计一颗中子星可用在振动模式中的能量为 ~10^{51} erg，所以所需的能量值看起来还算合理，不过在考虑到脉动源的演化史时，其他阻尼机制也有能发挥重要作用的可能性[4]。

尽管这种辐射的频率漂移特征会使人联想到第二类和第三类太阳暴，但它的产生机制似乎不太可能与这两类太阳暴相同。对于一颗白矮星或者中子星来说，其大气标高很小，预计扰动在传播时产生的频率漂移要比实际观测值高得多。正如上文所提到的，一种可能性更大的情况是脉冲辐射在穿过星际介质时被色散。

为了更好地理解这种奇特的新型射电源，显然还需要更多的观测证据。如果能够证明这种辐射的起源与本文中所设想的完全一致，那么经过进一步的研究人们或许可以了解到一些有关致密恒星行为以及高密度物质性质的重要线索。

我们要感谢马丁·赖尔爵士教授、鲍德温博士、朔伊尔博士和谢克沙夫特博士为此提出了有价值的意见，还要感谢科学研究理事会为此项研究提供了经费。我们中的一位作者（贝尔）要感谢北爱尔兰教育部，另一位作者（柯林斯）要感谢科学研究理事会提供了维持费用；皮尔金顿要感谢帝国化学工业公司提供研究基金。

（冯翀 翻译；蒋世仰 审稿）

Rotating Neutron Stars as the Origin of the Pulsating Radio Sources

T. Gold

Editor's Note

Three months after the publication above, Thomas Gold published his explanation of how pulsars function. The essential ingredients of his scheme are that pulsars are indeed neutron stars, that there must be intense magnetic fields associated with such structures and that the magnetic fields would generate intense radiation in the microwave (ultra-short radio waves) region of the spectrum. In Gold's view, the pulsation of these stars arose not from some kind of internal vibration as suggested by Hewish, but because the magnetic field generates "directional beams rotating like a lighthouse beacon". He made two predictions: the rate of rotation of a pulsar would decrease slowly but steadily with time and pulsars with a period of 1 second would be found to represent "the slow end of the distribution". Both predictions have been proved correct.

THE case that neutron stars are responsible for the recently discovered pulsating radio sources[1-6] appears to be a strong one. No other theoretically known astronomical object would possess such short and accurate periodicities as those observed, ranging from 1.33 to 0.25 s. Higher harmonics of a lower fundamental frequency that may be possessed by a white dwarf have been mentioned; but the detailed fine structure of several short pulses repeating in each repetition cycle makes any such explanation very unlikely. Since the distances are known approximately from interstellar dispersion of the different radio frequencies, it is clear that the emission per unit emitting volume must be very high; the size of the region emitting any one pulse can, after all, not be much larger than the distance light travels in the few milliseconds that represent the lengths of the individual pulses. No such concentrations of energy can be visualized except in the presence of an intense gravitational field.

The great precision of the constancy of the intrinsic period also suggests that we are dealing with a massive object, rather than merely with some plasma physical configuration. Accuracies of one part in 10^8 belong to the realm of celestial mechanics of massive objects, rather than to that of plasma physics.

It is a consequence of the virial theorem that the lowest mode of oscillation of a star must always have a period which is of the same order of magnitude as the period of the fastest rotation it may possess without rupture. The range of 1.5 s to 0.25 s represents periods that are all longer than the periods of the lowest modes of neutron stars. They would all be periods in which a neutron star could rotate without excessive flattening. It is doubtful that the fundamental frequency of pulsation of a neutron star could ever be so long (ref. 7 and unpublished work of A. G. W. Cameron). If the rotation period dictates the repetition rate, the fine structure of the observed pulses would represent directional beams rotating like a

旋转中子星作为脉动射电源的起源

戈尔德

编者按

在上一篇文章发表三个月后，托马斯·戈尔德发表了他对脉冲星工作原理的解释。他的模型包括三个要点：脉冲星实际上就是中子星；与此结构相关的是它一定具有很强的磁场；这种磁场会在光谱的微波（超短无线电波）波段产生很强的辐射。戈尔德认为：这些恒星的脉冲并不是由休伊什所假定的某种内部振荡产生的，而是因为磁场产生了“像灯塔中的信标一样不断旋转的定向波束”。他作了以下两项预言：脉冲星的旋转速率会随着时间的增长而有规律地缓慢下降；周期为 1 s 的那些脉冲星将代表着“分布范围中的慢端部分”。如今这两项预言都已被证实。

似乎有充足的理由相信，最近发现的脉动射电源[1-6]就是中子星。它们的脉动周期范围是 1.33 s 至 0.25 s，没有任何理论上已知的其他天体能有这样短而且精确的周期。有人曾指出，这是一颗白矮星所可能具有的较低基频的高次谐波；但是，在每个重复周期内的若干短脉冲存在复杂精细结构的事实使这类解释不太可能成立。源的距离可大致由不同射电频率的星际色散给出，很显然基于每单位发射体积的发射强度一定是非常高的；发出单个脉冲的发射区域大小无论如何也不可能远远超过光在几毫秒时间内所传播的距离，这个时间即单个脉冲的长度。这样高的能量聚集度是不可想象的，除非存在着一个强大的引力场。

高度稳定的内禀周期也表明：我们面对的是一个大质量的天体，而不仅仅是某种等离子体结构。一亿分之一的精度应归入大质量天体的天体力学领域，而非等离子体物理学。

由维里定理可以推出：一颗恒星的振荡有一个最低阶的模式，其周期通常与该恒星以最快速度转动而不致破裂的周期属于同一数量级。1.5 s 到 0.25 s 范围内的周期都要长于中子星的最低阶模式的周期。以这些周期旋转的中子星不会过度被压扁。而一颗中子星的脉动是否会有这样长的基频周期是值得怀疑的（参考文献 7 以及卡梅伦尚未发表的研究结果）。如果脉冲的重复频率是由转动周期决定的，那么观测到的脉冲的精细结构就代表了像灯塔中的信标一样不断旋转的定向波束。在不同的源中观测到了不同类型的精细结构，这只能归因于每颗恒星都各自有自己独

lighthouse beacon. The different types of fine structure observed in the different sources would then have to be attributed to the particular asymmetries of each star (the "sunspots", perhaps). In such a model, time variations in the intensity of emission will have no effect on the precise phase in the repetition period where each pulse appears; and this is indeed a striking observational fact. A fine structure of pulses could be generated within the repetition period, depending only on the distribution of emission regions around the circumference of the star. Similarly, a fine structure in polarization may be generated, for each region may produce a different polarization or be overlaid by a different Faraday-rotating medium. A single pulsating region, on the other hand, could scarcely generate a repetitive fine structure in polarization as seems to have been observed now[8].

There are as yet not really enough clues to identify the mechanism of radio emission. It could be a process deriving its energy from some source of internal energy of the star, and thus as difficult to analyse as solar activity. But there is another possibility, namely, that the emission derives its energy from the rotational energy of the star (very likely the principal remaining energy source), and is a result of relativistic effects in a co-rotating magnetosphere.

In the vicinity of a rotating star possessing a magnetic field there would normally be a co-rotating magnetosphere. Beyond some distance, external influences would dominate, and co-rotation would cease. In the case of a fast rotating neutron star with strong surface fields, the distance out to which co-rotation would be enforced may well be close to that at which co-rotation would imply motion at the speed of light. The mechanism by which the plasma will be restrained from reaching the velocity of light will be that of radiation of the relativistically moving plasma, creating a radiation reaction adequate to overcome the magnetic force. The properties of such a relativistic magnetosphere have not yet been explored, and indeed our understanding of relativistic magneto-hydrodynamics is very limited. In the present case the coupling to the electromagnetic radiation field would assume a major role in the bulk dynamical behaviour of the magnetosphere.

The evidence so far shows that pulses occupy about 1/30 of the time of each repetition period. This limits the region responsible to dimensions of the order of 1/30 of the circumference of the "velocity of light circle". In the radial direction equally, dimensions must be small; one would suspect small enough to make the pulse rise-times comparable with or larger than the flight time of light across the region that is responsible. This would imply that the radiation emanates from the plasma that is moving within 1 percent of the velocity of light. That is the region of velocity where radiation effects would in any case be expected to become important.

The axial asymmetry that is implied needs further comment. A magnetic field of a neutron star may well have a strength of 10^{12} gauss at the surface of the 10 km object. At the "velocity of light circle", the circumference of which for the observed periods would range from 4×10^{10} to 0.75×10^{10} cm, such a field will be down to values of the order of 10^3–10^4 gauss (decreasing with distance slower than the inverse cube law of an undisturbed dipole field. A field pulled out radially by the stress of the centrifugal force of a whirling plasma would decay as an inverse square law with radius). Asymmetries in the radiation could

特的非对称性（也许就像“太阳黑子”一样）。在这样的模型中，辐射强度随时间的变化将不会影响每个脉冲在重复周期内出现的精确相位——这确实是一个惊人的观测事实。一个重复周期内可能会出现由多个脉冲组成的精细结构，这只取决于该恒星周围的辐射区分布。同样，也可能会存在偏振的精细结构，因为每个区域都有可能产生不同偏振的辐射，或者被具有不同法拉第旋转的介质所覆盖。另一方面，单一的脉冲辐射区几乎不可能产生重复的偏振精细结构，目前的观测结果似乎与此相吻合[8]。

至今还没有足够的线索以确定射电辐射的机制。它可能是一个通过提取某种恒星内部能量源来获取自身能量的过程，因此和太阳活动一样很难进行分析。但是还有另外一种可能性，即辐射的能量来自恒星的转动能（转动能很可能是剩余能源中最主要的），而射电辐射是由共转磁层中的相对论效应产生的。

在一颗有磁场的旋转恒星周围，通常会形成共转的磁层。在一定距离之外，外部影响将占主导地位，因而共转会停止。对于一颗表面有强磁场并且旋转速度很高的中子星，其共转延伸的距离很可能会接近于共转速度为光速时的距离。磁层内等离子体不能达到光速的原因在于，以接近光速运动的等离子体所发出的辐射会产生足以抵抗磁力的辐射反作用力。这种相对论性磁层的性质还没有被研究过；实际上我们对相对论磁流体动力学的理解也是非常有限的。照现在的情况来看，等离子体与电磁辐射场的耦合可能是磁层动力学行为的主要根源。

从目前的证据来看，脉冲在每个重复周期内占大约 1/30 的时间。这使电磁辐射区域大小的数量级被限制在“光速圆周”周长的 1/30 左右。同样，在径向方向上尺寸也一定很小。我们可以认为要小到足以使脉冲上升时间等于或大于光穿过辐射区的时间。这意味着，产生辐射的等离子体的运动速度小于光速的 1%。对于有这样速度的区域，预计其辐射效应在任何情况下都会变得很重要。

这里暗含的轴向不对称性有待进一步讨论。一颗半径为 10 km 的中子星的表面磁场很可能会达到 10^{12} 高斯。对应于已观测到周期的“光速圆周”，其周长在 4×10^{10} cm 至 0.75×10^{10} cm 之间。在上述“光速圆周”处，磁场会降至 10^{3} 高斯 ~10^{4} 高斯的量级（对于标准的偶极磁场，场强随距离的负 3 次方的减小而下降。此处场强随距离的下降要慢于 –3 次方，因为一个旋转的等离子体在离心力的作用下会拖曳磁场沿径向向外运动，所以场强随半径的负 2 次方的减小而下降。）。辐射的非对称

arise either through the field or the plasma content being non-axially symmetric. A skew and non-dipole field may well result from the explosive event that gave rise to the neutron star; and the access to plasma of certain tubes of force may be dependent on surface inhomogeneities of the star where sufficiently hot or energetic plasma can be produced to lift itself away from the intense gravitational field (10–100 MeV for protons; much less for space charge neutralized electron-positron beams).

The observed distribution of amplitudes of pulses makes it very unlikely that a modulation mechanism can be responsible for the variability (unpublished results of P. A. G. Scheuer and observations made at Cornell's Arecibo Ionospheric Observatory) but rather the effect has to be understood in a variability of the emission mechanism. In that case the observed very sharp dependence of the instantaneous intensity on frequency (1 MHz change in the observation band gives a substantially different pulse amplitude) represents a very narrow-band emission mechanism, much narrower than synchrotron emission, for example. A coherent mechanism is then indicated, as is also necessary to account for the intensity of the emission per unit area that can be estimated from the lengths of the sub-pulses. Such a coherent mechanism would represent non-uniform static configurations of charges in the relativistically rotating region. Non-uniform distributions at rest in a magnetic field are more readily set up and maintained than in the case of high individual speeds of charges, and thus the configuration discussed here may be particularly favourable for the generation of a coherent radiation mechanism.

If this basic picture is the correct one it may be possible to find a slight, but steady, slowing down of the observed repetition frequencies. Also, one would then suspect that more sources exist with higher rather than lower repetition frequency, because the rotation rates of neutron stars are capable of going up to more than 100/s, and the observed periods would seem to represent the slow end of the distribution.

Work in this subject at Cornell is supported by a contract from the US Office of Naval Research.

(**218**, 731-732; 1968)

T. Gold: Center for Radiophysics and Space Research, Cornell University, Ithaca, New York.

Received May 20, 1968.

References:

1. Hewish, A., Bell, S. J., Pilkington, J. D. H., Scott, P. F., and Collins, R. A., *Nature*, **217**, 709 (1968).
2. Pilkington, J. D. H., Hewish, A., Bell, S. J., and Cole, T. W., *Nature*, **218**, 126 (1968).
3. Drake, F. D., Gundermann, E. J., Jauncey, D. L., Comella, J. M., Zeissig, G. A., and Craft, jun., H. D., *Science*, **160**, 503 (1968).
4. Drake, F. D., *Science* (in the press).
5. Drake, F. D., and Craft, jun., H. D., *Science*, **160**, 758 (1968).
6. Tanenbaum, B. S., Zeissig, G. A., and Drake, F. D., *Science* (in the press).
7. Thorne, K. S., and Ipser, J. R., *Ap. J.* (in the press).
8. Lyne, A. G., and Smith, F. G., *Nature*, 218, 124 (1968).

现象可能源自磁场或者等离子体的非轴对称性。倾斜的非偶极场极有可能是在产生中子星的超新星爆发过程中形成的；恒星表面的非均匀性或许就是出现特定力线束的等离子体的原因，在恒星表面可能会产生温度或能量足够高的等离子体，这些等离子体可以从表面的强引力场中逃逸（质子需要 10 MeV ~ 100 MeV；而空间电荷为中性的正负电子束所需要的能量则要低很多）。

观测到的脉冲幅度分布表明，幅度的变化不太可能由一种调制机制来解释（根据朔伊尔尚未公布的研究结果和康奈尔大学阿雷西博电离层天文台的观测结果），而必须用可变的发射机制理解这一效应。如果事实确实如此，那么所观察到的瞬时强度对频率的强烈依赖（观测频带发生 1 MHz 的改变会导致出现完全不同的脉冲幅度）就预示了一种非常窄频带的辐射机制，例如要比同步辐射窄很多。因为还必须符合能够从子脉冲长度估计得到的单位面积辐射强度，所以这应是一种相干辐射机制。这种相干机制将描绘出相对论性旋转区域内的非均匀静态电荷分布。因为磁场中静态电荷的非均匀分布比以不同速度高速运动的电荷的分布更容易建立和保持，所以这里讨论的非均匀分布也许对建立相干辐射机制非常有利。

如果上述基本构想是正确的话，就有可能发现观测到的重复频率会稳定地一点一点变慢。还可以由此猜测，具有较高频率的源要多于具有较低频率的源，因为中子星的旋转速度可以高达 100/s 以上。看来观测到的周期代表的是分布范围中的慢端部分。

在康奈尔大学进行的这项研究是与美国海军研究办公室签有协议的。

（岳友岭 翻译；于涌 审稿）

Evidence in Support of a Rotational Model for the Pulsar *PSR* 0833–45

V. Radhakrishnan *et al.*

Editor's Note

This paper, appearing within a few months of the discovery of pulsating stars (pulsars), is an almost immediate confirmation of Gold's theory that the apparent pulsation is caused by rapid rotation, sending a beam of radiation across the sky much as a lighthouse announces its presence. Radhakrishnan was trained as a radioastronomer in India and Australia; he was a son of the Nobel-Prize-winning scientist C. V. Raman and succeeded him as Director of the physical laboratory in Bangalore.

THE pulsar *PSR* 0833–45, discovered at the Molonglo Radio Observatory[1] and tentatively identified with the supernova remnant Vela *X*, has been observed with the Parkes 210 foot reflector at frequencies of 2,700 MHz and 1,720 MHz. We find that this pulsar is remarkable for its short pulse width, high and constant intensity, complete linear polarization and changing periodicity. Our observations of this object seem to rule out radial pulsations as the source of the radio emission and support a rotational hypothesis along the lines suggested by Gold[2] and discussed by Goldreich[3] as the most likely model.

The first observations were made on December 8, 9 and 10 using the wideband correlation receiver[4] operating at 2,700 MHz. The pulse energy, linear polarization and an accurate period were measured and are listed in Table 1. Attempts to obtain a pulse shape, however, soon indicated that the drift time of ~ 10 ms through the 500 MHz effective passband of the receiver was smoothing out the natural pulse beyond recognition. It seemed probable that the polarization characteristics of the pulse were also being smeared by the passband. Observations of the pulsar were therefore repeated on the nights of December 11, 12, 13 and 20 at 1,720 MHz with narrower bandwidths. The fast rise and short duration of the pulse as recorded by a 400 channel integrator[5] are illustrated in Fig. 1. Apart from occasional pulses of extraordinary strength, the amplitude variations were so low that an average of 100 pulses taken at any time during the several days this pulsar was observed on either frequency showed less than 10 percent variation in amplitude and even less in pulse shape. Large *et al.*[6] have remarked on similar behaviour at 408 MHz. Table 1 lists various parameters obtained from measurements at either one or both of the observing frequencies.

支持脉冲星*PSR* 0833–45的旋转模型的证据

拉达克里希南等

编者按

这篇文章发表于发现脉动星（现称脉冲星）之后的几个月内，它使戈尔德的理论立即得到印证，该理论认为脉动是由快速旋转引起的，脉冲星在旋转时发射出一束穿越空间的辐射，就像灯塔一样显示着自己的存在。本文的作者拉达克里希南曾在印度和澳大利亚接受过射电天文学的培训；他是诺贝尔奖获得者著名科学家拉曼的儿子，后来他继承父业成为班加罗尔物理实验室的负责人。

我们用帕克斯 210 英尺反射望远镜在 2,700 MHz 和 1,720 MHz 频率下对莫隆格勒射电天文台发现的脉冲星 *PSR* 0833–45[1]——现在暂时被证认为船帆座 – *X* 的超新星遗迹——进行了观测。我们发现这颗脉冲星具有脉冲宽度窄、强度高且稳定、完全线偏振和周期变动等值得注意的特征。我们对该天体的观测结果似乎排除了径向脉动是射电辐射起源的可能性，而支持由戈尔德提出[2]并被戈德赖希讨论过[3]的旋转模型。

最初的观测是在 12 月 8 日、9 日和 10 日用工作于 2,700 MHz 的宽带相关接收机[4]进行的。由测量得到的脉冲能量、线偏振和精确周期列于表 1 中。然而我们在试图获得脉冲波形的过程中很快发现：在通过接收机的 500 MHz 有效通带时，~ 10 ms 的漂移时间会将天然脉冲平滑到无法识别的程度。脉冲的偏振特征似乎也可能会被通带抹去。因此我们在 12 月 11 日、12 日、13 日和 20 日的夜晚用更窄的带宽在 1,720 MHz 频率下进行了重复的观测。用 400 通道积分器[5]记录脉冲的快速上升和短暂持续并示于图 1 中。除去偶然出现的强度异常的脉冲，在一般情况下振幅变动非常低：在我们用两种频率对该脉冲星进行观测的这几天中的任何时间里，取 100 个脉冲信号的平均值的结果显示仅有不到 10%的振幅变动，更不用说脉冲波形了。拉奇等人[6]曾讨论过在频率为 408 MHz 时也有类似情况。表 1 中列出了由在一种或两种频率下进行观测所得到的几个参数。

Table 1. Characteristics of *PSR* 0833–45

Heliocentric period (December 8–9, 1968)	$0.089208370 \pm 4 \times 10^{-9}$ s
Pulse width	~ 2 ms
Drift rate at 1,720 MHz	$-9{,}280 \pm 50$ MHz s^{-1}
$\int nedl$	63 cm^{-3} pc
Pulse energy at 1,720 MHz (both polarizations)	$(0.038 \pm 0.004) \times 10^{-26}$ J m^{-2} Hz^{-1}
Pulse energy at 2,700 MHz (both polarizations)	$(0.023 \pm 0.003) \times 10^{-26}$ J m^{-2} Hz^{-1}
Energy spectral index 1,720–2,700 MHz	–0.93
Percentage linear polarization 1,720 MHz	> 95 percent
Percentage linear polarization 2,700 MHz	> 60 percent (see text)
Percentage circular polarization 1,720 MHz	< 3 percent
Position angle of mean polarization 1,720 MHz	$115° \pm 5°$
Position angle of mean polarization 2,700 MHz	$77° \pm 7°$
Rotation measure	$+ 38 \pm 10$ rad m^{-2}
Intrinsic plane of polarization	≈ 50°
Mean longitudinal galactic field	≈ 0.8 μgauss

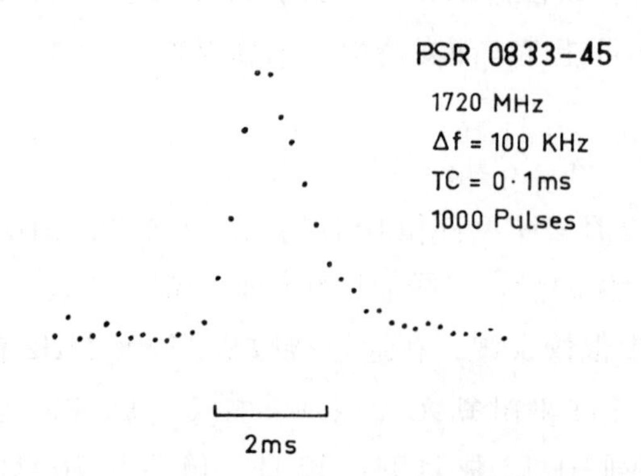

Fig. 1. A 1,000 pulse integration at 1,720 MHz of *PSR* 0833–45. The channel separation is approximately 0.22 ms.

The drift rate of the pulse at 1,720 MHz was measured by timing the arrival of the pulses in 100 kHz filters separated by 6 MHz. The interstellar dispersion derived from this measurement is somewhat higher than the value quoted by Large *et al.*[6] but not significantly so. The pulse energy at 1,720 MHz was obtained by integrating the observed pulses in two opposite polarizations. This is plotted in Fig. 2 together with the 2,700 MHz value and the 408 MHz energy[6]. The spectrum shows no break at the higher frequencies and in this respect *PSR* 0833–45 resembles *CP* 1133 (ref. 7).

表 1. *PSR* 0833–45 的特性

日心参照系中的周期（1968 年 12 月 8 日 ~ 9 日）	$0.089208370 \pm 4 \times 10^{-9}$ s
脉冲宽度	~ 2 ms
1,720 MHz 时的漂移率	$-9{,}280 \pm 50$ MHz s^{-1}
$\int n_e dl$	63 cm^{-3} pc
1,720 MHz 时的脉冲能量（两种偏振态）	$(0.038 \pm 0.004) \times 10^{-26}$ J m^{-2} Hz^{-1}
2,700 MHz 时的脉冲能量（两种偏振态）	$(0.023 \pm 0.003) \times 10^{-26}$ J m^{-2} Hz^{-1}
1,720 MHz ~ 2,700 MHz 的能谱指数	–0.93
1,720 MHz 时线偏振的百分比	>95%
2,700 MHz 时线偏振的百分比	>60%（参见正文）
1,720 MHz 时圆偏振的百分比	<3%
1,720 MHz 时平均偏振面的位置角	$115° \pm 5°$
2,700 MHz 时平均偏振面的位置角	$77° \pm 7°$
（法拉第）旋转量	$+38 \pm 10$ rad m^{-2}
固有偏振面	≈ 50°
星系场的平均纵向分量	≈ 0.8 μgauss

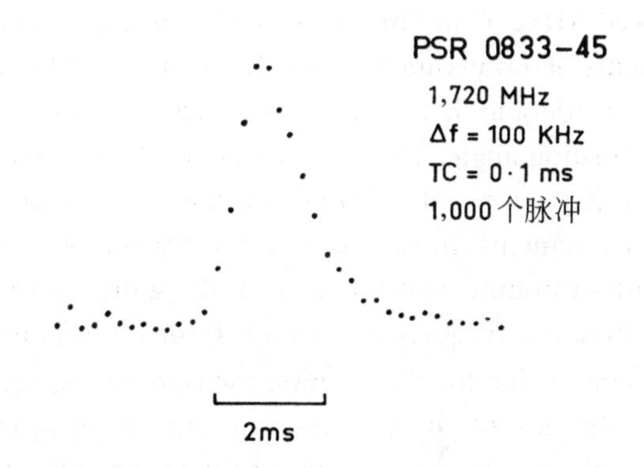

图 1. 在 1,720 MHz 下对 *PSR* 0833–45 的 1,000 个脉冲信号进行积分所得的结果。通道间隔约为 0.22 ms。

1,720 MHz 时的脉冲漂移率是用间隔为 6 MHz 的 100 kHz 滤波器记录脉冲到达时间测得的。从这一观测所推出的星际色散要比拉奇等人[6]提供的数值略微高一点，但差别不大。通过对观测到的两个相反偏振的脉冲进行积分可以得到 1,720 MHz 时的脉冲能量。所得结果与 2,700 MHz 和 408 MHz 时的能量值[6]一起示于图 2 中。能谱结果表明在较高频率处没有出现拐折，在这方面，*PSR* 0833–45 类似于 *CP* 1133（参考文献 7）。

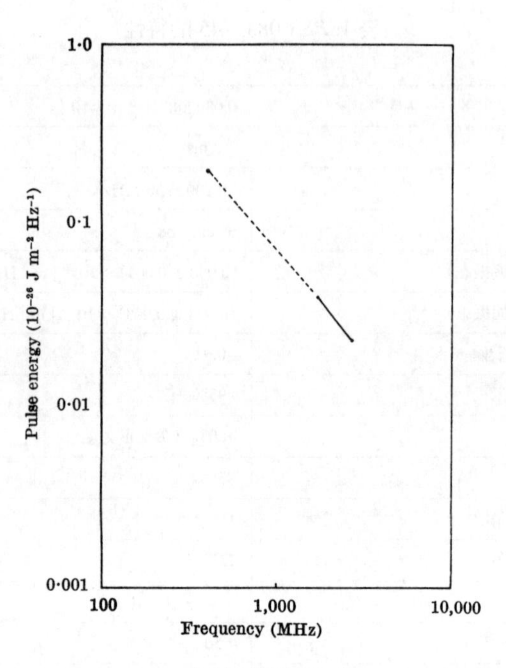

Fig. 2. The high frequency energy spectrum of *PSR* 0833–45. The spectral index is essentially constant at −0.9 between 408 MHz and 2,700 MHz.

The position angle at 1,720 MHz of the mean plane of linear polarization over the pulse was measured on the four nights of observation at this frequency and found to be the same. The internal agreement was considerably better than the error of 5 percent quoted in Table 1 for the absolute value of the position angle. The measurements of the position angle at 2,700 MHz on the previous three nights were also found to have remarkable internal consistency. Combining these two measurements on the assumption that the plane of polarization at these frequencies is indeed constant in time, and the same at the source, we have obtained a rotation of 38° ± n180° between these two frequencies. For $n = 0$ the corresponding rotation measure is +38 ± 10 rad m^{-2}. The mean value for the rotation measure in the region of Vela *X* obtained by Milne[8] is +46 rad m^{-2}. In view of the variations of rotation measure within the region the agreement is reasonable and provides strong support for the identification[1] of this pulsar with the supernova remnant. The average longitudinal component of the galactic magnetic field in this direction was then derived from the dispersion and the rotation measure[9]. A value of 0.8 μgauss was obtained which is intermediate between the values obtained by Smith[9,10] for three other pulsars in different directions in the galaxy.

The distribution of polarization across the pulse at 1,720 MHz is illustrated in Fig. 3. The Stokes parameter V has not been shown, for the circular polarization was found to be less than 3 percent. The whole pulse is almost completely linearly polarized (>95 percent) with the direction of the plane of polarization varying systematically across the main peak of the pulse. The Q axis was chosen as 115° to coincide with the polarization of the peak of the pulse. The plane of polarization rotates uniformly from a polarization angle (PA) near 140° on one edge of the pulse to less than PA 90° on the other, as sketched in Fig. 4. The

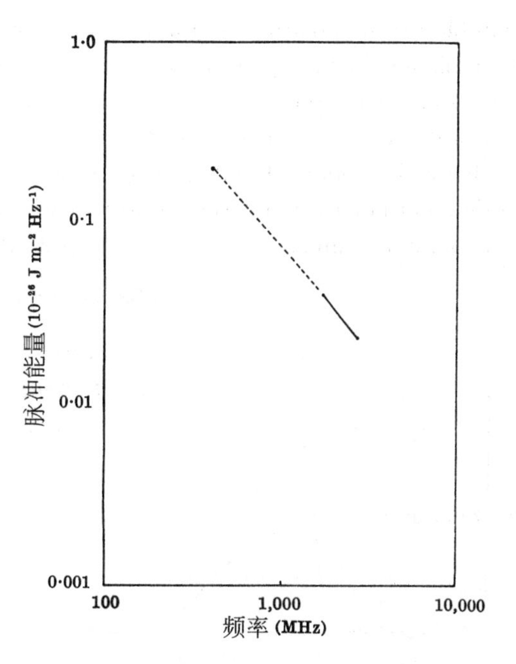

图 2. *PSR* 0833–45 的高频能谱。当频率在 408 MHz 到 2,700 MHz 之间时能谱指数基本上保持在常数 –0.9。

我们在 1,720 MHz 频率下观测了四个晚上，得到了对每个脉冲平均的线偏振面位置角，并发现数值是相同的。其内在一致性要比表 1 中所示的位置角绝对值的误差范围 5%好很多。前三个夜晚在 2,700 MHz 频率下对位置角进行观测所得的结果也具有很高的内在一致性。假定在上述频率下偏振面不随时间变化且源也不随时间变化，综合这两次观测的结果，我们得出在这两个频率之间有 38° ±n180°的转角。当 $n=0$时，相应的旋转量为 + 38 ± 10 rad m^{-2}。米尔恩 [8] 得到的船帆座 –*X* 区旋转量的平均值为 + 46 rad m^{-2}。考虑到在该区内旋转量存在变化，所以这两个结果之间的吻合还是比较好的，并且强有力地支持了这种脉冲星被证认为超新星遗迹的观点 [1]。因此，根据色散和旋转量就可以推导出在这一方向上星系磁场的平均纵向分量 [9]。得出的结果是 0.8 μgauss，此结果介于史密斯 [9,10] 由银河系中不同方向上的另外三颗脉冲星所得到的数值之间。

在 1,720 MHz 频率下脉冲中的偏振分布如图 3 所示。在图 3 中没有给出斯托克斯参数 *V*，因为我们发现圆偏振还不足 3%。整个脉冲几乎完全（>95%）是线偏振的，偏振面的方向从脉冲主峰的一侧到另一侧发生了系统的变化。*Q* 轴选取为 115° 以便与脉冲峰值处的偏振方向一致。偏振面从脉冲的一侧约 140° 偏振角（PA）处均匀地旋转到另一侧的小于 90° 偏振角处，如图 4 所示。因为辐射的振幅是从零开始增

magnetic field lines responsible for the polarization must vary smoothly over a similar range, for the amplitude of the radiation builds up from zero to a maximum and then falls off again. Detailed measurements across the pulse were not possible with the 2,700 MHz receiver because of the wide bandwidth as mentioned earlier. There were clear indications that the plane of polarization did change by about 40°–60° across the pulse, however. The smoothed pulse shapes obtained on different polarizations were consistent with the assumption that the polarization structure of the pulse was similar to that found at 1,720 MHz.

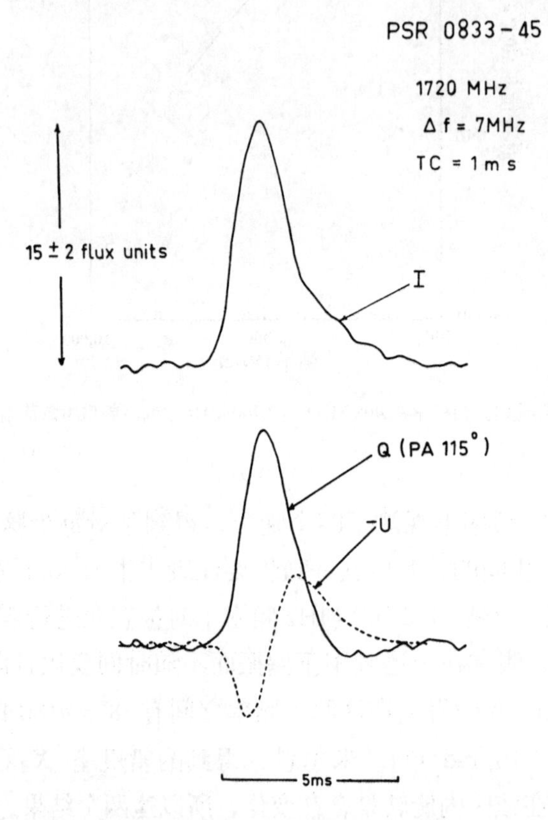

Fig. 3. The distribution of linear polarization across the pulse at 1,720 MHz. The distribution of Q and $-U$ represents a gradual change of approximately 45° in the position angle of polarization across the pulse. Within the errors of measurement the pulse is completely polarized.

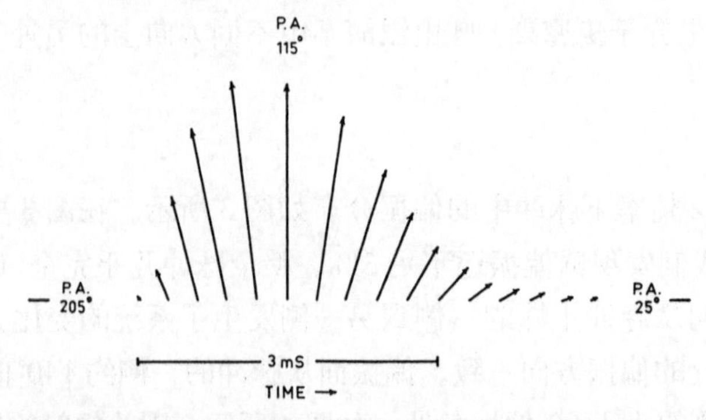

Fig. 4. Sketch showing the rocking of the plane of polarization during the pulse from *PSR* 0833–45.

加到最大值然后再次下降的，所以与偏振对应的磁力线也必然在类似的范围内平滑地变化。用2,700 MHz接收机不可能对脉冲进行更仔细的测量，原因在于前面已经提到过的带宽过大问题。然而仍然可以清晰地显示出偏振面在从脉冲一侧到另一侧时确实发生了约40°~60°的变化。在不同偏振下都得到了平滑的脉冲波形，这与2,700 MHz频率下的脉冲偏振结构类似于1,720 MHz下脉冲偏振结构的假定相吻合。

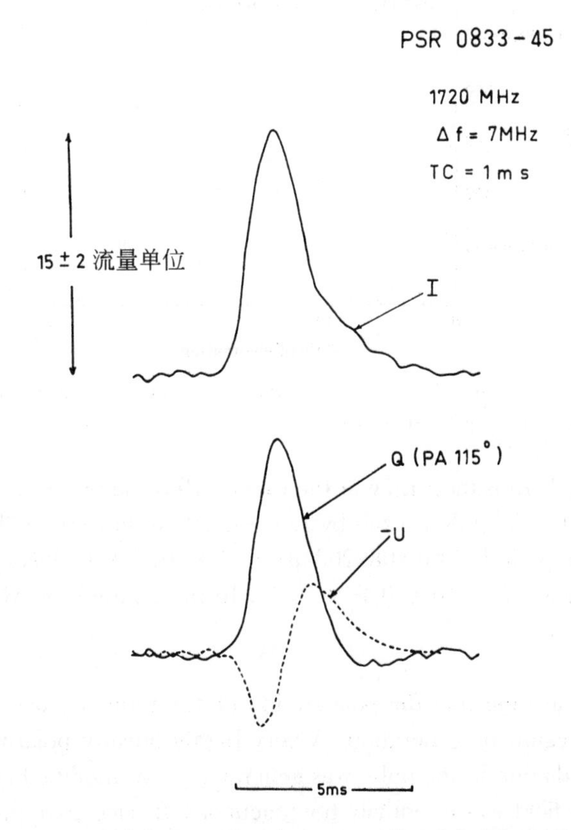

图3. 在1,720 MHz下一个脉冲期间的线偏振分布。Q和$-U$的分布状态表示出在一个脉冲期间偏振位置角逐渐改变了约45°。在测量误差范围之内该脉冲是完全偏振的。

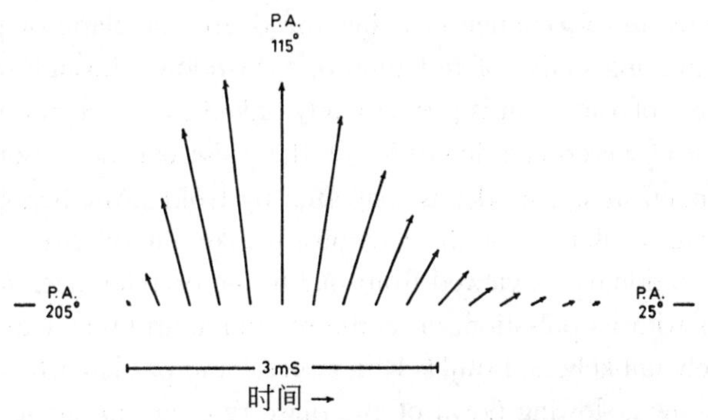

图4. 表示在*PSR* 0833–45的脉冲中存在偏振面摆动的示意图。

Mills[11] reports that the pulsar in the Crab nebula has been observed at Arecibo and found to be slowing down in periodicity by one part in 2,400 yr^{-1}. He further suggested that as *PSR* 0833–45 has the next shortest period known, it might also exhibit the same characteristic. Following up his suggestion we made several other measurements of the period and the results are illustrated in Fig. 5.

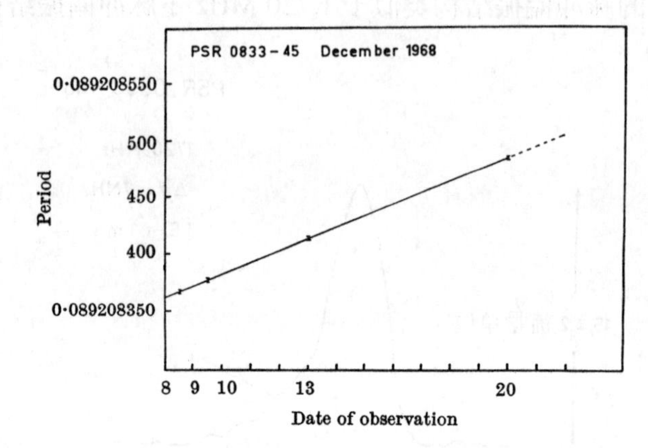

Fig. 5. The variation of the period of *PSR* 0833–45 over the 12 day observing span. The slope corresponds to an increase in the period of 10 ns or 1 part in 10^7 day^{-1}.

The period of *PSR* 0833–45 is increasing at the rate of 10 ns day^{-1}, or one part in 24,000 yr^{-1}. Assuming the position for the pulsar given by Large *et al.*[6] we have used all our measurements to give the heliocentric period $P = 0^s \cdot 089208483 \pm 2 \times 10^{-9} + R$ (Julian date = 2440210.0) where the rate of increase $R = (10 \pm 0.4) \times 10^{-9}$ s. In the notation of Moffet and Ekers[7] the stability $\Delta P/P \approx 1.2 \times 10^{-13}$.

It seems reasonable to assume that the polarization of the radiation from pulsars reflects the magnetic field in the region of generation. A very highly linearly polarized pulse must then imply that all of the radiation in the pulse was generated in, or modified in, passage through a small region where the field was essentially homogeneous. If successive pulses all have exactly the same polarization, then it must mean that the same locality in the magneto-sphere of the pulsar remains in view for as long as the polarization remains constant. We have observed *PSR* 0833–45 on December 11, 12, 13 and 20 at 1,720 MHz for many hours each night and never detected any measurable change in either the degree or plane of polarization. It must follow that either a pulsating source of radiation or a "window" through which we periodically glimpse a steady source of radiation is permanently linked to one region in the magnetic field of the star. All of the observed characteristics of the pulse can be reasonably understood by assuming a rotating neutron star model as suggested by Gold[2]. Any hypothesis associating the radio pulses with radial oscillations of the star must invoke one of three special conditions: (*a*) the magnetic field is stationary as viewed from our frame of reference, (*b*) the rotation of the pulsar is synchronous with its pulsation, or (*c*) the rotational and magnetic axes of the star are aligned. (*a*) is extremely unlikely, (*b*) is unlikely in general and particularly so on the basis of our observations which show a slowing down of the pulse rate but no change in the polarization, and (*c*) can be dismissed from symmetry considerations in view of the observed systematic sweeping of the plane of polarization across the pulse.

米尔斯[11]报道了他在阿雷西博观测到的蟹状星云脉冲星，他发现其周期在以每年 1 / 2,400 的速率加长。他进一步指出：由于 *PSR* 0833–45 具有目前已知的次短周期，或许它也会呈现出相同的特性。受其观点启示，我们又对 *PSR* 0833–45 的周期进行了几次观测，观测结果示于图 5 中。

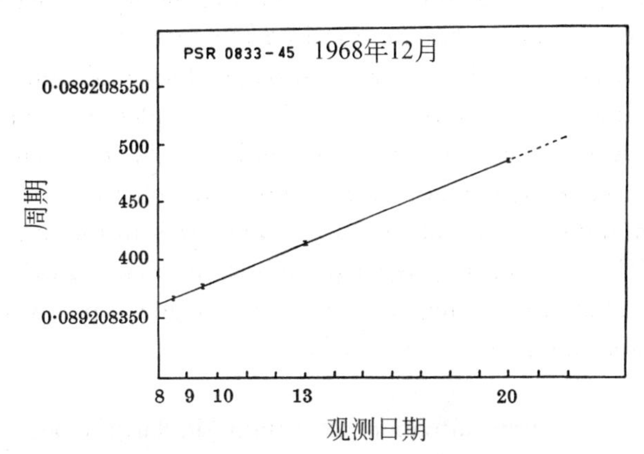

图 5. *PSR* 0833–45 的周期在 12 天观测期内显示出的变化。斜率所对应的周期增长为每天 10 ns 或 $1/10^7$。

PSR 0833–45 的周期在以每天 10 ns 的速率增加，即每年 1 / 24,000。假设该脉冲星位于拉奇等人[6]所给出的位置，则利用我们的全部观测数据可以得出：日心参照系中的周期 $P=0^s \cdot 089208483 \pm 2\times10^{-9} + R$（儒略日期 = 2440210.0)。其中，周期增长率 $R = (10\pm0.4)\times10^{-9}$ s。根据莫菲特和埃克斯[7]的记号，稳定度 $\Delta P/P\approx1.2\times10^{-13}$。

假定脉冲星辐射的偏振能反映其产生区域的磁场是合理的。那么脉冲的高度线偏振必定意味着脉冲中的所有辐射都产生自一个磁场基本均匀的小区域，或者在经过这样的区域时受到其影响。如果随后的脉冲都具有完全相同的偏振状态，那只能说明在偏振保持不变期间脉冲应该来自磁层中的同一部位。我们分别于 12 月 11 日、12 日、13 日和 20 日每晚在 1,720 MHz 下对 *PSR* 0833–45 进行了好几个小时的观测，未曾发现偏振度或偏振面有任何可观测到的变化。由此可以推定：辐射的脉动源，或者在我们周期性观测到稳定辐射源时所通过的“窗口”，是与恒星磁场的某个区域恒定地联系在一起的。利用戈尔德[2]提出的旋转中子星模型就可以合理地解释脉冲的所有观测特征。任何一种将射电脉冲与恒星径向振荡联系起来的假说都必须借助以下三个特殊条件中的一个：(*a*) 从我们的参照系看来该磁场是恒定的，(*b*) 脉冲星的旋转与它的脉动是同步的，或者 (*c*) 恒星的旋转轴和磁轴是在同一条直线上的。(*a*) 基本上没有成立的可能性；(*b*) 在一般情况下也不大可能成立，尤其是当我们在观测中发现脉冲速率在减慢而偏振状态并不改变时；基于对称性的要求 (*c*) 也可以被排除，因为已经观测到了偏振面从脉冲主峰一侧到另一侧的系统变化。

That the two most rapidly pulsating sources have shown an increase in period with time as predicted on Gold's model adds further support to the rotation hypothesis. Goldreich[3] has developed a simple axisymmetric version and shown that relativistic particles must be carried out along the field lines which extend beyond the velocity of light cylinder. If a pulsar is a non-axisymmetric version of this model as suggested, the beaming of the radiation in any given direction will be governed by the geometry linking the line of sight with the magnetic and rotational axes of the star. If this were indeed the case the average pulse shape would represent the profile of the magnetic "window" through which radiation can escape in a given direction, and the polarization of the pulse would be independent of frequency. Such a model would justify the tacit assumption in Faraday rotation interpretations that different frequencies are generated with the same polarization at the source. Any change of polarization across the pulse would then correspond to the rotation of the field lines in the duration of the pulse. A change of more the 45° in the plane of polarization such as seen in *PSR* 0833–45 implies that the radiation must emanate from a region close to the magnetic pole with the magnetic axis inclined at a considerable angle to the rotational axis.

We consider it relevant to draw attention to certain similarities between Jupiter and the pulsars. The low frequency radiation from both is characterized chiefly by unpredictability and fantastic complexity in structure and polarization. At the higher frequencies the polarization of Jupiter's radiation is chiefly linear, periodic and highly predictable. Over 180° of rotation the total intensity rises and falls due to beaming while the plane of polarization rocks[12] with the magnetic field. It is conceivable that at a high enough frequency the pulses from any pulsar will be as repetitive in character as they are in time.

If pulsars have satellites, and the satellites influence the activity of the pulsar in the manner of Io's influence on Jupiter[13], the radiation would exhibit many of the characteristics that are found at low frequencies. In spite of Jupiter's proximity and the time over which the low frequency radiation had been studied, it was only a statistical approach combined with a knowledge of Io's movements which finally led to the discovery of the modulation.

Goldreich[3] predicts a rate of change of period inversely proportional to the square of the period. The ratio of the rates of change of the Crab pulsar to the Vela pulsar is 10, in reasonable agreement with the predicted value of 7.5. The age of the Vela supernova on his model would then be $\sim 10^4$ yr.

We thank Dr. R. N. Manchester for help with the 1,720 MHz observations and Miss P. Beswick for help in the preparation of the manuscript.

(**221**, 443-446; 1969)

V. Radhakrishnan, D. J. Cooke, M. M. Komesaroff and D. Morris: Radiophysics Laboratory, CSIRO, Sydney, Australia.

正如戈尔德的模型所预言的那样，两个最快的脉动源都显示出周期随时间而增加的现象，这就为旋转假说提供了进一步的支持。戈德赖希[3]详尽阐述了一种简单的轴对称形式，并证明相对论性粒子一定是沿着延伸出光速圆柱面之外的磁力线运动的。如果一颗脉冲星被认为是处于此模型的非轴对称形式，那么在任意给定方向上的辐射束流都将由视线、该星体的磁轴及旋转轴这三者的几何关系所决定。如果事实确实如此，那么用平均脉冲波形将可以代表给定方向辐射逃逸时通过的磁“窗口”的轮廓，而脉冲的偏振状态将与频率无关。这一模型将证实在解释法拉第旋转时默认的假定，即认为不同的频率在源处以同一偏振产生。因此，偏振从脉冲主峰一侧到另一侧的任何变化都将与脉冲过程中磁力线的旋转相关联。如果偏振面的变化超过了45°，例如在*PSR* 0833–45中观察到的情况，则意味着辐射必定来自靠近磁极的区域，并且磁轴相对于旋转轴的倾斜角度不会太小。

我们认为有必要提请注意一下木星与脉冲星之间的某些相似之处。来自木星和脉冲星的低频辐射都以结构及偏振状态的不可预见性和高度复杂性为主要特征。而在较高频率下，木星辐射的偏振状态则基本上是线性的、周期性的和高度可预测的。在180°的转动中，总强度因束流而升高和降低，而偏振面会随着磁场的变化而摆动[12]。可以设想：当频率足够高时，来自任何脉冲星的脉冲在性质上和时间上都将是可重复的。

假设脉冲星有卫星，而且这个卫星能够像木卫一影响木星[13]那样影响脉冲星的活动，那么辐射就将呈现出在低频时观察到的许多特征。尽管木星近在咫尺，并且人们对其低频辐射的研究也已经经历了很长时间，但也只有在将统计方法与木卫一的运动知识相结合之后才最终导致了调制现象的发现。

戈德赖希[3]预言周期的变化率反比于周期的平方。蟹状星云脉冲星与船帆座脉冲星的变化率之比为10，与预估值7.5的吻合情况还是不错的。因此，根据他的模型，船帆座超新星的年龄为~10^4年。

感谢曼彻斯特博士帮助我们进行1,720 MHz频率下的观测，还要感谢贝斯威克小姐在准备手稿的过程中所提供的帮助。

（王耀杨 翻译；邓祖淦 审稿）

Received December 30, 1968.

References:

1. Large, M. I., Vaughan, A. E., and Mills, B. Y., *Nature*, **220**, 340 (1968).
2. Gold, T., *Nature*, **218**, 731 (1968).
3. Goldreich, P., *Proc. Ast. Soc. Austral.*, **1**, No. 5 (1969).
4. Batchelor, R. A., Brooks, J. W., and Cooper, B. F. C., *IEEE Trans. Antennas and Propagation*, **AP-16**, 228 (1968).
5. Radhakrishnan, V., Komesaroff, M. M., and Cooke, D. J., *Nature*, **218**, 229 (1968).
6. Large, M. I., Vaughan, A. E., and Wielebinski, R., *Nature*, **220**, 753 (1968).
7. Moffet, A. T., and Ekers, R. D., *Nature*, **218**, 227 (1968).
8. Milne, D. K., *Austral. J. Phys.*, **21**, 201 (1968).
9. Smith, F. G., *Nature*, **218**, 325 (1968).
10. Smith, F. G., *Nature*, **220**, 891 (1968).
11. Mills, B. Y., *Proc. Astro. Soc. Austral.*, **1**, No. 5 (1969).
12. Morris, D., and Berge, G. L., *Astrophys. J.*, **136**, 276 (1962).
13. Bigg, E. K., *Nature*, **203**, 1008 (1964).

Detection of Water in Interstellar Regions by Its Microwave Radiation

A. C. Cheung *et al.*

Editor's Note

As radio receivers became capable of observations at ever shorter wavelengths, new molecules were detected in space. A. C. Cheung and colleagues here report the discovery of radio emission from water in three regions in the sky. The brightness temperature of the emission from the Orion Nebula and the *W49* region was surprisingly high, which puzzles the authors because the energetically excited state of the molecules they observed has a natural lifetime of just 10 seconds. The observations therefore implied that the water molecules were being constantly excited. It turns out that collisions with molecular hydrogen cause this excitation—but molecular hydrogen had not yet been discovered in space.

MICROWAVE emission from the $6_{16} \rightarrow 5_{23}$ rotational transition of H_2O has been observed from the directions of Sgr *B*2, the Orion Nebula and the *W*49 source. This radiation, at 1.35 cm wavelength, was detected with the twenty foot radio telescope at the Hat Creek Observatory using techniques described earlier for the detection of the NH_3 spectrum[1]. In the case of Sgr *B*2, the H_2O emission is from the same direction in which considerable NH_3 is observed (unpublished work of A. C. C. *et al.*), although there is reason to believe the two molecular species may not be closely associated. Strong H_2O radiation producing an antenna temperature of 14°K is observed from the Orion Nebula (where no NH_3 was detected), and an antenna temperature at least as high as 55° was found for H_2O radiation from *W*49.

Fig. 1 shows the spectral intensity near the H_2O resonance in the approximate direction of Sgr *B*2, the antenna being pointed at the position α_{1950} =17 h 44 m 23 s. δ_{1950} = −28°25′. The observed antenna temperature is plotted as a function of Doppler velocity. The mean Doppler velocity of about 70 km s^{-1} for the spectral line observed is not very different from the 68 km s^{-1} mean velocity found for one of the OH emission and broad OH absorption features observed in this region[2], the 62 km s^{-1} Doppler velocity of a small nearly HII region[3], and the velocity of about 58 km s^{-1} found for NH_3 (unpublished work of A. C. C. *et al.*) observed in this direction. The results shown in Fig. 1 were obtained with filters producing a spectral resolution of about 1.3 MHz.

通过微波辐射探测到星际空间中存在水分子

Cheung 等

编者按

随着射电接收机能够观测的波长越来越短，星际空间中不断有新的分子被发现。在本文中，Cheung 及其同事在三个天区探测到了水的射电辐射。猎户座星云和 *W*49 辐射源的亮温度高得惊人，这使本文的作者们感到非常困惑，因为他们观测到的分子的能量激发态仅有 10 s 的自然寿命。因此，这些观测结果表明水分子是在不断地被激发。后来人们发现是氢分子的碰撞引发了这种激发过程——但当时人们还没有发现在星际空间中存在氢分子。

在人马座 *B*2、猎户座星云和 *W*49 源的方向上观测到了由水分子 $6_{16} \rightarrow 5_{23}$ 转动跃迁产生的微波辐射。这种波长为 1.35 cm 的射电辐射是在哈特克里克天文台用 20 英尺的射电望远镜观测到的，采用的技术与早先探测氨分子光谱的方法 [1] 相同。尽管有理由相信这两种分子之间并不存在密切的联系，但在人马座 *B*2 方向上，水分子的辐射和以前发现的大量氨分子的辐射（Cheung 等人尚未发表的研究结果）都来自同一个方向。从猎户座星云（在其中并未探测到氨分子）中观测到水分子强辐射产生的天线温度为 14 K；而从 *W*49 方向上观察到水分子辐射产生的天线温度至少可以达到 55 K。

图 1 展示出大致在人马座 *B*2 方向上水分子共振线附近的光谱强度。天线所指的位置是 $\alpha_{1950} = 17$ h 44 m 23 s，$\delta_{1950} = -28°25'$。图中给出了观测到的天线温度随多普勒速度变化的关系。被观测到的光谱线的平均多普勒速度约为 70 km/s，与在羟基发射线和在该区域宽广的羟基吸收特征中发现的平均多普勒速度 68 km/s[2]、在附近的一小块 HII 区（译者注：即电离氢区）内发现的多普勒速度 62 km/s[3] 以及在该方向上观测到的氨分子的多普勒速度 58 km/s（Cheung 等人尚未发表的研究结果）相比，并没有很大的差异。图 1 所示的结果是经过了能使光谱分辨率达到 1.3 MHz 左右的滤波器后得出的。

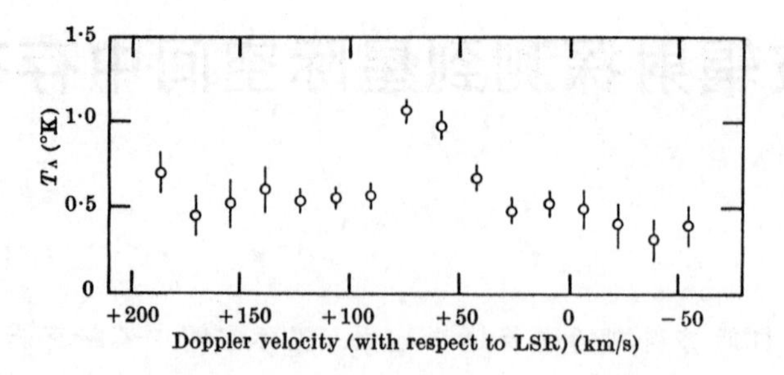

Fig. 1. Observed spectral intensity of $6_{16} \rightarrow 5_{23}$ H_2O rotational transition in the direction of Sgr *B*2 at $\alpha_{1950.0}$=17 h 44 m 23 s ± 6 s, $\delta_{1950.0}$ = 28° 24.9′ ± 1′.

Fig. 2 shows the antenna temperature as a function of Doppler velocities observed in the Orion Nebula at α_{1950} = 5 h 32 m 57 s ± 4 s and δ_{1950} = − 5° 25.5′±1.0′. In Orion, the radiation intensity was sufficiently high to make practical the use of filters producing a spectral resolution of about 350 kHz. In Fig. 2 the solid line represents the continuum temperature as it was measured with filters of width 2 MHz; the plotted points represent observations made with the filters having bandwidths of 350 kHz. These points are, in some places, closer together than the filter widths of 350 kHz because measurements were made during different runs in which the central frequencies of the filters were displaced in order to examine the structure of the spectral line, which is composite, representing at least two distinct Doppler velocities. The mean Doppler velocity of about +7 km s^{-1} found for the water line is somewhat different from the HII Doppler shifts in the range of −2 to −6 km s^{-1} found in the same direction[3,4]. OH radiation from this direction shows Doppler velocities near 20 km s^{-1} and 7 km s^{-1} (ref. 2); the latter coincides more closely with the H_2O velocity.

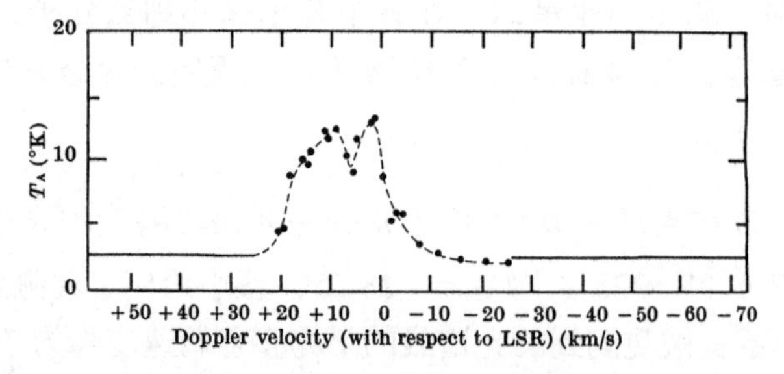

Fig. 2. Observed spectral intensity of $6_{16} \rightarrow 5_{23}$ H_2O rotational transition in the direction of the Orion Nebula. The position of peak emission is $\alpha_{1950.0}$ = 5 h 32 m 57 s ± 4 s, $\delta_{1950.0}$ = −5° 25.5′±1.0′.

Fig. 3 shows similar data for the still more intense source *W*49, at α_{1950} = 19 h 7 m 55 s±4 s and δ_{1950} = 9° 0.4′±1′. Over most of the spectral line, the narrow filters of width 350 kHz were used and several narrow and intense features were revealed. The measured antenna temperature was as high as 55° K, but allowance for atmospheric attenuation of this source, which was rather low in the sky during the observations, raises its effective antenna temperature to somewhat more than 70°. Presumably, narrower filters would give still higher antenna temperatures at some frequencies. The solid lines on either side of the diagram show the radiation intensity observed with filters about 2 MHz wide. Radiation present on the extreme right of Fig. 3 seems to be above the continuum level, showing that H_2O radiation also occurs

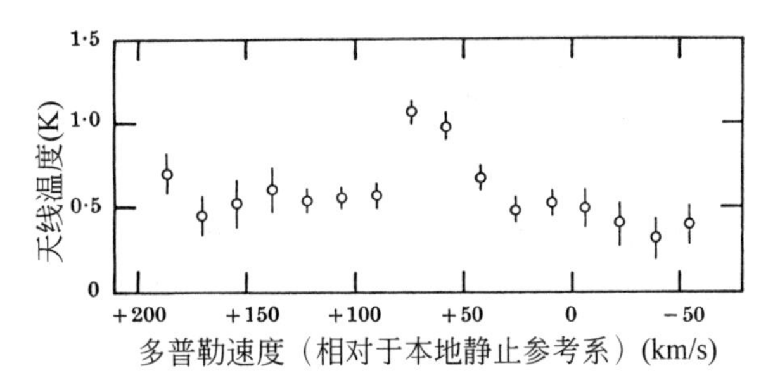

图 1. 在人马座 *B*2 方向（$\alpha_{1950.0}$ = 17 h 44 m 23 s ± 6 s，$\delta_{1950.0}$=−28° 24.9′ ± 1′）上观测到的水分子 $6_{16} \rightarrow 5_{23}$ 转动跃迁的光谱强度。

图 2 展示出在猎户座星云方向（α_{1950} = 5 h 32 m 57 s ± 4 s，δ_{1950} = −5° 25.5′ ± 1.0′）上观测到的天线温度随多普勒速度变化的关系曲线。在猎户星座，射电辐射的强度高到足以使用能使光谱分辨率达到约 350 kHz 的滤波器。在图 2 中，实线代表由带宽 2 MHz 的滤波器测得的连续谱温度；黑点则表示用带宽 350 kHz 的滤波器得到的观测值。黑点之间的距离在有些地方比滤波器的带宽 350 kHz 还小，这是因为为了探测光谱结构进行了多次测量，而在每次测量中滤波器的中心频率都会有所偏移的缘故。光谱结构表明它至少有两种不同的多普勒速度成分存在。从水分子谱线中观察到的平均多普勒速度约为 +7 km/s；它与在同一方向测得的 −2 km/s~−6 km/s 的 H II 区多普勒速度 [3,4] 有所不同。从这个方向上的羟基辐射测得的多普勒速度接近 20 km/s 和 7 km/s（参考文献 2）；后者更接近于水分子的多普勒速度。

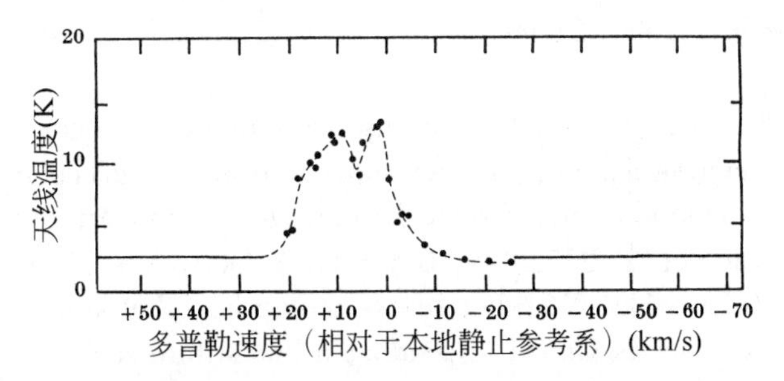

图 2. 在猎户座星云方向上观测到的水分子 $6_{16} \rightarrow 5_{23}$ 转动跃迁的光谱强度。辐射峰值的方位是 $\alpha_{1950.0}$ = 5 h 32 m 57 s ± 4 s，$\delta_{1950.0}$ = −5°25.5′ ± 1.0′。

图 3 中展示的是更强的 *W*49 源的类似数据，观测方向是 α_{1950} =19 h 7 m 55 s ± 4 s，δ_{1950} = 9° 0.4′ ± 1′。带宽为 350 kHz 的窄频带滤波器覆盖了谱线的大部分区域，揭示出一些很强的窄带特征。实际测出的天线温度为 55 K，但在补偿了该源的大气衰减（在观测时的夜空中非常低）之后，有效天线温度增加至稍稍大于 70 K。可以推测：在某些频率上，更窄的滤波器将给出更高的天线温度。位于这幅图两侧的实线所表示的是由带宽约为 2 MHz 的滤波器得到的辐射强度。图 3 中最右侧的辐射似乎高于连续谱，说明水分子的辐射也同样发生在宽频带滤波器的这一波段。在 *W*49 源中，羟

within the band of this last wide filter. The Doppler shift of OH in *W*49 ranges from -2 km s^{-1} to $+23$ km s^{-1} (ref. 2) and that of the HII region is $+6$ km s^{-1} (ref. 3). These are in the same general range, but by no means similar to the dominant values of Doppler shifts shown in Fig. 3.

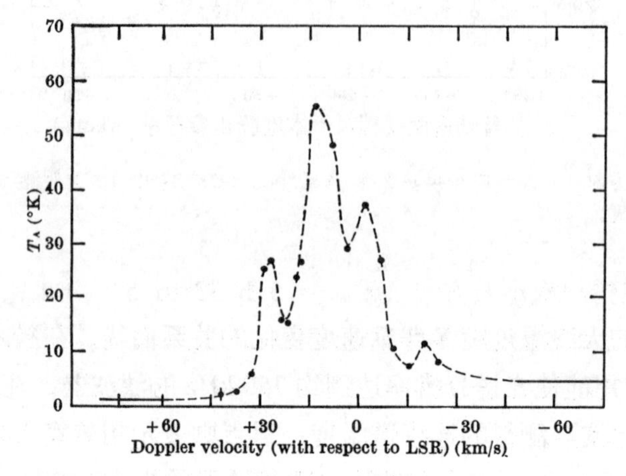

Fig. 3. Observed spectral intensity of H_2O transition in the direction of *W*49. The position of peak emission is $\alpha_{1950.0}$ = 19 h 07 m 55 s±5 s, and $\delta_{1950.0}$ = +09° 0.4′±1′.

In all three figures, the continuum temperatures should be scaled down by a factor of two, because the heterodyne detection system received continuum radiation in both side bands.

The Orion and *W*49 sources were not large enough in angular size noticeably to broaden the beam width of 8.8′ in drift scans across the sources. Thus we conclude that these radiating sources are not larger in angular diameter than 3′.

The radiation found is attributed to H_2O because its frequency coincides very closely to that found for H_2O in the laboratory, and no other known atomic or molecular species can explain the observations. The Doppler shifts plotted in the figures represent small departures from the laboratory frequency of 22,235.22 MHz for H_2O, for which 10 km s^{-1} corresponds to a frequency shift of 0.75 MHz. No other simple molecular of atomic species which can be expected in interstellar space is known to produce a line within several tens of MHz of this frequency except NH_3, the (3,1) inversion transition of which lies at 22,234.53 MHz. This is a rather weak NH_3 transition, but it is still important to examine carefully the possibility that the observed radiation might be caused by NH_3. A search was therefore made for other stronger transitions of NH_3. The (3,3) transition is somewhat more than an order of magnitude stronger than this (3,1) transition at excitation temperatures above 50° K, while the (1,1) transition is similarly more intense at temperatures below 50° K. Even at excitation temperatures considerably above 50°K, the (1,1) transition is always more than twice as intense as the (3,1) transition. A search for these two NH_3 transitions in the Orion Nebula was made. They were not found, with the upper detection limit of antenna temperature being set at less than 0.07°K for both the (1,1) and the (3,3) inversion transition. This limit is almost two orders of magnitude less than the intensity of radiation found at about 22,235 MHz, and hence we

基的多普勒速度在 –2 km/s~+23 km/s 之间（参考文献 2）；而 HII 区的多普勒速度为 +6 km/s（参考文献 3）。这些数值处于相同的范围内，但与图 3 中所示的多普勒速度的显著值并不相符。

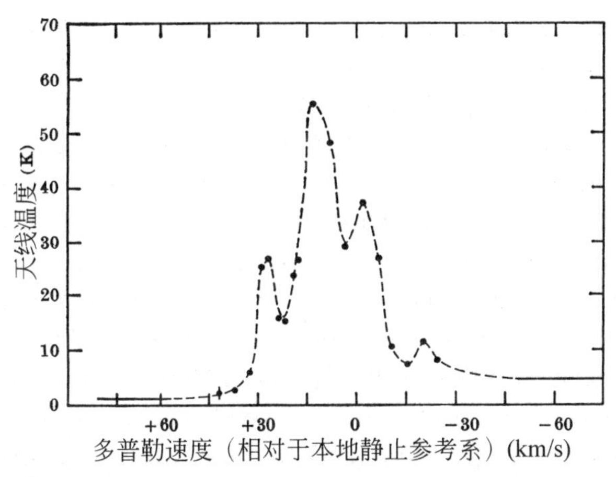

图 3. 在 *W*49 方向上观测到的水分子跃迁的光谱强度。辐射峰值的方位是 $\alpha_{1950.0}=19$ h 07 m 55 s±5 s，$\delta_{1950.0}=09°\ 0.4'\pm1'$。

因为外差式检波系统在波段的两侧都接受连续辐射，所以在这三幅图中，连续谱的温度都应下降为原来的一半。

由于猎户座星云和 *W*49 源的角直径不足够大，以致无法在对整个源的漂移扫描过程中将宽度为 8.8′ 的光束显著加宽。因此我们可以下结论认为这两个辐射源的角直径都不大于 3′。

认为被发现的辐射与水分子有关是因为其频率与实验室中的水分子频率非常接近，并且用其他已知原子或分子都解释不了这些观测现象。对水分子来说，10 km/s 的速度对应于 0.75 MHz 的频移，所以这些图中显示的多普勒位移与水分子在实验室中的频率 22,235.22 MHz 有小小的偏差。除位于 22,234.53 MHz 的氨分子（3,1）反演跃迁以外，在星际空间中再也没有其他简单分子能产生与这个频率相差几十兆赫之内的谱线了。虽然氨的这种跃迁很微弱，但仔细检查观测到的辐射是否有可能来自氨分子仍然很重要。因此需要搜索氨分子的其他一些较强的跃迁。当激发温度高于 50 K 时，（3,3）跃迁比（3,1）跃迁大致强一个量级以上；而在激发温度低于 50 K 时，（1,1）跃迁同样会更强。即使在激发温度远高于 50 K 时，（1,1）跃迁的发生概率也常常是(3,1)跃迁的两倍以上。我们在猎户座星云中搜索了氨的这两种跃迁。结果表明：直至天线温度的检测上限被设在小于 0.07 K 时，（1,1）和（3,3）反演跃迁都始终没有被探测到。这种限制条件比在 22,235 MHz 附近发现的辐射强度低大约两个量级，因此我们可以断定：我们已观测到的辐射并非由氨的（3,1）跃迁所引起，

conclude that the radiation we have observed cannot be caused by the (3,1) transition of NH_3, and can be explained only by H_2O.

In the case of Sgr *B*2, the situation is more complex. Rather strong NH_3 radiation was, in fact, observed from the same (or nearly the same) direction as the H_2O radiation. The (1,1), (2,2), (3,3) and (4,4) inversion transitions of NH_3 were all found with antenna temperatures between 0.5° K and 1.5° K (unpublished work of A. C. C. *et al.*). The possibility that the line observed at 22,235 MHz is caused by NH_3 may be ruled out by the following argument, however. The NH_3 rotational states (1,1), (2,2), (3,3) and (4,4) can be excited by collision, but are metastable, with exceedingly long radiation lifetimes. The (3,1) state can radiate in the infrared, however, by making a transition to the (2,1) state. The mean radiation lifetime of the upper state is about 50 s; thus unless the state is excited in a comparable time, it should be very much less populated than any of the states which were observed. The (4,3) state also radiates with a lifetime comparable with that of the (3,1) state. Thus a search was made for (4,3) level inversion radiation to find out whether or not some excitation mechanism was producing a significant population in the (4,3) level and, by inference, also producing an appreciable number of NH_3 molecules in the (3,1) state. The (4,3) inversion radiation was not observed, the limit of detection being an antenna temperature of 0.07° K. The (4,3) transition searched for is about an order of magnitude more intense than the (3,1) transition. If the observed antenna temperature of 1° K in Sgr *B*2 at 22,235 MHz were caused by the (3,1) transition, the (4,3) transition would have an antenna temperature of 10° K. No such temperature is observed, so we conclude that the observed radiation at 22,235 MHz must be caused by H_2O rather than this weak NH_3 transition.

In the case of *W*49, a search was made for the NH_3 (1,1) inversion transition. It was not found, and an upper limit of 0.5° could be established for the antenna temperature at its frequency.

There is, of course, H_2O in the Earth's atmosphere. One might wonder if it could produce the observed radiation. It can be eliminated as a possible source for several reasons. One is that the antenna beam is switched from one part of the sky to a neighbouring part, so that only radiation which varies very rapidly with angle is detected. A second reason is that the microwave resonant line was detected in the three particular fixed directions in space, and not nearby. Still a third is that an atmospheric water line would have occurred at the laboratory frequency, which corresponds to a Doppler shift of +15 km s^{-1} in Fig. 1 and −30 km s^{-1} in Fig. 2.

It has previously been suggested that the $6_{16} \rightarrow 5_{23}$ transition of H_2O could be of interest to radio astronomy[5], and recently a substantial proposal for its detection has been made by Snyder and Buhl at the AAS meeting, Dallas, December 1968. But it is surprising that the transition is as strong as observed, because it involves levels of rotational energy of 456 cm^{-1} which can radiate to lower states in about 10 s. They require moderately high temperatures and frequent excitations, either by collisions or by radiation. Rotational states of NH_3 which can similarly radiate have not been found, indicating that the NH_3 and H_2O have been detected in rather

它只能用水分子的存在来解释。

人马座 *B*2 的情况要复杂得多。事实上，在观测到水分子辐射的同一方向（或接近这一方向）上也发现了很强的氨分子辐射。在天线温度介于 0.5 K~1.5 K 之间时，氨的四种反演跃迁（1,1）、（2,2）、（3,3）和（4,4）都被探测到了（Cheung 等人尚未发表的研究结果）。然而，根据下述讨论也许可以排除在 22,235 MHz 观测到的谱线来源于氨分子的可能性。氨分子的转动态（1,1）、（2,2）、（3,3）和（4,4）会因为碰撞而激发，但它们在很长的辐射寿命之内是处于亚稳态的。氨分子可以从（3,1）态跃迁到（2,1）态并发出红外波段的辐射。其中高能态的平均辐射寿命约为 50 s。因此，除非该能态在 50 s 左右的时间内被激发，否则处于该能态的粒子数量会比处于观测到的任何其他能态的粒子数量少很多。（4,3）态的辐射寿命也与（3,1）态近似。因此有人对（4,3）能级的反演跃迁辐射进行了探测，力图判断是否存在某种可以产生大量（4,3）能级粒子的激发机制，并且还可以推知由这种机制同样会产生可观数量的（3,1）态氨分子。当天线温度的探测极限为 0.07 K 时，（4,3）态下的反演跃迁辐射并没有被观测到。所搜寻的（4,3）跃迁比（3,1）跃迁强大约一个数量级。如果在人马座 *B*2 方向上 22,235 MHz 处观测到的 1 K 天线温度是由（3,1）跃迁造成的，那么从（4,3）跃迁得到的天线温度将达到 10 K。鉴于人们从未观测到过如此高的温度，所以我们可以得出结论：在 22,235 MHz 处观测到的辐射一定来自水分子，而不可能来自微弱的氨分子跃迁。

人们在 *W*49 方向上对氨分子（1,1）的反演跃迁进行了搜索。但当在该频率处将天线温度的上限设定为 0.5 K 时，仍没有探测到这种跃迁。

当然，水分子在地球的大气中确实存在。有人怀疑可能是地球大气中的水分子产生了观测到的辐射。以下几条理由可以排除这种来源的可能性。原因一，因为天线射束从天空中的一个区域转向另一个邻近区域，所以只有随角度不同而迅速发生变化的辐射才会被探测到。原因二，微波共振线是在太空中的三个固定方向上探测到的，在这三个特殊方向附近并没有发现。原因三，大气中水分子谱线的出现位置应与实验室中水分子的频率相同，对应于图 1 中多普勒速度 +15 km/s 处和图 2 中多普勒速度 –30 km/s 处。

之前曾有人提出水分子的 $6_{16} \rightarrow 5_{23}$ 跃迁在射电天文学方面很有研究价值 [5]，最近在 1968 年 12 月于达拉斯举行的美国科学院会议上，斯奈德和布尔提出了一个与这项探测相关的实质性计划。观测到的跃迁出人意料地强，因为它涉及位于 456 cm^{-1} 处的转动能级，在大约 10 s 内即可实现向较低能级的跃迁并发出辐射。它们需要被较高的温度和频率所激发，通过碰撞或者辐射的方式。未曾探测到可发出类似辐射的氨分子转动态，这表明氨分子和水分子是在完全不同的区域内被检测到

different regions. Presumably the H_2O is present in rather special regions of higher than normal excitation. The matrix element for this H_2O line is appreciably less than that for the NH_3 inversion levels. Thus if there is thermal equilibrium between the 6_{16} and 5_{23} states, the population in these two levels, rather high above the ground state, must have a column density greater than that found for NH_3 (ref. 1), or about 10^{17} cm^{-2} in the Sgr *B*2 cloud.

In Orion, the actual microwave brightness temperature of the source would have to be at least as high as a few hundred degrees, and in *W*49 at least as great as about one thousand degrees. The actual temperatures may be much higher if these sources are smaller than the upper limit of 3′ of arc given, or if they are optically thin. The high intensity, very narrow, lines suggest that perhaps thermal equilibrium does not occur and that there may even be maser action. Further study of the distribution and condition of H_2O in interstellar space is clearly needed. If thermal equilibrium does in fact apply, this intense microwave radiation from H_2O and the existence of strong HDO transitions in the radio region should allow an interesting measurement of the hydrogen–deuterium abundance ratio.

We thank Professor Harold Weaver for giving us his positions of OH clouds in the Sagittarius region and for discussions, and Paul Rhodes for his help with the observations. This work is supported in part by NASA, the US Office of Naval Research and the US National Science Foundation.

(**221**, 626-628; 1969)

A. C. Cheung, D. M. Rank and C. H. Townes: Department of Physics, University of California, Berkeley.
D. D. Thornton and W. J. Welch: Radio Astronomy Laboratory and Department of Electrical Engineering, University of California, Berkeley.

Received January 13, 1969.

References:

1. Cheung, A. C., Rank, D. M., Townes, C. H., Thornton, D. D., and Welch, W. J., *Phys. Rev. Lett.*, **21**, 1701 (1968).
2. Weaver, H. F., Dieter, N. H., and Williams, D. R. W., *Ap. J. Suppl.*, **16**, 219 (1968).
3. Mezger, P. G., and Höglund, B., *Ap. J.*, 147, 490 (1967).
4. Gordon, M. A., and Meeks, M. L., *Ap. J.*, 152, 417 (1968).
5. Townes, C. H., in *Fourth IAU Symp., Manchester, 1955* (edit. by van de Hulst, H. C.) (Cambridge Univ. Press, 1957).

的。据推测，水分子存在于高于正常激发的特殊区域。导致出现水谱线的分子显然要少于对应氨反演能级的分子。因此，如果在 6_{16} 态和 5_{23} 态之间可以达到热平衡，那么在这两个比基态能量高很多的能态上的粒子数柱密度一定会高于氨分子的密度（参考文献 1），或者说在人马座 *B*2 云中约为 10^{17} cm^{-2}。

在猎户座星云中，源的实际微波亮温度至少可以达到几百度；而在 *W*49 中至少可以达到 1,000 度左右。如果这两个源的角直径比所给的上限 3′ 小，或者当它们为光学薄时，实际温度可能还会高出很多。这种强度很高并且宽度极窄的谱线说明可能没有达到热平衡，甚至还可能发生脉泽。进一步研究星际空间中水分子的分布和状态显然是有必要的。如果确实达到了热平衡，那么根据水分子的强微波辐射和存在于射电波段的强氘化水（HDO）跃迁就可以使我们进行一次关于氢－氘丰度比的有趣测量。

在此感谢哈罗德 · 韦弗教授为我们提供了人马座天区羟基云的具体位置并和我们进行讨论，还要感谢保罗 · 罗德为我们提供了观测上的帮助。美国国家航空航天局、美国海军研究办公室和美国国家科学基金会为这项工作提供了一定的支持。

（冯翀 翻译；沈志侠 审稿）

Galactic Nuclei as Collapsed Old Quasars

D. Lynden-Bell

Editor's Note

Shortly after the highly energetic and distant astrophysical objects called quasars were discovered, it was suggested that they might be powered somehow by gravitation, which ultimately is far more efficient at converting mass to energy than nuclear fusion. Here Donald Lynden-Bell develops this idea into what would eventually become the "standard model" of such active galactic nuclei. It is now generally accepted that all galaxies with central bulges (such as our own) contain supermassive black holes at their centres. When lots of gas is being fed into the black hole, it is known variously as a quasar, Seyfert or radio galaxy, depending on the total energy being generated and the wavelength of the peak radiation.

RYLE gives good evidence[1] that quasars evolve into powerful radio sources with two well separated radio components, one on each side of the dead or dying quasar. The energies involved in the total radio outbursts are calculated to be of the order of 10^{61} erg, and the optical variability of some quasars indicates that the outbursts probably originate in a volume no larger than the solar system. Now 10^{61} erg have a mass of 10^{40} g or nearly 10^{7} Suns. If this were to come from the conversion of hydrogen into helium, it can only represent the nuclear binding energy, which is 3/400 of the mass of hydrogen involved. Hence 10^{9} solar masses would be needed within a volume the size of the solar system, which we take to be 10^{15} cm (10 light h). But the gravitational binding energy of 10^{9} solar masses within 10^{15} cm is GM^2/r which is 10^{62} erg. Thus we are wrong to neglect gravity as an equal if not a dominant source of energy. This was suggested by Fowler and Hoyle[2], who at once asked whether the red-shifts can also have a gravitational origin. Greenstein and Schmidt[3], however, earlier showed that this is unlikely because the differential red-shift would wash out the lines. Attempts to avoid this difficulty have looked unconvincing, so I shall adopt the cosmological origin for quasar red-shifts. Even with this hypothesis the numbers of quasar-like objects are very large, or rather they were so in the past. I shall assume that the quasars were common for an initial epoch lasting 10^{9} yr, but that each one only remained bright for 10^{6} yr, and take Sandage's estimate (quoted in ref. 4) of 10^{7} quasar-like objects in the sky down to magnitude 22. This must represent a snapshot of the quasar era, so only one in a thousand would be bright. If these represent all the quasar-like objects that there are, then the density of dead ones should be $10^{7}\times10^{3} = 10^{10}$ per Hubble volume. The distance between neighbouring dead ones is then an average of 10^{-3} Hubble distances (10^{10} light yr) or 3 Mpc. From these statistics it seems probable that a dead quasar-like object inhabits the local group of galaxies and we must expect many nearer than the Virgo cluster and *M*87. If we restrict ourselves to old quasars bright at radio wavelengths, then Sandage reduces his estimate by a factor of

星系核是塌缩的老年类星体

林登 – 贝尔

编者按

在发现被叫作类星体的这种高能而遥远的天体后不久，就有人提出也许是引力在以某种方式为它们提供能量，在把质量转化为能量方面，从根本上说引力要比核聚变有效得多。在这里，唐纳德·林登 – 贝尔将这个想法发展到有可能最终成为这种活动星系核的“标准模型”的形式。现在大家普遍承认，所有中心有核球的星系（比如我们自己的星系）都会在它们的中心处包含超大质量的黑洞。当大量气体被摄入黑洞时，就会因其产生的总能量和辐射峰值波长的不同而被认为是类星体、塞弗特星系或射电星系等活动星系核。

赖尔给出了有力的证据 [1] 证明：类星体会演化为具有两个分得很开的射电子源的强射电源，它们分别位于这个死去或者正在死去的类星体的两侧。计算表明射电爆发的总能量为 10^{61} erg 量级，而一些类星体的光学光变表明这些爆发有可能源于一个不大于太阳系的空间体积中。10^{61} erg 的能量相当于 10^{40} g 或者将近 10^{7} 个太阳的质量。如果这些能量均来源于氢聚变为氦的反应，那么它就只能是原子核的结合能，其大小等于参加反应氢质量的 3/400。因而需要 10^{9} 倍太阳质量的物质处于和太阳系一样大的体积里，我们取太阳系的尺度为 10^{15} cm（10 光时）。但按公式 GM^2/r 计算，在 10^{15} cm 范围内，质量为 10^{9} 倍太阳的天体所具有的引力束缚能应该是 10^{62} erg。因此我们犯了忽视引力的错误——引力即使不是一个主要的能量来源，至少也是一个同等重要的能量来源。这一理论是由福勒和霍伊尔提出的 [2]，他们还曾问过红移会不会也是由引力造成的。然而，在这之前，格林斯坦和施密特 [3] 就曾指出这是不大可能的，因为不同的红移会把谱线抹掉。为避免这个困难所作的尝试看起来并不令人信服，所以我将采用类星体红移的宇宙学起源。即使在上述假定下，类星体的数目还是非常巨大，或者更确切地说它们在过去是这样的。我将假设在最初的一个持续 10^{9} 年的时期里类星体是常见的，但是每个类星体只能在 10^{6} 年内保持明亮，采用桑德奇的估计（见参考文献 4 中的引述），天空中应有 10^{7} 个亮于 22 星等的类似类星体的天体。这应该代表了类星体时期的一个快照，因此只有千分之一会是明亮的。如果这些代表了所有类似于类星体的天体，那么死亡类星体的密度应为每哈勃体积 10^{7} 个 $\times 10^{3} = 10^{10}$ 个。于是两个相邻的死亡类星体之间的平均距离就是 10^{-3} 哈勃距离（10^{10} 光年）或者 3 兆秒差距（Mpc）。从这些统计结果看来，很可能有一个死去的类似类星体的天体存在于本星系群中，并且我们必然可以预期有很多这样的天体距我们比室女座星系团和 *M*87 还近。如果我们仅限于研究那些在射电波段明亮的老年类星体，那么桑德奇的估计值就会减

200 so the average distance between dead quasars is around 20 Mpc. This is typical of the distance between clusters of galaxies.

If some 10^7–10^9 solar masses were involved in the quasar, releasing 10^7 solar masses as energy, then the dead quasar is likely still to be in the range 10^7–10^9 solar masses and to be bound still within a radius of the size of the solar system. Such an object is unlikely to exist for 10^{10} yr without burning out its nuclear fuel. There are no equilibria for burnt-out bodies of masses considerably in excess of a solar mass, however. Even uniform rotation hardly increases Chandrasekhar's critical mass of about 1.4 $M_\odot$ and non-uniform rotation always leads to the generation of magnetic fields and to angular momentum transport. For masses already of the size of the solar system such periods of angular momentum transport will not be very long. In a few thousand years the outer parts acquire a large fraction of the angular momentum and slow down in circular orbit while the more massive inner portion contracts and spins faster. This central portion will collapse and finally fall within its Schwarzschild radius and be lost from view. Nothing can ever pass outwards through the Schwarzschild sphere of radius $r=2\ GM/c^2$, which we shall call the Schwarzschild throat. We would be wrong to conclude that such massive objects in space-time should be unobservable, however. It is my thesis that we have been observing them indirectly for many years.

Effects of Collapsed Masses

As Schwarzschild throats are considerable centres of gravitation, we expect to find matter concentrated toward them. We therefore expect that the throats are to be found at the centres of massive aggregates of stars, and the centres of the nuclei of galaxies are the obvious choice. My first prediction is that when the light from the nucleus of a galaxy is predominantly starlight, the mass-to-light ratio of the nucleus should be anomalously large.

We may expect the collapsed bodies to have a broad spectrum of masses. True dead quasars may have 10^{10} or 10^{11} $M_\odot$ while normal galaxies like ours may have only 10^7 – 10^8 $M_\odot$ down their throats. A simple calculation shows that the last stable circular orbit has a diameter of $12\ GM/c^2 = 12m$ so we shall call the sphere of this diameter the Schwarzschild mouth. Simple calculations on circular orbits yield the following results, where M_7 is the mass of the collapsed body in units of 10^7 $M_\odot$, so that M_7 ranges from 1 to 10^4.

Circular velocity

$$V_c = [GM\ /\ (r - 2m)]^{1/2} \quad \text{where } r > 3m \tag{1}$$

Binding energy of a mass m^* in circular orbit

$$m^*\varepsilon = m^*c^2\ \{1-(r-2m)\ [r\ (r-3m)]^{-1/2}\} \tag{2}$$

少到原来的 1/200，于是死亡类星体之间的平均距离将约为 20 Mpc。这是星系团之间距离的典型值。

如果类星体的质量约为太阳质量的 $10^7 \sim 10^9$ 倍，并且释放了太阳质量的 10^7 倍作为能量，那么死亡类星体的质量很可能仍能达到太阳质量的 10^7 倍 ~ 10^9 倍，并且仍能被束缚于和太阳系同样半径大小的体积内。这样一个天体不大可能存在 10^{10} 年而不耗尽自身的核燃料。然而，一个质量远远超过太阳质量而耗尽了燃料的天体是没有平衡态的。即使均匀转动也很难增加量值约为 1.4 $M_\odot$（译者注：$M_\odot$代表太阳的质量）的钱德拉塞卡临界质量，而非均匀转动通常会导致磁场的产生以及角动量的转移。对于已经具有太阳系尺度的大质量天体，它们的角动量转移时间不会很长。在数千年期间，靠外的部分将获得角动量中的大部分并且减速为沿圆轨道运动；而质量较大的靠内的部分则会收缩并且旋转得更快。其中心部分将会塌缩并最终落到它的史瓦西半径之内而从视线中消失。任何物质都不能向外通过半径 $r=2\ GM/c^2$ 的史瓦西球面，我们在下文中将称之为史瓦西喉。然而，如果我们认为在时空中不可能观测到如此大质量的天体，那就错了。我的论述就是要说明我们已经间接观测到它们很多年了。

质量塌缩效应

因为史瓦西喉被认为是引力中心，所以我们预测将发现物质向它们聚集。于是我们可以预期这些喉将在大量恒星聚集区的中心被发现，而星系核的中心显然是一个不错的选择。我的第一个预测是：当来自一个星系核的光主要是星光时，这个星系核的质光比就会异常大。

我们可以预期那些塌缩后的天体会有很宽的质量范围。真正死亡的类星体可能有 $10^{10}\ M_\odot$或 $10^{11}\ M_\odot$的质量，而像银河系这样的普通星系在它们的喉内的质量可能只有 $10^7\ M_\odot \sim 10^8\ M_\odot$。简单的计算表明，最内面的稳定圆轨道的直径为 $12\ GM/c^2 = 12m$，因而我们将把直径这么大的一个球面称为史瓦西嘴。在对圆轨道进行简单的计算之后可以得到以下结果，如果 M_7 是以 $10^7\ M_\odot$为单位的塌缩天体的质量，则 M_7 的取值范围是 $1 \sim 10^4$。

圆周速度为：

$$V_c = [GM / (r - 2m)]^{1/2}，\text{其中 } r > 3m \tag{1}$$

圆轨道中质量为 m^* 的质点的束缚能为：

$$m^*\varepsilon = m^*c^2\ \{1 - (r - 2m)\ [r\ (r - 3m)]^{-1/2}\} \tag{2}$$

Angular momentum of circular orbit per unit mass

$$h = [mc^2r^2 \;/\; (r - 3m)]^{1/2} \tag{3}$$

The maximum binding energy in circular orbit is

$$m^*c^2\,(1 - 2\sqrt{2}\,/3) = 0.057\ m^*c^2 \sim m^*c^2/18 \tag{4}$$

which occurs at $r = 6m$ and $h = \sqrt{12}\,mc$ (equation (5)). This orbit is also the circular orbit of least angular momentum. The period of the circular orbit as seen from infinity is $(2\pi r/V_c)(1-[2m/r])^{-1/2} = 2\pi\ (r^3/GM)^{1/2}$ (equation (6)). The maximum wavelength change toward the blue visible from infinity is $\lambda_0/\lambda = 2^{1/2}$ for a stable circular orbit while for a stable parabolic orbit it is $\lambda_0/\lambda = 1+2^{-1/2}$. For parabolic orbits that will disappear down the Schwarzschild throat the greatest blueward change is seen from $r = 27m/8$ when $\lambda_0/\lambda = 1.77$. There is no possibility of synchrotron-like blue-shifts.

Numerical values are: $V_c = 200\ (M_7/r)^{1/2}$ km s^{-1} at r pc. $12m$ diameter of Schwarzschild mouth $= 1.6\times10^{13}\ M_7$ cm $\simeq M_7$, AU. Roche limit for a star of density ρ^* g cm^{-3} $(R) = 4.1\times10^{13}\ (M_7/\rho^*)^{1/3}$ cm $= 2.8\ (M_7/\rho^*)^{1/3}$ AU. Greatest swallowable angular momentum per unit mass $h_0=4mc=0.5\ M_7$ pc km s^{-1}. Once the initial very low angular momentum stars have been swallowed (those with $h < h_0$) the star swallowing rate rapidly declines to a negligible trickle of about 10^{-8} stars yr^{-1}. Because the Roche limit is near the Schwarzschild mouth we may likewise neglect the tearing apart of stars by tides. Spectroscopic velocity dispersion of the same energy as the circular orbit $\sigma^2 = \frac{1}{3}\,V_c^2$, that is, $\sigma=116\ (M_7/r)^{1/2}$ km s^{-1}. Period of circular orbit at r pc $= 3\times10^4\ (r^3/M_7)^{1/2}$ yr.

It is thus by no means impossible that there are collapsed masses in galactic nuclei. Considerable support for the notion comes from a detailed consideration of what happens when a cloud of gas collects in the galactic nucleus. (This was first considered by Salpeter, who also derived the $0.057\ F\,c^2$ power output[5] which is now described.)

Gas Swallowing

The total mass loss from all stars in a galaxy will be roughly 1 $M_\odot$ per year. A fraction of this accumulates in galactic nuclei, which are the centres of gravitational attraction. There is dissipation when gas clouds collide, due to shock waves that radiate the energy of collision. For a given angular momentum the orbit of least energy is circular, so we must expect gas to form a flat disk held out from the centre by circular motion. Such a differentially rotating system will evolve due to "friction" just as described earlier for a dying quasar. Nothing happens in the absence of friction so the energy is liberated via the friction. In the cosmic situation molecular viscosity is negligible and it is most probable that magnetic transport of angular momentum dominates over turbulent transport just as it does in Alfvén's theory of the primaeval solar nebula. To give a sensible model of the swallowing rate we must estimate the magnetic friction and investigate what happens to

圆轨道中单位质量的角动量为：

$$h = [mc^2r^2 / (r - 3m)]^{1/2} \tag{3}$$

圆轨道的最大束缚能为：

$$m^*c^2 (1 - 2\sqrt{2}/3) = 0.057\ m^*c^2 \sim m^*c^2/18 \tag{4}$$

这出现于 $r = 6m$ 和 $h=\sqrt{12}\,mc$（式(5)）的情况下。这个轨道也是角动量最小的圆轨道。从无穷远处观察到这个圆轨道的周期是 $(2\pi r/V_c)(1-[2m/r])^{-1/2} = 2\pi\ (r^3/GM)^{1/2}$（式（6））。从无穷远处可看到的最大的波长蓝移，对于稳定的圆轨道为 $\lambda_0/\lambda = 2^{1/2}$，对于稳定的抛物线轨道为 $\lambda_0/\lambda = 1+2^{-1/2}$。对于在史瓦西喉以内将会消失的抛物线轨道，最大的向蓝改变是在 $r = 27m/8$ 处观察到的 $\lambda_0/\lambda = 1.77$。类似于同步辐射的蓝移是不可能出现的。

具体的数值是：在 r 秒差距（pc）处，$V_c = 200\ (M_7/r)^{1/2}$ km/s。史瓦西嘴的直径 $12m = 1.6 \times 10^{13}\ M_7$ cm $\approx M_7$ 个天文单位（AU）。对于密度为 ρ^* g/cm^3 的恒星，洛希极限（R）$= 4.1 \times 10^{13}\ (M_7/\rho^*)^{1/3}$ cm $= 2.8\ (M_7/\rho^*)^{1/3}$ AU。每单位质量的最大可吞噬角动量 $h_0=4mc=0.5\ M_7$ pc km s^{-1}。一旦初始角动量非常低的恒星（即 $h<h_0$ 的恒星）被吞噬了以后，恒星吞噬率就会迅速下降到一个非常低的值，只有约 10^{-8} 恒星 / 年。因为洛希极限靠近史瓦西嘴，所以我们同样可以忽略由潮汐产生的恒星撕裂。和圆轨道有相同能量的光谱速度弥散度 $\sigma^2 = \frac{1}{3} V_c^2$，也就是 $\sigma = 116\ (M_7/r)^{1/2}$ km/s。位于 r pc 处的圆轨道的周期为 $3 \times 10^4\ (r^3/M_7)^{1/2}$ 年。

因此，在星系核中存在塌缩了的质量并非不可能。详细地考虑当一团气体云在星系核中聚集时将发生什么会有力地支持这一说法。(上述说法由萨尔皮特最先提出，他也推导出了 $0.057\ F c^2$ 的功率输出[5]，这正是现在所描述的结果。)

气体吞噬

一个星系中，所有恒星的总质量损失大约是 1 $M_\odot$/ 年。其中的一部分质量积聚在引力中心的星系核中。在气体云碰撞时，由于激波辐射了碰撞的能量，因此会产生耗散。对于给定的角动量，最小能量的轨道是圆形轨道，因此我们只能期望气体将形成一个扁平的盘，靠圆周运动来维持与中心的距离。这样一个较差自转系统会由于“摩擦”而不断演化，正如人们早期对一个正在死去的类星体所描述的那样。因为在没有摩擦的情况下什么都不会发生，所以能量是通过摩擦释放的。在宇宙环境中，分子的黏性是可以忽略的，而角动量的磁输运极有可能比湍动输运更起支配作用，这恰与在阿耳文的原初太阳星云理论中所遇到的情况类似。为了给出一个合理的吞噬率模型，我们必须估计磁摩擦并研究磁场所获得的能量会发生什么样的变

the energy acquired by the magnetic field. Before doing this let us assume a conservative swallowing rate of $10^{-3}\ M_{\odot}$ per year and work out the power available through magnetic friction. We assume that this mass flux is processed down through the circular orbits until it reaches the unstable orbit at $r = 6m$. We shall assume that the flux is swallowed by the throat without further energy loss. The power for a mass flow F is then $0.057\ F c^2 = 3.5\times10^{42}$ erg s^{-1} $\simeq 10^9\ L_{\odot}$ where the values are for $F = 10^{-3}\ M_{\odot}$ yr^{-1} and $L_{\odot}$ is the power of the energy output of the Sun. This sort of power emitted as light could just noticeably brighten a nucleus. If a fraction were emitted in the radio region the nucleus would be a radio source. Clearly it only requires a mass flux of $1\ M_{\odot}$ yr^{-1} and a conversion into light at 10 percent efficiency for the nucleus to equal the stellar light output from the whole galaxy. Can this be the explanation of the Seyfert galaxies?

We now return to the model for the magnetic transfer of angular momentum in a disk. As a magnetic field B is sheared by an initially perpendicular displacement. The component across the shear is left unchanged but the component down the shear is progressively amplified. We may therefore expect the magnetic field in a shearing medium to be progressively amplified until it somehow changes either the motion or its own configuration. The magnetic field first has a significant effect when it can bow upwards and downwards out of the differentially rotating disk leaving the material to flow down the field lines and so collect into clouds in the disk[6]. For significant cloudiness to result, the magnetic field pressure $B^2/8\pi$ must equal the turbulent and gas pressures ρc_s^2. Here ρ is the density and c_s^2 the combined velocity dispersion of microscopic and molecular motion. The thickness of the disk $2b$ is determined by the balance between the gravity of the central mass and the non-magnetic pressure, $GMr^{-3}z\rho = -c_s^2\, \partial\rho/\partial z$, which gives

$$\rho = \rho_0 \exp\left[-z^2/2b^2\right] \quad \text{where } b^2 = r^3 c_s^2/GM \tag{7}$$

The local shear rate is given by Oort's constant A

$$A = -\tfrac{1}{2} r\, \mathrm{d}(v/r)/\mathrm{d}r = \tfrac{3}{4}(GM/r^3)^{1/2} \tag{8}$$

Hence
$$\frac{B^2}{8\pi} \simeq \rho_0 c_s^2 = \frac{16}{9} A^2 b^2 \rho_0 = \left(\frac{16}{9\sqrt{2\pi}}\right) A^2 b \Sigma \tag{9}$$

Here Σ is the surface density of matter in the disk, $\Sigma = \sqrt{2\pi}\, b\, \rho_0$.

On multiplication by b^3, equation (9) tells us that the magnetic field energy in a volume b^3 is about equal to the kinetic energy of the shearing in the same region. Equation (9) may be obtained approximately from a dimensional analysis of a general shearing sheet of highly conducting material. As such, it claims general validity, so it is interesting to apply it to the gas in the neighbourhood of the Sun. The very reasonable result, $B = 4\times10^{-6}$ G, restores confidence in this rather inadequate treatment. We are now in a position to estimate the magnetic frictional force per unit length in a shearing disk. The Maxwell stress $B_x . B_y/(4\pi)$ acts over a thickness of about $2b$. Once the medium has broken up into clouds the magnetic field will be rather chaotic but the shearing will still give it some

化。在这之前，让我们先假设一个保守的吞噬率——$10^{-3}\ M_\odot$ / 年，并计算出通过磁摩擦可以产生的功率。我们假设这个质量流量沿着圆轨道不断流入，直到它到达 $r = 6m$ 处的不稳定轨道。我们假定这个质量流到史瓦西喉时被吞噬而没有进一步的能量损失。于是质量流量 F 产生的功率为 $0.057Fc^2 = 3.5 \times 10^{42}$ erg $s^{-1} \approx 10^9\ L_\odot$，其中：$F = 10^{-3}\ M_\odot$ / 年，$L_\odot$是太阳能量输出的功率。这种能量如以光的形式发射出来则刚好能使星系核显著增亮。如果有一部分能量是在射电波段发出的，那么这个星系核就会成为一个射电源。显然，为了使星系核辐射的光和整个星系的星光同样明亮，只需要质量流量达到 $1\ M_\odot$/年且转化为光的效率为 10% 就足够了。这是否可以作为对塞弗特星系的解释呢？

现在我们回到一个盘内角动量的磁转移模型中去。当磁场 B 被一个起初与之垂直的位移剪切时，垂直于剪切方向的分量保持不变而沿剪切方向的分量则逐渐放大。因此我们可以预期一个处于剪切介质中的磁场会被逐渐放大，直到它在某种程度上改变这种运动或者它自身的构型。在磁场可以向上和向下弯曲离开较差转动着的盘时，它会首先产生一个重要的效应使物质沿磁力线流动，从而在盘中聚集成云[6]。为产生大量的云，磁场压强 $B^2/8\pi$ 必须等于湍流和气体压强 ρc_s^2。这里 ρ 是密度，c_s^2 是微观运动和分子运动联合的速度弥散度。盘的厚度 $2b$ 由中心质量的引力和非磁场压强之间的平衡 $GMr^{-3}z\rho = -c_s^2\ \partial\rho/\partial z$ 确定，由此得到：

$$\rho = \rho_0 \exp[-z^2/2b^2]，其中\ b^2 = r^3 c_s^2/GM \tag{7}$$

局域剪切率由奥尔特常数 A 给出：

$$A = -\tfrac{1}{2} r\, \mathrm{d}(v/r)/\mathrm{d}r = \tfrac{3}{4}(GM/r^3)^{1/2} \tag{8}$$

因此
$$\frac{B^2}{8\pi} \approx \rho_0 c_s^2 = \frac{16}{9} A^2 b^2 \rho_0 = \left(\frac{16}{9\sqrt{2\pi}}\right) A^2 b \Sigma \tag{9}$$

这里 Σ 是盘中物质的面密度，$\Sigma = \sqrt{2\pi}\, b\, \rho_0$。

乘以 b^3 后，由式 (9) 可知，在体积 b^3 内的磁场能量差不多等于同一区域中的剪切动能。从对高导电率材料的一般剪切层的量纲分析中就可以近似得到式 (9)。就此而论，式 (9) 应该具有普遍的适用性，所以把它应用到太阳附近的气体中是很有意思的。由此得到了一个非常合理的结果：$B = 4 \times 10^{-6}$ 高斯 (G)，这使我们在这个不严格的处理中找回了信心。现在我们就可以估计一个剪切盘中每单位长度的磁摩擦了。麦克斯韦应力 $B_x . B_y/(4\pi)$ 作用于大约 $2b$ 的厚度上。一旦介质碎裂为云块，磁场就会变得非常混乱，但剪切仍会赋予介质某种系统性的倾向以反抗剪切。对于给

systematic tendency to oppose the shear. The greatest possible value of B_xB_y for given [**B**] is $B^2/2$, so we estimate $B^2/4$ as a typical value in a sense directed to oppose the shear. The force per unit length is then

$$bB^2/8\pi = [16/(9\sqrt{2\pi})]A^2b^2\Sigma \tag{10}$$

Unlike a normal viscous drag, the force here depends on the square of A, the rate of shear. This is reasonable because the shear itself is needed to build up the magnetic field which eventually opposes it. We shall use this estimate of the friction to make a model of the disk, but it is important to consider first where the energy goes. We must consider why the field is not further amplified by further shearing once it has reached the value estimated. Once the medium has split into clouds each cloud acts like a magnet. As the medium is sheared these magnets try to re-align themselves to take up a configuration of minimum energy. The intercloud medium, although dominated by the magnetic field, is still a highly conducting medium, however. In free space re-alignment of the magnets involves reconnexion of magnetic field lines through neutral points. This cannot happen in a force-free, strictly perfect, conductor. Rather one may show that neutral sheets develop with large sheet currents flowing through them. This situation does not really happen; any small resistivity causes a return to the reconnecting case and it is of the greatest interest to see how a plasma of particles behaves near a neutral point at which reconnexion takes place. (In the frame in which the neutral point is fixed this involves an electric field E. The mechanism of acceleration is basically that of Syrovatskii[7].) Tritton at this observatory has been studying with me the details of an exact model. To reconnect a finite flux in a finite time through a point at which $B = 0$, it is clear that the lines of force much move infinitely fast at the neutral point. They are not material lines and there is nothing wrong with this; even the normal formula for their velocity $v = c\,(\mathbf{E}\times\mathbf{B}/B^2)$ gives superluminous velocities when $|\mathbf{E}\times\mathbf{B}| > B^2$ (that is, for $|E| > |B|$ in gaussian units assuming they are perpendicular). The particles gyrate about their field lines until the lines move with the velocity of light; thereafter the electric field predominates, the magnetic field is too weak to change anything significantly and the particles are electrostatically accelerated. The potential drop follows directly from Faraday's law

$$\Phi = -\frac{1}{c}\,\mathrm{d}N/\mathrm{d}t$$

where $\mathrm{d}N/\mathrm{d}t$ is the rate of flux reconnexion. To calculate this e.m.f. we estimate the reconnexion rate as a flux of $2b^2B$ in a time of $(2A)^{-1}$. This gives an e.m.f. of $4b^2AB/c$ (equation (11)). We shall see that these e.m.f.s are of the order of 10^{12} V. Because I believe that this can be the primary source of dissipation for the highly conducting disk I deduce that the energy of dissipation can be converted directly into cosmic rays in the GeV range. We have now good reason to believe that galactic nuclei ought to be radio sources because such cosmic rays will clearly radiate by the synchrotron mechanism in the magnetic field. In summary, we expect from a shear $2A$ the generation of a magnetic field given by equation (2), a shearing force given by equation (3), and a power p per unit area dissipated into cosmic rays of energy up to $4b^2AB/c$ given by $p = 2AbB^2/8\pi = (32/(9\sqrt{2\pi}))\,A^3b^2\Sigma$

定的 [**B**]，B_xB_y 的最大可能值是 $B^2/2$，所以我们估计直接反抗剪切的典型值应该是 $B^2/4$。于是，单位长度上的力为：

$$bB^2/8\pi = [16/(9\sqrt{2\pi})]A^2b^2\sum \tag{10}$$

和普通的黏滞阻力不同，这里的力取决于剪切率 A 的平方。这是合理的，因为磁场的建立需要剪切，而建立后的磁场最终又会反抗剪切。我们将使用上面这种对摩擦的估计来构造一个盘模型，但重要的是首先要考虑这些能量的去向。我们必须思考以下问题，即为什么一旦达到了所估计的值，磁场便不再会被进一步的剪切继续放大？一旦介质分裂为云块，每个云块就会表现得像一块磁体。当介质受到剪切时，这些磁体将会试图重新排列以便占据一个能量最小的位形。对于云际介质，虽然它们被磁场所支配，但仍然是高导电率的介质。在自由空间中，磁体的重新排列会导致磁力线通过中性点的重联。这种情况不可能在无阻力的严格理想导体中发生。虽然可以证明中性片会在流经它们的大的片电流中产生，但这种情况并不会真的发生；因为任何小的电阻率都会导致再回到重联的情况，而了解由粒子组成的等离子体在发生重联的中性点附近有什么样的行为是极其有趣的。（在中性点固定的参考系里，可以产生一个电场 E。加速机制基本上就是瑟罗瓦茨基提出的机制[7]。）皇家格林尼治天文台的特里顿一直在和我一起研究一个精确模型的具体细节。为了在一个有限的时间内通过 $B = 0$ 的点使有限的磁通重联，显然需要磁力线在中性点处以无限快的速度移动。磁力线不是物质的线，因而以无限快的速度移动并没有什么不妥；甚至表征磁力线速度的标准公式 $v = c\,(\mathbf{E}\times\mathbf{B}/B^2)$ 也会在 $|\,\mathbf{E}\times\mathbf{B}\,| > B^2$（即，在高斯单位制中满足 $|E| > |B|$ 时，假设两者是垂直的）时给出视超光速的速度。粒子会一直绕着磁力线旋转直到磁力线的运动速度达到光速；随后电场开始起主导作用，而磁场变得很弱以至于不能使任何物理量发生显著的变化，这时粒子是通过静电场加速的。电位降可以直接由法拉第定律给出：

$$\Phi = -\frac{1}{c}\,\mathrm{d}N/\mathrm{d}t$$

其中 $\mathrm{d}N/\mathrm{d}t$ 是磁通的重联速率。为了计算这个电动势，我们可以根据在时间 $(2A)^{-1}$ 内重联的磁通为 $2b^2B$ 估算出重联速率。由此得到的电动势为 $4b^2AB/c$（式（11））。我们将会看到，这些电动势的量级为 10^{12} V。因为我认为这可能是高导电率盘能量的主要耗散源，所以推断耗散能量能够直接被转化为 GeV 量级的宇宙线。现在我们有充足的理由相信星系核应该是射电源，因为这些宇宙线显然将通过在磁场中的同步辐射机制进行辐射。综上所述，我们期望从剪切 $2A$ 将可以产生由式（2）给出的磁场，由式（3）给出的剪切力，和由 $p = 2AbB^2/8\pi = (32/(9\sqrt{2\pi}))\,A^3b^2\sum$（式（12））给出的每单位面积通过能量高达 $4b^2AB/c$ 的宇宙线耗散掉的功率 p。这些关系式在银

(equation (12)). These formulae have very general application in the astrophysics of the Galaxy, in the early history of the solar nebula and in the origin of peculiar A stars from binaries. There is, however, one caveat: the reconnexion energy only goes into cosmic rays if the density is low enough in the reconnecting region. If the particle being accelerated has a collision before it reaches the r.m.s. velocity, then it cannot run away to high energy; instead the reconnexion energy is dissipated by ohmic heating.

If m_A, m_p are the masses of the accelerated particle and the proton, then the condition for acceleration is

$$e\, m_A^{-1} E\, t_s > \left(\frac{3}{2}kTm_p^{-1}\right)^{1/2}$$

where t_s is the time between scatterings of the particles A. This time may be taken from Spitzer's book[8] to be approximately

$$t_s = m_A\, \rho^{-1}\, T^{3/2}\, m_A/(m_A+m_P) \text{ seconds}$$

T is the temperature and ρ is the density at the acceleration region. Putting in our expressions for the electric field $\Phi/2b$ we obtain for our disk in c.g.s. units the acceleration condition

$$\rho < 10^{-24}\left(\frac{2m_A}{m_A + m_p}\right) bABT$$

Notice the density for proton acceleration can be 918 times that for electron acceleration.

The Steady Model Disk

We look for a steady state of gas swallowing with a mass flux F. We shall assume that a small fraction of the rotational energy is converted into random motions so that $c_s \propto V_c$. Sensible values would be c_s = 10 km s^{-1} when V_c = 200 km s^{-1} so we write $c_s = (1/20)\, x\, V_c$, where x is of order unity but might be a weak function of r. The couple on the material inside r due to magnetic friction is from equations (10), (8) and (7)

$$g = 2\pi r^2 bB^2/8\pi = \sqrt{2}\,\pi\, c_s^2 r^2 \Sigma \tag{13}$$

Apart from the trickle of angular momentum $\sqrt{12}\, mcF$ into the singularity g must be balanced in a steady state by the inward flux of angular momentum carried by the material. Thus, using equations (3) and (5)

$$g = \sqrt{12}\, mcF + F\,[mc^2r^2(r-3m)^{-1}]^{1/2} \quad \text{for } r > 6m \tag{14}$$

that is

$$g \simeq F\,(GM\,r)^{1/2} \tag{15}$$

河系天体物理学中、在太阳星云的早期历史中以及在双星系统内特殊 A 型星的起源问题中都有着非常广泛的应用。但有一点需要说明：如果重联区内的密度低到一定程度，那么重联能量就只会以宇宙线的形式释放。如果被加速的粒子在达到均方根速度之前发生了一次碰撞，那么它就不能以很高的能量逃逸出去；这时，重联能量会通过欧姆加热的方式被耗散掉。

如果 m_A、m_p 分别表示被加速粒子和质子的质量，那么加速的条件是：

$$e\, m_A^{-1} E\, t_s > \left(\frac{3}{2}kTm_p^{-1}\right)^{1/2}$$

其中，t_s 是 A 粒子两次散射之间的时间。从斯皮策的著作 [8] 中可以了解到这个时间间隔大致为：

$$t_s = m_A\, \rho^{-1}\, T^{3/2}\, m_A/(m_A+m_P)\ \mathrm{s}$$

T 是温度，ρ 是加速区的密度。代入电场的表达式 $\Phi/2b$，则可以得到在我们的盘模型中，以厘米克秒单位表示的加速条件是：

$$\rho < 10^{-24}\left(\frac{2m_A}{m_A+m_p}\right) bABT$$

注意质子可加速的密度能达到电子可加速密度的 918 倍。

稳态模型盘

我们要寻求一个质量流量为 F 的气体吞噬的稳态。我们将假设转动能量中有一小部分转化成为随机运动，因而 $c_s \propto V_c$。当 $V_c = 200$ km/s 时，c_s 的合理取值为 $c_s = 10$ km/s，所以我们可以写出以下关系式：$c_s = (1/20)\ x\ V_c$，其中 x 的量级为 1，但有可能是微弱依赖于 r 的函数。由磁摩擦造成的 r 之内物质上的力偶可以从式（10)、(8）和（7）得到：

$$g = 2\pi r^2 bB^2/8\pi = \sqrt{2}\,\pi\, c_s^2 r^2 \Sigma \tag{13}$$

除了缓慢进入奇点的角动量流 $\sqrt{12}\,mcF$ 以外，在稳态中必须有被物质携带的角动量的内流来平衡 g。因此，根据式（3）和式（5）可得：

$$g = \sqrt{12}\ mcF + F\,[mc^2r^2(r-3m)^{-1}]^{1/2}，其中\ r > 6m \tag{14}$$

即

$$g \approx F\,(GM\,r)^{1/2} \tag{15}$$

Because $g \propto r^{1/2}$ we see from equation (13) that $\sum \propto r^{-3/2} c_s^{-2} \propto r^{-3/2} V_c^{-2} \propto r^{-1/2}$ and that the radial velocity

$$V_r = F\,(2\pi r \textstyle\sum)^{-1} \tag{16}$$

The power $2\pi r p(r)\,\delta r$ liberated into heat or cosmic rays in the region between r and $r + \delta r$ is $-F\mathrm{d}\varepsilon/\mathrm{d}r$ where $-\mathrm{d}\varepsilon/\mathrm{d}r$ is the derivative of the binding energy in circular orbit given by equation (2).

$$p(r) = \frac{Fc^2}{4\pi r}\frac{m\,(r - 6m)}{\left[r(r - 3m)\right]^{3/2}} \quad \text{for } r \geq 6m \tag{17}$$

$$p(r) \simeq FGM/(4\pi r^3) \qquad \text{for } r \gg m \tag{18}$$

For a given total mass and radius of the disk, a given Schwarzschild mass and choice of the parameter $x \simeq 1$, we have now a unique model of the disk including the cosmic ray power, the heating per unit area, the magnetic field and the mass flux. Rather than given the total mass and radius of the disk, we shall determine everything in terms of the mass flux. Any chosen maximum radius then determines the total mass within. We shall calculate two temperatures T_1 and T_2 as follows. T_1 is the temperature that the disk would have if it radiated as a black body just the power per unit area that is locally generated. Thus $T_1(r) = (P(r)/(2\sigma))^{1/4}$ where σ is Stefan's constant and the factor 2 arises because the disk has two sides. $T_2(r)$ is the ambient black body temperature at a distance r from a source the total power of which is the total power of one disk. Thus

$$T_2(r) = [P/(16\pi\sigma r^2)]^{1/4}$$

We now give the numerical values in terms of the following variables: F_{-3} the mass flux in units of $10^{-3}\,M_\odot$ per year, M_7 the Schwarzschild mass in units of $10^7\,M_\odot$, x the ratio of the turbulent velocity to $1/20\,V_c$. r_0, m_0 the running variable r in units of 1 pc and GM/c^2 respectively. From equation (15) couple $g = 4\times10^{48}\,r_0^{1/2}F_{-3}\,M_7^{1/2} = 2.6\times10^{45}\,m_0^{1/2}\,F_{-3}M_7$ g cm^2 s^{-1} ($m_0 \gg 6$). From equation (13) surface density $\sum = 0.16\,r_0^{-1/2}x^{-2}\,F_{-3}M_7^{-1/2} = 240\,m_0^{-1/2}\,x^{-2}\,F_{-3}M_7^{-1}$ g cm^{-2} ($m_0 \gg 6$). $\int_0^r 2\pi r \sum \mathrm{d}r = 6.7\times10^{36}\,r_0^{3/2}\,x^{-2}\,F_{-3}M_7^{-1/2}$ g $= 3.4\times10^3\,r_0^{3/2}$ solar masses. From equation (7) $b/r = 0.05x$. $\rho_0 = (2\pi)^{-1/2}\sum/b = 4.6\times10^{-19}\,r_0^{-3/2}x^{-3}F_{-3}M_7^{-1/2} = 1.4\times10^{-9}\,m_0^{-3/2}x^{-3}F_{-3}M_7^{-2}$ g cm^{-3} ($m_0 \gg 6$). From equation (13), $B = 3\times10^{-3}\,r_0^{-5/4}\,x^{-1/2}\,F_{-3}^{1/2}M_7^{1/4} = 2.7\times10^5\,m_0^{-5/4}\,x^{-1/2}\,F_{-3}^{1/2}M_7^{-1}$ G ($m_0 \gg 6$). From equation (18), $p(r) \simeq 0.22 r_0^{-3}F_{-3}M_7$ ergs cm^{-2} s^{-1}($r \gg 6m$). Notice that the power is strongly concentrated towards the centre, where we need the accurate relativistic formula (17)

$$p(r) = 0.22\,r_0^{-3}\left[\frac{1-2.6\times10^{-6}M_7 r_0^{-1}}{(1-1.3\times10^{-6}M_7 r_0^{-1})^{3/2}}\right]F_{-3}M_7$$

$$= 2.6\times10^{18}\left[\frac{1-6\,m_0^{-1}}{(1-3\,m_0^{-1})^{3/2}}\right]m_0^{-3}F_{-3}\,M_7^{-2}\ \text{erg cm}^{-2}\ \text{s}^{-1}$$

因为 $g \propto r^{1/2}$，所以我们从式（13）中可以得出 $\sum \propto r^{-3/2}c_s^{-2} \propto r^{-3/2} V_c^{-2} \propto r^{-1/2}$ 以及径向速度：

$$V_r = F(2\pi r \textstyle\sum)^{-1} \tag{16}$$

在 r 和 $r+\delta r$ 之间的区域，以热或宇宙线形式释放掉的功率 $2\pi rp(r)\delta r$ 为 $-F\mathrm{d}\varepsilon/\mathrm{d}r$，其中 $-\mathrm{d}\varepsilon/\mathrm{d}r$ 是由公式（2）给出的圆轨道束缚能的导数。

$$p(r) = \frac{Fc^2}{4\pi r}\frac{m(r-6m)}{\left[r(r-3m)\right]^{3/2}} \quad \text{其中 } r \geq 6m \tag{17}$$

$$p(r) \approx FGM/(4\pi r^3) \quad \text{其中 } r \gg m \tag{18}$$

对于一个给定的总质量和盘半径，在给定史瓦西质量并选定参数 $x \approx 1$ 的情况下，我们就会得到一个唯一的盘模型，其中包含宇宙线的功率、单位面积的加热、磁场以及质量流量。在没有给定总质量和盘半径的情况下，我们也可以根据质量流量得出所有其他的物理量。只要选定了最大半径就可以确定其内的总质量。下面我们将计算两个温度 T_1 和 T_2。T_1 是如果盘以黑体辐射释放出与局域单位面积产生的功率刚好相等时应该有的温度。于是 $T_1(r) = (P(r)/(2\sigma))^{1/4}$，其中 σ 是斯忒藩常数，出现因子 2 是因为盘有两个面。$T_2(r)$ 是总功率为一个盘的总功率而与源相距为 r 的周围黑体温度。所以：

$$T_2(r) = [P/(16\pi\sigma r^2)]^{1/4}$$

现在我们根据以下变量计算具体的数值：F_{-3} 是单位为 $10^{-3}\ M_\odot$/ 年的质量流量；史瓦西质量 M_7 的单位是 $10^7\ M_\odot$；x 是湍流速度与 1/20 V_c 的比值；r_0 和 m_0 分别是以 1 pc 和 GM/c^2 为单位的移动变量 r。由式（15）可以得到：力偶 $g = 4\times10^{48}\ r_0^{1/2}\ F_{-3}\ M_7^{1/2} = 2.6\times10^{45}\ m_0^{1/2}\ F_{-3}M_7$ g cm^2 s^{-1} $(m_0 \gg 6)$。由式（13）可以得到面密度 $\sum = 0.16\ r_0^{-1/2}x^{-2}\ F_{-3}M_7^{-1/2} = 240\ m_0^{-1/2}\ x^{-2}\ F_{-3}M_7^{-1}$ g cm^{-2} $(m_0 \gg 6)$，$\int_0^r 2\pi r\sum \mathrm{d}r = 6.7\times10^{36}\ r_0^{3/2}\ x^{-2}\ F_{-3}\ M_7^{-1/2}$ g $= 3.4\times10^3\ r_0^{3/2}$ 太阳质量。由式（7）得到：$b/r = 0.05x$; $\rho_0 = (2\pi)^{-1/2}\sum/b = 4.6\times10^{-19}\ r_0^{-3/2}x^{-3}F_{-3}M_7^{-1/2} = 1.4\times10^{-9}\ m_0^{-3/2}x^{-3}F_{-3}M_7^{-2}$ g cm^{-3} $(m_0 \gg 6)$。由式（13），$B = 3\times10^{-3}\ r_0^{-5/4}\ x^{-1/2}\ F_{-3}^{1/2}M_7^{1/4} = 2.7\times10^5\ m_0^{-5/4}\ x^{-1/2}\ F_{-3}^{1/2}M_7^{-1}$ G $(m_0 \gg 6)$。根据式（18），$p(r) \approx 0.22r_0^{-3}F_{-3}\ M_7$ erg cm^{-2} s^{-1} $(r \gg 6m)$。可以看出：功率具有向中心区高度聚集的趋势，在那里我们需要使用精确的相对论公式（17）：

$$p(r) = 0.22\ r_0^{-3}\left[\frac{1-2.6\times10^{-6}M_7r_0^{-1}}{(1-1.3\times10^{-6}M_7r_0^{-1})^{3/2}}\right]F_{-3}M_7$$

$$= 2.6\times10^{18}\left[\frac{1-6\ m_0^{-1}}{(1-3\ m_0^{-1})^{3/2}}\right]m_0^{-3}F_{-3}\ M_7^{-2}\ \text{erg cm}^{-2}\ \text{s}^{-1}$$

Total power $P = 0.057\, F c^2 = 3.2\times10^{42}\, F_{-3}$ erg s^{-1} ~ $10^9\, L_\odot$

$$T_1(r) = 6.7\, r_0^{-3/4}\left[\frac{1-2.6\times10^{-6}M_7 r_0^{-1}}{(1-1.3\times10^{-6}M_7 r_0^{-1})^{3/2}}\right]^{1/4} F_{-3}^{1/4}M_7^{1/4}$$

$$= 3.7\times10^5\, m_0^{-3/4}\left[\frac{1-6\, m_0^{-1}}{(1-3\, m_0^{-1})^{3/2}}\right]^{1/4} F_{-3}^{1/4}\, M_7^{-1/2}\ \text{K} \qquad (m_0\geq6)$$

$$T_2(r) = 100\, r_0^{-1/2}F_{-3}^{1/4} = 1.6\times10^5\, m_0^{-1/2}F_{-3}^{1/4}M_7^{-1/2}\ \text{K}$$

Maximum cosmic ray energy $e\Phi = 1.5\times10^{13}\, r_0^{-3/4}x^{3/2}F_{-3}^{1/2}M_7^{3/4} = 10^{18}\, m_0^{-3/4}\, x^{3/2}\, F_{-3}^{1/2}$ eV. Period of circular orbit (seen from infinity) $3\times10^4\, r_0^{3/2}\, M_7^{-1/2}$ yr $= 9.8\times10^{-6}\, m_0^{3/2}\, M_7$ yr. Circular velocity $V_c = 200\, r_0^{-1/2}\,(1-9.7\times10^{-7}\, M_7\, r_0^{-1})^{-1/2} = 3\times10^5\, m_0^{-1/2}\,(1-2\, m_0^{-1})^{-1/2}$ km s^{-1} ($m_0>6$). Radial velocity $V_r = 0.2\, r_0^{-1/2}\, x^2\, M_7^{1/2} = 3\times10^2\, m_0^{-1/2}\, x^2$ km s^{-1} ($m_0\gg6$). Condition for electron acceleration is $\rho < 3\times10^{-20}\, T_4\, r_0^{-7/4}\, x^{1/2}F_{-3}^{1/2}M_7^{3/4}$ and for proton acceleration $\rho < 3\times10^{-17}\, T_4$, etc., where T_4 is the temperature in the acceleration region in units of 10^4 K and ρ is the density. Acceleration will be between the clouds so at any region of the disk we should probably take $\rho_0/10$ for ρ (and T_4 ~ 1 unless the ambient temperature T_1 is greater.)

Diameter of Schwarzschild mouth $12m = 5.7\times10^{-6}\, M_7$ pc, that is, $m_0 = 12$. Diameter of region producing half the power $44m = 2.1\times10^{-5}\, M_7$ pc, that is, $m_0 = 44$. Rotational period at that radius $9\times10^{-4}\, M_7$ yr $= 0.32\, M_7$ days. Inward movement time r/V_r at that radius $1.6\times10^{-1}\, M_7\, x^{-2}$ yr ~ $59\, M_7$ days.

It should by now be clear that with different values of the parameters M_7 and F_{-3} these disks are capable of providing an explanation for a large fraction of the incredible phenomena of high energy astrophysics, including galactic nuclei, Seyfert galaxies, quasars and cosmic rays. The next section is therefore devoted to predicting the spectra.

Spectrum

The maximum temperature is at $m_0 = 7.05$ and is $T_1 = 6.6\times10^4\, F_{-3}^{1/4}\, M_7^{-1/2}$ K. The medium will be optically thick for $\Sigma = 90\, x^{-2}\, F_{-3}\, M_7^{-1}$ g cm^{-2}. The disk is in danger of becoming optically thin around $\Sigma = 1$, but there the temperature has fallen to T_2 ~ 700 K so dust will take over as a source of opacity (this may not happen for large M_7). Our standard model with the parameters all at unity will provide opacity out to about a parsec or so. Because all but the centre of our disk obeys a law $T = Ar^{-2a}$ with $a = 4$ in the outer parts where T_2 is relevant and $a = 8/3$ in the inner parts where T_1 is relevant, we study the radiation from disks with such power law temperature distributions. The total emission at frequency ν is given by

$$S_\nu = \int_0^\infty \frac{c}{4}\, u_\nu\,(T(r))\, 4\pi r dr = \frac{8\pi^2 h}{c^2}\int_0^\infty \frac{\nu^3\, r dr}{\exp(h\nu/kT) - 1}$$

Writing $x = h\nu/kT = h\nu r^{2a}(kA)$ we find

总功率 $P = 0.057\ F c^2 = 3.2\times10^{42}\ F_{-3}\ \text{erg s}^{-1} \sim 10^9\ L_\odot$

$$T_1(r) = 6.7\ r_0^{-3/4}\left[\frac{1-2.6\times10^{-6}M_7 r_0^{-1}}{(1-1.3\times10^{-6}M_7 r_0^{-1})^{3/2}}\right]^{1/4} F_{-3}^{\ 1/4}M_7^{\ 1/4}$$

$$= 3.7\times10^5 m_0^{-3/4}\left[\frac{1-6\ m_0^{-1}}{(1-3\ m_0^{-1})^{3/2}}\right]^{1/4} F_{-3}^{\ 1/4}\ M_7^{-1/2}\ \text{K} \qquad (m_0\geq6)$$

$$T_2(r) = 100\ r_0^{-1/2}F_{-3}^{\ 1/4} = 1.6\times10^5\ m_0^{-1/2}F_{-3}^{\ 1/4}M_7^{-1/2}\ \text{K}$$

宇宙线能量的最大值为 $e\Phi = 1.5\times10^{13}\ r_0^{-3/4}x^{3/2}F_{-3}^{\ 1/2}M_7^{3/4} = 10^{18}\ m_0^{-3/4}\ x^{3/2}\ F_{-3}^{\ 1/2}$ eV。圆轨道周期（在无穷远处看）为 $3\times10^4\ r_0^{3/2}\ M_7^{-1/2}$ 年 $= 9.8\times10^{-6}\ m_0^{3/2}\ M_7$ 年。圆周速度为 $V_c = 200\ r_0^{-1/2}\ (1-9.7\times10^{-7}\ M_7\ r_0^{-1})^{-1/2} = 3\times10^5\ m_0^{-1/2}\ (1-2\ m_0^{-1})^{-1/2}$ km/s $(m_0>6)$。径向速度为 $V_r = 0.2\ r_0^{-1/2}\ x^2\ M_7^{1/2} = 3\times10^2\ m_0^{-1/2}\ x^2$ km/s $(m_0\gg6)$。电子加速的条件是 $\rho < 3\times10^{-20}\ T_4\ r_0^{-7/4}\ x^{1/2}F_{-3}^{\ 1/2}M_7^{3/4}$；质子加速的条件是 $\rho < 3\times10^{-17}\ T_4$，等等。其中 T_4 是加速区的温度，单位是 10^4 K；ρ 代表密度。因为加速将发生在云之间，所以在盘中的任何区域我们或许应该把 ρ 的值取为 $\rho_0/10$（且 $T_4\sim1$，除非周围的温度 T_1 比较高。）

史瓦西嘴的直径为 $12m = 5.7\times10^{-6}\ M_7$ pc，即 $m_0 = 12$。产生一半功率的区域直径为 $44m = 2.1\times10^{-5}\ M_7$ pc，即 $m_0 = 44$。在该半径处的转动周期为 $9\times10^{-4}\ M_7$ 年 $= 0.32\ M_7$ 天。由该半径处向内运动的时间 r/V_r 为 $1.6\times10^{-1}\ M_7\ x^{-2}$ 年 $\sim 59\ M_7$ 天。

现在应该已经很清楚：利用参数 M_7 和 F_{-3} 的不同取值，这些盘可以解释高能天体物理中的很多难以置信的现象，包括星系核、塞弗特星系、类星体和宇宙线。下一节将专门讨论对其光谱的预测。

光 谱

当 $m_0 = 7.05$ 时温度达到最大值，即 $T_1 = 6.6\times10^4\ F_{-3}^{\ 1/4}\ M_7^{-1/2}$ K。当面密度 $\sum = 90\ x^{-2}\ F_{-3}\ M_7^{-1}\ \text{g cm}^{-2}$ 时，介质是光学厚的。当大致有 $\sum = 1$ 时，盘有变成光学薄的危险，但这时温度已经下降到 $T_2\sim700$ K，因而尘埃将代之成为不透明度的主要来源（如果 M_7 很大，这种情况可能不会发生）。在我们的标准模型中，如果取所有这些参数均为 1，根据我们的标准模型可以得出达到 1 pc 左右的不透明度。因为在我们的盘中，除中心以外的所有区域都服从 $T = Ar^{-2a}$ 的分布律，在与 T_2 相应的靠外区域，$a = 4$；在与 T_1 有关的靠内区域，$a = 8/3$。我们在研究来自盘的辐射时采用的就是这种幂律的温度分布。频率为 v 的总辐射由下式给出：

$$S_v = \int_0^\infty \frac{c}{4}\ u_v\ (T(r))\ 4\pi r\text{d}r = \frac{8\pi^2 h}{c^2}\int_0^\infty \frac{v^3 r\text{d}r}{\exp(hv/kT)-1}$$

令 $x = hv/kT = hvr^{2a}(kA)$，则有：

$$S_\nu = \frac{4\pi^2 h}{c^2}\left(\frac{kA}{h}\right)^a \int_0^\infty \frac{a\,x^{a-1}}{e^x - 1}\mathrm{d}x\ \nu^{3-a}$$

where

$$\int_0^\infty \frac{a\,x^{a-1}}{e^x - 1}\mathrm{d}x = a\,\Gamma(a)\zeta(a)$$

Thus for $a = 8/3$ we have $S_\nu \propto \nu^{1/3}$ while for $a = 4$ we have $S_\nu \propto \nu^{-1}$. Before trying to use these formulae it is important to find out at what radius the main contributions to S_ν arise. For $a < 1$ the main contributions come from radii close to those for which $h\nu \sim kT$. We may therefore deduce that for our standard model $S_\nu \propto \nu^{-1}$ when $100\ \mathrm{K} < h\nu/k < 3{,}000\ \mathrm{K}$ and $S_\nu \propto \nu^{1/3}$ when $3\times10^4\ \mathrm{K} < h\nu/k < 10^5\ \mathrm{K}$, and that for frequencies corresponding to temperatures of 10^5 K or greater the system shines like a black body of 10^5 K. In practice it is known that at least for Seyfert galaxies the reddening by dust takes a large fraction of the energy out of the ultraviolet and replaces it in the infrared. Because I have no theory for the amount of dust at each radius I cannot predict the final optical spectrum in detail. But because dust evaporates at a thousand degrees or so it would seem likely that the radiation should be peaked on the red side of the corresponding frequency. Fig. 1 shows the details of the emitted "black body" radiation. It is clear that fluorescence from the ultraviolet will mean that the optical spectrum should be full of emission lines. Those arising from where the disk has ambient temperatures near 2×10^3 K come from regions $r_0 \simeq 10^{-3}$ where the circular velocities are $V_c \simeq 6\times10^3\ \mathrm{km\ s^{-1}}$. These emission lines should therefore be very broad. We may expect that the real disk is not steady, although exact periodicity due to a source in orbit is unlikely. Rather, we should expect variations on a time scale given by $2r/V_r$ at $m = 22$ (120 days) because that is the time scale in which the material flux can vary over the region in which most of the flux is emitted. Using our theory in the most straight-forward way it is clear that more power is used in accelerating protons than is used in accelerating electrons. Protons can be accelerated in a density 918 times as great as that in which the electron acceleration can operate. Density in our model behaves like $r^{-3/2}$ while total power between $3/2\ r$ and $1/2\ r$ behaves like r^{-1}. We deduce that the proton power is about 10^2 times the electron power. The steady state spectrum is easily determined. From our accelerator we expect energy proportional to potential drop. Because particles start from all points the energy spectrum ejected by the accelerator is uniform up to the maximum energy, $e\Phi \sim 10^{13}$ eV.

$$S_\nu = \frac{4\pi^2 h}{c^2}\left(\frac{kA}{h}\right)^a \int_0^\infty \frac{a\, x^{a-1}}{e^x - 1}\mathrm{d}x\, \nu^{3-a}$$

其中

$$\int_0^\infty \frac{a\, x^{a-1}}{e^x - 1}\mathrm{d}x = a\,\Gamma(a)\zeta(a)。$$

于是，当 $a = 8/3$ 时，我们得到 $S_\nu \propto \nu^{1/3}$；而当 $a = 4$ 时，我们有 $S_\nu \propto \nu^{-1}$。在使用这些公式之前，找出对 S_ν 的主要贡献到底出现在多大半径处是很重要的。当 $a < 1$ 时，主要贡献来自满足条件 $h\nu \sim kT$ 的半径附近。于是我们可以推断出：在我们的标准模型中，当 $100\ \mathrm{K} < h\nu/k < 3{,}000\ \mathrm{K}$ 时，$S_\nu \propto \nu^{-1}$；而当 $3\times10^4\ \mathrm{K} < h\nu/k < 10^5\ \mathrm{K}$ 时，$S_\nu \propto \nu^{1/3}$。此外还可以得出：对于与 10^5 K 或更高温度相应的频率，这个系统的发光和一个 10^5 K 的黑体一样。实际上，人们已经知道至少在塞弗特星系中，尘埃造成的红化将很大一部分紫外的能量转移到了红外。因为我没有理论来描述每个半径处的尘埃数量，所以不能详细预言最终的光谱。不过因为尘埃会在 1,000 度左右蒸发，所以辐射很可能应在相应频率的红边达到峰值。图 1 给出了发射的“黑体”辐射的细节。显然来自紫外的荧光意味着光谱中充满了发射线。那些由盘周围温度接近 2×10^3 K 的区域发出的发射线来自 $r_0 \approx 10^{-3}$ 处,其对应的圆周速度为 $V_c \approx 6\times10^3$ km/s。因而这些发射线会非常宽。我们可以预期真实的盘是不会处于稳态的，尽管一个在轨道里运动的源不大可能产生严格的周期性。更准确地说，我们预期会在 $m = 22$ 处发生时标为 $2r/V_r$（120 天）的光变，因为这刚好是物质流能够在其发射大部分流量的区域内发生变化的时标。应用我们的理论所得到的一个最直接的结果显然是：更多的能量被用于加速质子而不是电子。质子被加速所要求的密度是电子被加速所需密度的 918 倍。在我们的模型中，密度的变化如同 $r^{-3/2}$，而总功率在 $3/2\ r$ 和 $1/2\ r$ 之间的变化如同 r^{-1}。我们推测质子的功率大约是电子功率的 10^2 倍。稳态的谱容易被确定。从我们的加速器来看，我们期望能量与电位降成比例。因为粒子是从各个点开始运动的，所以被加速器喷射出的能谱在直到最高能量 $e\Phi \sim 10^{13}$ eV 的范围内都是均匀的。

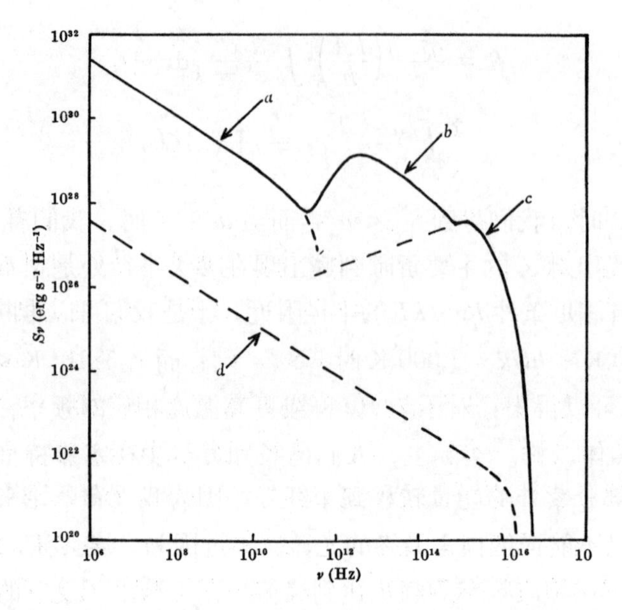

Fig. 1. The emitted spectrum of disk and synchrotron radiation for the standard model. The flux from Sagittarius *A* is weaker by a factor 100, indicating only 1 percent efficiency of the proton synchrotron. *a*, Proton synchrotron; *b*, outer disk; *c*, central disk; *d*, electron synchrotron.

Each particle of energy E radiates at a rate proportional to E^2. If there were constant monoenergetic injection into the medium, the flux of particles downwards in energy would be constant. Thus the number per unit area at any E less than the injection energy would follow the law $N(E) = KE^{-2}$ for $E < E_{max} \equiv E_m$. This law is only slightly modified by our uniform injection at all energies up to E_{max}. It is

$$N(E) = K\,(1 - E/E_m)E^{-2} \quad E < E_m$$

where K is related to the total power of injection power per unit area p by

$$K = \frac{2p}{E_m}\frac{9\,m_A^{\,2}c^7}{4e^2}\,B^{-2}$$

Because the power law is near E^{-2} except close to E_{max}, we expect the γ of synchrotron radiation theory to be close to two, and the corresponding spectrum to be close to $S_\nu \propto \nu^{-\alpha}$ with $\alpha = \frac{1}{2}$. It is possible to work out a better approximation using the δ function approximation to the frequency spectrum of a single electron[9]. For our disk model the flux is

$$S_\nu = \int_{r_A}^{\infty}\int_{0}^{E_m} \frac{2p}{E_m}\,\delta\,(\nu - \nu_m)\left(1 - \frac{E}{E_m}\right) dE\; 2\pi r\; dr$$

where r_A is the least radius at which the particles can be accelerated and

$$\nu_m = 0.07\left(\frac{2}{3}\right)^{1/2}\frac{e}{m_A^{\,3}c^5}BE^2$$

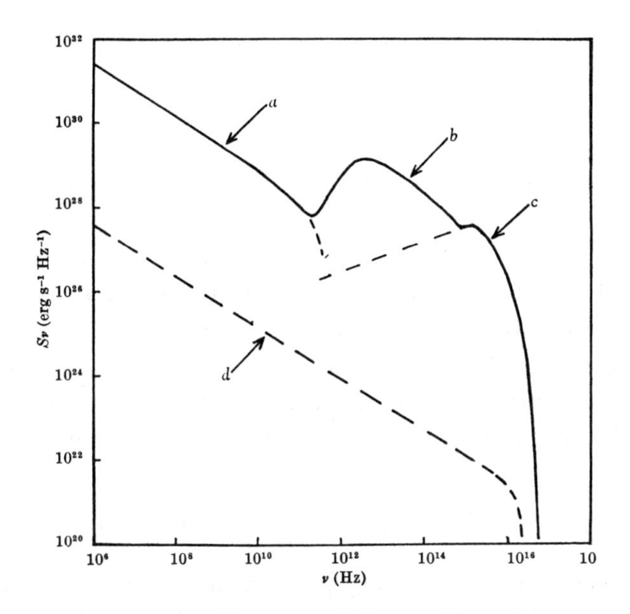

图 1. 标准模型中盘和同步辐射的发射谱。来自人马座 A 的流量减弱为原来的 1/100，表明质子同步辐射的效率只有 1%。a，质子同步辐射；b，外盘；c，中心盘；d，电子同步辐射。

每个能量为 E 的粒子都会以正比于 E^2 的速率发射辐射。如果有持续的单一能量注入介质中，那么能量更低的粒子的流量应该保持不变。因此，对于任何低于注入能量的能量 E，单位面积的粒子数都应遵循 $N(E) = KE^{-2}$ 的规律，其中 $E < E_{max} \equiv E_m$。在直到 E_{max} 的所有能量都均匀注入时，这个规律只需进行微小的修正，即：

$$N(E) = K\,(1 - E/E_m)E^{-2} \quad E < E_m$$

其中 K 与单位面积中注入的总功率 p 的关系是：

$$K = \frac{2p}{E_m}\frac{9\,m_A^{\,2}\,c^7}{4e^2}\,B^{-2}$$

因为除了在 E_{max} 附近以外，幂律都接近于 E^{-2}，所以我们可以预期同步辐射理论中的 γ 应该接近于 2，而相应的谱接近于 $S_\nu \propto \nu^{-\alpha}$，其中 $\alpha = \frac{1}{2}$。用 δ 函数对单个电子频谱的近似很可能会是一种更好的近似[9]。将其应用于我们的盘模型，得到的流量为：

$$S_\nu = \int_{r_A}^{\infty}\int_{0}^{E_m} \frac{2p}{E_m}\,\delta\,(\nu - \nu_m)\left(1 - \frac{E}{E_m}\right)\mathrm{d}E\; 2\pi r\,\mathrm{d}r$$

其中 r_A 是粒子可以被加速的最小半径，且

$$\nu_m = 0.07\left(\frac{2}{3}\right)^{1/2}\frac{e}{m_A^{\,3}\,c^5}BE^2$$

Using our power laws for $p(r)$, $E_m(r) \equiv e\Phi$, $B(r)$, we obtain

$$S_\nu = 3.8\times 10^{28} \left(\frac{m_A}{m_p}\right)^{12/11} \nu_9^{-7/11}$$

$$\left[1 - \frac{1}{11}\left(14 - 3\left(\frac{\nu_9}{\nu_A}\right)^{1/2}\right)\left(\frac{\nu_9}{\nu_A}\right)^{3/22}\right] x^{-10/11} F_{-3}^{5/11} M_7^{4/11}$$

where this formula holds for $\nu < \nu_A \equiv \nu_m(r_A)$ and ν_9 is ν in units of GHz. This formula has assumed that all the power dissipated goes into fast protons or electrons as the case may be. In regions where both electrons and protons are accelerated the power should obviously be divided by two. In practice it is probable that S_ν should be reduced by some efficiency factor because probably only a fraction of the total reconnexion energy really gets into fast particles. The radius of the source at frequency ν is about

$$r_0 = 0.7\, \nu_9^{-4/11} \left(\frac{m_A}{m_p}\right)^{-12/11} x^{10/11} F_{-3}^{6/11} M_7^{7/11}, \quad \nu < \nu_A$$

Notice that the fast electrons can only be produced much further out than the protons but that they nevertheless produce radiation to much higher frequencies.

Comparison with Observations

In the Galaxy it is not clear that the circular velocity near the centre falls below 200 km s^{-1}, but the OH observations do suggest velocities as low as 100 km s^{-1} within 70 pc of the centre. This indicates a nuclear mass of $M_7 \sim 3$ for the central singularity. The size and flux from Sagittarius *A* are in rough accord with our estimate of the synchrotron spectrum. An infrared flux found at 100 μm could be due to dust from an ultraviolet source radiating $10^9\, L_\odot$, so the flux of mass into the throat must be around $F_{-3} = 1$. The general level of activity observed at radio wavelengths close to the nucleus indicates that high energy phenomena are involved[10].

The Magellanic clouds have no nucleus. *M*31 has a strong radio source rather larger but weaker than that found in the Galaxy[11]. Code has discovered strong ultraviolet emission from the nucleus, which Kinman finds to have a large mass-to-light ratio and to contain some 10^8 solar masses. This suggests a small mass flux into the centre of *M*31 and only a very small ultraviolet disk about the Schwarzschild mouth. *M*32 has a nucleus which is not a radio source but the system is very deficient in gas. We suggest that this system has a Schwarzschild mass but the Galaxy has run out of gas and left it hungry. *M*82 has had a recent violent radio explosion and an infrared nucleus with a small bright radio source in the centre. I suspect $M_7 \sim 3$ and $F_{-3} \sim 10$ but that F_{-3} was larger in the recent past. *M*81 has a very small flux but an intense radio source at its nucleus. I suggest that it is intermediate between the Galaxy and *M*31. *M*87 is the nearest really bright radio galaxy. Luckily the velocity dispersion in its nucleus has been measured[12]. We can therefore measure M_7 with some pretence of accuracy to be about $4\times 10^{10}\, M_\odot$, that is, $M_7 = 4\times 10^3$. Over 10^{10} yr it would take an F_{-3} of 4×10^3 to build such an object. This is probably the nearest old dead radio-bright quasar. It is only a shadow of its former self, as F_{-3} has declined severely as the gas has run out. Its electron synchrotron still produces copious

把我们的幂律代入 $p(r)$、$E_m(r) \equiv e\Phi$ 和 $B(r)$，得到：

$$S_\nu = 3.8\times10^{28}\left(\frac{m_A}{m_p}\right)^{12/11}\nu_9^{-7/11}$$

$$\left[1-\frac{1}{11}\left(14-3\left(\frac{\nu_9}{\nu_A}\right)^{1/2}\right)\left(\frac{\nu_9}{\nu_A}\right)^{3/22}\right]x^{-10/11}F_{-3}^{5/11}M_7^{4/11}$$

上式的成立条件是 $\nu < \nu_A \equiv \nu_m(r_A)$，$\nu_9$ 代表以 GHz 为单位的 ν。该式假设所有耗散的功率都转移给了快质子或快电子。在电子和质子都被加速的区域，功率显然应该除以 2。实际上，S_ν 的值很可能应乘以某个效率因子使其适当减小，因为在总的重联能量中很可能只有一部分真正转移到了快粒子中。频率为 ν 的源的半径大约是：

$$r_0 = 0.7\,\nu_9^{-4/11}\left(\frac{m_A}{m_p}\right)^{-12/11}x^{10/11}\,F_{-3}^{6/11}\,M_7^{7/11},\ \nu < \nu_A$$

注意：快电子只能在比产生快质子靠外很多的地方产生，但尽管如此，它们仍会产生频率高得多的辐射。

与观测结果的比较

在银河系中，中心附近的圆周速度是否会小于 200 km/s 尚不清楚，然而对羟基（OH）的观测结果确实表明：在距中心 70 pc 的范围之内，速度低至 100 km/s。这表明中心奇异性的核质量是 $M_7 \sim 3$。人马座 A 的尺度和辐射流量大致与我们对同步辐射谱的估计一致。100 μm 处观测到的红外流量可能来源于一个辐射功率为 $10^9\,L_\odot$的紫外源中的尘埃，因此进入喉的质量流量应该大致为 F_{-3}=1。在射电波段观测到的星系核附近的整体活动性水平表明有高能现象产生 [10]。

麦哲伦云没有星系核。M31 有一个强射电源，比在银河系中发现的射电源更大但更弱 [11]。科德已经发现了来自这个星系核的强紫外辐射，欣曼发现它有大的质光比并且其质量约为太阳质量的 10^8 倍。这表明只有少量的质量流入了 M31 的中心，并且在史瓦西嘴周围只有一个非常小的紫外盘。M32 有一个不是射电源的星系核，但是这个系统十分缺乏气体。我们认为该系统有一个史瓦西黑洞，但是这个星系中的气体已经耗尽从而使它无气体可吞噬。最近，M82 发生了一次强烈的射电爆发，其中心有一个带有小而明亮的射电源的红外核。我认为 $M_7 \sim 3$ 以及 $F_{-3} \sim 10$，但 F_{-3} 在不久之前比现在还要大。M81 的流量非常小，但在其中心处有一个很强的射电源。我认为它介于银河系和 M31 之间。M87 是真正最近邻的亮射电星系。幸而其核的速度弥散度已经被测定出来了 [12]。因此我们可以假定其 M_7 的量值已经被精确测定为 $4\times10^{10}\,M_\odot$，即 $M_7 = 4\times10^3$。以 4×10^3 的 F_{-3} 构建这样一个天体需要超过 10^{10} 年的时间。这可能就是距离我们最近的老死了的射电亮类星体。它只是自己过去的一个影子，

X-rays[13], however. *NGC* 4151 is a Seyfert galaxy, and M_7 need not be greater than 3, or more likely 30, but F_{-3} is high because there is still much gas in the central regions. Seyfert galaxies that are active have $F_{-3} = 10^3$ but probably are only active at this flux level for one-hundredth of the time. The breadths of the wings of the Balmer lines are 6,000 km s^{-1}— I suggest that these are Doppler widths[14]. *NGC* 4151 is a strong infrared source.

Quasars

When F_{-3} achieves large values ~10^6 or 10^7 (that is, 10^{3-4} $M_\odot$ yr^{-1}) the mass of the Schwarzschild throat rapidly build up to 10^9–10^{10} $M_\odot$. When galaxies first formed there was this amount of gaseous material in them. Large proto-galaxies rapidly achieved large Schwarzschild throats and greedily swallowed gas. It is clear that the right energy is available and by making M close to 10^{10} $M_\odot$ we lower the densities close to the Schwarzschild throat. This allows the radio phenomena to occur closer to the singularity where more of the power is.

Note added in proof. Low's recent observations of the galactic centre at 100 μm (reported at the Cambridge conference on infrared astronomy, 1969) suggest that F_{-3} for the galaxy is nearer 10^{-2} than 1. A dust model by Rees can explain the infrared observations assuming a single central source of visible or ultraviolet light. The light pressure from such sources will expel dusty material from the nuclei of Seyfert galaxies causing the observed outflow as suggested by Weymann. Such a mechanism could cut off the flux F_{-3} and therefore produce the changes in the emitted flux. The light pressure may drive the dust out of the nucleus so that no dust is ever swallowed. This could leave a great enhancement of dust in the surroundings of the nucleus corresponding to the dust content of all material swallowed in the past. A violently active outburst of such a nucleus would then be associated with the expulsion of great swathes of dust such as those seen across several radio galaxies.

The proton synchrotron radiation discussed here is probably replaced in practice by synchrotron radiation from electron secondaries and X and γ-ray bremsstrahlung corresponding to a sizable fraction of the power input into fast protons.

I thank the Astronomer Royal for discussions, and Drs. Pagel, Bingham, Tritton, Rowan-Robinson, Weymann and Osterbrock for further help and encouragement.

(**223**, 690-694; 1969)

D. Lynden-Bell: Royal Greenwich Observatory, Herstmonceux Castle, Sussex.

Received July 8, 1969.

因为在其气体耗尽之后 F_{-3} 已经急剧减小。然而，它的电子同步辐射仍在产生大量的 X 射线[13]。*NGC* 4151 是一个塞弗特星系，M_7 不一定要大于 3，或者可能性更大的 30，但 F_{-3} 很大，因为在中心区域仍存在着大量的气体。对于活动的塞弗特星系，有 $F_{-3}=10^3$，在这个流量水平保持活跃的时间很可能只占总时间的 1%。巴耳末线的线翼宽度为 6,000 km/s——我认为这正是多普勒宽度[14]。*NGC* 4151 是一个强红外源。

类 星 体

当 F_{-3} 达到 $\sim 10^6$ 或 10^7（即 $10^3 \sim 10^4\ M_\odot$/ 年）的高值时，史瓦西喉的质量就会迅速增大到 $10^9\ M_\odot \sim 10^{10}\ M_\odot$。在星系刚开始形成时，其中就有如此大量的气体。巨大的原初星系很快形成巨大的史瓦西喉并贪婪地吞噬着气体。显然该过程可以产生足够的能量，并且通过使 M 接近于 $10^{10}\ M_\odot$，我们可以降低史瓦西喉附近的密度。这使得射电现象能够发生在更靠近奇点的地方，那里也是有更高功率的地方。

附加说明。洛最近在 100 μm 波段对银河系中心进行了观测（报告于 1969 年的剑桥红外天文学会议），结果表明银河系的 F_{-3} 更接近于 10^{-2} 而不是 1。瑞斯提出的尘埃模型可以在假设存在单一的可见光或紫外光中心源的前提下来解释这些红外观测结果。来自这些源的光压将把含尘埃的物质从塞弗特星系的星系核中排出，导致观测到的外流，正如魏曼所提出的一样。这样一种机制可以截断质量流 F_{-3} 并因此产生辐射流量的变化。光压可能会驱使尘埃离开星系核，因此没有尘埃再被吞噬。这将导致星系核周围的尘埃相对于过去被吞噬的所有物质中的尘埃量而言大大增加。因此，这样一个星系核的剧烈爆发活动将伴随着大量尘埃的排出，正如人们在对几个射电星系进行观测时所看到的那样。

这里所讨论的质子同步辐射在实践中很可能被替换为来自电子的次级同步辐射，而 X 射线和 γ 射线波段的轫致辐射提供了相当一部分注入于快质子的功率。

我要感谢皇家天文学家们对此进行了多次讨论，也要感谢帕格尔博士、宾厄姆博士、特里顿博士、罗恩 - 鲁滨逊博士、魏曼博士和奥斯特布罗克博士给了我进一步的帮助和鼓励。

（钱磊 金世超 翻译；邓祖淦 审稿）

References:

1. Ryle, M., *Highlights of Astronomy* (edit. by Perek, L.) (D. Reidel, 1968).
2. Hoyle, F., Fowler, W. A., Burbidge, G., and Burbidge, E. M., *Astrophys. J.*, **139**, 909 (1964).
3. Greenstein, J. L., and Schmidt, M., *Astrophys. J.*, **140**, 1 (1964).
4. Schmidt, M., *Texas Conf. Relativistic Astrophys.* (edit. by Maran, S. P., and Cameron, A. G. W.) (1968).
5. Salpeter, E. E., *Astrophys. J.*, **140**, 796 (1964).
6. Parker, E. N., *Astrophys. J.*, **149**, 517 (1967).
7. Syrovatskii, S. I., *IAU Symp. No. 31, Radio Astronomy and the Galactic System* (edit. by Van Woerden, H.), 133 (Academic Press, 1967).
8. Spitzer, L., *Physics of Fully Ionised Gases* (Interscience, 1955).
9. Ginzburg, V. L., and Syrovatskii, S. I., *Ann. Rev. Astron. Astrophys.*, **3**, 297 (1965).
10. Lequeux, J., *Astrophys. J.*, **149**, 393 (1967).
11. Poolley, G. G., *Mon. Not. Roy. Astron. Soc.*, **144**, 101 (1969).
12. Brandt, J. C., and Rosen, R. G., *Astrophys. J. Lett.*, **156**, L59 (1969).
13. Byram, E. T., Chubb, T. A., and Friedman, H., *Science*, **152**, 66 (1966).
14. Woltjer, L., *Astrophys. J.*, **130**, 38 (1959).

Superfluidity in Neutron Stars

G. Baym *et al.*

Editor's Note

Physicists had come to understand that matter in neutron stars must exist at densities as high as 10^{15} grams per cubic cm. Some suggested that the interiors of such stars might well be superfluid (flowing without viscosity), and Gordon Baym and colleagues here comment on the likely consequences for these stars' properties. Superfluids, they note, tend to expel magnetic fields from their interiors, but their analyses suggested this could be so slow in stars as to be insignificant. The researchers suggest that interactions between superfluid neutrons and a star's surrounding magnetosphere might account for the observed slowing down of a number of pulsars—thought to be neutron stars—although more detailed calculations would be required to confirm that.

MATTER in the interior of a typical neutron star is a mixture of three degenerate interacting quantum liquids—neutrons, protons and electrons, the latter two having a density at most a few percent that of the neutrons[1]. The mixture, bounded on the inside by a superdense core of hadrons, muons and so on, and most likely by a solid mantle on the outside[2], is of density between 5×10^{13} and 10^{15} g cm^{-3}. As was first pointed out by Migdal[3], and more recently discussed by other[4-8], there are quite possibly superfluid states in this interior. Here we discuss certain general features of such states and the extent to which they influence the properties of the star.

As Ginzburg[7] has observed, the electrons are very unlikely to be superconducting, because they form a highly degenerate relativistic plasma in which effects of Coulomb interactions, relative to the kinetic energy, are of the order $e^2/\hbar c$. In particular, their superconducting transition temperature is vanishingly small on the scale of typical neutron star temperatures; the electrons are a weakly interacting normal system which can be treated microscopically by perturbation theory. On the other hand, strong interaction forces make neutron superfluidity and proton superconductivity serious possibilities. (Precise criteria for these possibilities, as well as their interrelation, can be furnished only on the basis of reliable microscopic calculations, which have yet to be performed.)

At first sight, one would expect that superconductivity of the protons would have a drastic effect on the magnetic properties of a neutron star, because superconductors exhibit some form of the Meissner effect—either a complete or an incomplete expulsion of magnetic flux from superconducting region[9]. As we shall now show, however, the enormous electrical conductivity of the normal state implies that the characteristic times for flux expulsion from macroscopic regions are typically comparable with the age of the universe.

中子星内部的超流

贝姆等

编者按

物理学家们逐渐认识到中子星内部的物质密度必须达到10^{15} g/cm^3。有人认为，这类恒星的内部有可能会处于超流（无黏滞性流动）状态。戈登·贝姆和他的同事们在本文中讨论了超流对这些恒星的性质可能产生的影响。他们注意到超流体有将磁场排除出其内部的倾向，但他们通过分析表明这个过程是非常缓慢的，以至于对恒星而言并不重要。这几位研究者指出：超流中子与这类恒星周围磁层之间的相互作用也许能解释为什么会观测到若干脉冲星（被认为是中子星）的转速在减慢，尽管为确认这一点尚需要更详细的计算。

典型中子星的内部物质是三种相互作用的简并量子液体的混合物——中子、质子和电子。后两者的密度至多是中子密度的百分之几[1]。该混合物被约束在由强子和μ介子等组成的高密度星体内核之中，其外部很可能有固态的幔[2]，混合物的密度在5×10^{13} g cm^{-3}~10^{15} g cm^{-3}之间。米格达尔[3]最早提出在这种星体内部很可能存在超流态，最近有不少人对这种可能性进行了讨论[4-8]。我们将在下文中论述这些超流态的一些基本特征，以及它们在何种程度上影响了中子星的性质。

正如京茨堡[7]所述，电子不太可能是超导的，因为电子形成了高度简并的相对论性等离子体，并且在等离子体中库仑相互作用与动能之比的量级是$e^2/\hbar c$。尤其是，它们的超导转变温度与典型中子星的温度相比是极其小的；这些电子组成了一个正常的弱相互作用系统，在微观上可以用微扰理论进行处理。另一方面，强相互作用力使存在中子超流和质子超导的可能性变得非常大。（只有基于可靠的微观计算才能给出存在中子超流和质子超导的精确判据以及两者之间的相互关系，但至今还没有人进行过这样的计算。）

初看起来，我们会认为质子超导将对中子星的磁性质产生重大影响，因为超导体会表现出迈斯纳效应的某种形式——超导区域对磁场的完全排斥或部分排斥[9]。然而，正如我们马上要说明的，因为正常态物质的电导率非常大，所以磁通量被排斥出宏观区域的特征时标通常会与宇宙年龄的量级相当。

The characteristic time, τ_D, for flux diffusion in normal, that is, non-superconducting, neutron star matter is $\sim 4\pi\sigma R^2/c^2$, where σ is the electrical conductivity of the normal state, R is the scale of variation of the magnetic field, typically the radius of the star, and c is the speed of light. Electrical conduction is primarily by the highly relativistic degenerate electrons, and thus $\sigma \approx n_e e^2 \tau_{tr} c/\hbar k_f$, where n_e is the electron number density, k_f the electron (and proton) Fermi wavenumber, and τ_{tr} is the transport relaxation time. From a detailed calculation, to be reported separately, we find that

$$\tau_{tr}^{-1} \lesssim \frac{\pi^2}{12}\left(\frac{e^2}{\hbar c}\right)^2\left(\frac{T}{T_p}\right)^2\frac{ck_f^2}{k_{FT}}$$

Here T_p is the proton Fermi Temperature, $k_{FT} \approx (4k_f m_n e^2/\pi\hbar^2)^{1/2}$ is the proton Fermi–Thomas wavenumber and m_n the nucleon mass. For $k_f \sim 0.7\times10^{13}$ cm^{-1} (corresponding to 2×10^{13} g of protons cm^{-3}) and $T \sim 10^8$ K, one has $T_p \sim 1.2\times10^{11}$ K, and $\tau_{tr} \sim 6\times10^{-14}$ s. Thus for $R \sim 10$ km the flux diffusion time is $\sim 10^{22}$ s (a striking manifestation of the proton degeneracy). In other words, the magnetic flux is rigidly tied to normal neutron star matter.

Suppose now it is energetically favourable for a given macroscopic region to become superconducting, with a condensation energy per unit volume $H_c^2/8\pi$, where H_c is the thermodynamic critical field. An elementary calculation shows that characteristic time to expel magnetic induction B from an initially normal region is $\tau_{nucl} \sim \tau_D(B^2/H_c^2)$; for typical values of B and H_c, τ_{nucl} is $\sim 10^{15}$ s. Nucleation of superconductivity thus cannot be accompanied by expulsion of flux. Superconductivity, if it occurs, occurs at constant B. Put another way, on a macroscopic scale, the magnetic flux at any point does not depend on whether the matter there is normal or superconducting.

There are two possible ways in which magnetic flux can penetrate superconductors[9]: in the mixed state in which one has a periodic array of quantified vortices of supercurrents, aligned parallel to the field; or in the intermediate state, where one has, on a fine scale, alternating regions of normal material containing flux, and superconducting material exhibiting a complete Meissner effect. Which of these two situations one expects depends on the ratio of the proton coherence length ξ_p to the penetration depth λ_p; for $\xi_p/\lambda_p<\sqrt{2}$ (Type II superconductivity) the vortex state is energetically preferable, while for $\xi_p/\lambda_p>\sqrt{2}$ (Type I) the intermediate state results. For s-state pairing the coherence length is given by $\xi_p \approx (2/\pi k_f)(\varepsilon_p/\Delta_p)$, where $\varepsilon_p=\hbar^2k_f^2/2m_n$, and Δ_p is the proton superconducting energy gap. When $\lambda_p\gg\xi_p$, as we anticipate for proton superconductivity, λ_p is given by the London formula $\lambda_p^2 = m_n c^2/4\pi n_e e^2$, and

$$\frac{\xi_p}{\sqrt{2}\,\lambda_p} = \left(\frac{8}{3\pi^2}\frac{e^2k_f}{m_n c^2}\right)^{1/2}\frac{\varepsilon_p}{\Delta_p} \sim n_c^{1/6}\frac{\varepsilon_p}{\Delta_p}$$

For $k_f = 0.7\times10^{13}$ cm^{-1} one finds $\lambda_p = 6.7\times10^{-12}$ cm, and $\xi_p/\sqrt{2}\,\lambda_p<1$ for $\Delta_p/\varepsilon_p\geq1.0\times10^{-2}$. Thus the protons are most likely a Type II superconductor, which, because of the finite B, will be in the vortex state. The magnetic flux associated with each vortex is

在正常的、或称非超导的中子星物质中，磁通量扩散的特征时标 τ_D 为 ~ $4\pi\sigma R^2/c^2$。其中 σ 为正常态物质的电导率；R 为磁场的变化尺度，一般是指这类恒星的半径；c 为光速。因为导电主要依靠高度相对论性简并的电子，所以 $\sigma \approx n_e e^2 \tau_{tr} c/\hbar k_f$，其中 n_e 是电子数密度，k_f 是电子（也是质子）的费米波数，τ_{tr} 是输运弛豫时间。经过复杂的计算，具体过程将在其他论文中给出，我们得到：

$$\tau_{tr}^{-1} \backsimeq \frac{\pi^2}{12}\left(\frac{e^2}{\hbar c}\right)^2\left(\frac{T}{T_p}\right)^2\frac{ck_f^2}{k_{FT}}$$

此处 T_p 是质子的费米温度，$k_{FT} \approx (4k_f m_n e^2/\pi\hbar^2)^{1/2}$ 是质子的费米－托马斯波数，m_n 是核子质量。由 $k_f \sim 0.7\times10^{13}\ cm^{-1}$（对应于每 cm^3 2×10^{13} g 质子）和 $T \sim 10^8$ K 可以得到 $T_p \sim 1.2\times10^{11}$ K 和 $\tau_{tr} \sim 6\times10^{-14}$ s。对于 $R \sim 10$ km，磁通量扩散时标为 ~10^{22} s（可以看出质子简并的显著影响）。换句话说，就是磁通量与正常态的中子星物质是严格绑定的。

现在假设一个每单位体积凝聚能为 $H_c^2/8\pi$ 的给定宏观区域（式中 H_c 是热力学临界场强）转变为超导态在能量上是有利的。初步的计算表明，从一个起先为正常的区域排斥出磁感应强度 B 的特征时标为 $\tau_{nucl} \sim \tau_D(B^2/H_c^2)$；代入 B 和 H_c 的典型值，得到 τ_{nucl} 为 ~10^{15} s。因而超导成核过程不可能同时伴随着磁通量被排斥出超导区。如果超导能够产生，则一定是发生在 B 保持恒定的情况下。换一种说法就是：在宏观尺度上，任何一点的磁通量与该点处的物质是处于正常态还是处于超导态无关。

磁通量穿过超导体的可能方式有两种[9]：在混合态中，通过与磁场平行的超流量子涡旋的周期排列；在中间态中，通过小尺度上有磁通量的普通物质区和表现出完全迈斯纳效应的超导体区的交替出现。这两种情况中到底哪一种会发生取决于质子相干长度 ξ_p 与穿透深度 λ_p 的比值。当 $\xi_p/\lambda_p<\sqrt{2}$（II 型超导）时，涡旋态（译者注：即混合态）在能量上更有利；而当 $\xi_p/\lambda_p>\sqrt{2}$（I 型超导）时，中间态是有利的。对于 s 态配对超导态，相干长度由公式 $\xi_p \approx (2/\pi k_f)(\varepsilon_p/\Delta_p)$ 给出，其中 $\varepsilon_p=\hbar^2k_f^2/2m_n$，$\Delta_p$ 是质子超导能隙。在 $\lambda_p \gg \xi_p$ 的情况下，当我们预期会发生质子超导时，λ_p 由伦敦公式 $\lambda_p^2 = m_n c^2/4\pi n_e e^2$ 和下式给出：

$$\frac{\xi_p}{\sqrt{2}\,\lambda_p} = \left(\frac{8}{3\pi^2}\,\frac{e^2k_f}{m_n c^2}\right)^{1/2}\frac{\varepsilon_p}{\Delta_p} \sim n_c^{1/6}\,\frac{\varepsilon_p}{\Delta_p}$$

由 $k_f = 0.7\times10^{13}\ cm^{-1}$ 可以得到 $\lambda_p = 6.7\times10^{-12}$ cm；还可以得到：当 $\Delta_p/\varepsilon_p \geq 1.0\times10^{-2}$ 时，$\xi_p/\sqrt{2}\lambda_p<1$。因而质子极有可能是 II 型超导体。由于 B 是有限的，所以这种超

$\varphi_0 = hc/2e \approx 2\times10^{-7}$ G cm^2 and the number of vortices per unit area is B/φ_0, where B is the locally averaged magnetic induction. For $B \sim 10^{12}$ G (ref. 10), the vortex lattice, which is triangular, has a lattice constant $\sim 5\times10^{-10}$ cm, large compared with the interparticle spacing. The vortices have normal cores of radius ξ_p, while the magnetic induction falls off in a distance λ_p from the core centre.

The critical magnetic fields (H, not B) between which the vortex state is thermodynamically stable are given by $H_{c1} \approx (\varphi_0/4\pi\lambda_p^2)$ ln λ_p/ξ_p and $H_{c2} \approx \varphi_0/2\pi\xi_p^2$; for $\varepsilon_p/\Delta_p \sim 20$ and $k_f = 0.7\times10^{13}$ cm^{-1} one has $H_{c1}\sim10^{15}$ G and $H_{c2} \sim 3\times10^{16}$ G. We note further that the time for expulsion of flux from the vortex state is substantially greater than that for a normal region, τ_D.

In conditions of constant B, the transition from the normal to the superconducting state is first order. The maximum B for which the superconducting state is thermodynamically preferable to the normal state is that for which the free energy density of the vortices ($=BH_{c1}/4\pi$ for $B \ll H_{c1}$) is equal to the superconducting condensation energy density, $H_c^2/8\pi$ ($=m_n k_f \Delta_p^2/2\pi^2\hbar^2$). This critical B in fact equals H_{c2}; this is far greater than values of B one expects in neutron stars.

Because the flux in the superconducting regions is frozen, superconductivity in neutron stars is expected to have far less effect on magnetic properties of pulsars than previously believed[8]. We note also that in the rotation of a neutron star the electrons must corotate with the protons quite independently of whether the protons are superconducting or not, because any appreciable differential rotation would give rise to inordinately large magnetic fields.

Neutron superfluidity due to *s*-state pairing is expected[11,12] to occur for neutron densities between 0.6 and 2×10^{14} g cm^{-2}. In this regime the neutron Fermi wavenumber k_n varied from 1.0 to 1.5×10^{13} cm^{-1} and the neutron energy gap Δ_n is of the order of 1 MeV, falling with increasing density. The angular velocities of neutron stars lie well above Ω_{c1}, the minimum angular velocity at which it is energetically favourable to create vortices in the neutron superfluid, but well below Ω_{c2}, the angular velocity above which bulk superfluidity is destroyed by rotation. Thus, as Ginzburg and Kirzhnits[4] have noted, the neutron superfluid will contain an array of vortex lines parallel to the rotation axis, of sufficient number that on a macroscopic scale the neutrons will appear to be rotating as a rigid body and thus have their classical moment of inertia. The critical angular velocities are given by $\Omega_{c1} \approx (\hbar/2m_n R^2)$ ln (R/ξ_n) and $\Omega_{c2} \sim \hbar/2m_n\xi_n^2$, where R is a typical dimension of the superfluid, and ξ_n, the neutron coherence length, is $\approx (2/\pi\, k_n)(\varepsilon_n/\Delta_n)$, where $\varepsilon_n = \hbar^2 k_n^2/2m_n$. Taking $R \sim 10$ km, $k_n \sim 1.4\times10^{13}$ cm and $\Delta_n = 1$ MeV, one finds $\xi_n \sim 1.6\times10^{-12}$ cm, $\Omega_{c1} \sim 10^{-14}$ s^{-1} and $\Omega_{c2} \sim 10^{20}$ s^{-1}. The angular momentum per neutron pair per vortex is $\hbar$, and thus the density of vortices per unit area is $\sim 2m_n\Omega/\pi\hbar \sim 2\times10^5$ cm^{-2} for $\Omega \sim 2\times10^2$ s^{-1}, appropriate for the Crab pulsar *NP* 0532.

As pointed out by several authors[10,13,14], the slowing down of a neutron star results from the loss of energy via electromagnetic coupling of the charged particles in the star to the

导体将处于涡旋态。与每个涡旋相关联的磁通量为 $\varphi_0 = hc/2e \approx 2\times10^{-7}$ G cm^2，而每单位面积的涡旋数量为 B/φ_0，其中 B 是局域平均的磁感应强度。当 $B \sim 10^{12}$ G（参考文献 10）时，涡旋晶格呈三角形，晶格常数约为 5×10^{-10} cm，比粒子间隔要大。这种涡旋具有半径为 ξ_p 的正常核，同时磁感应强度在距离核中心 λ_p 处迅速降低。

涡旋态在两个临界磁场强度（H，而非 B）之间是热力学稳定的，它们分别是 $H_{c1}\approx(\varphi_0/4\pi\lambda_p^2)\ln\lambda_p/\xi_p$ 和 $H_{c2}\approx\varphi_0/2\pi\xi_p^2$；当 $\varepsilon_p/\Delta_p\sim20$ 和 $k_f=0.7\times10^{13}\ cm^{-1}$ 时，可得出 $H_{c1}\sim10^{15}$ G 及 $H_{c2}\sim3\times10^{16}$ G。我们还注意到，把磁通量排斥出涡旋态的时间要比排斥出正常态区域的时间 τ_D 长很多。

对于 B 保持不变的情况，从正常态到超导态的转变是一阶相变。使超导态在热力学上比正常态更稳定的 B 有一个上限，此时涡旋的自由能密度（$=BH_{c1}/4\pi$，当 $B\ll H_{c1}$ 时）应与超导态的凝聚能密度 $H_c^2/8\pi$ $(=m_nk_f\Delta_p^2/2\pi^2\hbar^2)$ 相等。这个临界 B 值实际上就等于 H_{c2}，远高于预期的中子星中的 B 值。

由于在超导区磁通量是被冻结的，因而可以预测中子星内的超导对于脉冲星磁性质的影响要远远小于之前所认为的值[8]。我们还注意到：当中子星旋转时，无论质子是否处于超导态，电子都必须与质子共转，因为任何明显的较差转动都会产生非常大的磁场。

可以预测由 s 态配对导致的中子超流[11,12]会发生在中子物质密度为 $0.6\times10^{14}\ g\ cm^{-2}$ ~$2\times10^{14}\ g\ cm^{-2}$ 时。在这个范围内，中子的费米波数 k_n 为 $1.0\times10^{13}\ cm^{-1}$~$1.5\times10^{13}\ cm^{-1}$，而中子能隙 Δ_n 可达 1 MeV 量级并随密度的增加而降低。中子星转动的角速度应远大于 Ω_{c1}，Ω_{c1} 是使中子超流体产生涡旋在能量上更有利的最小角速度；同时中子星转动的角速度要远远小于 Ω_{c2}，大于此值时整体超流将会被旋转破坏。因此，正如京茨堡和基尔日尼茨[4]所指出的：中子超流体将包含与旋转轴平行的涡线阵列，因为涡线的数量足够多，所以在宏观尺度上中子物质将像刚体一样旋转，因而其转动惯量也具有经典的形式。这两个临界角速度可由 $\Omega_{c1}\approx(\hbar/2m_nR^2)\ln(R/\xi_n)$ 和 $\Omega_{c2}\sim\hbar/2m_n\xi_n^2$ 给出。其中 R 是超流体的典型尺度；ξ_n 是中子的相干长度，其值可由 $\xi_n\approx(2/\pi k_n)(\varepsilon_n/\Delta_n)$ 得到，上式中的 $\varepsilon_n=\hbar^2k_n^2/2m_n$。假设 $R\sim10$ km、$k_n\sim1.4\times10^{13}$ cm 和 $\Delta_n=1$ MeV，则可以得到 $\xi_n\sim1.6\times10^{-12}$ cm、$\Omega_{c1}\sim10^{-14}\ s^{-1}$ 以及 $\Omega_{c2}\sim10^{20}\ s^{-1}$。每涡旋每中子对的角动量为 $\hbar$，所以当 $\Omega\sim2\times10^2\ s^{-1}$ 时，单位面积上的涡旋密度为 $\sim2m_n\Omega/\pi\hbar\sim2\times10^5\ cm^{-2}$，这对于蟹状星云脉冲星 *NP* 0532 来说是合适的。

正如一些作者所指出的[10,13,14]，中子星转速的减慢是由这类恒星中带电粒子与周

surrounding magnetosphere; the pulsar clock is determined by the rotation of the charged particles in the star. The principal energy source, however, is the rotational energy of the neutrons. One is immediately led to ask how the neutrons can transfer their rotational energy to the charged particles. If both the neutron and proton fluids are normal, any relative motion is quickly damped out by strong interactions in a characteristic time $\tau \approx 9\hbar^3/[32\sigma_{np}m_n (k_B T)^2] \sim 10^{-17}$ s in bulk matter (σ_{np} is a typical neutron-proton scattering cross-section and k_B is Boltzmann's constant). Should the protons be superfluid, coupling between the charged particles and the neutrons is still provided by the interaction of the electrons with the neutron magnetic moments. Such processes cause charged particle–neutron relative motion in bulk to relax in a time $\sim 10^{-11}$ s, a factor 10^6 greater than when the protons are normal, but still microscopic.

If the neutrons are superfluid, the situation changes drastically. Now the only effective scattering of the charged particles by the neutrons takes place in the normal cores of the neutron vortices. This increases the relaxation times by a factor $\Omega_{c2}/\Omega \sim 10^{18}$, which is a measure of the relative volume of the vortex cores. In the case of simultaneous proton and neutron superfluidity the coupling time reaches years, not altogether negligible on the evolutionary time scale of a neutron star. Any change in the angular momentum of the neutron superfluid must occur via the creation or destruction of vortices, either in the bulk of the superfluid or at the interface of the superfluid and normal regions. Estimating the rates for these processes remains, as in the case of laboratory superfluid helium, an unsolved problem. The observed decrease in the slow-down rate of the Vela pulsar (*PSR* 0833–45), following its sudden speed-up, is conceivably due to such a relaxation between the solid crust and the fluid interior. (The details of the calculations will be published later.)

We thank Professor John Bardeen for useful comments and the Aspen Center for Physics for hospitality during the preparation of this communication. This research is supported by the US National Science Foundation, the Army Research Office (Durham), and the US Air Force Office of Scientific Research.

(**224**, 673-674; 1969)

Gordon Baym, Christopher Pethick* and David Pines: Department of Physics, University of Illinois, Urbana, Illinois 61801.

Received September 2, 1969.

References:

1. Nemeth, J., and Sprung, D. W. L., *Phys. Rev.*, 176, 1496 (1968).
2. Ruderman, M., *Nature*, **218**, 1128 (1968).
3. Migdal, A. B., *Zh. Exp. Theor. Fiz.*, 37, 249 (1959).
4. Ginzburg, V. L., and Kirzhnits, D. A., *Zh. Exp. Theor. Phys.*, 47, 2006 (1964). English trans. in *Sov. Phys. JETP*, **20**, 1346 (1965).

* on leave of absence from Magdalen College, Oxford.

围磁层发生电磁耦合而导致的能量损失造成的；脉冲星的时钟取决于这类恒星中带电粒子的转动。但主要的能量来源仍是中子的转动能。于是我们马上会问：中子如何才能把转动能传递给带电粒子？如果中子流体和质子流体均处于正常态，那么任何相对运动都会因在大块物质中特征时标为 $\tau \approx 9\hbar^3/[32\sigma_{np}m_n(k_BT)^2] \sim 10^{-17}$ s（σ_{np} 是典型的中子－质子散射截面，k_B 为玻尔兹曼常数）的强相互作用而很快耗散掉。即使质子是超流体，带电粒子与中子的耦合仍由电子与中子磁矩的相互作用提供。这种作用使带电粒子与中子间大块相对运动的弛豫时间为 $\sim 10^{-11}$ s。这比质子处于正常态时大 6 个数量级，但仍然很小。

如果中子是超流体，情况就会发生显著的变化。此时，中子对带电粒子的有效散射仅发生在中子涡旋的正常态核心处。这将使弛豫时间增加 $\Omega_{c2}/\Omega \sim 10^{18}$ 倍，这个值也是涡旋核心相对体积的一种量度。在质子超流和中子超流同时存在的情况下，耦合时标将达到若干年，这相对于中子星演化的时间尺度来说是不能完全忽略不计的。中子超流体角动量的任何变化都必须通过生成或破坏涡旋来实现，无论是在超流体的内部还是在超流区与正常区的交界处。与在实验室超流氦研究中遇到的问题一样，怎样估计这些过程的速率也是一个尚未解决的问题。已经观测到船帆座脉冲星（*PSR* 0833–45）在突然加速之后慢下来的速率有所下降，这显然是由固体壳和液体核的弛豫过程造成的。（计算细节将在以后发表。）

感谢约翰 · 巴丁教授为我们提出了有价值的建议以及阿斯彭物理中心在准备此次会谈期间的热情招待。这项研究得到了以下单位的资助：美国国家科学基金会、陆军研究办公室（位于达勒姆）和美国空军科学研究办公室。

（岳友岭 翻译；蒋世仰 审稿）

5. Wolf, R. A., *Astrophys. J.*, **145**, 834 (1966).
6. Ruderman, M., in *Proc. Fifth Eastern Theor. Phys. Conf.* (edit. by Feldman, D.) (W. A. Benjamin, New York, 1967).
7. Ginzburg, V. L., *Uspekhi Fiz. Nauk*, **97**, 601 (1969).
8. Ginzburg, V. L., and Kirzhnits, D. A., *Nature*, **220**, 148 (1968).
9. De Gennes, P.-G., *Superconductivity of Metals and Alloys* (W. A. Benjamin, New York, 1966).
10. Gold, T., *Nature*, **218**, 731 (1968).
11. Ruderman, M., *J. de Physique* (in the press).
12. Kennedy, R., Wilets, L., and Henley, E. M., *Phys. Rev.*, **133**, B1131 (1964).
13. Pacini, F., *Nature*, **219**, 145 (1968).
14. Goldreich, P., and Julian, W. H., *Astrophys. J.*, **157**, 869 (1969).

Implications of the "Wave Field" Theory of the Continuum from the Crab Nebula

M. J. Rees

Editor's Note

By 1971 it was well accepted that pulsars were rotating neutron stars. But the mechanism that actually generates the pulse was not known. British astrophysicist Martin Rees here explores the observational consequences of a particular model known at the time as the "wave field theory". The model required that radiation from some parts of the nebula had to be circularly polarized. But subsequent observations showed an insufficient amount of polarized light, meaning that the model was incorrect.

THE Crab Nebula almost certainly derives a continuing power input from a rotating neutron star associated with the pulsar NP 0532. Studies by Ostriker and Gunn[1] and others of the "oblique rotator" model suggest that the rotational braking of pulsars may be primarily due to emission of electromagnetic waves at the rotation frequency of ~30 Hz (or low harmonics thereof). Indeed, it had been suggested[2-4], even before pulsars were discovered, that neutron stars might emit by this mechanism. In the case of the Crab Nebula, this view naturally suggests the further speculation that the continuum emission from the "amorphous mass" may arise from relativistic electrons moving in a low frequency wave, rather than the usual synchrotron process. This "wave field" model is especially attractive when one recalls the difficulty of accounting for the presence of a large-scale ordered magnetic field in the nebula. In this article I shall show that the "wave field" theory of the Crab Nebula leads to some distinctive predictions, especially regarding the polarization of the continuum. It may therefore be possible to test its validity and thus, indirectly, to infer something about the energetic link between the nebula and the pulsar.

The rate of emission by an isotropic distribution of relativistic electrons moving in an electromagnetic field is determined by the electromagnetic energy density. It is therefore convenient to express the intensity of the 30 Hz wave in terms of the magnetic field for which $H^2/8\pi$ equals the wave energy density. We find

$$H_{\text{eq}} \simeq 1.8 \times 10^{-4} \left(\frac{L_{30\ \text{Hz}}}{5 \times 10^{38}\ \text{erg s}^{-1}} \right)^{\frac{1}{2}} \left(\frac{r}{10^{18}\ \text{cm}} \right)^{-1} (1-\varepsilon)^{-\frac{1}{2}} \text{G} \tag{1}$$

This approximate relation assumes that the wave propagates with speed c, but ignores the fact that strict dipole emission would be twice as intense along the rotation axis as in the equatorial plane. ε is the effective albedo at the boundary of the nebula, and the factor

蟹状星云连续谱“波场”理论的推论

瑞斯

编者按

到 1971 年时，脉冲星是旋转中子星这一观点已经得到了人们的普遍认可。但产生脉冲的真实机制尚不清楚。在本文中英国天体物理学家马丁·瑞斯探讨了那时被称为“波场理论”的一个特定模型在观测上的推论。这个模型要求来自星云中某些部分的辐射必须是圆偏振的。但是随后的观测结果显示圆偏振光的量不够，因此这个模型还是被推翻了。

几乎可以肯定：蟹状星云从一颗与脉冲星NP 0532 相关的旋转中子星中获得了持续的能量注入。奥斯特里克和冈恩[1]以及其他人对“倾斜转子”模型的研究表明，脉冲星的转动变慢可能主要是因为在 ~30 Hz 转动频率（或者它的低谐频）处的电磁波发射。确实，甚至在脉冲星被发现以前就已经有人提出[2-4]中子星有可能会以这种机制发射辐射。就蟹状星云而言，这一观点很自然地使人联想到：来自那些“不定形物质”（译者注：指星云中脉冲星周围的等离子体）的连续谱辐射可能来源于在一个低频波中运动的相对论性电子，而非通常的同步辐射过程。如果我们回想起在解释星云中存在大尺度有序磁场时所遇到的困难，就会对这个“波场”模型格外感兴趣。在这篇文章中，我将向大家说明由蟹状星云的“波场”理论可以作出哪些与众不同的预言，尤其是在关于连续谱的偏振方面。这样就为今后检验它的正确性提供了可能，并由此间接地得出星云与脉冲星之间存在能量上的联系的推论。

由在电磁场中运动的相对论性电子的各向同性分布而引起的辐射率取决于电磁能量密度。因此可以很方便地用波的能量密度等于 $H^2/8\pi$ 的磁场表达出 30 Hz 波的强度。我们发现：

$$H_{\mathrm{eq}} \simeq 1.8\times10^{-4}\left(\frac{L_{30\ \mathrm{Hz}}}{5\times10^{38}\,\mathrm{erg\ s^{-1}}}\right)^{\frac{1}{2}}\left(\frac{r}{10^{18}\,\mathrm{cm}}\right)^{-1}(1-\varepsilon)^{-\frac{1}{2}}\,\mathrm{G} \qquad (1)$$

这个近似关系式假设波以速度 c 传播，但它忽略了这样一个事实，即对严格的偶极辐射而言，其沿旋转轴的强度应该是赤道面中的两倍。ε 是星云边界处的有效

$(1-\varepsilon)^{-\frac{1}{2}}$ allows for possible reflexions. For the Crab all the terms in brackets will be ~1. Thus the wave energy density is comparable with that of the weakest magnetic field (~10^{-4} G) permitted by energetic and dynamical considerations. It is also useful to define an "equivalent electron gyrofrequency" $\Omega/2\pi \simeq 3\times10^6 H_{eq}$ Hz, because the parameter $f = \Omega/\omega$, where ω is the wave frequency, determines both the character of the orbits of electrons exposed to the wave and the radiation which these particles emit.

Throughout the Crab Nebula, we expect $f \gtrsim 10$ (taking $\omega/2\pi \simeq 30$ Hz). In this situation a relativistic particle of Lorentz factor γ emits at frequencies $\sim\gamma^2\Omega$, as in the case of synchrotron radiation, and not $\sim\gamma^2\omega$, as for standard inverse Compton emission[5-7]. Observable features of this "synchro-Compton" emission—and, in particular, its polarization—will be considered here. For the moment, however, it is sufficient to note that, in general terms, the usual results of synchrotron theory still hold, so that the standard inferences of the electron density and spectrum in the nebula remain applicable.

Some of the rotational energy lost by an "oblique rotator" would accelerate relativistic particles in the vicinity of the speed of light cylinder[1], but for our considerations to be relevant it is essential that a substantial fraction (~10% at the very least) should escape into the nebula as 30 Hz radiation. It is interesting to investigate the eventual fate of this wave energy. When the wave reaches the boundary of the nebula, only a fraction $\sim v_{exp}/c$ of its energy is used in pushing against the external medium, v_{exp} being the expansion velocity of the boundary. The bulk of the energy must be either reflected or absorbed. As we shall see, the high linear polarization of the continuum from the nebula implies that the 30 Hz radiation must be ordered rather than random, and this precludes more than ~50% reflexion (so that, in equation (1), $\varepsilon \lesssim 0.5$). This means that most of the wave energy must be deposited in a thin "skin" at the boundary and at the inner edges of the filaments. The densities are so low that there is no possibility of this energy being radiated thermally, so it seems inevitable that relativistic particles, probably chiefly electrons, will be generated. Thus the pulsar would be almost 100% efficient as an accelerator—whatever fraction of the energy escapes into the wave zone as 30 Hz emission will produce fast particles in the outer parts of the nebula.

Even though the wave field simulates a stationary magnetic field as regards the radiation emitted by relativistic electrons (except, as we shall see, in the important respect of circular polarization) the particle orbits are very different. An electron's general motion is a superposition of a uniform translation velocity and a periodic relativistic oscillation with Lorentz factor $\sim f$. The form of this oscillation depends on the polarization properties of the low frequency wave: for linear polarization it is a "figure of eight" in a plane perpendicular to the H-vector, and for pure circular polarization it is a circle. All particles exposed to the wave must therefore be relativistic with $\gamma \gtrsim f$. Because the orbits of particles with $\gamma \gg f$ are basically straight lines, the wave field is ineffective for confining particles. But even a very weak magnetic field, which is negligible as regards the emission mechanism, could confine the particles adequately if it were sufficiently tangled. Alternatively the particles would be "mirrored" at the boundary by the external interstellar field and by the

反射率，而因子 $(1-\varepsilon)^{-\frac{1}{2}}$ 是允许有反射的。对蟹状星云来说，括号里的所有项都将为 ~1。因此波的能量密度就会与能量和动力学所允许的最弱磁场（$\sim10^{-4}$ G）的能量密度相当。定义一个“等价电子回旋频率” $\Omega/2\pi \simeq 3\times10^{6} H_{\rm eq}$ Hz 也是很有用的，因为参数 $f=\Omega/\omega$（其中 ω 为波的频率）决定了波中电子轨道的性质以及这些粒子发出的辐射。

对于整个蟹状星云，我们预言 $f\gtrsim10$（取 $\omega/2\pi\simeq30$ Hz）。在这种情况下，洛伦兹因子为 γ 的相对论性粒子会在频率 $\sim\gamma^2\Omega$ 下发射，而不会在频率 $\sim\gamma^2\omega$ 下发射；前者与同步辐射时的情形相同，后者则相当于标准的逆康普顿辐射 [5-7]。本文将讨论这种“同步－康普顿”辐射的观测特征，尤其是它的偏振。不过，就现在的情况而言，注意到以下这一点就足够了：在一般情况下，同步辐射理论中的常见结论仍然有效，因此关于星云中电子密度和光谱的标准推论还可以沿用。

在一个“倾斜转子”所损失的转动能中，有一部分将加速光速圆柱附近的相对论性粒子 [1]，但为了使我们的考虑有意义，其中必须有很大一部分（最少 ~10%）以 30 Hz 的辐射逃逸到星云中去。研究波能量的最终去向是很有趣的。当波到达星云的边界时，在它的能量中仅有 $\sim v_{\rm exp}/c$ 的部分被用于推动外面的介质，其中 $v_{\rm exp}$ 是边界的膨胀速度。大部分能量不是被反射就是被吸收。正如我们将要看到的，这个星云连续谱的高线偏振度意味着 30 Hz 辐射必定是有序的而不是随机的，这就排除了反射会高于 ~50% 的可能性（因而在式（1）中，$\varepsilon\lesssim0.5$）。这意味着波的大部分能量会积聚在边界处和丝状结构内边缘处的一层薄“外皮”里。这里密度太低，以至于这些能量不可能转化为热辐射，如此看来，产生相对论性粒子（可能主要为电子）就成为一种不可避免的趋势了。于是脉冲星将会和加速器一样达到接近于 100% 的效率——不管以 30 Hz 辐射逃逸到充满波的区域的能量比例有多大，都将在星云的靠外部分产生快速粒子。

尽管这种波场能够模拟一个与相对论性电子所发射的辐射相关的静态磁场（除了我们将要看到的，在圆偏振这一重要方面），但各个粒子的轨道迥然不同。电子的一般运动是一个均匀的平动速度和一个洛伦兹因子为 $\sim f$ 的周期性相对论振动的叠加。这种振动的形式取决于低频波的偏振性质：对线偏振而言，它在垂直于 H 矢量的一个平面内呈“8 字形”；对于纯粹的圆偏振，它是一个圆。因此，波中的所有粒子都必须是满足 $\gamma\gtrsim f$ 的相对论性粒子。因为 $\gamma\gg f$ 的粒子的轨道基本上是直线，所以这种波场不能有效地限制粒子。然而，即使是一个在涉及辐射机制时可以忽略的非常弱的磁场也足以把粒子限制住，只要它能被充分缠结。或者，粒子可能会在边界处被外部星际场和丝状结构“镜面反射”，如果这些结构含有磁场的话。

filaments if these contain magnetic fields.

Self-consistency demands that the plasma density within the nebula should be low enough to allow the 30 Hz radiation to propagate. At first sight one might suspect that the formal plasma frequency $9\times10^3 n_e^{\frac{1}{2}}$ Hz would have to be below 30 Hz, which would lead to the exceedingly stringent condition that the electron density n_e throughout the nebula be $\lesssim 10^{-5}$ cm^{-3}. In the case of a "strong" wave ($f \gtrsim 1$), however, this condition can be relaxed somewhat. For relativistic electrons with differential number spectrum $n(\gamma)$ the propagation condition is[8]

$$\int n(\gamma)\frac{\log\gamma}{\gamma}\mathrm{d}\gamma \lesssim 10^{-5}\ \mathrm{cm}^{-3} \tag{2}$$

Because all the particles exposed to the wave have $\gamma \gtrsim f$ the propagation condition is $n_e \lesssim 10^{-5} f$ (ref. 1). Relation (2) is satisfied, with a factor $\gtrsim 10$ to spare, by the particles with $\gamma \gtrsim 100$ whose density can be directly inferred from observations of the continuum emission from the Crab at frequencies above a few MHz. As relation (2) is obviously not fulfilled by the general interstellar medium, the 30 Hz waves cannot propagate beyond the boundary of the nebula—indeed, the observable nebula would, in this picture, be delineated by the region which has been evacuated sufficiently for the wave to propagate. Also, the waves would be unable to penetrate the filaments in the nebula.

It would be agreeable if the "wave field" theory led to a clear-cut prediction of the surface brightness distribution over the Crab Nebula, but unfortunately this is not the case. The r^{-1} dependence of H_{eq} means that a particle radiates more efficiently near the centre, but there would not necessarily be an enhanced surface brightness around the pulsar because this effect is counteracted by the tendency of particles to be excluded, by the force of the wave, from regions where $f \gtrsim \gamma^{\frac{1}{2}}$.

A distinctive consequence of the "wave field" theory is that most of the continuum originates in the half of the nebula farthest from the observer. This is because the synchro-Compton emission is due to electrons coming towards the observer, and the radiation rate per electron is larger for particles moving at large angles to the wave propagation direction. One might, in principle, detect such an effect by determining the Faraday rotation due to those filaments whose radial velocities indicate that they lie on the back side of the nebula.

What polarization should the Crab Nebula possess if the continuum is synchro-Compton emission rather than straightforward* synchrotron radiation? It is easy to see that the

* If, as was suggested most recently by Hoyle[9], the field in the Crab Nebula were attached to a spinning neutron star and had been amplified by being tightly wound, the emission would again not be straightforward synchrotron radiation because the scale of the field reversals would be small compared with the relativistic electron gyroradii. In any case, the particle density in the nebula would not be high enough to carry the currents associated with the rapid field reversals, so a magnetohydrodynamic treatment may be inappropriate in this situation.

自洽性要求星云内的等离子体密度应该足够低以至于 30 Hz 辐射可以传播。乍一看人们也许会表示怀疑，如果要使等离子体的正常频率 $9\times10^3 n_e^{\frac{1}{2}}$ Hz 下降到 30 Hz 以下，就会产生非常严格的条件，即整个星云的电子密度 n_e 必须 $\lesssim 10^{-5}\ \mathrm{cm}^{-3}$。然而，在“强”波 ($f\gtrsim1$) 的情况下，这个条件可以稍微放松一点。对于有差异性数量谱 $n(\gamma)$ 的相对论性电子，其传播条件为[8]：

$$\int n(\gamma)\frac{\log\gamma}{\gamma}\mathrm{d}\gamma \lesssim 10^{-5}\ \mathrm{cm}^{-3} \tag{2}$$

因为波中的所有粒子都满足 $\gamma\gtrsim f$，所以传播条件是 $n_e\lesssim10^{-5} f$（参考文献 1）。对于 $\gamma\gtrsim$ 100 的粒子，其密度可以直接从在高于几兆 Hz 频率观测蟹状星云连续谱辐射的结果中推测出来，这些粒子满足关系式（2），并富余一个 $\gtrsim10$ 的因子。因为一般的星际介质显然不满足关系式（2），所以 30 Hz 的波不能在星云边界之外传播——确实，在这个图景里，可观测的星云轮廓是由足够稀疏使得波可以传播的区域划定的。同时，波应该不能穿透星云中的丝状结构。

如果这个“波场”理论能够明确地预言整个蟹状星云的表面亮度分布，那当然再好不过；但不幸的是，情况并非如此。H_{eq} 对 r^{-1} 的依赖意味着粒子在靠近中心的地方辐射效率更高，但这并不一定能够说明脉冲星周围的表面亮度一定会增强，因为该效应被波的作用力将粒子从 $f\gtrsim\gamma^{\frac{1}{2}}$ 的区域驱逐出去的趋势抵消了。

“波场”理论的一个与众不同的推论是：连续谱辐射中的大部分源自星云离观测者最远的那一半。这是因为同步－康普顿辐射由朝向观察者运动的电子所产生，并且每个电子的辐射率对于相对波传播方向成大角度的粒子来说是比较大的。原则上，人们可以通过测量由丝状结构造成的法拉第旋转来探测这个效应，这些丝状结构的径向速度表明它们位于星云的背面。

如果连续谱是同步－康普顿辐射而不是直接的 * 同步辐射，那么蟹状星云的偏振应该是什么样的呢？很容易看到，在一个相干线偏振波中运动并满足 $\gamma\gg f$ 的电子

* 如果，正如霍伊尔[9]最近提出的，蟹状星云中的场附属于一颗自旋的中子星并且由于缠绕得很紧而被放大，那么其所发出的辐射同样不会是直接的同步辐射，因为场反转的尺度与相对论性电子的回旋半径相比是很小的。在任何情况下，星云中的粒子密度应该都不会高到足以承载与快速场反转有关的电流的程度，因此在这种情况下用磁流体动力学进行处理是不合适的。

synchro-Compton emission from an electron with $\gamma \gg f$ moving in a coherent linearly polarized wave will itself be highly linearly polarized in the direction of the projected E-vector of the low frequency wave. (Remember that the electron is deflected both by the E and the B-field of the wave.) When the low frequency wave is circularly polarized, the polarization of the synchro-Compton emission is best estimated by transforming to the "guiding centre" frame. In this frame an electron executes a circular orbit, with Lorentz factor f, in a plane which (except when the particle is traveling nearly in the propagation direction of the wave) is almost perpendicular to the direction of its mean velocity v. This motion would give rise to synchrotron-type radiation concentrated in a "fan" at angles $\pi/2 \pm O(f^{-1})$ to v. This radiation would be circularly polarized in opposite senses on the two sides of the orbital plane. In the transformation from the guiding centre frame to the source frame, however, the factor $(1-v/c \cos \theta)$ in the Doppler formula favours the emission from the forward hemisphere by a factor $\sim(1+f^{-1})/(1-f^{-1})$, which leads to a net polarization $\sim f^{-1}$. A more refined calculation using the methods of Roberts and Komesaroff[10] leads to the estimate $\sim 0.6\, f^{-1}$. Therefore, synchro-Compton emission by electrons of all energies can possess circular polarization of order f^{-1}. The synchro-Compton continuum from any distribution of electrons moving in a circularly polarized low frequency wave should therefore be circularly polarized to this extent (the precise degree of polarization depending on the slope of the electron spectrum). This contrasts with the γ^{-1} dependence expected for synchrotron radiation[11], which leads to an undetectably small predicted degree of circular polarization in most astronomical objects even at radio frequencies.

The electromagnetic waves emitted by an ideal spinning magnetic dipole *in vacuo* would, in the equatorial plane, be completely linearly polarized, the electric vector lying perpendicular to the plane. At higher latitudes the waves would be elliptically polarized, and along the rotation axis the polarization would be purely circular. In order to compare the polarization predicted by the "oblique rotator" model with the observations of the Crab Nebula, we must assume an orientation for the dipole. In certain pulsar models the existence of an interpulse, as seen in NP 0532, indicates that the observer lies close to the equatorial plane. Guided by this, let us suppose that the rotation axis of NP 0532 lies in the plane of the sky. Then, provided that relation (2) is satisfied by a large enough margin for the effects of the medium to be negligible, the synchro-Compton radiation from the equatorial plane should be highly linearly polarized parallel to the rotation axis. The direction of linear polarization will be similar at other latitudes (and even along lines of sight intercepting the rotation axis there will be a linearly polarized contribution). Both the optical and the radio observations show fairly uniform linear polarization in a NW–SE direction over the inner part of the Crab Nebula. This would therefore imply that the pulsar's rotation axis also pointed in this direction.

The observed continuum from a point in the nebula at latitude θ relative to the pulsar should therefore display $\sim 60\, f^{-1} \sin \theta$ percent circular polarization. If the pulsar's rotational axis lies in the plane of the sky, all points along a single line of sight through the nebula would contribute the same degree of circular polarization (the latitude-dependence being

所发出的同步－康普顿辐射本身就会是高度线偏振的，偏振的方向是低频波投影的 E 矢量方向。（记住：电子是在波的 E 场和 B 场的共同作用下偏转的。）当低频波为圆偏振时，最好通过变换至“导向中心”参考系来估算同步－康普顿辐射的偏振。在这个参考系中，电子是在一个几乎垂直于其平均速度 v 的平面内作圆轨道运动（除非当粒子近似沿着波的传播方向运动时）的，洛伦兹因子为 f。这种运动将产生集中在一个与 v 成 $\pi/2 \pm O(f^{-1})$ 角范围内的“扇形区域”中的同步辐射型辐射。这个辐射在轨道面的两边应该是方向相反的圆偏振。然而，在从导向中心参考系到源参考系的变换中，多普勒公式里的 $(1-v/c\cos\theta)$ 因子使得来自前面半球的辐射多出一个 $\sim(1+f^{-1})/(1-f^{-1})$ 因子，这就导致了一个净的 $\sim f^{-1}$ 的偏振。采用罗伯茨和科梅萨罗夫[10]的方法对上述过程进行更为精确的计算，得到的结果是 $\sim 0.6\, f^{-1}$。因此，各种能量电子的同步－康普顿辐射可能会有量级为 f^{-1} 的圆偏振。由此可知，来自在一个圆偏振低频波中运动的任意分布电子的同步－康普顿连续谱都应该有这么高的圆偏振（偏振度的具体数值取决于电子谱的斜率）。这与同步辐射对 γ^{-1} 的依赖[11]相对，因而在大多数天体中即使是在射电频率下也只能预测出有探测不到的极小的圆偏振度。

由在真空中快速旋转的理想磁偶极子发出的电磁波在赤道面内应该是完全线偏振的，其电矢量垂直于赤道面。在纬度更高的地方，波应该是椭圆偏振的，而在沿旋转轴的方向，偏振应为纯粹的圆偏振。为了将由“倾斜转子”模型预言的偏振与对蟹状星云的观测结果进行比较，我们必须假设偶极子的取向。在某些脉冲星模型中，存在中间脉冲即说明观测者靠近赤道面，正如在 NP 0532 中所看到的情况。以此为依据，让我们假设 NP 0532 的旋转轴位于天球切面中。然后，只要一个介质效应可以忽略的足够大的边缘能够满足关系式（2），那么来自赤道面的同步－康普顿辐射应该在与旋转轴平行的方向上是高度线偏振的。在其他纬度处，线偏振的方向类似（甚至在沿着与旋转轴相交的视线方向上也会有线偏振的贡献）。光学和射电观测结果都显示出在蟹状星云的靠内部分沿西北—东南方向有相当均匀的线偏振。因此这可能表明脉冲星的旋转轴也指向这个方向。

因此，从星云中相对于脉冲星而言纬度为 θ 的一个点观测到的连续谱应该显示出百分比为 $\sim 60\, f^{-1}\sin\theta$ 的圆偏振。如果脉冲星的旋转轴位于天球切面中，则在一条穿过星云的单一视线上的所有点都应对圆偏振度有相同的贡献（纬度依赖性被 f 对

cancelled by the r^{-1} dependence of f). The predicted degree of circular polarization would vary over the nebula, being proportional to the angular distance from the line of sight to the projected equatorial plane (assumed to be a line running NE–SW through the pulsar). The circular polarization should be greatest in the outermost parts of the nebula in the NW and SE directions, and should amount to a few percent. It would have opposite senses on the two sides, so the net circular polarization of the whole nebula could be very slight.

(Towards the outer parts of the nebula, one would expect a gradual increase in the density of low-γ electrons, which may cause the refractive index μ at 30 Hz (which is related to the integral in relation (2)) to drop significantly below unity. When $\mu<1$ the $\langle E^2\rangle/\langle B^2\rangle$ ratio of an electromagnetic wave exceeds its vacuum value, and the foregoing discussions of the polarization of synchro-Compton radiation are somewhat modified[5,6]. In particular, even when the low frequency wave is unpolarized, the synchro-Compton radiation will display linear polarization perpendicular to the projected propagation direction of the wave. Also, when μ differs from unity the low frequency wave will be refracted, and can no longer be assumed to propagate radially outward from the pulsar.)

In conclusion, the simple "wave field" model can account naturally for the high linear polarization in the Crab Nebula, and for the required highly efficient acceleration of relativistic electrons. The direction of the linear polarization determines the orientation of the rotation axis, so any independent evidence of the pulsar's alignment would provide a test of the theory. The simple model also indicates that the NW and SE parts of the nebula should display circular polarization—at radio, optical and X-ray wavelengths—amounting to a few percent. The estimated circular polarization would be somewhat reduced if the magnetic field of the pulsar were more complicated than a simple dipole, or if allowance were made for deflexion of the low frequency waves by irregularities within the nebula. Provided, however, that the low frequency waves are sufficiently well ordered to account for the observed linear polarization, it does not seem possible to reduce the predicted circular polarization by more than a factor ~2. Also, of course, circular polarization cannot, like linear polarization, be smeared out by differential Faraday rotation or similar effects. Therefore, if it were found that no parts of the Crab Nebula displayed even ~ 1% circular polarization, this would indicate that the 30 Hz waves certainly do not penetrate beyond the region of the wisps and would be strong evidence against the "wave field" model for the nebula. It would also suggest that the popular "oblique rotator" magnetic dipole model for NP 0532 would require some reappraisal.

I acknowledge valuable discussions with Drs. S. A. Bonometto, J. E. Gunn, F. C. Michel, J. P. Ostriker and V. L. Trimble.

(**230**, 55-57; 1971)

M. J. Rees: Institute of Theoretical Astronomy, Madingley Road, Cambridge.

Received December 15, 1970.

r^{-1} 的依赖所抵消）。预测出的圆偏振度在星云内会有所不同，它与从视线到投影赤道面（假设为一条沿东北—西南方向穿过脉冲星的直线）的角距离成正比。圆偏振度应该在星云最外部的西北和东南方向上最高，并且应该能达到百分之几。它在两侧有相反的偏振方向，所以整个星云的净圆偏振可能会非常小。

（对于星云靠外的部分，我们可以预期低 γ 电子的密度会逐渐增加，这有可能导致 30 Hz 处的折射率 μ（与关系式（2）中的积分有关）降低到显著小于 1。当 μ<1 时，电磁波的 $\langle E^2\rangle/\langle B^2\rangle$ 值将超过其真空值，因而要对前面所讨论的同步 – 康普顿辐射的偏振情况进行一些修正 [5,6]。尤其是，即使在低频波无偏振时，同步 – 康普顿辐射也会显示出垂直于投影的波传播方向的线偏振。同样，当 μ 偏离 1 时，低频波将会被折射，因而不能再假设它是沿径向从脉冲星向外传播的。）

总之，利用简单的“波场”模型可以很自然地解释蟹状星云中的高线偏振度以及所要求的相对论性电子的高效加速。线偏振的方向决定了旋转轴的取向，所以任何关于脉冲星取向的独立证据都可以作为对这个理论的一种检验。这个简单的模型还表明，星云的西北和东南部分应该显示出百分之几的圆偏振，在射电、光学和 X 射线波段都是如此。如果脉冲星的磁场比一个简单的偶极子更复杂，或者如果考虑到低频波会因星云内的不规则结构而发生偏折，那么估算出的圆偏振度就会有所降低。然而，如果低频波的有序度非常高以至于可以用来解释观测到的线偏振，那么预测出的圆偏振就不太可能会降低至 50%以下。此外，圆偏振当然不会像线偏振那样被较差的法拉第旋转或与之类似的效应抹掉。因此，如果发现在蟹状星云中没有任何一个部分能显示出哪怕 ~1% 的圆偏振，这就可以说明，30 Hz 的波必然没有穿透波束区并且将可以作为反驳星云“波场”模型的有力证据。这还将表明，我们必须对得到广泛认可的 NP 0532 的“倾斜转子”磁偶极模型进行重新评估。

感谢博诺梅托博士、冈恩博士、米歇尔博士、奥斯特里克博士和特林布尔博士和我进行了很多次有价值的讨论。

（钱磊 翻译；蒋世仰 审稿）

References:

1. Ostriker, J. P., and Gunn, J. E., *Astrophys. J.*, **157**, 1395 (1969).
2. Hoyle, F., Narlikar, J. V., and Wheeler, J. A., *Nature*, **203**, 914 (1964).
3. Wheeler, J. A., *Ann. Rev. Astron. Astrophys.*, 4, 393 (1966).
4. Pacini, F., *Nature*, **216**, 567 (1967).
5. Rees, M. J., *IAU Symposium No. 46*, "*The Crab Nebula*" (edit. by Davies, R. D., and Smith, F. G.) (Dordrecht, D. Reidel, in the press).
6. Rees, M. J., *Nature*, **229**, 312 (1971).
7. Gunn, J. E., and Ostriker, J. P., *Astrophys. J.* (in the press).
8. Zheleznakov, V. V., *Sov. Astron.*, **11**, 33 (1967).
9. Hoyle, F., *Nature*, **223**, 936 (1969).
10. Roberts, J. A., and Komesaroff, M. M., *Icarus*, 4, 127 (1965).
11. Legg, M. P. C., and Westfold, K. C., *Astrophys. J.*, **154**, 499 (1968).

Carbon Compounds in Apollo 12 Lunar Samples

B. Nagy *et al.*

Editor's Note

Samples of lunar "soil" (mineral powder, or regolith) were brought back by Apollo 11, the first manned mission to the Moon's surface. The Apollo 12 mission was more systematic in collecting rocks and regolith. Here geochemist Bartholomew Nagy and his coworkers, including the Nobel laureate chemist Harold Urey who pioneered research on the chemical origin of life, report that the Apollo 12 samples contain traces of organic (carbon-based) compounds, including small traces of amino acids (potentially contaminants from handling). While some carbon on the lunar surface originates in the stream of charged particles coming from the Sun (the solar wind), carbon deeper in the rock samples was puzzling. There was no suggestion, however, of an origin involving life.

SMALL quantities of carbon compounds have been found in three samples of surface fines (12001, 34, 12033, 8 and 12042, 13), in one sample from the bottom of a 15 cm deep trench dug by the astronauts (12023, 8), in two core samples from 9 and 39 cm below the lunar surface (12025, 60 and 12028, 207, respectively) and in the interior of a lunar rock (12022, 81). The locations where these samples were collected, the site of the lunar descent module, and the basic topography of the collection area are shown in Fig. 1. We identified hydrocarbons (in the parts per billion range, except CH_4) using a pyrolysis instrument containing a helium atmosphere at 700°C, connected to a tandem gas chromatograph-mass spectrometer. The gaseous components were analysed by vacuum pyrolysis at 681°C and 1,084°C. Some gaseous components were separated with high resolution mass spectrometry. A search for amino-acids was carried out using Soxhlet extraction with hot water and ion exchange chromatography. In addition, an extensive examination with transmitted light and scanning electron microscopy was made in an attempt to establish the locales of the organic molecules and to shed some light on their possible origin(s) and history. It is difficult to interpret the results unequivocally, which is not surprising in view of the difficulties encountered with the explanations of the Apollo 11 and 12 findings. This is illustrated by the apparent disparity in age of lunar fines and rocks[1,2].

阿波罗12号月球样品中的含碳化合物

纳吉等

编者按

人类搭乘阿波罗 11 号首次登上了月球，并带回了一些月球表面的“土壤”（矿物粉末或表土）。在阿波罗 12 号登月任务中，采集岩石和表土样品的工作就更加系统化了。地球化学家巴塞洛缪·纳吉及其同事（包括诺贝尔化学奖获得者哈罗德·尤里）率先发起了对生命化学起源的研究，他们在本文中报告称：在阿波罗 12 号采集的样品中含有痕量的有机物（碳基化合物），包括超痕量的氨基酸（可能源自处理样品时的污染）。虽然可以认为月球表面上的碳是由来自太阳的带电粒子流（太阳风）产生的，但岩石深处的碳实在令人费解。然而，这一研究并没有带给我们关于生命起源的更多提示。

目前已经在以下这些样品中发现了少量的含碳化合物：三份表面微粒土样品（12001, 34；12033, 8 和 12042, 13），一份宇航员从 15 cm 深的沟槽底部挖出的样品（12023, 8），两份在月球表面以下 9 cm 和 39 cm 处采得的土芯样品（分别为 12025, 60 和 12028, 207）以及一份取自月球岩石内部的样品（12022，81）。上述样品的采集地点、登月舱的降落位置以及采集区域的基本地貌见图 1。我们是利用工作在 700℃氦气氛下并与气相色谱 – 质谱联用仪相连的热解装置鉴定出碳氢化合物（除 CH_4 外都在十亿分之一或称 p.p.b. 量级）的。对气体组分的分析采用 681℃和 1,084℃真空热解。有些气体组分是用高分辨质谱进行分离的。在寻找氨基酸的过程中我们使用了以热水为溶剂的索氏提取法和离子交换色谱法。此外，我们还用透射光显微镜和扫描电子显微镜进行了大范围的检查，目的在于确定有机分子所处的位置并尝试说明它们的来源和历史。很难毫不含糊地对研究结果加以解释，从解释阿波罗 11 号和阿波罗 12 号发现物时所遇到的种种困难来考虑，这并不令人惊奇。可以用月球上的微粒土和岩石在年代上的明显不同 [1,2] 来说明两者之间的差异。

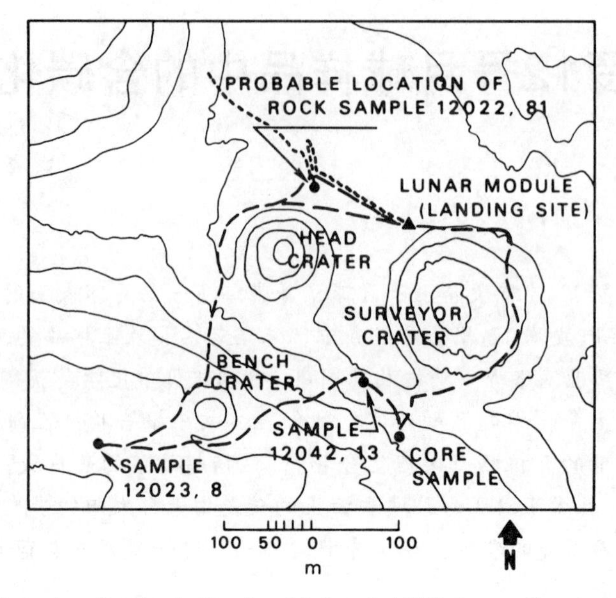

Fig. 1. Location of samples referred to in Figs. 2 and 3. Sample 12033, 8 was collected north of Head Crater on the second Extra Vehicular Traverse. The location of sample 12001, 34 was not accurately noted by NASA, although it was indicated that it was collected near the landing module (modified from ref. 10). - - -, First Eva traverse; – – –, second Eva traverse; ——, elevation contour lines at 5 feet intervals.

Vacuum Pyrolysis

Lunar samples were pyrolysed *in vacuo* in a quartz tube designed to fit the gas inlet system of the mass spectrometer. This system was modified so that all evolved gases could be directed into the ion source of the mass spectrometer on reaching the desired pyrolysis temperatures. Before pyrolysis, the quartz tube containing the sample was degassed for 20 min by the mass spectrometer pumping system. A portable furnace was then placed around the quartz tube. The pyrolysis temperatures were kept constant during the mass spectral analyses.

Total carbon based on p.p.m. of carbon from CH_4, CO, and CO_2 was determined for Apollo 12 fines, core material and one rock chip. Before mass spectrometric determinations were made, calibration curves were calculated from various quantities of these gases. To prevent ambiguity with respect to the amount of CO in the lunar samples, we made high resolution determinations at $m/e = 28$, to distinguish the relative intensities of CO, N_2, and C_2H_4.

All lunar samples were vacuum pyrolysed at 681°C. After pumping out the gases from the mass spectrometer, the samples were subsequently pyrolysed at 1,084°C and analysed as before. The contribution of carbon to the gases is shown in Table 1. The core samples and the fines showed CO_2 in almost equal quantities from the 681°C and 1,084°C pyrolysis. In addition, the total carbon from CO_2 was approximately the same in surface fines and

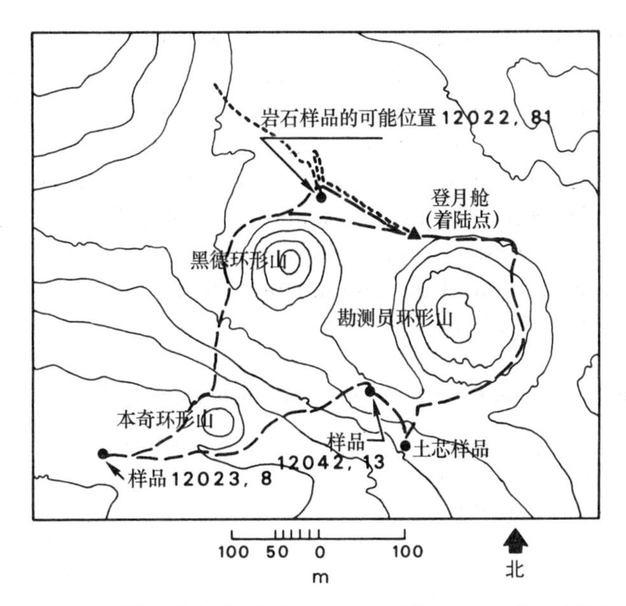

图 1. 在图 2 和图 3 中所涉及的样品的位置。样品 12033，8 是在第二次舱外活动路线上的黑德环形山北部采集的。美国国家航空航天局没有准确地注明样品 12001, 34 的位置，仅指出是在着陆舱附近采集到的（根据参考文献 10 进行了修改）。- - -，第一次舱外活动路线；– – –，第二次舱外活动路线；——，5 英尺间隔的等高线。

真空热解

将月球样品置于为适应质谱的气体进样系统而设计的石英管内并在真空中热解。调节系统以保证生成的所有气体在到达所需的热解温度时都能被导入到质谱仪的离子源中。在热解前，用质谱仪的真空系统对盛放样品的石英管进行 20 min 的除气处理。接着在这个石英管的周围安插一个可移动的加热炉。在质谱分析过程中热解温度保持恒定。

阿波罗 12 号采集的微粒土、土芯物质和岩屑样品中的总碳量来自 CH_4、CO 和 CO_2，以 p.p.m.（译者注：百万分之一）表示。在进行质谱检测前，我们根据对不同量上述气体的计算作出相应的校准曲线。为了避免在判断月球样品中的 CO 含量时出现不确定性，我们在 m/e = 28 处进行了高分辨测定，以便区分出 CO、N_2 和 C_2H_4 的相对强度。

所有的月球样品都在 681℃下进行真空热解。当质谱仪抽空所有气体后，样品接着在 1,084℃下热解，并按前述方法进行分析。表 1 中列出了碳在气体中的含量。根据 681℃和 1,084℃的热解结果，土芯样品和微粒土中的 CO_2 量几乎相同。此外，在表面微粒土样品和两份土芯样品中，来自 CO_2 的总碳量是大致相同的。不过，岩

in the two core samples. The rock chip, however, yielded 80% of the total CO_2 present when pyrolysed at 681°C and contained almost three times more CO_2 than any other sample. The CO content was roughly half of the amount found in all other samples. In all other samples pyrolysis at 1,084°C released 90% of the CO. CH_4 was produced in approximately equal amounts at both pyrolysis temperatures in all samples. The rock chip produced the most CH_4. Some of the CH_4 could be attributed to contamination, because the Ottawa quartz sand control from the LRL cabinet where the lunar samples were processed yielded about one-third the amount of CH_4 found in the lunar samples when pyrolysed under identical conditions.

Table 1. Vacuum Pyrolysis, Combined Results for 681°C and 1,084°C

Sample	p.p.m. C from CO, CO_2 and CH_4
12028, 207 (core 39 cm below lunar surface)	96.2
12025, 60 (core 9 cm below lunar surface)	111.2
12042, 13	140.8
12001, 34	115.2
12023, 8	109.2
12033, 8	108.6
12022, 81 (rock chip from interior of lunar rock)	89.6

All of the Apollo 12 lunar samples (except the rock chip) received in this laboratory were stored under nitrogen at the LRL. Pyrolysis at 681°C with high resolution determinations at m/e =28 showed a large amount of nitrogen for all samples. The nitrogen content was approximately 40% greater than the CO content at 681°C, even though the samples were placed under high vacuum for 20 min before heating. The subsequent pyrolysis at 1,084°C showed a considerable decrease in the nitrogen content to less than one-third that of CO.

Helium Pyrolysis

The lunar samples were pyrolysed in a stream of He at 700°C for 7.5 min in a modified Hamilton pyrolysis unit[3-5]. The He stream (3–5 ml./min) from the fused quartz pyrolysis tube was connected to a Perkin–Elmer model 226 gas chromatograph. It also served as the carrier gas for this unit. In turn the gas chromatograph was connected by means of a Watson–Biemann molecular separator to a Hitachi RMU-6E mass spectrometer. A small trap, cooled with liquid nitrogen, was placed between the pyrolyser unit and the gas chromatograph so that, following pyrolysis, the trapped pyrolysis products could be introduced as a single sample into the gas chromatograph by means of a small portable oven which quickly heated the trap to 250°–300°C. The gas chromatograph contained a 50 feet × 0.02 inch internal diameter polyphenyl ether capillary column and a hydrogen flame detector.

Because it has been reported that similar aromatic hydrocarbons can be synthesized when methane is heated to 1,000°C over silica gel[6], we passed methane over pre-pyrolysed lunar fines (12001, 34) at 700°C for 7.5 min. No evidence of any synthesis was observed (Fig. 3*E*). It could be argued that the initial pyrolysis, which removed the carbon compounds,

屑样品在 681℃的热解中产生了总 CO_2 量的 80%，并且所含的 CO_2 量近乎是任何其他样品的 3 倍。CO 的含量大致为所有其他样品中含量的一半。所有其他样品在 1,084℃热解时都释放出 90%的 CO。全部样品在两个热解温度下都产生了近乎等量的 CH_4。岩屑样品产生的 CH_4 最多。可以把一部分 CH_4 归因于污染，因为来自专门处理月球样品的月球物质回收实验所处理间的渥太华石英砂对照样品在同样条件下被热解时，也产生了约为在月球样品中发现量的 1/3 的 CH_4。

表 1. 真空热解，综合 681℃和 1,084℃下的结果

样品	来自 CO、CO_2 和 CH_4 的 p.p.m. 碳量
12028, 207（月球表面以下 39 cm 处的土芯）	96.2
12025, 60（月球表面以下 9 cm 处的土芯）	111.2
12042, 13	140.8
12001, 34	115.2
12023, 8	109.2
12033, 8	108.6
12022, 81（月球岩石内部的岩屑）	89.6

本实验室接收的所有由阿波罗 12 号采集的月球样品（除岩屑外）都曾被月球物质回收实验所保存于氮气中。在 m/e=28 处对 681℃下热解产物进行高分辨测定所得的结果表明：所有样品都含有大量的氮气。在 681℃时氮气的量比 CO 的量多出约 40%，尽管在加热前我们已将样品置于高真空中长达 20 min。随后在 1,084℃下的热解结果显示，氮气含量大幅下降至不足 CO 量的 1/3。

氦 热 解

将月球样品置于改进的哈密顿热解装置中，在 700℃下于氦气流中热解 7.5 min[3-5]。来自熔凝石英热解管的氦气流（3 ml/min ~ 5 ml/min）被接入到一台珀金－埃尔默 226 型气相色谱仪上。氦气也是这个装置中的载气。接着将气相色谱仪经由一台沃森－比曼分子分离器连接到日立 RMU-6E 质谱仪上。热解器和气相色谱仪之间有一个用液氮来冷却的小型冷阱，这样，在热解之后收集到的热解产物就可以各自作为单独的样品通过一台能迅速将冷阱加热至 250℃ ~ 300℃的小型可移动加热器而引入到气相色谱仪中。该气相色谱仪中装有一根内径为 50 英尺 × 0.02 英寸的聚苯醚毛细管柱和一台氢焰检测器。

由于已经有人报道过将甲烷置于硅胶上加热到 1,000℃就可以生成类似芳香族化合物的烃类物质[6]，我们将甲烷在 700℃时通过预热解过的月球微粒土（12001, 34）长达 7.5 min。没有观测到有任何合成现象出现（图 3*E*）。有人可能会说：初始热解

destroyed any "catalytic" effect the fines might have had before heating. This argument is, however, tenuous because of the lack of experimental evidence. Another He pyrolysis control was run using a CO–H_2 (1/2.5, v/v) gas mixture over pre-pyrolysed lunar fines (12023, 8) at 700°C for 7.5 min, and did not show any of the hydrocarbons found in the lunar samples.

He pyrolysis of lunar fines, the two core samples, and a rock chip (Figs. 2 and 3) showed the presence of primarily CO, CO_2, CH_4 and H_2. Many organic compounds were present in the p.p.b. range, together with lesser amounts of Ne and possibly COS. (See captions for Figs. 2 and 3.) It may be possible, however, that some of the higher molecular weight hydrocarbons were artefacts resulting from the He pyrolysis technique. In a control experiment benzene was pyrolysed in He at 600°C and yielded traces of, predominantly, biphenyl with lesser amounts of naphthalene. Pre-Cambrian rock samples which had given He pyrolysis products similar to those found in the lunar samples were subjected to vacuum pyrolysis at 500°C and 600°C using a liquid nitrogen cold trap and showed only ions of m/e=76, 78, 91, 92, 154, 177 which indicate the presence of benzene, toluene, alkyl benzenes, biphenyl, and so on, plus smaller molecular weight hydrocarbons below m/e=78. It was also determined in a separate experiment with this cold trap vacuum pyrolysis technique that biphenyl can be destroyed by pyrolysis at temperatures of 700°C and above. The results showed degradation and intramolecular rearrangement to products such as naphthalene.

过程去除了含碳化合物，因而破坏了微粒土在加热前可能会具有的“催化”效应。不过因为缺乏实验证据，所以这种说法是空洞无力的。我们还进行了另一组氦热解的对照实验，即将 CO–H_2（体积比为 1/2.5）气体混合物在 700℃时通过预热解过的月球微粒土样品（12023,8）长达 7.5 min，结果在月球样品中并未发现有任何烃类物质。

月球表面微粒土、两份土芯样品和一片岩屑的氦热解结果（图 2 和图 3）显示主要产物为 CO、CO_2、CH_4 和 H_2。很多有机物的存在量在 p.p.b. 级的范围之内，还有更微量的 Ne 和可能存在的 COS。（参见图 2 和图 3 的说明文字。）不过，某些分子量较高的烃类物质有可能是由氦热解技术导致的人工产物。在对照实验中，将苯在 600℃氦气氛下热解，产生的痕量物质主要为联苯，并含有少量萘。前寒武纪岩石样品的氦热解产物与在月球样品中所发现的类似；将前寒武纪岩石样品在 500℃和 600℃下进行真空热解并由液氮冷阱引入质谱，结果只发现了 m/e = 76、78、91、92、154 和 177 的离子，表明存在苯、甲苯、烷基苯和联苯等物质以及低于 m/e = 78 的较小分子量的烃。在另一次单独的实验中，我们还用这种冷阱真空热解技术证明：在 700℃或更高温度下的热解中，联苯是会被破坏的。这些结果表明联苯通过降解和分子内重排生成了诸如萘等产物。

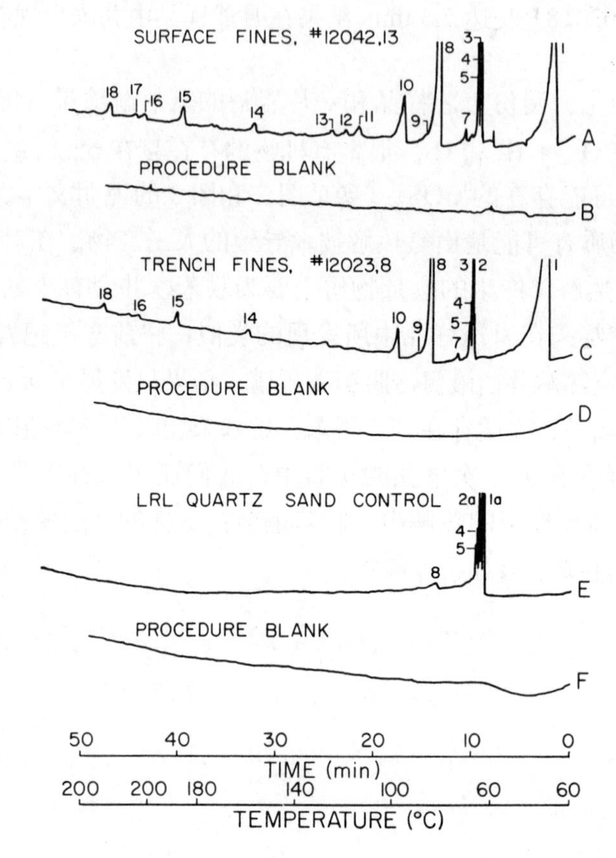

Fig. 2. Gas chromatograms of the pyrolysis products of *A*, surface lunar fines (sample 12042, 13), *C*, lunar trench fines (sample 12023, 8), and *E*, LRL Ottawa quartz sand control. Peak identities, all traces: 1, H_2, CH_4, CO; 2, C_2H_4, C_2H_6; 3, CO_2, C_3H_6; 4, C_4H_8; 5, C_4H_6; 6, C_5H_8 (small peak preceding peak 7 in trace *A* only); 7, C_5H_6; 8, benzene; 9, thiophene; 10, toluene; 11, C_2 alkyl benzene;12, o-xylene; 13, styrene; 14, indene; 15, naphthalene; 16 and 17, methyl naphthalenes; 18, biphenyl. For trace *E* only: 1a, CH_4, C_3H_8; 2a, CO_2, C_2H_4. Additionally, on trace *A*: 1, also contains Ne; 2, C_2F_4; and on trace *C*: 2, CO_2. Compounds 9, 11 and 13–18 may well be artefacts synthesized in the pyrolyser as Nagy and Preti *et al.* pointed out[12,13].

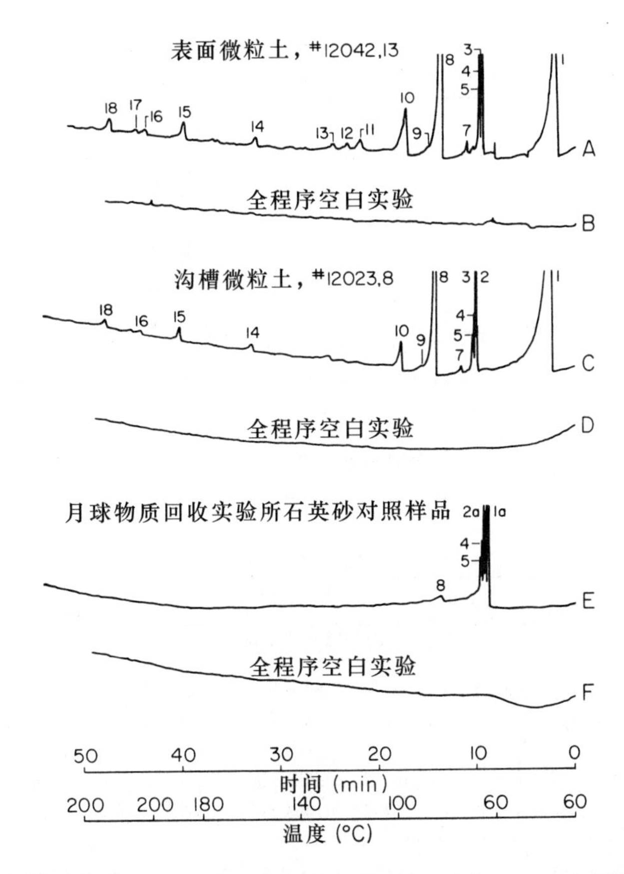

图 2. 热解产物的气相色谱图。*A*，月球表面微粒土（样品12042, 13）；*C*，月球沟槽中的微粒土（样品12023, 8）；*E*，月球物质回收实验所的渥太华石英砂对照样品。谱峰代号如下，对所有迹线：1，H_2、CH_4、CO；2，C_2H_4、C_2H_6；3，CO_2、C_3H_6；4，C_4H_8；5，C_4H_6；6，C_5H_8（7号峰前面的小峰，仅出现在迹线*A*中）；7，C_5H_6；8，苯；9，噻吩；10，甲苯；11，取代基为二碳的烷基苯；12，邻二甲苯；13，苯乙烯；14，茚；15，萘；16和17，甲基萘；18，联苯。迹线*E*特有的峰：1a，CH_4、C_3H_8；2a，CO_2、C_2H_4。此外，对于迹线*A*：1，还含有Ne；2，C_2F_4。对于迹线*C*：2，CO_2。正如纳吉和普雷蒂等人所指出的那样，化合物 9、11以及13～18很可能是在热解器中生成的人工产物[12,13]。

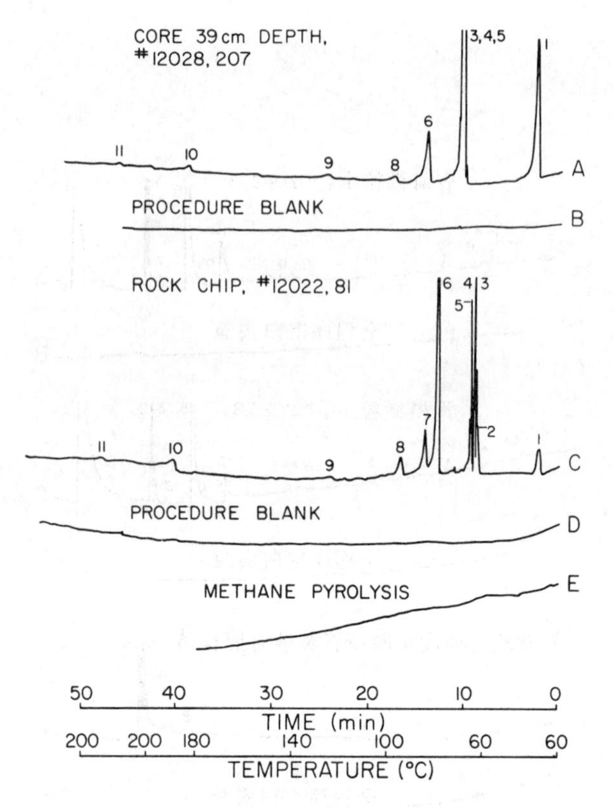

Fig. 3. Gas chromatograms of the pyrolysis products of (*A*) lunar core, 39 cm depth (sample 12028, 207); (*C*) lunar rock chip (sample 12022, 81), and (*E*) CH_4 pyrolysis over pre-pyrolysed lunar fines (sample 12001, 34). Peak identities, all traces: 1, H_2, CH_4, CO; 2, CH_4; 3, C_2H_4; 4, C_3H_6; 5, C_4H_8; 6, benzene; 7, thiophene; 8, toluene; 9, styrene; 10, naphthalene; 11, biphenyl. Additionally, on trace *A*: 3 also contains C_2F_4, $C_2H_2F_2$; 4, $C_4F_4H_2$; 5, C_3F_6; 6, C_4H_6; and on trace *C*: 1, CO absent; 3, CH_4, CO, CO_2; 4, CO_2, Ne, possibly COS; 5, CO_2, C_3H_6. Compounds 7 and 9–11 can be artefacts.

Trace amounts of fluorocarbon compounds were also observed in some of the pyrolysed samples, but these were probably "Teflon" contaminations. A "Teflon" control pyrolysis, using "Teflon" supplied and used by the NASA Manned Spacecraft Center, showed these same fluorocarbon compounds. An Ottawa quartz sand control, which was exposed in the same cabinet at the LRL where the lunar samples were processed, was also He pyrolysed; Fig. 2*E* shows its gas chromatogram. Note that it shows no compounds beyond the small benzene peak. It has been reported (R. B. Erb, personal communication) that the container in which this sample was shipped was cleaned with benzene.

Amino-acid Analyses

Three samples of lunar fines, 12001, 34 (4.1703 g), 12023, 8 (9.0000 g) and 12033, 8 (9.0015 g), were analysed for their amino-acid content by hot water Soxhlet extraction and by ion exchange column chromatographic techniques[7]. Complete parallel procedure blanks were run at the same time. The water used for the extraction was triple distilled and all glassware was cleaned with acid. Before the ion exchange chromatography the extracts

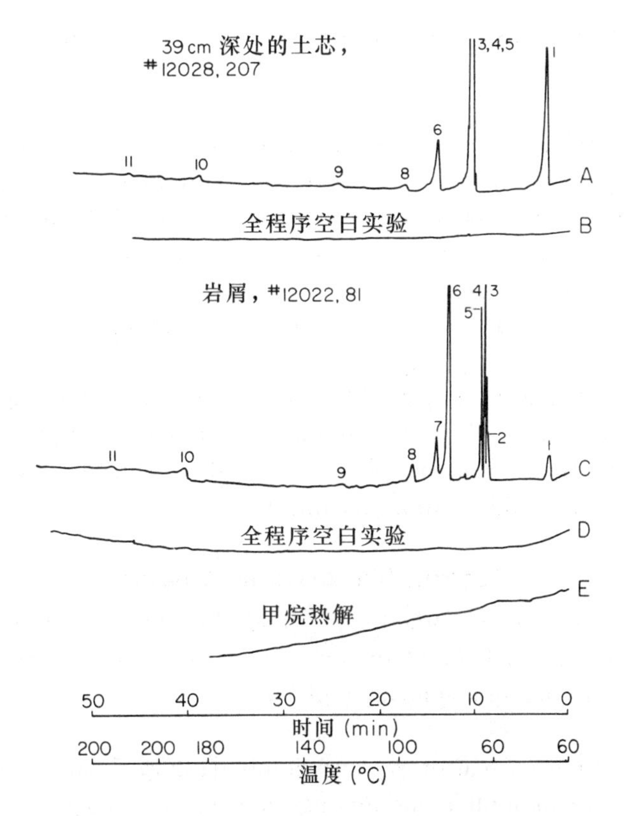

图 3. 热解产物的气相色谱图。(*A*) 月球土芯样品，来自地面以下 39 cm 深处（样品 12028, 207）；(*C*) 月球上的岩屑（样品 12022, 81）；(*E*) 在经预热解的月球微粒土（样品 12001, 34）上进行的 CH_4 热解。谱峰代号如下，对所有迹线：1, H_2、CH_4、CO；2, CH_4；3, C_2H_4；4, C_3H_6；5, C_4H_8；6, 苯；7, 噻吩；8, 甲苯；9, 苯乙烯；10, 萘；11, 联苯。此外，对于迹线 *A*：3 中还含有 C_2F_4 和 $C_2H_2F_2$；4, $C_4F_4H_2$；5, C_3F_6；6, C_4H_6。对于迹线 *C*：1, 没有 CO；3, CH_4、CO、CO_2；4, CO_2、Ne，可能有 COS；5, CO_2、C_3H_6。化合物 7 和 9 ~ 11 可能是人工产物。

在一些热解后的样品中还发现了痕量的碳氟化合物，不过它们很可能是来自“特氟隆”（译者注：聚四氟乙烯塑料）的污染。我们用美国国家航空航天局载人航天器中心提供和使用的“特氟隆”进行对照热解实验，结果发现了相同的碳氟化合物。我们对与月球样品同在一个月球物质回收实验所处理间的一份渥太华石英砂对照样品也进行了氦热解实验；其气相色谱结果见图 2*E*。请注意在小的苯峰之后该图中并没有出现其他化合物。据说（厄尔布，个人交流）在运输过程中盛放该样品的容器曾用苯清洗过。

氨基酸分析

我们用热水索氏提取法和离子交换柱层析技术对三份在月球上采集的微粒土样品——12001, 34（4.1703 g）、12023, 8（9.0000 g）和 12033, 8（9.0015 g）进行了氨基酸含量的分析[7]。同时还进行了全程序空白实验。提取所用的水是经过三次蒸馏的，且每一件玻璃器皿都用酸清洗过。在用离子交换色谱法进行分析之前，我们

and the corresponding blanks were checked for bacterial contamination. The cultures showed no growth.

The analyses of the three lunar samples indicated the presence of only traces of amino-acids in the low p.p.b. range. After subtracting the amino-acids in the blanks from the lunar sample chromatograms, samples 12001, 34 and 12023, 8 showed traces of urea, aspartic acid, threonine, serine, glycine, alanine and ornithine. The highest concentration of amino-acids occurred in sample 12001, 34. NASA reported that this sample was likely to be contaminated. The amino-acid distribution found is similar to that expected from hand contamination[8]. In sample 12023, 8 the total amount of amino-acids is approximately 1/20 of that in sample 12001, 34 and the distribution again suggests hand contamination. In sample 12033, 8 the amino-acid content appeared to be zero. Urea, however, was present even in this sample. It seems, therefore, that these samples did not contain the amino-acids found in Apollo 11 fines[3].

Electron Microscopic Studies

All samples of lunar fines were studied before and after pyrolysis at 700°C in He and at 1,084°C in vacuum, using a Leitz Ortholux microscope with a transmitted light and oil immersion objective having a magnification of 950.

In the fines, silicate spheres, "tear drop", and dumbbell shaped particles (with and without large inclusions) were commonly found among the irregularly shaped mineral matter. The spheres and dumbbells resembled the spheres and dumbbells found in the Apollo 11 fines but there were some subtle differences. In all of the untreated fines which we examined, many of the spheres and dumbbells appeared to be more irregular than the almost perfect spheres found in the Apollo 11 fines. Some particles had relatively large inclusions, while other spheres had no inclusions at all; others had an appearance similar to the heat treated (1,200°C) Apollo 11 spheres. Only the untreated sample 12042, 13 contained very few large inclusions, although numerous smaller inclusions were also observed in this sample. In the same sample, an elongated dumbbell (Fig. 4*A*) and a bead enclosed in an irregularly shaped glass fragment were found. In the untreated sample 12033, 8, reported to be an ash layer[9], there was an elongated and twisted dumbbell, which gave the appearance of having started to elongate and melt on heating. There was also a small hole at one end of this particle.

检查了提取物和相应空白样品的细菌污染情况。没有发现培养菌表现出生长的迹象。

这三份月球样品的分析结果表明：只有 p.p.b. 级的氨基酸存在。在从月球样品色谱图里扣除空白样中的氨基酸后，发现样品 12001, 34 和 12023, 8 中含有痕量的尿素、天冬氨酸、苏氨酸、丝氨酸、甘氨酸、丙氨酸和鸟氨酸。氨基酸浓度最高的样品是 12001, 34。美国国家航空航天局报道称这份样品很可能遭到了污染。已发现的氨基酸分布与根据手触污染预测到的情况类似[8]。样品 12023, 8 中的氨基酸总量大概是样品 12001, 34 的 1/20，而其分布也暗示着存在手触污染。样品 12033, 8 的氨基酸含量似乎为 0。不过，即使在样品 12033, 8 中也有尿素存在。因此，事实似乎表明这些样品并不含有在阿波罗 11 号微粒土样品中所发现的氨基酸[3]。

电子显微镜研究

我们用一台带有放大倍数为 950 的油浸物镜的莱茨 – 奥托卢克斯透射光显微镜对所有月球微粒土样品在 700℃氦热解和在 1,084℃真空热解之前以及之后都进行了研究。

在微粒土样品中，我们经常能从形状不规则的矿物质里找到硅酸盐球、“滴斑”和哑铃形微粒（有或者没有大的内含物）。这些球状物和哑铃状物类似于在阿波罗 11 号微粒土样品中所发现的球状物和哑铃状物，但有一些细微的差别。在我们检测过的所有未经处理的微粒土中，很多球状物和哑铃状物似乎比在阿波罗 11 号微粒土中所发现的近乎完美的球体显得更不规则些。有些微粒中含有较大的内含物，而另一些球状物完全不含内含物；其余则具有与在经过热处理（1,200℃）的阿波罗 11 号微粒土中发现的球体相类似的外观。只有未经处理的 12042，13 样品含有极少的大内含物，但在该样品中仍然可以观察到大量的较小内含物。在同一个样品中我们看到了一个拉长的哑铃（图 4*A*）和一粒被封闭于一个形状不规则的玻璃碎片中的熔珠。据报道样品 12033，8 属于火山灰层[9]，在未经处理的该样品中，有一个长而扭曲的哑铃状物，看上去像是在加热时要开始伸长和熔化的样子。在该微粒的一端还有一个小洞。

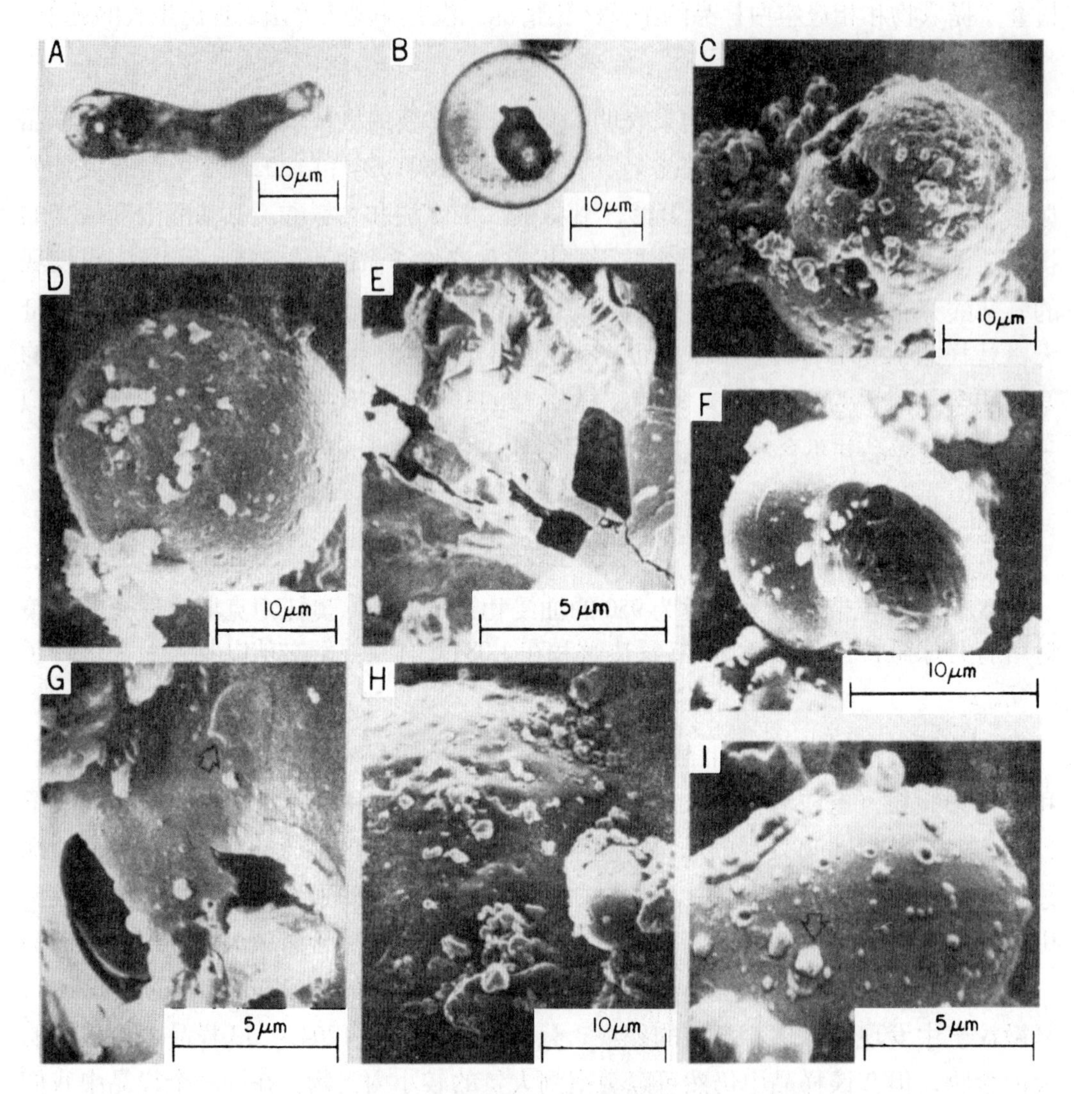

Fig. 4. *A* and *B*, Transmitted light photomicrographs of "twisted" glass dumbbell and glass sphere with opening, respectively; *C* and *D*, scanning electron micrographs of beads after pyrolysis at 1,084°C; *E*, freshly broken surface of interior rock chip (note absence of vesicles); *F*, partially broken glass bead showing cavity near the surface; *G*, a portion of a thin walled glass bead with broken openings (note arrow pointing to a flat object, probably produced by collision with a molten particle); *H*, part of a glass bead showing what appear to be several blow holes; *I*, one end of a glass dumbbell showing holes produced by hypervelocity impacts. Note arrow pointing to larger and slower velocity particle which caused imprint but did not penetrate the surface.

Morphologically, the core samples were no different from the surface fines, with the possible exception that many of the glass beads were covered with attached, fine, particulate matter. The glass beads from the core samples also showed numerous well defined cavities and inclusions. The lunar rock showed no vesicles or holes. The morphological effects of pyrolysis on all the samples at 700°C and 1,084°C are summarized in Table 2.

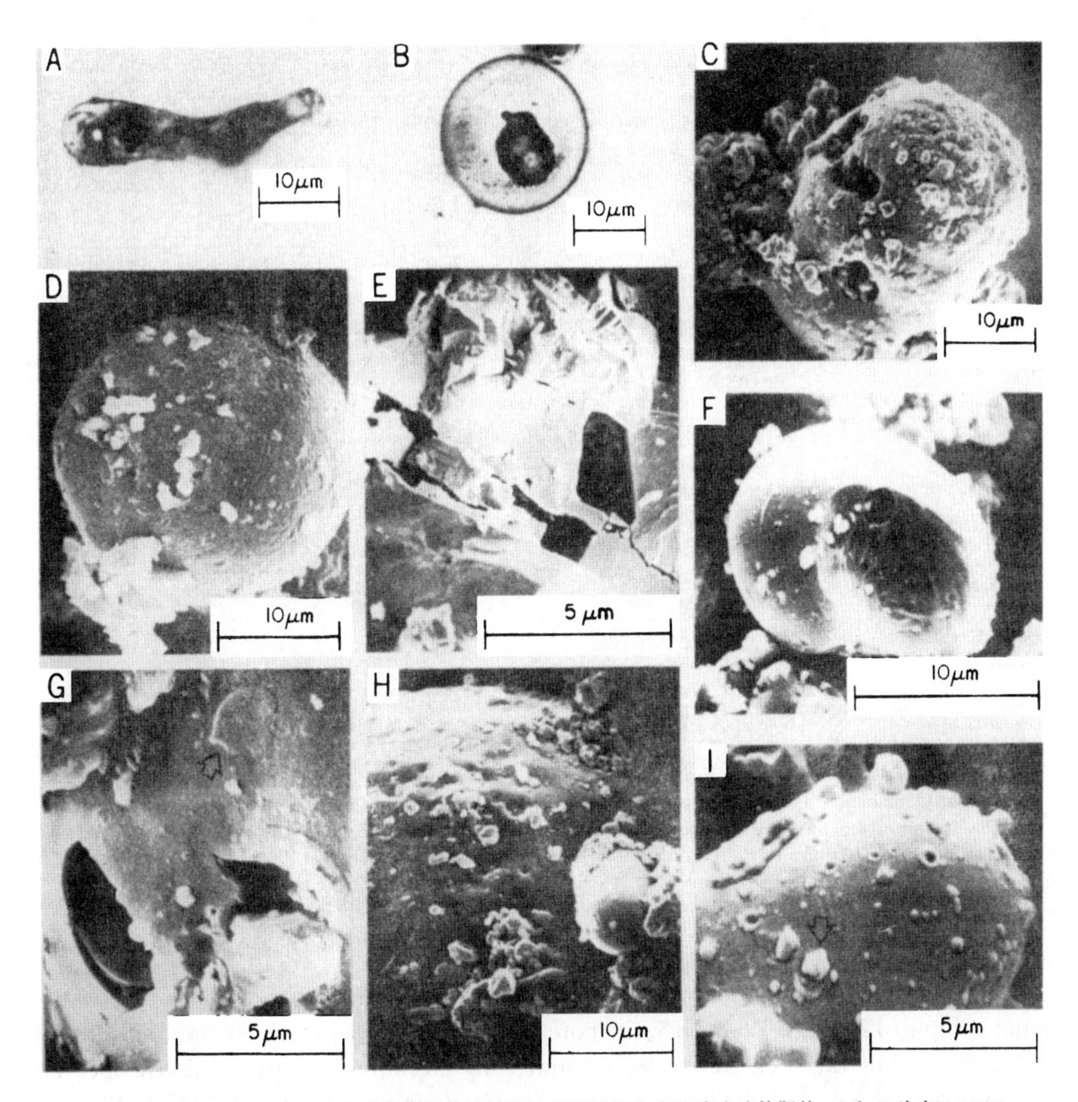

图 4. *A* 和 *B* 分别是用透射光显微镜拍摄的“扭曲”玻璃哑铃和有孔玻璃球的照片；*C* 和 *D* 为在 1,084℃热解后对熔珠拍摄的扫描电子显微镜照片；*E*，岩石内部的岩屑新切开的断面（请注意没有气泡）；*F*，部分破裂的玻璃熔珠在表面附近出现空穴；*G*，有破口的薄壁玻璃熔珠的一部分（请注意箭头所指的片状物，很可能产生自与一个熔融微粒的碰撞）；*H*，玻璃熔珠的一部分，看上去好像有若干个气孔；*I*，玻璃哑铃的一端，上面有由超速撞击导致的小孔。请注意图中箭头所指的是一个只留下印痕但未穿透表面的大且速度慢的微粒。

从形态学角度来看，可能除了很多玻璃熔珠表面会有附着于其上的细小颗粒状物质以外，土芯样品和表面微粒土样品并没有什么不同。来自土芯样品的玻璃熔珠还呈现出大量轮廓分明的空穴和内含物。在月球岩石样品中没有发现气泡或小孔。表 2 总结了 700℃和 1,084℃热解对所有样品形态的影响。

Table 2. Morphological Features and Differences Observed under Transmitted Light

Sample No.	Untreated lunar material	Pyrolysed at 700°C in He	Pyrolysed at 1,084°C in vacuum
12001, 34 fines (*a*)	Numerous distorted spheres and dumbbells with large central and/or small inclusions. Some spheres without inclusions have the appearance of recrystallized structures	Inclusions unchanged; some well defined blow holes	No inclusions; openings but no well defined blow holes
12023, 8 fines (*b*)	As above	Same as above	No inclusions; well defined blow holes; one thick irregular glass fragment containing a bead 21 μm in diameter, with inclusion
12033, 8 fines (*c*)	As above	No inclusions; no well defined blow holes	No blow holes; beads almost unrecognizable
12042, 13 fines (*d*)	As above, with the exception of few large central inclusions, numerous small inclusions	Inclusions unchanged, many small blow holes	No inclusions; beads deformed; irregular and well defined holes are visible
12025, 60 fines. Core, 9 cm below surface (*e*)	Spheres covered with fine particulate matter; broken and deformed beads; inclusions and well defined blow holes	Small blow holes, inclusions	No inclusions; no blow holes; beads deformed
12028, 207 fines. Core, 39 cm below surface (*f*)	As above; one well formed 77 μm elongated dumbbell with inclusion	Same as above	Same as above
Rock 12022, 81 (*g*)	No inclusions; no blow holes	No inclusions; no blow holes	Not examined

(*a*) Documented fines collected near Lunar Module. (*b*) Documented fines from Lunar Environment Sample Container; collected from the bottom of 15 cm trench inside north rim of Sharp Crater. (*c*) This sample is regarded to be an ash layer; documented and dug near north rim of Head Crater. (*d*) Documented collection in an area of "wrinkled texture" near rim of Surveyor Crater. (*e*) Top of double core tube No. 2010, collected on rim of 10 m diameter crater south of Halo Crater. (*f*) Bottom of double core tube No. 2012 (same collection site). (*g*) Inside chip from documented olivine dolerite. This rock was partially buried and was located approximately 120 m north-west of the Lunar Module.
Sample locations (*a*)—(*g*) are taken from ref. 10.

Lunar samples 12023, 8, 12042, 13, the core sample from a depth of 39 cm, and a freshly broken surface of the rock chip, were examined by a scanning electron microscope. The core sample and sample 12023, 8 were also examined after pyrolysis at 1,084°C. The fines were mounted on the electron microscope pedestal with a drop of absolute methanol. The rock chip was mounted with double adhesive tape. All sample preparations for electron microscopy were performed in the clean room of the Naval Weapons Center at China Lake, California. The pyrolysed fines showed much less tendency to adhere to the bare pedestal surface than did the untreated samples. The samples were coated with an ~500 Å thick layer of gold-palladium alloy by vacuum deposition[3] and then inserted in the scanning electron microscope.

The pyrolysis at 1,084°C seems to have led to the development of rough surfaces on the beads (Fig. 4*C* and *D*). Smooth beads were difficult to find and larger holes were more noticeable in the beads with rough surfaces than in the unpyrolysed beads. These effects might indicate softening of the glass accompanied by degassing at the pyrolysis temperature.

表 2. 在透射光下观测到的形态特征与差异

样品编号	未经热处理的月球物质	700℃氦热解	1,084℃真空热解
12001, 34 微粒土（*a*）	大量扭曲的球状物或哑铃状物，中心有大的和 / 或小的内含物。某些没有内含物的球状物具有重结晶结构的外观	内含物未发生变化；出现了一些轮廓分明的气孔	无内含物；有开口但没有轮廓分明的气孔
12023, 8 微粒土（*b*）	如上	同上	无内含物；有轮廓分明的气孔；一块有内含物并包含 21 μm 直径熔珠的不规则厚玻璃体
12033, 8 微粒土（*c*）	如上	无内含物；无轮廓分明的气孔	无气孔；有几乎不可辨认的熔珠
12042, 13 微粒土（*d*）	如上，不同之处是中心几乎没有大的内含物，有大量小的内含物	内含物未发生变化，出现了很多小气孔	无内含物；熔珠变形；可以看到不规则且轮廓分明的小孔
12025, 60 微粒土。表面以下 9 cm 处的土芯样品（*e*）	表面覆盖着细小颗粒物的球；破碎和变形的熔珠；内含物和轮廓分明的气孔	小气孔，内含物	无内含物；无气孔；熔珠变形
12028, 207 微粒土。表面以下 39 cm 处的土芯样品（*f*）	如上；一个形状完好、有内含物的 77 μm 拉长哑铃状物	同上	同上
岩石 12022, 81（*g*）	无内含物；无气孔	无内含物；无气孔	未检测

（*a*）在登月舱附近采集到的有记录的微粒土。（*b*）来自月球环境标本容器的有记录的微粒土；从夏普环形山内北部边缘处深 15 cm 的沟底采集。（*c*）该样品被认为属于火山灰层；从黑德环形山北部边缘附近被挖掘出来并有记录。（*d*）在勘测员环形山附近的“褶皱构造”区域采集到的有记录的样品。（*e*）第 2010 号双层岩心管的顶部，采集地点是在黑洛环形山南方一直径为 10 m 的环形山的边缘。（*f*）第 2012 号双层岩心管的底部（同一采集地点）。（*g*）从有记录的橄榄石粗玄岩内部取得的岩屑。这块被部分掩埋的岩石位于登月舱西北约 120 m 处。

样品采集地点（*a*）~（*g*）引自参考文献 10。

我们用扫描电子显微镜检测了月球样品 12023，8、12042，13、来自深 39 cm 处的土芯样品以及岩屑新切开的断面。还检测了 1,084℃热解后的土芯样品和 12023，8 样品。用一滴无水甲醇将微粒固定在电子显微镜的样品台上。岩屑用双面胶带固定。所有用电子显微镜检测的样品都是在加州中国湖美国海军武器中心的无尘室中制备的。经热解处理的微粒远比未经热解处理的微粒更不容易黏在空样品台的表面。利用真空沉积法 [3] 在样品表面镀上一层厚约 500 Å 的金钯合金，然后将其置入扫描电子显微镜中。

1,084℃热解似乎会导致熔珠粗糙表面的扩展（图 4*C* 和 *D*）。在热解之后的样品中很难发现表面光滑的熔珠，而且粗糙表面熔珠中较大的孔也比在未经热处理的熔珠中见到的更醒目。这些结果或许可以说明在热解温度下的脱气过程同时伴随着玻璃的软化。

The freshly broken surface of the lunar rock showed no vesicles, only massive crystalline structure with fissures which were possibly arranged along cleavage planes (Fig. 4*E*); conchoidal fractures were also present. Other untreated samples showed a number of interesting features. From the particle illustrated in Fig. 4*F*, a chip was removed, probably by impact, resulting in a conchoidal surface and revealing an internal cavity very near to the original surface of this glass bead. Fig. 4*G* shows part of a thin walled glass bead with irregular shaped broken holes and a plaque (note arrow), probably resulting from a completely molten, impacting object. Fig. 4*H* is particularly noteworthy because it contains numerous small openings which resemble blow holes. The end of a dumbbell shown in Fig. 4*I* is covered with what appear to be impact craters with lips, which might have been caused by impacts from hypervelocity particles of about 0.1–0.2 μm in size. Lower velocity and larger impacting particles, ~1.0 μm in diameter, show impact prints but remained attached to this dumbbell without penetrating its surface (note arrow).

There seem to be subtle differences between the morphologies of the Apollo 11 and 12 lunar fines. The abundance of what appear to be degassing blow holes, distorted particles and so on may suggest a different thermal history at the Apollo 12 landing site. It appears that the carbon compounds are present in some of the enclosures of the fines, and there seem to be carbon compounds dissolved in the lunar rock (interior) that we examined. This lunar rock contained no vesicles. One can account for the carbon in the Apollo 12 fines partially from solar wind emplacement[11]. It is, however, difficult to account for the carbon in the interior of the rock by this mechanism. Degassing of the Moon could be one mechanism which might account for the carbon in the interior of igneous lunar rocks.

We thank Drs. T. Timothy Myoda, Marjorie Lou and Barbara Rowley and Mrs. Johanne C. Dickinson, Roberta deFiore, Urmi Patel and Janet Greenquist for help with the experiments. This work was sponsored by a NASA contract.

(**232**, 94-98; 1971)

Bartholomew Nagy, Judith E. Modzeleski, Vincent E. Modzeleski, M. A. Jabbar Mohammad, Lois Anne Nagy and Ward M. Scott: Department of Geosciences, University of Arizona, Tucson, Arizona 85721.
Charles M. Drew, Joseph E. Thomas and Reba Ward: Naval Weapons Center, China Lake, California 93555.
Paul B. Hamilton: Alfred I. duPont Institute, Wilmington, Delaware 19899.
Harold C. Urey: Department of Chemistry, University of California at San Diego, La Jolla, California 92037.

Received May 25, 1971.

References:

1. Silver, L. T., *Geol. Soc. Amer. Abstract*, **2**, 684 (1970).
2. Wasserburg, G. J., and Papanastassiou, D. A., *Geol. Soc. Amer. Abstract*, **2**, 715 (1970).
3. Nagy, B., Drew, C. M., Hamilton, P. B., Modzeleski, V. E., Murphy, Sr. M. E., Scott, W. M., Urey, H. C., and Young, M., *Science*, **167**, 770 (1970).
4. Nagy, B., Scott, W. M., Modzeleski, V. E., Nagy, L. A., Drew, C. M., McEwan, W. S., Thomas, J. E., Hamilton, P. B., and Urey, H. C., *Nature*, **225**, 1028 (1970).
5. Scott, W. M., Modzeleski, V. E., and Nagy, B., *Nature*, **225**, 1129 (1970).
6. Oró, J., and Han, J., *J. Gas Chromatog.*, **5**, 480 (1967).
7. Hamilton, P. B., *Anal. Chem.*, **35**, 2055 (1963).

在月球岩石的新断面上没有发现气泡，只有大量带有裂纹的晶体结构，这些裂纹有可能是沿解理面排布的（图 4*E*）；此外还有贝壳状断口。其他未经处理的样品呈现出很多值得关注的特征。如图 4*F* 所示的微粒，很可能是由于撞击使一块碎片脱离，从而生成了贝壳状表面并在非常接近于这一玻璃熔珠原始表面的部位出现了一个内部空穴。图 4*G* 是一块薄壁玻璃熔珠的一部分，上面有形状不规则的破洞和一个斑痕(注意箭头所指)，可能是由一个完全熔融的冲击物造成的。图 4*H* 格外引人注意，因为它含有大量类似于气孔的小开口。图 4*I* 显示的是一个哑铃状物的末端，它的表面上有一些带凸边的冲击坑，可能是由约 0.1 μm ~ 0.2 μm 大小的高速粒子冲撞而产生的。由速度较低并且体积较大（直径约 1.0 μm）的粒子所造成的撞击印痕只附着于该哑铃状物之上而未能穿透其表面（注意箭头所指）。

看来，在阿波罗 11 号和 12 号月球微粒土样品的形态之间存在着细微的差别。排气孔和变形微粒等的大量存在也许可以说明在阿波罗 12 号着陆点处的物质有不同的受热过程。看来含碳化合物在某些微粒土样品的外壳中是存在的，而且在我们检测过的月球岩石（内部）中似乎也存在着融于其中的含碳化合物。这块来自月球的岩石没有气泡。一种解释是阿波罗 12 号微粒土样品中的碳有一部分来自太阳风的入侵 [11]。但是用这种机制很难解释岩石内部的碳。月球的排气过程也许可以作为解释月球火成岩内部含碳的一种机制。

感谢蒂莫西 · 迈奥达博士、玛乔丽 · 洛乌博士和芭芭拉 · 罗利博士以及约翰妮 · 迪金森、罗伯塔 · 德菲奥里、乌尔米 · 帕特尔和珍妮特 · 格林奎斯特在实验中为我们提供的帮助。本项研究得到了美国国家航空航天局的资助。

（王耀杨 翻译；周江 审稿）

8. Hamilton, P. B., *Nature*, **205**, 284 (1965).

9. *NASA Lunar Sample Information Catalog Apollo 12*, 121 (1970).

10. *NASA Apollo 12 Preliminary Science Report*, sample description on pp. 137-144, map on p. 33 (1970).

11. Moore, C. B., Larimer, J. W., Lewis, C. F., Delles, F. M., and Gooley, R. C., *Geol. Soc. Amer. Abstract*, **2**, 628 (1970).

12. Nagy, B., *Geotimes*, **15**, 18 (1970).

13. Preti, G., Murphy, R. C., and Biemann, K., *Apollo 12 Lunar Science Conference*, Houston, Texas (1970).

Detection of Radio Emission from Cygnus X–1

L. L. E. Braes and G. K. Miley

Editor's Note

Cygnus X–1 was one of the first X-ray emitting objects found when a rocket with an X-ray detector was launched in 1964. But X-ray detectors in those days had very crude spatial resolution. Luc Braes and George Miley here report variable radio emission coming from that region of the sky, and show that it stems from the radio counterpart to Cyg X–1. This enabled identification of Cyg X–1's orbit, and we now know that Cyg X–1 is a black hole of about 9 solar masses co-orbiting with a blue supergiant star. They are separated by only about 20 percent of the distance between the Earth and Sun, and the wind from the star feeds a disk of material around the black hole, which the black hole is accreting.

THE X-ray source Cyg X–1 is known to be highly variable and recently has been found to pulsate with a period of 73 ms (ref. 1). Here we report the detection of its radio counterpart at a frequency of 1,415 MHz.

Two sets of observations at 1,415 MHz were made with the Westerbork synthesis radio telescope of a 1°×1° field centred near the position of Cyg X–1. The bandwidth was 4 MHz. Baselines ranging in length from 36 to 1,471 m were used, resulting in a synthesized beam of half-power diameter 23″ in right ascension and 40″ in declination. The first observations on February 28, 1971, from 10 h 29 min to 15 h 02 min UT showed no source stronger than 0.005 flux units near the X-ray position. The field was observed again during April 28–29 from 23 h 04 min to 11 h 06 min UT, and in the map resulting from these observations there is an unresolved source within the X-ray error box.

The flux density of the radio source is 0.021±0.004 f.u. and its position, determined by a least-squares fit to the antenna pattern, is in 1950 coordinates: α=19 h 56 min 28.9 s±0.2 s, δ=35° 03′56″± 3″. This can be compared with the X-ray position determined by the Uhuru satellite which is α=19 h 56 min 25 s ± 15 s, δ=35° 03′25″± 1′ (unpublished work of Tananbaum *et al.*). Because of the close agreement in position and because this radio source is so strongly variable, it is almost certainly associated with Cyg X–1. A search at the radio position may therefore enable the X-ray source to be identified optically. A likely candidate is the ninth magnitude star AGK2 +35° 1,910 only 1″ from our position, but photometric and spectroscopic observations are needed to confirm the identification. Further study of the radio emission is also of great importance. Simultaneous measurements over a range of frequencies should reveal whether its radio spectrum is non-thermal as in the case of Sco X–1 (refs. 2 and 3), and observations with high time resolution are essential to determine whether the radio source is a pulsar.

探测到天鹅座X–1的射电辐射

布拉埃斯，米利

编者按

天鹅座 X–1 是 1964 年一枚载有 X 射线探测器的火箭升空后所发现的第一批 X 射线源之一。但是在当时，X 射线探测器的空间分辨率很差。卢克 · 布拉埃斯和乔治 · 米利在本文中报道了来自该天区的随时间变化的射电辐射，并证明其来源于天鹅座 X–1 的射电对应体。这使得人们可以确定天鹅座 X–1 的轨道。现在我们知道，天鹅座 X–1 是一个大约 9 倍太阳质量的黑洞，它与一颗蓝超巨星互相绕转。两者之间的距离只有日地距离的 20% 左右，蓝超巨星的星风被吸积并在黑洞周围形成吸积盘。

大家都知道，天鹅座 X–1 是一个变化剧烈的 X 射线源，最近发现它的脉动周期为 73 ms（参考文献 1）。本文报道我们在频率 1,415 MHz 处探测到了它的射电对应体。

我们用韦斯特博克综合孔径射电望远镜在 1,415 MHz 处进行了两组观测，观测区域是大致以天鹅座 X–1 为中心的 1° × 1° 天区。观测带宽为 4 MHz。基线长度从 36 m 到 1,471 m，因而综合波束的半功率宽为赤经方向 23″，赤纬方向 40″。我们在 1971 年 2 月 28 日进行了第一次观测，时间是世界时 10 h 29 min 至 15 h 02 min，结果并没有发现在 X 射线源位置附近有强于 0.005 流量单位的源。我们在世界时 4 月 28 日 ~ 29 日 23 h 04 min 至 11 h 06 min 又对该区域进行了一次观测。观测成图后发现在 X 射线源位置的误差框范围内有一个不可分辨的源。

该射电源的流量密度为 0.021 ± 0.004 流量单位。用最小二乘法拟合天线方向图得到它的位置在 1950 历元：α = 19 h 56 min 28.9 s ± 0.2 s, δ = 35° 03′ 56″ ± 3″。这一结果与由乌呼鲁卫星得到的 X 射线位置 α = 19 h 56 min 25 s ± 15 s, δ = 35° 03′ 25″ ± 1′（塔南鲍姆等人尚未发表的研究结果）大致相当。由于两者的位置吻合并且该射电源的流量变化剧烈，几乎可以肯定它就是天鹅座 X–1 的对应体。因此在射电位置处进行搜索或许可以从光学上识别出该 X 射线源。九等星 AGK2 +35° 1,910 就是一个可能的候选者，它仅与我们现在的位置相差 1″，但还需要通过测光和光谱观测进行确认。对射电辐射的进一步研究也是非常重要的。在一定频率范围内进行同时观测应能确定它的射电谱是否就是类似于天蝎座 X–1 的非热谱（参考文献 2 和 3）。高时间分辨率的观测对于确定这个射电源是不是脉冲星是至关重要的。

During the preparation of this article we learned from Drs. R. M. Hjellming and C. M. Wade that the Cyg X-1 radio source has been detected independently at a frequency of 2,695 MHz with the NRAO interferometer at Green Bank. Their position is in good agreement with that given here.

The Westerbork Radio Observatory is operated by the Netherlands Foundation for Radio Astronomy with the financial support of the Netherlands Organization for the Advancement of Pure Research (Z.W.O.).

(**232**, 246; 1971)

L. L. E. Braes and G. K. Miley: Leiden Observatory, Leiden 2401.

Received June 28; revised July 9, 1971.

References:

1. Oda, M., Gorenstein, P., Gursky, H., Kellogg, E., Schreier, E., Tananbaum, H., and Giacconi, R., *Astrophys. J. Lett.*, **166**, L1 (1971).
2. Hjellming, R. M., and Wade, C. M., *Astrophys. J. Lett.*, **164**, L1 (1971).
3. Braes, L. L. E., and Miley, G. K., *Astron. Astrophys.* (in the press).

在准备这篇文章的过程中，我们从耶尔明博士和韦德博士那里得知，其他人也独立地探测到了天鹅座 X–1 射电源，他们使用了位于格林班克的美国国家射电天文台的干涉仪，观测频率为 2,695 MHz。他们测得的位置与本文中给出的位置非常吻合。

韦斯特博克射电天文台由荷兰射电天文学基金会负责管理，经费上的支持来自荷兰纯理论研究提升组织。

（岳友岭 孟洁 翻译；肖伟科 审稿）

The Solar Spoon

F. W. W. Dilke and D. O. Gough

Editor's Note

The nuclear fusion reactions taking place in the Sun's core release photons and neutrinos. The photons take about a million years to reach the surface of the Sun, then arrive 8 minutes after that. But the neutrinos, because they interact very weakly with matter, escape from the Sun's interior almost at once. By the early 1970s it was clear that fewer neutrinos were being observed than theory predicted. F. W. W. Dilke and Douglas Gough here propose a model for the Sun in which it undergoes periodic (every few hundred million years) mixing due to convection. This would, for a period of about ten million years, lead to a lower temperature at the surface of the Sun, and fewer neutrinos coming from it. The neutrino problem has since been solved in another way: some of the neutrinos change their type on the way from the solar core to the Earth. The kind of convective instability proposed here has now almost certainly been ruled out.

SOLAR models are normally constructed by evolving for a time τ a stellar model of 1 $M_\odot$ which initially had uniform chemical composition and was situated either somewhere on the Hayashi track or on the zero age main sequence. The principal direct observational data are the present luminosity L, effective temperature T_e, upper bound to the neutrino flux F_ν, and some knowledge of the chemical composition of the surface. The luminosity, effective temperature and chemical composition are essentially surface boundary conditions; the neutrino flux is an integral to which only regions very close to the center contribute. Conditions can be imposed on the past history of the Sun by the theory of stellar evolution and by the requirement that our ideas about the history of the interstellar medium are not contradicted. But there remains considerable freedom in our choice of the initial abundances by weight, Y and Z, of helium and of heavier elements; further freedom is provided by our lack of knowledge of the precise value of τ, though limits can be provided by geological evidence and by measures of certain isotope ratios in the Sun.

There may be serious errors in the microscopic physics providing the equation of state, the opacity and the nuclear reaction rates, and our understanding of the macroscopic physics describing the fluid motions within the Sun is incomplete. In particular, the mixing length formalism normally used to describe the outer convection zone is unlikely to be accurate, but even if it were the mixing length l itself remains a free parameter (or, to be more precise, free function) of the theory. Varying l causes the relatively diffuse outer layers of the model to readjust so that the heat can still be transported. This has little effect on the structure of the central regions so L is hardly affected. Broadly speaking, therefore, l determines T_e, but in practice the converse is the case: the theoretical solar model serves to calibrate the mixing length theory for use in models of other stars.

太阳的调羹

迪尔克，高夫

编者按

发生在太阳核心的核聚变反应会释放出光子和中微子。光子从太阳核心到太阳表面约需 10^6 年的时间，之后再要 8 分钟才到达地球。但中微子很快就能从太阳内部逃逸出来，因为它们与物质间的相互作用极其微弱。20 世纪 70 年代早期，观测到的太阳中微子数目比理论预测值少的事实就已经明确了。在本文中，迪尔克和道格拉斯·高夫提出了一个太阳模型，在这个模型中太阳由于对流而经历周期性（每几亿年一次）的混合。由此将导致太阳在大约 10^7 年的时期里具有较低的表面温度，并且从太阳来的中微子也会减少。后来，中微子问题从另一途径得到了解决，即某些中微子在从太阳核心到地球的途中改变了它们的类型。本文中提出的这类对流不稳定性解释现在已经基本上被排除了。

太阳模型通常是从一个初始化学组成均匀并处在林忠四郎线上或者零龄主序位置上的、质量为一个太阳质量（$M_{\odot}$）的恒星模型经时间 τ 的演化而构建的。主要的直接可观测数据有：当前光度 L、有效温度 T_e、中微子流量上限 F_ν 以及一些表面化学组成方面的信息。光度、有效温度和化学组成实质上是在表面处的边界条件；中微子流量是一个积分，只有非常接近中心的区域才对它有贡献。我们可以根据恒星演化理论附加上一些与太阳过去历史有关的条件，但要求不能与我们关于星际介质历史的概念相抵触。即使如此，我们仍然在选择氦和更重元素的初始（按重量计的）丰度 Y 和 Z 上保有相当大的自由度；虽然地质学证据和太阳中某些同位素比可以给出一些限制，可是，由于我们对 τ 的精确值缺乏了解导致出现了进一步的自由度。

微观物理学所提供的状态方程、不透明度和核反应率可能存在严重的误差，而我们对用来描述太阳内部流体运动的宏观物理学的理解也是不全面的。尤其是，通常用于描述外对流区的混合长的方式未必准确，其实即使是混合长 l 本身也仍然只是一个理论上的自由参量（或者，更准确地说是一个自由函数）。改变 l 可以使模型相对弥散的外层重新调整从而热量仍然能被传输。这对中心区域的结构几乎没有影响，所以 L 几乎没有受到影响。因此，一般来说是 l 决定了 T_e，但在实际情况中却恰好相反：理论太阳模型被用于标定混合长理论以用于其他恒星模型。

Early measurements of F_v (ref. 1) were considerably lower than solar models predicted at the time, and now it cannot be said with confidence that any neutrinos have been detected at all (Davis, R., jun., Wolfendale, A. W., and Young, E. C. M., in preparation). This has led to a re-examination of the microphysics and the general tendency has been to depress F_v, although the latest predictions[2-4] are still greater than the experimental upper bound. It is clearly worth re-examining the macrophysics too. In particular it is normally assumed that the solar core remains unmixed because the Schwarzschild criterion for convective stability[5,6] is satisfied. Although Öpik has suggested that diffusion of material builds up a shell with high opacity around the core which eventually becomes convectively unstable and mixes[7], his rough description of the diffusion process differs considerably from the results of more sophisticated calculations[8], and if one is to believe the latter it is unlikely that the mechanism, in unmodified form, can operate in a time shorter than the age of the Sun. The possibility that there may be a strong Eckman circulation[9,10] has led to a consideration of the effect on F_v of continuous mixing in the core[11-15]. The results depend upon the timescale assumed for mixing and suggest that F_v can be reduced, but it is not clear whether the reduction is adequate. In any case it seems unlikely that sufficiently rapid Eckman circulation exists in the Sun at the present time[16]. Here we consider the stability of the solar core and find a process which may lead to intermittent mixing; indeed Fowler[17] has pointed out that such mixing may be even more efficient at reducing F_v.

Instability of the Solar Core

In the core of a main sequence star the thermonuclear generation of energy has negligible effect on the Schwarzschild criterion, because the timescale for generating energy is very much greater than the growth time of convective modes, and so is always ignored when testing for convection. But energy generation might destabilize gravity modes in much the same way as Eddington[18] thought pulsations might be driven in Cepheids. We consider a grossly oversimplified model: a plain-parallel stratified fluid layer of thickness d within which energy is being generated at a rate ε per unit mass. For the moment we assume that the only significant reactions are $p(p, \beta^+ \nu)\,D\,(p, \gamma)^3He$ and $^3He(^3He, 2p)^4He$ of the *pp*I chain which release energy at rates ε_{11} and ε_{33} respectively. Because it is unlikely that the solar core is in rapid rotation now[16] we will ignore rotation altogether, and to simplify matters still further the layer will be presumed so thin that the Boussinesq approximation can be employed to describe buoyancy driven motions[19] and gravity fluctuations will be ignored.

Because diffusion of material and momentum and the change of the H and He concentrations by nuclear burning are insignificant over the length and timescales considered in this section, the material derivatives of X and X_3, the mass fractions of H and ^{3}He, will be taken to be zero and viscous stresses will be neglected. In cartesian coordinates (x, y, z) with z vertical, the linearized equations can then be written

$$\rho\frac{\partial \mathbf{u}}{\partial t} = -\nabla p' + \mathbf{g}\rho' \tag{1}$$

F_v 的早期测量值（参考文献 1）显著低于那时候太阳模型的预期值，但现在尚不能非常肯定地说所有中微子都被探测到了（小戴维斯、沃尔芬德尔和扬，完稿中）。这使一些人重新核查微观物理学，总的趋势是压低 F_v 值，然而最新的预测值[2-4]仍然高于观测值的上限。显然，重新核查宏观物理学也是一件值得做的事。尤其是，因为太阳核心满足对流稳定性的史瓦西判据[5,6]，所以宏观物理学通常假设太阳核心保持在非混合的状态。尽管奥皮克曾经提出过：物质的扩散会围绕着核心形成一个不透明度很高的壳层，核心区最终会变得对流不稳定，并发生混合[7]；但他对扩散过程的粗略描述与更仔细的计算得出的结果显著不同[8]。如果我们相信详细的计算结果，那么在不经过修正的情况下，奥皮克机制不太可能在短于太阳年龄的时间内起作用。存在一个强埃克曼环流[9,10]的可能性使得核心的连续混合对 F_v 造成的影响成为一件值得考虑的事[11-15]。这些结果依赖于所取的混合时标，它们表明 F_v 值可以降低，但并不清楚降低的量是否足够。无论如何，当前太阳中都不太可能存在足够快速的埃克曼环流[16]。在本文中我们考虑了太阳核心的稳定性并且发现了一个会导致间歇性混合的过程；实际上福勒[17]已经指出过这种混合可能对降低 F_v 值更为有效。

太阳核心的不稳定性

在一颗主序星的核心，热核反应产能对史瓦西判据的影响可以忽略不计。这是因为产能的时标要远大于对流模式增长的时间，所以当考虑对流模式的时候通常忽略产能。但是能量产生可能会使重力模式不稳定，就像爱丁顿[18]认为在造父变星内部脉动有可能会被驱动一样。我们考虑一个极其简化的模型：一层简单的平行分层液体，厚度为 d，它内部每单位质量的产能率为 ε。我们暂且假设主要的核反应只有 ppI 链中的 $p(p,\ \beta^+\nu)D(p,\ \gamma)^3He$ 和 $^3He(^3He,\ 2p)^4He$ 反应，其产能率分别为 ε_{11} 和 ε_{33}。因为太阳核心区目前不太可能在快速旋转[16]，所以我们将旋转一并忽略掉。为了使模型进一步简化，我们假设流体层非常薄，可以用博欣内斯克近似来描述浮力驱动运动[19]，并且忽略重力的起伏。

由于物质和动量的扩散以及由核燃烧引起的 H 和 He 浓度的变化在本节所考虑的尺度和时标内是微不足道的，所以 X 和 X_3 的拉格朗日导数、H 和 ^{3}He 的质量分数都被取为零，黏滞应力也将被忽略。在笛卡尔坐标系 (x, y, z) 下，当 z 为垂直轴时，线性化方程可以被写作：

$$\rho\frac{\partial \mathbf{u}}{\partial \boldsymbol{t}} = -\nabla p' + \mathbf{g}\rho' \tag{1}$$

$$\operatorname{div} \mathbf{u} = 0 \tag{2}$$

$$C_p\left(\frac{\partial T'}{\partial t} - \beta w\right) = \varepsilon' + \frac{K}{\rho}\nabla^2 T' \tag{3}$$

$$\frac{\partial T'}{\partial t} + w\frac{dX}{dz} = 0 \qquad \frac{\partial X'_3}{\partial t} + w\frac{dX_3}{dz} = 0 \tag{4}$$

where $\mathbf{u}$ is the fluid velocity and w its vertical component; p, ρ, T are pressure, density and temperature, K is the radiative conductivity $4acT^3/3\kappa\rho$, C_p is the specific heat at constant pressure, β is the superadiabatic temperature gradient and $\mathbf{g}$ is the gravitational acceleration. In these equations unprimed variables refer to the equilibrium state and primed variables to the perturbations. The equations must be supplemented by an equation of state; the perfect gas law for a fully ionized gas, in Boussinesq approximation, will be assumed so that

$$\frac{\rho'}{\rho} = -\frac{T'}{T} - \lambda\frac{X'}{X} \tag{5}$$

where
$$\lambda = -\left(\frac{\partial \ln\rho}{\partial \ln X}\right)_{p,T} \backsimeq \frac{5X}{5X+3}$$

With the help of this equation the perturbed energy generation rate is given by

$$\frac{\varepsilon'}{\varepsilon} = \eta\frac{T'}{T} + \nu\frac{X'}{X} + \nu_3\frac{X'_3}{X_3} \tag{6}$$

where

$$\eta = \frac{\varepsilon_{11}}{\varepsilon}\left(\frac{\partial \ln \varepsilon_{11}}{\partial \ln T}\right)_{\rho,X,X_3} + \frac{\varepsilon_{33}}{\varepsilon}\left(\frac{\partial \ln \varepsilon_{33}}{\partial \ln T}\right)_{\rho,X,X_3} - \left(\frac{\partial \ln \varepsilon}{\partial \ln \rho}\right)_{T,X,X_3}$$

$$\nu = \frac{\varepsilon_{11}}{\varepsilon}\left(\frac{\partial \ln \varepsilon_{11}}{\partial \ln X}\right)_{\rho,X,X_3} - \lambda\left(\frac{\partial \ln \varepsilon}{\partial \ln \rho}\right)_{T,X,X_3}$$

and
$$\nu_3 = \frac{\varepsilon_{33}}{\varepsilon}\left(\frac{\partial \ln \varepsilon_{33}}{\partial \ln X_3}\right)_{\rho,T,X}$$

As is customary in initial studies of convective instability, the simplest boundary conditions will be assumed:

$$w = 0,\ T' = 0 \text{ at } z = 0, d \tag{7}$$

After the usual manipulations[20] it can be shown that the space and time dependence of w, ρ', T', X' and X'_3 can be written in the separated form

$$e^{qwt}\, e^{i(k_x x + k_y y)} \sin\left(\frac{m\pi z}{d}\right)$$

where m is an integer. If $w^2 = \dfrac{k_x^2 + k_y^2}{k^2}\dfrac{|\mathbf{g}|}{d}$ where $k^2 = k_x^2 + k_y^2 + \left(\dfrac{m\pi}{d}\right)^2$ equations (1) to (6) imply the dispersion relation

$$\mathrm{div}\,\mathbf{u} = 0 \tag{2}$$

$$C_p\left(\frac{\partial T'}{\partial t} - \beta w\right) = \varepsilon' + \frac{K}{\rho}\nabla^2 T' \tag{3}$$

$$\frac{\partial T'}{\partial t} + w\frac{\mathrm{d}X}{\mathrm{d}z} = 0 \qquad \frac{\partial X'_3}{\partial t} + w\frac{\mathrm{d}X_3}{\mathrm{d}z} = 0 \tag{4}$$

其中 $\mathbf{u}$ 为流体速度，w 是它的垂直分量；p、ρ、T 分别是压强、密度和温度，K 代表辐射传导系数 $4acT^3/3\kappa\rho$，C_p 为定压比热，β 为超绝热温度梯度，$\mathbf{g}$ 为重力加速度。在这些方程中，无撇号的变量表示平衡状态下的量，带撇号的变量表示扰动量。上述方程组还必须补充一个状态方程；在博欣内斯克近似下，可以取完全电离气体的理想气体状态方程，因而：

$$\frac{\rho'}{\rho} = -\frac{T'}{T} - \lambda\frac{X'}{X} \tag{5}$$

其中
$$\lambda = -\left(\frac{\partial \ln\rho}{\partial \ln X}\right)_{p,T} \backsimeq \frac{5X}{5X+3}$$

借助这一方程，扰动态下的产能率就可以表示为：

$$\frac{\varepsilon'}{\varepsilon} = \eta\frac{T'}{T} + \nu\frac{X'}{X} + \nu_3\frac{X'_3}{X_3} \tag{6}$$

其中

$$\eta = \frac{\varepsilon_{11}}{\varepsilon}\left(\frac{\partial \ln \varepsilon_{11}}{\partial \ln T}\right)_{\rho,X,X_3} + \frac{\varepsilon_{33}}{\varepsilon}\left(\frac{\partial \ln \varepsilon_{33}}{\partial \ln T}\right)_{\rho,X,X_3} - \left(\frac{\partial \ln \varepsilon}{\partial \ln \rho}\right)_{T,X,X_3}$$

$$\nu = \frac{\varepsilon_{11}}{\varepsilon}\left(\frac{\partial \ln \varepsilon_{11}}{\partial \ln X}\right)_{\rho,X,X_3} - \lambda\left(\frac{\partial \ln \varepsilon}{\partial \ln \rho}\right)_{T,X,X_3}$$

和
$$\nu_3 = \frac{\varepsilon_{33}}{\varepsilon}\left(\frac{\partial \ln \varepsilon_{33}}{\partial \ln X_3}\right)_{\rho,T,X}$$

习惯上，在开始研究对流不稳定性时，首先取最简单的边界条件：

$$\text{在 } z = 0 \text{ 和 } d \text{ 处，} w = 0,\ T' = 0 \tag{7}$$

在经过常规处理 [20] 之后发现，量 w、ρ'、T'、X' 和 X'_3 对空间和时间的依赖关系可以写成分离的形式：

$$e^{qwt}\, e^{i(k_x x + k_y y)} \sin\left(\frac{m\pi z}{d}\right)$$

其中 m 是一个整数。如果 $w^2 = \dfrac{k_x^2 + k_y^2}{k^2}\dfrac{|\mathbf{g}|}{d}$，式中 $k^2 = k_x^2 + k_y^2 + \left(\dfrac{m\pi}{d}\right)^2$，则方程(1)到(6)中含有色散关系：

$$q^3 + E\alpha q^2 + \left[\lambda \frac{d}{H_x} - (\nabla - \nabla_{ad}) \frac{d}{H_p}\right] q + E\left[(\nu + \lambda\alpha)\frac{d}{H_x} + \nu_3 \frac{d}{H_3}\right] = 0 \tag{8}$$

where H_p, H_x and H_3 are the scale heights of p, X and X_3 in the equilibrium configuration, $\alpha = \frac{KTk^2}{\rho\varepsilon} - \eta$ and $E = \frac{\varepsilon}{wC_pT}$. The superadiabatic temperature gradient β has been written as $(\nabla - \nabla_{ad})\frac{T}{H_p}$ where $\nabla = \frac{\mathrm{dln}\,T}{\mathrm{dln}p}$ and $\nabla_{ad} = \left(\frac{\partial \ln T}{\partial \ln p}\right)_{ad}$. Note that provided $k_x^2 + k_y^2 \lll k^2$, E is essentially the ratio of the free-fall time to the Kelvin–Helmholtz time for the core and is therefore small. Because on the main sequence the luminosity is largely balanced by the energy generated with the core, $\left(\frac{KT}{d}\right)d^2 \backsimeq \rho\varepsilon d^3$, where d represents the scale of the core, and the parameter $\frac{KTk^2}{\rho\varepsilon}$ is approximately $(kd)^2$; for the lower modes this is of order unity.

The dispersion relation (8) admits three roots. Granted that $E \ll 1$ the motion is nearly adiabatic and these roots may be written approximately:

$$q_{1,2} \backsimeq \pm i\left[\lambda \frac{d}{H_x} - (\nabla - \nabla_{ad})\frac{d}{H_p}\right]^{\frac{1}{2}} + \frac{1}{2}E\left[\alpha(\nabla - \nabla_{ad})H_x + \left(\nu + \nu_3 \frac{H_x}{H_3}\right)H_p\right] / \left[\lambda H_p - (\nabla - \nabla_{ad})H_x\right] \tag{9}$$

which correspond to the two g modes, and

$$q_3 \backsimeq -E\left(\lambda\alpha + \nu + \nu_3 \frac{H_x}{H_3}\right)H_p / \left[\lambda H_p - (\nabla - \nabla_{ad})H_x\right] \tag{10}$$

which corresponds to the secular mode. The f and p modes have been filtered out by the Boussinesq approximation. If

$$\nabla - \nabla_{ad} > \lambda \frac{H_p}{H_x} \tag{11}$$

the leading term of either q_1 or q_2 is real and positive, and provided this exceeds the small second term the fluid layer is unstable to direct convective motions. This criterion was obtained by Ledoux[21]. But we pointed out above that solar models are stable to even the Schwarzschild criterion: $\nabla - \nabla_{ad} < 0$, and the g modes are oscillatory. This does not, however, imply that the core is stable; if

$$\left(\nu + \nu_3 \frac{H_x}{H_3}\right)H_p > \alpha(\nabla_{ad} - \nabla)H_x \tag{12}$$

the amplitude of the oscillations grows in a time comparable to the Kelvin–Helmholtz time for the core.

If the ^{3}He is in nuclear equilibrium with the hydrogen, $\frac{X_3}{3X} = \sqrt{\frac{\lambda_{pp}}{2\lambda_{33}}}$ where

$$q^3+E\alpha q^2+\left[\lambda\frac{d}{H_x}-(\nabla-\nabla_{\mathrm{ad}})\frac{d}{H_p}\right]q+E\left[(\nu+\lambda\alpha)\frac{d}{H_x}+\nu_3\frac{d}{H_3}\right]=0 \tag{8}$$

其中 H_p、H_x 和 H_3 分别是平衡位形下 p、X 和 X_3 的标高，$\alpha=\frac{KTk^2}{\rho\varepsilon}-\eta$ 及 $E=\frac{\varepsilon}{wC_pT}$。超绝热温度梯度β被写成：$(\nabla-\nabla_{\mathrm{ad}})\frac{T}{H_p}$，其中 $\nabla=\frac{\mathrm{dln}\,T}{\mathrm{dln}p}$ 和 $\nabla_{\mathrm{ad}}=\left(\frac{\partial\ln T}{\partial\ln p}\right)_{\mathrm{ad}}$。注意：如果 $k_x^2+k_y^2\lll k^2$，则 E 实际上是核心自由下落时间与开尔文–亥姆霍兹时间之比，因此很小。因为在主序时，核心的产能大致与光度相抵，$\left(\frac{KT}{d}\right)d^2\backsimeq\rho\varepsilon d^3$，其中 d 代表核心的尺度，而参量 $\frac{KTk^2}{\rho\varepsilon}$ 近似等于 $(kd)^2$；对于更低的模式这个值的数量级是 1。

色散关系式（8）有 3 个根。假定 $E\ll1$，则这一运动过程基本上是绝热的，这几个根可以被近似写成：

$$q_{1,2}\backsimeq\pm i\left[\lambda\frac{d}{H_x}-(\nabla-\nabla_{\mathrm{ad}})\frac{d}{H_p}\right]^{\frac{1}{2}}+\frac{1}{2}E\left[\alpha(\nabla-\nabla_{\mathrm{ad}})H_x+\left(\nu+\nu_3\frac{H_x}{H_3}\right)H_p\right]\Big/\left[\lambda H_p-(\nabla-\nabla_{\mathrm{ad}})H_x\right] \tag{9}$$

q_1、q_2 对应于两个 g 模式，和

$$q_3\backsimeq-E\left(\lambda\alpha+\nu+\nu_3\frac{H_x}{H_3}\right)H_p\Big/\left[\lambda H_p-(\nabla-\nabla_{\mathrm{ad}})H_x\right] \tag{10}$$

q_3 对应于长期模式。f 模式和 p 模式已经被博欣内斯克近似筛除。如果

$$\nabla-\nabla_{\mathrm{ad}}>\lambda\frac{H_p}{H_x} \tag{11}$$

则 q_1 或 q_2 的主项是正实数，这时只要其超过较小的第二项，流体层就是不稳定的并将导致对流运动。这一判据是由勒杜归纳得出的[21]。但是我们在上文中指出，即使在史瓦西判据 $\nabla-\nabla_{\mathrm{ad}}<0$ 和 g 模式为振荡模式的情况下，太阳模型仍是稳定的。然而，假如

$$\left(\nu+\nu_3\frac{H_x}{H_3}\right)H_p>\alpha(\nabla_{\mathrm{ad}}-\nabla)H_x \tag{12}$$

核心区振荡幅度增长的时间与核心的开尔文–亥姆霍兹时间相当的话，则并不意味着核心是稳定的。

如果 ^{3}He 与 H 在核反应中是平衡的，则有 $\frac{X_3}{3X}=\sqrt{\frac{\lambda_{pp}}{2\lambda_{33}}}$，其中 λ_{33} 由

λ_{33} is defined in terms of the reaction rate r_{33} of the $^3He(^3He,2p)^4He$ reaction by $r_{33} = \lambda_{33}\left(\frac{X_3}{3}\right)^2$ and λ_{pp} is defined similarly. $\frac{\lambda_{pp}}{\lambda_{33}}$ is a function of temperature alone, and $\frac{\mathrm{dln}(\lambda_{pp}/\lambda_{33})}{\mathrm{dln}\,T} \equiv -2b \approx -12$ at the centre of the Sun. Thus if the core is approximated by a polytrope of index n, $\frac{1}{H_3} = \frac{1}{H_x} + \frac{b}{(n+1)H_p}$ and the criterion (12) for overstability may be written

$$\frac{1}{H_x} > \frac{1}{\nu+\nu_3}\left[\alpha(\nabla_{\mathrm{ad}} - \nabla) - \frac{b\nu_3}{n+1}\right]\frac{1}{H_p} \tag{13}$$

The analysis presented above very closely resembles the work of Defouw[22]. The thermodynamics of the overstability is identical, though the physical processes responsible for the energy source term ε are quite different. The existence of gradients in composition adds an extra term to the stability criterion: because in a hydrogen burning core $\nu>0$, $\nu_3>0$, and X and X_3 do not decrease upwards, the perturbations in composition help to enhance the energy generation rate in a downward displaced fluid element relative to its surroundings and so add to the tendency towards overstability. Eddington's discussion of the overstability of spherically symmetric oscillations of Cepheids is similar, but in purely radial motion any displaced element always has the same fluid in its immediate environment and again the composition gradient does not directly enter into the stability criterion.

The existence of the composition gradients is also responsible for the secular mode whose growth rate is given by equation (10). Imagine, for example, a fluid element to experience a positive temperature fluctuation. It will rise to find a new hydrostatic equilibrium position which, because condition (11) is not satisfied, is stable on a convective timescale. Thus it will have the same pressure and density as its surroundings. But if $\mathrm{d}X/\mathrm{d}z > 0$, it will have a lower H abundance than its surroundings, and hence a higher temperature. Thus there will be a difference in the energy generation rate; if the increase resulting from the higher temperature, offset by thermal diffusion, exceeds the decrease resulting from the lower concentration of H and the inevitably lower concentration of ^{3}He, the temperature excess will be accentuated and the perturbation will grow.

An estimate of the growth rates of g modes in the Sun cannot be made by merely substituting typical values into equation (9) because the perturbations will not be confined solely to the energy generating core. Outside the core the only surviving nonadiabatic effect is the radiative damping. Auré[23] has estimated that in early type stars damping in the envelope by far exceeds the excitation (in the semiconvective zone) beneath, primarily because the amplitudes of the perturbations increase rapidly with radius. But the comparatively deep outer convection zone in the Sun makes an important difference. When $\nabla - \nabla_{\mathrm{ad}} > 0$ radiative heat transfer tends to destabilize the oscillation, just as it does p modes[24], and in any case the amplitude then decreases with radius. Further, one might expect the interaction with the unsteady direct convective motion in this zone to destroy

$^3He(^3He,2p)^4He$ 反应的反应率 r_{33} 定义为 $r_{33}=\lambda_{33}\left(\frac{X_3}{3}\right)^2$，$\lambda_{pp}$ 的定义与此类似。$\frac{\lambda_{pp}}{\lambda_{33}}$ 只是温度的函数，并且在太阳核心处有 $\frac{\mathrm{dln}(\lambda_{pp}/\lambda_{33})}{\mathrm{dln}\,T}\equiv -2b\approx -12$。因此，如果核心被近似为一个多方指数为 n 的多方球，则 $\frac{1}{H_3}=\frac{1}{H_x}+\frac{b}{(n+1)H_p}$ 并且超稳定性判据（12）可以写作：

$$\frac{1}{H_x}>\frac{1}{\nu+\nu_3}\left[\alpha(\nabla_{\mathrm{ad}}-\nabla)-\frac{b\nu_3}{n+1}\right]\frac{1}{H_p} \tag{13}$$

上述分析过程非常类似于迪佛的研究工作[22]。尽管与能源项 ε 有关的物理过程完全不同，但超稳定性热力学是一样的。组成成分梯度的存在为稳定性判据增加了一个新的项：因为在氢燃烧的核心区 $\nu>0$，$\nu_3>0$，且 X 和 X_3 不会越向上越小，组成的扰动有助于提高相对于其周围环境有向下位移的流体元的产能率，因而增加了趋于超稳定态的倾向。爱丁顿对造父变星球对称振动超稳定性的讨论与此类似，但在纯径向运动中，任何发生位移的流体元紧挨着的环境总是同样的流体，从而组分梯度也不直接进入稳定性判据。

组分梯度的存在也是造成其增长率由等式（10）给出的长期模式的原因。例如，试设想一个流体元经历温度正起伏的情况。它将上升以寻求一个新的流体静力学平衡位置。这时由于条件（11）不被满足，这个位置在对流时间尺度内是稳定的，因而流体元将具有与周围环境等同的压强和密度。但是，如果 $\mathrm{d}X/\mathrm{d}z>0$，那么该流体元中的 H 丰度将比周围环境中的低，所以温度将比周围的高。因此，产能率就会有所不同；如果由较高温度导致的被热扩散抵消后的增长超过了由较低 H 浓度以及不可避免的较低 ^3He 浓度引起的降低，那么温度升高将会加剧，扰动也会增长。

不能只通过把典型值代入等式（9）来估计太阳 g 模式的增长率，因为扰动不仅仅局限于产生能量的核心。在核心外，唯一存在的非绝热效应就是辐射阻尼，奥雷[23]曾估计早型星包层的阻尼作用远远超过了下层（在半对流区内）的激发，这主要是因为扰动的幅度随半径增加得很快。但是，在太阳中相对较深的外对流区却存在一个重要的不同。当 $\nabla-\nabla_{\mathrm{ad}}>0$ 时，正如在 p 模式的情况一样[24]，辐射热传递倾向于使振动失去稳定性，并且在任何情况下振幅都会随着半径而降低。此外，我们可以预期在此区域中与非稳定直接对流运动之间的相互作用将破坏该运动与核心内运动之间可能存在的任意相关。值得注意的是：假如辐射阻尼足以使等式（9）和（10）

whatever correlation the motion might have with the motion in the core. It is worth noting that if radiative damping is sufficient to make the effective value of α in equations (9) and (10) positive, the secular mode must be stable.

An approximate stability criterion for the Sun was derived by evaluating the work integral[25] for nonradial modes using the adiabatic eigenfunction for the fundamental g mode associated with the surface harmonic with $l=2$ published by Cowling[26] for the polytrope of index 3, and making the Boussinesq approximation in calculating the divergence of the radiative heat flux. The polytrope approximates the Sun reasonably well beneath the convection zone. In view of the more extensive analysis by, for example, Smeyers[27] of nonradial modes in other stars, one might expect the $g_1(l=2)$ oscillation to be a strong contender for the most unstable mode. Bearing in mind the discussion in the previous paragraph, the integrals were truncated at 70% of the total radius of the polytrope, the position of the base of the convection zone in the solar model considered in the next section. The analogue of criterion (13) for instability so obtained is

$$\frac{1}{H_x} > \left[(30-\eta)(\nabla_{\rm ad}-\nabla) - \frac{\nu_3 b}{n+1}\right]\frac{1}{(\nu+\nu_3)H_p} \tag{14}$$

where now one must enter typical values. In the solar core $\nu \simeq \frac{1}{2}$, $\nu_3 \simeq 1$, $\eta \simeq 9$, $b \simeq 6$ and $\nabla_{\rm ad} - \nabla \simeq 0.1$, and the criterion becomes

$$H_x < 2.5\, H_p \tag{15}$$

The Main Sequence Evolution of the Sun

The early main sequence Sun is roughly chemically homogeneous (H_x is infinite). The core is stable and no mixing occurs, except possibly at the very beginning when X_3 is far from its equilibrium value. A spatial variation in composition builds up until criterion (14) is satisfied and the core becomes unstable to oscillatory g modes.

What happens when large amplitude motions occur? If the core is mixed until the chemical composition and specific entropy are uniform, local nuclear equilibrium is upset: in the central regions X_3 increases proportionately more than X (because $H_3 < H_x$ before mixing) and the pp chain is dominated by the temperature sensitive ^{3}He burning reactions. The rate at which energy is generated close to the center is increased by the redistribution of fuel; if, for example, an equilibrium abundance of ^{3}He in the inner 20% by mass is thoroughly mixed, the central abundance is increased by a factor of about 5 in a polytrope model and if the temperature were to stay constant the rate of the ^{3}He–^{3}He reaction would increase by a factor of 25. The consequent increased heat flux can no longer be stably transported by radiation alone, the Schwarzschild criterion for stability is violated and a mechanism to drive the motions originally postulated for mixing is thus provided. The core, though stable to infinitesimal direct convective modes, is unstable to finite amplitude convection.

中的 α 有效值为正的话，那么长期模式必然是稳定的。

太阳的一个近似稳定性判据可以通过以下方式导出：用基本 g 模式下的绝热本征函数，结合考林[26]针对指数等于 3 的多方球发表的 $l=2$ 的面调和函数计算出非径向模式下的功积分[25]，并利用博欣内斯克近似值计算热辐射流的散度。对于太阳对流区以下的区域用多方球近似是合理的。考虑到更广泛的分析，例如斯梅耳斯[27]对其他恒星在非径向模式下所作的分析，我们可以预期，$g_1(l=2)$ 振动模式将是最不稳定模式的一个可能性很大的候选者。别忘了我们在前一段中的讨论，积分在多方球总半径的 70% 处被截断，这正是在下一节我们将考虑的太阳模型中对流区基部所在的位置。因此，不稳定状态的判据（13）可以类似地表达为：

$$\frac{1}{H_x} > \left[(30-\eta)(\nabla_{ad}-\nabla) - \frac{\nu_3 b}{n+1}\right]\frac{1}{(\nu+\nu_3)H_p} \tag{14}$$

现在需要将一些典型值代入式中。在太阳核心，$\nu \simeq \frac{1}{2}$，$\nu_3 \simeq 1$，$\eta \simeq 9$，$b \simeq 6$ 和 $\nabla_{ad} - \nabla \simeq 0.1$，因而上述判据变为：

$$H_x < 2.5\ H_p \tag{15}$$

太阳在主序阶段的演化

早期主序阶段的太阳在化学上大致是均匀的（H_x 为无穷大）。除了在非常早期 X_3 可能会远远偏离它的平衡值外，其核心是稳定的并且没有混合发生。直到满足判据（14）且核心对于振荡 g 模式变得不稳定的时候，组分在空间的变化才得以建立起来。

大幅度运动出现时会发生什么？如果在化学组成和比熵达到均衡之前，核心一直在混合，那么局域核反应平衡就会被打破：在中心区域，X_3 按比例的增长比 X 要大（因为混合之前 $H_3<H_x$），而 pp（质子－质子）链取决于对温度敏感的 ^3He 燃烧反应。中心区附近的产能率随着燃料的重新分配而不断增加；比如，如果在质量占 20% 的内部区域，具有平衡丰度的 ^3He 被彻底混合，则在多方球模型中，中心丰度会增加到原来的 5 倍左右，如果温度保持不变，那么 ^3He–^3He 反应的速率将增加到原来的 25 倍。仅仅通过辐射便不再能够稳定地传输随之增长的热流量了，史瓦西稳定性判据不再被满足，从而提供了那种一开始假设的驱动物质混合运动的机制。虽然核心对于极微小的直接对流模式来说是稳定的，但是对于有限幅度对流来说却是不稳定的。

The combination of overstability to infinitesimal disturbances and instability to finite amplitude convection has been studied in other contexts, for example, convection in a rotating layer of Boussinesq fluid[28] and thermosolutal convection[29]. It is thought that overstable oscillations first appear which ultimately grow until they are of sufficient amplitude to trigger convection. Laboratory experiments, though not conclusive, support this idea[30-32]. The combination of the subsequent persistent convective motion and diffusion finally destroys or substantially weakens the stabilizing mechanism. We note that in the circumstances considered here, although diffusion of material was ignored when deriving the linear stability criterion, it was tacitly invoked later to render the solar core chemically homogeneous.

What are the evolutionary consequences for the Sun? As a result of transporting fuel to hotter regions of the star there is a tendency for the total energy generation rate to increase. But because the Sun is thermally stable this is adequately offset by an expansion of the core, converting thermal energy into gravitational energy, reducing the temperature and density and hence the energy generation rate. The timescale for this process, the Kelvin–Helmholtz time for the core, is shorter than the Kelvin–Helmholtz time for the whole star by a factor of about ten. So the envelope, which adjusts almost immediately to stay in hydrostatic equilibrium, at first expands almost adiabatically and cools. This causes a reduction in the effective temperature and surface luminosity. Meanwhile, as the ^3He abundance relaxes to equilibrium the temperature sensitivity of the energy generation is reduced and the convective core disappears. The bulk of the envelope has not yet had time to adjust thermally and the central temperature still remains anomalously low. Finally, in about 10^7 yr the whole star reaches thermal equilibrium; the central temperature, effective temperature and luminosity have increased somewhat, but not to their previous values, for the Sun is now chemically less inhomogeneous than it was before mixing began and takes on an appearance similar to when it was in a less evolved state. Normal main sequence evolution without mixing follows until condition (14) is satisfied once more, and the process is repeated.

To check this argument quantitatively, we modified a stellar evolution computer program provided by B. Paczyński. The only nuclear reactions considered were those of the pp chain, using published data for cross-sections[15] and for the rate of electron capture by ^7Be (ref. 33).

The time which elapses between successive mixing phases is that required to build up a composition gradient sufficient to make the core overstable. This is a nuclear burning time and, taking criterion (14) literally, was found to be of the order of 10^9 yr. We emphasize that this value is not accurate because the coefficient in (14) was obtained by taking the difference between two numbers each of which was obtained from the structure of a particular g mode in a simple polytrope. We are now studying the stability of nonradial modes in a more realistic solar model. Because the Sun is now about 40% more luminous than it was when it arrived on the main sequence, evolution is occurring more rapidly. Consequently, provided the structure is not changing so much as to significantly alter the

对无限小扰动的超稳定性和对有限幅度对流的不稳定性已在其他情况下被联合研究过：比如，在旋转的博欣内斯克流体层中的对流[28]和热溶质对流[29]。人们认为超稳态下的振动首先出现，它们最终会发展到具有足够的振幅以触发对流。实验室中的一些结果，虽然不具备结论性，但是支持这一想法[30-32]。随后持续的对流运动与扩散过程相结合，最终破坏或者极大地削弱了这种稳定机制。我们注意到：在本文所讨论的情况下，尽管在推导线性稳定性判据时忽略了物质扩散，但后来还是默许用它来说明太阳核心在化学组成上的均一性。

太阳最终会演化成什么样子？由于燃料向恒星中更高温度处的输运，因而总产能率趋向于不断增加。但因为太阳是热稳定的，所以这一效应会被核心区的膨胀适当抵消，从而将热能转化为重力势能，降低温度和密度从而降低产能率。这一过程的时标，即核心的开尔文－亥姆霍兹时间，还不到整个恒星的开尔文－亥姆霍兹时间的 1/10。因此，几乎立即调节到处于流体静力学平衡的包层将首先几乎绝热地膨胀并冷却。这就导致了有效温度和表面光度的降低。同时，随着 ^{3}He 丰度逐渐通过弛豫趋于平衡，产能的温度敏感度会下降，对流核心也随之消失。而这时整个包层还没来得及进行温度上的调节，中心温度仍然维持在异常低的状态。最后，经过大约 10^{7} 年，整个恒星达到了热平衡；中心温度、有效温度和光度都有所上升，但是达不到它们以前的值，因为太阳此时在化学组成上的均一性比混合开始之前高，并且呈现出类似于处在更早演化阶段的状态。随后发生没有混合的正常主序演化过程，直到条件（14）再被满足，上述过程又得以重复。

为了从数值上检验这一观点，我们对帕金斯基提供的关于恒星演化的计算机程序进行了修改。只考虑 pp 链的核反应，使用已发表的截面数据[15]和 ^{7}Be 电子俘获率的数据（参考文献 33）。

两个相继的混合期之间经过的时间就是建立足以使核心区变成超稳定的组分梯度所需的时间。这就是核燃烧的时间，从判据（14）的表面意义上看，这一时间的数量级为 10^{9} 年。我们要说明的是这个值并不准确，因为判据（14）中的系数是由简单多方球中两个特殊 g 模式结构分别给出的值的差得到的。我们现在正在研究的是一个更实际太阳模型中的非径向模式的稳定性。因为太阳目前的光度大约比它刚进入主序阶段时高了 40%，所以它会更快地演化。因此，只要结构的变化不足以大到使稳定性判据发生显著的改变，那么两个相继的混合期之间的时间间隔就会缩短。

stability criterion, the period between successive mixing phases is decreasing. In the next section we argue that it is now about 2.5×10^8 yr.

To simulate the gross effects of earlier mixing a solar model was evolved for 4.45×10^9 yr with a core of mass m_c artificially mixed. Normal evolution was allowed for a further 2.5×10^8 yr, then the entropy and chemical composition of the core were instantaneously mixed and normal evolution was allowed to continue again. We found that a superadiabatic core did develop after the sudden mixing, though it did not extend to m_c. The velocity of the ensuing motion was sufficient to advect ^{7}Be away from the central region in a time somewhat shorter than the half life against electron capture. We assumed that mixing of ^{3}He and ^{7}Be occurs only in the superadiabatic region, because the amplitudes of at least the linear direct convective modes decay rapidly in the subadiabatic region surrounding the core[34], though finite amplitude motion may penetrate further[35]. It can be seen in Fig. 1 that with $m_c = 0.25\ M_\odot$ the minimum luminosity is some 5% lower than the luminosity before mixing. At this point, although much of the interior is in a relatively expanded state the radius at the photosphere is somewhat smaller than normal. The cooling of the outer envelope which results from the smaller heat input from below allows the temperature and pressure to fall and the surface layers to contract. Later, the luminosity rises as the thermal wave produced by the burst of energy generated immediately after mixing reaches the surface, and finally the star settles down with a luminosity somewhat lower and an effective temperature slightly higher than it had before mixing began.

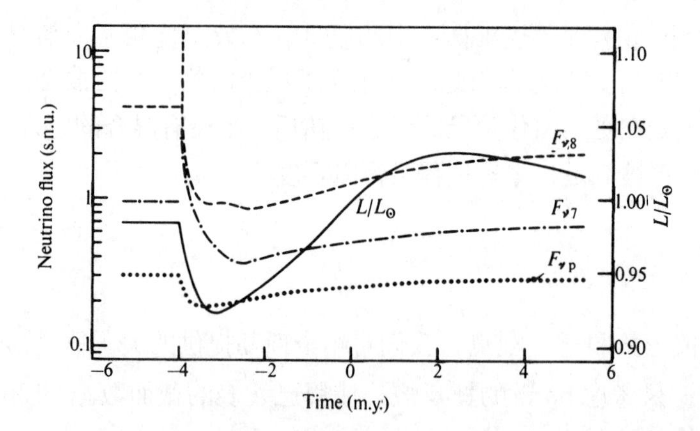

Fig. 1. Time variations of the surface luminosity of the Sun, in units of its present value, and of the neutrino fluxes $F_{\upsilon p}$, $F_{\upsilon 7}$ and $F_{\upsilon 8}$ measured in s.n.u. (The neutrino capture cross-sections were taken from Bahcall[36].) The first minimum in $F_{\upsilon 8}$ occurs soon after the convective core begins to recede, the second when the central temperature is also a minimum. Time is measured in units of m.y.

Finally, we consider the neutrino flux. The mixing of ^{3}He towards the center initially increases the rate of ^{7}Be production and the ^{7}Be neutrino flux $F_{\upsilon 7}$ rises. But mixing of ^{7}Be away from the centre reduces the rate of the temperature sensitive proton capture and the ^{8}B neutrino flux $F_{\upsilon 8}$ drops rapidly. As the central temperature subsequently falls $F_{\upsilon 7}$ attains a maximum and then decreases, and $F_{\upsilon 8}$ drops still further; temperature is the dominant

在下一节中我们将论证这个时间段的长度大致为 2.5×10^8 年。

为了模拟出更早期混合的总效应，我们让一个具有人为混合过的质量为 m_c 的核心的太阳模型演化 4.45×10^9 年。再继续进行正常的演化 2.5×10^8 年，然后核心区的熵和化学组成被迅速混合，随后再按正常演化继续进行。我们发现超绝热的核心在经过突然混合之后确实扩展了，虽然它还没有扩展到 m_c。结果产生的运动速度足以把 ^{7}Be 在比电子俘获的半减期更短的时间内输送出核心区。我们假设 ^{3}He 和 ^{7}Be 的混合只发生在超绝热区，因为：虽然有限幅度的运动还可以进一步穿透 [35]，但至少线性直接对流模式的幅度会在核心周围的亚绝热区内迅速衰减掉 [34]。从图 1 中可以看出：当 $m_c=0.25\ M_\odot$ 时，其最低光度比混合前的光度低了大约 5%。在这一点上，尽管大部分内部区域处于相对膨胀的状态，但是其光球层的半径却比正常的要小。由于从下层传输来的热量比正常时变少导致了外包层的冷却，从而使温度和压强下降，因而表面层也将收缩。之后，由紧接着混合后立即导致的能量暴发所产生的热波到达表面，光度随之上升。最终，恒星稳定在一个光度比混合开始前略低、而有效温度比混合开始前略高的状态。

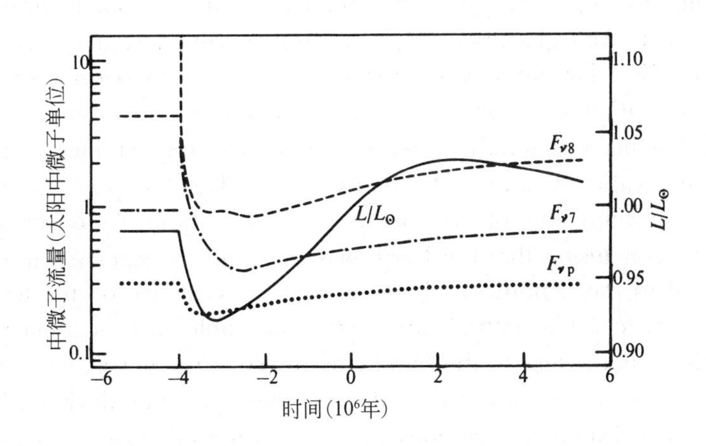

图1. 太阳表面光度的时变，以当前值为单位；以及中微子流量 $F_{\nu p}$、$F_{\nu 7}$ 和 $F_{\nu 8}$ 的时变，以太阳中微子单位为单位（中微子俘获截面取自巴考[36]）。$F_{\nu 8}$ 的首个极小值紧跟在对流核开始缩小之后出现，当中心区温度也达到极小值时出现第二个极小值。时间的单位为 10^6 年。

最后，我们来考虑中微子流量。^{3}He 向中心区的混合使 ^{7}Be 的生成率开始增加，也使 ^{7}Be 中微子流量 $F_{\nu 7}$ 增大。但是 ^{7}Be 离开中心区的混合却使对温度敏感的质子俘获率下降并使 ^{8}B 中微子流量 $F_{\nu 8}$ 也迅速下降。随着中心温度随后的下降，$F_{\nu 7}$ 在达到一个极大值之后也开始下降，$F_{\nu 8}$ 下降的幅度更大；温度对 pep（质子－电子－质子）

influence on the pep neutrino flux $F_{\nu p}$ which also decreases. The ^{7}Be and ^{8}B fluxes reach a minimum at the same time as the central temperature, and then gradually rise as the Sun relaxes back to thermal equilibrium. The precise details of the evolution depend on how much of the core is mixed; the model illustrated in Fig. 1 is typical. Because many of the recent improvements in the microphysics have not been included in the calculation, $F_{\nu 8}$ in Fig. 1 has been scaled by a constant factor to make its value prior to mixing similar to the latest published theoretical results. R. Rood and D. N. Schramm (private communication) report similar results. The minimum value of the total neutrino flux is of the same order as the currently quoted experimental upper bound. It appears, therefore, that if it can be demonstrated that the Sun has recently been mixed, the observed neutrino flux may no longer present a serious problem. The possibility still remains, however, that more refined measurements will lower the upper bound still further. No doubt a model can be found for which $F_{\nu 8}$ (which is very sensitive to comparatively small adjustments of the model) is negligible, but it is unlikely that the sum of $F_{\nu p}$ and $F_{\nu 7}$ can be reduced appreciably below about 0.7 s.n.u.

Solar Chronology and the Terrestrial Climate

The most recent of the Earth's major ice ages were separated by approximately 2.5×10^8 yr (ref. 37) which is of the same order as our rough estimate of the interval between successive mixing phases. The view that reductions in the solar luminosity L were responsible for the Earth's glaciations[7,38,39] is not universally accepted[37], but this is chiefly because evidence for solar luminosity variations has not been considered convincing. If major ice ages are initiated by fluctuations in L they must be a transient response, for it is believed that L has been generally increasing approximately exponentially from its zero age main sequence value of about 70% of its value today. Accepting the view[7,40] that an ice age is the equilibrium state of the climate when L is about 5% lower than its present value leads to the conclusion that the Earth was continuously glaciated until about 10^9 yr ago. It is essential to the hypothesis, therefore, that at least one of the terrestrial factors controlling the climate has a natural timescale comparable with or longer than the time taken for L to drop subsequent to the Sun being mixed. Theoretical estimates are difficult to make but there is some empirical evidence suggesting that during the Tertiary the climate underwent substantial variations on a timescale of several million years[41-44]. It seems most likely that these variations were produced either by solar variations, which would at least confirm the conclusion that the climate is strongly influenced by the Sun, or by some factor of confluence of factors intrinsic to the Earth, which would establish the existence of a sufficiently long natural timescale.

It is also necessary to ask whether the 5% drop in L is enough to produce an ice age. Theoretical discussions[7,40,45] suggest it is, but because these are deficient at least so far as they do not predict the long term stability of the climate, they cannot be relied on. The supporting empirical evidence is that the relatively short period low amplitude fluctuations in the insolation at a particular latitude arising from precession and planetary

反应产生的中微子流量 F_{vp} 起主导作用，所以 F_{vp} 也不断下降。^7Be 和 ^8B 流量与中心温度同时达到最小值，而在太阳通过弛豫回到热平衡的过程中，它们又逐渐增大。演化的精确细节取决于核心的混合程度；图 1 中给出的模型具有代表性。因为在计算中没有考虑到微观物理学的许多最新进展，所以我们在绘制图 1 时将 F_{v8} 乘上了一个常数因子，以使它在混合之前的数值接近于最新发表的理论值。鲁德和施拉姆也给出了相似的结果（个人交流）。总中微子流量的最小值与目前引用的实验值的上限在同一个数量级。因此，假如可以证明太阳在近期发生过混合，那么观测到的中微子流量也许就不再是一个严重的问题了。然而，更精密的测量仍有可能使上限值进一步降低。毫无疑问，一个 F_{v8}（其对模型中相当微小的调节十分敏感）可以忽略的模型是能够找到的，但是要使 F_{vp} 和 F_{v7} 的总和降到显著低于 0.7 太阳中微子单位则不太可能。

太阳年代学和地球气候

地球上离现在最近的两次主要冰期之间大约相隔 2.5×10^8 年（参考文献 37），这与我们对两个相继混合期之间的时间间隔的粗略估计是同一量级。太阳光度 L 的减少使地球上出现冰期的观点 [7,38,39] 并不被大家广泛接受 [37]，但这主要是由于证明太阳光度变化的证据尚不令人信服。如果说主要的冰期是由 L 的波动引起的，那么这种现象必定不会长久，因为一般认为从太阳零龄主序开始到现在，L 总是大致呈指数增长，其初始值是现在值的 70% 左右。如果接受这一观点 [7,40]，即冰期是当 L 比其现在值低大约 5% 时的一种气候上的平衡态，就可以导出直到 10^9 年以前地球一直处于冰河时期的结论。因此，下述假设是非常重要的：在控制气候的各种地球内部因素中至少有一个所具有的自然时标相当于或者长于在太阳混合之后 L 出现下降所用的时间。很难作出理论上的估计，但是一些经验上的证据表明：在第三纪时，气候经历了几百万年时间尺度上的重大变迁 [41-44]。这些变化似乎极有可能或者是由太阳变化引起的，从而至少证实了太阳会强烈影响气候；或者是由地球内在的某些因素联合形成的某个因素引起的，这要求承认存在一个充分长的自然时标。

另一个必须回答的问题是，L 下降 5% 是否足以产生冰期。虽然理论上的讨论 [7,40,45] 都支持这一点，但因为至少迄今为止它们仍是有缺陷的——它们不能预言气候的长期稳定性，所以并不可靠。一个支持此观点的经验上的证据是：已经发现在由地球运动岁差和行星摄动而引起的日射能量在某一特定纬度处的短周期低幅波动

perturbations of the Earth's motion have been correlated with the fluctuations in the extent of glaciation over the last 3×10^5 yr (refs. 46–49). Because the amplitudes of the fluctuations in the total annual insolation are typically only about 1%, one can at least say that it is not implausible that a 5% fluctuation is sufficient to bring about a major ice age.

Accepting this picture leads one to expect the beginning of the Pleistocene epoch to be located near the luminosity minimum at about 3 m.y. BP, when glaciation in midlatitudes began[50] (Fig. 1). The beginning of the depression of L, at about 4 m.y. BP, is then concomitant with the estimated onset of the Antarctic glaciation[51]. Because the beginning of the Pleistocene epoch was less than a solar Kelvin–Helmholtz time ago, the Sun is not at present in thermal equilibrium. This accounts at least in part for the discrepancy between the theoretically predicted neutrino flux and the upper bound measured by Davis and his collaborators.

After the core is mixed the Sun's evolution temporarily reverses its direction in the H–R diagram, envelope convection becomes more vigorous and presumably solar activity and the strength of the solar wind increase. The flux of high energy particles incident on the Earth and hence the production of radioactive nuclei in the atmosphere is therefore modified. One might hope that calibrating ^{14}C dating against other methods, for example, would show this, but during the comparatively short times over which this is possible other perturbing influences, such as the changes in the geomagnetic field, dominate[52]. Alternatively one might expect to unravel a direct record of the cosmic ray flux by analysing nuclear tracks in rocks, but this has not yet been possible[53].

We thank Professor N. H. Baker, Dr. A. Delany, Mr. D. C. Heggie, Professor K. H. Prendergast, Dr. D. N. Schramm, Dr. N. J. Shackleton, Professor C. T. Shaw, Professor E. A. Spiegel, Dr. N. O. Weiss, Professor A. Wolfendale, and Professor L. Woltjer for useful discussions, and Dr. J. R. Gribbin and Dr. E. Phillips for their help in rewriting the paper. We are grateful to Dr. B. Paczyński for giving us a copy of his stellar evolution program and to Dr. P. P. Eggleton for showing us how to use it. F. W. W. D. acknowledges the tenure of an SRC research studentship.

(**240**, 262-264&293-294; 1972)

F. W. W. Dilke and D. O. Gough: Institute of Theoretical Astronomy, University of Cambridge.

Received June 28; revised October 6, 1972.

References:

1. Davis, jun., R., Harmer, D. S., and Hoffman, K. C., *Phys. Rev. Lett.*, **20**, 1205 (1968).
2. Bahcall, J. N., and Ulrich, R. K., *Astrophys. J. Lett.*, **160**, L57 (1970).
3. Abraham, Z., and Iben, I., *Astrophys. J.*, **170**, 157 (1971).
4. Ezer, D., and Cameron, A. G. W., *Astrophys. Space Sci.*, **10**, 52 (1971).
5. Reye, T., *Die Wirbelstürme, Tornados und Wettersäulen* (Gesenius, Halle, 1872).

与过去 3×10^5 年之中冰川范围的波动之间存在关联（参考文献 46~49）。因为全年总日照能量的波动幅度通常只有大约 1%，所以我们至少可以认为 5% 的浮动足以诱发一次大的冰期的观点有一定道理。

接受这一理论框架将导致我们预期：更新世的开始时间接近于距今大约 3×10^6 年前的光度极小值时期，那时中纬度地区已开始进入冰期 [50]（图 1）。L 在距今大约 4×10^6 年前开始下降，伴随着的是预期的南极冰期的开始 [51]。由于更新世的开始时间没有达到太阳的开尔文 – 亥姆霍兹时间，因此太阳到现在还没有达到热平衡。这至少可以部分地解释在中微子流量理论估计值和戴维斯及其合作者测量出的上限值之间存在的差异。

在核心被混合之后，太阳的演化在赫罗图上暂时变为反向，包层对流区变得更加活跃，太阳活动性和太阳风的强度可能也会增加。因而入射到地球的高能粒子流和大气层中放射性核的生成量也发生了改变。例如，我们可以期望用定标的 ^{14}C 年龄测量与其他方法相比对来证明以上结论，但是其他扰动因素，如地磁场的变化，也可能会在较短的时间内占据主导作用 [52]。作为另一种可选择的方案，我们希望能通过分析岩石中的核径迹来解释宇宙射线流的直接记录结果，但是这种做法到目前为止还是不可能的 [53]。

感谢贝克教授、德拉尼博士、赫吉先生、普伦德加斯特教授、施拉姆博士、沙克尔顿博士、肖教授、施皮格尔教授、韦斯博士、沃尔芬德尔教授和沃尔特耶尔教授与我们进行了有益的讨论，还要感谢格里宾博士和菲利普斯博士在我们改写这篇文章时所给予的帮助。帕金斯基博士为我们提供了他的恒星演化程序的拷贝，埃格尔顿博士给我们演示了如何使用这一程序，在此也对他们表示感谢。迪尔克要感谢英国科学研究理事会一直为他提供研究生奖学金。

（孟洁 翻译；邓祖淦 于涌 审稿）

6. Schwarzschild, K., *Göttinger Nachrichten*, **1**, 41 (1906).
7. Öpik, E. J., *Contrib. Armagh Obs.*, No. 9 (1953).
8. Chapman, S., and Cowling, T. G., *The Mathematical Theory of Nonuniform Gases*, third edition (Cambridge University Press, 1970).
9. Howard, L. N., Moore, D., and Spiegel, E. A., *Nature*, **214**, 1297 (1967).
10. Bretherton, F. P., and Spiegel, E. A., *Astrophys. J. Lett.*, **153**, L77 (1968).
11. Ezer, D., and Cameron, A. G. W., *Astrophys. Lett.*, **1**, 177 (1968).
12. Bahcall, J. N., Bahcall, N. A., and Ulrich, R. K., *Astrophys. Lett.*, **2**, 91 (1968).
13. Iben, I., *Astrophys. J. Lett.*, **155**, L101 (1969).
14. Schatzman, E., *Astrophys. Lett.*, **3**, 139 (1969).
15. Shaviv, G., and Salpeter, E. E., *Astrophys. J.*, **165**, 171 (1971).
16. Gough, D. O., and Spiegel, E. A., *Comm. Astrophys. Space Phys.*, 4 (1972).
17. Fowler, W. A., *Nature*, **238**, 24 (1972).
18. Eddington, A. S., *The Internal Constitution of the Stars* (Cambridge University Press, 1926).
19. Spiegel, E. A., and Veronis, G., *Astrophys. J.*, **131**, 442 (1960); (correction: **135**, 655, 1962).
20. Chandrasekhar, S., *Hydrodynamic and Hydromagnetic Stability* (Oxford University Press, 1961).
21. Ledoux, P., *Astrophys. J.*, **105**, 305 (1947).
22. Defouw, R. J., *Astrophys. J.*, **160**, 659 (1970).
23. Auré, J.-L., *Astron. Astrophys.*, **11**, 345 (1971).
24. Spiegel, E. A., *Astrophys. J.*, **139**, 959 (1964).
25. Ledoux, P., and Walraven, Th., *Handbuch der Physik*, **51**, 353 (Springer, Berlin, 1958).
26. Cowling, T. G., *Mon. Not. Roy. Astron. Soc.*, **101**, 367 (1941).
27. Smeyers, P., *Bull. Soc. Roy. Sci. Liège*, **36**, 357 (1967).
28. *J.F.M.*, **5**, 401 (1959); Veronis, G., *J.F.M.*, **24**, 545 (1966).
29. Veronis, G., *J. Marine Res.*, **23**, 1 (1965).
30. Turner, J. S., *J.F.M.*, **33**, 183 (1968).
31. Shirtcliffe, T. G. L., *J.F.M.*, **35**, 677 (1969).
32. Rossby, H. T., *J.F.M.*, **36**, 309 (1969).
33. Bahcall, J. N., and Moeller, C. P., *Astrophys. J.*, **155**, 511 (1969).
34. Saslaw, W. C., and Schwarzschild, M., *Astrophys. J.*, **142**, 1468 (1965).
35. Latour, J. (reported by Spiegel, E. A.), *Ann. Rev. Astron. Astrophys.*, **10** (1972).
36. *Phys. Rev. Lett.*, **12**, 300 (1964); Bahcall, J. N., *Phys. Rev.*, **135**, B137 (1964).
37. Holmes, A., *Principles of Physical Geology* (Nelson, London, 1965).
38. Dutton, C. E., *Amer. J. Sci.*, third series, **27**, 1 (1884).
39. Öpik, E. J., *Mon. Not. Roy. Astron. Soc.*, **110**, 49 (1950).
40. Sellers, W. D., *J. Appl. Meteor.*, **8**, 392 (1969).
41. Dorman, F. H., *J. Geol.*, 74, 49 (1966).
42. Devereux, I., *New Zealand J. Sci.*, **10**, 988 (1967).
43. Douglas, R. G., and Savin, S. M., *Initial Reports of the Deep Sea Drilling Project*, **6**, 1123 (1971).
44. Wolfe, J. A., *Palaeogeogr. Palaeoclim. Palaeoecol.*, **9**, 27 (1971).
45. Eriksson, E., *Meteor. Mon.*, **8**, 68 (1968).
46. Milankovitch, M., *Théorie Mathématique des Phénomènes Thermiques Produits par la Radiation Solaire* (Gauthier-Villars, Paris, 1920).
47. Emiliani, C., *J. Geol.*, **63**, 538 (1955).
48. van Woerkom, A. J. J., *Climatic Change* (edit. by Shapley), 147 (Harvard, Mass., 1953).
49. van den Heuvel, E. P. J., *Geophys. J. Roy. Astron. Soc.*, **11**, 323 (1966).
50. McDougall, I., and Stipp, J. J., *Nature*, **219**, 51 (1968).
51. Hays, J. D., and Opdyke, N. D., *Science*, **158**, 1001 (1967).
52. Grey, D. C., *J. Geophys. Res., Sp. Phys.*, 74, 6333 (1969).
53. Fleischer, R. L., Price, P. B., and Walker, R. M., *Science*, **149**, 383 (1965).